There is a wide variety of optical instruments with which the human eye forms an integral part of the overall system. This book provides a comprehensive description of the construction and image formation in such visual optical instruments, ranging from simple magnifiers, through microscopes and telescopes, to more sophisticated interferometric and diffractive devices. Throughout, details of the eye's interaction with a particular optical instrument, the visual ergonomics of the system, are emphasised.

The book begins with a section on the general theory of image formation and a description of basic optical components. The various optical instruments that can be adequately described using geometrical optics are then discussed, followed by a section on diffraction and interference and the instruments based on these effects. There are separate sections devoted to ophthalmic instruments and aberration theory, with a final section covering visual ergonomics in depth.

Dealing with basic physical principles, as well as engineering and design issues, and containing many problems and solutions, this book will be of great use to undergraduate and graduate students of optometry, optical design, optical engineering, and visual science, and to professionals working in these and related fields.

The eye and visual optical instruments

The eye and
visual optical instruments

GEORGE SMITH
University of Melbourne

DAVID A. ATCHISON
Queensland University of Technology

PUBLISHED BY THE PRESS SYNDICATE OF THE UNIVERSITY OF CAMBRIDGE
The Pitt Building, Trumpington Street, Cambridge CB2 1RP, United Kingdom

CAMBRIDGE UNIVERSITY PRESS
The Edinburgh Building, Cambridge CB2 2RU, United Kingdom
40 West 20th Street, New York, NY 10011-4211, USA
10 Stamford Road, Oakleigh, Melbourne 3166, Australia

First published 1997

Printed in the United States of America

Typeset in Times Roman

Library of Congress Cataloging-in-Publication Data

Smith, George, 1941 Oct. 19–
The eye and visual optical instruments / George Smith, David A.
Atchison.

p.   cm.

Includes index.

ISBN 0-521-47252-0 (hbk.). – ISBN 0-521-47820-0 (pbk.)

1. Physiological optics.   2. Optical instruments.   3. Optics.
I. Atchison, David A.   II. Title.

QP475.S576   1996
681$'$. 4–dc20   96-23142
                                          CIP

A catalog record for this book is available from the British Library.

ISBN 0-521-47252-0 hardback
ISBN 0-521-47820-0 paperback

Contents

Preface

The purpose of this book is to present a thorough description of the construction and image formation of visual optical instruments, ranging from simple magnifiers, through microscopes and telescopes, to the more sophisticated instruments based upon interference and diffraction. There are many other types of optical instruments, such as spectrophotometers and laser systems, but these are not visual optical instruments; that is, they are not used with the eye as an essential component in the imaging process. The only instrument that we include in this book that may not be regarded as a visual optical instrument is the camera. However, we have included the camera, because while one can take a photograph without any "eye" input, the eye is often used to aim the camera and the final image is usually viewed by the eye, either directly or with a projection system.

There are many other textbooks on optics but most of these only briefly discuss visual optical instruments and even more briefly discuss any visual ergonomic aspects of these instruments. We believe that the major strength of this book is its emphasis on the detail of the construction and image formation and most importantly the visual ergonomic aspects. Visual ergonomics is the study or application of the properties of the eye to human performance. In this context, visual ergonomics involves the following factors that may affect vision through an optical instrument: the aberrations of the eye, depth-of-field of the eye, the role of the pupil of the eye, the amplitude of accommodation, refractive errors, visual acuity and the coordination of the two eyes in binocular vision.

Apart from a more comprehensive treatment of visual optical instruments and the inclusion of visual ergonomics, this book includes other topics not normally covered in standard texts. Perhaps the most important is a discussion (Chapter 10) on defocus and focussing techniques.

The book is divided into six parts and a set of appendices, with the chapters in each part having a common theme. Part I covers the general theory of image formation and the description of the optical components that make up a system. Part II is dedicated to individual optical instruments that can be adequately described using geometrical optics, usually with one instrument per chapter. Part III describes two important aspects of physical optics (interference and diffraction), some interesting visual optical instruments based upon these effects and the importance of diffraction to image formation. Part IV is set aside for specific ophthalmic instruments. Part V covers general aberration and image quality theory as well as the aberrations and image quality of the eye. Part VI is dedicated to the visual ergonomics of visual optical instruments.

Various professionals – vision scientists, optometrists, ophthalmologists, microscopists, astronomers and those in surveillance professions such as the police and military – regularly use visual optical instruments. This book will be useful to these professionals, particularly those who need to understand the workings of these instruments in order to understand their limitations. However, visual optical instruments have a wider range of uses and by a wider community.

Some typical instrument uses are listed in the following table:

Instrument	Chapter No.	Uses
Ophthalmic lenses	14	correction of refractive errors of the eye
Simple magnifiers and eyepieces	15	inspection of fine detail, down to about 0.01 mm in size, fine mechanisms, e.g. watches, electronic components low vision magnifiers
Microscopes	16	inspection of very fine detail from about 0.1 mm down to the wavelength of light which is approximately 0.0005 mm components in other instruments such as the slit lamp or the bio-microscope
Telescopes	17	magnification of distant objects (e.g. astronomical bodies) components in many other instruments such as binoculars, spectrometers, focimeters low vision magnifiers or field expanders viewfinders (usually Galilean) in cameras and security doors
Macroscopes	18	magnification of objects at a close distance, but not so close that a microscope can be used
Relay systems, e.g. periscopes, endoscopes and fibrescopes	19	transmission of images over some distances and around corners, e.g. inspection of internal organs and inside machines
Angle and distance measuring instruments (e.g. sextants and rangefinders)	20	measurement of angular distance measurement or estimate of distance
Cameras	21	recording of scenes on photographic material or for electronic recording, e.g. video camera
Projection systems	22	projection of photographic slides or other suitable objects, usually at a high magnification
Collimators	23	production of images at optical infinity, i.e. simulation of very distant scenes or targets checking of infinity settings on many instruments such as in eyepieces, telescopes and camera lenses
Photometers and colorimeters	24	measurement of the light level or colour of a target
Interferometers	25	testing of visual acuity by producing sinusoidal fringes on the retina that by-pass the optics
Diffraction and diffractive devices	26	Fresnel zone plate forms of bifocal ophthalmic lenses speckle patterns for the measurement of refractive error effect of diffraction on image quality
Focimeters	27	measurement of the vertex power of ophthalmic lenses
Radiuscopes and keratometers	28	measurement of the radius of curvature of surfaces
Ophthalmoscopes	29	detailed inspection of the retina
The Badal optometer	30	presentation of stimuli of constant size at different distances
Optometers	31	measurement of the level of accommodation or refractive error
Binocular vision testing instruments	32	measurement of binocular vision

We have included worked examples in the book for two reasons. One is that worked examples help to give some concrete interpretation to what may at first appear to be abstract quantities. The other reason is that many calculations, such as ray tracing, can be done by computer. The worked examples can be used to check the computer program.

Acknowledgements

This book was started by George Smith, who owes his interest in optics to Professor H. H. Hopkins FRS (deceased), his doctorate supervisor while at the Reading University (1968–1972). In those early days of the book, before being joined by David Atchison, he was also encouraged by Professor V. V. Rao (deceased) at the Physics Department of the Regional Engineering College, Warangal, Andra Pradesh, India.

Symbols, signs and other conventions

Symbols

The following is a list of general symbols used in this book. A list of other symbols specific to each chapter is given at the end of the chapter. Where possible the lower case (small) characters are used for distance and the upper case (capitals) are used for the corresponding reciprocals. The most common exceptions are the radius of curvature of a surface and the diameter of the pupil. The surface curvature is denoted by C not R and the pupil diameter is denoted by the symbol D.

As a general rule, a symbol that is not followed by a prime symbol (′) indicates an object space quantity and a symbol followed by a prime symbol indicates an image space quantity.

λ	wavelength in vacuum
λ_d	= 587.6 nm: wavelength of the helium yellow spectral line
λ_F	= 486.1 nm: wavelength of the hydrogen blue spectral line
λ_C	= 656.3 nm: wavelength of the hydrogen red spectral line
r	radius of curvature
C	surface curvature ($C = 1/r$)
F	equivalent power of a system

Object and image space quantities (Note that symbols for points are written in the *Poetica Chancery* font)

o, o'	object and image positions (axial case)
Q, Q'	object and image positions (off-axis case)
$\mathcal{F}, \mathcal{F}'$	front and back focal points
$\mathcal{P}, \mathcal{P}'$	front and back principal points
$\mathcal{N}, \mathcal{N}'$	front and back nodal points
$\mathcal{V}, \mathcal{V}'$	front and back vertex points
$\mathcal{E}, \mathcal{E}'$	centres of entrance and exit pupils
l, l'	object and image distances from respective principal planes
L, L'	corresponding (reduced) vergences
l, l'_v	object and image distances from respective vertex planes
x, x'	object and image distances from respective focal points
$\bar{l}, \bar{l}'$	distances of pupils from respective principal planes
$\bar{L}, \bar{L}'$	corresponding vergences
$\bar{l}_v, \bar{l}'_v$	distances of pupils from respective surface vertices
$\bar{L}_v, \bar{L}'_v$	corresponding vergences
η, η'	object and image sizes
$\bar{\rho}, \bar{\rho}'$	radii of entrance and exit pupils
D, D'	diameters of entrance and exit pupils
u, u'	paraxial ray angles (also angles of paraxial marginal ray)
h, h'	paraxial ray heights (also heights for paraxial marginal ray)
$\bar{u}, \bar{u}'$	paraxial pupil ray angles
$\bar{h}, \bar{h}'$	paraxial pupil ray heights

Greek alphabet

α A	alpha	η H	eta	ν N	nu	τ T	tau
β B	beta	$\theta\Theta$	theta	$\xi\Xi$	xi	$\upsilon\Upsilon$	upsilon
$\gamma\Gamma$	gamma	ι I	iota	o O	omicron	$\phi\Phi$	phi
$\delta\Delta$	delta	κ K	kappa	$\pi\Pi$	pi	χ X	chi
ϵ E	epsilon	$\lambda\Lambda$	lamda	ρ P	rho	$\psi\Psi$	psi
ζ Z	zeta	μ M	mu	$\sigma\Sigma$	sigma	$\omega\Omega$	omega

Sign convention

The mathematical theory of optical systems requires a sign convention, particularly for ray tracing. The choice of a sign convention is arbitrary but it must be consistent. In this book we use the standard cartesian and trigonometric sign conventions. That is, distances to the left of a surface or lens or below the optical axis are negative and those to the right or above are positive. Angles which are due to an anti-clockwise rotation of the ray from the optical axis are positive and those due to a clockwise rotation are negative. This sign convention is explained more fully in Chapter 2.

Distance notation and sign of distance

Distances are denoted by either a single lower case letter such as l or two upper case letters, e.g. $\mathcal{VF}$. In this example, $\mathcal{V}$ and $\mathcal{F}$ are both points, usually on the optical axis, and thus $\mathcal{VF}$ denotes the distance from $\mathcal{V}$ to $\mathcal{F}$. If $\mathcal{F}$ is to the right of $\mathcal{V}$, this distance is positive, and if $\mathcal{F}$ is to the left of $\mathcal{V}$, then the distance is negative.

Notation for refractive index on diagrams

The refractive indices are denoted by the symbol n, n', μ or a number written inside a circle or ellipse.

Units and their abbreviations

metre	m
centimetre	cm
millimetre	mm
micrometre	μm
nanometre	nm
second	s
prism dioptre	Δ
hertz	Hz
Kelvin	K
radian	rad
degrees	°
cycles per degree	c/deg
cycles per radian	c/rad
Joule	J
exposure	lux.s
lumen	lm

lux (lumens per squared metre) lm/m^2
candela   cd
candelas per squared metre cd/m^2
steradian   st
Watt   W
Joule-seconds   J.s

References and bibliography

From time to time in various chapters, we have cited other published work, and this material is fully referenced at the end of the respective chapter. In some chapters we have also supplied references for alternative or background reading. The cited material is marked with an asterisk (∗).

Part I

General theory

1

Introduction

1.0   Introduction

Optics is the study of light whereas visual optics is the study of the optical properties of the eye and sight. Ancient civilizations such as those of Greece were familiar with some of the properties of light, for example the laws of reflection. However, the Greeks misunderstood the nature of sight and the optical principles of the eye. They believed that light was emitted by the eye and only produced a visual response when the emitted rays struck an object. Many centuries passed before it was realized that light passes from the object to the eye and not from the eye to the object.

We will see later in this book, when we come to look at the optics of the eye, that the ability to sense the visual word around us is limited by the optical properties of the eye and its defects. For example before the advent of optical instruments, the smallest creature that could be seen with the unaided eye was about 0.05 mm in length and the mountains of the moon were unknown. Of particular frustration must have been the deterioration of eyesight with age. For example, as we age, the closest point of clear sight recedes, making it more and more difficult to do some things that we enjoy or need to do, such as reading and fine craft work. The discovery or invention of optical instruments enabled these restrictions to be overcome and allowed mankind to discover a world that was much more complex than ever envisaged, from the discovery of micro flora and fauna to countless galaxies far out in space.

The development of visual optical instruments took place over many centuries and the earliest instruments were developed without any knowledge of how they worked. The first visual optical instrument was probably the spectacle lens which appeared in Europe about 1200 A.D., although it is possible that spherical balls or beads of glass had been used as magnifying lenses well before that. The telescope was developed towards the end of the sixteenth and the early years of the seventeenth century. The invention of the telescope (1609) and microscope (1610) has been accredited to Galileo, but it is possible that they had been built and used by others before then. Time and the discovery of the laws of optics have enabled numerous other optical instruments to be developed since that time. The early instruments were crude, and without an understanding of the laws of optics, it was difficult to design and build instruments that gave good

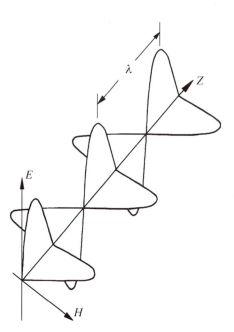

Fig. 1.1: The transverse electric and magnetic components of electromagnetic wave motion.

quality images. Now that we understand these laws, we can design and build optical instruments that give images of the highest possible quality.

While the use of visual optical instruments allows us to see far beyond the limitations of our eyes alone, these instruments also have limitations which are set by the laws of optics and the properties of light, in particular the wavelength of light. We will see that it is not possible to see objects smaller than the wavelength of the illuminating beam of radiation. As a rule, an object can only be "seen" by radiation whose wavelength is less than the object dimension. Therefore to "see" smaller and smaller detail, we must use shorter and shorter wavelengths, for example the X-ray microscope. However, since our eyes cannot respond to X-rays, we need to convert the X-ray image into a visible image.

The aim of this book is to describe the optical properties and functions of a wide range of visual optical instruments. To appreciate these aspects fully, we need to understand the nature of light and some of its basic properties. We will use the remainder of this chapter to cover this material, starting with the nature of light.

1.1 Electromagnetic radiation

Light is only a very small part of the electromagnetic radiation spectrum. Away from the source, electromagnetic radiation is a transverse wave motion composed of an electric (E) and a magnetic (H) field. These two fields are mutually perpendicular and also perpendicular to the direction of propagation, as shown in Figure 1.1. For this reason, electromagnetic radiation is sometimes called **transverse electromagnetic radiation** or wave motion.

Generally in optics, only the electric field component is important. This is because the interaction of electromagnetic radiation with matter involves interaction of the radiation with the electrons in the material, and whereas the electric field interacts with all electrons, the magnetic field only interacts with fast moving electrons. The electrons in optical materials are usually moving sufficiently slowly that their speed can be neglected and therefore they are only affected by the electric field component of the radiation.

Thus for our purposes, the transverse electromagnetic radiation can be sufficiently described in terms of the electric wave motion alone. For plane wave motion in a vacuum, at a distance z from some arbitrary origin and at a time t, the electric field $E(z, t)$ can be described by the equation

$$E(z, t) = E_0 \cos[2\pi(z/\lambda + \nu t + \delta)] \qquad (1.1)$$

and in the complex algebra notation, it can be expressed in the form

$$E(z, t) = \text{real part of } E_0 e^{[i2\pi(z/\lambda + \nu t + \delta)]}$$

where E_0 is the **amplitude** of the electric field, λ is the **wavelength**, ν is the temporal frequency and δ is an arbitrary phase factor. However, we usually omit the reference to the real part and simply write

$$E(z, t) = E_0 e^{[i2\pi(z/\lambda + \nu t + \delta)]} \qquad (1.1a)$$

where the real part is assumed. The direction of the electric vector, or field or plane containing it, is called the direction of the plane of **polarization**.

Physical detectors of light, such as a light meter or the eye, cannot detect the instantaneous electric field. Instead they detect the time averaged square of the field. If we square the instantaneous electric field function $E(z, t)$ given by equation (1.1), carry out a temporal summation by integration with respect to t and finally determine the average value for an infinite integration time, the final value is simply E_0^2. This quantity is often called the **intensity** as opposed to the amplitude. One advantage of the complex representation above is that the intensity is equivalent to the product of the complex field and its complex conjugate. That is,

$$\text{intensity} = E(z, t)E^*(z, t) = E_0^2 \qquad (1.2)$$

where $E(z, t)$ is the complex electric field given by equation (1.1a) and $E^*(z, t)$ is the complex conjugate of $E(z, t)$, which is the same function as $E(z, t)$ except that the complex quantity $i[= \sqrt{(-1)}]$ is replaced by $-i$.

In a vacuum, the wavelength λ and frequency ν are connected by the following equation:

$$\lambda \nu = c \qquad (1.3)$$

where c is the speed or velocity of propagation of the electromagnetic radiation in a vacuum. Its value is given in the summary of symbols at the end of the chapter.

In the visible part of the spectrum, the wavelength ranges from about 400 to 780 nm, with corresponding frequencies of 7.25×10^{14} to 3.72×10^{14} Hz. The wavelengths and frequencies of the different components of the electromagnetic spectrum are shown in Figure 1.2.

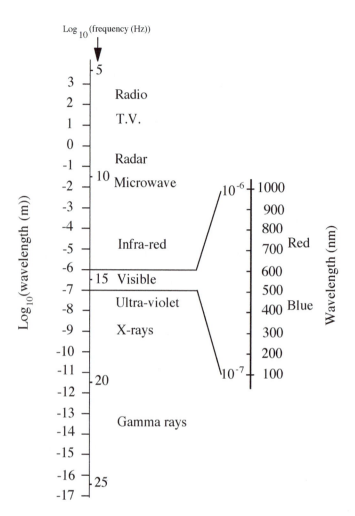

Fig. 1.2: The electromagnetic spectrum.

1.1.1 Particle or quantum theory

While light has the properties of a wave motion, under some circumstances, it behaves like a stream of particles. This behaviour is embodied in the quantum theory. In the quantum theory description of electromagnetic radiation, the radiation is quantized into discrete energy packets called photons. The energy E of a photon is given by the simple equation

$$E = h\nu = hc/\lambda \tag{1.4}$$

where h is Planck's constant. Its value is given in the summary of symbols at the end of the chapter.

1.2 Refractive index and dispersion

The velocity of propagation of electromagnetic radiation through a medium depends upon how strongly it interacts with the charged particles in the medium. The refractive index (n) is a measure of the propagation velocity through the

Table 1.1. The approximate refractive indices of some common materials

Material	Index
Air (15° and 76 cm Hg)	1.00028
Benzene	1.50
Dense flint	1.625
Diamond	2.419
Perspex	1.490
Sapphire	1.77
Sodium chloride	1.54
Water	1.333
White ophthalmic crown glass	1.523
Silicon	4.000 (approx.)

medium. It is defined as

$$n = \frac{\text{velocity in a vacuum } (c)}{\text{velocity in a material } (v)} = \frac{c}{v} \qquad (1.5)$$

For any medium, the velocity v is less than that in a vacuum. Therefore, the refractive index of any medium other than a vacuum is greater than 1.0. The above index is the **absolute refractive index**. However, most of the time we use the **relative refractive index**, that is the index relative to air and not to vacuum. Since the refractive index of dry air under normal conditions is close to unity ($n = 1.0003$), the absolute indices are about 0.03% higher than the corresponding relative indices. Because this difference is small, we often take the index of air as 1.0 and do not make a distinction between the relative and absolute index. Typical values of the refractive index of some common materials are given in Table 1.1.

1.2.1　Dependency of wavelength and frequency on refractive index

Because there is a change of propagation speed or velocity when light enters a medium, there is a corresponding change in wavelength; therefore we can denote the wavelength dependency as $\lambda(n)$. For a medium with a refractive index n, the wavelength $\lambda(n)$ and frequency ν are connected by the equation

$$\lambda(n)\nu = v \qquad (1.6)$$

which has the same form as equation (1.3). It follows from this equation that

$$\lambda(n)\nu = c/n = \lambda\nu/n$$

where λ is the wavelength in vacuum. Therefore

$$\lambda(n) = \lambda/n \qquad (1.7)$$

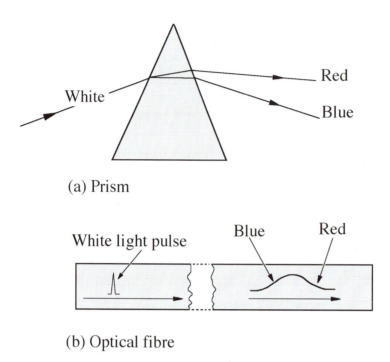

Fig. 1.3:   Examples of the effect of dispersion.

(a) Prism

(b) Optical fibre

However, while wavelength changes with refractive index, frequency does not change. Because of the constancy of frequency, it is common in many circumstances to specify a particular part of the electromagnetic radiation by its frequency rather than by its wavelength.

1.2.2   The dependence of refractive index on wavelength (dispersion)

The refractive index varies with wavelength and, as a general rule, the refractive index decreases with increase in wavelength. The dependency of refractive index on wavelength is called **dispersion**.

The term dispersion is used because under many circumstances, the variation of refractive index with wavelength leads to a white light beam being broken up into its spectral colours. For example, if a beam of light passes through a prism, the beam is deviated through an angle which increases with increase in the refractive index of the prism. Now if a beam of white light passes through a prism as shown in Figure 1.3a, rays of different wavelengths are deviated by different amounts. Since the refractive index for blue light is higher than for red, the blue wavelengths are deviated through a greater angle. Thus the beam is dispersed into an angular distribution. The **rainbow** is due to a similar process in raindrops. A second example is the following. If a very short pulse of white light is sent into a long length of material such as an optical fibre as shown in Figure 1.3b, different wavelengths will travel at different velocities, thus lengthening or dispersing the pulse in the direction of travel. Because the index increases with a decrease in wavelength, red light is at the front of the pulse and the blue light is at the rear.

Because of dispersion, any stated refractive index should also be accompanied by the corresponding wavelength. For example it is common to specify the refractive index of optical glass at the wavelength $\lambda = 587.6$ nm, which is the yellow spectral line of helium. Sometimes average values, say over the visible spectrum, are used instead. The values given in Table 1.1 are mostly average values.

For glass and other common transparent optical materials, the variation of index can be accurately determined by a number of simple mathematical formulae. One due to Cauchy (1836) is of the form

$$n(\lambda) = A + B/\lambda^2 + C/\lambda^4 \tag{1.8a}$$

A second, known as the Hartmann equation (see Longhurst 1973, 500; Smith 1990, 164), is

$$n(\lambda) = n_o + A/(\lambda - \lambda_o)^{1.2} \tag{1.8b}$$

The values of the coefficients A, B, C, n_o and λ_o depend on the actual material and must be determined experimentally. A third equation is Sellmeir's dispersion formula (Born and Wolf, 1989, 97). This is commonly used by the manufacturers of optical glass, e.g. Schott (1992), and is as follows:

$$n^2(\lambda) - 1 = \frac{B_1\lambda^2}{\lambda^2 - C_1} + \frac{B_2\lambda^2}{\lambda^2 - C_2} + \frac{B_3\lambda^2}{\lambda^2 - C_3} \tag{1.8c}$$

where the values of B_1, B_2, B_3, C_1, C_2 and C_3 depend upon the particular glass type, which is given by Schott for each glass in their glass catalog. The accuracy of this equation is claimed to be ± 0.00001 over the spectral range at which the refractive indices are specified. Another dispersion equation that has been used is

$$n^2(\lambda) = a_0 + a_1\lambda^2 + a_2/\lambda^2 + a_3/\lambda^4 + a_4/\lambda^6 + a_5/\lambda^8 \tag{1.8d}$$

1.2.2.1   Quantification of dispersion

Like the refractive index, the dispersion of a material is a very important property of that material and expressing it as a single numerical value gives optical workers some immediate idea of how the refractive index of a material varies with wavelength. There are two common ways of quantifying dispersion. One is the Abbe V-value, often denoted by the symbol V_d. This is defined by the equation

$$V_d = \frac{(n_d - 1)}{(n_F - n_C)} \tag{1.9}$$

where

n_d = is the refractive index at $\lambda = \lambda_d$

n_F = is the refractive index at $\lambda = \lambda_F$

and

$$n_C = \text{is the refractive index at } \lambda = \lambda_C$$

where λ_d, λ_F and λ_C are wavelengths of certain spectral lines of gaseous elements, with λ_d being in the middle of the visible spectrum (yellow) and the other two being towards the edges; λ_F is blue and λ_C is red. Their values are given in the symbols section at the front of the book. The V-value is sometimes defined for a different set of wavelengths, but we will use the above definition throughout this book and for simplicity mostly refer to it as the V-value and not as the V_d-value. A second way is in terms of the difference $(n_F - n_C)$ and this is called the **principal dispersion**.

The magnitude of the dispersion depends upon the type of material and as a general rule, the higher the refractive index the greater the dispersion. For most optical glasses, the values of V_d are in the range of about 25 to 65, with the trend that the higher the refractive index the lower the V-value. Water has a V-value of about 55. Note that with the definition given by equation (1.9), the higher the dispersion, the lower the value of V_d. The corresponding values of the principal dispersion are in the range of about 0.03 to 0.008.

1.2.3   Gradient index materials

The refractive index of most materials encountered in conventional optics is nominally constant throughout the bulk of the material. However, there are some materials in which the refractive index changes within the material, sometimes in a regular manner, and these are called gradient index materials. Two common naturally occurring examples are the atmosphere and the crystalline lens of the eye. Perhaps the most important man-made example is the gradient index optical fibre which is used in telecommunication transmissions. Gradient index fibres are briefly referred to again in Chapter 19.

1.3   Waves and rays

As mentioned in Section 1.1, there are two theories or descriptions of the nature of light. These are the wave theory and the particle theory. Both are valid but only when correctly applied. As a general rule, while light is travelling, the wave theory is used, but on interaction with materials, particularly when there is some absorption, the particle theory is usually applicable.

1.3.1   The wave theory

If a point source of light is radiating light in all directions in an isotropic medium, the radiation field, for each wavelength in the source, can be pictured as a set of spherical **wavefronts** expanding outwards. The wavefronts are often regarded as the crests or troughs of the waves but any other phase of the wave may be used to define the wavefronts. Thus in a set of wavefronts, each neighbouring pair of wavefronts is one wavelength apart. In a more general situation, the wavefronts will be more complex in shape, for example if the propagation velocity or refractive index varies within the medium.

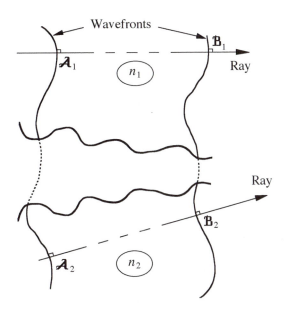

Fig. 1.4: Wavefronts and rays. The rays are the paths of normals to the wavefronts.

1.3.2 Rays

Light rays are a useful concept to be used to trace the paths of the beams of radiation. Rays are imaginary lines drawn perpendicular to the wavefronts or may be interpreted as the expected paths travelled by particles (quanta or photons) of radiation. However, rays as paths of the wave normals is a much more useful interpretation. In a general situation, the wavefronts may not be spherical and instead may be of any shape, as shown in Figure 1.4. Two typical rays are shown in this diagram.

Let A_1A_2 and B_1B_2 be two successive wavefronts and A_1B_1 and A_2B_2 be two rays joining these wavefronts. We also let the two rays pass through regions with different refractive indices, denoted by n_1 and n_2 respectively. Since two neighbouring wavefronts in a set of wavefronts are one wavelength apart, it follows from equation (1.7) that the distances A_1B_1 and A_2B_2 are wavelengths in the respective media and thus

$$A_1B_1 = \lambda/n_1 \quad \text{and} \quad A_2B_2 = \lambda/n_2$$

where λ is the vacuum wavelength. Therefore

$$n_1A_1B_1 = n_2A_2B_2 \tag{1.10a}$$

The quantity

$$\text{refractive index} \times \text{physical distance}$$

is called the **optical path length** or **optical distance**. Thus the distances $n_1A_1B_1$ and $n_2A_2B_2$ are the optical path lengths of the physical distances A_1B_1 and A_2B_2. We often denote optical path lengths or distances by the squared brackets [ ].

Thus

$$[\mathcal{A}_1\mathcal{B}_1] = [\mathcal{A}_2\mathcal{B}_2] \tag{1.10b}$$

It follows from this equation that the optical path length of two rays joining any two wavefronts in a beam must be equal.

1.3.3 *Waves, rays, geometrical optics and physical optics*

In investigating optical systems, sometimes we can use the ray approach and at other times we must use the wave theory approach. **Geometrical optics** is mainly concerned with the ray approach. In homogeneous isotropic media, rays travel in straight lines and only change direction when they encounter a medium of different refractive index. In gradient index media, rays continually change direction and therefore follow curved paths.

A description of the propagation of light that is based upon the wave model is called **physical optics**. Geometrical optics is usually the simpler of the two approaches and therefore we often use geometrical optics as often as possible. However, many observed phenomena cannot be explained by geometrical optics and we have to fall back on physical optics to explain these events. Typical examples are interference and diffraction.

1.4   Beams and paths of rays

A ray is a purely mathematical quantity and in a real situation, we should think of light in the form of beams. Using the mathematical concept of rays, we can regard the beam as being made of rays. If we wish to follow the path of a beam of light through space we can look at the propagation of individual rays in the beam.

The propagation of light through an optical system can be studied by ray tracing techniques. These involve the laws of geometrical optics, which assume that rays travel in straight lines and are only deviated by a change in refractive index. In media of constant index, this only occurs at the boundary with a medium of different index. However, in gradient index media such as the atmosphere, the crystalline lens of the eye and gradient index optical fibres, there is a continuous change in index and hence a continuous change in ray direction. In this book, we will only consider materials of constant refractive index.

Thus in media of constant index, ray tracing involves applying simple algebraic formulae to follow the ray path to the point of contact with the next boundary or surface. Application of **Snell's law** at this interface is used to find the new direction of the refracted or reflected ray. **Refraction** is the process of a ray crossing a boundary between two media of different refractive indices. **Reflection** is the process by which the ray "bounces" back off the boundary. Detailed ray tracing techniques, how they are used to analyse the properties of optical systems and the formation of images by these systems are discussed in detail in the next two chapters. Here we will finish off this chapter looking at some of the fundamental laws of ray propagation, particularly rules for determining the new direction of a ray when it meets a boundary separating two media of different refractive index.

We can begin by considering the situation shown in Figure 1.5a, in which a ray is to go from $\mathcal{A}$ to $\mathcal{B}$. Which path will it take? The answer is that the

Fig. 1.5: Fermat's principle of least travel time.

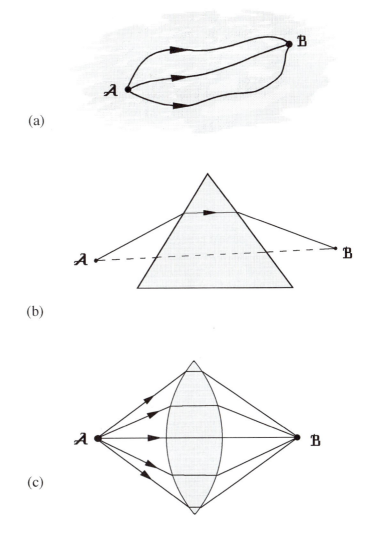

(a)

(b)

(c)

light ray takes the path with the least travel time. This statement is Fermat's principle.

1.4.1 Fermat's principle

Fermat's principle may be described in a number of different ways, for example

(a) the path followed is such that the time of travel is least or
(b) the path followed by a ray is such that it has the minimum optical path length or travel time of all neighbouring paths.

Now the shortest physical path is not always the fastest path. Consider the situation shown in Figure 1.5b, where a ray going from $\mathcal{A}$ to $\mathcal{B}$ through a prism follows the path as shown. This ray path will be confirmed in Chapter 8. In this case, the quickest path is not along the straight line joining $\mathcal{A}$ to $\mathcal{B}$. While this path is physically shorter, if the ray took this path, it would spend more time in a medium of higher refractive index and therefore would travel more slowly over a longer distance. By taking the longer route shown in the diagram, it takes a longer physical path but its travel time is shorter because it spends less time travelling more slowly.

In Figure 1.5b, there is only one path from $\mathcal{A}$ to $\mathcal{B}$ with the least travel time and that path is as shown in the diagram. But in other optical arrangements, there may be more than one path with the least travel time. Figure 1.5c shows a lens perfectly imaging a point $\mathcal{A}$ to a point $\mathcal{B}$. Since all rays from $\mathcal{A}$ travel to $\mathcal{B}$, they must all have the same travel time.

In Section 1.5, we will use Fermat's principle to derive Snell's law, which is the fundamental law of ray tracing.

1.4.2   *Optical path length and travel time*

In Section 1.3.2, the term "optical path length" was introduced. We will now show that this quantity is related to travel time and therefore is relevant to Fermat's principle. Consider a ray travelling a distance d with a speed v of propagation. The time t of travel is then

$$t = d/v$$

but from equation (1.5)

$$n = c/v$$

Therefore it follows that

$$t = nd/c = [d]/c \qquad\qquad (1.11)$$

where $nd = [d]$ is the optical path length for the distance d. Therefore we can conclude that the optical path length is a measure of travel time, and hence a route with the minimum travel time is the route of shortest optical path.

1.4.3   *Principle of reversibility*

The principle of reversibility simply states that the path of a ray is independent of its direction of travel; that is a ray travelling from one point, say $\mathcal{A}$, to another, say $\mathcal{B}$, along a certain path would follow the same path if going in the other direction (i.e. from $\mathcal{B}$ to $\mathcal{A}$).

1.5   Laws of refraction and reflection

The process of following the path of a ray through an optical system involves finding the point of intersection with each surface in turn, finding the new direction after it has been refracted or reflected by that surface and repeating the procedure at the next surface until the ray exits the system. Figure 1.6 shows a simple case of a ray meeting a plane boundary between two media with refractive indices n and n' as shown. We consider the more general case of a curved boundary in the next chapter. We can identify three components to the ray: the incident ray, the refracted ray and the reflected ray. In general,

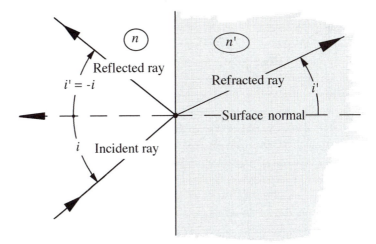

Fig. 1.6: Refraction and reflection at a plane surface: introduction to Snell's law.

when a ray meets such a boundary, part of the energy in the ray is reflected and part is refracted. For normal incidence, we discuss the distribution of this energy in Chapter 34. At this stage, we are only concerned with the directions of the reflected and refracted rays and not with how much energy they carry. The laws of refraction and reflection can be used as rules for determining the new direction.

Referring to Figure 1.6, the basic laws of refraction and reflection can be stated as follows.

(a) The incident ray, the refracted or reflected ray and the normal to the surface are coplanar.

(b) The angles of incidence (i) and refraction (i') or reflection (i') are given by **Snell's law**, which states

$$n' \sin(i') = n \sin(i) \tag{1.12}$$

We should note that since the sine of an angle increases with the angle, it follows from Snell's law that for the refracted ray, if $n' > n$ then $i' < i$.

1.5.1 *Proof of Snell's law*

Snell's law can be easily derived by either ray theory or wave theory. Let us look at both of these approaches.

1.5.1.1 *Ray theory*

To prove Snell's law using the ray approach, we use Fermat's principle. Consider the case as shown in Figure 1.7, where a ray has to travel from the point A to the point C. The path chosen must have a minimum travel time. Let a possible path be the path ABC where B is the point of intersection on the boundary. The total time of travel t can be expressed as

$$t = \frac{\text{distance}}{\text{velocity}} = \frac{\sqrt{[h^2 + x^2]}}{v} + \frac{\sqrt{[d^2 + (s - x)^2]}}{v'}$$

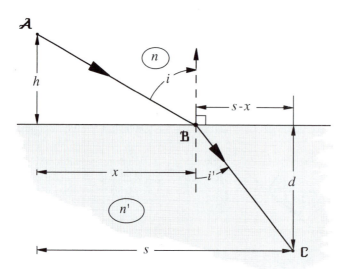

Fig. 1.7:   Derivation of
Snell's law using ray theory
and Fermat's principle.

where v is the propagation velocity from $\mathcal{A}$ to $\mathcal{B}$ and v' is the velocity from $\mathcal{B}$ to $\mathcal{C}$. Now from equation (1.5), since $v = c/n$ and $v' = c/n'$,

$$t = \frac{n\sqrt{[h^2 + x^2]} + n'\sqrt{[d^2 + (s - x)^2]}}{c}$$

The path which has minimum time is the one for which

$$dt/dx = 0$$

Differentiating with respect to x leads to

$$\frac{nx}{\sqrt{[h^2 + x^2]}} - \frac{n'(s - x)}{\sqrt{[d^2 + (s - x)^2]}} = 0$$

But since

$$\sin(i) = x/\sqrt{[h^2 + x^2]}$$

and

$$\sin(i') = (s - x)/\sqrt{[d^2 + (s - x)^2]}$$

the above equation reduces to

$$n\sin(i) = n'\sin(i')$$

which is Snell's law.

Snell's law can also be applied to the case of reflection. In this case, as shown in Figure 1.6, since the angle of reflection is equal to the angle of incidence

Fig. 1.8: Derivation of Snell's law using wave theory and plane waves.

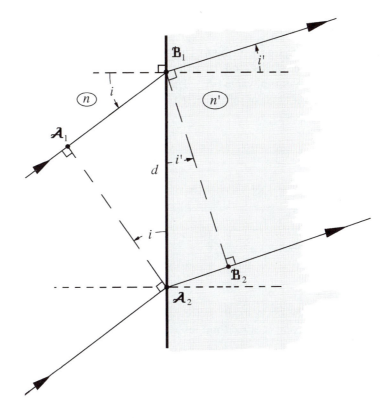

except that it is of the opposite sign, one must make the rule, that on reflection

$$n' = -n$$

Thus on reflection, the refractive index changes sign. We will pursue reflection further in Chapter 4.

1.5.1.2 Wave theory

Let us now look at a much simpler proof using the wave theory of light. Figure 1.8 shows two plane wavefronts $\mathcal{A}_1\mathcal{A}_2$ and $\mathcal{B}_1\mathcal{B}_2$, one just before and one just after refraction. The lines $\mathcal{A}_1\mathcal{B}_1$ and $\mathcal{A}_2\mathcal{B}_2$ are two rays joining these wavefronts and therefore from Section 1.3.2, their optical paths must be equal. Therefore

$$[\mathcal{A}_1\mathcal{B}_1] = [\mathcal{A}_2\mathcal{B}_2], \quad \text{that is} \quad n\mathcal{A}_1\mathcal{B}_1 = n'\mathcal{A}_2\mathcal{B}_2$$

Now since

$$\sin(i) = \mathcal{A}_1\mathcal{B}_1/d \quad \text{and} \quad \sin(i') = \mathcal{A}_2\mathcal{B}_2/d$$

it follows that

$$n' \sin(i') = n \sin(i)$$

which is Snell's law.

1.5.2   Critical angle

When a ray is moving from a medium of higher to a medium of lower refractive index, the ray will be totally (and internally) reflected if the angle of incidence (i) is greater than a critical value (i_{crit}). This value is called the **critical angle** and occurs when the angle of refraction (i') is $90°$. From Snell's law, it follows that

$$n \sin(i_{crit}) = n' \sin(90°) \tag{1.13}$$

that is

$$\sin(i_{crit}) = n'/n$$

For all angles in which

$$i > i_{crit}$$

all the energy in the ray is totally reflected and no energy enters the second medium. This situation is called **total internal reflection**.

1.5.2.1   Waveguides

A number of devices use the principle of total internal reflection to channel light over a certain distance. For example if a ray of light enters a parallel slab of material and strikes one of the faces at an angle greater than the critical angle, the ray will be totally reflected back into the slab and strike the opposite wall at the same angle, and once again be totally reflected. Thus the ray is constrained within the slab and travels down the slab, bouncing back and forth off the walls until it reaches the exit end. Optical fibres, which we discuss in Chapter 19, are based upon this principle but this is not the only application. Biological applications also exist and it is believed that the photoreceptors (light sensitive cells) of the eye use this principle.

1.5.3   Deviation of a ray

On refraction or reflection, the ray is deviated from its original direction. The angle through which the ray is effectively bent is called the **angle of deviation**. This angle is most important in prisms and the study of **aberrations**, which we define and discuss in the next chapter and in Chapter 5.

Exercises and problems

1.1   Calculate the energy in a photon of wavelength 555 nm in air.

ANSWER: 3.58×10^{-19} J

1.2   Calculate the frequency of light of wavelength 555 nm in air.

ANSWER: 5.40×10^{14} Hz

1.3   For radiation of wavelength 555 nm in a vacuum, calculate the wavelength in a medium of refractive index 1.523.

        ANSWER: 364.4 nm

1.4   Calculate the critical angle for light passing from water ($n = 1.333$) to air.

        ANSWER: 48.6°

1.5   Draw a ray travelling from the left and incident on the left vertical entrance face of a horizontal slab that has horizontal top and bottom surfaces, at an angle of incidence of θ. Let this ray enter the slab and be incident on the top surface. If the slab material has a refractive index of 1.333 and the ray has to be just totally internally reflected at the top face, calculate the maximum value of the angle θ.

        ANSWER: $\theta_{max} = 61.8°$

Summary of main symbols and equations

λ, λ_o	wavelengths in vacuum
$\lambda(n)$	wavelength in a medium with a refractive index n
$\lambda_d, \lambda_F, \lambda_C$	special wavelengths whose values are given in the symbols section at the front of the book
n_d	the refractive index at $\lambda = \lambda_d$
n_F	the refractive index at $\lambda = \lambda_F$
n_C	the refractive index at $\lambda = \lambda_C$
$E(z, t)$	electric field at a point z and time t
t	time
E_o	amplitude of electric field
ν	temporal frequency
δ	phase factor
c	velocity of light in a vacuum (2.99792×10^8 m/s)
v, v'	velocity of light in a medium
E	energy in a photon
h	Planck's constant (6.62620×10^{-34} J. s)
i, i'	angles of incidence and refraction or reflection
i_{crit}	critical angle of incidence
V_d	the V-value or a measure of the dispersion of an optical material

Section 1.2: Refractive index and dispersion

$$n = \frac{\text{velocity in a vacuum } (c)}{\text{velocity in a material } (v)} \tag{1.5}$$

$$V_d = \frac{(n_d - 1)}{(n_F - n_C)} \tag{1.9}$$

Section 1.5: Laws of refraction and reflection

$$n' \sin(i') = n \sin(i) \quad \text{(Snell's law)} \qquad (1.12)$$

$$\sin(i_{\text{crit}}) = n'/n \qquad (1.13)$$

References and bibliography

*Born M. and Wolf E. (1989). *Principles of Optics*, 6th (corrected) ed., Pergamon Press, Oxford.

*Cauchy A. L. (1836). Mémoire sur la dispersion de la lumiére. *Nouv. exerc. de math.* in Oeuvres Complètes d'Augustin Cauchy, 2nd Series, Vol. 10, Gauthier-Villars, Paris, 1895.

*Longhurst R. S. (1973). *Geometrical and Physical Optics*, 3rd ed., Longman, London.

*Schott. (September 1992). *Optical Glass Catalog*, Schott Glaswerke, Mainz, Germany.

*Smith W. J. (1990). *Modern Optical Engineering*, 2nd ed., McGraw-Hill, New York.

2
Image formation and ray tracing

2.0   Introduction

In this chapter, we will introduce the concept of an image forming system in its most general sense. By tracing rays from an object through the system, using Snell's law at each surface, we will show how to find the image of that object. When we decide to ray trace, there are two types of rays that we can choose, (a) **finite** or **real** rays and (b) **paraxial** rays. A finite or real ray is a general exact ray, and a paraxial ray is a special type of finite ray that is traced very close to the **optical axis**. One distinct advantage of paraxial rays is that their ray trace equations are much simpler than finite ray trace equations and hence are easier to apply. In this chapter, we will look at each of these two types and use the paraxial rays to develop a concept of the "ideal" image.

In the next chapter, Chapter 3, we will use the behaviour of paraxial rays to explore some of the properties of both simple and more complex optical systems. We will show that given the details of these properties, we can often find the ideal image positions and sizes without recourse to any type of ray tracing.

2.1   Image formation

We define an imaging optical system as a system consisting of any number of refracting or reflecting surfaces. Usually the surfaces will be spherical and we will assume that the centres of curvature of each of the spherical surfaces lie on a single line called the **optical axis**. Such a system is depicted schematically in Figure 2.1, but without any individual surfaces shown. The purpose of this system is to produce an image of a specified object, as shown in the diagram. The object may be two or three dimensional. We say that the object is in **object space** and the image is in **image space**.

We can think of the imaging of extended objects in terms of a more fundamental process in which the object is broken up into an infinite number of points and the system images each of these points separately. The final image is then the collection of the image points. Thus the optical system collects some of the light from each point in the object space, passes the light through the system by controlled refraction and/or reflection and focusses the light onto a point in image space.

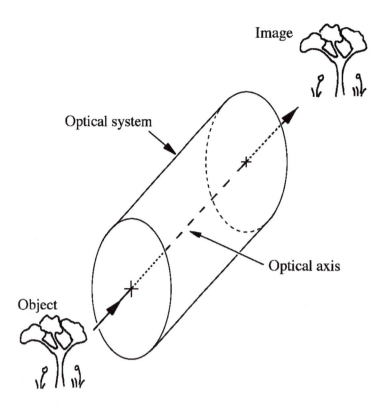

Image

Optical system

Optical axis

Object

Fig. 2.1: The formation of an image by an optical system.

We can describe this process in terms of either waves or rays. Figure 2.2a shows an object point at Q emitting waves with typical wavefronts as shown. For an isotropic medium, these wavefronts will be spherical. Ideally, they should maintain their spherical shape as they pass through the system and on exiting the system should still be spherical and concentric with their centres of curvature at the desired image point Q'. In the ray model of image formation, as shown in the same diagram, the system collects a beam of rays from the point Q, passes the beam through the system and "focuses" the beam onto the point Q'. That is, all the rays in the beam should be concurrent at Q'. The point Q' is the image of the point Q and these points are called **conjugate points**.

2.1.1 *Real and virtual images*

In Figure 2.2a, the image Q' of Q is formed after the last surface of the optical system. If a screen were placed in the plane of the image, the image would be seen formed on the screen. Such an image is called a **real image**. In many situations, the wavefronts emerging from the system are expanding from a point Q' to the left of the last surface or the rays are diverging from this point as shown in Figure 2.2b. In this case, the image cannot be formed on a screen and the image is said to be **virtual**. As we will see later, visual optical systems usually must produce a virtual image or an image at infinity, if the image is to be seen clearly by the eye.

Fig. 2.2a: The formation
of real images using wave
and ray theory.

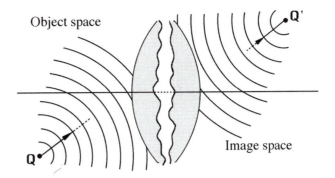

Object space

Q'

Image space

Q

Wave model

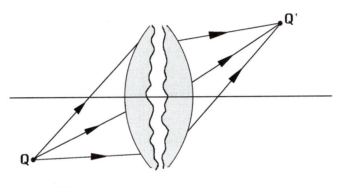

Q'

Q

Ray model

2.1.2   *The ideal image*

Ideally, the image of an extended object should have the same shape as the object, but may be different in size and orientation. However in Chapter 3, we will see that for three dimensional objects, the image cannot be the same shape as the object because the change in size (**magnification**) perpendicular to the axis is not the same as that along the axis. On the other hand, if we restrict the objects to being in a plane perpendicular to the optical axis, then we can design systems that will produce near perfect imagery.

We could define the perfect image as one in which the light distribution in the image has exactly the same form as that in the object. In real systems, these two light distributions are different because of (a) **aberrations** and (b) **diffraction**. Let us look further at aberrations.

In the real system, on leaving the system, the wavefronts, as shown in Figure 2.2a or b, would no longer be spherical and the exiting rays would not be concurrent at any point Q'. This is due to a phenomenon called **aberration**. Aberrations cause the exiting wavefronts to be distorted from their initial spherical shape and cause the rays to depart from their ideal paths. Thus in the presence of aberrations, light from a point object is not imaged as a point but instead as a spread out patch of light. As a result, the relative light distribution in the image

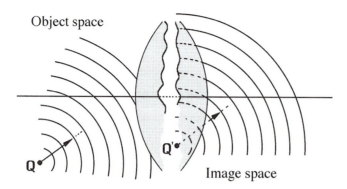

Object space

Image space

Fig. 2.2b: Same as
Fig. 2.2a but here the
images are virtual.

Wave model

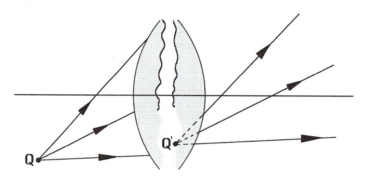

Ray model

space is no longer identical to that in the object space. In Chapter 5, we will
look at some of the factors that affect the aberrations of an optical system.

   Diffraction is due to the wave nature of light and is a phenomenon in which
light tends to "bend around corners", and thus light at the edge of a beam tends
to be deflected out of the beam. It also spreads out the light in the image of a
point and the amount of spread increases with a decreasing width of the beam
and an increase in wavelength. However, while an optical designer can reduce
the effects of aberrations by increasing the sophistication of the optical system,
diffraction is an inherent property of the wave nature of light and there is little
an optical designer can do to reduce it, except increase the beam diameter or
decrease the wavelength where possible. We discuss diffraction in greater depth
in Chapter 26.

   Thus a real system does not act like the ideal system. However, it is usually
the aim of optical designers to design a system that is as close as possible to
the ideal. In order to do this, they must know the properties of the ideal system.
While we can use either the wave or the ray theory to study aberrated or ideal
image formation, the ray approach is simpler in the earlier stages of the study.
Therefore from here on, we will mostly use the ray model.

   Any study of the construction, properties and performance of optical systems
is perhaps best begun by investigating the properties of the equivalent ideal (or
perfect) system. The most important property of an ideal system is that it is
aberration free; that is, imagery is point to point. Point to point imagery means

that all the rays leaving any object point and passing through the system are concurrent at the corresponding image point.

One further requirement of the ideal system is that objects that lie on a plane perpendicular to the optical axis are imaged onto a plane that is also perpendicular to the optical axis. We will show that if imagery is formed according to the rules of **paraxial optics**, this criterion is satisfied. Thus paraxial ray theory is very useful in defining the ideal system or ideal image formation. Paraxial theory is introduced and discussed in Section 2.4.

2.2 Ray tracing: General principles

Optical systems may consist of refracting and reflecting components, but this chapter is only concerned with refracting components. Reflection will be studied in Chapter 4. The most common refracting optical component is a single spherical refracting surface. A lens may be regarded as being constructed of two or more refracting surfaces and an optical system is often constructed from a number of surfaces or lenses (but may contain reflecting components such as mirrors). Therefore we will begin our study of image formation by looking at the refracting properties of a single spherical refracting surface and then extend the process to a system of any number of refracting surfaces.

We will begin by looking at the rules for tracing finite or exact rays through a single refracting surface. Such rays are generally aberrated. We will then proceed to look at rules for tracing a special kind of finite ray, a ray that is very close to the optical axis and free of aberration. This special kind of ray is called a **paraxial ray**. Apart from being aberration free, paraxial rays are useful for two reasons. Firstly, the paraxial ray tracing equations are much simpler than the general finite ray trace procedures and secondly, the ideal properties discussed in the preceding section are easily determined using these rays.

Ray tracing falls within the realm of geometric optics. According to geometric optics, light travels in straight lines or follows paths which are only deviated by reflections or refractions due to a change in refractive index. These paths are known as rays. We use the term "geometric optics" because the rays can be traced using conventional geometry and trigonometry. Ray tracing thus reduces to constructing an appropriate straight line to represent the ray, locating the point of intersection with the next surface, refracting the ray by applying Snell's law and representing the refracted ray by another straight line.

2.2.1 Direction of ray tracing

It is conventional to trace rays from left to right, that is have the object on the left of the optical system and the image nominally on the right. The word "nominally" is used because the image is not always formed on the righthand side of the optical system. As stated in Section 2.1.1 and shown in Figure 2.2a, the image is only physically formed on the right if it is a real image. On the other hand, if it is a virtual image, it appears to be formed either inside the optical system or on its left, as discussed in Section 2.1.1 and shown in Figure 2.2b. Because the ray tracing equations to be derived in the following sections are derived on the assumption than the rays will be going from left to right, care has to be taken in using them to trace from right to left. However, rarely is it necessary to trace rays from right to left.

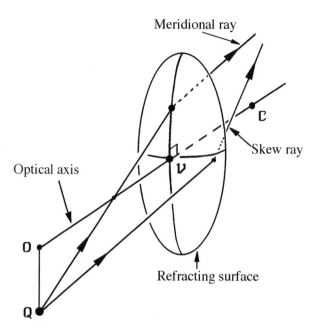

Fig. 2.3: Refraction at a single surface showing the distinction between meridional and skew rays.

2.2.2 *Number of rays to be traced to locate an image*

We will use ray tracing processes to locate the image of an object point. Since in the ideal system all rays that form an image are concurrent at the image point, only two rays should be needed to be traced and their intersection point in image space is the image point. However, as we will see in the next section, because of aberrations, finite rays are not concurrent at a single point in image space. In contrast, we will see that paraxial rays are concurrent and therefore only two paraxial rays need to be traced to locate an image point. Only by looking at the behaviour of finite rays can we fully appreciate paraxial rays.

2.2.3 *Meridional and skew rays*

Consider refraction by a single spherical surface as shown in Figure 2.3. The point v is the geometrical centre of the surface and the centre of the curvature of this surface is at c on the **optical axis**. The optical axis is usually defined for a system of surfaces in which the centres of curvatures of all the surfaces lie on the same line and this line is defined as the optical axis. For a single surface, this axis is undefined but for this special case, we will define it as the line $v c$. The point o lying on this axis is known as the axial object point.

Now consider a typical off-axis point Q, also shown in this diagram. We wish to examine the paths of rays from Q that are refracted by the surface. For ray tracing purposes, we consider two types of rays, (a) **meridional** rays and (b) **skew** rays. Meridional rays only lie in the plane defined by the optical axis and the off-axis line oQ. Before and after refraction, these rays lie in this same plane. Skew rays are those rays that do not lie in this plane, either before or after refraction. Furthermore, for a general optical system, no single plane contains the path of a skew ray. Thus meridional rays lie on a two dimensional surface but skew rays need to be described by three dimensional geometry and algebra.

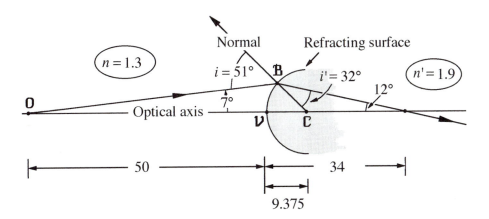

Fig. 2.4:   Refraction at a single surface for Example 2.1. The radius of curvature of the refracting surface is 9.375 units.

As a result, the tracing of meridional rays is much easier than the tracing of skew rays. Skew ray tracing, because of its complexity, is beyond the scope of this book. Procedures for skew ray tracing can be found in texts such as Welford (1986). Fortunately, for rotationally symmetric systems we can determine all of the ideal properties of an optical system from the behaviour of meridional rays, and we will now look at procedures for tracing this type of ray.

2.3 Ray tracing: Finite rays

In this section, we will restrict ourselves to the use of finite rays to determine the image position of an object point on the optical axis. By only considering axial object points we avoid the need to trace skew finite rays. Figure 2.4 shows an object point o to the left of a single refracting surface. Let us find the image of this point object formed by the surface. We will present two ray trace techniques, one using a graphical construction and the second based upon algebraic equations. We will begin with the graphical technique.

2.3.1 A graphical technique

Rays can be traced graphically by use of a ruler, protractor, compass and Snell's law [equation (1.12)], that is

$$n' \sin(i') = n \sin(i) \qquad (2.1)$$

To see how this is done, let us consider a particular example.

Example 2.1: Consider the case shown in Figure 2.4, where a point object at o on the optical axis is 50 units from a spherical surface with a radius of curvature of 9.375 units. All rays traced from a point on the optical axis are effectively meridional rays. Let us follow the path of a typical ray refracted by this surface, using simple trigonometrical techniques. Let us take a ray from o , inclined at a convenient angle to the axis (here taken as 7°), and draw this ray to the surface and then construct the normal at the point of intersection. The normal is the line drawn from the centre of curvature c through the point $\mathcal{B}$ of intersection. Using a protractor, we can measure the angle of incidence

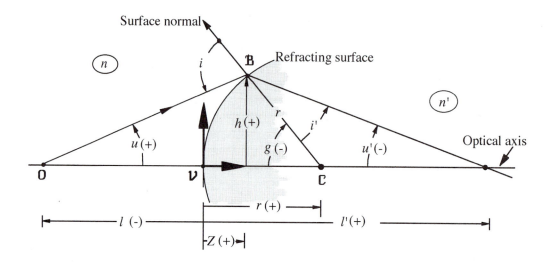

Fig. 2.5: Diagram for the development of finite and paraxial refraction equations at a surface with sign convention.

i. In this example, the measured angle of incidence is 51°. We then calculate the angle of refraction i' using Snell's law, that is equation (2.1), where in the example, as shown in the diagram, $n = 1.3$ and $n' = 1.9$. This gives the refraction angle $i' = 32°$. We then draw a line from the point B and inclined at 32° to the normal. This line represents the refracted ray which crosses the optical axis at a distance of 34 units from the surface vertex and makes an angle of 12° to the axis.

If we have more than one surface, we can repeat the procedure at each surface in turn. However, this method is limited by the precision with which one can read and set angles using a protractor. This would not be better than about ±0.5° and the errors will accumulate with increasing number of surfaces. For more accurate ray tracing, we can trace rays using algebraic or numerical methods and the accuracy of the numerical methods is only limited by the number of decimal places used in the calculations.

2.3.2 *Algebraic or numerical ray tracing*

Before one can investigate the problem algebraically, it is first necessary to establish a **sign convention**. We will use Figure 2.5 for this purpose. Throughout this book, the common cartesian and trigonometric sign conventions will be used with the axis origin placed at the vertex V of the surface (or lens) under consideration. This leads to the following rules for the sign convention:

(a)   Distances to the right and above the axis origin are positive and those to the left and below are negative.

(b)   The angle between a ray and the optical axis will be positive, if on rotating a line from the optical axis to the ray by the quicker of the two routes, the rotation is anti-clockwise. Otherwise the angle will be negative. This is consistent with the normal trigonometric convention.

(c)   For surface radii of curvature, if the centre of curvature c is to the right of the surface vertex V, the radius of curvature is positive, and negative otherwise.

Thus in the diagram, the angles u' and g, and the distance l are negative and the angle u, the ray height h, the distance l' and radius of curvature r are positive.

The signs of the angles i and i' are not explicitly defined by these rules. However, if we look at the angles of the rays relative to the surface normal at $\mathcal{B}$, the angles are anti-clockwise and therefore to be consistent with the above rules, we will take the angles as positive if as shown in the diagram.

Using this sign convention and Figure 2.5, we will now develop a procedure for accurate numerical finite ray tracing.

Step 1: Calculation of the angle of incidence i. In Figure 2.5, the sine rule applied to triangle $O\mathcal{B}C$ gives

$$\frac{r}{\sin(u)} = \frac{r - l}{\sin(i)}$$

noting that l is negative and r is positive in this diagram. Solving for angle i gives

$$i = \arcsin\{(r - l)[\sin(u)/r]\} \tag{2.2a}$$

Step 2: Calculation of the angle of refraction i'. The angle of refraction i' is then given by Snell's law [equation (2.1)].

Step 3: Calculation of the angle g. Now using triangle $O\mathcal{B}C$ again and the rule that the external angle is the sum of the two internally opposite angles we have

$$i = u - g \tag{2.2b}$$

where the negative sign occurs on g because it is numerically negative in the diagram. Therefore

$$g = u - i \tag{2.2c}$$

Step 4: Calculation of the angle u'. Now using triangles $O'\mathcal{B}C$ and the rule that the external angle is the sum of the two internally opposite angles we have

$$-g = i' - u'$$

where the negative sign occurs on g and u' because they are numerically negative in the diagram. From these equations, we have

$$u' = i' + g \tag{2.2d}$$

Step 5: Calculation of the distance l'. Finally, applying the sine rule once again, but this time to triangle $O'\mathcal{B}C$, we have

$$\frac{l' - r}{\sin(i')} = -\frac{r}{\sin(u')}$$

The minus sign occurs on the righthand side because u', as shown in the diagram, is negative. This equation gives

$$l' = r\{1 - [\sin(i')/\sin(u')]\} \qquad (2.2e)$$

Step 6 (arbitrary) : Calculation of the intersection height with the surface. The intersection height h of the ray with the surface can be found from the equation

$$h = -r\sin(g) \qquad (2.2f)$$

The minus sign occurs on the righthand side because g, as shown in Figure 2.5, is negative.

Example 2.2: Let us use these equations to check the trigonometric ray trace done in Example 2.1 and Figure 2.4 in the previous section.

Solution: Taking $l = -50$ units, $r = 9.375$ units and $u = 7°$, from equation (2.2a) we have first

$$i = 50.5°$$

Using this value, we have in turn

from equation (2.1), $\quad i' = 31.9°$
from equation (2.2c), $\quad g = -43.5°$
and from equation (2.2d), $\quad u' = -11.6°$

Equation (2.2e) then gives

$$l' = 33.9 \text{ units}$$

The trigonometric ray trace results of Example 2.1 and shown in Figure 2.4 compare well with these values.

If we want the intersection height h at the surface, equation (2.2f) gives

$$h = 6.46 \text{ units}$$

Using either of these techniques, we could trace a number of rays from o, each with a different intersection height h with the surface. If we did this, for the case shown in Figure 2.4, we would end up with a situation shown schematically in Figure 2.6a. We first note that the refracted rays are not concurrent at a single point and the distance l' decreases with increase in ray height h. This is due to an aberration called **spherical aberration**. There are a number of other aberrations, but these will not be discussed until Chapter 5. For the situation already discussed in Examples 2.1 and 2.2, the algebraic ray tracing method was used to calculate l' for various values of ray height h and the results are plotted in Figure 2.6b. Of course we cannot calculate the value of l' for $h = 0$. However, we find that in the limit h approaches zero, the distance l' approaches

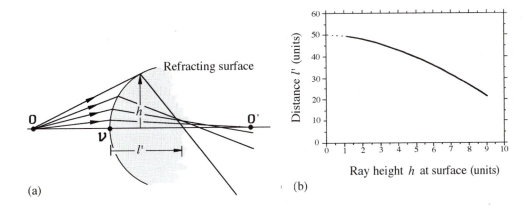

Fig. 2.6: Refraction of finite or real rays at a single surface, showing the effect of spherical aberration. Fig. 2.6a is not to scale.

a finite limit, in this case 50 units, and this limit defines the intersection point o', shown in Figure 2.6a for rays traced limitingly close to the axis. These rays, which travel limitingly close to the optical axis, are called **paraxial rays**. Rays traced in the vicinity of the edge of the surface, in contrast, are called **marginal rays**. The point o' is the **paraxial image** of o. We will investigate the properties of paraxial rays in the next section.

Ray tracing through a system of surfaces

The above procedure can readily be applied to tracing a finite or exact ray through a system of any number of surfaces. After refraction, the ray is "transferred" to the next surface, where the next refraction takes place. At this surface the new angle u is the old angle u' and the new distance l is given in terms of the old distance l' and the distance d to the next surface by the equation

$$l = l' - d \tag{2.2g}$$

and this is a "transfer" equation.

2.4   Ray tracing: Paraxial rays

The ideal properties of a particular optical system can be obtained by finding the positions of the paraxial images of various object points, for example the position of the image of an axial object at infinity. We will show that paraxial rays from any object point are concurrent in image space and thus aberration free. The concurrency property of paraxial rays means that to locate the image point, only two rays need be traced from any object point. The intersection point of these two rays in image space gives the location of the image point. For an axially symmetric system and an object point on the optical axis, the optical axis can be regarded as one of these rays and thus in practice only one ray need be traced. To prove this concurrency property of paraxial rays, we must first develop equations and procedures for tracing paraxial rays and examine their properties. The first of these is the **paraxial refraction equation**, which as the name implies is for refracting a paraxial ray at a surface.

2.4.1   *The paraxial refraction equation*

To begin the development of the laws governing the behaviour of paraxial rays, we return to Figure 2.5. In this case, we need to find the position of the paraxial image o' of an axial object point o for a system consisting of a single refracting surface. The object is at a distance l from the surface vertex $\mathcal{V}$. To locate the image position o', only one ray has to be traced, the ray $o\mathcal{B}o'$. The location where this ray crosses the optical axis (i.e. meets the second ray) is the position of the image at o'. As shown in this diagram, this ray makes an angle i with the surface normal at $\mathcal{B}$ and an angle i' after refraction. If the object space medium has a refractive index n and the image space has an index n', then angles i and i' are related by Snell's law, which is given by equation (2.1). This equation applies to all finite or real rays, but if we make the assumption that the ray is close to the axis, then the angles u, i, i', u' and g will be small, and for small angles one can make use of the approximation

$$\sin(x) \approx x \approx \tan(x) \tag{2.3}$$

For this approximation to be valid, the angle x must be in radians. Making this approximation, which will be explained more fully towards the end of this subsection, is the first step in developing the paraxial optics laws of ray tracing. Thus in the paraxial approximation, Snell's law, equation (2.1), can be written

$$n'i' = ni$$

If we now eliminate i and i' from this equation using equations (2.2b) and (2.2d), we have

$$n'(u' - g) = n(u - g) \tag{2.4}$$

or

$$n'u' - nu = g(n' - n)$$

Now, the angle g is related to the ray height h and surface radius of curvature r by equation (2.2f), but in the paraxial approximation defined by equation (2.3), this can be written

$$g = -h/r$$

Using the curvature C instead of radius of curvature r, where C is defined as

$$C = 1/r$$

it follows that

$$g = -hC$$

Thus equation (2.4) can be written in the form

$$n'u' - nu = -hC(n' - n) \qquad (2.5a)$$

which is usually referred to as the **paraxial refraction equation**.

The quantity $C(n' - n)$ on the righthand side of the paraxial refraction equation is known as the **power** of the surface and will be denoted by the symbol F. We can now write the above paraxial refraction equation in the alternative form

$$n'u' - nu = -hF \qquad (2.5b)$$

where

$$F = C(n' - n) \qquad \text{surface power} \qquad (2.6)$$

Returning to Figure 2.5 and given only the position of the object point o, this equation cannot be used to find the image o', until a specific ray is chosen. It will be proved later that any paraxial ray traced from o will give the same image point at o'. Hence the choice of ray is arbitrary. From the diagram, it can be seen that u, h and l are exactly related by the equation

$$h/(l - z) = -\tan(u) \qquad (2.7)$$

where the negative sign in front of $\tan(u)$ is necessary because h and u are positive and l is negative in the diagram. In the paraxial approximation, equation (2.3), the tangent of an angle is also replaced by the angle itself, and thus

$$h = -u(l - z) \qquad (2.7a)$$

So far, the paraxial approximation has not been fully explained. A statement or rule of this approximation is as follows:

In all expansions of power series functions, only the first order terms are taken; higher orders are ignored.

This explains the rule for replacing sines and tangents of angles by the angles. The sine function is expandable in the following power series

$$\sin(x) = x - (x^3/3!) + (x^5/5!) + \cdots \qquad (2.8a)$$

and the tangent function is expandable in a similar power series

$$\tan(x) = x + (x^3/3) + (2x^5/15) + \cdots \qquad (2.8b)$$

where, in both cases, x must be in radians. Thus if terms higher than the first in x are ignored, it is clear that both $\sin(x)$ and $\tan(x)$ are identical to the angle x in radians.

Definition of the paraxial region: *This is the region close to the optical axis where expanding functions as a power series but only up to the first order terms, with the higher order terms being neglected, does not lead to a significant error.*

If one now turns to the quantity z in equations (2.7) and (2.7a), it can be expressed as a power series in h and C. The surface in Figure 2.5 is spherical (or circular in cross-section) and can be represented in cross-section by the equation

$$h^2 + (z - r)^2 = r^2$$

Solving for z leads to the following

$$z = r\{1 - \sqrt{[1 - (h^2/r^2)]}\}$$

Using the binomial expansion and replacing the radius of curvature r by $C (= 1/r)$, we can express z as

$$z = Ch^2/2 + \text{terms of } h^4 \text{ and higher} \tag{2.9}$$

This equation shows that the lowest order term is a h^2 term, which is second order in h. Now in the paraxial approximation, all terms in the expansion of second order or higher are neglected and therefore z is taken as zero. Thus equation (2.7a) can now be expressed simply as

$$h = -ul \tag{2.10a}$$

and similarly

$$h = -u'l' \tag{2.10b}$$

One consequence of the above approximation is that the height h of intersection of the ray with the surface can now equally be regarded as the height of intersection at the vertex plane, which is the plane tangent to the surface vertex at v.

Equation (2.10a) can now be used to select a particular paraxial ray leaving the object point. Since the value of l is usually known, one has only to choose an arbitrary value of either u or h and then find the value of the other quantity from equation (2.10a). Thus a typical paraxial ray trace involves firstly selecting a ray using (2.10a) and then applying equation (2.5a or b) to find the corresponding value of u'. The distance l' of the image point o' from the surface vertex v is then given by equation (2.10b). It must be stressed that in these calculations, the correct use of the sign convention is most important.

2.4.1.1   *Units*

Distances

In the metric system of units, the distances are most commonly expressed in millimetres (mm), centimetres (cm) or metres (m). Thus the corresponding units for curvature and power would be mm^{-1}, cm^{-1} or m^{-1}. In ophthalmic optics, the unit of power is the **dioptre**, denoted by the symbol D, which is equivalent to mm^{-1}. However, we will not use the symbol D to denote the dioptre in this book, because D is used for other purposes.

Angles

Since the paraxial approximation

$$\sin(x) = \tan(x) = x$$

requires the angle x to be in radians, one may initially assume that the angles u and u' can be interpreted also to be in radians. However, this is only true for very small angles. Paraxial equations and concepts are used well outside the paraxial region, and in these cases, the angles are large. For example, let us suppose in Figure 2.5 that the object space ray angle u has a value of 2. If this were a true radian measure, the corresponding angle in degrees would be $114.59°$ and since this value is greater than $90°$, the ray should be sloping backwards or to the left. This is obviously incorrect, showing the limitation of interpreting the values of the paraxial angles u and u' as radians for large angles. The physical interpretation of paraxial quantities is discussed further in Section 2.4.4.

Example 2.3: Let us use the paraxial refraction equation to find the paraxial image in the situation shown in Figure 2.4 and suppose the distances are in millimetres. Thus in this example

$$n = 1.3, \qquad C = 1/9.375 = +0.106667 \text{ mm}^{-1},$$

$$l = -50 \text{ mm} \quad \text{and} \quad n' = 1.9$$

Solution: Choice of ray. Let us choose a value of $u = +1$. From equation (2.10a), the corresponding value of h will be $+50.0$ mm.
   Refraction at the surface. From equation (2.6), the surface power is

$$F = +0.106667 \times (1.9 - 1.3) = 0.064000 \text{ mm}^{-1}$$

The paraxial refraction equation (2.5b) can now be written as

$$1.9 \times u' - 1.3 \times (+1) = -50.0 \times 0.064000$$

that is

$$u' = -1.0000$$

   Image distance. The distance l' of the image from the surface vertex is then given by equation (2.10b)

$$l' = -h/u' = -50.0/(-1.0000)$$

that is

$$l' = +50.0 \text{ mm}$$

Thus the paraxial image of the point o is formed 50 mm to the right of the surface and this result confirms the limiting value of $+50.0$ quoted in Example 2.2.

It is necessary to comment here on the above choice of the value of $u(+1)$, which is of course well outside the region of the paraxial approximation. The value of $+1$ was taken for numerical simplicity. However, any paraxial ray, traced from the same axial object point, will give the same axial image position, whether the ray is within or beyond the paraxial region. The reader can check this by repeating the calculation with any other arbitrary values of u.

We can also prove this algebraically. The proof relies on combining the refraction equation (2.5b) and equations (2.10a and b). Dividing both sides of equation (2.5b) by h gives

$$\frac{n'u'}{h} - \frac{nu}{h} = -F$$

Now since $l = -h/u$ from equation (2.10a), and $l' = -h/u'$ from equation (2.10b), the above equation can be written

$$\frac{n'}{l'} - \frac{n}{l} = F \tag{2.11}$$

This equation is a special form of the paraxial refraction equation, which is now independent of u, u' and h. Therefore it is also independent of any particular paraxial ray traced from the object point at o and hence the final value of l' is independent of the particular ray chosen. Thus for a single refracting surface, all the paraxial rays from an axial object point are concurrent at the same axial image point. Thus we have just shown that the surface is perfect for axial imagery with paraxial rays. If we now have a system consisting of a number of surfaces, the image formed by any particular surface becomes the object for the next surface. Since we have just shown that the paraxial imagery by one surface is perfect, it then follows that the imagery by any number of surfaces is also perfect, within the paraxial approximation.

For systems consisting of more than one refracting surface, the paraxial refraction equation (2.5b) alone is insufficient for ray tracing. Consider a two surface system as shown in Figure 2.7. The ray refracted at the first surface intersects the second surface at a height h'. The refraction equation can then be used to find the final image space ray angle u'' by applying the refraction equation again at the second surface, but the refraction equation at this surface involves the ray height h', and this must be found before the final angle u'' can be determined. The equation for this calculation is the paraxial transfer equation.

2.4.2   *The paraxial transfer equation*

From Figure 2.7, it may be deduced that

$$h' - h = [d - (z_1 - z_2)]\tan(u')$$

which is an exact equation, if h and h' are in fact the ray heights at the surfaces. Now according to the rules of paraxial optics, (a) both z_1 and z_2 are zero as

Fig. 2.7: Diagram for the
derivation of the paraxial
transfer equation.

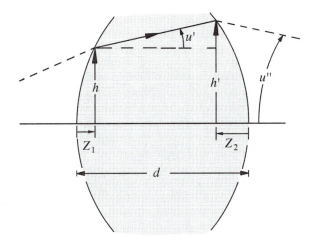

discussed in relation to equations (2.9) and (2.10a and b), since they are at least second order in ray height, and (b) $\tan(u')$ can be replaced by u'. Hence we can now write the above equation in the paraxial form

$$h' = h + u'd \tag{2.12}$$

This equation is known as the **paraxial transfer equation**.

The two equations, the paraxial refraction equation (2.5a or b) and the paraxial transfer equation (2.12), are a very useful pair of equations for tracing paraxial rays through an optical system consisting of any number of refracting surfaces. They are ideally suited for programmable calculators or computers because they can be built into a repeating loop.

2.4.3 Ray tracing through a system of surfaces

We will now show how to use the paraxial refraction and transfer equations to find the axial image position for any particular optical system, but we will change the notation slightly, to make the equations more suitable for tracing a paraxial ray through a multi-surface system.

In general, we will be given the distance (l_1) of the object o from the front surface or vertex $\mathcal{V}$ of the system consisting of k surfaces as shown in Figure 2.8. Before we can use the paraxial refraction and transfer equations, we must choose a particular ray, specified in terms of the angle u_1 the ray makes with the axis and the intersection height h_1 with the first surface. These quantities are related by equation (2.10a), now written as

$$h_1 = -u_1 l_1 \tag{2.13a}$$

If u'_k is the final angle in image space and h_k is the ray intersection height with the last surface, then the distance l'_k of the image o' from the last surface vertex $\mathcal{V}'$ is given by the equation (2.10b), here written as

$$l'_k = -h_k / u'_k \tag{2.13b}$$

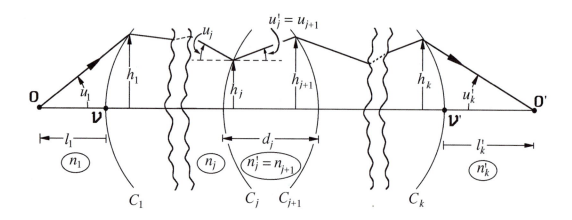

Once we have chosen the ray, we apply the paraxial refraction equation (2.5b) at each surface in turn. At the jth surface, this equation can be written as

$$n'_j u'_j - n_j u_j = -h_j F_j \tag{2.14}$$

where

$$F_j = C_j(n'_j - n_j) \tag{2.15}$$

$$u_{j+1} = u'_j \tag{2.16a}$$

and

$$n_{j+1} = n'_j \tag{2.16b}$$

Between surfaces, we must apply the paraxial transfer equation (2.12), here written as

$$h_{j+1} = h_j + u'_j d_j \tag{2.17}$$

Some of these symbols are also explained in Figure 2.8.

To show how these equations are used to trace a ray through a general system, let us apply them to a particular example.

> **Example 2.4:** Consider the Gullstrand number 2 accommodated eye (Appendix 3) shown in Figure 2.9, and suppose the position of the conjugate of the retina has to be found, if the retina is taken to be 23.896 mm from the corneal vertex, that is at a distance of 16.696 mm from the vertex v of the lens. This conjugate is the "near point of accommodation" for this eye. In other words, the retinal point o is the object and we want to know the position o' of its image. In this eye, the refractive index of the vitreous and aqueous is 4/3, which for computational purposes will be taken as 1.3333.

Fig. 2.8: Diagram showing the notation for tracing a paraxial ray through a system of k surfaces. C_j and C_{j+1} are the surface curvatures of surface j and $j+1$, respectively.

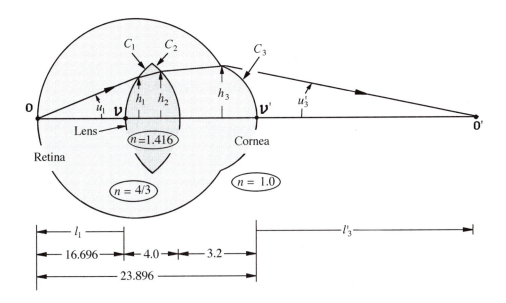

Fig. 2.9: Application of the paraxial refraction and transfer equations to ray tracing through the Gullstrand number 2 schematic eye (Example 2.4), in order to determine the position of the axial image position o' of the axial retinal point o. $C_1 = +0.2 \, \text{mm}^{-1}$, $C_2 = -0.2 \, \text{mm}^{-1}$ and $C_3 = -0.1282 \, \text{mm}^{-1}$. Other units are also in millimetres.

Solution: *Choice of ray.* The choice of paraxial ray is arbitrary. Let us take $u_1 = +0.1$, and since $l_1 = -16.696$ mm, it then follows from equation (2.13a) that $h_1 = +1.6696$ mm.

Refraction at surface 1. The paraxial refraction equation (2.14) can be used to find u'_1, with equation (2.15) used to find the surface power at the first surface. At this surface, $n'_1 = 1.4160$, $n_1 = 1.3333$ and $C_1 = 0.20 \, \text{mm}^{-1}$, therefore

$$F_1 = 0.20 \times (1.4160 - 1.3333) = 0.01654 \, \text{mm}^{-1}$$

At this surface, $u_1 = +0.1$, $h_1 = 1.6696$ mm, and so from equation (2.14)

$$1.4160 \times u'_1 - 1.3333 \times (+0.1) = -1.6696 \times 0.01654$$

Thus

$$u'_1 = +0.074657$$

Transfer to surface 2. Using this value of u'_1 and the paraxial transfer equation (2.17), the ray height h_2 at the second surface can be found. In this case, $h_1 = 1.6696$ mm, $u'_1 = +0.074657$ and $d_1 = 4.0$ mm so we have

$$h_2 = 1.6696 + 0.074657 \times 4.0 = 1.968229 \, \text{mm}$$

Refraction at surface 2. The refraction equation (2.14) can now be applied once again, this time at the second surface. At the second surface, the power, from equation (2.15), is

$$F_2 = (-0.20) \times (1.3333 - 1.4160) = 0.01654 \text{ mm}^{-1}$$

With $u_2 = u_1' = +0.074657$, the paraxial refraction equation (2.14) at this surface is

$$1.3333 \times u_2' - 1.4160 \times 0.074657 = -1.968229 \times 0.01654$$

and thus

$$u_2' = +0.054872$$

Transfer to surface 3. Transferring to find the height h_3 at the corneal surface, using equation (2.17), gives

$$h_3 = 1.968229 + 0.054872 \times 3.2 = 2.143819 \text{ mm}$$

Refraction at surface 3. At this surface, from equation (2.15)

$$F_3 = (-0.1282) \times (1.0 - 1.3333) = 0.0427291 \text{ mm}^{-1}$$

With $u_3 = u_2' = +0.054872$, refracting at this surface with equation (2.14) gives

$$1.0 \times u_3' - 1.3333 \times 0.054872 = -2.143819 \times 0.0427291$$

and finally

$$u_3' = -0.018443$$

Image distance. The above ray trace results have been put in tabular form in Table 2.1. The image distance l_3' as shown in Figure 2.9 is now found from the equation (2.13b), with $k = 3$, that is

$$l_3' = -h_3/u_3' = -2.143819/(-0.018443) = +116.24 \text{ mm}$$

Thus the near point of accommodation of the eye is 116.2 mm in front of the corneal vertex and thus the eye forms a real image of the retina at this distance.

Note: Because paraxial ray tracing is an iterative process, errors due to any rounding off at each step will build up progressively through the calculation. Therefore intermediate calculations should be carried out with at least two more decimal places than that set by the precision of the data or required precision of the final answer.

2.4.4   *Interpretation of paraxial ray heights and angles*

The ray height h at a surface may be equally interpreted as the ray height at the vertex plane or at the refracting surface. This is because within the paraxial

Table 2.1. *Ray trace results for Example 2.4*

Surface	u	h (mm)
	+0.1	
1		+1.669600
	+0.074657	
2		+1.968229
	+0.054872	
3		+2.143819
	−0.018443	

approximation, these two ray heights are equal.

The ray angle u is a little more difficult to interpret. However, if we recall the derivation of the paraxial transfer equation (2.12), and assume the ray height is measured at the surface vertices, then the paraxial angle u is the tangent of the real angle.

2.4.5 A linear property of paraxial rays

The above paraxial raytrace equations can be expressed in a **matrix algebra** form and this leads to an optical system being represented by a set of 2×2 refraction matrices (one for each surface) and 2×2 transfer matrices. This matrix representation is discussed fully in Section A1.3 in Appendix 1.

In that section, we use the matrix representation to show that for any paraxial ray, refracted by an optical system, the height h at any plane in the system and the angle u on either side of the surface are related to the height h' at any other plane and the angle u' on either side, by the simple pair of linear equations

$$u' = \mathfrak{A}u + \mathfrak{B}h \qquad\qquad (2.18a)$$

$$h' = \mathfrak{C}u + \mathfrak{D}h \qquad\qquad (2.18b)$$

where for any particular system, the quantities $\mathfrak{A}$, $\mathfrak{B}$, $\mathfrak{C}$ and $\mathfrak{D}$ have numerical values that depend only on the positions of the two surfaces, the respective side at which the two angles are measured and the system parameters. These equations apply to any optical system. The values of $\mathfrak{A}$, $\mathfrak{B}$, $\mathfrak{C}$ and $\mathfrak{D}$ are not independent because we show in Appendix 1 that they must satisfy the relationship

$$\mathfrak{A}\mathfrak{D} - \mathfrak{B}\mathfrak{C} = n/n' \qquad\qquad (2.19)$$

where n and n' are the refractive indices in the respective spaces.

In any particular case, the numerical values of $\mathfrak{A}$, $\mathfrak{B}$, $\mathfrak{C}$ and $\mathfrak{D}$ can be found by a number of different methods, and in Appendix 1 we describe three. We show that we can find the values of these quantities by tracing no particular ray, or one or two rays. Let us look at their values for the object and image planes.

2.4.5.1 *Special case of object and image planes*

Consider the ray from the axial point o on the object plane that passes through the axial point o' on the image plane. For this ray

$$h = h' = 0$$

and therefore equations (2.18) become

$$u' = \mathfrak{A}u + \mathfrak{B}0$$

$$0 = \mathfrak{C}u + \mathfrak{D}0$$

that is

$$u' = \mathfrak{A}u \qquad\qquad (2.20a)$$

$$0 = \mathfrak{C}u \qquad\qquad (2.20b)$$

for any value of u (or u'). These equations can only be satisfied if

$$\mathfrak{C} = 0$$

Thus for the object and image planes, equations (2.18a and b) and (2.19) become

$$u' = \mathfrak{A}u + \mathfrak{B}h \qquad\qquad (2.21a)$$

$$h' = \mathfrak{D}h \qquad\qquad (2.21b)$$

and

$$\mathfrak{A}\mathfrak{D} = n/n' \qquad\qquad (2.22)$$

These equations can be used to confirm readily that all paraxial rays from an axial object point and imaged by a system are all concurrent at one point on the optical axis in image space: the paraxial image point. This can be done as follows. For any ray from the axial object point, the ray height h at the object plane must be zero and it follows from equation (2.21b) that the ray height h' at the image plane must be zero. We will now use these equations to examine some properties of off-axis imagery.

2.5 Off-axis imagery

We have just demonstrated that all paraxial rays coming from an axial object point o, and traced through the system, are concurrent at one point, the paraxial image point o'. If we take an off-axis object point ϱ which lies on a plane passing through o and perpendicular to the optic axis, what is the nature of its image and where is it formed? We will now show that all meridional paraxial rays from any point ϱ on the perpendicular object plane through o are concurrent at some point ϱ' on the perpendicular image plane through o'.

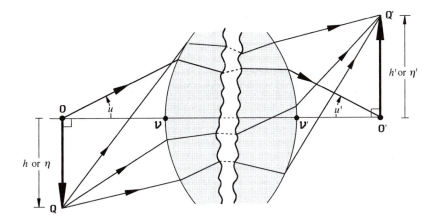

Fig. 2.10: Concurrency of paraxial rays from an off-axis image point and the position of the image.

Figure 2.10 shows an off-axis point Q on the perpendicular plane through O. The point Q is at a height h. The intersection height h' with the image plane is given by equation (2.21b) and this equation gives the same height h' for all values of the angle of rays leaving the object plane. Therefore all meridional rays leaving an off-axis object point on a perpendicular object plane at O are concurrent at a point on the perpendicular image plane through O', as shown in the diagram.

Thus for meridional rays, we have proved that all points on a plane perpendicular to the optical axis in object space are imaged as points on a plane also perpendicular to the optical axis in image space.

2.5.1 Image size and transverse (or lateral) magnification

In equation (2.21b), the quantities h and h' can be regarded as the heights of an extended object and its image, respectively. Thus the quantity $\mathfrak{M}$ in that equation, which is the ratio (h'/h), is also the transverse magnification M for that pair of conjugate planes and we often need to know its value. Since, in general, we use the symbols h and h' as paraxial ray heights, let us replace them by the symbols η and η' when they refer, as they do here, to object and image heights, as shown in Figure 2.10. Now, we define **transverse magnification M** as

$$M = \eta'/\eta \tag{2.23}$$

and it follows from the above that

$$M = \mathfrak{M} \tag{2.23a}$$

which shows that the magnification of the image is independent of the size of the object and therefore all parts of the image have the same magnification. This is another property of "ideal" imagery and we will see in Chapter 5 that for real images the magnification varies across the image, leading to a distortion of the image.

We will now show that we can express the transverse magnification M in terms of the paraxial angles u and u' of the ray that is used to locate the position of the paraxial image O' of O. This ray and the angles are shown in Figure 2.10.

If we use equation (2.22) to eliminate $\mathcal{B}$ from equation (2.23a), we have

$$M = n/(n'\mathcal{A})$$

For the ray coming from the axial object point o, equations (2.20) are applicable and we can use equation (2.20a) to eliminate $\mathcal{A}$ from this equation, to get

$$M = \frac{nu}{n'u'} \qquad\qquad (2.23b)$$

which shows that the transverse magnification can be found from the same ray trace used to locate the axial image position.

Exercises and problems

2.1   By using a compass, ruler and protractor, trace an exact ray from an axial point 10 cm to the left of a surface of radius of curvature +10 cm, refract at this surface and trace to where the ray crosses the optical axis. The ray should make an angle of 10° to the axis in object space. Take the object and image space refractive indices as 1.4 and 1.9, respectively.
      Repeat with a ray making an angle 20° to the axis.

2.2   Repeat the task in problem 1 for the 10° ray but this time using exact ray trace equations. Plot the path of the ray on the same diagram.

2.3   Taking the same situation as given in problems 2.1 and 2.2 but only for the 10° ray, add a second surface that has a radius of curvature −15 cm situated at a distance of 7.5 cm from the first and to the right and take the refractive index to the right of this surface as 1.0. Now continue the ray trace and find where the ray crosses the axis after refraction by the second surface.

       ANSWER: Ray crosses axis at a distance −333.1 cm from last surface

2.4   By any means find the surface power (F) in the following cases:

	C	n	n'	Answers (F)
(a)	0.1	1.0	1.5	0.05
(b)	0.3	1.0	1.6	0.18
(c)	−0.5	1.2	1.7	−0.25

2.5   Use the paraxial refraction equation to find the positions (l') of the paraxial images o' in the following cases.

	C	n	n'	l'	Answers (l')
(a)	0.1	1.3	1.5	−10	−13.6
(b)	−0.3	1.0	1.52	−100	−9.15
(c)	0.03	1.5	1.8	infinity	+200.0

2.6   Spherical glass beads are mixed in paint to increase the reflectivity of white lines on the road and white material used in projection screens. The purpose of the beads is to reflect the light towards the source no matter what the direction of the source. Such materials are called retro-reflectors. For a source at infinity, calculate the optimum refractive index of the glass beads if they are to act as perfect retro-reflectors.

ANSWER: 2.0

2.7 Given the optical system in the following table:

	Refractive indices	Surface curvatures (cm^{-1})	Surface separations (cm)
Object space	1.00		
		+0.01000	
	1.55		5.000
		−0.05000	
	1.75		3.000
		+0.05000	
Image space	1.33		

use the paraxial ray trace equations to find the position and magnification of an object placed 30 cm to the left of the first surface.

ANSWERS: The image is formed −24.40 cm from the last surface; therefore the image is virtual. The transverse magnification is 0.529.

Summary of main symbols and equations

i, i' angles of incidence and refraction relative to surface normal
j refers to the jth surface in an optical system
k number of surfaces in an optical system
g angle of normal relative to the optical axis
d surface separation
c centre of curvature of a spherical surface

Section 2.4: Paraxial ray tracing

$$n'u' - nu = -hF \tag{2.5b}$$

$$F = C(n' - n) \tag{2.6}$$

$$h = -ul \quad \text{and} \quad h = -u'l' \tag{2.10a,b}$$

$$\frac{n'}{l'} - \frac{n}{l} = F \tag{2.11}$$

$$h' = h + u'd \tag{2.12}$$

Section 2.4.3: Ray tracing through a system of surfaces

$$h_1 = -u_1 l_1 \tag{2.13a}$$

$$l'_k = -h_k/u'_k \tag{2.13b}$$

$$n'_j u'_j - n_j u_j = -h_j F_j \tag{2.14}$$

$$F_j = C_j(n'_j - n_j) \tag{2.15}$$

$$u_{j+1} = u'_j \tag{2.16a}$$

$$n_{j+1} = n'_j \tag{2.16b}$$

$$h_{j+1} = h_j + u'_j d_j \tag{2.17}$$

Section 2.5: Off-axis imagery

$$M = \eta'/\eta \quad \text{(definition)} \tag{2.23}$$

$$M = \frac{nu}{n'u'} \tag{2.23b}$$

References and bibliography

*Welford W.T. (1986). *Aberrations of Optical Systems*. Adam Hilger, Bristol.

3

Paraxial theory of refracting systems

3.0 Introduction

In Chapter 2, we showed that by paraxial ray tracing, we could find the position and magnification of any object for an "ideal" optical system of any complexity. We also showed that all points on an object plane perpendicular to the optical axis are imaged as points on a plane perpendicular to the optical axis in image space and that the magnification of the object is independent of object size; that is there is no distortion of the image.

In this chapter, we will use paraxial theory to explore the properties of a number of optical systems with varying complexity. We will also show that by using some of these optical properties, we can find the positions of images and their magnification without any need to ray trace.

Since paraxial theory will be used extensively in this chapter, let us first briefly review the paraxial ray trace equations.

3.1 Paraxial ray tracing: Review

Tracing paraxial rays through an optical system involves the use of the paraxial refraction and transfer equations.

3.1.1 A single surface

For a single surface shown in Figure 3.1a, we need only the following refraction equation [equation (2.5b)]:

$$n'u' - nu = -hF \tag{3.1}$$

where

$$F = C(n' - n) \tag{3.2}$$

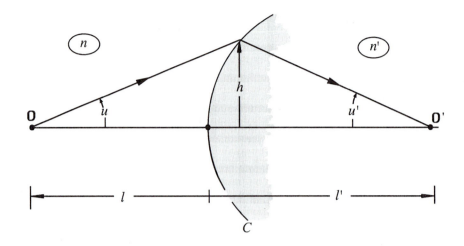

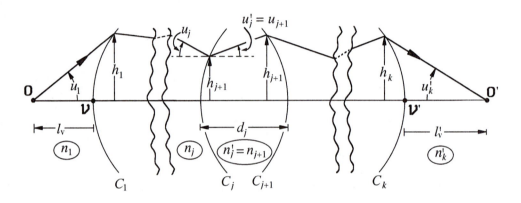

is the power F of the surface [equation (2.6)]. The object distance l and image distance l' from the surface are connected to the ray height h and angles u and u' by equations (2.10)

$$h = -ul \tag{3.3a}$$

and

$$h = -u'l' \tag{3.3b}$$

Fig. 3.1: (a) Refraction at a single surface, showing the meaning of symbols. C is the surface curvature. (b) Diagram showing the notation for tracing a paraxial ray through a system of k surfaces. C_j and C_{j+1} are the surface curvatures of surfaces j and $j + 1$, respectively.

In any particular problem, we are usually given the object distance l. We then must choose a particular ray using equation (3.3a), but we are free to choose a suitable combination of values of u and h which satisfies this equation. After application of the refraction equation (3.1) to find the value of u', equation (3.3b) gives the image distance l' from the values of h and u'. A numerical example has been given in Example 2.3.

The choice of the actual ray to trace is not a problem, because in the previous chapter we have shown that as long as equation (3.3a) is satisfied, any paraxial ray will give the same image distance. This follows from equation (2.11),

that is

$$\frac{n'}{l'} - \frac{n}{l} = F \tag{3.4}$$

which is known as the **lens equation** even though it is only applied here to a single surface.

3.1.2 A general optical system of k surfaces

For a system of more than one surface, after the refraction at each surface, we also must apply the paraxial transfer equation which transfers the ray to the next surface. Thus we repeatedly use the refraction and transfer equations, here written in a more suitable notation for use in a multi-surface system. From Section 2.4.3, the equations [equations (2.14) to (2.16)] for paraxial refraction at the jth surface are

$$n'_j u'_j - n_j u_j = -h_j F_j \tag{3.5}$$

where

$$F_j = C_j (n'_j - n_j) \tag{3.6}$$

$$u_{j+1} = u'_j \tag{3.7a}$$

and

$$n_{j+1} = n'_j \tag{3.7b}$$

This situation is shown in Figure 3.1b.

Between surfaces, we must apply the paraxial transfer equation (2.17),

$$h_{j+1} = h_j + u'_j d_j \tag{3.8}$$

The object and image distances from the front and back surface vertices are related to the initial and final ray angles and heights by equations (2.13)

$$h_1 = -u_1 l_{\text{v}} \tag{3.9a}$$

$$l'_{\text{v}} = -h_k/u'_k \tag{3.9b}$$

where l_1 and l'_k in equations (2.13) have been replaced by l_{v} and l'_{v} respectively. Once again, equation (3.9a) is used to generate the ray from the object distance, and at the end of the ray trace, equation (3.9b) gives the final image distance l'_{v} from the last surface. Example 2.4 in the preceding chapter shows how these equations are applied in practice.

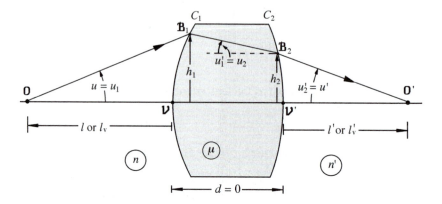

Fig. 3.2: Refraction by a thin lens. C_1 and C_2 are the surface curvatures.

3.2   The thin lens

Let us now examine the refracting properties of a lens consisting of two surfaces and of negligible thickness. Such a lens is known as a **thin lens**. This is an unreal example, but it does however reveal some interesting results which are useful when examining the properties of thick lenses and later the properties of general optical systems.

Figure 3.2 shows a thin lens with an arbitrary axial object point at o and its corresponding paraxial image point at o'. The path oB_1B_2o' shows the route of a typical paraxial ray. Let the refractive indices be n, μ and n', as shown in the diagram, and the first and second surface curvatures be C_1 and C_2, respectively. The object and image space paraxial angles are u_1 and u'_2 as shown, and let $u'_1 = u_2$ be the ray angle inside the lens. Substituting the required values in the paraxial ray trace equations given in Section 3.1.2, at the first surface, we have

$$\mu u'_1 - n u_1 = -h_1 F_1 \tag{3.10}$$

and

$$F_1 = C_1(\mu - n) \tag{3.10a}$$

which is the power of the first surface. Refracting at the second surface gives

$$n' u'_2 - \mu u_2 = -h_2 F_2 \tag{3.11}$$

where

$$F_2 = C_2(n' - \mu) \tag{3.11a}$$

is the power of the second surface. Transferring the ray between the first and second surfaces, we have

$$h_2 = h_1 + u'_1 d$$

However, since the lens is negligibly thin, the distance d is zero and thus

$$h_2 = h_1 = h$$

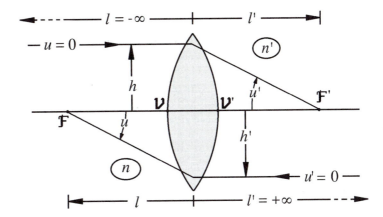

Fig. 3.3: Physical significance of the power F and focal points $\mathcal{F}$ and $\mathcal{F}'$ of a thin lens.

Therefore adding the two refraction equations (3.10) and (3.11), with $h_2 = h_1 = h$ and noting that $u'_1 = u_2$, leads to the following

$$n'u'_2 - nu_1 = -h(F_1 + F_2)$$

If we write

$$F = (F_1 + F_2) \tag{3.12}$$

and write $u_1 = u$ and $u'_2 = u'$, this refraction equation can be written as

$$n'u' - nu = -hF \tag{3.13}$$

which is now identical to the refraction equation (3.1) for a single surface with F now being the power of the thin lens. This power is the sum of the surface powers F_1 and F_2, which are given by equations (3.10a) and (3.11a). The power F of this lens can be alternatively expressed explicitly in terms of refractive indices and surface curvatures as

$$F = C_1(\mu - n) + C_2(n' - \mu) \tag{3.14}$$

For a thin lens in air, $n = n' = 1$, and thus

$$F = (C_1 - C_2)(\mu - 1) \quad \text{(thin lens in air)} \tag{3.14a}$$

If we now divide both sides of equation (3.13) by h and recall equations (3.9a and b), this refraction equation can be expressed as

$$\frac{n'}{l'} - \frac{n}{l} = F \tag{3.15}$$

where the distance symbols l_v and l'_v in equations (3.9) are now written here as l and l'. These distances are shown in Figure 3.2 for this special case of a thin lens. This equation is the same as equation (3.4) for a single surface and is known as the **lens equation**.

3.2.1   Physical interpretation of the power F

It is now a convenient point to pause and give some consideration to the physical interpretation of the power F. If one traces a ray from infinity on the left from an axial object (i.e. $l = -\infty$) as shown in Figure 3.3, this ray intersects the axis in image space at the point $\mathcal{F}'$, which is called the **back focal point**. The plane perpendicular to the optical axis and passing through the back focal point is called the **back focal plane**. It can be seen from equation (3.15) and the diagram that when $l = -\infty$

$$l' = n'/F = \mathcal{V}'\mathcal{F}'$$

Thus

$$F = n'/\mathcal{V}'\mathcal{F}'$$

where $\mathcal{V}'\mathcal{F}'$ is the distance from the back vertex $\mathcal{V}'$ of the lens to the back focal point $\mathcal{F}'$ and in this diagram, this distance is positive.

Similarly, a ray traced backwards from infinity on the right, from an axial image point (i.e. $l' = \infty$), as shown in Figure 3.3, intersects the optical axis in object space at the point $\mathcal{F}$, known as the **front focal point**. The plane perpendicular to the optical axis and passing through this point is called the **front focal plane**. It can now be seen from equation (3.15) and the diagram that

$$l = -n/F = \mathcal{V}\mathcal{F}$$

and thus

$$F = -n/\mathcal{V}\mathcal{F}$$

where $\mathcal{V}\mathcal{F}$ is the distance from front vertex $\mathcal{V}$ of the lens to the front focal point $\mathcal{F}$. In the diagram, this distance is negative.

In summary

$$F = \frac{n'}{\mathcal{V}'\mathcal{F}'} = -\frac{n}{\mathcal{V}\mathcal{F}} \tag{3.16}$$

where for the thin lens, the vertex points $\mathcal{V}$ and $\mathcal{V}'$ coincide.

The distances $\mathcal{V}\mathcal{F}$ and $\mathcal{V}'\mathcal{F}'$ are known as **vertex focal lengths**. There are a number of different types of focal lengths and these will be discussed in more detail later in this chapter when the general optical system is examined. The points $\mathcal{F}'$ and $\mathcal{F}$ shown in Figure 3.3 are the conjugates of the axial object and image points at infinity, respectively. Thus the power F is a reciprocal measure of the vertex focal lengths of the thin lens, which are measurable quantities.

3.2.2   Extension to more complex systems

The above example of the thin lens has shown that the refraction equation (3.1) for a single surface can also be applied to a thin lens, equation (3.13), providing the power is appropriately defined in each case. The question can now be asked,

"Can the same refraction equation be applied to a more complex system?" The answer is yes and a convenient and instructive approach to a proof is firstly to show that it applies to simple but more complex systems than a thin lens, systems such as a thick lens and a system of two thin lenses. We will then show that it applies to a completely general multi-surface system. Let us now look at the refracting properties of a thick lens.

3.3   The thick lens

We have shown that the paraxial refraction equation for a single surface, equation (3.1), also applies to a complete thin lens, equation (3.13), where the power F for the thin lens is the sum of the surface powers. We will now show that it can also be applied to a thick lens, providing the power F and the ray height h are appropriately defined. We begin by defining the power F and the ray height h, starting with the power.

3.3.1   The equivalent power of a thick lens

Consider the thick lens shown in Figure 3.4a and the ray traced from infinity on the left. The ray may typically follow the path as shown and crosses the optical axis in image space at the back focal point $\mathcal{F}'$. The dashed lines represent the extensions of the object and image space rays. At the point of intersection of these two extensions, the normal to the optical axis is drawn. The point where this normal meets the optical axis is denoted by the symbol $\mathcal{P}'$.

Using the thin lens theory developed in the previous section, the extended ray paths show that this thick lens could be replaced by a thin lens placed in the plane through $\mathcal{P}'$, perpendicular to the axis and having a power

$$F = \frac{n'}{\mathcal{P}'\mathcal{F}'} \tag{3.17}$$

and this equation is analogous to the first part of equation (3.16), where $\mathcal{P}'$ replaces $\mathcal{V}'$. An equation for this power, in terms of the thick lens constructional parameters, can be derived by tracing the above ray using the paraxial refraction and transfer equations.

We trace this ray using the same procedure as in the previous section, except that here the incident ray angle u_1 is zero and the lens now has a finite thickness. In Figure 3.4a, at the first surface $u_1 = 0$ and therefore the paraxial refraction equation gives

$$\mu u_1' = -h_1 F_1$$

where F_1 is the front surface power given by equation (3.10a). Now transferring the ray to the second surface, we have

$$h_2 = h_1 + u_1'd$$

that is

$$h_2 = h_1(1 - dF_1/\mu) \tag{3.18}$$

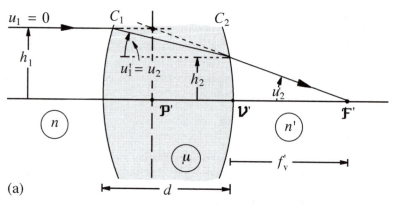

Fig. 3.4: Ray tracing through a thick lens to find its equivalent power: (a) left to right and (b) right to left.

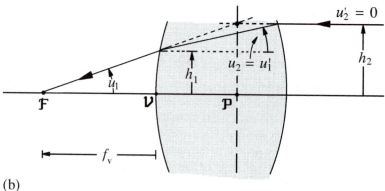

Refracting at the second surface, with $u'_1 = u_2$

$$n'u'_2 - \mu u_2 = -h_2 F_2$$

where F_2 is given by equation (3.11a). On replacing $\mu u'_1 (= \mu u_2)$ and h_2 by expressions from above, we have

$$u'_2 = -\frac{h_1[F_1 + F_2 - (F_1 F_2 d/\mu)]}{n'} \tag{3.19}$$

Now the distance

$$\mathcal{P}'\mathcal{F}' = -\frac{h_1}{u'_2} = \frac{n'}{[F_1 + F_2 - (F_1 F_2 d/\mu)]}$$

Therefore using equation (3.17), the "equivalent thin lens" power F of a thick lens is given by the equation

$$F = F_1 + F_2 - (F_1 F_2 d/\mu) \quad \text{(thick lens)} \tag{3.20}$$

where F_1 and F_2 are surface powers given by equations (3.10a) and (3.11a), respectively.

We now trace a ray from infinity on the right as shown schematically in Figure 3.4b following the same procedure as before and this ray crosses the optical axis in object space at the front focal point $\mathcal{F}$. This time we denote the point on the optical axis, for which a normal to the optical axis passes through the ray extension intersection point, by the symbol $\mathcal{P}$. In this case, the thick lens can be replaced by an equivalent thin lens in this plane having a power given by the righthand expression in equation (3.16), but in which $\mathcal{P}$ has replaced $\mathcal{V}$, that is

$$-\frac{n}{\mathcal{P}\mathcal{F}}$$

If this ray is traced algebraically as before, an expression for this power in terms of the construction parameters will be found to be identical to that given by equation (3.20). Hence the "equivalent thin lens" power of a thick lens is the same irrespective of the direction of the ray. Thus one can write

$$F = -\frac{n}{\mathcal{P}\mathcal{F}} = \frac{n'}{\mathcal{P}'\mathcal{F}'} \tag{3.21}$$

Since this is the power of an equivalent thin lens, it is called the **equivalent power**. The points $\mathcal{P}$ and $\mathcal{P}'$ are called the **front** and **back principal points**.

3.3.2 *Vertex focal lengths and vertex powers of a thick lens*

The distances $\mathcal{V}\mathcal{F}$ and $\mathcal{V}'\mathcal{F}'$ as shown in Figure 3.4 are also of some interest. They are called the **front** and **back vertex focal lengths**, respectively, and denoted by the symbols f_v and f_v', respectively. Firstly from Figure 3.4a, it follows that

$$f_v' = \mathcal{V}'\mathcal{F}' = -h_2/u_2'$$

Now, using equations (3.18), (3.19) and (3.20) it follows that

$$f_v' = \mathcal{V}'\mathcal{F}' = \frac{n'[1 - (dF_1/\mu)]}{F} \tag{3.22}$$

Similarly denoting

$$f_v = \mathcal{V}\mathcal{F}$$

it can be shown that

$$f_v = \mathcal{V}\mathcal{F} = \frac{n[(dF_2/\mu) - 1]}{F} \tag{3.23}$$

In ophthalmic optics, the vertex powers F_v and F_v' are used more frequently than the vertex focal lengths. Firstly let us look at the back vertex power F_v'. This is defined as

$$F_v' = n'/f_v' \quad \text{(back vertex power)} \tag{3.24a}$$

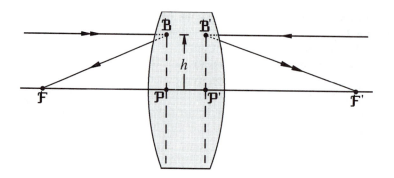

Fig. 3.5: The formation of the principal planes of a thick lens and proof that they have unit positive transverse magnification.

and thus from equation (3.22)

$$F'_v = \frac{F}{[1 - (dF_1/\mu)]} \qquad (3.24b)$$

Similarly, the front vertex power F_v is defined as

$$F_v = -n/f_v \quad \text{(front vertex power)} \qquad (3.25a)$$

and from equation (3.23)

$$F_v = \frac{F}{[1 - (dF_2/\mu)]} \qquad (3.25b)$$

Occasionally, an equation giving the equivalent power in terms of the vertex powers is very useful. Re-arranging equations (3.24b) and (3.25b) gives

$$F = [1 - (dF_1/\mu)]F'_v \qquad (3.26a)$$

and

$$F = [1 - (dF_2/\mu)]F_v \qquad (3.26b)$$

3.3.3 The principal planes and their properties

The planes perpendicular to the optical axis and drawn through the principal points $\mathcal{P}$ and $\mathcal{P}'$ are called the **front** and **back principal planes**, respectively. We can easily show that these planes are conjugate planes and with the special property that they have positive unit magnification. We can prove this with the aid of Figure 3.5, where we have taken the two rays from Figures 3.4a and b, placed them on the same diagram and chosen the incident ray heights to be equal. From this diagram, it is clear that the two rays are concurrent on the front and back principal planes and at the same height, denoted by h. Therefore the two points $\mathcal{B}$ and $\mathcal{B}'$ are conjugate points and since these two points are equidistant from the axis and are on the same side of the axis, the transverse magnification of the two planes is $+1$.

3.3.3.1 The positions of the principal planes

The positions of the principal planes or points relative to the respective vertex points are worth noting, that is the distances $\mathcal{VP}$ and $\mathcal{V'P'}$ in Figure 3.4. Firstly to find $\mathcal{V'P'}$ we can proceed as follows. From Figure 3.4a

$$\mathcal{P'F'} = \mathcal{P'V'} + \mathcal{V'F'}$$

That is

$$\mathcal{V'P'} = -\mathcal{P'V'} = \mathcal{V'F'} - \mathcal{P'F'}$$

From equations (3.22) and (3.21) it follows that

$$\mathcal{V'P'} = \frac{n'[1 - (dF_1/\mu)]}{F} - \frac{n'}{F}$$

that is

$$\mathcal{V'P'} = -\frac{n'dF_1}{\mu F} \tag{3.27a}$$

It can easily be shown by a similar procedure, using Figure 3.4b and equations (3.23) and (3.21), that

$$\mathcal{VP} = \frac{ndF_2}{\mu F} \tag{3.27b}$$

3.3.4 The paraxial refraction and lens equations

We have just shown that for a ray coming from infinity, we can replace a thick lens by an "equivalent thin lens" placed at the appropriate principal plane. We will now use the above results to prove that we can apply the paraxial refraction equation for a thin lens, equation (3.13), to a thick lens and show that the power in this equation is the equivalent power defined by equation (3.20) and that the ray height h is the ray height at the principal planes.

Let us trace a ray from a finite distance through the thick lens, as shown in Figure 3.6. At the first surface, we have

$$\mu u_1' - n u_1 = -h_1 F_1 \tag{3.28a}$$

and at the second surface

$$n' u_2' - \mu u_2 = -h_2 F_2 \tag{3.28b}$$

If we add equations (3.28a) and (3.28b) with $u_1' = u_2$, we have

$$n' u_2' - n u_1 = -(h_1 F_1 + h_2 F_2) \tag{3.29}$$

If we extend the object and image space rays to the principal planes as shown in the diagram, we know that the two rays must intersect these planes at the same

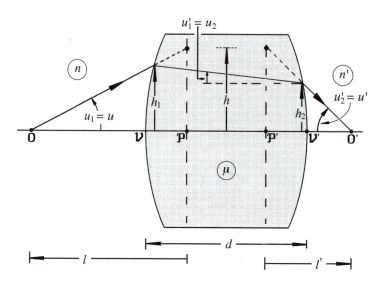

Fig. 3.6: Diagram to show
the concept of the
equivalent thin lens and
used to prove that the
paraxial refraction equation
(3.13) applies to the thick
lens.

height h, because the principal planes have positive unit magnification. Thus on applying the paraxial transfer equation to the object space ray, we have

$$h = h_1 + u_1 \mathcal{VP}$$

Using equation (3.27b) to eliminate the distance $\mathcal{VP}$ from this equation, we can get

$$h_1 = h - [u_1 n d F_2/(\mu F)]$$

Similarly, if we extend the image space ray back to the back principal plane we have

$$h_2 = h + [u_2' n' d F_1/(\mu F)]$$

On substituting these equations into equation (3.29) and after some manipulation, we have

$$(n'u_2' - nu_1)(\mu F + dF_1 F_2) = -h(F_1 + F_2)\mu F$$

Using equation (3.20), and writing $u_2' = u'$ and $u_1 = u$, we can readily reduce this equation to

$$n'u' - nu = -hF \qquad\qquad (3.30)$$

which is the paraxial refraction equation we aimed to establish. Therefore we have shown that this equation is applicable to a thick lens providing we define the power F according to equation (3.20) and the ray height h is taken as the ray height at the principal planes.

3.4   Two lens systems

3.4.1   Two thin lenses

It can easily be shown that the equations and principles developed for a thick lens apply equally to two thin lenses. The surface powers F_1 and F_2 of the thick lens now become the thin lens powers, and the refractive index μ of the medium of the thick lens becomes the index between the two thin lenses. For two thin lenses in air, this requires the index μ to be replaced by 1.0.

3.4.2   Two thick lenses

Systems containing two lenses occur frequently in optics. However, because the lenses are real, they have thicknesses that must often be taken into account and in these cases, the "two thin lenses" equations apply, providing the "thin lens" separation d is taken as the distance between the back principal plane of the first thick lens and the front principal plane of the second thick lens. This principle can be extended to combining any pair of optical systems of any complexity and this is done in Section 3.7.

3.5   Three component systems

By using the same procedure as applied to the thick lens, it can be shown that the equivalent power F of a thick lens with three surfaces or a system of three thin lenses is given by the equation

$$F = F_1 + F_2 + F_3 - (d_1 F_1 F_2 / \mu_1) - (d_2 F_2 F_3 / \mu_2) \tag{3.31}$$

$$- [(d_1/\mu_1) + (d_2/\mu_2)]F_1 F_3 + [d_1 d_2 F_1 F_2 F_3 / (\mu_1 \mu_2)]$$

where F_1, F_2 and F_3 are the surface powers of the thick lens or the lens powers of the thin lenses, d_1 and d_2 are the surface or lens separations and μ_1 and μ_2 are the refractive indices in between the components, with the subscripts numbered left to right.

If we now compare the equations for power (3.20) for a two component system and the above equation for a three component system, it is probably obvious that as the number of components increases, the equation for the equivalent power becomes increasingly complex and would very soon be unmanageable. Therefore in the general case, we need a different approach to looking at the properties of a general optical system and determining its equivalent power.

3.6   The general lens system

3.6.1   The paraxial refraction equation

We will now show that the refraction equation (3.1) applied to a single surface and equation (3.13) applied to a thin lens and equation (3.30) applied to a thick lens; that is the equation

$$n'u' - nu = -hF \tag{3.32}$$

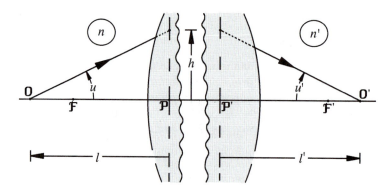

Fig. 3.7: A general system, the focal points, principal points and principal planes.

also applies to a general optical system consisting of any number of surfaces providing the following conditions are satisfied.

(i)   The system has both focal points ($\mathcal{F}$ and $\mathcal{F}'$) and principal planes (at the principal points $\mathcal{P}$ and $\mathcal{P}'$) at finite distances.

(ii) The height h is the "extended" ray height at these planes, as shown in Figure 3.7.

(iii) The system has a non-zero equivalent power F which is defined in terms of the principal to focal point distances as

$$F = -\frac{n}{\mathcal{P}\mathcal{F}} = \frac{n'}{\mathcal{P}'\mathcal{F}'} \qquad (3.33)$$

which is the same definition as for a thick lens, given by equation (3.21). Systems which have zero power are called **afocal** systems. Telescopes are common examples of afocal systems and these are discussed in Chapter 17.

To prove that the refraction equation (3.32) applies to a general optical system, we need to recall the discussion in Section 2.4.5. In that section, we stated that for any paraxial ray refracted by an optical system, the height h at any plane in the system and the angle u on either side of the surface are related to the height h' at any other plane and the angle u' on either side by equations (2.18). In this discussion, equation (2.18a) is relevant, that is

$$u' = \mathfrak{A}u + \mathfrak{B}h \qquad (3.34)$$

where for any particular system, the numerical values of the quantities $\mathfrak{A}$, $\mathfrak{B}$, $\mathfrak{C}$ and $\mathfrak{D}$ depend upon the positions of the two surfaces, the respective sides at which the two angles are measured and the system parameters. These equations apply to any optical system, even those of zero equivalent power.

Firstly we note that equation (3.34) has the same form as equation (3.32) and thus the proof involves replacing the quantities $\mathfrak{A}$ and $\mathfrak{B}$ by appropriate expressions in terms of some system parameters.

Thus we will use equation (3.34) to prove equation (3.32). We take the above two planes as the front and back principal planes and angles u and u' to be in object and image spaces, respectively. Given this condition, we will now

Fig. 3.8: Ray tracing to find equivalent power and positions of the focal and principal points of a general system.

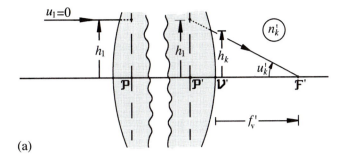

(a)

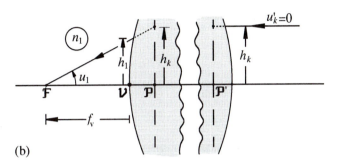

(b)

proceed to find expressions for the values of $\mathfrak{A}$ and $\mathfrak{B}$ and we need the results of two ray traces to do this.

Let us take the first ray as one coming from an axial point at infinity on the left, as shown in Figure 3.8a. This ray will have the angle $u = u_1 = 0$ and height $h = h_1$ at the principal planes and an angle $u' = u'_k$ in image space. In this case, equation (3.34) then becomes

$$u'_k = \mathfrak{A}0 + \mathfrak{B}h_1 \qquad (3.35)$$

Now this ray must pass through the back focal point $\mathcal{F}'$, as shown in the diagram, and so

$$h_1/u'_k = -\mathcal{P}'\mathcal{F}'$$

and using equation (3.33), it follows that

$$h_1/u'_k = -n'/F$$

Therefore, from equation (3.35)

$$\mathfrak{B} = -F/n'$$

With this expression for $\mathfrak{B}$, equation (3.34) now becomes

$$u' = \mathfrak{A}u + (-F/n')h \qquad (3.36)$$

A suitable second ray is a ray parallel to the optical axis in image space, that is the ray with the image space angle $u' = u'_k = 0$. Let this ray have a height $h' = h_k$ at the principal planes, as shown in Figure 3.8b. If the angle in object space is $u = u_1$, then equation (3.36) now can be written as

$$0 = \mathfrak{A}u_1 + (-F/n')h_k \qquad (3.37)$$

This ray must pass through the front focal point $\mathcal{F}$. Now from the diagram

$$-h_k/u_1 = \mathcal{PF}$$

Using equation (3.33), it follows that

$$h_k/u_1 = n/F$$

If we substitute for this expression in equation (3.37), it follows that

$$\mathfrak{A} = n/n'$$

If we substitute this expression for $\mathfrak{A}$ into equation (3.36) and after some minor manipulation, we have

$$n'u' - nu = -hF$$

which is identical with the refraction equation (3.32) and thus the proof is completed.

3.6.2   Calculation of the positions of principal and focal points and equivalent power

Before we can apply equation (3.32) in numerical calculations, we need to know (a) the positions of the principal planes, which are required for the calculation of the value of h, and (b) the value of the equivalent power F. For a general optical system, there is no suitable practical equation that expresses the equivalent power of the system in terms of the system parameters such as surface curvatures or powers, refractive indices and surface or component separations. On inspection of equations (3.20) and (3.31), it is clear that the equation for the equivalent power rapidly increases in complexity as we increase the number of surfaces or components. Therefore, the most convenient alternative is to trace a suitable ray numerically. Suitable equations for this ray trace are given in Section 3.1.2 and here we will use the same symbol notation, which is shown in Figure 3.1b.

We can find the position of the back principal plane and the equivalent power from the results of a ray traced from infinity at the left (i.e. with an object space angle u_1 of 0) with an arbitrary height h_1 at the first surface. This ray is shown in Figure 3.8a. This ray trace will give the numerical value of the ray height h_k at the last surface and the image space angle u'_k. The distance $\mathcal{P}'\mathcal{F}'$ between the back principal point $\mathcal{P}'$ and back focal point $\mathcal{F}'$ is then given by the equation

$$\mathcal{P}'\mathcal{F}' = -h_1/u'_k \qquad (3.38a)$$

This ray trace also gives the distance $\mathcal{V}'\mathcal{F}'$ of the back focal point $\mathcal{F}'$ from the back vertex $\mathcal{V}'$. It is given by the equation

$$\mathcal{V}'\mathcal{F}' = -h_k/u'_k \tag{3.38b}$$

It now follows from these two equations that

$$\mathcal{V}'\mathcal{P}' = (h_1 - h_k)/u'_k \tag{3.39}$$

which gives the position of the back principal point $\mathcal{P}'$ relative to the back vertex $\mathcal{V}'$ of the system.

This ray trace also gives the equivalent power F. Equation (3.33) expresses the equivalent power in terms of the distance $\mathcal{P}'\mathcal{F}'$. In this case, this equation should be written as

$$F = n'_k/\mathcal{P}'\mathcal{F}'$$

and using equation (3.38a), the equivalent power is given in terms of the ray trace results by the equation

$$F = -n'_k u'_k / h_1 \tag{3.40}$$

Since the value of h_1 is arbitrary, there is some advantage in setting its value as 1 and if we do this, then we have

$$F = -n'_k u'_k \quad (h_1 = 1) \tag{3.40a}$$

To find the position of the front principal point $\mathcal{P}$, we can trace a ray backwards from image space where it is parallel to the axis, as shown in Figure 3.8b. This ray trace will give the same equivalent power as the above left to right ray trace. In tracing from right to left, we use the same ray trace equations as used for the left to right ray trace, but we must note the following.

(i)   In the paraxial refraction equation (3.5), at any surface, we will know the value of u'_j and will be calculating the value of u_j.

(ii) When we transfer with equation (3.8), we will know the value of h_{j+1} and be calculating the value of h_j.

We start with the image space ray angle $u'_k = 0$ and choose some value for the ray height h_k at the last surface. This ray trace will finally give the object space angle u_1 and ray height h_1 at the first surface. It then follows from the diagram that

$$\mathcal{P}\mathcal{F} = -h_k/u_1 \tag{3.41a}$$

and

$$\mathcal{V}\mathcal{F} = -h_1/u_1 \tag{3.41b}$$

These two equations give the position of the front principal point $\mathcal{P}$ relative to the front vertex point $\mathcal{V}$

$$\mathcal{VP} = (h_k - h_1)/u_1 \tag{3.42}$$

Ray tracing backwards can be troublesome and it is often easier to trace from one direction. An alternative method for finding the positions of the principal and focal points is described in Appendix 1.

We will now give a numerical example to show how we find the equivalent power and positions of the focal and principal points in a particular case. Example 2.4 showed how to use these equations for an object point at a finite distance. In that example, we only needed to trace one ray, a ray from left to right. Now we will repeat the procedure by tracing two rays from infinity (one from the left and one from the right), to find the equivalent power and positions of the focal points and principal planes of the same eye.

> **Example 3.1:** Find the equivalent power and positions of the focal points and principal planes of the Gullstrand number 2 accommodated schematic eye. The data are given in Appendix 3.
>
> **Solution:** In the following two ray traces, we will use equations (3.5) to (3.8).

Left to right ray trace

Choice of ray. A ray from infinity on the left has the angle $u_1 = 0$ in object space. The ray height h_1 at the first surface is arbitrary, but a value of $h_1 = 1$ is most convenient and we will use this value here.
Refraction at surface 1. The power F_1 at this surface is

$$F_1 = 0.128205 \times (1.3333 - 1.0) = +0.0427307 \text{ mm}^{-1}$$

Substituting in the paraxial refraction equation

$$1.3333 \times u_1' - 1.0 \times (0) = -1.0 \times 0.0427307$$

Thus

$$u_1' = -0.0320512$$

Transfer to surface 2. Using this value of u_1' and the paraxial transfer equation, the ray height h_2 at the second surface can be found. In this case

$$h_1 = 1.0 \text{ mm}, \quad u_1' = -0.0320512, \quad d_1 = 3.2 \text{ mm}$$

and hence

$$h_2 = 1.0 + (-0.0320512) \times 3.2 = 0.897436 \text{ mm}$$

Refraction at surface 2. The refraction equation can now be applied once again, this time at the second surface $u_2 = u'_1$. At this surface, the power is

$$F_2 = (0.20) \times (1.4160 - 1.3333) = 0.0165334 \text{mm}^{-1}$$

and the paraxial refraction equation is

$$1.416 \times u'_2 - 1.3333 \times (-0.0320512) = -0.897436 \times 0.0165533$$

and thus

$$u'_2 = -0.0406586$$

Transfer to surface 3. Transferring to find the height h_3 at the corneal surface gives

$$h_3 = 0.897436 + (-0.0406586) \times 4.0$$

that is

$$h_3 = 0.734801 \text{ mm}$$

Refraction at surface 3. At this surface, the power is

$$F_3 = (-0.2) \times (1.3333 - 1.416) = +0.01654 \text{ mm}^{-1}$$

Refracting gives

$$1.3333 \times u'_3 - 1.416 \times (-0.0406586) = -0.734801 \times 0.01654$$

and finally

$$u'_3 = -0.0522911$$

The results of this ray trace are listed in Table 3.1.

Equivalent power. Since we have used an initial ray height of $h_1 = 1.0$, we can use equation (3.40a) to find the equivalent power F, where n'_k and u'_k in that equation are n'_3 and u'_3. Thus

$$F = -n'_3 u'_3 = -1.3333 \times (-0.0522911) = 0.069721 \text{ mm}^{-1}$$

Back principal and focal points. From equation (3.38b), we have

$$\mathcal{V}'\mathcal{F}' = -0.734801/(-0.0522911) = 14.052 \text{ mm}$$

and from equation (3.39), we have

$$\mathcal{V}'\mathcal{P}' = (1 - 0.734801)/(-0.0522911) = -5.072 \text{ mm}$$

Table 3.1. Ray trace results for Example 3.1

Surface	Left to right		Right to left	
	u	h (mm)	u	h (mm)
	$+0.000000$		0.069728	
1		$+1.000000$		0.875738
	-0.032051		0.024231	
2		$+0.897436$		0.953277
	-0.040659		0.011681	
3		$+0.734802$		1.000000
	-0.052291		0.000000	

Right to left ray trace

Here we choose a ray with an angle $u'_3 = 0$ in image space and arbitrarily choose a height h_3 at the last surface as $h_3 = 1$. The paraxial refraction equation at this surface then is

$$n'_3 u'_3 - n_3 u_3 = -h_3 F_3$$

that is

$$1.3333 \times 0 - 1.416 \times u_3 = -1 \times 0.01654$$

Solving for u_3, we get

$$u_3 = +0.0116808 = u'_2$$

After refraction, we use the transfer equation

$$h_3 = h_2 + u'_2 d_2$$

that is

$$1.0 = h_2 + 0.0116808 \times 4$$

Solving for h_2 gives

$$h_2 = 0.953277 \text{ mm}$$

If we continue the refraction and transfer calculations, we finally have

$$u_1 = +0.069728 \quad \text{and} \quad h_1 = 0.875738 \text{ mm}$$

These final values and the intermediate values of the ray trace are listed in Table 3.1. Substituting the relevant values in equations (3.41b) and

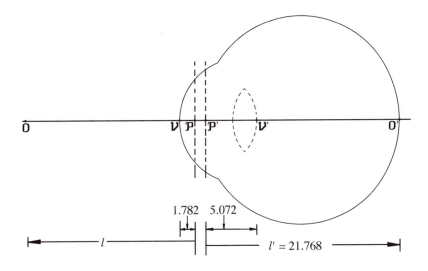

Fig. 3.9: Example of application of the paraxial refraction equation to an optical system as a whole (the Gullstrand number 2 accommodated eye). The diagram is not to scale and the distances are in millimetres.

(3.42), we have

$$\mathcal{VF} = -h_1/u_1 = -0.875738/0.069728 = -12.559 \text{ mm}$$

$$\mathcal{VP} = (h_3 - h_1)/u_1 = (1.0 - 0.875738)/0.069728 = 1.782 \text{ mm}$$

The accommodated Gullstrand eye and the positions of its principal points are shown in Figure 3.9.

3.6.3   The lens equation

It has been established in Section 3.6.1 that the paraxial refraction equation, equation (3.32), applies to any general optical system that has a non-zero power and principal planes at finite distances. It follows that the lens equation

$$\frac{n'}{l'} - \frac{n}{l} = F \tag{3.43}$$

is also valid under the same conditions, where the ray angles (u and u') and height h in equation (3.32) are related to distances l and l' by the equations

$$l = -h/u \quad \text{and} \quad l' = -h/u' \tag{3.44}$$

where l and l' are measured from the front and back principal planes, respectively, *and not from the front and back surface vertices.*

These distances are shown in Figure 3.7. At this stage, it is appropriate to introduce the following useful variants of the above equation.

$$l = \frac{nl'}{(n' - l'F)} \quad \text{and} \quad l' = \frac{n'l}{(n + lF)} \tag{3.45a,b}$$

Example 3.2: Let us use the results of Example 3.1 and determine the position of the conjugate of the retina of the Gullstrand number 2 accommodated eye. The problem was solved earlier in Example 2.4 by direct ray tracing. Here we will solve the problem using the equivalent power and the positions of the principal planes.

Solution: The Gullstrand number 2 accommodated eye with its principal planes is shown in Figure 3.9. The relevant dimensional data are taken from Appendix 3 and the positions of the principal points are taken from results of Example 3.1. The values are shown on this diagram and from this diagram we have

$$l' = 21.768 \text{ mm}$$

We can now use equation (3.45a) to give

$$l = -118.056 \text{ mm}$$

Now

$$l = -(o\mathcal{V} + \mathcal{VP}) = -(o\mathcal{V} + 1.782)$$

Therefore

$$\mathcal{V}o = -116.27 \text{ mm}$$

The value found in Example 2.4 was -116.24 mm. The difference is due to truncation or rounding errors.

3.6.4   Interpretation of the equivalent power

An interpretation of the power was discussed in Section 3.2.1. In the light of equation (3.43), let us look at this once again. Recalling equation (3.43), the object distance l is the radius of curvature of the wavefront from the axial object point o as it arrives at the front principal plane and l' is the radius of curvature of the exiting wavefront at the back principal plane. Thus the equivalent power F is a measure of the change in the radius of curvature of the incident and refracted wavefronts.

3.6.5   Off-axis or extended image formation

In Chapter 2, we showed that for a general optical system, all object points on a plane perpendicular to the optical axis are imaged on a plane in image space which is also perpendicular to the optical axis. We also showed that the transverse magnification of an extended object was independent of position in the object plane or its size. We will look a little further at the magnification.

3.6.5.1   Transverse (or lateral) magnification

Referring to Figure 3.10 and Chapter 2, we defined the transverse magnification M, in terms of the object size η and image size η', as

$$M = \eta'/\eta \tag{3.46}$$

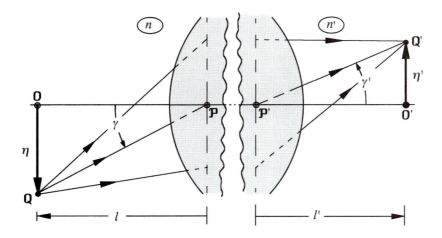

and this is also called the **lateral magnification**. We also showed that the transverse magnification could be expressed in terms of the object and image space angles u and u' of a ray traced from the axial object, by the equation (2.23b), that is

$$M = \frac{nu}{n'u'} \tag{3.47}$$

We will now give an alternative proof and other forms of this equation.

Figure 3.10 shows an object OQ of height η and its image $O'Q'$ with a height η'. The diagram shows a paraxial beam forming the point image at Q'. Because all paraxial rays in the beam are concurrent at the image point Q', we can follow the path of any of these rays to find its position. A convenient ray is the ray $Q\mathcal{P}\mathcal{P}'Q'$. In the triangle $OQ\mathcal{P}$

$$\eta/l = \tan(\gamma)$$

However, if we assume the paraxial approximation is still valid, the tangent of the angle γ is simply the angle γ itself. Therefore we can write

$$\eta = \gamma l$$

Similarly in triangle $O'Q'\mathcal{P}'$

$$\eta' = \gamma'l'$$

and thus

$$\frac{\eta'}{\eta} = \frac{\gamma'l'}{\gamma l} \tag{3.48}$$

We now apply the paraxial refraction equation (3.32) to the ray $Q\mathcal{P}\mathcal{P}'Q'$. Since the ray height is zero at the principal planes, we have $h = 0$ and thus

$$n'\gamma' = n\gamma$$

and therefore

$$\frac{\gamma'}{\gamma} = \frac{n}{n'}$$

Finally if this equation is used to eliminate γ'/γ from equation (3.48), we have the following useful equation for transverse magnification

$$M = \frac{nl'}{n'l} \tag{3.49}$$

If we replace l and l' by u and u' using the equations (3.44), we obtain equation (3.47).

The transverse magnification can be expressed in a number of alternative forms that are occasionally very useful: for example, equations that express the magnification in terms of only one of l or l' and the equivalent power F. Recalling the lens equation (3.43) and using equation (3.49), we can show that

$$M = \frac{(n' - l'F)}{n'} \tag{3.50a}$$

and

$$M = \frac{n}{(n + lF)} \tag{3.50b}$$

Effect of a change in conjugate position on image size or magnification

Equations (3.50) show that the magnification is a function of object or image positions. Sometimes we need to know how the magnification changes for a given change in object or image position. Let M_1 and M_2 be the transverse magnifications for two distinct pairs of conjugate planes denoted by the subscripts 1 and 2. Starting with equations (3.50a) and (3.50b) in turn, we can easily derive the following equations

$$(M_1 - M_2) = (l'_2 - l'_1)F/n' \tag{3.51a}$$

and

$$(1/M_1 - 1/M_2) = (l_1 - l_2)F/n \tag{3.51b}$$

which give the change in magnification in terms of a shift in position of the conjugate planes.

If the equivalent power F is zero, these equations show that

$$M_1 = M_2$$

That is, the magnification is constant and independent of the positions of the conjugates. This result is very useful in understanding the optical properties of afocal systems such as telescopes (Chapter 17).

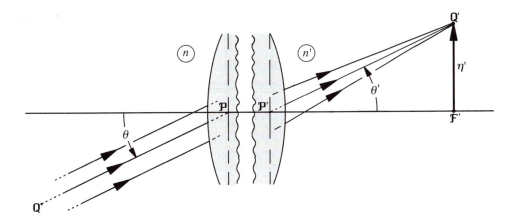

Fig. 3.11: Image size of an object at infinity.

For a general system of non-zero power, if either conjugate is at infinity, none of the above equations for transverse magnification is valid. In fact in these cases, the magnification is either zero or infinite. For an object that is at a great distance (effectively at infinity), its size is best specified as an angular dimension and its image is formed in the back focal plane. For such distant objects, we may need to know the focal plane image size.

3.6.5.2 *Image size of an object at infinity*

If an object is at infinity, its image is formed in the back focal plane of the optical system. Given its angular subtense θ, its image size η' can be found by application of a simple equation. Referring to Figure 3.11

$$\eta' = \theta' \mathcal{P}' \mathcal{F}'$$

Applying the paraxial refraction equation (3.32) to the ray $Q\mathcal{P}\mathcal{P}'Q'$, we have

$$n'\theta' = n\theta$$

Now from equation (3.33) we have

$$F = n'/\mathcal{P}'\mathcal{F}'$$

and after combining these three equations, we finally have

$$\eta' = \theta n/F \tag{3.52}$$

3.6.6 *Longitudinal magnification*

Occasionally of interest is the effect of a change in object position on image position, in a direction along the optical axis. The sensitivity of image position to a change in object position is called the **longitudinal magnification** and is denoted here by the symbol M_{L}. With reference to Figure 3.12, we define this

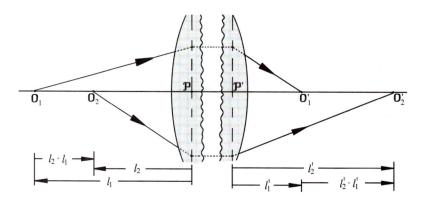

Fig. 3.12: Longitudinal magnification.

as

$$M_L = \frac{l_2' - l_1'}{l_2 - l_1} \tag{3.53}$$

We can use equations (3.51a) and (3.51b) to show that

$$M_L = n' M_1 M_2 / n \tag{3.54}$$

If we now define M_{L0} as the limiting longitudinal magnification, in the limit $l_1 \Rightarrow l_2$, it could be written in terms of infinitessimally small increments δl and $\delta l'$, in the form

$$M_{L0} = \delta l' / \delta l \tag{3.55}$$

In the limit that $l_1 \Rightarrow l_2$, $M_1 = M_2 = M$ and thus, in this limit

$$M_{L0} = (n'/n) M^2 \tag{3.56}$$

3.6.7 The cardinal points

Optical systems have six **cardinal points**. Four of these have already been discussed, and these are the front and back focal points ($\mathcal{F}$ and $\mathcal{F}'$, respectively) and front and back principal points ($\mathcal{P}$ and $\mathcal{P}'$, respectively). The other two points not mentioned so far are the **nodal points**. However before these are introduced, it will be instructive briefly to go over once again the key features of focal and principal points.

3.6.7.1 Focal points ($\mathcal{F}$ and $\mathcal{F}'$)

Focal points are defined as the conjugates of infinity. The front focal point is the conjugate of the axial object at infinity to the right and the back focal point is the conjugate of the axial point at infinity to the left. For simple lenses, they can be located by application of the standard equations introduced previously. For the more general systems, they are best found by ray tracing, as explained in Section 3.6.2. The planes through the focal points and perpendicular to the optical axis are called **focal planes**.

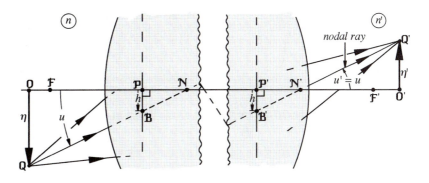

Fig. 3.13: Definition of the nodal points N and N'.

3.6.7.2 Principal points (P and P')

Principal points are the intersection points of the optical axis with the principal planes. Principal planes are the planes at which the paraxial ray height h must be measured when using equation (3.32) and from which the distances l and l' must be measured when using equation (3.43) or any of its derivatives.

One very interesting property of principal planes is that they are conjugate planes with positive unit magnification. This was shown in Section 3.3.3 for a thick lens, but the same argument is readily extended to a general optical system. As an alternative confirmation, it is left as an exercise for the reader to prove the conjugacy property using either of the equations (3.45a or b) and the magnification property using either of the equations (3.50a or b).

Position of principal planes

In the diagrams shown so far, for example Figure 3.12, the back principal point P' is shown to the right of the front principal point P and both are shown inside the system. However, this is not always so, depending upon the nature of the optical system. For example in numerical problem 3.4 given at the end of this chapter, P' is to the left of P. It is also possible that the principal planes can be outside the system and both on the same side. Examples of these situations are given in Chapter 6.

3.6.7.3 Nodal points (N and N')

Nodal points are the remaining two cardinal points. From any off-axis object point, there is one ray in the beam which could pass through the system without any angular deviation; that is the ray subtends the same angle to the axis in both object and image spaces. We can call this ray the **nodal ray** and it is not usually the central ray of the beam. The points where this ray appears to cross the axis in object and image space are called the **nodal points** and their positions are denoted by the symbols N and N', respectively. This situation is shown schematically in Figure 3.13. The positions of the nodal points relative to the principal points can be found from a very simple equation, which can be derived as follows.

Figure 3.13 shows an image forming beam and several of its rays passing through an optical system. In this diagram, the triangles PNB and $P'N'B'$ are identical, because they have two angles (the angle u and the right angle) and corresponding sides that are equal. The corresponding sides, $\mathcal{PB}$ and $\mathcal{P'B'}$, are equal because they are on the principal planes which have positive unit magnification. Hence

$$\mathcal{PN} = \mathcal{P'N'}$$

and the distances $\mathcal{PN}$ and $\mathcal{P'N'}$ are given by the expression

$$\mathcal{PN} = \mathcal{P'N'} = -h/u \qquad (3.57)$$

Applying the paraxial refraction equation (3.32) to the nodal ray, we have $u' = u$ and so the equation can be written as

$$n'u - nu = -hF$$

Dividing by u gives

$$(n' - n) = -F(h/u)$$

and using equation (3.57) to eliminate the term (h/u), we finally have

$$\mathcal{PN} = \mathcal{P'N'} = \frac{(n' - n)}{F} \qquad (3.58)$$

When the object and image space indices are equal, it is obvious that the principal and nodal points coincide, and this of course is the case for any optical system in air. However in the case of the eye, the object and image space refractive indices are unequal and hence the nodal points and principal points do not coincide. In the Gullstrand number 1 schematic relaxed eye, the separation is about 5.73 mm (see Appendix 3).

Two simple examples of optical systems with separated principal and nodal points are a single refracting and a single reflecting spherical surface. In these cases, both principal points are at the surface vertices and the two nodal points are at the centres of curvature. The reflecting example is discussed in Chapters 4 and 7.

3.6.8   Focal lengths

The general system has a number of different focal lengths. Considering the generalized system shown in Figure 3.8, the different focal lengths can be listed as follows:

Front vertex focal length (f_v)	$\mathcal{VF}$
Back vertex focal length (f_v')	$\mathcal{V'F'}$
Front equivalent focal length	$\mathcal{PF}$
Back equivalent focal length	$\mathcal{P'F'}$

The front and back vertex focal lengths were introduced in Section 3.3.2 in the discussion of the thick lens, and in that case, simple equations exist for their calculation from lens parameter data. In the general case, no similar simple equations exist but their values can be determined by ray tracing, as described in Section 3.6.2.

The distances $\mathcal{PF}$ and $\mathcal{P'F'}$ have already been introduced in the previous sections and an important equation involving these quantities is equation (3.33), that is

$$F = -\frac{n}{\mathcal{PF}} = \frac{n'}{\mathcal{P'F'}}$$

If the object and image space refractive indices n and n' are equal, it is obvious that

$$\mathcal{PF} = -\mathcal{P'F'}$$

If the refractive indices are equal and also unity (i.e. in air) then we can use the term **equivalent focal length** (f) to denote the distances $\mathcal{PF}$ and $\mathcal{P'F'}$ and thus

$$\mathcal{P'F'} = -\mathcal{PF} = 1/F = f \text{(in air)} \tag{3.59}$$

3.6.8.1 Nodal points and focal lengths

Using Figure 3.13, it can be easily seen that

$$\mathcal{FN} = \mathcal{FP} + \mathcal{PN}$$

and using equations (3.33) and (3.58), it follows that

$$\mathcal{FN} = \mathcal{P'F'} = n'/F \tag{3.60a}$$

Similarly, it can be shown that

$$\mathcal{N'F'} = \mathcal{FP} = n/F \tag{3.60b}$$

Let us apply these equations to a schematic eye.

> **Example 3.3:** Let us look at the values of the distances $\mathcal{FN}$, $\mathcal{P'F'}$, $\mathcal{N'F'}$ and $\mathcal{FP}$ for the Gullstrand number 1 relaxed schematic eye given in Appendix 3. From the data given in this appendix
>
> $$\mathcal{FN} = \mathcal{P'F'} = 22.784 \text{ mm}$$
>
> and
>
> $$\mathcal{N'F'} = \mathcal{FP} = 17.054 \text{ mm}$$

3.6.8.2 Vertex powers (F_v and F_v')

For the general optical system, the front and back vertex powers (F_v and F_v', respectively) are defined in terms of the vertex focal lengths, exactly as for a thick lens, that is by equations (3.25a) and (3.24a), respectively. Thus for the general optical system

$$F_v = -n/f_v \qquad (3.61a)$$

and

$$F_v' = n'/f_v' \qquad (3.61b)$$

For the general case, unlike the thick lens or two thin lenses, there is no simple convenient equation for these vertex powers in terms of the system constructional data, but their values can be found by the ray tracing procedure described in Section 3.6.2.

3.6.9   *Vergences*

In ophthalmic optics, it is common to express many distances in terms of their reciprocals rather than the distances themselves. These reciprocals are called vergences and defined as follows:

$$\text{(a) object vergences: } L = n/l \quad \text{and} \quad L_v = n/l_v \qquad (3.62a)$$

$$\text{(b) image vergences: } L' = n'/l' \quad \text{and} \quad L_v' = n'/l_v' \qquad (3.62b)$$

where L and L' are the vergences at the principal planes and L_v and L_v' are the vergences at the front and back surface vertices. In terms of L and L', the lens equation (3.43) becomes

$$L' - L = F \qquad (3.63)$$

and the transverse magnification given by equation (3.49) becomes

$$M(= \eta'/\eta) = L/L' \qquad (3.64)$$

Note: In some textbooks, these vergences are called **reduced vergences**. The adjective "reduced" has not been used here because the term is unnecessary and may be confusing. For example, if one calls the quantity $1/l$ a vergence and the quantity n/l a reduced vergence, this may be confusing because the magnitude of the reduced vergence is larger than the ordinary vergence. The use of the adjective "reduced" implies the magnitude should be less, not greater.

3.6.10   *Newton's equation*

The general refraction equation (3.32) and the lens equation (3.43) are not always easily applied to a general optical system since they require a knowledge of the positions of the principal points and these may not be known or easily found. On the other hand, focal points (especially for positive power lenses) are far more readily determined, particularly by laboratory techniques. Newton's

Fig. 3.14: Newton's
equation.

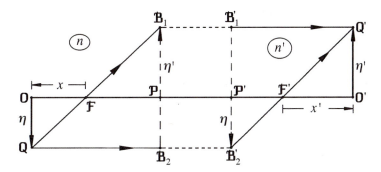

equation, an alternative to the above equations, uses the focal points as reference points instead of the principal points.

Newton's equation can be derived using Figure 3.14, which shows only the object and image, the focal points and principal planes of the system. In this diagram an object at O of size η is imaged with a size η' at O'. The rays $Q\,\mathcal{B}_1\mathcal{B}_1'Q'$ and $Q\mathcal{B}_2\mathcal{B}_2'Q'$ are two rays forming this image. Applying the rules of similar triangles, it follows that

$$\frac{-\eta}{-x} = \frac{\eta'}{\mathcal{FP}} \quad \text{(triangles } QO\mathcal{F} \text{ and } \mathcal{B}_1\mathcal{PF}) \tag{3.65a}$$

The negative signs are used because η and x are negative as drawn in the diagram. Similarly

$$\frac{\eta'}{x'} = \frac{-\eta}{\mathcal{P}'\mathcal{F}'} \quad \text{(triangles } Q'O'\mathcal{F}' \text{ and } \mathcal{B}_2'\mathcal{P}'\mathcal{F}') \tag{3.65b}$$

Combining these two equations leads to the following

$$\frac{\mathcal{FP}}{x} = \frac{-x'}{\mathcal{P}'\mathcal{F}'}$$

Therefore

$$xx' = -\mathcal{FP} \times \mathcal{P}'\mathcal{F}'$$

However from equation (3.33),

$$\mathcal{FP} = n/F \quad \text{and} \quad \mathcal{P}'\mathcal{F}' = n'/F$$

and therefore finally

$$xx' = -\frac{nn'}{F^2} \tag{3.66}$$

which is Newton's equation.

3.6.10.1   Transverse magnification

Equations (3.49), (3.50a and b) give the transverse magnifications in terms of distances measured from the principal points. Let us find corresponding equations for the distances measured from the focal points. Referring to Figure 3.14 and equation (3.65a), it follows that

$$\frac{\eta'}{\eta} = \frac{\mathcal{FP}}{x}$$

Now from equation (3.33)

$$\mathcal{FP} = n/F$$

and the equation (3.46) for the definition of the transverse magnification can now be written

$$M = \frac{n}{xF} \tag{3.67a}$$

Similarly, starting with equation (3.65b), it can be shown that

$$M = -\frac{x'F}{n'} \tag{3.67b}$$

3.6.10.2   Longitudinal magnification

Equation (3.53) for longitudinal magnification in terms of distances from the principal planes is still applicable here, because that equation expresses the longitudinal magnification in terms of shifts of the conjugate planes and not their absolute positions. Therefore we could write it in terms of the distances from the focal points rather than the principal points and since

$$x_2 - x_1 = l_2 - l_1 \quad \text{and} \quad x_2' - x_1' = l_2' - l_1'$$

we could write equation (3.53) as

$$M_L = \frac{x_2' - x_1'}{x_2 - x_1} \tag{3.68}$$

Equation (3.56) for the limiting longitudinal magnification is also applicable when the distances are measured from the focal and not principal points, because this equation does not make any reference to these distances. However, we can express it in terms of x and x' by using equations (3.67a and b) by the following alternatives

$$M_{L0} = \frac{nn'}{x^2F^2} = \frac{x'^2F^2}{nn'} = -\frac{x'}{x} \tag{3.69}$$

3.7 Cascaded systems

A complex system can be thought of as made up of two or more simpler systems cascaded together. For example, a single thin or thick lens is the combination of two simple systems, each consisting of a single surface. A system of two thin lenses is clearly the result of cascading two systems, each consisting of a single thin lens. We can extend the principle to regard a system of two thick lenses as the combination of two systems each consisting of a thick lens, and so on, to a system of any complexity. Under special circumstances, cascading of two systems can lead to some interesting results. We will explore these in the following sub-section.

3.7.1 Combining two systems

3.7.1.1 Equivalent power

If we know the cardinal point positions and equivalent powers of the two systems, we can find the equivalent power of the new system without the need for any ray tracing. If F_1 and F_2 are the equivalent powers of the two systems, the equivalent power F of the new system is given by equation (3.20) providing the distance d is now the distance between the back principal point of the first system and the front principal point of the second system and μ is the refractive index of the medium separating the two systems. A useful alternative equation expresses the system separation in terms of the distance between the focal points rather than the distance between the principal points. If d_f is the distance between the back focal point of the first system and the front focal point of the second system then

$$d = \mathcal{P}'_1\mathcal{F}'_1 + d_f + \mathcal{F}_2\mathcal{P}_2 \tag{3.70}$$

where the subscript "1" refers to the first system and "2" refers to the second system. By substituting this expression for the d in equation (3.20) and using equation (3.33) to replace the distances $\mathcal{P}'_1\mathcal{F}'_1$ and $\mathcal{F}_2\mathcal{P}_2(= -\mathcal{P}_2\mathcal{F}_2)$ by the respective equivalent powers, equation (3.20) reduces to

$$F = -d_f F_1 F_2/\mu \tag{3.71}$$

Special case 1: The back focal point of the first system coincides with the front principal plane of the second. In this case

$$d = \mathcal{P}'_1\mathcal{F}'_1$$

and equation (3.20) reduces to

$$F = F_1 \tag{3.72a}$$

Similarly, if the back principal point of the first system is coincident with the front focal point of the second system, it follows that

$$F = F_2 \tag{3.72b}$$

Special case 2: The back focal point of the first system coincides with the front focal point of the second. In this case, the distance d_f between focal points is zero and therefore from equation (3.71), the equivalent power is zero. This result applies to the construction of two lens afocal systems such as telescopes (Chapter 17).

3.8 Afocal systems

It is possible for an optical system to have zero power, and equations (3.51a and b) have hinted that such systems may still have some magnification. Such systems are termed **afocal** and one sub-class (telescopes) is discussed in Chapter 17. Afocal systems have unique properties in that they do not have any cardinal points. Firstly, if we recall equations (3.33), it is clear that for afocal systems, the distance $\mathcal{P}'\mathcal{F}'$ must be infinite, implying that either $\mathcal{P}'$ or $\mathcal{F}'$ or both must be at infinity. Secondly, the position of the back focal point is where the ray from infinity crosses the axis. The power of the lens is given by equation (3.40), but since the power is zero, the angle u'_k must be zero, implying that the ray travels parallel to the axis so that the back focal point is at infinity. Thirdly, the principal plane is the plane of intersection of the extended object and image space rays, and since these rays are parallel to each other, they never meet (or do so at infinity); therefore the back principal plane is at infinity. The same argument applies to the front focal and principal points. Therefore afocal systems have no cardinal points. This property finally implies that we cannot apply the paraxial refraction equation (3.32) and the lens equation (3.43) to an afocal system for the system as a whole. Also we cannot apply any derivatives of the lens equation, if the equations require a knowledge of the absolute object or image distances from the principal planes. However, equations (3.51a and b) can still be applied since they only require a knowledge of the shift in object or image planes and do not require a knowledge of the absolute positions of these from the principal planes. For afocal systems, the alternative is to apply the ray trace equations through the system, component by component, i.e. surface by surface or thin lens by thin lens.

3.9 Gaussian optics

Gaussian optics can be suitably defined as follows. The Gaussian optical system is the idealization of a real optical system, based upon the properties of paraxial rays, but extended beyond the paraxial region. Thus strictly speaking, paraxial rays traced outside the paraxial region should be called Gaussian rays.

The focal (i.e. one with a non-zero equivalent power) Gaussian optical system can be described by the following characteristics:

(a)   the equivalent power and
(b)   the positions of the six cardinal points.

It must be remembered that cardinal points are only defined using paraxial optics. Therefore they are only meaningful within the paraxial region. Using them outside the paraxial region is Gaussian optics.

Apart from these fundamental quantities, the Gaussian system obeys the following rules.

(a) Aberrations are neglected.
(b) All points on the object plane at o which is perpendicular to the optical axis have their images in a plane which is also perpendicular to the optical axis and passing through the paraxial image o' of o.
(c) The position of an image Q' of an object at Q on the above object plane is given by the transverse magnification equation

$$M = \frac{o'Q'}{oQ} = \frac{\eta'}{\eta} = \frac{nu}{n'u'} = \frac{nl'}{n'l} \tag{3.73}$$

(d) It follows from the above that the Gaussian image of an object in the perpendicular object plane is a replica of the object and is only different in size.

Exercises and problems

3.1 Calculate the surface powers and thin lens power of the following thin lens:

$$C_1 = 0.2 \text{ cm}^{-1}, \quad C_2 = -0.1 \text{ cm}^{-1}, n = 1.336, \quad \mu = 1.7,$$

$$n' = 1.336$$

ANSWERS:

$$F_1 = 0.0728 \text{ cm}^{-1}, \quad F_2 = 0.0364 \text{ cm}^{-1}, \quad F = 0.1092 \text{ cm}^{-1}$$

3.2 By any means, find the equivalent power of the following simple lenses, where n is the refractive index of the object medium, μ that of the lens and n' that of the image space medium, d is the lens thickness and C_1 and C_2 are the front and back surface curvatures, respectively.

	n	μ	n'	C_1	C_2	d	Answers (F)
(a)	1.5	1.8	1.3	−0.1	0.5	0	−0.28
(b)	1.0	1.7	1.0	−0.1	0.2	1	−0.2157

3.3 Using the paraxial refraction and transfer equations, find the equivalent power and back focal length of the lenses defined by the following data.

$$n_0 = 1.0, \quad n_1 = 1.3374, \quad n_2 = 1.42, \quad n_3 = 1.336$$

$$C_1 = 0.1282, \quad C_2 = 0.098, \quad C_3 = -0.1669$$

$$d_1 = 3.5, \quad d_2 = 4.0$$

The subscript 0 on the refractive index is for the object space, and so on; the 1 on C is the curvature of the first surface, and so on; and the 1 on d is the thickness in the first medium after the object space, and so on.

ANSWERS: $F = 0.060874, \quad f'_v = 16.34$

3.4 A two component system consists of two thin lenses of powers $F_1 = 0.01 \text{ mm}^{-1}$ and $F_2 = -0.02 \text{ mm}^{-1}$ which are separated by a distance of 5 mm.

(a) Calculate the power of the system and hence its equivalent focal length.

(b) By paraxial ray tracing, find the positions of the back and front focal points and hence the positions of the principal planes.

(c) Accurately sketch or graph the system, showing the positions of the focal and principal points.

(d) For an object placed 100 mm in front of the first lens, find the position of its image and its magnification.

ANSWERS:

(a) $F = -0.009$ mm^{-1} and $f = -111.1$ mm
(b) $v'\mathcal{F}' = -105.5$ m, $v\mathcal{F} = +122.2$ mm, $v\mathcal{P} = 11.1$ mm, $v'\mathcal{P}' = 5.5$ mm
(d) $v'o' = -50$ mm, $M = +0.5$

3.5 For a single thick lens with a refractive index of 1.65, curvatures $C_1 = +0.02$ mm^{-1} and $C_2 = +0.01$ mm^{-1} and a thickness of 3 mm, calculate the following.

(a) the surface powers
(b) the equivalent power
(c) the back vertex power

ANSWERS: (a) $F_1 = 13.0$ m^{-1}, $F_2 = -6.5$ m^{-1},   (b) $F = 6.65$ m^{-1},
(c) $F'_v = 6.81$ m^{-1}

3.6 Given a thick lens with a front surface power of $+6$ m^{-1}, thickness 2.5 mm, refractive index of 1.55 and equivalent power of 4 m^{-1}, calculate the back vertex power.

ANSWER: $F'_v = 4.04$ m^{-1}

3.7 If a lens with a back vertex power of 32 m^{-1} has a thickness of 10 mm, a refractive index of 1.523 and a flat back surface, calculate its equivalent power and error in assuming the equivalent power is the same as the back vertex power.

ANSWERS: $F = 26.44$ m^{-1} and percentage error $= 21.0\%$

3.8 Calculate the refractive indices of a lens, in air, in the shape of a sphere under two conditions: (a) the back vertex focal length is zero and (b) the back vertex focal length is positive and equal to the radius.

ANSWERS: (a) $\mu = 2$, (b) $\mu = 4/3$

3.9 Below is a diagram showing the positions of the cardinal points of a certain optical system

where $\mathcal{P}\mathcal{N} = 7$ mm and $\mathcal{P}'\mathcal{F}' = 80$ mm. If the refractive index of the image space is 1.7, calculate

(a) the equivalent power
(b) the refractive index of the object medium
(c) the distance $\mathcal{F}\mathcal{P}$

ANSWERS: (a) $F = 0.02125$ mm^{-1}, (b) $n = 1.55$, (c) $\mathcal{F}\mathcal{P} = 73$ mm

3.10 Calculate the separation of two thin lenses of powers 10 m^{-1} and 20 m^{-1}, if the equivalent power of the combination is 25 m^{-1}.

ANSWER: $d = 0.025$ m

3.11 Given a thin lens with a power of 5.0 cm^{-1}, and an object 2.5 cm to the left of the lens, find the transverse magnification M.

ANSWER: $M = -0.0870$

3.12 Calculate the power in air of a thin intra-ocular lens, assuming it has a power of 19 m^{-1} when in the eye. Take the aqueous and vitreous indices as 1.336 and the intra-ocular lens index as 1.5.

ANSWER: $F = 57.9 \text{ m}^{-1}$

3.13 Given an optical system with an equivalent power of 20 m^{-1} and in which the refractive index of the object space is 1.333 and that of the image space is 1.550, calculate the image distance from the back focal point, if the object is placed 5 cm from the front focal point and towards the lens system. Also calculate the transverse magnification.

ANSWERS: Distance = 10.3 cm in a direction towards the optical system; magnification = 1.333

3.14 For an optical system with an equivalent power of 60 m^{-1} and an image space refractive index of 1.336, calculate the position of the image of an object 10 cm to the left of the front focal point. Express the image position as a distance from the back focal point.

ANSWER: 0.371 cm beyond the back focal point

Summary of main symbols and equations

μ	refractive index of a single lens
u_j, u'_j	ray angles on left and right of the jth surface
u, u_1	paraxial angle in object space
u', u'_k	paraxial angle in image space
h, h'	general paraxial ray heights
h_1, h_k	special cases: h_1 (first surface), h_k (last surface)
d	surface separation
d_f	separation of focal points of cascaded systems
F_n	equivalent power of the nth single surface, lens or system
f_n	corresponding equivalent focal length
M_L	longitudinal magnification
M_{L0}	longitudinal magnification in the limit $l_1 \Rightarrow l_2$
θ	angular size of a distant object

Section 3.1.1: Single surface paraxial equations

$$F = C(n' - n) \quad \text{surface power} \tag{3.2}$$

Section 3.1.2: The paraxial equations and the general system

$$n'_j u'_j - n_j u_j = -h_j F_j \tag{3.5}$$

$$F_j = C_j(n'_j - n_j) \tag{3.6}$$

$$u_{j+1} = u'_j \quad \text{and} \quad n_{j+1} = n'_j \tag{3.7a,b}$$

$$h_{j+1} = h_j + u'_j d_j \qquad (3.8)$$

$$h_1 = -u_1 l_v \qquad (3.9a)$$

$$l'_v = -h_k/u'_k \qquad (3.9b)$$

Section 3.2: The thin lens

$$F_1 = C_1(\mu - n) \qquad (3.10a)$$

$$F_2 = C_2(n' - \mu) \qquad (3.11a)$$

$$F = F_1 + F_2 \qquad (3.12)$$

$$n'u' - nu = -hF \qquad (3.13)$$

$$F = (C_1 - C_2)(\mu - 1) \qquad \text{in air} \qquad (3.14a)$$

Section 3.3: The thick lens

$$F = F_1 + F_2 - (F_1 F_2 d/\mu) \qquad (3.20)$$

$$f'_v = \mathcal{V}'\mathcal{F}' = \frac{n'[1 - (dF_1/\mu)]}{F} \qquad (3.22)$$

$$f_v = \mathcal{V}\mathcal{F} = \frac{n[(dF_2/\mu) - 1]}{F} \qquad (3.23)$$

$$F'_v = \frac{n'}{f'_v} = \frac{F}{[1 - (dF_1/\mu)]} \qquad (3.24a,b)$$

$$F_v = -\frac{n}{f_v} = \frac{F}{[1 - (dF_2/\mu)]} \qquad (3.25a,b)$$

$$\mathcal{V}'\mathcal{P}' = -\frac{n'dF_1}{\mu F} \qquad \mathcal{V}\mathcal{P} = \frac{ndF_2}{\mu F} \qquad (3.27a,b)$$

Section 3.4.1 Two thin lenses

The equations are identical to the above with (i) the surface powers becoming thin lens powers and (ii) the lens refractive index μ above becoming the index between the two lenses. Thus for two lenses in air, $\mu = 1$.

Section 3.4.2 Two thick lenses

The two thin lens equations apply to two thick lenses providing the lens separation d is taken as the distance from the back principal plane of the first lens to the front principal plane of the second lens.

Section 3.6: The general system

$$n'u' - nu = -hF \tag{3.32}$$

$$F = -\frac{n}{\mathcal{PF}} = \frac{n'}{\mathcal{P'F'}} \tag{3.33}$$

Section 3.6.3: The lens equation

$$\frac{n'}{l'} - \frac{n}{l} = F \tag{3.43}$$

$$l = \frac{nl'}{(n' - l'F)} \quad \text{and} \quad l' = \frac{n'l}{(n + lF)} \tag{3.45a,b}$$

Section 3.6.5.1: Off-axis image formation and image magnification and sizes

$$M = \frac{\eta'}{\eta} = \frac{nu}{n'u'} = \frac{nl'}{n'l} \tag{3.46,47,49}$$

$$M = \frac{(n' - l'F)}{n'} = \frac{n}{(n + lF)} \tag{3.50a,b}$$

$$\eta' = \frac{\theta n}{F} \quad \text{(object at infinity)} \tag{3.52}$$

Section 3.6.9: Vergences

$$L = n/l \quad \text{and} \quad L_v = n/l_v \tag{3.62a}$$

$$L' = n'/l' \quad \text{and} \quad L'_v = n'/l'_v \tag{3.62b}$$

$$L' - L = F \tag{3.63}$$

$$M(= \eta'/\eta) = L/L' \tag{3.64}$$

Section 3.6.10: Newton's equation

$$xx' = -\frac{nn'}{F^2} \quad \text{Newton's equation} \tag{3.66}$$

$$M = \frac{n}{xF} = -\frac{x'F}{n'} \tag{3.67a,b}$$

$$M_{L0} = \frac{nn'}{x^2 F^2} = \frac{x'^2 F^2}{nn'} = -\frac{x'}{x} \tag{3.69}$$

4

Paraxial theory of reflecting optics

4.0 Introduction

This chapter is a brief introduction to the paraxial theory of reflecting optics. The term "mirror" has not been used because although all refracting surfaces also act as reflecting surfaces, they cannot be classified as mirrors. Here mirrors are defined as reflecting surfaces where there is no transmission of rays.

Although reflecting optics do not have a very large role to play in the optics of the eye or visual and ophthalmic instruments, they are very useful and important in some cases. The reflections from the four refracting surfaces of the eye are most useful as they can be used to measure the radii of curvature of these surfaces. Measurement of the radius of curvature of the cornea is a special case and is called keratometry. Reflections from the refracting surfaces of the eye are known as **Purkinje** images.

Many optical situations involving reflections also involve some refraction. For example in the measurement of the radius of curvature of the front surface of the crystalline lens of the eye, the beam is refracted by the cornea, reflected from the front surface of the lens and then refracted by the cornea once again. Optical systems that are a mix of refracting and reflecting elements are called **catadioptric** systems (see Section 4.2). Those that are purely reflecting are called **catoptric**. However, very few optical systems are catoptric.

Reflecting components are often used instead of refracting components because they can be made with a smaller mass and they have no chromatic aberration. They are also useful with high energy beams, where the smallest amount of absorption would damage a lens. A mirror can be made more durable to such beams by suitable choice of mirror substrate or by cooling of the substrate.

It will now be shown that the paraxial ray tracing equations developed for refracting components or systems in Chapters 2 and 3 can be equally applied to reflecting components and catadioptric systems, provided that we use modified forms of the refractive indices and surface separations.

4.1 The paraxial reflection ray trace equations

Let us look at the optics of a plane reflecting surface. Consider a ray incident on the surface as shown in Figure 4.1. The ray has an angle of incidence i and an angle of reflectance i'. Snell's law, equation (1.12), is

$$n' \sin(i') = n \sin(i)$$

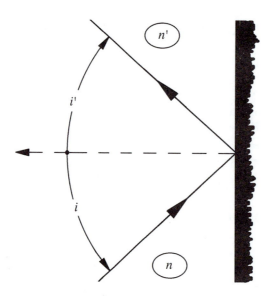

and can be used to analyse this problem further. Now the angle of incidence is equal to the angle of reflectance or at least numerically so. Thus

$$|i'| = |i|$$

However by the sign convention used so far, i' and i must be of opposite sign, and as drawn, i is positive and i' is negative. Thus

$$i' = -i$$

For this result to be compatible with Snell's law above, the numerical value of the refractive index n' must be of opposite sign to that of n, that is

$$n' = -n$$

These results lead to the conclusion that Snell's law of refraction applies to reflection provided that on reflection, the refractive index changes sign. Now since the paraxial refraction equations developed in the previous two chapters were based upon Snell's law, it can be concluded that the paraxial refraction equations can be applied to reflecting cases, if the refractive index changes sign on reflection.

The concept of negative refractive index may be somewhat new because we are accustomed to all refractive indices being positive. However, let us go back to the definition of refractive index [equation (1.5)] , that is

$$n = \frac{\text{velocity } (c) \text{ of light in a vacuum}}{\text{velocity } (v) \text{ of light in the medium}}$$

When the ray is reflected and travelling backwards, it is travelling in a negative distance direction according to the sign convention used here. Therefore, the velocity (v) would have a negative magnitude and thus so must the refractive

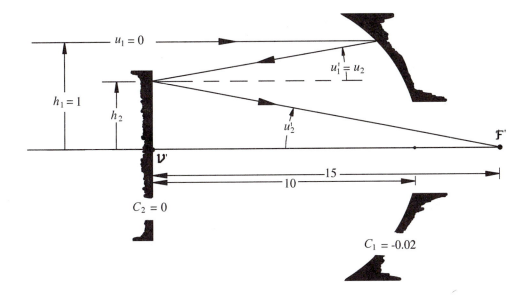

Fig. 4.2: Example of paraxial ray tracing in a simple reflecting system (not to scale). See Examples 4.1 and 4.2.

index. According to this interpretation, it follows that the refractive index must be negative whenever the ray is travelling backwards, i.e. when travelling from right to left.

On reflection, the ray will travel in the opposite direction. If the ray was initially travelling from the left to right, distances travelled would be positive. After the first reflection, these distances will now be negative and so we could also conclude that on reflection, the sign of surface separation changes sign; i.e. it follows the same rule as refractive index. Therefore, the sign of the numerical value of d in the transfer equation, equation (3.8), is negative when the ray is moving from right to left.

Thus the paraxial refraction and transfer equations and all equations derived from these fundamental equations are valid for reflecting cases provided that when the ray is reflected and moving from right to left, both refractive indices and distances become negative and remain negative until the ray is further reflected and then moves once more from left to right.

Example 4.1: Consider the simple reflecting system shown in Figure 4.2. Let us find the equivalent power of the system by tracing the ray shown in the diagram.

Solution: The ray tracing is done using equations (3.5) to (3.8), but with the above rules for the changes of the sign of the refractive index and surface separation on reflection.

The "refraction" equation at the curved surface gives

$$n_1' u_1' - n_1 u_1 = -h_1 F_1$$

where

$$F_1 = C_1(n_1' - n_1)$$

and using the rules outlined above, since $n_1 = 1$, n_1' must be -1. Also $u_1 = 0$ and $h_1 = 1$. Therefore

$$(-1)u_1' - 1 \times 0 = -1 \times (-0.02) \times (-1 - 1)$$

that is

$$u_1' = +0.04 = u_2$$

We now apply the transfer equation

$$h_2 = h_1 + u_1' d_1$$

to transfer this ray to the plane surface. Now we have $d_1 = -10$ and therefore

$$h_2 = 1 + 0.04 \times (-10)$$

Thus

$$h_2 = +0.60$$

Now applying the "refraction" equation at the second reflecting surface, we have

$$n_2' u_2' - n_2 u_2 = -h_2 F_2$$

where

$$F_2 = C_2(n_2' - n_2)$$

Here

$$n_2' = +1, n_2 = -1, u_2 = u_1' = +0.04, h_2 = +0.60 \text{ and } C_2 = 0$$

and we have

$$1 \times u_2' - (-1) \times 0.04 = -0.60 \times 0 \times [+1 - (-1)]$$

which gives

$$u_2' = -0.04$$

Finally the equivalent power is, from equation (3.40), with $k = 2$

$$F = -n_2' u_2'/h_1 = -(+1)(-0.04)/(1) = +0.04 \, \text{unit}^{-1}$$

4.1.1 Surface power of a reflecting spherical surface

The equation for the power of a refracting surface is given by equation (3.6), that is

$$F = C(n' - n) \tag{4.1}$$

Applying the rule that on reflection, the refractive index changes sign, that is $n' = -n$, the power of a reflecting surface reduces to

$$F = -2Cn \tag{4.2}$$

We can express this equation in terms of the radius of curvature r, thus

$$F = -2n/r \tag{4.3}$$

and in the special case in air (i.e. $n = 1$)

$$F = -2/r \quad \text{(in air)} \tag{4.3a}$$

The focal length f in air (defined as the distance $\mathcal{P}'\mathcal{F}'$) is then simply given by equation (3.33) and thus

$$f = r/2 \quad \text{(in air)} \tag{4.4}$$

4.1.2 The paraxial "reflection" equation at a single surface

Using the rules given above, for a reflection at a single surface, the paraxial refraction equation given by equation (3.1), that is

$$n'u' - nu = -hF$$

becomes

$$u' + u = -2hC \tag{4.5}$$

4.1.3 The lens equation of a single surface

For a reflecting surface, the lens equation (3.4) with $n' = -n$ becomes

$$-(n/l') - (n/l) = F = -(2n/r)$$

and this reduces to

$$\frac{1}{l'} + \frac{1}{l} = \frac{2}{r} \tag{4.6}$$

and this should now be called the "mirror equation" instead of the lens equation.

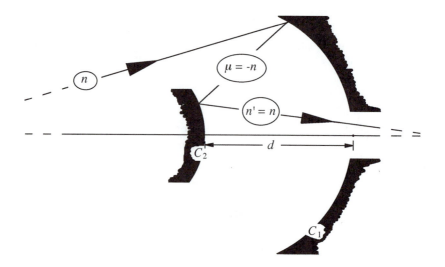

4.1.4   The equivalent power of a two-surface reflecting system

Fig. 4.3:   A general
two-surface reflecting
system, showing symbols
and change in signs.

Equation (3.20) for the equivalent power of a thick lens is

$$F = F_1 + F_2 - (F_1 F_2 d / \mu) \tag{4.7}$$

This equation can now be used to find an equation for the power of a reflecting system consisting of two surfaces, by appropriate definition of the powers F_1 and F_2. In the reflecting case, they remain surface powers. For the system shown in Figure 4.3 and recalling equation (4.1) for surface powers, we use

$$\mu = -n, \quad n' = n$$

Thus

$$F_1 = C_1[(-n) - n] = -2nC_1$$

$$F_2 = C_2[n - (-n)] = +2nC_2$$

and

$$d = -|d|$$

Substituting these quantities into equation (4.7) leads to the equation

$$F = 2n(-C_1 + C_2 + 2C_1 C_2 |d|) \tag{4.8}$$

where $|d|$ is of course now to be taken as positive.

> **Example 4.2:** Let us solve the problem given in Example 4.1 (Figure 4.2), by direct substitution in equation (4.8).

Table 4.1. System construction for Example 4.3

Surface	n	C	d	
	1.0			
1		0.129870		
	1.376		0.5	cornea
2		0.147059		
	1.336		3.1	aqueous
3		0.1		
	−1.336		−3.1	aqueous
4		0.147059		
	−1.376		−0.5	cornea
5		0.129870		
	−1.0			

Note: These data are for a reflection from the front surface of the lens of the Gullstrand number 1 relaxed schematic eye. The relevant section of the eye is shown in Figure 4.4.

Solution: In equation (4.8), we put $n = 1, C_1 = -0.02, C_2 = 0$ and $d = 10$. Therefore

$$F = 2 \times 1 \times [-(-0.02) + 0 + (-0.02) \times 0 \times 10]$$

$$= +0.04 \, \text{unit}^{-1}$$

which is the same value found in Example 4.1 by ray tracing.

4.2 Catadioptric systems

As stated in the introduction, catadioptric systems are optical systems containing both refracting and reflecting elements. A simple example in visual optics is the phenomenon of a Purkinje image. This is an image of an object reflected from one of the four refracting surfaces of the eye.

If we know the optical construction of an eye, we can easily locate the position of a Purkinje image and its size by ray tracing. This ray tracing is made very easy if we construct a suitable data table. The following example shows how this is done.

Example 4.3: Find the position of the Purkinje image formed by reflection from the front surface of the crystalline lens, using the Gullstrand number 1 relaxed schematic eye. Take the object at infinity.

Solution: The data for this eye are given in Appendix 3. The first step in solving this problem is to assemble a system construction data table that can be used for the ray tracing. This data table is given in Table 4.1 and the optical system is shown in Figure 4.4. In constructing this table, we used the rule that on reflection both refractive indices and distances changed sign, but not surface curvatures.

This data was used to trace a ray from an axial object point at infinity, using equations (3.5) to (3.8). Table 4.2 contains the ray trace results

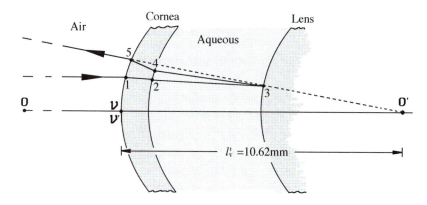

Fig. 4.4: A catadioptric system, a reflection off the front surface of the crystalline lens of the eye (surface 3), for Example 4.3. The data are given in Table 4.1 and ray trace results are given in Table 4.2. The eye is the Gullstrand number 1 relaxed eye.

Table 4.2. A ray trace through the catadioptric system given in Table 4.1

	u	h
	0.000000	
1		1.000000
	− 0.035488	
2		0.982256
	− 0.032225	
3		0.882357
	− 0.144246	
4		1.329520
	− 0.145737	
5		1.402388
	− 0.132053	

Note: The ray trace is also shown in Figure 4.4.

and this ray is shown in Figure 4.4. The image distance measured from the corneal vertex is given by equation (3.9b)

$$l'_v = -h_k/u'_k \text{ where } k = 5$$

Therefore from Table 4.2

$$l'_v = -(1.402388)/(-0.132053) = +10.62 \text{ mm}$$

Thus the image is formed inside the eye at a distance of 10.62 mm from the corneal vertex.

Exercises and problems

4.1 Calculate the equivalent power of a spherical mirror surface that has a radius of curvature of +6.7 cm.

ANSWER: $F = -0.299 \text{ cm}^{-1}$

4.2   Taking the radius of curvature of the front surface of the cornea as 7.8 mm, calculate the image position of an object placed 50 cm in front of the cornea and reflected in the cornea.

        ANSWER: 3.870 mm behind the cornea

4.3   Calculate the equivalent power of a reflecting system consisting of two mirrors, $r_1 = +8$ cm and $r_2 = +5$ cm, separated by a distance of 6 cm.

        ANSWER: $F = 0.75$ cm^{-1}

4.4   Construct a data table similar to Example 4.2 that can be used for ray tracing to find the position and magnification of the Purkinje image formed by a reflection of the back surface of the lens of the Gullstrand number 2 accommodated schematic eye (see Appendix 3 for the data of this eye).

Summary of main symbols and equations

$n'_j = n_{j+1}$ refractive index in j^{th} space
$u'_j = u_{j+1}$ paraxial angle in j^{th} space
C_j surface curvature of jth surface
d surface separation (change sign on reflection)
i, i' angles of incidence and refraction relative to the normal
h_j paraxial ray height at j^{th} surface
F_j surface power of jth surface

Section 4.1.1: Surface power of a reflecting spherical surface

$$F = -2n/r \qquad\qquad (4.3)$$

Section 4.1.3: The "lens equation" of a single surface

$$\frac{1}{l'} + \frac{1}{l} = \frac{2}{r} \qquad\qquad (4.6)$$

5

Non-Gaussian optics: Introduction to aberrations

5.0   Introduction

Paraxial optics only gives a guide to the image formation by real optical systems. Real imagery is different from the ideal or Gaussian model because of the effects of **aberrations** and **diffraction**. Both of these cause the light distribution in the image space, and most importantly in the Gaussian image plane, to be different from that in the object space or plane. Whereas diffraction can only be explained in terms of physical optics, aberrations can be discussed in terms of either geometrical or physical optics. As a general rule, geometrical optics only adequately describes the image plane light level distribution on a coarse scale and this is only accurate in highly aberrated systems. On the other hand, physical optics is more accurate than geometrical optics and so better describes the light level distribution on a fine scale, which is particularly important when the aberrations are small or zero. However, physical optical calculations are usually more complex and difficult and therefore we prefer to use the simpler geometric optical approach as often as possible.

Aberrations may be defined as the factors which cause the departure of real rays from the paths predicted by Gaussian optics. They may be investigated by following the paths of real rays through an optical system, using some suitable ray tracing procedure (e.g. the one described in Section 2.3 of Chapter 2) and comparing their paths with the paths of equivalent paraxial rays.

5.0.1   *Aberration of a beam*

Beams, not single rays, form images and therefore the quality of an image depends upon the combined aberrations of all the rays in the beam. Figure 5.1 shows an ideal beam from a general off-axis point Q on the object plane at o and passing through a general optical system and focussing to a point at Q' on the Gaussian image plane of o'. This general optical system is only represented by its front vertex plane at v and back vertex plane at v'. The **central** ray of this beam crosses the optical axis at ε and ε' in object space and image space, respectively. Now the aberrations of the beam depend upon the path of the beam

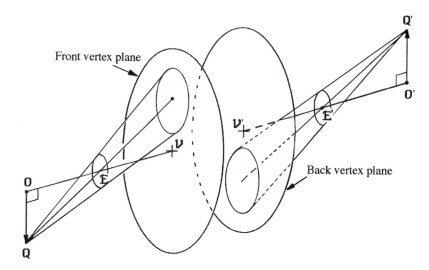

Fig. 5.1: The ideal image formation by a beam.

through the optical system, that is, depend upon where the beam strikes the first surface, which in turn depends upon where the central ray of the beam crosses the optical axis in object (or image) space.

One of the most important effects of aberrations is that a point object is no longer imaged as a point, because in general, the real rays from any object point are no longer concurrent at any point in image space. There are two important exceptions to this rule, but in these cases, the point images are no longer at the positions predicted by Gaussian optics. When the rays are not concurrent, the light in the beam is focussed over a region in the neighbourhood of the paraxial image point Q'. Whether the rays are concurrent or not, this spread of light is called the **point spread function**.

5.0.1.1   *The point spread function*

The point spread function is defined as the light distribution in the image of a point, measured in a plane at or close to the Gaussian image plane. This distribution depends upon both the level of aberration in the beam and diffraction, both of which depend upon the wavelength. If one neglects diffraction, the point spread function is simply the density of the ray intersections with the image plane. This pattern is known as the **spot diagram**. However, for small aberration levels, diffraction affects the point spread function more than the geometrical optics ray aberrations, but diffraction effects will be neglected in this chapter. Therefore in this chapter, only the geometrical optics point spread function will be considered. A deeper discussion which includes a mathematical theory of the point spread function is presented in Chapter 34, where the effects of both aberrations and diffraction will be quantitatively investigated.

The form of the point spread function depends upon the amount and types of aberrations present, and examples of these distributions will be given in Sections 5.1 and 5.2 for the different aberration types. The shape of the point spread function also depends upon the cross-sectional shape of the image forming beam. In the following discussion, we will assume that this cross-section is circular in object space, as hinted in Figure 5.1.

5.0.2 Quantification of aberrations

There are a number of methods for specifying the levels of the aberration of an optical system. For example, the width of the geometrical point spread function is one measure of the aberration in a beam. At a more fundamental level, we can look at the aberrations of the rays that make up the beam.

5.0.2.1 Aberrations of a ray

When a ray is aberrated, there are several ways in which the level of aberration can be quantified and these depend upon whether the ray is from a point on the axis or a point off the axis and whether it is a meridional or skew ray.

Rays from an axial point

For a point on the axis, the rays are all meridional. Figure 5.2a shows a beam arising from a point O on axis and a particular real ray of this beam, the ray $O\mathcal{B}\mathcal{B}'\mathcal{H}$. According to Gaussian theory, this ray should meet the Gaussian image plane at the point O'. Instead it intersects the axis at the point G and the Gaussian image plane at the point $\mathcal{H}$. The path $O\mathcal{B}O'$ indicates the route of the corresponding paraxial ray. The aberration of the real ray may be specified in terms of any of the following quantities

Longitudinal aberration	$O'G$	
Transverse aberration	$O'\mathcal{H}$	(5.1a)
Wave (path length) aberration	$[O\varepsilon\varepsilon'O'] - [O\mathcal{B}O']$	

where the square brackets [ ] refer to optical path lengths. The wave aberration is the difference in optical path length between that of the central ray of the beam and that of any other ray of the beam. In this case, the central ray of the beam travels along the optical axis. The transverse aberrations can be used to assemble the spot diagram, which is a geometrical optical measure of the point spread function. For any ray, these aberrations must in some way be connected and these connections are discussed in Chapter 33.

Rays from an off-axis point

Real rays from an off-axis object point may be either meridional or skew, and in general the rays will be skew. Since skew rays do not intersect the optical axis or the central ray of the beam, longitudinal aberration is not applicable to skew rays. For an off-axis point Q as shown in Figure 5.2b, the transverse aberration is now $Q'\mathcal{H}$ but this has two mutually perpendicular components that we will denote by the coordinates $(\delta\xi', \delta\eta')$. Therefore the transverse aberration, as defined above [equation (5.1a)] for an axial ray, is defined here to include the two components of the transverse aberration. The wave aberration is the same as defined above but now the central ray of the beam does not follow the optical axis, but instead the path $Q\varepsilon\varepsilon'Q'$. In summary, we can quantify the aberrations

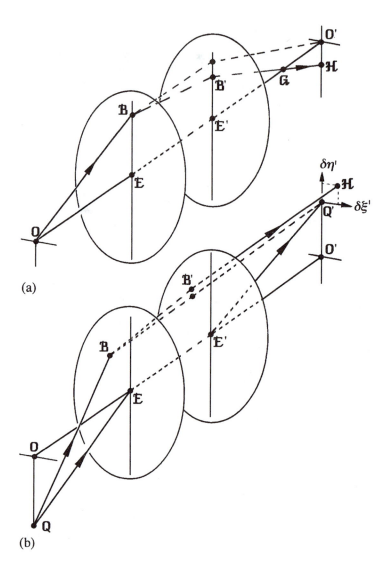

Fig. 5.2: Different measures of aberration levels of a ray (a) from an axial object point: longitudinal ($o'\mathcal{G}$), transverse ($o'\mathcal{H}$), wave ($[o\varepsilon\varepsilon'o'] - [o\mathcal{B}o']$) and (b) from an off-axis object point: transverse ($\varrho'\mathcal{H}$) and wave ($[\varrho\varepsilon\varepsilon'\varrho'] - [\varrho\mathcal{B}\varrho']$).

(a)

(b)

of rays from off-axis points in terms of the following quantities:

Transverse aberration	$\varrho'\mathcal{H}$	
Two perpendicular components	$(\delta\xi', \delta\eta')$	(5.1b)
Path length (wave) aberration	$[\varrho\varepsilon\varepsilon'\varrho'] - [\varrho\mathcal{B}\varrho']$	

There are other methods of quantifying the level of aberrations of rays and beams and some of these will be introduced in the following sections, when we discuss the individual aberrations.

5.0.3 Theory of aberrations

The mathematical theory of aberrations shows that aberrations may be classified into distinct aberration types which lead to distinct patterns for the longitudinal, transverse and wave aberrations and the point spread function distributions. We will look at these aberration types in the next two sections and their longitudinal and transverse aberration patterns. The theory also shows that, in general, the aberrations of a ray increase in magnitude (a) the farther the ray is from the central ray of the beam, and (b) with distance of the object point ϱ from the axis.

For low levels of aberrations, there are seven aberration types and these aberrations are called **Seidel** or **primary** aberrations. For higher levels of aberrations, these and other aberration types are present in the beam. The "other" aberrations are known as **higher order aberrations**, but we will not examine the form of these higher order aberrations in this book.

This aberration theory also explains how the aberration of a ray can be calculated from the path of the ray and the constructional parameters of the system (refractive indices, surface curvatures and surface separations). An example of this approach is given in Section 5.3, where we show how to predict the aberration for an axial image point, imaged by a spherical surface. This example also shows how the higher order aberrations arise. ˙

We will begin the aberration classification by dividing them into two groups:

(a)   monochromatic aberrations, which are those that occur in monochromatic light (and also polychromatic light)
(b)   chromatic aberrations, which are those that occur only in polychromatic light and are due to the dispersion of material.

The development of the theory of these seven aberrations assumes that the optical system is rotationally symmetric. A rotationally symmetric system must be constructed with rotationally symmetric surfaces such as spherical surfaces and the centres of curvature of all surfaces must be colinear. We will also assume rotational symmetry in the following discussion.

5.1   Monochromatic aberrations

The mathematical theory of aberrations, mentioned briefly in Section 5.0.3 and to be discussed further in Chapter 33, shows that the aberration level of any ray depends upon the distance of the ray from the central ray in the beam and the distance of the object point ϱ from the optical axis point o. This theory shows that the aberration of a ray can be expanded as a power series in these quantities and the different terms in the expansion correspond to distinct aberration types. Thus the different types of aberrations can be distinguished by how the aberration level depends on the distance between the ray and the central ray of the beam and the distance of the off-axis object point from the axis. In this chapter, we will use the same monochromatic aberration types as identified by this theory but group them according to whether they

(a)   occur both on- and off-axis or only off-axis and
(b)   lead to point to point imagery or not.

This mathematical theory predicts the existence of five monochromatic aberrations and these are laid out in Table 5.1 according to this classification.

Table 5.1. The seven primary aberrations

Monochromatic aberrations	
On- and off-axis aberration	
Do not give point to point imagery	Spherical aberration
Off-axis only aberrations	
Do not give point to point imagery	Coma
	Astigmatism
Give point to point imagery	Petzval curvature
	Distortion
Chromatic aberrations	
On- and off-axis aberration	Longitudinal chromatic aberration
Off-axis only aberration	Transverse chromatic aberration

Aberrations that do not give point to point imagery are aberrations in which the rays from a point in object space are not concurrent at any point in image space; that is, a point object does not form a point image. With these aberrations, the image of a point is a spread of light (the point spread function) in any image plane, whether it be the Gaussian image plane or not. There are three aberrations of this type. These aberrations can be distinguished by their effect on ray paths or by the shape of their point spread functions (or spot diagrams).

There are two aberrations that give point to point imagery, but in these cases, the image point is not formed in the position predicted by Gaussian optics.

We will now look at these aberrations in turn, observe the particular ray patterns that are characteristic of each aberration and note the forms of the point spread function (or the equivalent spot diagram).

5.1.1   On- and off-axis aberration

There is only one monochromatic aberration of the on- and off-axis aberration type. This is spherical aberration and because it occurs on axis in a rotationally symmetric system, it also must be rotationally symmetric.

Spherical aberration

In Chapter 2, we observed the effect of spherical aberration on rays from an axial object point and refracted by a single surface. Here we will examine this aberration a little further.

The best way of describing the effect of spherical aberration is to consider the effect of the aberration on a beam of rays from an axial object point. Figure 2.6 shows a beam from a point o and typical ray paths, for a positive power surface with positive spherical aberration, but a more complex system with positive spherical aberration would show the same ray pattern. Ideally, all the rays should pass through o', the paraxial image of o. However, the effect of positive spherical aberration is to make the rays cross the axis closer to the surface as the ray height at the surface increases. The aberration is rotationally symmetric so we only show the ray paths in one section.

We can interpret the aberration as being due to an excess or lack of peripheral or marginal refractive power; that is, we can quantify the aberration of a particular ray as

$$\text{spherical aberration} = \text{peripheral power} - \text{central (paraxial) power}$$

(5.2)

The magnitude of this quantity will depend upon the ray height in the beam. The farther the ray is from the centre of the beam, the greater the aberration. Thus for a positive power system, in which (real) rays cross the axis closer to the lens as the ray height in the beam increases (the case in the diagram), the spherical aberration is positive. For a positive power system with negative spherical aberration, peripheral rays must cross the axis farther away from the system than the paraxial image point o'. The effect of positive and negative spherical aberration on the ray paths in a negative power system can be left to the reader to investigate and is given as an exercise at the end of this chapter (problem 5.1).

An imaginary line that would bound the beam in image space is called the **caustic curve**.

The point spread function and spot diagram. For a rotationally symmetric optical system, the ray patterns shown in Figure 2.6 must also be rotationally symmetric. Therefore the point spread function has rotational symmetry. From the transverse aberrations in the diagram, the ray density is highest in the centre and therefore the point spread function has a maximum value in the centre and decreases towards the edge.

5.1.2 Off-axis only aberrations

There are four monochromatic aberrations that occur only off-axis. Two of these (coma and astigmatism) do not give point to point imagery and two (field curvature and distortion) do.

5.1.2.1 Aberrations that do not give point to point imagery Coma

The coma aberration is best described in terms of the ray intersection pattern with the image plane for rays in object space that form annular cones with increasing diameter, as shown in Figure 5.3. In the object space beam cross-section, these rays form a set of concentric circles. The ray intersection pattern on the Gaussian image plane is also a set of circles, but they are no longer concentric, as shown in the diagram. One interesting property of the ray pattern is that as a ray moves around the circle in object space, the corresponding ray in image space moves around the circle at twice the speed.

The point spread function and spot diagram. From the discussion in the preceding paragraph, it is clear that the coma point spread function is composed of a set of circles, which are progressively displaced as the circles increase or decrease in size, as shown in Figure 5.3. In this case, the point spread function in the Gaussian image plane has a similar shape to a comma but without the curl on the tail. The light level is not uniform and is maximum at the pointed end and lowest at the thicker and rounded other end. The angle at the apex of the coma flare is $60°$.

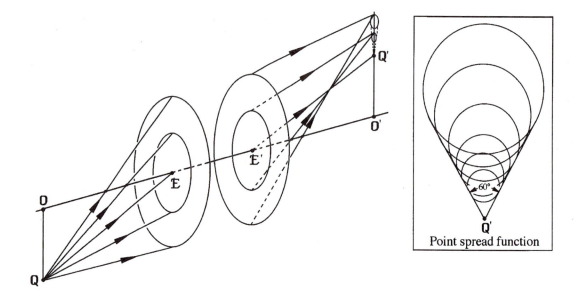

Point spread function

The amount of coma in the beam may be quantified in terms of the dimensions of the coma flare. The coma flare may point away from the axis, as shown in Figure 5.3, or may point towards the axis. The direction depends upon the sign of the aberration in a particular case.

Fig. 5.3:   Coma aberration.

Astigmatism

The astigmatism aberration can be described with reference to Figure 5.4. In this diagram, the beam can be thought of as composed of a set of fans of rays in one section or a set of fans of rays in a section perpendicular to the first. One section is the **tangential** (T) section. This section is in the direction of the plane containing the off-axis object OQ and the optical axis. The other section is called the **sagittal** (S) section. These sections do not have a common focus. The focus of the tangential fans of the beam is called the tangential (T) focus, and that of the sagittal fans is the sagittal (S) focus. Both of these sections focus to lines which are perpendicular to each other and at different distances from the Gaussian image plane, as shown in the diagram.

As the point Q moves over the object plane, the positions of the sagittal and tangential image lines map out two curved surfaces, as shown in Figure 5.5. In the vicinity of the axial image point O', these surfaces are close to being spherical. Sometimes this pattern is called a "teacup and saucer" diagram, because if the tangential surface looks like a teacup, the sagittal surface looks like the saucer.

The level of astigmatism in the beam forming the image Q' may be quantified by the distance between the tangential and sagittal image lines. This distance is called the **interval of sturm**, and it is clear that the magnitude of this interval depends upon the distance of the object point Q from the axis. The level of aberration for the image as a whole may be quantified by the radii of curvatures of the sagittal and tangential surfaces shown in Figure 5.5. Equations for these radii are given in Chapter 33.

Fig. 5.4: Astigmatism aberration.

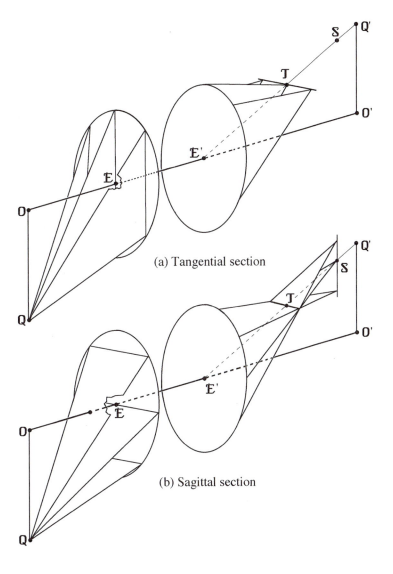

(a) Tangential section

(b) Sagittal section

The point spread function. In this situation, we do not think of this aberration as having a unique point spread function, as we often observe the light distribution in any plane between the tangential and sagittal focuses and the Gaussian image plane. The shape of the point spread function in these planes is indicated in Figure 5.6. It is clear from this diagram that the tangential image line is tangential to a circle, perpendicular to and centred, on the optical axis. On the other hand, the sagittal image line is along a radius of this circle. The tangential and sagittal image lines are the point spread functions in the tangential and sagittal image planes, respectively. In other image planes, the point spread function has other shapes. For example, as the observation plane moves from the tangential to the sagittal plane, the shape of the point spread function changes from a line to an ellipse with its major axis in the direction of the tangential image line, to a circle, to another ellipse with its major axis in the direction of the sagittal image line and finally to a line at the sagittal image. This

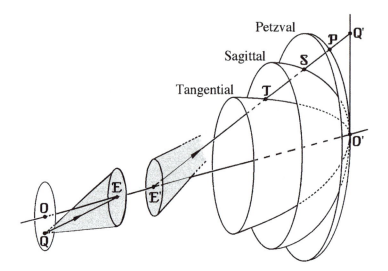

Fig. 5.5: The tangential, sagittal and Petzval surfaces.

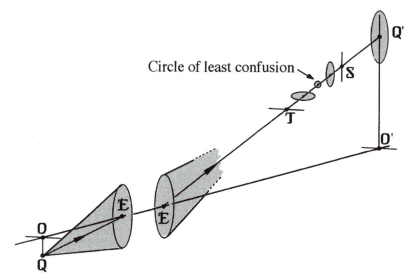

Fig. 5.6: The changing shape of the point spread function in the presence of astigmatism as a function of the position of the image plane.

is shown in the diagram and the ray density is uniform in all these distributions and hence the light level is uniform across the different point spread functions. The smallest circle that encloses the beam at its narrowest point is called the **circle of least confusion** and for a beam with a circular cross-section it is the circle shown in the diagram.

5.1.2.2 Aberrations that give point to point imagery

There are two aberrations in which the rays from a point object are concurrent at a single point in the image space, but this image point is not at the expected Gaussian image position.

Fig. 5.7: Field curvature.

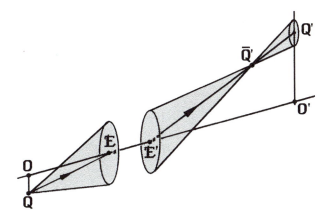

Field curvature

Neglecting all other aberrations, a point Q in the object plane, as shown in Figure 5.7, is not imaged at the Gaussian image point Q'. Instead it is imaged at a point $\bar{Q}'$ which lies on a curved surface that passes through the Gaussian axial image point O' as shown in Figure 5.5. In the region close to the optical axis, the image surface is almost spherical. This surface is called the **Petzval** surface.

The level of a field curvature in the beam may be quantified by the distance between the image position $\bar{Q}'$ and the Gaussian image plane. For the image as a whole, it may be quantified by the radius of curvature of the Petzval surface.

The point spread function. In the plane containing the point $\bar{Q}'$, the point spread function is a point, but in other planes it is an evenly illuminated circular disc, whose diameter increases with distance from this plane. On the (flat) Gaussian image surface, the point spread function at Q' will be a circular blurred image of Q because it is defocused at this plane. The level of defocus will increase with distance of the point Q off-axis. In some optical systems, the object or image recording surface is curved to compensate for this effect.

For any real aberrated system, one can calculate the shapes of the tangential, sagittal and Petzval surfaces and, as indicated, close to the axis these surfaces can be regarded as spherical. A typical situation is shown in Figure 5.5. The tangential surface is farther than the sagittal surface from the Petzval surface, and close to the axis, the distances are in a ratio of 3:1. As astigmatism decreases, the tangential and sagittal surfaces approach the Petzval surface, and in the limit of zero astigmatism, they coincide with the Petzval surface. Depending upon the sign of the aberration, the tangential and sagittal surfaces may be on either side of the Petzval surface. In the diagram, they are on the side nearer the optical system, but if the numerical value of astigmatism had the opposite sign, they would be on the side away from the optical system.

Distortion

When the distortion aberration is present, the point Q is imaged on the Gaussian image plane but not at the expected Gaussian position. Instead the image $\bar{Q}'$ is either farther away or closer to the optical axis than the expected position Q'.

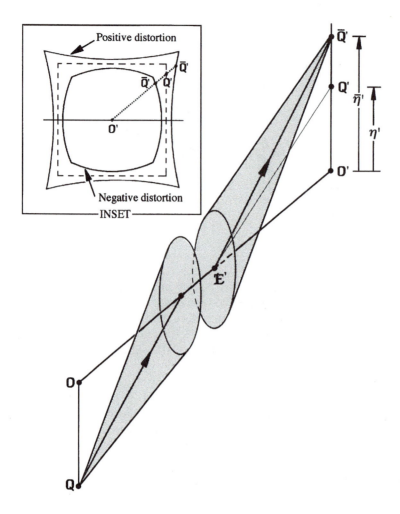

Fig. 5.8: Distortion aberration and, in the inset, the effect of distortion on the image of a square. The dashed line is the Gaussian image of the square.

Figure 5.8 shows the case where the actual image point $\overline{Q}'$ is farther away from the axis.

The magnitude of the distortion of the beam is usually quantified in terms of the fractional distance between the real image point $\overline{Q}'$ and the Gaussian image point Q', that is, by the equation

$$\text{fractional distortion} = \frac{o'\overline{Q}' - o'Q'}{o'Q'} = \frac{\overline{\eta}' - \eta'}{\eta'} \tag{5.3}$$

For low aberration levels, the magnitude of this value is proportional to the square of the Gaussian image height η'.

The value of the fractional distortion may be positive (**pincushion**) or negative (**barrel**) distortion. The effect of these two types of distortion is well perceived by looking at the imagery of a square, as shown in the inset of Figure 5.8.

The point spread function. In the Gaussian image plane, the point spread function is a point.

5.1.3 Laboratory tests for distinguishing among the different aberrations

If we wish to examine a particular optical system for the presence of aberrations, a convenient laboratory test is the **star test**. In this test, we examine the image of a point source of light. Thus we observe the point spread function. In real optical systems, all the above aberrations will usually be present simultaneously, but since the level of each aberration depends by different amounts on such factors as conjugate position, beam width and distance of the point from the axis, the conditions can sometimes be manipulated to increase the level of a particular aberration and at the same time reduce the others. Whatever the situation, we may wish to identify the dominant aberrations present. The shape of the point spread function is an initial guide to the dominant aberration. The observation image plane can be moved backwards and forwards to detect the presence of astigmatism because its point spread function changes shape with distance from the Gaussian image plane.

A very useful technique for differentiating among the different aberrations is to observe the change in the shape and size of the point spread function as the diameter of the beam is increased or decreased. From the foregoing description of the formation of the point spread function for each aberration, it is easy to deduce how the point spread function changes in size and shape as the diameter of the beam is increased or decreased. This exercise is left to the reader.

Field curvature and distortion usually cannot be detected by the star test because their point spread functions are either points or simply blurred images of points. If they are present, these aberrations are best examined by observing the image of a square grid pattern centred on the axis. A defocus in the peripheral or central field would indicate field curvature, and departures from straightness of the grid lines would indicate distortion.

5.2 Chromatic aberrations

Chromatic aberrations arise because the refractive index of optical materials, whether they be glass, plastic or water, varies with wavelength. This effect is called dispersion and was first mentioned in Section 1.2. Figure 5.9 shows typical variations for two types of optical glass, one with a low refractive index and one with a high index. In optical systems, dispersion causes the imaging properties (e.g. power) of the component lenses to vary with wavelength.

Dispersion produces two types of chromatic aberrations, and these are listed and differentiated in Table 5.1. These aberrations are classified by using the same scheme as used for monochromatic aberrations.

5.2.1 On- and off-axis aberration

Longitudinal chromatic aberration

While longitudinal chromatic aberration occurs on- and off-axis, it is best described by its effects on the image of on-axis object points. In Figure 5.10, a white light beam arises from an axial object point at o. If the system is not corrected for longitudinal chromatic aberration, the rays in the beam are dispersed by the system, with a typical red beam focussing at o'_{red}, a yellow beam focussing at o'_{yellow} and a blue beam at o'_{blue}. These three points are corresponding wavelength dependent images of o.

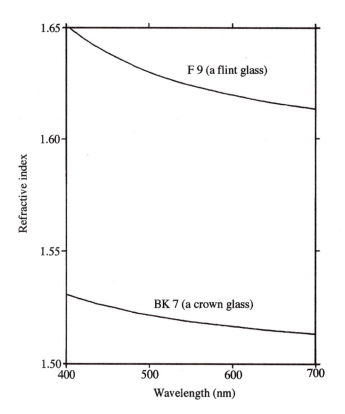

Fig. 5.9: Variation of refractive index with wavelength (dispersion) for two samples of Schott (1992) glass.

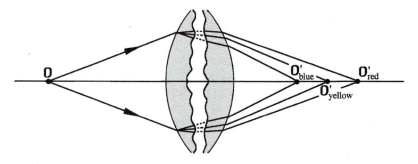

Fig. 5.10: Longitudinal chromatic aberration.

The longitudinal distance between the red and blue images is one measure of the longitudinal chromatic aberration.

The point spread function. For a white light source, the point spread function in the Gaussian image plane is circular and should show some colour differences across the patch, with the yellow rays concentrating at the centre and blue and red towards the edge, so that the point spread function should have a yellowish centre and a purplish edge. In other planes, different colours may concentrate at the centre and at the edge, giving different variations in colour across the point spread function.

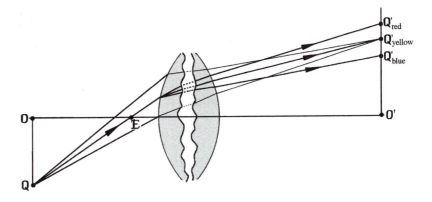

5.2.2 Off-axis only aberration

Transverse chromatic aberration

For off-axis beams, the effect of the dispersion is to change the transverse magnification with wavelength. Thus an off-axis object point Q is imaged at different distances from the axis, as shown in Figure 5.11. For a lens with no correction for chromatic aberrration, the red image Q'_{red} will be farther from the axis than the yellow image Q'_{yellow}, and the blue image Q'_{blue} will be closer, as shown in the diagram.

Transverse chromatic aberration (TCA) may be quantified in terms of the distance between the red and blue images, or as a fraction change in image size, according to the equation

$$\text{TCA} = \frac{O'Q'_{red} - O'Q'_{blue}}{O'Q'_{yellow}} \tag{5.4}$$

The point spread function. For a white light source, in the paraxial image plane, the point spread function is a line of light that shows the colours of the spectrum.

Equivalent prismatic effect

For a simple lens, these chromatic aberrations can be thought of as being due to the lens having a prismatic form which increases with the distance from the optical axis. The prismatic effect of a simple lens is discussed further in Chapter 6 and the ability of a prism to disperse polychromatic light is discussed in more detail in Chapter 8.

5.3 Spherical aberration at a surface

In this section, we will investigate the level of spherical aberration at a spherical and also at a non-spherical surface, and use the results to investigate some of the factors that affect the aberration level, such as the position of the conjugates and surface shape. We will then proceed to look at surfaces that are free of spherical aberration.

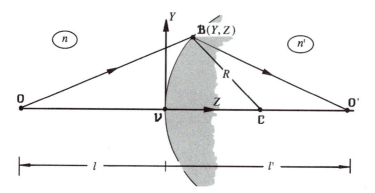

Fig. 5.12: Spherical wave (W) aberration at a refracting surface. $W = [ovo'] - [oBo']$.

5.3.1   Spherical aberration at a spherical surface

A convenient way of examining the level of spherical aberration at a spherical surface is to use the concept of wave aberration that was introduced and defined in Section 5.0.2.1 by equation (5.1a). Figure 5.12 shows an object point o on axis at a distance l from the vertex v of a spherical refracting surface. The paraxial image is at o', which is at a distance l' from the vertex. If all the rays from o that are refracted by the surface are to be concurrent at o', Fermat's principle states that the optical path length $[oBo']$ must be the same for all positions of B or implies that

$$[oBo'] = n'l' - nl \tag{5.5}$$

for perfect imagery. However, we now know that spherical refracting surfaces have spherical aberration and the real rays from o are not concurrent at o'. In this case, equation (5.5) is not satisfied and a measure of the aberration of a ray is the wave aberration. The wave aberration, denoted by the symbol W, of the ray oB is the optical path difference between it and the central ray of the beam; that is

$$W = (n'l' - nl) - [oBo'] \tag{5.6}$$

From the diagram,

$$W = (n'l' - nl) - \{n'\sqrt{[Y^2 + (l' - Z)^2]} \tag{5.7}$$

$$+ n\sqrt{[Y^2 + (Z - l)^2]}\}$$

where Y and Z are related by the following equation of the spherical surface

$$Z^2 - 2rZ + Y^2 = 0 \tag{5.8}$$

That is,

$$Z = r - \sqrt{(r^2 - Y^2)} \tag{5.9}$$

Using the binomial expansion, we can express this equation for Z as a power series in Y, that is,

$$Z = \frac{Y^2}{2r} + \frac{Y^4}{8r^3} + \text{higher order terms} \tag{5.9a}$$

At this stage, we will neglect the higher order terms in this expansion and we can use this equation to eliminate Z from equation (5.7). If we do this and once again use the binomial expansion to eliminate the square roots from equation (5.7) and again omit terms of order higher than the Y^4 term, we finally have

$$W = -\frac{Y^2}{2}\left\{\frac{n'}{l'} - \frac{n}{l} - \frac{(n'-n)}{r}\right\} - \frac{Y^4}{8r^2}\left\{\frac{n'}{l'} - \frac{n}{l} - \frac{(n'-n)}{r}\right\}$$

$$+ \frac{Y^4}{8}\left\{\frac{n'}{l'}\left(\frac{1}{l'} - \frac{1}{r}\right)^2 - \frac{n}{l}\left(-\frac{1}{l} + \frac{1}{r}\right)^2\right\} \tag{5.10}$$

Now since the point o' is the paraxial image of o, the lens equation (2.11), with F defined by equation (2.6), gives

$$\frac{n'}{l'} - \frac{n}{l} = F = C(n'-n) = \frac{(n'-n)}{r}$$

and therefore the above wave aberration W reduces to

$$W = w_4 Y^4 \tag{5.11}$$

where

$$w_4 = \left\{\frac{n'}{l'}\left(\frac{1}{l'} - \frac{1}{r}\right)^2 - \frac{n}{l}\left(-\frac{1}{l} + \frac{1}{r}\right)^2\right\} \Big/ 8 \tag{5.11a}$$

This equation shows that the wave aberration form of the spherical aberration varies at least as the fourth power in ray height Y at the surface. We say "at least" because in the binomial expansions above, we neglected terms of order higher than Y^4. If we had included these terms, we would find that the wave aberration is of the form

$$W = w_4 Y^4 + w_6 Y^6 + w_8 Y^8 + \cdots \tag{5.12}$$

where w_4, w_6, and so on, are the coefficients of the terms. The $w_4 Y^4$ term is the primary spherical aberration and the other terms are the **higher order** spherical aberration.

Equation (5.11) also shows that spherical aberration at a surface

(a)   depends upon the position of the conjugates,
(b)   is non-linearly related to the surface radius of curvature (r) or curvature ($C = 1/r$) (in fact it is quadratic in curvature), and
(c)   also occurs if r is infinite, that is, at a flat surface.

In some texts, spherical aberration is attributed to the general spherical nature of refracting and reflecting surfaces. However, this is misleading because we have just seen that a flat surface has spherical aberration. Let us investigate the spherical aberration at a flat surface a little further.

5.3.2   Spherical aberration at a flat surface

For a flat surface, the value of r is infinite and equation (5.11) can be written as

$$W = \frac{Y^4}{8} \left\{ \frac{n'}{l'^3} - \frac{n}{l^3} \right\}$$

Now from the lens equation (2.11) with $F = 0$,

$$l' = (n'/n)l$$

and so we have

$$W = \frac{Y^4}{8} \frac{n}{l^3} \frac{(n^2 - n'^2)}{n'^2} \tag{5.13}$$

which expresses the primary spherical aberration of a flat refracting surface in terms of (a) the ray intersection distance Y from the axis or normal to the surface and (b) the object distance l. Let us now look at the transverse and longitudinal forms of primary spherical aberration.

Figure 5.13a shows a slab of material and a point at the bottom of the slab at the position o. If we view the point o from immediately above, it appears to be at o' and this corresponds to the paraxial image position. Applying the lens equation (2.11) (with $F = 0$) to this situation, we have the apparent depth l'_o

$$l'_o = n'l/n \tag{5.14}$$

Figure 5.13b shows a non-paraxial or real ray being traced from o, to the point $\mathcal{B}$ on the upper surface, out of the slab and extended back to the paraxial image plane at o'. We will find an expression for the longitudinal aberration $(l' - l'_o)$ of the real ray. First we find the apparent depth l', as a function of the distance Y from the normal. We begin by applying Snell's law, equation (1.12), that is,

$$n' \sin(i') = n \sin(i)$$

From simple trigonometry

$$\sin(i) = \frac{Y}{\sqrt{(l^2 + Y^2)}} \quad \text{and} \quad \sin(i') = \frac{Y}{\sqrt{(l'^2 + Y^2)}}$$

Thus from Snell's law above

$$\frac{nY}{\sqrt{(l^2 + Y^2)}} = \frac{n'Y}{\sqrt{(l'^2 + Y^2)}}$$

Fig. 5.13:   Example of
spherical aberration at a
plane surface and the
longitudinal and transverse
aberrations.

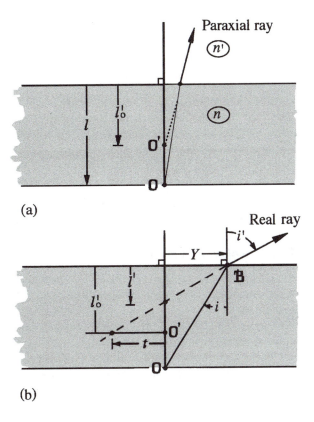

Therefore

$$n^2(l'^2 + Y^2) = n'^2(l^2 + Y^2)$$

That is

$$l' = \frac{n'l}{n}\sqrt{\left\{1 + \frac{(n'^2 - n^2)}{n'^2}\frac{Y^2}{l^2}\right\}}$$

If we take the special case $Y = 0$, that is the paraxial case, the apparent depth reduces to that given by equation (5.14). If we expand the above square root using the binomial theorem and form the difference $(l' - l'_o)$, which is the longitudinal aberration, then

$$l' - l'_o = \frac{Y^2(n'^2 - n^2)}{2nn'l} + \text{terms of order } Y^4 \text{and higher} \qquad (5.15)$$

This equation shows that for small values of Y, the longitudinal aberration is quadratic in Y.

We can use simple algebra and the binomial theorem again to show that the transverse aberration (t in Figure 5.13b) is cubic in Y. To do this, we begin with the relation

$$t/(l' - l'_o) = Y/l'$$

and solve for t in terms of Y and finally we get

$$t = \frac{Y^3(n'^2 - n^2)}{2n'^2l^2} + \text{terms of order } Y^5 \text{and higher} \tag{5.16}$$

The distance Y is the distance from the surface normal through o and o' to the point $\mathcal{B}$, which would be the optical axis in the general case. Therefore the results show that the longitudinal spherical aberration [equation (5.15)] and transverse spherical aberration [equation (5.16)] for a plane surface are quadratic and cubic in the distance of the ray from the optical axis respectively. In comparison, we have seen that the wave form of the aberration [equation (15.13)] is quartic in this distance. These results apply also to a surface of any shape.

5.3.3 *Non-spherical or aspheric surfaces*

Detailed examination of equation (5.11) shows that, in general, the spherical aberration of a spherical surface is not and cannot be made zero, except for the special case where the object and image are at the centre of curvature of the surface (i.e. $l = l' = r$). We could now ask the question, does any other type of surface have zero spherical aberration for other conjugates? The answer is yes. Therefore why do we use spherical surfaces if they have spherical aberration? The answer to this question is that spherical surfaces are much easier and therefore much cheaper to make than non-spherical surfaces. In the design of optical systems there is a balance between manufacturing cost and optical quality.

Surfaces that are not spherical are called **aspheric** surfaces. A common type of rotationally symmetric aspherical surface is the conicoid. Others are possible, but let us firstly look at the conicoid.

5.3.3.1 *Conicoid surfaces*

The conicoid surface is formed by rotation of a conic around one of the axes. If the axis of rotation is the Z-axis, the general equation of a conicoid, which passes through the point $(0, 0, 0)$, is

$$p^2 + (1 + Q)Z^2 - 2rZ = 0 \tag{5.17}$$

as shown in Figure 5.14, where

$$p^2 = X^2 + Y^2 \tag{5.18}$$

r is the radius of curvature at the vertex $(0, 0, 0)$ and Q is the surface asphericity and defines the type of surface according to the rules:

$$
\begin{array}{rcl}
Q > & 0 & \text{ellipsoid, with the major axis being in the } X\text{-}Y \text{ plane} \\
Q = & 0 & \text{sphere} \\
0 > Q > & -1 & \text{ellipsoid, with the } Z\text{-axis being the major axis} \\
Q = & -1 & \text{paraboloid} \\
Q < & -1 & \text{hyperboloid}
\end{array}
$$

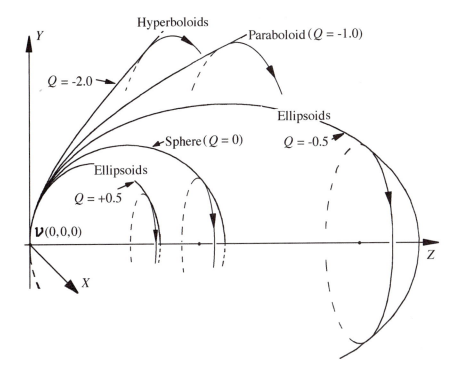

Fig. 5.14: Formation of a rotationally symmetric conicoid and effect of numerical value of asphericity Q on the shape of a conic. All curves have same curvature at vertex $\mathcal{V}$.

The cross-section in the X-Y plane is a circle and the effect of Q on surface shape is shown in Figure 5.14.

Equation (5.17) can be expressed explicitly in terms of Z, by solving the equation as a quadratic in Z. Of the two solutions, the correct solution is

$$Z = \frac{r - \sqrt{[r^2 - (1+Q)p^2]}}{(1+Q)} \tag{5.19a}$$

For some purposes, this is not a suitable form since it is indeterminate if r is infinite or $Q = -1$. An alternative and more suitable form that copes with these cases can be found by taking the other solution of the quadratic, that is, the one with the plus sign in front of the square root sign (call this Z_2) and solving for Z_1 (the correct one) above as follows. The quadratic in Z given by equation (5.17) can be written as

$$Z^2 - (2Zr)/(1+Q) + [p^2/(1+Q)]$$

The incorrect solution Z_2 of this equation is

$$Z_2 = \frac{r + \sqrt{[r^2 - (1+Q)p^2]}}{(1+Q)} \tag{5.19b}$$

Now the two solutions for Z must satisfy the condition

$$Z_1 Z_2 = p^2/(1+Q) \tag{5.20}$$

Solving for the correct solution $Z = Z_1$, we have

$$Z = \frac{p^2}{r + \sqrt{[r^2 - (1 + Q)p^2]}} \qquad (5.21)$$

and this is always determinate. This equation was used to calculate the shape of the curves shown in Figure 5.14.

Shortly, we will show that the spherical aberration of a conicoid surface depends upon the numerical value of the asphericity Q, but before we do that we will briefly discuss other types of aspheric surfaces.

5.3.3.2 *More general aspheric surfaces*

In many situations, the required surface shape cannot be represented by a conicoid, that is of the form given by equation (5.17). For example, we will see shortly that a refracting surface free of spherical aberration for conjugates at a finite distance is not a conicoid. However, these surfaces can be represented by a base conicoid with some modifications to surface contours. These modified surfaces are called **figured conicoids**.

Figured conicoids

Taking the conicoid form given by equation (5.21), here written as

$$Z_{\text{conicoid}} = \frac{p^2}{r + \sqrt{[r^2 - (1 + Q)p^2]}} \qquad (5.22)$$

a figured conicoid is expressed in the form

$$Z = Z_{\text{conicoid}} + f_4 p^4 + f_6 p^6 + f_8 p^8 + \text{etc.} \qquad (5.23)$$

where the coefficients f_4, f_6, and so on, are called figuring coefficients. The individual values of the figuring coefficients will depend upon the aspheric surface being modelled.

An exact representation of a non-conicoid aspheric by a figured conicoid would usually require an infinite number of terms in equation (5.23). In practice, the power series must be terminated at a finite number of terms, which leaves some residual error. The number of figuring coefficients required in practice will depend upon the maximum permissible error.

5.3.4 *Spherical aberration at a conicoid surface*

Let us now look at the effect of the asphericity value on the spherical aberration of a conicoid. We can find an expression for the primary wave spherical aberration W by proceeding as we did in Section 5.3.1, but this time using either equation (5.19a) or (5.21) expanded in the binomial form, instead of equation (5.9) to eliminate Z. If we do this, we arrive at the following equation for the wave primary spherical aberration coefficient w_4.

$$w_4 = \left\{ \frac{n'}{l'} \left(\frac{1}{l'} - \frac{1}{r} \right)^2 - \frac{n}{l} \left(-\frac{1}{l} + \frac{1}{r} \right)^2 \right\} \Big/ 8 \qquad (5.24)$$

$$+ Q(n' - n)/(8r^3)$$

This equation shows that the primary spherical wave aberration value is proportional to the asphericity value Q and thus for any position of the conjugates or surface radii of curvature (r), we can find some value of asphericity Q which will give zero primary spherical aberration. This result leads us to speculate that a given conicoid may have zero total [i.e. primary (i.e. Seidel) plus higher order] and not just zero primary spherical aberration alone for some particular pair of conjugates. Let us explore this idea further.

5.3.5 Spherical aberration free surfaces

Equation (5.24) indicates that by some suitable choice of asphericity Q, we can have an axial image free of primary spherical aberration. However, we must at this stage remind ourselves of the difference between the total and primary spherical aberration. While we can always find some value of Q to set the primary spherical aberration to zero, the other and higher order terms may not be zero and we would have to examine every higher order spherical aberration term. Fortunately in some particular cases, there are simpler ways to solve this problem.

5.3.5.1 A refracting surface free of spherical aberration

Figure 5.12 shows a refracting surface imaging a point o to o'. We wish to find the surface shape that provides spherical aberration free imaging. We can use Fermat's principle to help solve the problem. According to this principle, all ray paths from o to o' through $\mathcal{B}$ must have the same optical path length for all positions of $\mathcal{B}$, that is,

$$n' \sqrt{[Y^2 + (l' - Z)^2]} + n \sqrt{[Y^2 + (Z - l)^2]} = n'l' - nl \qquad (5.25)$$

This equation cannot be reduced to the conicoid form given by equation (5.17), unless $n = n'$, which is a trivial case in optics, since there would be no refraction. Curves of the form given by equation (5.25) are known as a Descartes or Cartesian oval (*Encyclopaedia of Mathematics* 58). However, a figured conicoid could satisfy this condition, with the value of the figuring coefficients depending upon the positions of the conjugates and the number of terms depending upon the desired accuracy of fit.

Thus while the general refracting surface, free of spherical aberration, is not a conicoid, it is a conicoid for the special case of the object at infinity. We will demonstrate this now.

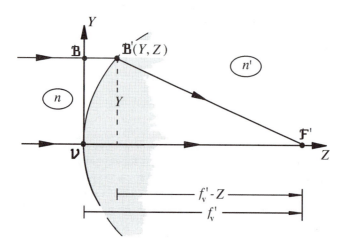

Fig. 5.15: Refracting surface free of spherical aberration for an object at infinity. This surface is an ellipsoid if $n < n'$ and a hyperboloid if $n' < n$.

Object at infinity

To find the shape of a spherical aberration free refracting surface for the object at infinity, we could solve equation (5.25), by setting the object distance l as infinite, but the method would be tedious. Fortunately, there is a simpler and more direct approach.

Consider the situation shown in Figure 5.15. The axial object is at infinity and is being imaged at the back focal point $\mathcal{F}'$ of the surface with a back vertex focal length f_v'. If the surface is free of spherical aberration, a ray striking the surface at any point $\mathcal{B}'$, as shown in the diagram, will be refracted and pass through the back focal point $\mathcal{F}'$. The vertex plane through $\mathcal{VB}$ can be regarded as a wavefront and therefore, for any point $\mathcal{B}$, the ray must pass through $\mathcal{F}'$ and Fermat's principle gives

$$n\mathcal{BB}' + n'\mathcal{B}'\mathcal{F}' = n'\mathcal{VF}'$$

That is

$$nZ + n'\sqrt{[Y^2 + (f_v' - Z)^2]} = n'f_v'$$

Simplifying leads to

$$\frac{n'^2}{f_v'(n'^2 - nn')}\left\{Y^2 + Z^2\frac{(n'^2 - n^2)}{n'^2}\right\} - 2Z = 0$$

This is equivalent to equation (5.17) if

$$r = f_v'(n - n')/n' \tag{5.26a}$$

and

$$Q = -n^2/n'^2 \tag{5.26b}$$

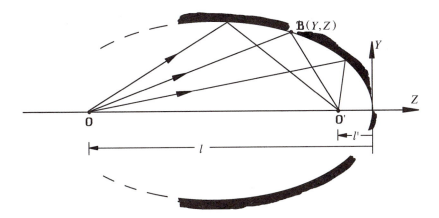

Fig. 5.16: Reflector free of spherical aberration for an object at a finite distance. This surface is ellipsoidal.

Thus equation (5.26b) gives the asphericity of a refracting surface free of spherical aberration for an object at infinity. If $n < n'$ it must be elliptical and if $n' < n$ it must be hyperbolic. If we take the example of the front surface of the cornea, which has a typical refractive index of 1.376, it would be free of spherical aberration for an object at infinity, if

$$Q = -0.53$$

Studies of corneal shape (Kiely et al. 1982) have shown that the average corneal asphericity is about

$$Q = -0.26$$

so the cornea is under-corrected for spherical aberration when viewing distant objects.

These equations show that the required asphericity to give zero spherical aberration (total not just primary), for an object at infinity, is independent of the surface radius of curvature.

Thus for a refracting surface to be free of spherical aberration, the surface shape cannot be a conicoid, unless the object (or image) is at infinity and in this case, the spherical aberration free surface is either an ellipsoid or a hyperboloid depending upon the relative values of the two refractive indices. Now let us look at reflecting surfaces free of spherical aberration.

5.3.5.2   A reflecting surface free of spherical aberration

Figure 5.16 shows a reflecting surface imaging the point o to o' free of spherical aberration. Once again we can use Fermat's principle, which states that the (optical) path length for all rays must be constant, that is,

$$\sqrt{[Y^2 + (Z - l')^2]} + \sqrt{[Y^2 + (Z - l)^2]} = -(l' + l) \qquad (5.27)$$

The minus sign on the righthand side is there because the distances l and l' have negative values in the diagram. This equation is similar to the one [equation (5.25)] for a refracting surface free of spherical aberration, which is not a

conicoid. However, if we manipulate equation (5.27) to eliminate the square
root signs, we finally have the equation

$$Y^2 + \frac{4ll'}{(l+l')^2}Z^2 - \frac{4ll'}{(l+l')}Z = 0 \tag{5.28}$$

which has the same form as equation (5.17) and therefore a conicoid, with

$$(1+Q) = \frac{4ll'}{(l+l')^2} \quad \text{and} \quad r = \frac{2ll'}{(l+l')} \tag{5.29}$$

that is,

$$Q = \frac{4ll'}{(l+l')^2} - 1 \tag{5.29a}$$

Therefore a reflecting surface free of spherical aberration must be a conicoid,
but what type of conicoid? Let us determine the type.

l and l' with the same sign (i.e. typically concave surfaces)

For a reflector with a real object and image, l and l' must be of the same sign.
Therefore the above expression for $(1 + Q)$ must always be positive and thus
the value of Q must always be numerically greater than -1, which means that
the surface must be an ellipsoid.

The result that a spherical aberration free reflector must be ellipsoidal in
shape could have been deduced without any recourse to mathematics. By one
definition of an ellipse, it is the locus of a point moving so that the sum of the
path lengths of the straight lines joining two points via the locus is constant.
A simple drawing of an ellipse can be made by taking a piece of loose string
pinned at its ends, placing a pencil tightly against the string and drawing a
curve.

l and l' with opposite sign (i.e. typically convex surfaces)

If l and l' have opposite sign then the value of Q must be numerically less than
-1 and therefore the surface must be hyperbolic.

Now let us look at the special case of the object at infinity and a concave
reflector.

Object at infinity and a concave reflector

We can find the surface shape of a concave reflector free of spherical aberration
for an object (or image) at infinity by three methods, as follows:

(1)   *As a special case of the above general result.* If we put $l = \infty$ in equation
      (5.29a), we have $Q = -1$. Therefore the ideal surface is parabolic.
(2)   *From first principles.* Let us consider a reflector with the object on axis
      and at infinity (see Figure 5.17). What is the surface shape that would
      give zero spherical aberration? The general ray from a point $\mathcal{B}$ on a plane

Fig. 5.17: Reflector free of spherical aberration for an object at infinity. This surface is a paraboloid.

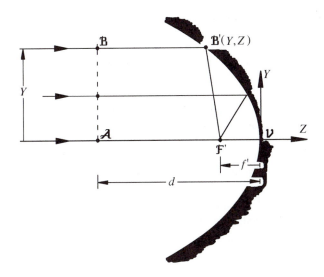

wavefront and meeting the reflector at $\mathcal{B}'$ must pass through the back focal point $\mathcal{F}'$ for all positions of $\mathcal{B}$ or height Y. From Fermat's principle, this requires that the optical path length along the ray be the same for all positions of $\mathcal{B}$. Thus

$$[\mathcal{B}\mathcal{B}'\mathcal{F}'] = [\mathcal{A}\mathcal{V}\mathcal{F}']$$

that is,

$$[\mathcal{B}\mathcal{B}'] + [\mathcal{B}'\mathcal{F}'] = [\mathcal{A}\mathcal{V}] + [\mathcal{V}\mathcal{F}']$$

or

$$d + Z + \sqrt{[Y^2 + (Z - f')^2]} = d - f'$$

The above equation easily simplifies to give

$$Z = Y^2/(4f') \tag{5.30}$$

which is the equation of a parabola.

(3) *As a special case of the refracting surface [equation (5.26b)].* This problem could have been analysed directly from the result of the refracting surface. The required asphericity for a refracting surface to be free of spherical aberration is given by equation (5.26b). To find the value of asphericity Q_o for a reflecting surface, we only have to substitute $n = 1$ and $n' = -1$ into this equation and we get

$$Q_o = -1$$

which is the asphericity of a paraboloid.

5.4 Some factors which affect aberration level

The level of aberrations of rays in a beam depends upon a number of factors. From Sections 5.1 and 5.2, we qualitatively saw that the level of aberration of a ray depends upon

- the distance of the ray from the central ray in the beam
- the distance of the object point from the optical axis

From Section 5.3 on the study of spherical aberration at a single refracting surface, we also observed that the aberration of the ray depends upon

- position of the conjugates
- radius of curvature (and hence curvature) of the surface
- surface shape (e.g. asphericity)
- refractive indices

Equations were given for calculating the exact and approximate (primary) levels of spherical aberration. As a general rule, the other aberrations also depend upon these factors and equations exist for their calculation and these will be presented and discussed in Chapter 33. The aberration level also depends upon other factors, which we will now describe.

5.4.1 *Effect of shape of lens*

For a lens of given power, the aberrations depend upon the distribution of curvature between the two surfaces. In Section 5.3, we saw that primary spherical aberration at a refracting surface depends upon the inverse of the square of the curvature. As a result, the primary spherical aberration of a thin lens is also quadratic in either of the surface curvatures. Therefore, there is an optimum lens shape which minimizes the aberration level and this optimum shape depends upon the positions of the conjugates and refractive index. For example, if we use a single thin lens with one conjugate at infinity, the spherical aberration cannot be made zero, only minimized, and the minimum aberration occurs with a lens that has most of the power on one surface, and this surface faces the more distant conjugate. This is discussed a little further in the next chapter and in greater depth in Chapter 33.

5.4.2 *Effect of the path of the beam through the system*

By using two examples, we will now show that the aberrations experienced by a beam from an object point and the amount and type of a particular aberration in the beam depend upon the particular path of the beam through an optical system and other aberrations in the beam from some other object point.

Let us take the example of the aberrations of a simple thin lens. While this is an unrealistic example it allows us to gain some insight into the aberration behaviour of more complex systems. Let us look at the example shown in Figure 5.18. In (a), a thin beam passes through the centre of the lens and since all rays pass through the centre of a thin lens without any deviation, real and paraxial rays will follow the same path. While this beam may have some

Fig. 5.18: Effect of beam
path through a simple thin
lens on the distortion
aberration in the presence
of spherical aberration.

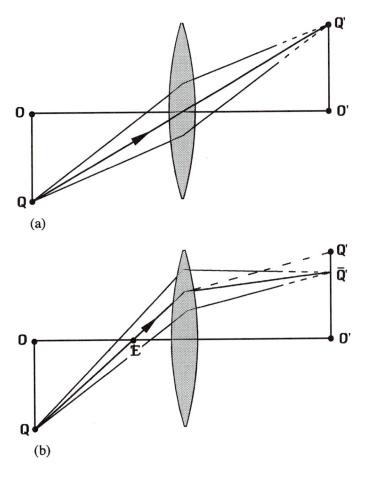

spherical aberration, the image is formed at the expected Gaussian image point Q' by the central ray of the beam. Therefore there will be no distortion. However, if we let the beam pass through the upper part of the lens as shown in (b), the central ray of the beam will pass through the axial point ε. If rays from ε suffer some spherical aberration, the central ray will not meet the image plane at the Gaussian image point Q'. If the spherical aberration is positive, the central ray will meet the Gaussian image plane at a point $\overline{Q}'$ closer to the axis than Q' and the beam will now have some negative distortion (see Figure 5.8). Now if the beam passed through the bottom part of the lens and the spherical aberration suffered by the central ray were still positive, the beam would be focussed higher than Q' and thus have positive distortion. Thus the particular path of the beam through the lens affects the level and sign of distortion, which in turn depend upon the level of spherical aberration for the object plane at the point ε where the central ray of the beam crosses the optical axis.

Now let us consider a lens with longitudinal chromatic aberration and let this lens image an off-axis point Q as shown in Figure 5.19. If a thin beam passes through the centre of the lens as shown in (a), the central ray is undeviated for all wavelengths and the beam will have zero transverse chromatic aberration. Now let the beam pass through the upper part of the lens as shown in (b) and let the central ray of the beam pass through the point ε on the optical axis. Since

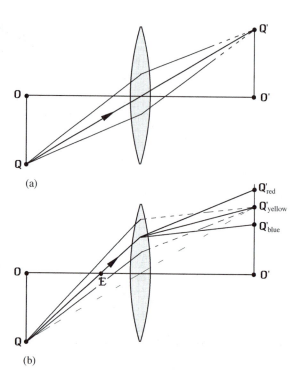

Fig. 5.19:   Effect of beam path through a simple thin lens on transverse chromatic aberration.

the lens has some longitudinal chromatic aberration, rays passing through the point ε will be deviated by different amounts for different wavelengths and these rays will meet the image plane at different heights, as shown in the diagram, leading to transverse chromatic aberration. Thus the particular path of the beam through the lens affects the level and sign of transverse chromatic aberration.

These two simple examples show that the particular path of the beam through a system affects the type and level of aberration picked up by the beam and that the aberrations are inter-dependent.

5.5   Primary and higher order aberrations

The aberrations discussed so far in this chapter are called primary (or sometimes Seidel) aberrations. The sum of these aberrations is not the total aberration in the beam, because the "higher" order aberrations have been neglected.

To understand the concept of higher order aberrations, let us recall equations (5.11) and (5.11a) for the spherical aberration at a spherical surface. In their derivation, we neglected terms of order higher than Y^4. These neglected terms, which are shown in equation (5.12), are the higher order spherical aberration terms. The same argument also applies to the other aberrations and thus there are also higher order coma, astigmatism, field curvature and distortion.

5.6   Aberrations and Gaussian optics

Since the cardinal points are defined in terms of paraxial optics, one should take care in applying them to a real (aberrated) system, because real rays are

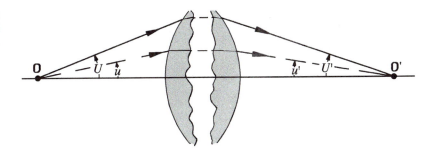

Fig. 5.20: The sine condition (free of coma and spherical aberration).

not bound by the rules of paraxial optics. For example, a focal point is defined as the point of intersection of a paraxial ray, from an axial point at infinity, with the axis in the conjugate space. In paraxial optics, these points are unique; that is, they are independent of the initial ray height. However when one traces real rays, the previous discussion of spherical aberration shows that the point of intersection with the axis moves along the axis and therefore its position depends upon the height of the ray at the surface. Similarly, since principal and nodal points are defined in terms of paraxial optics, they are only applicable for rays inclined at small angles to the optical axis.

The effect of finite or real (non-Gaussian) optics on the shape of the principal "planes" is also of some importance. In Gaussian optics, these surfaces are regarded as planes as the term *principal planes* implies. However, in real systems they are curved, as the following example will show.

5.6.1 The sine condition

A system free of spherical aberration is free of coma when the sine condition is satisfied. Referring to Figure 5.20, the sine condition relates the angles made by both paraxial and real rays. In this diagram, the broken lines represent a paraxial ray and the full lines represent a real ray. The real ray meets the axis at o' because the beam from o is free of spherical aberration. The sine condition states that this system will be free of coma if the following condition is satisfied

$$\frac{\sin(U')}{u'} = \frac{\sin(U)}{u} \tag{5.31}$$

where u and u' are the paraxial angles, as shown in the diagram, and U and U' are the corresponding real ray angles.

By considering a special case, it will now be shown that if the sine condition holds, then the principal "planes" (or at least one of them) must be curved. Consider the case of the object at infinity, as shown in Figure 5.21a. In this case the sine condition changes to

$$\frac{Y}{h} = \frac{\sin(U')}{u'}$$

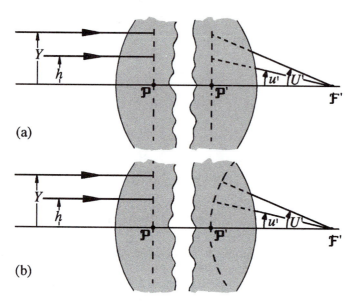

Using equation (3.38a), but with the notation in this section,

$$-u'/h = 1/\mathcal{P}'\mathcal{F}'$$

which leads to

$$\sin(U') = -Y/\mathcal{P}'\mathcal{F}'$$

However, this cannot be so, since it is obvious that

$$Y/\mathcal{P}'\mathcal{F}' = -\tan(U')$$

However, if we now assume that the back principal "plane" is curved and centred on the back focal point $\mathcal{F}'$, as shown in Figure 5.21b, it follows that we can write

$$\sin(U') = -Y/\mathcal{P}'\mathcal{F}' \tag{5.32}$$

Therefore, for the sine condition to be satisfied, that is, in a system free of spherical aberration and coma, the principal "planes" (or at least one of them) must be curved. A more detailed proof of equation (5.32) has been given by Born and Wolf (1989).

It should be noted that the amount of spherical aberration in a beam depends upon the position of the conjugates and the previous discussions indicate that it can only be made zero for one object plane position. We demonstrated this for a single surface in Section 5.3, but it is also true for more complex systems. Therefore the sine condition also only applies to one particular position of the object plane.

Exercises and problems

5.1 Sketch the paths of a bundle of rays from an axial object point at infinity and refracted by a negative power simple lens with (a) negative spherical aberration and (b) positive spherical aberration.

5.2 Look again at Section 5.3.2 and the spherical aberration at a flat surface. Imagine that you are looking into a pool of water with your eye close to the surface. If the pool depth is constant, determine how the apparent depth varies with distance from your eye. Does the apparent depth ever become zero?

5.3 Calculate the asphericity Q for a refracting surface free of spherical aberration for an object at infinity, if the incident index is 1.0 and the final index is 1.7.

ANSWER: $Q = -0.346$

5.4 (i) Calculate the asphericity of an elliptical reflector, designed to image a point source at a distance of 10 cm to a point at a distance of 100 cm, and image the beam free of spherical aberration.

(ii) What is the vertex radius of curvature of this reflector?

ANSWERS: (a)$Q = -0.699$, (b) $r = \pm18.2$ cm

Summary of main symbols and equations

Q	surface asphericity
U, U'	real ray angles in object space and image space, respectively
Y	real ray height at a surface
Z	distance along optical axis
f_n	figuring coefficient
$\mathcal{B}$	intersection of a ray with a surface
$\varepsilon, \varepsilon'$	axis crossing points of central ray in an off-axis beam
$\overline{Q}'$	real image position, in contrast to the Gaussian position Q'
W	wave (or path length) aberration
t	transverse spherical aberration

Section 5.3.3.1: Conicoids

$$p^2 + (1 + Q)Z^2 - 2rZ = 0 \text{ (conicoid)} \tag{5.17}$$

$$p^2 = X^2 + Y^2 \tag{5.18}$$

$$Z_{\text{conicoid}} = \frac{p^2}{r + \sqrt{[r^2 - (1 + Q)p^2]}} \tag{5.22}$$

Section 5.3.5.1: A refracting surface free of spherical aberration (infinite conjugate)

$$Q = -n^2/n'^2 \tag{5.26b}$$

References and bibliography

*Born M. and Wolf E. (1989). *Principles of Optics*. 6th (corrected) ed. Pergamon Press, Oxford.

Encyclopaedia of Mathematics. (1989). Vol. 3. Kluwer Academic Publishers, Dordrecht, The Netherlands.

*Kiely P.M., Smith G., and Carney L.G. (1982). The mean shape of the human cornea. *Opt. Acta* 29(8), 1027–1040.

*Schott (September 1992). *Optical Glass Catalog*. Schott Glaswerke, Mainz, Germany.

6

Simple lens types, lens systems and image formation

6.0 Introduction

In this chapter, we will look further at the optical properties of single or simple lenses, some special lenses and some interesting examples of more complex lens systems.

We study the properties of single lenses to learn more about how they image beams. Such knowledge also helps us to understand the properties of more complex optical systems because these more complex systems are composed of single lenses and an understanding of the role of each single lens helps us to understand the operation of the system as a whole.

There are a wide range of simple lenses. We will initially classify them according to whether they are rotationally symmetric or non-rotationally symmetric and start with the symmetric lenses.

In this chapter, we will assume the lenses are in air, unless it is specifically stated otherwise and there will be such cases. There are some interesting situations where the lenses are not in air and we will look at some examples in Section 6.4.

6.1 Rotationally symmetric simple lenses

Rotationally symmetric lenses are constructed with surfaces that are rotationally symmetric and the axes are co-linear. These lenses may be made with spherical or aspheric surfaces. Non-rotationally symmetric lenses will be discussed in the next major section, Section 6.2.

6.1.1 Spherical lenses

Most lenses are constructed using spherical surfaces with co-linear centres of curvature. The line joining the centres is the optical axis. Lenses are usually made with spherical surfaces because of the low manufacturing cost relative to that of other surface forms.

$\Gamma < -1$ $\qquad$ $\Gamma = -1$ $\qquad$ $\Gamma = 0$ $\qquad$ $\Gamma = +1$ $\qquad$ $\Gamma > +1$

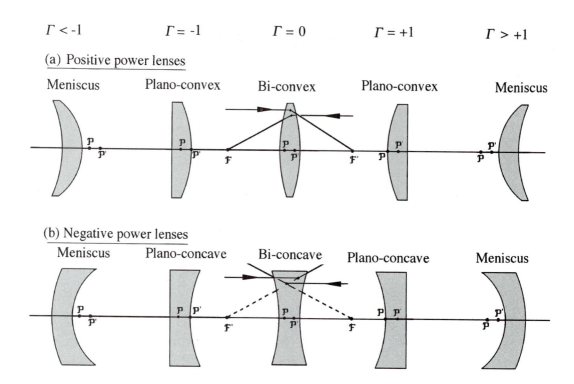

Fig. 6.1: Positive and negative power lenses, their different shapes, the shape factor Γ, the positions of the focal points and the effect of shape on the positions of the principal planes.

6.1.1.1 Terminology

Spherical lenses can be firstly categorized according to their equivalent power. If the equivalent power is positive, they are called positive power, positive, plus, convex or converging lenses. If the lenses have negative power, they are called negative power, negative, minus, concave or diverging lenses. Let us look at the meaning of these terms for both single lenses and more complex optical systems.

Positive/negative power

The term "positive" or "negative" power relates to the equivalent power of a lens or system and since by far the majority of lenses have a unique and non-zero value for this power, this is a useful way of classifying the lens. The exceptions are those lenses whose power is zero (afocal lenses or systems) and a few whose power is variable (see Section 6.3). Therefore, the terms **positive** or **negative** power lenses are preferred names since these are applicable to most lenses, no matter how complex. A positive or negative power simple lens can have a variety of forms or shapes and some of these are shown in Figure 6.1. It can be seen from this diagram that the positive power lenses are thicker in the middle and the negative lenses are thicker at the edge.

Convex/concave

Positive power lenses are sometimes called convex lenses and negative power lenses are sometimes called concave lenses. These terms are due to the most

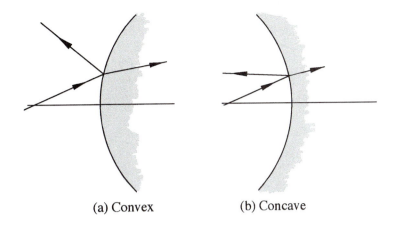

(a) Convex   (b) Concave

common shapes of these lenses. Convex and concave surfaces are best defined in
terms of their apparent shape for an observer travelling along an incident ray. The
centre of a convex surface bows towards the observer as shown in Figure 6.2a.
In contrast, the centre of a concave surface bows away from the observer as
shown in Figure 6.2b. These definitions are readily applicable to mirrors, since
mirrors are single surfaces and only viewed from one side. However, when
applied to lenses, complications arise. For example, it can be seen from Figure
6.1 that the left-most meniscus positive power lens has two concave surfaces
for a ray approaching from the left. Yet this lens is often called a convex lens.
Similarly, the left-most meniscus negative power lens has two convex surfaces
for a ray approaching from the left and this type of lens is often called a concave
lens. Therefore, the use of the terms convex and concave when applied to simple
lenses is perhaps unfortunate since a positive power lens may have one or two
concave surfaces and a negative power lens also may have one or two convex
surfaces, as seen from the direction of the incident ray. For more complex
optical systems, that is those made up of a number of lenses, the terms convex
and concave are even less meaningful. For example, a complex lens system is
usually a combination of both positive and negative simple lenses.

However for simple lenses, the terms convex and concave are so commonly
used that we will have to use the terms occasionally in this book, but the use
will be kept to a minimum.

Converging/diverging

The terms converging and diverging have been applied to lenses because, under
common conditions, a positive power lens converges the beam leaving the lens
and a negative power lens diverges the beam. This is shown in Figure 6.3a. These
terms are also not good terms for the following reasons. Firstly the above rules
are not always true. For example, Figure 6.3b shows a diverging beam leaving
a positive lens and a converging beam leaving a negative lens. Secondly, we
should be aware that even for the positive lens in Figure 6.3a, the beam finally
diverges after passing through the beam focus. Thus both positive and negative
lenses diverge the beam eventually. However, the main argument against using
these terms is that the above definition does not apply to more complex lens
systems. For example, an optical system consisting of two positive power lenses

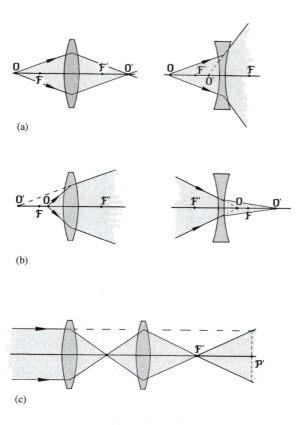

Fig. 6.3: Positive and
negative power lenses, the
convergence and divergence
of beams and the formation
of real and virtual images.

separated by more than the sum of their focal lengths has a negative equivalent
power. However, a beam from infinity is focused beyond the last lens, as shown
in Figure 6.3c, and is initially converged by the lens system. Thus this system
has a negative equivalent power, yet the beam leaving the system is initially
converging.

6.1.1.2 Quantification of lens shape

Simple lenses are sometimes categorized according to the "shape" of the lens.
This is the shape of the lens in a cross-section through the optical axis as shown
in Figure 6.1. For any given power F of a thin lens, the choice of curvatures C_1
(front surface) and C_2 (back surface) is arbitrary within the constraints of the
lens power [equation (3.14a)], that is

$$F = (C_1 - C_2)(\mu - 1) \tag{6.1}$$

where μ is the refractive index of the lens material. If one of the two curvatures
is chosen arbitrarily, the other is fixed by the above equation. Apart from the
constraints on the lens power, no radius of curvature can be less than the desired
aperture radius ρ of the lens, that is

$$|r_1| \quad \text{and} \quad |r_2| > \rho \tag{6.2a}$$

or

$$|C_1| \quad \text{and} \quad |C_2| < 1/\rho \tag{6.2b}$$

If the lens is regarded as thin, its shape is sometimes quantified by the shape factor Γ defined as follows

$$\Gamma = \frac{C_1 + C_2}{C_1 - C_2} \tag{6.3}$$

The curvatures C_1 and C_2 can be expressed explicitly in terms of the shape factor Γ, power F and refractive indices by simultaneously solving equations (6.1) and (6.3). The solutions are

$$C_1 = \frac{F(\Gamma + 1)}{2(\mu - 1)} \qquad C_2 = \frac{F(\Gamma - 1)}{2(\mu - 1)} \tag{6.4}$$

Figure 6.1 shows the effect of the value of Γ on the actual lens shape. This diagram also gives the particular names of some lens shapes.

The above shape factor Γ is commonly used in the mathematical theory of aberrations of thin lenses (Chapter 33). However in ophthalmic optics, it is more usual to specify the shape of a lens in terms of the back surface power F_2. This is given by equation (3.11a). Because the lens is in air, we have $n' = 1$ and therefore

$$F_2 = C_2(1 - \mu) \tag{6.5}$$

If one uses equation (6.4) to eliminate C_2 from equation (6.5), the shape factor Γ and the back surface power F_2 are related by the equation

$$\Gamma = 1 - 2(F_2/F) \tag{6.6}$$

or

$$F_2 = F(1 - \Gamma)/2 \tag{6.7}$$

In using these quantities to describe lenses, we must note that the above methods of specifying the shape of a lens are only valid for a thin lens.

6.1.1.3 Cardinal point positions

Focal points

The positions of the focal points of positive and negative power simple lenses are shown in Figure 6.1. It should be noted that for negative power lenses, the back focal point is on the object side of the lens and the front focal point is on the image side of the lens.

Principal points

The principal points of a thin lens coincide with the lens. When the lens is thick, the positions of the principal points depend significantly on the shape of the lens. Figure 6.1 shows how the shape affects the position of the principal planes. It is clear from these diagrams that as the lenses become more meniscus, the principal points move farther to the side with the surface of higher curvature, and for heavy meniscus shapes, may be outside the lens, as shown in the diagram.

Nodal points

For a lens in air, the nodal points coincide with the principal points. Otherwise the distance between the nodal and principal points is given by equation (3.58). For an actual lens system, the position of the nodal points can be found independently by a laboratory method described in Chapter 11.

6.1.1.4 Aberrations and lens shape

The aberration levels of a simple lens depend upon a number of factors, two of which are lens shape and conjugate positions. Thus in choosing a lens, one should determine the shape that gives the best imagery and that depends upon the positions of the conjugates and the most important aberrations.

For example, if a simple positive power lens is used to image a small object, placed at its front focal point, at infinity, it should have zero spherical aberration. A lens specifically designed for this task is called a **collimator** (a discussion of collimators is given in Chapter 23). However, the spherical aberration of a thin lens cannot be made zero. It can only be minimized. For minimum primary spherical aberration, the optimum shape of a thin lens is independent of lens power but dependent on refractive index. For a lens with an index of 1.5, the optimum lens shape factor is $\Gamma = -0.714$. With a power of 1 m^{-1} the front surface curvature would be 0.286 m^{-1} and the back surface curvature would be -1.714 m^{-1}, and these curvatures are in the exact ratio of 1:6. Figure 6.4a shows such a thin lens acting as a collimator. The front surface is almost flat and slightly convex. Note that the more curved surface faces the more distant conjugate.

On the other hand, in the design of spectacle lenses, the most important aberrations are astigmatism and field curvature. For the elimination of primary astigmatism, there are either no, one or two shapes, which depend upon lens power, refractive index, conjugate position and the distance of the lens from the eye. For an object at infinity, thin lenses with a refractive index of 1.5 and placed 27 mm from the **centre of rotation** of the eye, primary astigmatism is zero at two shape factors for powers in the range approximately -22 m^{-1} to $+7$ m^{-1}. Outside this power range, astigmatism can only be minimized. Thus for a thin lens with a power of $+10$ m^{-1}, primary astigmatism can only be minimized. For a refractive index of 1.5, its shape that minimizes this aberration is $\Gamma = +2.646$ and hence has front and back surface curvatures of 0.03646 mm^{-1} and 0.01646 mm^{-1}, respectively. This lens is shown in Figure 6.4b. For a lens with a power of -10 m^{-1}, primary astigmatism can be eliminated for two shape factors; $\Gamma = -1.436$ and $\Gamma = -3.855$. These are drawn to scale in Figure 6.4c.

Fig. 6.4: Typical lens shapes for the reduction of certain aberrations for some applications: (a) for a collimator – spherical aberration; (b) and (c) spectacle lenses – astigmatism. See Section 6.1.1.4 for details.

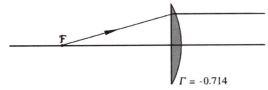

$\Gamma = -0.714$

(a) Collimator: Minimum spherical aberration

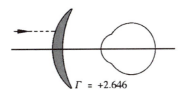

$\Gamma = +2.646$

(b) Positive power spectacle lens: Minimum astigmatism

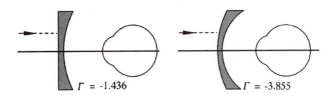

$\Gamma = -1.436$ $\qquad$ $\Gamma = -3.855$

(c) Negative power spectacle lenses: Zero astigmatism

In practice the shape of $\Gamma = -1.436$ would be the preferred option because of the smaller surface powers. This lens has front and back surface curvatures of 0.00436 mm^{-1} and 0.02436 mm^{-1}, respectively. Note that the front surface is almost flat. A more detailed and quantitative treatment of the effect of shape on the aberrations of a thin lens and the equations used for the above calculations is given in Chapter 33. The background for the above results is given in Examples 33.1 and 33.2 of that chapter.

6.1.1.5 Maximum aperture radius of a lens (in air)

Equations (6.2) state that the radius of curvature of both surfaces of a lens must be greater than the aperture radius ρ. If the lens is to be manufactured, this condition imposes a limit on the range of the shape factor Γ. This condition can also be used to limit the power F, given a desired aperture radius. The lens shape that has the maximum radii of curvature for both surfaces is an equiconvex (or equiconcave) lens. For an equiconvex (or equiconcave) lens, let us write the curvatures as

$$C = C_1 = -C_2$$

Suppose this lens is made from a material with a refractive index of 1.5, which is a typical value of low index materials, particularly plastic. Then we have from equation (6.1), with $\mu = 1.5$,

$$F = 2C(1.5 - 1)$$

or

$$C = F$$

Now recalling equation (6.2b) and replacing C by F, we have finally

$$|F\rho| < 1 \quad \text{(equi-convex/concave)} \tag{6.8a}$$

Note that this discussion and conclusions only apply to thin lenses and lenses made with a material with a refractive index of 1.5. Nevertheless, equation (6.8) gives a good guide to the practicability of making a thick lens with a power F and aperture radius ρ. If the above condition cannot be met, the lens can be split into two lenses, say of approximately equal powers.

By a similar argument, it can be shown that if the lens is to be plano-convex or plano-concave in shape, then

$$|F\rho| < 0.5 \quad \text{(plano-convex/concave)} \tag{6.8b}$$

In some cases, these restrictions can be overcome using a lens with one or both surfaces made aspheric.

6.1.2   Aspheric surfaced lenses

In Chapter 5, we have seen that the spherical aberration of a spherical surface can be eliminated by making the surface aspheric. If one of the conjugates is at infinity, the surface is conicoid. If both conjugates are at finite distances then the surface is a cartesian oval (Section 5.3.5.1) instead of a conicoid. In both cases, the spherical aberration is reduced because the local surface power reduces with distance from the surface vertex. This means that the local surface curvature reduces and the surface flattens towards the periphery, as shown in Figure 6.5a.

Thus lenses made with these types of aspheric surfaces either can be made thinner for the same aperture radius (Figure 6.5b), or, if they have the same thickness, can be made with wider aperture radii (Figure 6.5c). Such aspherized lenses also can be made with a wider aperture than predicted by equation (6.8a or b).

6.1.3   Fresnel lenses

In a number of instances, a single lens with a large aperture radius and a high equivalent power is required. Typical examples are **condenser** lenses. This requirement usually leads to very thick and hence heavy lenses. Providing image quality is not of major importance; the thickness and weight can be significantly reduced by using an equivalent Fresnel lens. The Fresnel lens equivalent of a plano-convex lens is shown in Figure 6.6a. The spherical refracting surface is constructed as a series of concentric zones as shown in the diagram and the

Fig. 6.5: Effect of surface asphericity on (a) surface shape and (b) thickness or (c) possible aperture diameter of a simple lens.

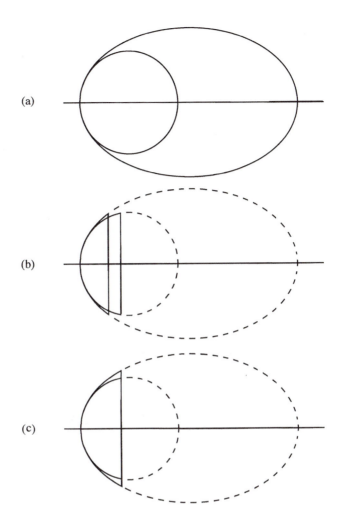

(a)

(b)

(c)

curvature of each zone is identical to that of the base sphere. It is obvious from the diagram that such lenses are relatively thin and the narrower the zones, the thinner the lens. It would be possible to construct a Fresnel equivalent of any other shaped lens, for example a meniscus, but this is not usually done.

The main disadvantage of Fresnel lenses is that a significant amount of light is scattered at the junction between adjacent zones, and this leads to poor image quality and thus reduces their use when good image quality is required. Their major use is as condenser lenses (see Section 6.5 and Chapter 22), e.g. in overhead projectors, and they are used as collimators in some signal lights.

The main advantages of Fresnel lenses are their light weight, as mentioned above, especially when large diameters are needed and that they can be so easily moulded using plastic. The moulding process significantly reduces their cost compared to that of full thick spherical lenses.

In practice, it is difficult to machine the concentric curved sections. Therefore they are usually made flat instead. In this cross-section, they can be regarded as prismatic elements (i.e. in the shape of a prism) with a progressively changing

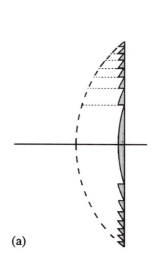

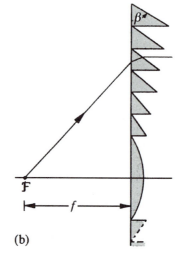

Fig. 6.6: (a) Schematic cross-section construction of a Fresnel lens with a positive power. (b) A design of a Fresnel lens used as a collimator. This lens has a refractive index of 1.75, focal length of 50 mm and the central zone has a radius of curvature of 37.5 mm.

(a)   (b)

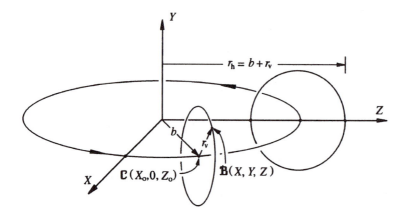

Fig. 6.7: Generation of a toroidal surface.

apex angle β. Figure 6.6b shows an example of a such a Fresnel lens acting as a collimator. In this case, the apex angle is chosen to make the final refracted ray parallel to the axis. Using this construction, a Fresnel lens can be made with a much wider aperture radius than indicated by the restriction specified by equation (6.8a or b). In fact the example shown in the diagram is closer to representing an aspheric surfaced lens corrected for spherical aberration.

6.2 Toric lenses

Toric lenses are non-rotationally symmetric lenses with at least one surface of the toric form. A toric surface is a surface of revolution usually formed when an arc (most commonly a circle) is revolved about an axis, which in the case of the circle does not pass through the circle centre. An example using a circle is shown in Figure 6.7. Here a circle of vertical radius r_v in the vertical plane is rotated about the vertical Y-axis, a distance b from the centre of the circle. An equation for the surface can be found as follows.

6.2.1 *Mathematical representation of toroidal surfaces*

In Figure 6.7, the centre of the circle is at $c(X_0, 0, Z_0)$ which is at a distance b from the Y-axis or axes origin $(0, 0, 0)$. A point $\mathcal{B}(X, Y, Z)$ on this circle must satisfy the equation

$$(X - X_0)^2 + Y^2 + (Z - Z_0)^2 = r_v^2 \tag{6.9}$$

where r_v is the radius of the circle. However, this equation is also the equation of a sphere centred on c. Therefore we need to restrict the values of X, Y and Z to lie on the vertical circle. This can be done by adding the condition that the normal to the plane of the circle centred on c is perpendicular to the Y-axis. This condition can be written in terms of the vector product, as

$$[(X_0, 0, Z_0)\mathbf{x}(X - X_0, Y, Z - Z_0)]\cdot(0, 1, 0) = 0$$

This reduces to the simple condition

$$X_0 Z - X Z_0 = 0 \tag{6.10}$$

Now since the point c $(X_0, 0, Z_0)$ lies on a circle of radius b, we have

$$X_0^2 + Z_0^2 = b^2 \tag{6.11}$$

If we now expand equation (6.9), it becomes

$$X^2 - 2XX_0 + X_0^2 + Y^2 + Z^2 - 2ZZ_0 + Z_0^2 = r_v^2$$

and on replacing $X_0^2 + Z_0^2$ by b^2 from equation (6.11), this equation reduces to

$$X^2 - 2XX_0 + Y^2 + Z^2 - 2ZZ_0 = r_v^2 - b^2 \tag{6.12}$$

We now have to eliminate X_0 and Z_0 from this equation. This can be done by solving equations (6.10) and (6.11) as a pair of simultaneous equations in these quantities. The solutions are

$$X_0 = \pm bX/\sqrt{(X^2 + Z^2)}$$

and

$$Z_0 = \pm bZ/\sqrt{(X^2 + Z^2)}$$

After substituting these expressions for X_0 and Z_0 in equation (6.12), and after some simplification, we have finally

$$Y^2 + (X^2 + Z^2) \pm 2b\sqrt{(X^2 + Z^2)} = r_v^2 - b^2$$

If we now replace the radius b by the radius r_h of the outer vertex of the toroid (as shown in Figure 6.7), that is

$$b = r_h - r_v$$

we have

$$Y^2 + (X^2 + Z^2) \tag{6.13}$$

$$\pm 2(r_h - r_v)\sqrt{(X^2 + Z^2)} = r_v^2 - (r_h - r_v)^2$$

This equation has also been presented by Wray (1981), but with an alternative derivation.

This toric surface is centred on the Y-axis and if we wish to place the X-Y-Z axes origin $(0, 0, 0)$ on the surface, with the positive Z-axis being the optical axis, then we must carry out a shift of the form

$$Z \quad \text{becomes or is replaced by } Z - r_h$$

With this translation of the toroid, equation (6.13) becomes

$$Y^2 + [X^2 + (Z - r_h)^2] \tag{6.14}$$

$$\pm 2(r_h - r_v)\sqrt{[X^2 + (Z - r_h)^2]} = r_v^2 - (r_h - r_v)^2$$

This equation now describes a toroid which lies entirely on the positive Z-axis and the cross-section in the Y–Z plane is shown in the diagram.

The "$\pm$" sign in these equations may lead to some ambiguity but is necessary to cover the complete toric. The plus and minus signs cover separate sections of the toric. Example calculations show that the region of the toric near the origin $(0, 0, 0)$ requires the plus sign; that is this portion of the toric is described by the equation

$$Y^2 + [X^2 + (Z - r_h)^2] \tag{6.15}$$

$$+ 2(r_h - r_v)\sqrt{[X^2 + (Z - r_h)^2]} = r_v^2 - (r_h - r_v)^2$$

and the region of the toric farthest from the origin requires the minus sign.

The toric can be expressed in other mathematical forms; for example Fry and Loshin (1975) have given the alternative equation

$$Z^2 - 2(r_v + b)Z + Y^2 + X^2 + 2r_v b - 2b\sqrt{(r_v^2 - Y^2)} = 0 \tag{6.16}$$

but without any derivation. This equation also describes a toroid with the axis origin at the surface, but only describes the outer half of the surface of the toroid.

Some typical constructions of toric surfaces are shown in Figure 6.8. In Figure 6.8a, the axis of rotation is well outside the circle (i.e. $r_h > 2r_v$) and the surface formed in this case is a conventional toroid (i.e. doughnut or tyre shape). If we take the inner surface of this toroid, we have the capstan form of

Fig. 6.8: Formation of
various toric surfaces.

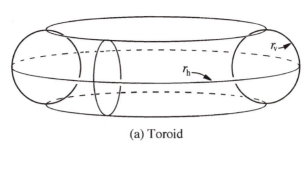

(a) Toroid

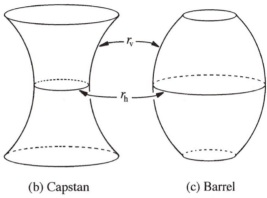

(b) Capstan   (c) Barrel

toroid shown in Figure 6.8b. The surface in Figure 6.8c is formed when the axis
of revolution is inside the circle (i.e. $r_h < 2r_v$).

An aspheric toroid could be formed by revolving a conic section, such as an
ellipse, parabola or hyperbola, around the Y-axis.

Toroidal surfaces have two principal radii of curvature; here they are the
radius of revolution r_h and the radius of the generating arc r_v (commonly a
circle). Thus the toroid has two principal sections which are perpendicular, pass
through the optical axis and contain the maximum and minimum curvatures.
In Figure 6.8, these two principal sections are horizontal and vertical. However
for a general toric surface, the principal sections may have any orientation, but
they are always mutually perpendicular.

6.2.1.1 Special case of the cylindrical surface

A special case of a toroidal surface is the **cylindrical** surface, which is formed
when one of the principal radii is infinite. A cylindrical lens, with one flat surface, is shown in Figure 6.9 for both positive and negative power. The direction
of zero power is called the **cylinder axis**.

6.2.2 The power of a toric lens

Once a lens has at least one toroidal surface, the equivalent power of the lens
now varies with meridian. There will be one meridian in which the power is
largest and one meridian in which the power is smallest. If only one surface

Direction of cylinder axis

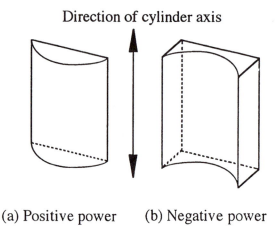

Fig. 6.9: A special toric surface: the cylindrical surface.

(a) Positive power   (b) Negative power

is toric, these principal meridians are the same as the principal sections of that surface. If both surfaces are toroidal with principal sections in different orientations, the resulting power distribution is analysed using the theory presented in Section 6.2.5.

It is conventional to express the equivalent power F of a toric lens as a function of azimuth θ, in the form

$$F(\theta) = F_s + F_c \sin^2(\theta - \alpha) \tag{6.17}$$

where the lens is regarded as being made up of a pure spherical component F_s and a pure cylindrical component F_c with the cylinder axis orientated at an angle α to the horizontal section or meridian. Thus along the cylinder axis, the power of the cylindrical component is zero and the powers in the two principal sections are F_s along α° and $F_s + F_c$ along $\alpha^\circ + 90^\circ$.

However, equation (6.17) is not strictly valid for any other angles than α and $\alpha + 90^\circ$. This is because paraxial rays traced in meridians at other angles are not concurrent after refraction. This in turn is due to the fact that the normals to the surface along these meridian sections do not lie in a single plane; that is they are not coplanar.

6.2.3 Ophthalmic notation of toric lenses

Toric lenses (often called sphero-cylindrical lenses in ophthalmic optics) are commonly used in ophthalmic optics to correct for axial astigmatism in the eye, which is usually caused by a toroidally shaped cornea. In this case, the clinician must determine the patient's refractive error in the two principal meridians of the pupil and the orientation of one of these.

Using the above representation of power [equation (6.17)] in ophthalmic optics, it is conventional practice to specify the toric lens prescription by the notation

$$F_s / F_c \times \alpha^\circ$$

For example, the following prescription

$$+4/-2 \times 66°$$

describes a lens with principal powers of $4\,\mathrm{m}^{-1}$ along $66°$ and $+4-2 = +2\,\mathrm{m}^{-1}$ along $156°$. However, this representation is not the only one that is used. The same prescription could be written in the form

$$+2/+2 \times 156°$$

The first of these $(+4/-2)$ is known as the minus cylinder form and the second $(+2/+2)$ is known as the plus cylinder form.

6.2.4 Toric astigmatism and the astigmatic aberration

A meridional change in power has the same effect on the image of a point source as the regular astigmatic aberration described in Chapter 5; that is a point object forms two image lines, mutually perpendicular and separated in space. The positions of these image lines can be found by applying the conventional paraxial equations in turn for the two principal sections. However, whereas the regular astigmatism is only present off-axis, the above astigmatic effect is present on-axis and is not an aberration. For this reason, the conventional astigmatic aberration, described in Chapter 5, is sometimes referred to as oblique astigmatism in ophthalmic optics applications.

6.2.5 Crossed torics or crossed cylinders (zero separation)

In ophthalmic practice, it is common to superimpose two toric surfaces or two toric lenses or optical systems, for example a toric lens and the "toric" eye. We need to know how to determine the powers and axis of such a combination.

We will now show that the combination of two toric systems is also a system with a cylindrical and a spherical power. We will derive equations for these powers and the new cylindrical axis. We firstly assume that the two torics are thin and in contact, with principal powers F_{s1}/F_{c1} and F_{s2}/F_{c2} with axes α_1 and α_2, respectively. We also assume that the power of the combination is their sum. We will now prove that this sum is identical to that of a single toric, that is

$$F_s + F_c \sin^2(\theta - \alpha) = F_{s1} + F_{c1} \sin^2(\theta - \alpha_1) \tag{6.18}$$

$$+ F_{s2} + F_{c2} \sin^2(\theta - \alpha_2)$$

To prove this equation, we firstly recall the trigonometric identity

$$2\sin^2(x) = 1 - \cos(2x)$$

and use it to expand the $\sin^2$ terms in the righthand side of equation (6.18). This gives us the expression

$$F_{s1} + (F_{c1}/2)\{1 - \cos[2(\theta - \alpha_1)]\} + F_{s2} \tag{6.19}$$

$$+ (F_{c2}/2)\{1 - \cos[2(\theta - \alpha_2)]\}$$

If we now expand the cosine terms using the trigonometric identity

$$\cos(A - B) = \cos(A)\cos(B) + \sin(A)\sin(B)$$

the resulting expansion of equation (6.19) has the same form as the identical expansion of the expression of the lefthand side of equation (6.18), which proves that a combination of two torics in contact is equivalent to a single toric. It remains to determine the value of the resulting powers and cylinder axis.

These can be found by equating terms in the two expansions and we get

$$\tan(2\alpha) = \frac{F_{c1}\ \sin(2\alpha_1) + F_{c2}\ \sin(2\alpha_2)}{F_{c1}\ \cos(2\alpha_1) + F_{c2}\ \cos(2\alpha_2)} \tag{6.20}$$

$$F_c = [F_{c1}\ \cos(2\alpha_1) + F_{c2}\ \cos(2\alpha_2)]/\cos(2\alpha) \tag{6.21a}$$

or

$$F_c = [F_{c1}\ \sin(2\alpha_1) + F_{c2}\ \sin(2\alpha_2)]/\sin(2\alpha) \tag{6.21b}$$

and

$$F_s = F_{s1} + F_{s2} + [(F_{c1} + F_{c2} - F_c)/2] \tag{6.22}$$

Thus the equations (6.20) to (6.22) enable us to find the resulting spherical and cylindrical powers of a pair of crossed toric systems and the resulting axis orientation.

6.2.6   *Paraxial ray tracing through toric surfaces*

To trace a paraxial ray through a toric surface, we need to trace the ray in terms of two pairs of angles and heights, say in two mutually perpendicular planes. These can be conveniently the X–Z and Y–Z planes. Let these pairs be (u, x) and (v, y) where u and v are the angles and x and y are the paraxial heights. In terms of these variables, the paraxial refraction equations for a toric surface are

$$n'u' - nu = -x[F_s + F_c\ \sin^2(\alpha)] + y[F_c\ \sin(\alpha)\cos(\alpha)] \tag{6.23}$$

$$n'v' - nv = +x[F_c\ \sin(\alpha)\cos(\alpha)] - y[F_s + F_c\ \cos^2(\alpha)]$$

and this pair of equations is equivalent to the paraxial refraction equation (2.5b) for a spherical surface.

The paraxial transfer equations are

$$x' = x + du'$$ (6.24)

$$y' = y + dv'$$

and these are equivalent to the paraxial transfer equation (2.12) for a spherical surface.

6.3 Variable power lenses

There is a need for variable power optical systems and the zoom optical systems have been extensively developed to meet this demand. However, zoom systems are both optically and mechanically complex devices. The system must consist of a number of separate lenses which must be moved relative to each other in precise motions to achieve the different powers. In certain circumstances there is a need for much simpler variable power systems, especially in ophthalmic optics.

Currently there are two such types of variable power lenses. In one type, the power varies over the surface of the lens surface, here to be called **progressive addition lenses**. In the other type, the lens consists of two thick elements that have complex shapes and the power is varied by sliding the two elements sideways relative to each other. This type will be called the **Alvarez lens** after its developer.

6.3.1 *Progressive addition lenses*

The purpose of the progressive addition type of variable power ophthalmic lens is to provide a variation in lens power in the vertical meridian, and is intended to be an alternative to bi- or tri-focal lenses for presbyopes. The power increases from top to bottom, with the top for distant viewing and the bottom for near work.

The change in power, increasing from top to bottom, is achieved by smoothly increasing the curvature. This leads to a decrease in instantaneous radius of curvature down the vertical meridian cross-section. To prevent astigmatism along this line, the radii in horizontal sections must be the same as that in the vertical section. If taken to the extreme, such a surface would look like an elephant's trunk. However, this surface would have excessive aberrations, namely astigmatism, in the periphery.

The astigmatism can be reduced by changing the horizontal curvatures and shapes of the peripheral parts of the lens. For example, this could be done by replacing the circular horizontal arcs by suitable conic sections. However, other considerations will lead to the use of more complex shapes.

A review of progressive addition lenses has been given by Sullivan and Fowler (1988).

6.3.2 *The Alvarez lens*

Alvarez (1978) described a variable power lens consisting of two thick elements with complex but complementary surface shapes. In one position, which will be called the zero position, the thickness of the combination is constant all over the surface. However, if these elements are sheared sideways, equally and in opposite directions, the thickness resembles a thick spherical lens. This will be made clearer in the mathematics to follow.

Alvarez suggested that the surface shape of one element should have a form such that the thickness $d_1(X, Y)$ is

$$d_1(X, Y) = k + aXY^2 + (aX^3/3) + bX$$

and for the other element

$$d_2(X, Y) = k - aXY^2 - (aX^3/3) - bX$$

It can easily be checked that the thickness $d(X, Y)$ of the combination is constant and equal to $2k$. If now the first element is moved sideways in the X direction by an amount $-h$ and the other by an amount $+h$, the combined thickness $d(X, Y)$ should now be written $d(X, Y, h)$, which is now

$$d(X, Y, h) = k + a(X - h)Y^2 + a[(X - h)^3/3] + b(X - h)$$

$$+ k - a(X + h)Y^2 - a[(X + h)^3/3] - b(X + h)$$

This reduces to

$$d(X, Y, h) = 2k - 2ah(X^2 + Y^2) - 2[ah^3/3] - 2bh \qquad (6.25)$$

Now we have shown in Chapter 2 through equation (2.9) that in the vicinity of the vertex, the sag of a spherical surface or the thickness of a spherical lens varies as the square of the distance from the vertex. With the notation used here, equation (2.9) has the form

$$Z = C(X^2 + Y^2)/2 + \text{higher order terms} \qquad (6.26)$$

If we compare this equation with equation (6.25), it is clear that equation (6.25) describes a spherical sag in the vicinity of the vertex $(0, 0)$ with a curvature C given by the equation

$$C = -4ah$$

Thus the central power F_A of the Alvarez lens would be given by the equation

$$F_A = -4ah(\mu - 1) \qquad (6.27)$$

These equations show that the shift or shear h induces an effective curvature C and thus a refractive power. It also shows that the induced curvature or power is linear in shift h.

The remaining part of the righthand side of equation (6.25) is a constant for any shift and therefore represents the minimum sag of the surface or lens as well as the minimum or maximum thickness. Thus it does not contribute to the power of the lens.

6.4   Simple lenses not in air

Most lenses that we are familiar with are used in air and therefore the equations given so far in this chapter are suitable for most situations. However there are some interesting exceptions, for example the natural lens of the eye or its artificial substitute (the intra-ocular lens) and some "air lenses". For a lens immersed in a medium other than air, the power will be different from that of the lens in air. The closer the external medium refractive index to that of the lens material, the lower the power will be. If the external medium has a higher index, the power of the lens will have opposite sign to that in air. Let us look at some of these situations, starting with the intra-ocular lens.

6.4.1   Intra-ocular lenses

The intra-ocular lens is usually made of polymethlymethacrylate, has a refractive index (μ) of about 1.49 and has a power of about 19 m^{-1} in the eye. This power will vary from eye to eye. If the lens were regarded as thin, then the thin lens power would be

$$F = (C_1 - C_2)(\mu - n)$$

where n is the index of the aqueous/vitreous which has a refractive index of about 1.336. For a lens with a power of 19 m^{-1} in the eye

$$19 = (C_1 - C_2)(1.49 - 1.336)$$

that is

$$(C_1 - C_2) = 123.38 \text{ m}^{-1}$$

The power of this lens when placed in air would be

$$F = (C_1 - C_2)(1.49 - 1.0) = 123.38 \times 0.49 = 60.5 \text{ m}^{-1}$$

which is about three times the value in the eye.

6.4.2   Air lenses

A spherical bubble of air in water is an example of an air lens. If it were a spherical water lens in air, the power would be positive and therefore an air bubble in water would have a negative power.

Let us look at another example of the application of an air lens. Let us suppose we need to form a real image of an object, as shown in Figure 6.10a. We could achieve the same result with the arrangement shown in Figure 6.10b. We could interpret this situation as using two very thick plano-convex lenses. However, we could interpret the situation in an alternative way. We could say that we are using a glass medium in the object and image space and using an "air" lens as the refracting element. In this diagram, the air lens has the shape of a conventional negative simple lens, but actually has a positive power.

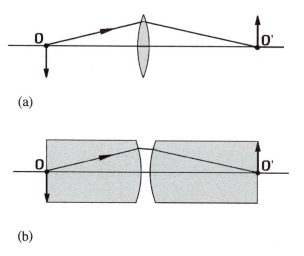

Fig. 6.10:   A positive
power conventional lens
replaced by an "air" lens.

(a)

(b)

6.5   Systems of lenses

Many important and common optical systems consist of more than one simple lens. There are three main reasons for the extra complexity.

One reason is that the power/aperture ratio may be too large for a single lens. In Section 6.1.1.5, we looked at the physical limitations in making a high power large diameter single lens, and equations (6.8a and b) give some indication of the limitation in special cases using the product of the equivalent power and aperture radius. If the product of the power and the aperture radius is too high, we can split the single lens into two or more separate simple lenses of equal power. A common example of this situation is in the design of condenser lenses (see Chapter 22). In practice, many condensers consist of two lenses, for this reason.

A second reason is that the desired aberration level may not be achievable with a single simple lens. For example, it is not possible to eliminate either spherical or chromatic aberration in a single simple lens. However, if we replace the lens by a combination of a positive and a negative lens, both of these aberrations can be made zero. Such a configuration is called an **achromatic** doublet. This is possible because the positive lens contributes positive spherical and chromatic aberration and the negative lens has aberrations of the opposite sign. The effective design of achromatic doublets requires an appropriate choice of lens shapes, refractive indices and dispersions, in order to balance the aberration contributions from each component. For the maximum degree of freedom in the design, the two lenses are separated by a small distance as shown in Figure 6.11a. This configuration allows each lens to be bent separately, leading maximum control over the aberration level. However, the most common form of an achromatic doublet is the cemented type in which the two are joined together as shown in Figure 6.11b. In this arrangement, the requirement for a common curvature at the contact surface restricts the ability to manipulate the aberration level, but has the advantage that it is more robust. The actual design of these doublets is discussed further in Chapter 33.

The third reason for the use of complex systems is that the required Gaussian properties cannot always be satisfied by a simple single lens. For example, suppose we need a symmetric long focal length (f) lens but with a short overall

Fig. 6.11: Various forms of achromatic doublets.

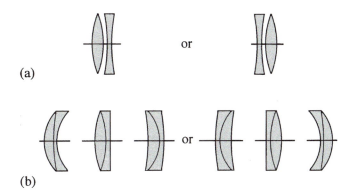

(a)

(b)

Fig. 6.12: A symmetric "telephoto" lens system: $d = 220.1$ mm, positive powers are both $3.481\mathrm{m}^{-1}$, negative power is -11.486 m^{-1} and $\mathcal{V}'\mathcal{F}' = 58.8$ mm.

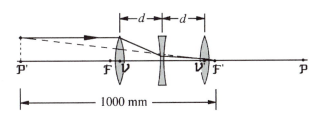

length and short back focal length (f_{v}'). Such a requirement cannot be achieved with a single lens but can with a three lens system. Figure 6.12 shows a symmetric triplet lens system that has principal planes outside the system. This has the advantage that it has a long equivalent focal length but a shorter distance from the front element to the back focal point. Such a design is a type of **telephoto** lens and the **telephoto ratio** is defined as

$$\text{telephoto ratio} = \frac{\text{system length} + f_{\mathrm{v}}'}{f} \tag{6.28}$$

The system shown in the diagram has an equivalent focal length of 1 m and telephoto ratio of 0.5.

A second example is the **anamorphic** lens. This is a complex lens system designed to have different (equivalent) focal lengths and hence magnifications in two meridians that are 90° apart. Anamorphic lenses are widely used in cinematography. Standard motion picture film is 35 mm film, which has a 35×24 mm format. By using an anamorphic camera lens, a much wider format scene can be photographed. The different focal lengths in the two meridians cause a compression of the image in the meridian with the shorter focal length, which is normally along the horizontal. The film is projected on the screen with a similar anamorphic lens, which decompresses the image and therefore expands it in the horizontal direction. An example of a schematic anamorphic lens, having equivalent focal lengths of $f(0°) = 50$ mm and $f(90°) = 100$ mm in the two perpendicular meridians, a back focal length of 40 mm and a lens separation of 30 mm, is shown in Figure 6.13. This lens system would give a magnification ratio of 1 : 2 between these two meridians for an object plane at infinity. The powers of the two lenses were found by simultaneously solving

(a) $f = 50$ mm, $0°$

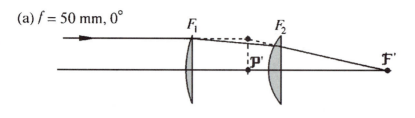

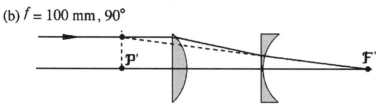

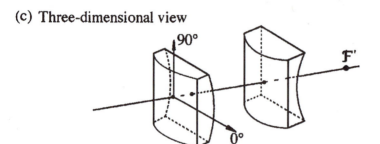

Fig. 6.13: An example of an anamorphic lens system. The powers are horizontal section ($0°$) : $F_1 = 6.7$ m^{-1} and $F_2 = 16.7$ m^{-1} and vertical section ($90°$) : $F_1 = 20.0$ m^{-1} and $F_2 = -25.0$ m^{-1}.

|←—30 mm—→|←40mm→|

(b) $f = 100$ mm, $90°$

(c) **Three-dimensional view**

equations (3.20) and (3.26a or b) with $\mu = 1$. For the above specification, the powers are

$$F_1(0°) = +6.66 \text{ m}^{-1} \quad \text{and} \quad F_2(0°) = +16.66 \text{ m}^{-1}$$

and

$$F_1(90°) = +20.00 \text{ m}^{-1} \quad \text{and} \quad F_2(90°) = -25.00 \text{ m}^{-1}$$

These are shown in the diagram.

There are many other types of more complex systems, for example telescopes and microscopes. We will look at these and others in Part II of this book.

6.6   Image formation

The main purpose of an optical system is to produce an image of an object with the image in a suitable position and having an appropriate size or magnification. In a complex optical system, each individual lens can be thought of as producing an image which then becomes the object for the next lens.

Images so formed may be either **real** or **virtual**. Real images can be projected onto a screen. They are formed beyond the optical system on the image side.

Fig. 6.14: Rays for graphical ray tracing, used to determine the position and size of images.

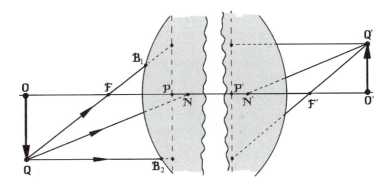

Virtual images are formed within the system or on the object side and thus cannot be projected onto a screen. With most visual optical instruments, the image is usually virtual since it must be in front of the eye, except in cases of **hyperopia**. In a few cases, where real images are formed [such as in the **indirect ophthalmoscope** (Chapter 29)], the image can only be seen if the eye is placed sufficiently far back to put the image within the **accommodation range** of the eye. For a discussion of hyperopia and accommodation range, see Chapter 13.

A positive power thin lens produces a real image if the object is farther from the lens than the front focal point. Otherwise, a virtual image is produced. Negative power thin lenses produce a real image if the object is between the lens and the front focal point and produce a virtual image otherwise. These image formations are explained further in Figure 6.3a and b.

6.6.1 *Graphical ray tracing*

If the construction parameters of an optical system are known, the position and magnification of any object can be found by applying the paraxial ray trace equations described in Chapter 2. On the other hand, if the positions of the cardinal points are known, graphical or sketching techniques can be used to find the image position and size. It has already been established that only two paraxial rays need to be traced to locate an image. Consider the situation in Figure 6.14, where the image Q' of Q must be located. In this general situation, three rays can be easily traced. These are as follows.

(a)   The ray $Q\mathcal{N}\mathcal{N}'Q'$ from Q and passing or appearing to pass through the nodal points, which must make the same angle to the axis in both object and image space. For a thin lens in air, the nodal points are at the lens centre.

(b)   The ray $Q\mathcal{F}B_1$, which passes or appears to pass through the front focal point $\mathcal{F}$, intersects the principal planes at the same height and is parallel to the optical axis in image space.

(c)   The ray $QB_2\mathcal{F}'$, which initially travels parallel to the optical axis, intersects the principal planes at the same height and passes through the back focal point $\mathcal{F}'$ after refraction.

All three rays will be or appear to be concurrent at the image point Q'. Only two (any two) of these rays need to be traced but a third ray is useful as a check.

Exercises and problems

6.1 Find the shape factor of a thin lens with the following construction:

$$C_1 = 0.235 \text{ mm}^{-1}, C_2 = 0.100 \text{ mm}^{-1}$$

ANSWER: $\Gamma = 2.48$

6.2 Calculate the surface curvatures of a thin lens in air, if its power is 10 m^{-1} and it has a shape factor of -2. Use a refractive index of 1.5.

ANSWERS: $C_1 = -10 \text{ m}^{-1}$ and $C_2 = -30 \text{ m}^{-1}$

6.3 Calculate the approximate maximum possible aperture radius of a plano-convex 40 m^{-1} lens.

ANSWER: 12.5 mm

6.4 Sketch the shape of a Fresnel lens with a negative power.
6.5 Calculate the resultant sphero-cylinder powers and angle produced by a combination of the following two thin sphero-cylindrical lenses in contact.

$$\text{lens 1: } F_s = 10 \text{ m}^{-1}, \quad F_c = -5 \text{ m}^{-1} \quad \text{and } \alpha = 30°$$

$$\text{lens 2: } F_s = 5 \text{ m}^{-1}, \quad F_c = -1 \text{ m}^{-1} \quad \text{and } \alpha = 70°$$

ANSWERS: $F_s = 14.63 \text{ m}^{-1}, F_c = -5.27 \text{ m}^{-1}$ and $\alpha = 35.39°$

Summary of main symbols and equations

C, C_1, C_2	surface curvatures
ρ	aperture radius of a surface or lens
X, Y, Z	cartesian co-ordinates; Z-axis is the optical axis
d	lens thickness
F_1, F_2	front and back surface powers

Section 6.1.1: Spherical lenses

$$\Gamma = \frac{C_1 + C_2}{C_1 - C_2} \tag{6.3}$$

Section 6.2: Astigmatic lenses and toric surfaces

r_h, r_v	radii of curvature in horizontal and vertical sections for toric surfaces
$\alpha, \alpha_1, \alpha_2$	axis directions of cylindrical components
F_s, F_c	spherical and cylindrical power of a surface or lens
θ	azimuth angle with usual trigonometric sign convention

References and bibliography

*Alvarez L.W. (1978). Development of variable-focus lenses and a new refractor. *J. Am. Optom. Assoc.* 49(1), 24–29.

*Fry G.A. and Loshin D.S. (1975). Ray tracing through a toric refracting surface. *Am. J. Optom. Physiol. Opt.* 52, 258–262.

Long W.F. (1976). A matrix formulation for decentering problems. *Am. J. Optom. Phys. Opt.* 53(1), 27–33.

*Sullivan C.M. and Fowler C.W. (1988). Progressive addition and variable focus lenses: A review. *Ophthal. Physiol. Opt.* 8, 402–413.

*Wray L. (1981). The toroidal surface. *Optician* 181 (4691); 29 May, 14–21.

7

Mirror types and image formation

7.0 Introduction

This chapter deals with the properties of mirrors or reflecting surfaces. Two types of mirrors will be looked at: (a) plane mirrors and (b) curved mirrors. So far, we have tended to concentrate on the paraxial properties of optical systems and therefore looked at such phenomena as image formation in terms of paraxial rays, because these rays are aberration free. When we discuss the properties of plane mirrors, we need not restrict ourselves to paraxial rays because plane mirrors are aberration free and therefore we can use either paraxial or finite rays. However, when we discuss the properties of curved mirrors, these are not free of aberration and therefore we must return to paraxial optics.

Optical systems can be constructed solely of mirrors or reflecting elements and such systems are called **catoptric** systems. Systems that consist of refracting and reflecting elements are called **catadioptric** systems. In Chapter 4, we showed how to use paraxial ray tracing to study the properties and image formation in such systems. In this chapter, we will look at the properties of single mirrors and simple mirror systems, starting with plane mirrors.

7.1 Plane mirrors

In this section, we will investigate the properties of plane mirrors. These properties are useful in their own right, but are also very useful in understanding the properties of systems of plane mirrors or reflecting surfaces, such as occur in some reflecting prisms, which are discussed in Chapter 8.

The optics of plane mirrors can be analysed using Snell's law. For reflection, Snell's law reduces to the statement that the angle of reflection is equal to the angle of incidence. A single ray from the object and representing the central ray of the beam, reflected from the surface using Snell's law, gives us the direction of the reflected beam. However, in many situations, this is not sufficient as we also need to know the orientation of the reflected image. It is well known that images formed in mirrors are inverted in some manner. We will now look at the conditions of this inversion.

7.1.1 *Image formation and orientation*

The optics of the image formed by a reflection from plane mirrors is, in one sense, simple. In Figure 7.1a, Q and A are points on an extended object which

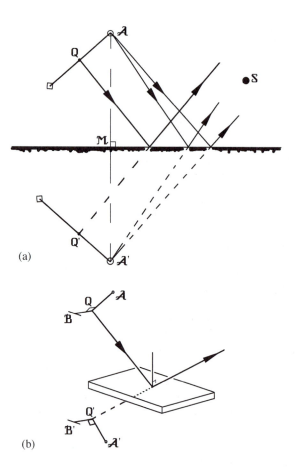

Fig. 7.1: Reflection from a plane mirror, showing image inversion in the plane of incidence.

(a)

(b)

are imaged in the mirror at Q' and A'. We can locate the position of either of these image points by tracing two rays from a point and reflecting them from the surface using Snell's law. The two reflected rays will appear to orginate from the image point. As an example, we have drawn two rays from A. It is easy to show, using simple geometry and trigonometry, that this image point must be an equal distance behind the mirror and formed on the normal drawn from the object point to the mirror. That is, the line AA' is normal to the mirror and the distances AM and $A'M$ are equal. Thus the image of any point is formed at an equal distance on the other side of the mirror and hence the object must be imaged with unit magnification.

Now, if we observed the original object and its image from a point such as S, we would observe that the image is inverted relative to the object. Thus we may be tempted to conclude that the image formed in the mirror is inverted in some manner. This brings us to a fundamental rule for image formation by reflection from a plane mirror.

7.1.1.1   A rule for determining the image orientation

From the above discussion, we can say that

> *Objects lying in the plane of incidence are inverted in this plane on reflection.*

Fig. 7.2: Two
interpretations of image
inversion after more than
one reflection.

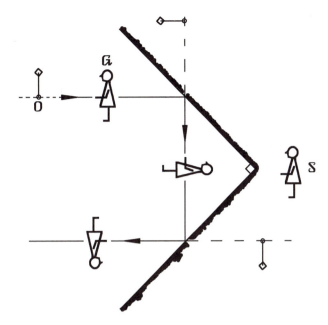

The plane of incidence is the plane containing the incident ray, the reflected ray and the normal to the surface. Furthermore, objects lying perpendicular to the plane of incidence, that is the line QB shown in Figure 7.1b, are not inverted on reflection. This diagram shows the image formation more clearly. This situation applies to all image-forming beams reflected by plane mirrors.

It also follows that after an odd number of reflections in the same plane, the image will be inverted and after an even number of reflections, the image will be erect. However, with mirrors, the notions "inverted" and "erect" depend upon the observing conditions. To demonstrate this, let us look at the example shown in Figure 7.2, where an object is reflected twice by two perpendicular mirrors.

Let us suppose we have an observer initially at G looking backwards towards the orginal object at O. Now regard this observer as travelling along a representative ray, for example a ray from the central point on the object, as shown in Figure 7.2. Further, regard the ray as a moving wire and the observer as firmly attached to the wire and moving along with it. After the first reflection, the observer would note that the image is inverted in the plane of incidence. After the second reflection, the image is once again inverted as seen by the observer, but after two inversions it appears to be erect for our observer. However, the image is inverted relative to the original object as seen by an external observer at S. Therefore while the above rule always applies, we have to be very careful when we interpret it in a given situation, because we have just seen that the concept of image "inversion" depends upon the status of the observer.

In order to appreciate further the effect of observing conditions on the perception of image orientation, let us look at the case of our own image in the mirror.

7.1.2 The image of our own reflection

While the above rule helps us to determine image orientation in any situation, image formation by mirrors is often confusing because of the subjective perception that mirrors invert left to right but not top to bottom. When we look at

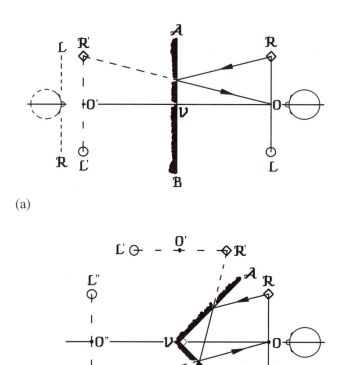

Fig. 7.3: Comparison of image formation between a simple plane mirror and two mirrors inclined at 90°.

(a)

(b)

ourselves in a mirror, we have the perception that the mirror inverts us left for right but not top to bottom. This seems paradoxical, but it is purely due to our construct of handedness. To prove this, consider the situation shown in Figure 7.3a. Here, imagine that we are looking at ourselves in the mirror. We are at o and our mirror image is at o'. It is clear from this diagram that the mirror has not inverted us left for right, because the righthand side of our body is still on the righthand side of the image as we see it. However, if we now imagine ourselves as in the mirror as shown in the diagram and looking out of the mirror at our original self, our right hand is now on our left, and vice versa. This is the reason for the apparent left- to righthand inversion of our mirror image.

To demonstrate further that the mirror has not inverted left for right, let us consider holding a page of print up against a mirror and reading the mirror image. Obviously, the print appears to be back to front. However, if we held up the same print on a transparent sheet, and looked at the sheet from behind and through the sheet at the image, we would see that the original print is identical in orientation to the mirror image; that is both are back to front. Therefore the mirror has not inverted the print left to right.

However, we can invert left for right using two mirrors inclined at 90° to each other. Let us look at ourselves in such a mirror system, as shown in Figure 7.3b. Let us follow a ray that leaves the point $\mathcal{R}$ and is reflected by the mirror $\mathcal{AV}$ onto the mirror $\mathcal{BV}$ back to the observer. This ray path shows that the mirror $\mathcal{AV}$ forms an intermediate image at o' and the second mirror $\mathcal{BV}$ forms a final image at o''. It is clear from this diagram that the image is now inverted relative

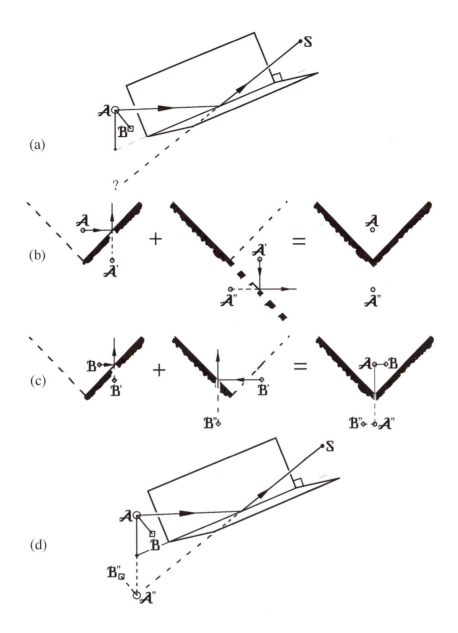

Fig. 7.4: Image formation in a roof reflector.

to the object, in the plane of the object. This mirror arrangement does clearly invert left for right.

This type of construction is used in some reflecting prisms that contain a roof, and examples of these are given in Chapter 8. Let us look at this roof construction a little further.

7.1.2.1   The roof reflector

A roof reflecting system is shown in Figure 7.4a and consists of two plane mirrors accurately aligned at 90° to each other. If we place an object as shown, its image will be seen in the mirror system from some observation point s. What is the orientation of this image? Because image formation by plane mirrors is aberration free, all rays that can form the image will be concurrent at the image point. It follows that rays that would be reflected from "mirror" extensions would also be concurrent at the same image point. Most of the rays that form the image move in three dimensions and the path of these rays is difficult to picture and draw on this diagram. Therefore to examine the image formation, we should choose the most convenient rays, preferably those that are restricted to two dimensions.

Let us examine the image formation by firstly determining the image position of the point $\mathcal{A}$. To find this image position, we must first find the image position in one mirror and then find the image of this image formed in the other mirror. Figure 7.4b shows the point $\mathcal{A}$ imaged at $\mathcal{A}'$ by one mirror, the righthand side mirror. To find the image of this image formed by the second (the lefthand side) mirror, let us trace a ray to this "extended" second mirror. It is clear that the second mirror images $\mathcal{A}'$ to $\mathcal{A}''$. Therefore an observer at the point s will see the image of $\mathcal{A}$ at $\mathcal{A}''$ at a point immediately below the junction of the two mirrors and the same distance below as $\mathcal{A}$ is above. Thus the final image of $\mathcal{A}$ is formed as if it were reflected from a single plane mirror, which in this diagram would be horizontal.

Let us now look at the image position of the point $\mathcal{B}$. Its imagery is shown in Figure 7.4c. By following the formation of its images in the two mirrors in turn, it is clear that this image is finally at $\mathcal{B}''$. Thus the final image, as seen by the observer at s, is oriented as shown in Figure 7.4d.

Now some of the rays from the points $\mathcal{A}$ or $\mathcal{B}$ will hit the lefthand side mirror first and then be reflected from the righthand side mirror. We can easily show that this ray path will give the same image position as rays traced to the righthand mirror first. Therefore the image orientation is independent of which mirror we first take.

In conclusion, the final image formed by the two mirrors is inverted top to bottom and left to right; that is the image has been rotated through $180°$. This is a very useful property of this mirror construction. We should also note that this roof reflector does not invert an image that is parallel to the line of the roof.

7.1.3   Rotation of a mirror

The main application of plane mirrors in general optics is to deviate and invert beams or images. Mirrors are replaced by prisms in some situations (see Chapter 8). We need to know how the reflecting surface orientation affects the deviation angle of the beam. Figure 4.1 shows a beam being reflected from a flat surface. If this surface is rotated through an angle $\Delta\theta$, it can be easily demonstrated that the reflected beam is rotated through twice this angle; that is the beam is rotated through an angle

$$\Delta i' = 2\Delta\theta \tag{7.1}$$

One application of this result is the use of reflecting surfaces to deviate beams. For example, a mirror tilted at $45°$ to a beam will deviate the beam through $90°$. This particular example is used frequently in prisms (Chapter 8).

On the other hand, this result has the disadvantage that any angular error ϵ in aligning the mirror leads to double the angular error, that is 2ϵ, in beam deviation. In the above example, if the mirror is set at $46°$, instead of $45°$, the beam will be deviated by $92°$ instead of $90°$. That is, the beam deviation error is $2°$. This result is particularly relevant to the manufacturing tolerances of beam deviating prisms (see Chapter 8).

7.2 Spherical mirrors

Curved mirrors, like lenses, produce aberrations in incident rays. Therefore we should begin by looking at the paraxial properties of mirrors. The paraxial optical theory of reflecting surfaces was established in Chapter 4 and seen to be a special case of refracting optics with two modification rules as follows. While the ray is travelling backwards because of a reflection (i.e. right to left),

(1)   refractive indices are negative and
(2)   the surface separations are also negative.

These rules can be applied to any optical system containing curved or plane mirrors or reflecting surfaces, and, as an example, have been applied to a system of two mirrors in Chapter 4.

The most common curved mirror has a spherical surface as this is the easiest and hence cheapest type to manufacture. Other less common types are the parabolic and the elliptical mirrors. We have seen in Chapter 5 that these types of conicoid surfaces are free of spherical aberration under special circumstances and therefore are useful when we need a beam free of spherical aberration. In this section, we will look at the simple spherical mirror and look at conicoid mirrors in Section 7.3.

Spherical mirrors can be classified according to whether they are convex or concave, and the definition of these terms should be clear from Figures 7.5a and b.

7.2.1   Power

The equivalent or surface power of a single spherical mirror is given by equations (4.2) and (4.3), that is

$$F = -2nC = -2n/r \tag{7.2}$$

with the usual sign convention. This equation applies to both convex and concave mirrors. Therefore the power of a convex mirror appears to depend upon the sign of the incident refractive index n (i.e. the direction of the incident ray) and the surface curvature. Figure 7.5 shows both convex and concave mirrors in both orientations. For a ray travelling left to right, the sign of n and the curvature of a convex mirror are both positive and therefore the power of the convex mirror in this orientation would be negative. Thus the power of the concave mirror, in the same orientation, is positive. If the ray is travelling from right to left, the sign of n is now negative and the curvature of the convex mirror is also now negative, so once again the power of the convex mirror is negative. Therefore it follows that the power of the concave mirror is once again positive. Thus the power of a mirror is independent of the direction of the ray, just as it is for a lens, the power of the convex mirror being negative and that of the concave mirror being positive.

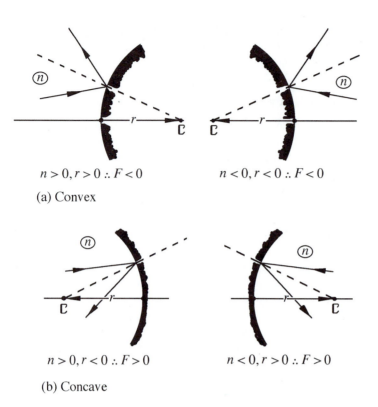

Fig. 7.5: Definitions of convex and concave mirrors and the sign of the powers.

$n > 0, r > 0 \therefore F < 0$ 　　　　 $n < 0, r < 0 \therefore F < 0$

(a) Convex

$n > 0, r < 0 \therefore F > 0$ 　　　　 $n < 0, r > 0 \therefore F > 0$

(b) Concave

7.2.2 The "mirror" equation

In Chapter 4, we derived the following equation (4.6)

$$\frac{1}{l'} + \frac{1}{l} = \frac{2}{r} \tag{7.3}$$

which is the "reflecting" equivalent of the lens equation and now should be called the "mirror" equation.

7.2.3 Cardinal point positions

The position of the back principal point $\mathscr{P}'$ must be at the surface vertex $\mathscr{V}$ as shown in Figure 7.6. The position of the back focal point $\mathscr{F}'$ is given by equation (3.33), that is

$$\mathscr{P}'\mathscr{F}' = n'/F$$

but since $n' = -n$

$$\mathscr{P}'\mathscr{F}' = -n/F$$

Using equation (7.2) to eliminate F from this equation gives

Fig. 7.6: Positions of
cardinal points for single
spherical mirrors.

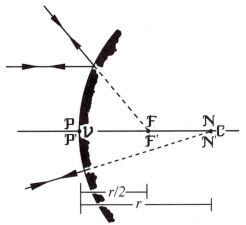

(a) Convex mirror

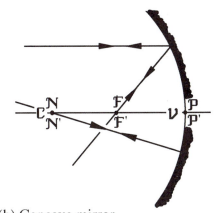

(b) Concave mirror

$$\mathcal{P}'\mathcal{F}' = \frac{1}{2C} = \frac{r}{2} \tag{7.4}$$

Therefore for both convex and concave mirrors, the back focal point $\mathcal{F}'$ is midway between the centre of curvature C and the surface as shown in Figure 7.6. Now for a mirror in air, the equivalent focal length f is equal to the distance $\mathcal{P}'\mathcal{F}'$. Thus we have

$$f = r/2 \tag{7.5}$$

The position of the back nodal point $\mathcal{N}'$ is given by the principal to nodal point distance, equation (3.58), that is

$$\mathcal{P}'\mathcal{N}'(= \mathcal{P}\mathcal{N}) = \frac{(n'-n)}{F}$$

Since $n' = -n$ and $F = -2n/r$ from equation (7.2),

$$\mathcal{P}'\mathcal{N}'(= \mathcal{P}\mathcal{N}) = r \tag{7.6}$$

Thus the back nodal point is at the centre of curvature c of the mirror.

With single mirrors, the object and image space are identical; therefore the image and object space cardinal points must also be identical. Therefore the front cardinal points coincide with the back cardinal points. The positions of the six cardinal points are shown in Figure 7.6.

7.2.4   Image formation

To find the position and size of the image formed by a mirror, we can use the mirror equation, equation (7.3). This equation can be transformed to give

$$l' = \frac{lr}{(2l - r)} \tag{7.7}$$

The position of the image depends upon the magnitude and sign of the object distance l and the nature of the surface. An analysis of equation (7.7) will show that for a real object (that is, the object distance is negative according to our sign convention)

- a convex mirror always gives a virtual image, that is behind the surface vertex
- a concave mirror gives a real image providing the object is farther from the mirror than the focal points and a virtual image if the object lies between the focal points and the mirror.

7.2.4.1   Magnification

The transverse magnification M of the image is given by the standard equations (3.46) and (3.49)

$$M = \frac{\eta'}{\eta} \tag{7.8a}$$

and

$$M = \frac{nl'}{n'l} \tag{7.8b}$$

Since $n' = -n$

$$M = -\frac{l'}{l} \tag{7.9}$$

Replacing l' by the corresponding expression from (7.7)

$$M = \frac{r}{(r - 2l)} \tag{7.10}$$

Fig. 7.7: Definition of
relative magnification of a
curved mirror.

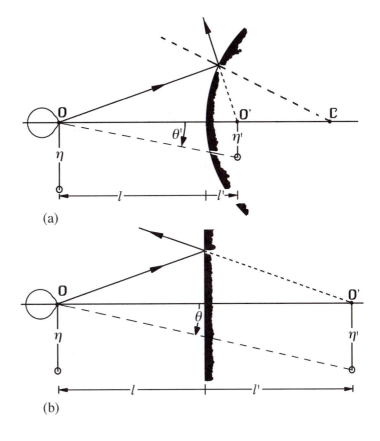

(a)

(b)

7.2.4.2   *Relative angular magnification of a curved mirror as seen from the object*

Curved mirrors are sometimes used as viewing systems to provide either (a) a wide field-of-view with an associated reduction in image size or (b) some increase in image size (magnification > 1) with an associated reduction in field-of-view. The effect of radius of curvature on image size can be examined by considering the **relative magnification** M_r defined as

$$M_r = \frac{\text{angular size } (\theta') \text{ of image in curved mirror}}{\substack{\text{angular size } (\theta) \text{ of image in plane mirror} \\ \text{at the same position}}} \qquad (7.11)$$

with the point of observation being the object position in both cases.

A simple equation for this quantity can be derived as follows. Firstly by referring to Figure 7.7a,

$$\theta' = \frac{\eta'}{(l' - l)}$$

where θ' is negative in the diagram. By using equations (7.8a) to (7.10), we can show that

$$\theta' = \frac{r\eta}{2l(l-r)}$$

It follows from Figure 7.7b that

$$\theta = -\frac{\eta}{2l}$$

where like θ', θ must be negative in the diagram. Substituting the above expressions for θ and θ' in equation (7.11) leads to the equation

$$M_r = \frac{r}{(r-l)} \tag{7.12}$$

An analysis of this equation shows that

- concave mirrors give positive magnification greater than unity when the object is between the centre of curvature c (or nodal points) and the mirror, and an inverted and reduced image if the object point is beyond the centre of curvature.
- convex mirrors always give an erect but reduced image.

7.2.4.3   *Graphical ray tracing*

The position and size of an image can be found by a number of techniques. For single mirrors, the equations (7.3) and (7.10) or paraxial ray tracing from first principles may be used. For more complex systems, paraxial ray tracing is the only practical alternative. In some single mirror cases when only general trends are being investigated without any particular numerical values being available or important, graphical ray tracing is a very useful tool in the investigation.

It has already been established that only two paraxial rays need to be traced to locate an image. Consider the situation in Figure 7.8, where the image Q' of Q is to be located. In this general situation three rays can be easily traced. These are as follows.

(a)   The ray $Q\mathcal{N}$ ($\mathcal{N}'$ or c), from Q and passing or appearing to pass through the nodal points ($\mathcal{N}$ or $\mathcal{N}'$) or centre of curvature c
(b)   The ray $Q\mathcal{B}$, which is parallel to the axis and after reflection passes or appears to pass through the focal point $\mathcal{F}'$ (or $\mathcal{F}$)
(c)   The ray $Q\mathcal{F}$, which appears to pass through the front focal point $\mathcal{F}$ and which on reflection is parallel to the axis.

All three rays will be or appear to be concurrent at the image point Q'. Only two (any two) of these rays need to be traced.

In tracing the above rays, we should not forget that the rays should be traced close to the axis in order to resemble paraxial rays. If they are too far from the axis, they will be outside the paraxial region, suffer some aberration and not be concurrent in the image space.

Fig. 7.8: Graphical ray
tracing used to find position
and size of image formed
by a spherical mirror.

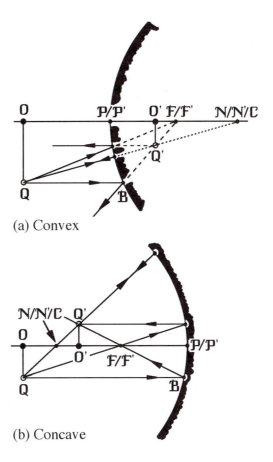

(a) Convex

(b) Concave

7.3 Conicoid mirrors

In Chapter 5, it was shown that certain types of conicoid mirrors are free of spherical aberration for a pair of axial conjugates and the value of the asphericity will depend upon the positions of these conjugates. It was shown that concave mirrors free of spherical aberration were ellipsoidal and convex mirrors free of spherical aberration were hyperbolic in shape. Let us examine these types of conicoids a little further.

A conicoid mirror, rotationally symmetric about the optical (Z-) axis, may be expressed in the general conicoid form by equation (5.17), that is

$$p^2 + (1+Q)Z^2 - 2rZ = 0 \tag{7.13}$$

where

$$p^2 = X^2 + Y^2$$

r = radius of curvature at the vertex and Q is the asphericity

The value of Q defines the type of conicoid according to the following rules:

$$
\begin{aligned}
Q &> 0 \quad \text{an ellipsoid with the major axis being} \\
& \text{perpendicular to the optical } (Z\text{-}) \text{ axis} \\
Q &= 0 \quad \text{a sphere} \\
Q &= -1 \quad \text{a paraboloid} \\
-1 < Q &< 0 \quad \text{an ellipsoid with the optical axis being} \\
& \text{the major axis}
\end{aligned}
\tag{7.14}
$$

The effect of the value of Q on the surface shape is shown in Figure 5.14.

Ellipsoidal mirrors

An alternative mathematical expression for an ellipsoid is

$$
\frac{(Z-a)^2}{a^2} + \frac{p^2}{b^2} = 1
\tag{7.15}
$$

where $2a$ is the length of the ellipsoid along the optical axis and $2b$ is the maximum width in the X–Y plane. This ellipsoid has vertices at the points $Z = 0$ and $Z = 2a$. The values of a and b are related to the vertex radius of curvature r and asphericity Q. By comparison of equation (7.13) with equation (7.15), we have

$$
r = b^2/a
\tag{7.16a}
$$

and

$$
Q = (b/a)^2 - 1
\tag{7.16b}
$$

Sometimes the asphericity of an ellipse is given in terms of the eccentricity e. For an ellipsoid, this is defined as

$$
e^2 = 1 - (b/a)^2
\tag{7.17}
$$

and therefore it follows that

$$
Q = -e^2
\tag{7.18}
$$

Paraboloidal mirrors

A paraboloid mirror can be regarded as a limiting case of the ellipsoidal mirror in the limit that the value of a is infinite. A paraboloid has a Q value of -1. Substituting this value of Q in equation (7.13) gives

$$
Z = p^2/(2r)
\tag{7.19}
$$

For a paraboloid, the parameters a and b are not meaningful and therefore we cannot define e in terms of a and b as we do for the ellipsoids and hyperboloids. However, if we make use of equation (7.18) with $Q = -1$, then for a paraboloid, e^2 would have a value of 1.

Hyperboloidal mirrors

An alternative mathematical expression for a hyperboloid is

$$\frac{(Z+a)^2}{a^2} - \frac{p^2}{b^2} = 1 \tag{7.20}$$

This equation describes two hyperboloids which have vertices at $Z = 0$ and $Z = -2a$. The values of a and b are related to the vertex radius of curvature r and asphericity Q. By comparison of equation (7.13) with equation (7.20), we have

$$r = b^2/a \tag{7.21a}$$

and

$$Q = -(b/a)^2 - 1 \tag{7.21b}$$

Sometimes, the asphericity of a hyperboloid is given in terms of the eccentricity e. For a hyperboloid this is defined as

$$e^2 = 1 + (b/a)^2 \tag{7.22}$$

and therefore it follows that once again

$$Q = -e^2 \tag{7.23}$$

7.3.1 Cardinal points

The cardinal points are paraxial quantities and therefore have to be found using paraxial rays. In the paraxial region, conicoids are identical to spherical surfaces with the same vertex radius of curvature. This can be demonstrated by firstly expressing equation (7.13) explicitly in Z and then expanding the resulting square root as a polynomial in p^2, using the binomial theorem. This expansion would show that the first term is a p^2 term and its coefficient does not contain the Q value. Thus all members of the family of conicoids, specified by equation (7.13), all have identical first terms in their polynomial expansions and it is the coefficient of this term which contains the vertex curvature. Therefore the positions of the cardinal points of a conicoid are the same as for a spherical mirror with the same vertex radius of curvature r (or curvature C). These positions are given in Section 7.2.3.

7.3.2 Conjugates free of spherical aberration

For a conicoid with an asphericity Q and vertex radius of curvature r, the positions of the conjugates free of spherical aberration can be found by solving equations (5.29) for l and l'. Solving these equations leads to the following distances

$$\text{conjugate distances} = \frac{r}{1 \pm \sqrt{(-Q)}} = \frac{r}{1 \pm e} \tag{7.24}$$

These distances are measured from the surface vertex and Q and e are related by equation (7.18) or (7.23). Let us now look at the individual types of conicoids.

Ellipsoidal mirrors

For an ellipsoidal mirror, the rules given by equation (7.14) allow Q to be in the range -1 upwards in a positive direction. However, from equation (7.24), it is clear that Q must be negative; otherwise there are no conjugates along the Z-axis free of spherical aberration.

The above equations for the position of the conjugates free of spherical aberration can be expressed alternatively in terms of a and b. If we substitute for r and Q from equations (7.16a and b) into equations (7.24), the distances of these two points from the vertex are

$$\text{conjugate distances} = a \pm \sqrt{(a^2 - b^2)} \qquad (7.25)$$

or by replacing b from equation (7.17), we have

$$\text{conjugate distances} = a(1 \pm e) \qquad (7.26)$$

For conjugate pairs at other distances, there will be some spherical aberration.

Paraboloidal mirrors

For a paraboloidal mirror, $Q = -1$ and equation (7.24) leads to the conjugates being at distances

$$r/2 \quad \text{and} \quad \infty \qquad (7.27)$$

From equation (7.5), the focal length of a mirror is $r/2$; therefore a paraboloidal mirror is free of spherical aberration for the conjugate points, which are the focal point(s) and infinity.

Hyperboloidal mirrors

For a hyperboloidal mirror, the rules given by equation (7.14) require Q to be in the range -1 downwards in a negative direction. Now since equation (7.24) requires Q to be always negative, it follows that any hyperbolic mirror is free of spherical aberration for some pair of conjugates.

Equation (7.24) can be expressed alternatively in terms of a and b. If we substitute for r and Q from equations (7.21a and b) into equations (7.24), the distances of the conjugates from the surface vertex are

$$\text{conjugate distances} = -a \pm \sqrt{(a^2 + b^2)} \qquad (7.28)$$

or by replacing b from equation (7.22), we have

$$\text{conjugate distances} = -a(e + 1) \quad \text{and} \quad a(e - 1) \qquad (7.29)$$

For conjugate pairs at other distances, there will be some spherical aberration.

7.4 Mirror systems

Pure mirror systems are very rare in optics. Perhaps the most common application of mirror systems is in astronomical telescopes. In these cases, only the objective is a mirror system and the eyepiece is a conventional refracting system. The structure of the reflecting objectives is shown later in Chapter 17.

Exercises and problems

7.1 If a plane mirror rotates through an angle of $5°$, through how many degrees does the reflected ray rotate?

ANSWER: $10°$

7.2 What is the power of a spherical mirror with a radius of curvature of $+100\,\text{mm}$?

ANSWER: $F = -0.02\,\text{mm}^{-1}$ or $-20\,\text{m}^{-1}$

7.3 Taking the radius of curvature of the front surface of the cornea as $7.8\,\text{mm}$, calculate the image position of an object placed $50\,\text{cm}$ in front of the cornea and reflected in the cornea.

ANSWER: $3.870\,\text{mm}$ behind the cornea

7.4 For a concave mirror of radius $10\,\text{cm}$, calculate the equivalent focal length. On an accurately drawn diagram, show this mirror and the positions of the cardinal points.

ANSWER: $f = 5\,\text{cm}$

7.5 Calculate the "relative" magnification of a concave mirror with a radius of curvature of $(-)3\,\text{m}$, viewed from a distance of (a) $10\,\text{m}$ and (b) $2\,\text{m}$.

ANSWERS: $M_\text{r} = $ (a) $- 0.428$, (b) $+ 3$

Summary of main symbols and equations

$X, Y \& Z$	cartesian co-ordinates
Q	surface asphericity
e	eccentricity
a, b	semi-major and semi-minor axes of an ellipse (a is horizontal and b is vertical)
M_r	angular magnification of a spherical mirror relative to a plane mirror seen from the object position

Section 7.2.1: Spherical mirrors

$$F = -2nC = -2n/r \quad \text{power of a surface} \tag{7.2}$$

$$\frac{1}{l'} + \frac{1}{l} = \frac{2}{r} \tag{7.3}$$

Section 7.2.4: Image formation

$$M_{\rm r} = \frac{\text{angular size of image in curved mirror } (\theta')}{\text{angular size of image in plane mirror } (\theta)} \atop \text{at the same position} \qquad (7.11)$$

$$M_{\rm r} = \frac{r}{(r - l)} \qquad (7.12)$$

8

Prisms

8.0 Introduction

In many visual optical systems, prisms play an important role in the formation of the final image. They are used to control or change the direction of the image forming beam and are often used to change the orientation of the image. In these operations, the prism may only refract the beam, may refract and reflect it or may only reflect it but one or more times. Therefore in discussing the structure and properties of prisms, it is convenient to classify prisms according to whether they (a) are purely refracting, (b) combine refraction with internal reflections or (c) are purely reflecting.

The optical construction of most prisms is relatively simple but the ray paths inside a prism, which reflect the beam one or more times, can be complex and some of the properties of the prism are not readily obvious from looking at the ray paths. However these properties are sometimes made clearer by examining the **unfolded** version. Unfolding a prism involves reflecting each reflecting surface in the next reflecting surface along the ray path. Ideally, this process will become clearer when examples are discussed later in this chapter.

8.1 Refracting-only prisms

Refracting prisms occur frequently in visual optical and ophthalmic instruments. They can be used to deviate beams or images, as aids to focussing to form double or split images and often to compensate for eye problems such as phorias and squint.

Just as lenses may be examined using finite (or real) ray tracing on one hand or paraxial theory on the other, so can refracting prisms, but before we examine the paraxial properties of prisms, we will examine their effects on finite rays. Let us look firstly at the angle of deviation of a ray passing through a prism.

8.1.1 Angle of deviation

A typical prism is shown in Figure 8.1. A ray is incident at $\mathcal{B}$ on the lefthand side face, is refracted, strikes the righthand side face at $\mathcal{B}'$ and is refracted again. Of general interest is the angle θ of deviation of the ray. Let us solve this problem by following the ray through the prism.

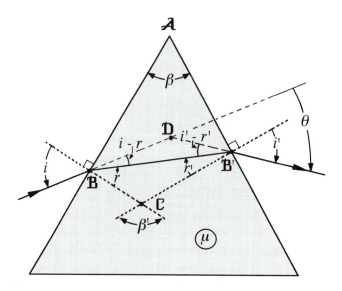

Firstly, using Snell's law with the prism in air, we have at $\mathcal{B}$

$$r = \sin^{-1}[\sin(i)/\mu] \qquad (8.1a)$$

The next step is to find the angle r', the angle of incidence at the righthand side face. In triangle $\mathcal{BB'C'}$

$$r' + r + \beta' = 180°$$

but since the angles at $\mathcal{B}$ and $\mathcal{B}'$ are right angles in the quadrilateral $\mathcal{AB'CB}$, we have

$$\beta + \beta' = 180°$$

Thus

$$r' = \beta - r \qquad (8.1b)$$

Once again applying Snell's law, this time at $\mathcal{B'}$ and since the prism is in air,

$$i' = \sin^{-1}[\mu \sin(r')] \qquad (8.1c)$$

Now, from triangle $\mathcal{BB'D}$ and using the rule that the external angle is the sum of the two internally opposite angles, we have

$$\theta = (i - r) + (i' - r')$$

but, from equation (8.1b)

$$r' + r = \beta$$

Therefore, finally, the angle of deviation (θ) is given by the equation

$$\theta = i + i' - \beta \qquad (8.1\text{d})$$

Using equations (8.1a to d), we can show that

$$\theta = i + \sin^{-1}\{\sin(\beta)\sqrt{[\mu^2 - \sin^2(i)]} - \sin(i)\cos(\beta)\} - \beta \qquad (8.2)$$

These equations can be used to calculate the angle of deviation for a ray with an angle of incidence i on a prism with apex angle β and a refractive index μ. If the resulting angle of deviation θ were plotted against i, it would be seen to pass through a minimum value. This **minimum angle of deviation** depends upon the refractive index and apex angle of the prism and it is easy to derive an equation for this minimum angle. This is done below.

8.1.1.1 *Minimum angle of deviation*

An equation for the minimum angle of deviation can be derived as follows. The amount of mathematics required for the derivation can be reduced considerably by using the reversibility principle of rays and the symmetry of the prism about a bisecting line through the prism apex. Because of these factors, if the deviation angle θ has only one stationary value, then the corresponding angle of incidence i and final refraction angle i' are equal.

This point of stationarity is a minimum. Thus when $i = i' =$ say $i_{\min}$, θ is a minimum, say $\theta_{\min}$, and from equation (8.1d)

$$\theta_{\min} = 2i_{\min} - \beta$$

Therefore

$$\sin[(\theta_{\min} + \beta)/2] = \sin(i_{\min}) = \mu \sin(r)$$

but from equation (8.1b) with $r = r'$

$$2r = \beta$$

Therefore finally

$$\sin[(\theta_{\min} + \beta)/2] = \mu \sin(\beta/2) \qquad (8.3\text{a})$$

or

$$\sin(i_{\min}) = \mu \sin(\beta/2) \qquad (8.3\text{b})$$

A proof from first principles can be found in a number of other texts, for example Born and Wolf (1989). Their proof does not assume that $i = i'$ at the minimum but begins by differentiating equation (8.1d) and seeking the condition that $d\theta/di = 0$.

8.1.2 The paraxial approximation or thin prism

If the apex angle β of a prism and the angle of incidence are small, all of the other angles will also be small. If these are small enough, such that all sines of angles can be replaced by the angles themselves, the above mathematics for the deviation of a ray by a prism can be greatly simplified. This simplification is equivalent to a paraxial approximation. Returning to Figure 8.1 and once again tracing the ray through the prism, but this time replacing all sines of angles by angles, we get the following equations in turn

$$i = \mu r \quad \text{Snell's law applied to the first surface} \tag{8.4a}$$

$$r' = \beta - r \tag{8.4b}$$

$$i' = \mu r' \quad \text{Snell's law applied to the second surface} \tag{8.4c}$$

$$\theta = i + i' - \beta \tag{8.4d}$$

which are equivalent to equations (8.1a to d) and we can solve these equations to get

$$\theta = \beta(\mu - 1) \tag{8.5}$$

This equation shows that in the paraxial approximation, the deviation angle θ is independent of the angle of incidence i.

8.1.2.1 The power of a thin prism and the prism dioptre

We can see that the ability of a prism to deviate a finite ray is a complex function of the refractive index, apex angle and angle of incidence. However, if the prism is thin and the angle of incidence is small, we can reduce the problem to the paraxial case. In the paraxial approximation, the angle of deviation is given by equation (8.5) and is independent of the angle of incidence. Thus the angle of deviation could be used as a measure of the "refractive power" of the prism, but in ophthalmic optics it is not. Instead the refractive power is specified in terms of the corresponding transverse displacement in centimetres, at a distance of 1.0 m. Referring to Figure 8.2 and denoting the displacement by the symbol y, we have

$$y = 100 \tan(\theta) \text{ cm} \tag{8.6}$$

but since θ is small, one can use the approximation $\tan(\theta) = \theta$, and equation (8.6) becomes

$$y = 100\theta \text{ cm}$$

Fig. 8.2: Thin prism and the prism dioptre. The power of the prism in prism dioptres (Δ) is the distance y in centimetres.

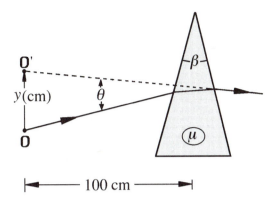

where θ must now be in radians. If we now replace θ by the prism constructional parameters β and μ from equation (8.5), we have

$$y = 100\beta(\mu - 1)\,\text{cm}$$

where β must also be in radians. This displacement is called the **power** of the prism (here denoted by the symbol F_p). Although it has the unit of centimetre/metre, the unit is more commonly known as the **prism dioptre** and usually is given the symbol Δ. Thus

$$F_p = 100\beta(\mu - 1)\,\Delta \tag{8.7}$$

In terms of the deviation angle θ, we can write this power also in the form

$$F_p = 100\theta\,\Delta \tag{8.7a}$$

It should be noted that the prism dioptre is only valid within the paraxial approximation, in the sense that the deviation distance is assumed to be independent of the angle of incidence of the incident ray.

Sign of the power

Unlike the power of a lens, the power of a prism is not assigned a positive or negative value. Instead the direction of the deviation is denoted by the orientation of the base of the prism. In ophthalmic optics, when a prism is placed before an eye, terms such as "base in/out/up/down" are used to denote the direction of the base.

Direction of deviation of the image

It is clear from the preceding discussion that a prism in air deviates the beam or ray towards the base. The result is that the image is deviated towards the apex, as shown in Figure 8.2.

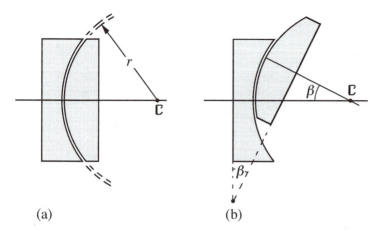

Fig. 8.3: Variable power prism using cylindrical lenses.

(a)   (b)

8.1.3   Variable power prisms

Just as there is a need for variable power lenses (Chapter 6), there is a need for variable power prisms and there are various solutions to the problem of their construction.

One solution is to take two plano-cylindrical lenses with equal and opposite power. These are placed with their curved surfaces in contact as shown in Figure 8.3a, with their flat surfaces parallel. This construction acts as a plane block of parallel sided glass with no prismatic power. If one of the lenses is rotated through an angle β about the centre of curvature of the curved surfaces, the initially parallel sides are tilted and the combination now acts as a prism, as shown in Figure 8.3b. The region of overlap forms a prism with an apex angle β. The limitation of this method is that as the prism angle increases, the effective aperture of the prism decreases. Another solution is the Risley prism arrangement.

The Risley prisms

The Risley prisms are a pair of conventional prisms placed close together as shown in Figure 8.4a, and a variable prismatic effect is achieved by rotating them in opposite directions.

Figure 8.4b shows the effect of each prism's acting independently on the image of the object o when the prisms are aligned as shown in Figure 8.4a. The image of o will appear to be deviated downwards to o'_A by prism A and upwards to o'_B by prism B. We can let the amount of deviation be the power F_p of each prism. For thin prisms in contact, the final image deviation is the vector sum of the deviations of each prism. In this case, the deviations are equal and opposite and therefore there is no net deviation.

If the prisms are now rotated by the same amount but in opposite directions about the optical axis as shown, prismatic power is induced in the horizontal direction but not in the vertical direction. Let us see how this happens.

If the prisms are rotated in opposite directions through an angle γ, the two images will move along the arc of a circle of radius F_p in opposite directions, through an angle γ to the new positions shown in Figure 8.4c. If we add the deviations, the vertical deviations are equal and opposite and therefore cancel.

Fig. 8.4: The Risley variable power prism.

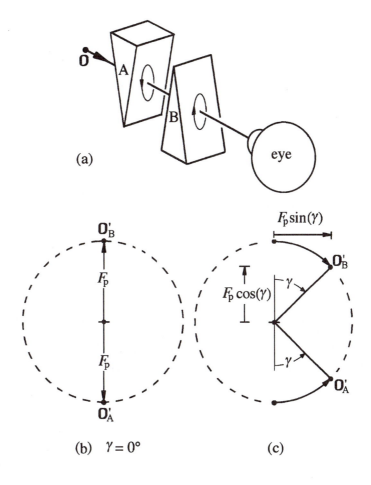

(a)

(b) $\gamma = 0°$ (c)

The horizontal deviations are in the same direction and therefore are additive and do not cancel out.

It therefore follows from Figure 8.4c that in the horizontal direction, each prism deviates the image by an amount

$$F_p \sin(\gamma)$$

and therefore the total deviation is

$$2F_p \sin(\gamma)$$

and this is equivalent to the combined prismatic power of the Risley prisms. Therefore if we denote the power of the Risley prisms as $F_{p(ris)}$ then

$$F_{p(ris)} = 2F_p \sin(\gamma)\ \Delta \tag{8.8}$$

The power of each of the prisms in the pair is about 10Δ, giving a range of powers from 0 to about 20Δ.

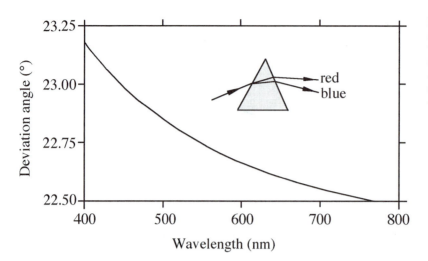

Fig. 8.5: Deviation angle θ as a function of wavelength of a prism (dispersion) made with Schott BK7 glass. The apex angle is 30° and the angle of incidence is 60°.

8.1.4 Dispersion of a prism

Both equation (8.2) for the real rays and equation (8.5) for the thin prism or paraxial case show that the deviation of a ray by a prism depends upon the refractive index of the prism. Now the refractive index varies with wavelength and this variation has already been discussed in Chapters 1 and 5. Since the refractive index decreases with increase in wavelength, the angle of deviation should decrease with increase in wavelength; that is red light is deviated less than blue light. When a small white light source is viewed through a prism, a normal spectrum is observed. We can call this effect **transverse colour fringing** or **colour fringing**. Figure 8.5 shows the dependence of deviation angle on wavelength in a typical case. In this example, over the visible spectrum, the angular spread is about one degree. Now the angular resolving power of the eye is about 1 minute of arc and therefore such a spectrum should be clearly visible. This chromatic fringing can be reduced by using an **achromatic prism**, which is analogous to the achromatic lens referred to in Section 6.5. The theory of this special prism is as follows.

8.1.4.1 The thin achromatic prism

The change in angle of deviation θ with wavelength can be reduced by combining two thin prisms in contact, as shown in Figure 8.6, with the prisms being made of materials with different dispersions. In the thin prism approximation, for two thin prisms in contact, the angle of deviation θ will be the sum of the deviations due to each prism separately and thus from equation (8.5), that is

$$\theta = \beta_1(\mu_1 - 1) - \beta_2(\mu_2 - 1)$$

where μ_1 and μ_2 are the refractive indices. We can reduce the chromatic dispersion of this combination by requiring the deviation to be the same for two wavelengths at opposite ends of the visible spectrum. This effectively "folds" the dispersion about some intermediate wavelength and therefore reduces the

Fig. 8.6: An achromatic thin prism. The refracted rays for λ_F and λ_C are parallel and hence have the same deviation angle θ.

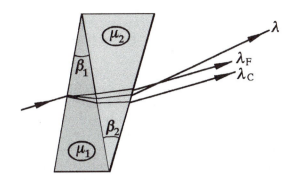

angular dispersion. Let us choose the two wavelengths to be the hydrogen blue line λ_F and the hydrogen red line λ_C, which were introduced in Section 1.2.2.1. At the blue wavelength λ_F the deviation angle is

$$\theta_F = \beta_1(\mu_{1F} - 1) - \beta_2(\mu_{2F} - 1)$$

and at the red wavelength λ_C

$$\theta_C = \beta_1(\mu_{1C} - 1) - \beta_2(\mu_{2C} - 1)$$

where μ_{1F}, μ_{2F}, μ_{1C} and μ_{2C} are the corresponding refractive indices at the F and C lines. It follows that these two angles of deviations are equal if

$$\beta_1(\mu_{1F} - \mu_{1C}) - \beta_2(\mu_{2F} - \mu_{2C}) = 0$$

Replacing the differences $\mu_F - \mu_C$ by their V_d values [equation (1.9)], we have

$$\frac{\beta_1(\mu_{1d} - 1)}{V_{1d}} = \frac{\beta_2(\mu_{2d} - 1)}{V_{2d}} \tag{8.9}$$

This is the condition that the wavelengths λ_F and λ_C will be deviated by the same amount. However some colour fringing remains because this condition only sets the deviation angle to be the same at the F and C wavelengths, but it will be different at other wavelengths. This residual colour fringing is called the **secondary spectrum**.

8.1.5 Applications of refracting prisms

8.1.5.1 Compensation for phorias or squints

In ophthalmic optics, prisms are used to measure **phorias** and **tropias** and in some cases compensate for them. A **heterophoria** is a mis-alignment of the two eyes when the stimulus for the two eyes to look at the same object is removed (e.g. by covering one eye), and a **heterotropia** (or squint) is a mis-alignment of the two eyes even when there is a stimulus present. The mis-alignment of the two eyes leads to double vision unless the vision in one eye is suppressed by the brain. When a phoria or squint is small, a thin prism with a small prismatic power can be used to provide an optical alignment of the two eyes. However, large phorias and squints require prisms with large prismatic power, which, because of their thickness, may lead to objectionable transverse colour fringing.

8.1.5.2   *Image doubling or splitting*

Image doubling or splitting principles are used in a number of visual optical instruments such as rangefinders or focussing systems, and in ophthalmic instruments such as the keratometer. The doubling can be achieved by using refracting prisms. Thick glass plates can also be used for doubling and while these are not prisms, they are worth a brief discussion at this point. Providing the image is not at infinity, a thick plate can be used for doubling, if the plate intersects only half of the image forming beam. The beam that by-passes the plate forms an undeviated image. The portion of the beam that passes through the plate can be variably displaced by rotating the plate about an axis perpendicular to the beam axis. Thus two images are formed.

Referring to Figure 8.7a, if a plate of thickness d and refractive index μ is tilted through an angle i, relative to the direction of the image, the beam is displaced transversely by an amount t. By using the diagram, it can easily be shown that

$$t = d[\sin(i) - \cos(i)\tan(r)] \tag{8.10}$$

where from Snell's law

$$\sin(r) = \sin(i)/\mu$$

The maximum value of the displacement t is the thickness of the plate, which occurs when the plate is parallel to the direction of the incident beam.

If the ray shown in Figure 8.7a arises from an object at o, a distance w as shown in Figure 8.7b, this object and the deviated image at o' will also be transversely separated by a distance t and their angular separation ψ is simply

$$\psi = t/w \tag{8.11}$$

It can be seen from the above equation that there is no effective doubling if the image is at infinity (that is $w = \infty$) and that the amount of doubling can be varied by rotating the glass plate. We will now proceed to discuss a very simple prism image doubling technique, using the Fresnel biprism.

Fresnel biprism

The Fresnel biprism may be thought of as consisting of two thin prisms cemented together at their base as shown in Figure 8.8. This biprism is placed in the beam as shown. If the object at o or primary image is at a distance z from the biprism, two images are formed. One at o'_1 is due to refraction and hence deviation of the rays by the top half of the biprism, and the other image at o'_2 is formed similarly by the bottom part of the biprism. If the prism apex angle β is small, the biprism can be investigated using the paraxial approximation.

Suppose that the eye is at a distance w from the image as shown in Figure 8.8. Without the biprism in position, the image is observed at the point o. With

Fig. 8.7: Deviation of images using a tilted thick glass plate.

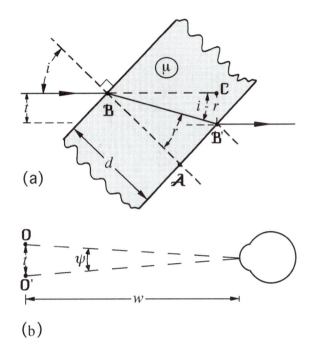

(a)

(b)

Fig. 8.8: The Fresnel biprism used to produce variable doubling of an image.

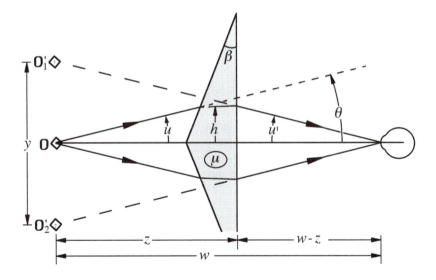

the biprism in position, the upper image of o is displaced to o'_1 and its angular displacement u' can be found as follows. Firstly

$$\theta = u + u'$$

where the sign convention is neglected and u and u' are both taken to be positive. Now in the paraxial approximation, from the diagram

$$u = h/z \quad \text{and} \quad u' = h/(w - z)$$

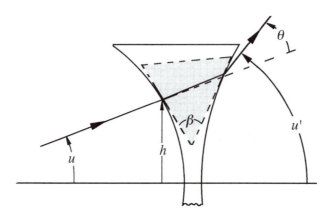

where z and w are also taken to be both positive and w is the distance from the point o to the eye. By eliminating h and solving for u', we have

$$u' = \theta z/w$$

Also from the diagram, it is clear that the physical displacement $y/2$ of each image is given by the equation

$$y/2 = wu'$$

that is

$$y = 2\beta z(\mu - 1) \tag{8.12}$$

which is the physical displacement of the images o_1' and o_2'.

The displacement y can be varied by moving the prism backwards and forwards along the optical axis of the beam, that is by varying z, and this equation shows that when the prism and object are coincident (i.e. $z = 0$), the image displacement is zero.

8.1.6   Lenses as prisms and the prismatic effect

Lenses may be regarded as prisms of variable power, in which the prismatic power changes with distance from the axis. As an example, consider the situation as shown in Figure 8.9. A ray from the left at an angle u to the axis meets the lens at a height h. The lens at this point can be represented as a prism, whose sides are the tangents to the surfaces of the lens at the points where the ray intersects them, as shown in the diagram. The apex angle of this prism is denoted by β. In the paraxial approximation, the ray will be deviated by an amount θ, which is equal to the difference in angles u and u'. That is

$$\theta = u' - u$$

with the usual sign convention. Now from the paraxial refraction equation

(3.13), in air we have

$$u' - u = -hF$$

therefore

$$\theta = -hF \tag{8.13}$$

Using this angle of θ and the definition of prismatic power in Section 8.1.2.1, it follows that the equivalent prismatic power here is

$$F_p = 100hF\Delta \tag{8.14}$$

The negative sign has been dropped in equation (8.14) because prismatic powers do not have a sign and therefore the sign of the lens power F could also be neglected. The prismatic power, given by equation (8.14), is the **prismatic effect** of the lens.

8.2   Refracting and reflecting prisms

There are a number of prisms that combine refraction with internal reflection. The beam enters and leaves the prism at oblique incidence and thus some dispersion (that is colour fringing) may result. However, under some circumstances, no net dispersion occurs if the dispersions occurring at the input and output faces are equal and opposite. The actual conditions for zero dispersion depend upon the prism construction and the ray paths.

Some of these prisms are used to deviate the image forming beam or change the orientation of an image, by inverting it in one or more planes. A single reflection inverts the image in the plane of incidence. Two successive reflections from planes at 90° to each other are equivalent to an image rotation of 180°, but reflections from a roof are also sometimes used. A roof is equivalent to two reflecting surfaces inclined at precisely 90° to each other. The optical principle of the roof system has been described in Section 7.1.2.1, using two plane mirrors to form the roof. Several examples of roofed prisms will be given later in this chapter.

We will now look at a number of these prisms, and a summary of their properties is given in Table 8.1.

8.2.1   Single prisms

8.2.1.1   The Dove prism (Figure 8.10)

The Dove prism is typically a right-angle (45°–90°–45°) prism. However, it may be constructed with other angles. This prism has only one reflection and therefore inverts the image in the plane of incidence but does not deviate the image. With this prism, the beam is not incident normally on the entrance and exit faces and hence there will be some refraction. Unless the beam is collimated, the resulting refraction will introduce extra aberrations into the beam, particularly astigmatism and chromatic aberration.

Table 8.1. Summary of combined refracting/reflecting prisms and reflecting-only prisms

Prism	Image orientation	Beam deviation (°)	Beam translation
Refracting and reflecting prisms			
Dove	L $\Leftrightarrow$ R only	0	
Dove + roof	erect	0	
NAP	erect	approx.90	
Trihedral	—	180	
Double Dove	erect	0	
Reflecting-only prisms			
Right-angle	L $\Leftrightarrow$ R only	90	
Amici (roof)	erect	90	
Penta	—	90	
Penta + roof	erect	90	
Schmidt	erect	45	
Porro	L $\Leftrightarrow$ R		
	U $\Leftrightarrow$ D	0	yes
Pechan	L $\Leftrightarrow$ R		
	U $\Leftrightarrow$ D	0	

Dove prisms often have the apex cut off as shown in Figure 8.10. The reason is that this part of the prism is not used and removing reduces its weight and size. For example if we have a collimated beam as shown in Figure 8.10b, rays that strike the input surface at a height greater than some value ϕ will not hit the bottom reflecting surface. Instead they will strike the exit face first. For a prism made of a material with a refractive index μ and with angles $(\theta, 180° - 2\theta, \theta)$, we can show that the relationship between ϕ and base length L is given by the equation

$$L/\phi = 1/\tan(\theta) + \tan(\theta + i') \tag{8.15}$$

where i' is given by Snell's law, that is

$$\mu \sin(i') = \sin(i) = \sin(90° - \theta)$$

For $\theta = 45°$ and a refractive index of 1.5, $i' = 28.13°$ and therefore this ratio is 4.30. A few calculations with different refractive indices will show that the higher the index, the lower the ratio and hence the shorter the prism.

The unfolded prism and field-of-view

The unfolded version of the Dove prism is shown in Figure 8.10c, but with the top intact. The unfolded version is formed by reflecting the exit face in the bottom reflecting surface. It is clear that the unfolded version is equivalent to a plane parallel sided block of glass. Blocks of glass induce spherical aberration into the beam and tilted blocks induce astigmatism and chromatic transverse aberration for objects that are at a finite distance. The reason for astigmatism

Fig. 8.10: Dove prism.
The angles are only typical.

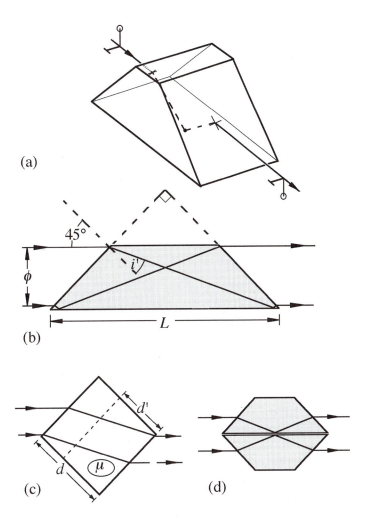

is not readily obvious, but it is easy to understand why an object at a finite distance suffers some chromatic aberration. It is due to the fact that tilted plates produce a transverse shift of the image for objects at a finite distance and the amount of shift depends upon the refractive index (Section 8.1.5.2), which in turn depends upon the wavelength.

The field-of-view can be determined by examining the unfolded version and imagining that we are looking into the unfolded version from the exit side. The view is equivalent to looking through two apertures, the exit face and the apparent image of the entrance face, which is at a distance d', where

$$d' = d/\mu$$

from the exit face. These apertures have the same width; therefore the closer they are, the wider the field-of-view. Therefore the higher the refractive index, the wider the field-of-view. The field-of-view can be increased by using two Dove prisms base to base as shown in Figure 8.10d. This arrangement doubles the field-of-view in the direction perpendicular to the base.

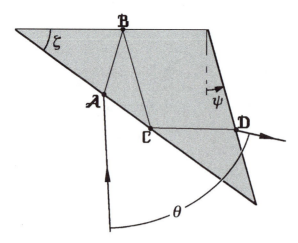

Fig. 8.11:   The NAP prism
and a typical ray path.

If the prism rotates about its optical axis (parallel to the reflecting edge), the image viewed through the prism is seen to rotate at twice the rate of the prism rotation, and in the same direction. If the prism is held fixed and a rotating object is viewed through it, the image will appear to rotate in the opposite direction and at the same speed.

The Dove prism with roof

The Dove prism has only one reflecting surface and if this surface is replaced by a 90° roof, this prism now produces an image that is effectively rotated through 180°. However, this type of prism has the serious disadvantage that the presence of the roof significantly reduces the field-of-view.

8.2.1.2   The NAP prism (Figure 8.11)

The NAP prism is shown in Figure 8.11 and has been commercially used in a pair of spectacles called the "NAP glasses". Its purpose is to produce an erect image with the beam deviated through about 90°. A typical ray path is shown in the diagram. There are a range of angles ψ and ζ, which will give an angle of deviation θ of 90°. These angles will depend upon the direction of the incident ray and the refractive index of the prism. Smith et al. (1990) have analysed the properties of this prism and determined the conditions for which the angle of deviation is 90°. Figure 8.12 shows the relationship between ψ and ζ for a deviation of 90° using a refractive index of 1.5 and the incident ray normal to the top face, which also coincides with the exit ray being parallel to this top face.

For this prism to work effectively, it is desirable that total internal reflection takes place at $\mathcal{B}$ and $\mathcal{C}$. Total internal reflection is easy to achieve at $\mathcal{C}$ but not at $\mathcal{B}$; therefore the surface at $\mathcal{B}$ must be coated with a highly reflecting layer such as aluminium or silver. Total internal reflection at $\mathcal{C}$ can be achieved by a suitable choice of angles ψ and ζ; in fact there is a range of combinations, but there are constraints on the choice because certain combinations of these variables will give deviations very different from 90°. For the conditions stated in the preceding paragraph, the total internal reflection occurs for angles of ψ less than about 45.5°.

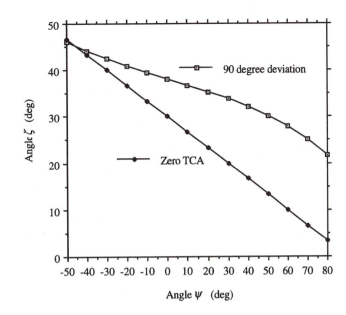

Fig. 8.12: Relationship between the angle ψ and ζ of the NAP prism shown in Fig. 8.11 for a refractive index of 1.5 for two conditions. (a) A deviation angle θ of 90° for the exit ray parallel to the top face. The value of ψ corresponding to total internal reflection just occurring at c is about 45.5°. (b) No colour fringing or transverse chromatic aberration (TCA); i.e. the unfolded prism apex angle is 0°. Note that for (b) the angle of deviation is not 90°, except for $\psi = -45°$.

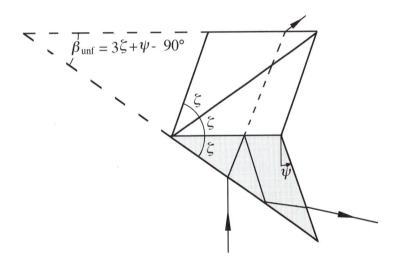

Fig. 8.13: The "unfolded" NAP prism.

Now because there is refraction at two points $\mathcal{A}$ and $\mathcal{D}$, there will be some transverse chromatic aberration or colour fringing. We can examine this colour fringing by looking at the refraction through the equivalent "unfolded" prism.

The unfolded prism

The unfolded NAP prism is shown in Figure 8.13 and appears to be equivalent to a simple refracting prism, whose apex angle β_{unf} is given by the equation

$$\beta_{unf} = 3\zeta + \psi - 90° \tag{8.16}$$

It is clear that this prism is equivalent to a plane parallel sided thick plate if

$\beta_{unf} = 0°$, that is

$$3\zeta + \psi = 90° \qquad\qquad\qquad (8.17)$$

Once this condition is satisfied, the deviation of the beam is independent of the initial direction and the refractive index, and the original NAP prism is now free of transverse chromatic aberration. Equation (8.17) is plotted in Figure 8.12. While this condition ensures no transverse chromatic aberration, it does not ensure that the angle of deviation is still 90°. This only occurs where the two curves shown in this diagram intersect or appear to do so for ψ between $-50°$ and $-40°$. If we let $\psi = -45°$, then it follows from equation (8.17) that $\zeta = 45°$, which is equivalent to a 45°–90°–45° prism.

A deeper discussion of the NAP prism has been given by Smith et al. (1990), who state that the angles for commercially available prisms are $\psi = 10°$ and $\zeta = 30°$ and that this construction gives an angle of deviation of about 115° and an unfolded apex angle of β_{unf} of 10°, which is equivalent to a prismatic power of about 8.7Δ for a refractive index of 1.5. Smith et al. investigated the field-of-view and showed that while the above 45°–90°–45° prism has zero colour fringing, it has about half the field-of-view of the commercial design described above and suffers from secondary imaging.

8.2.1.3   *Trihedral retro-reflector or corner cube (Figure 8.14)*

Retro-reflection is the reflection of a beam back in the direction that it came from. Thus the purpose of retro-reflecting components or materials is to reflect an incident beam back in the direction of incidence, and independent of the direction of the incident ray. Thus on a macroscopic scale, Snell's law appears to be disobeyed, although on the microscopic level the law is obeyed. Light beams can be reflected in the direction of incidence by using the trihedral prism described below. Single trihedral prisms or prism arrays are used extensively to increase the brightness of illuminated signs, particularly safety and warning signs. Glass spheres also function as retro-reflectors, but less efficiently because of the lower reflectance at the rear surface.

The construction of the trihedral prism is shown in Figure 8.14. The apex angles of the three sloping sides are all 90° and each side is inclined at 90° to the adjacent sides. Therefore on looking into the prism from the input/output face, the walls of the prism have the shape of a (90°) corner, hence the alternative name "corner cube" prism. The trihedral prism has the property that any ray entering the front face will emerge from this face in a direction parallel to the direction of the incident ray, though slightly displaced. Thus an incident beam filling the front face aperture is reflected back on itself.

Apart from being used as a retro-reflector, this prism can be used as a mirror that rotates the image through 180°. An observer looking into this prism will see his or her face upside down and left to right, that is, rotated through 180°.

Unfolded prism

If this prism were unfolded, the input and output faces would be seen to be parallel. Therefore, the unfolded prism appears to be equivalent to a plane parallel slab, and therefore the angle of deviation is independent of the direction of the incident ray.

Fig. 8.14:   The trihedral or
corner cube retro-reflecting
prism.

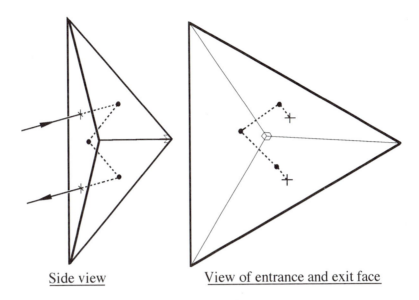

Side view   View of entrance and exit face

8.2.2   Prism systems

8.2.2.1   Double Dove prisms

An intermediate inverted image in some instruments can be reverted to the correct orientation by rotating it through 180°. An image can be rotated through
180° by using two Dove prisms in line but rotated relatively by 90°. Alternatively, a single Dove prism with a roof would perform the same operation. These
prism arrangements do not displace or deviate the line of sight. However, both
of these prism arrangements will give a reduced field-of-view, the first due to
the extra length and the second due to the roof.

If the incident beam is not collimated, the prism will introduce some aberrations, mainly astigmatism and chromatic aberration. Therefore, the Dove prism
system is not suitable for erecting images in telescopes and microscopes.

8.3   Reflecting-only prisms

These are prisms in which the central ray of the beam is incident normally
on the entrance and exit surfaces of the prism. Therefore there is no oblique
refraction and hence there is no induced chromatic aberration or colour fringing.
Reflections take place internally, and ideally these reflections should be total
internal reflections. However, if the angle of incidence is less than the critical
angle, the reflecting surface must be mirror coated.

Reflecting prisms are a very common component of visual optical instruments. They are mostly used to erect or rotate images. For example in the
astronomical or Keplerian telescope, the image is inverted, which is acceptable
in astronomy but is undesirable for terrestrial purposes. Thus the images in terrestrial telescopes are usually erect and the erection is most commonly achieved
with prisms. Monocular microscopes also have inverted images. However, inverted images in binocular stereo-microscopes may lead to a reverse stereo-
effect (Chapter 37); therefore stereo-microscopes contain an image erecting
system which is usually based upon prisms.

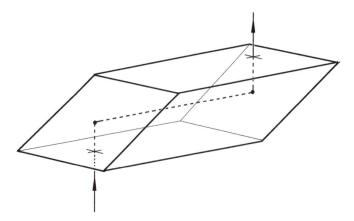

Apart from image rotation, reflecting prisms are sometimes used to deviate a beam through an angle, most commonly 90°. Some prisms deviate, invert or rotate the image, for example the penta prism in single lens reflex (SLR) cameras (Chapter 21). The above image rotations, inversions and beam deviations can also be done with mirror systems. However, reflecting prisms are often preferred for the following reasons.

(a) The reflecting surfaces of a prism are much more durable than that of a mirror. A prism usually reflects by total internal reflection and to work efficiently all that is required is that the reflecting surfaces be kept clean. On the other hand, mirrors are either front or rear surface coated with a metal such as silver or aluminium or in some cases dielectric materials. For high quality image formation, these films are placed on the front surface. Rear surface coatings are not advisable since there is always a small but often significant reflection from the front uncoated surface. However a front surface mirror coating is much more vulnerable to damage on cleaning or from atmospheric attack. Partial protection can be given by coating with a layer of silicon dioxide. The efficiency and durability of a metal coating greatly depend upon the original surface (the substrate) quality, cleanliness and the conditions during the coating process.

(b) Many reflection geometries use multi-reflections from a number of surfaces accurately aligned to each other at set angles. A prism, though not as easy to construct, is much more robust than an equivalent assembly of individual mirrors.

We will now look at some of the most common prisms and prism systems and a summary is given in Table 8.1.

8.3.1   Single prisms

8.3.1.1   Rhomboidal prism (Figure 8.15)

This prism displaces a beam axis but does not deviate the beam or change the orientation of the image. There is no inversion or rotation of the image because there are an even number (two) of reflections in the same plane of incidence.

Applications: It is used in binocular instruments to allow a change in inter-pupillary distance. This will be discussed further in Chapter 37.

Fig. 8.16: Right-angle
prism. This is a
45°–90°–45° prism.

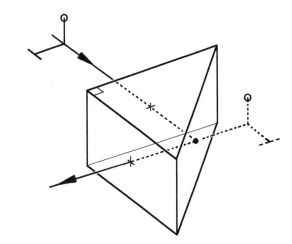

Fig. 8.17: The Amici
prism.

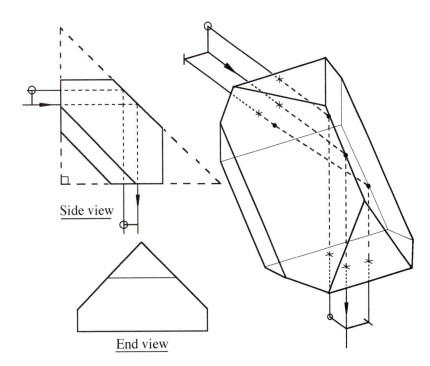

Side view

End view

8.3.1.2   Right-angle prism (Figure 8.16)

The right-angle prism is used for deviating the beam through 90° and since
there is only one reflection, the image is inverted within the plane of incidence.

8.3.1.3   Amici prism (Figure 8.17)

The Amici prism is a modification of the right-angle prism described above. In
this case, the single reflecting surface is replaced by a 90° roof and the beam is
as before deviated through 90° and the image is now rotated through 180°.

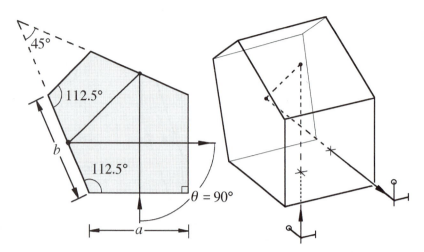

Fig. 8.18:   The penta prism.

8.3.1.4   Penta prism (Figure 8.18)

The right-angle prism shown in Figure 8.16 deviates the beam through 90° and also inverts it within the plane of incidence. If this inversion is undesirable, the penta prism can be used instead. This prism has two reflections with the same plane of incidence and hence does not change the orientation of the image. With this prism, the angle of incidence is less than the critical angle and hence the reflecting surfaces must be coated with a reflecting layer and a protective overcoat.

From the penta prism angles shown in Figure 8.18, it follows that the dimensions a and b are related by the equation

$$b = a/\cos(22.5°) = 1.0824a \qquad (8.18)$$

The unfolded prism

The optical properties of the unfolded prism can be further examined by "unfolding" it. This involves reflecting the images of each surface in the preceding surface. Using this technique to "unfold" the penta prism, it can be shown to be equivalent to a plane parallel block of glass, as shown in Figure 8.19.

Using simple trigonometry and geometry, it can be shown that the geometrical path length d of the ray inside the prism is given by the equation

$$d = a[3 + \tan(22.5°)] = 3.414a \qquad (8.19)$$

which is the length of the unfolded block.

A very useful and important property of any prism used in a visual optical system is its field-of-view. The field-of-view through this prism is limited by the apparent size of the incident face as seen through the prism. Using Figure 8.19, the incident face (at o) has an apparent position at o'. Using the lens equation, it can be shown that this is at a distance d' from the exit face given by the equation

$$d' = d/\mu \qquad (8.20)$$

Fig. 8.19: The "unfolded" penta prism.

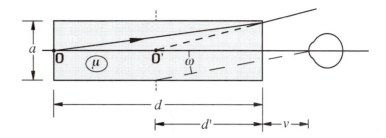

where μ is the refractive index of the prism material, but there is no change in size. If the eye is at a distance v from the exit face, as shown in the diagram, the field-of-view would be the same as that looking through an aperture of width a, from a distance $d' + v$. Thus we can describe the field-of-view in terms of the angular radius ω of the field-of-view, which is given by the equation

$$\tan(\omega) = 0.5a/(d' + v) = 0.5\mu a/(3.414a + \mu v) \tag{8.21}$$

This equation shows that if the eye is placed in contact with the prism, that is $v = 0$, the field-of-view is

$$\tan(\omega) = \mu/3.414$$

which is independent of the size of the prism and thus is the maximum possible field-of-view radius. If we take a numerical example, with a typical refractive index of 1.5, then from the above equation

$$\omega_{\text{max}} = 23.7° \tag{8.22}$$

Thus while the size of the field-of-view is independent of the prism size when the eye is in contact with the exit face, equation (8.21) shows that the size has some effect when the eye is set back from this face.

Sensitivity to tilt

The "unfolded" prism shown in Figure 8.19 demonstrates that the prism can be regarded as a thick parallel sided block of glass. Therefore the angle of deviation θ of a ray passing through the prism is independent of incidence angle and will be 90° for all incidence angles. It also shows that the prism will be free of transverse chromatic aberration.

Applications: This prism is often used to deviate a beam through 90°, without any change in orientation of the image. A simple mirror or right-angle mirror would flip the image in the plane of incidence. Also because it is insensitive to tilt, its alignment is not critical. In contrast a mirror or a right-angle prism would have to be accurately aligned.

8.3.1.5 Penta prism with a roof (Figure 8.20)

If the penta prism as described in Figure 8.18 has the top or first reflecting surface replaced by a 90° roof, the prism will have the construction shown in

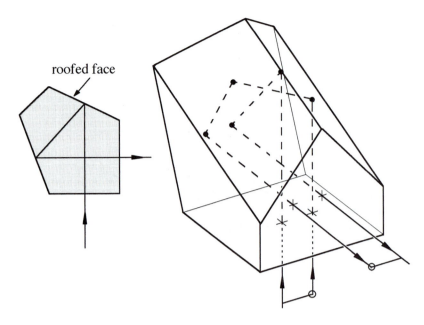

Fig. 8.20:   Penta prism
with roof.

roofed face

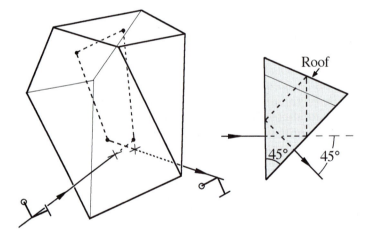

Fig. 8.21:   The Schmidt
prism.

Roof

45°   45°

Figure 8.20. This prism will now invert the image left to right as shown, but
does not invert the image in a direction parallel to the roof edge.

Applications: These prisms are widely used in single lens reflex cameras and
this particular application is described in detail in Chapter 21.

8.3.1.6   Schmidt prism (Figure 8.21)

The Schmidt prism is a roof prism and deviates the beam through 45°. It has
reflections from the two faces which are inclined at 45° and therefore the incidence angle is greater than the critical angle for a refractive index of 1.414.
This prism rotates the image through 180°.

Fig. 8.22: Porro prism type 1.

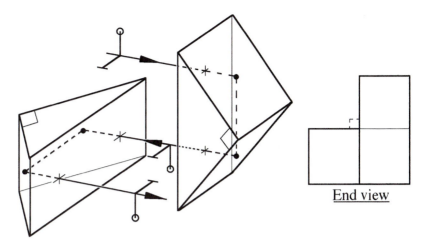

End view

8.3.2 Prism systems

8.3.2.1 Porro prism – type 1 (Figure 8.22)

The Porro prism assembly consists of two right-angle prisms with their hypotenuse faces in contact and rotated by 90° relative to each other, as shown in the diagram.

There are two internal reflections in each prism, both in the plane of incidence, and hence the image orientation along the beam is not changed. However relative to the original direction, the beam is inverted in the plane of incidence. The reason for this has been discussed in Chapter 7. After traversing the two prisms, the image is rotated through 180° in two stages. It is obvious that with this arrangement the optical axis is displaced though not deviated. The advantages of this system are the ease and cheapness of manufacture. The main disadvantage of this arrangement is that the prisms have to be aligned very accurately (90°) to each other. An angular mounting error ϵ from 90° will lead to a rotation of 2ϵ in the final image. The prisms must be mounted very securely as any small knock can displace the prisms relative to each other.

Applications: These prisms are widely used in prism binoculars to provide an erect image.

8.3.2.2 Porro prism – type 2 (Figure 8.23)

The type 2 Porro prism can be thought of as consisting of four 45°–90°–45° prisms as shown in the diagram. It has the same effect as the type 1 Porro prism, that is it rotates the image through 180° without any deviation but with some translation. This type of Porro prism is claimed to be more compact than the type 1 construction.

Applications: This prism is used in stereo microscopes.

8.3.2.3 The Pechan prism (Figure 8.24)

The Pechan prism is composed of two dissimilar prisms, one of which is the Schmidt prism shown in Figure 8.21. The two single prisms are in contact and there is an internal reflection at the boundary.

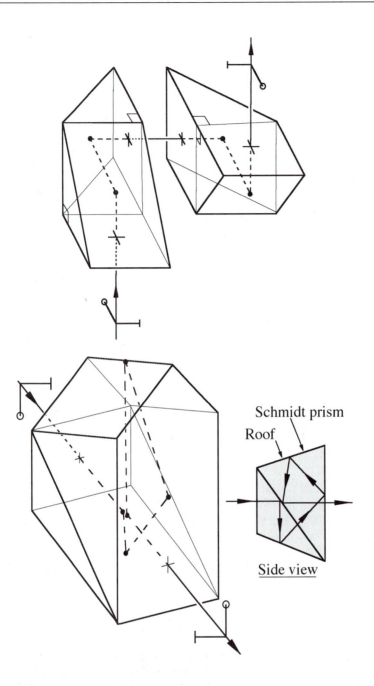

Fig. 8.23:   Porro prism
type 2.

Fig. 8.24:   The Pechan
prism with a roof. Note that
one of the prisms in this
construction is the Schmidt
prism.

Schmidt prism

Roof

Side view

The main advantage of this prism is that it erects an inverted image (that is rotates the image through 180°) without any deviation of the optical axis. More detailed design details of this prism are given in MIL-HNDBK-141.

Applications: Pechan prisms are sometimes used in binoculars and tele-scopes.

8.3.2.4   *The optical trombone (Figure 8.25)*

The optical trombone system is a combination of two right-angle prisms and what appears to be a Dove prism but is not used in the same manner. The beam

Fig. 8.25: The optical trombone.

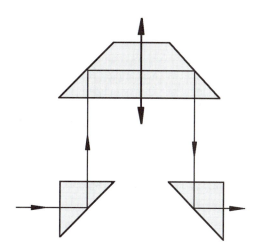

is not displaced and since there are an even number of reflections in the same plane, there is no change of image orientation.

Applications: This system can be used to move the focus of the beam longitudinally and hence provide a method of adjustable focussing.

8.4 Effects of prisms on image forming beams

8.4.1 Aberrations

Light beams passing through a prism suffer conventional aberrations. If the beam is refracted and deviated by a prism, usually the most important aberration is chromatic aberration, and this is in fact used in some refracting prisms to analyse the spectral content of light sources. On the other hand, if the unfolded prism is equivalent to a thick plate, the beam will suffer the same aberrations as a conventional thick plate.

The aberrations of a thick plate are discussed in Chapter 33. Where there is oblique refraction, on entering and leaving the prism for objects or images at finite distances, the oblique refraction induces some astigmatism and chromatic aberration. There is always some spherical aberration induced in the beam and the amount of this aberration depends upon the width of the beam, the refractive index of the prism material and the length of the prism. Equations for calculating the level of these aberrations are given in Chapter 33.

8.4.2 Angular width of the beam (or numerical aperture)

When a beam is reflected from an internal surface of a prism, all of the rays should be totally and internally reflected. This will be so only if the angle of incidence is greater than the critical angle. In many cases, this is not so and the reflecting surface must be coated with a highly reflected layer, for example silver or aluminium.

Total internal reflection occurs when the angle of refraction just exceeds 90°. In the case of a prism, the ray usually moves from glass of index μ to air and the critical angle is given by equation (1.13), that is

$$i_{\text{crit}} = \sin^{-1}(1/\mu) \tag{8.23}$$

Now many prisms are designed around the reflecting surfaces being inclined at 45° to the incident beam. In this case the difference between the angle i_{crit} and 45° limits the width of the cone of rays (numerical aperture) that will be totally internally reflected. Consider as an example the situation shown in Figure 8.26. A cone of rays of half-angle width α is incident normally on the surface of a prism. Inside the prism the extreme ray makes an angle α' with the beam axis and α and α' are related by Snell's law, that is here

$$\sin(\alpha) = \mu \sin(\alpha') \qquad (8.24)$$

Now from the diagram, all the rays in the beam will be totally and internally reflected if

$$45° - \alpha' = i_{crit} \qquad (8.25)$$

that is

$$\alpha' = 45° - i_{crit} \qquad (8.26)$$

Example 8.1: Calculate the maximum beam width for a beam entering the prism shown in Figure 8.26, if the prism material has a refractive index of 1.5.

Solution: From equation (8.23), we have firstly

$$i_{crit} = \sin^{-1}(1/1.5) = 41.8°$$

Substituting this value in equation (8.26) gives

$$\alpha' = 3.19°$$

and, finally, solving equation (8.24), we have

$$\alpha = 4.79°$$

We can note that because of Snell's law, the product of the refractive index and the sine of the angle of incidence of the ray has the same value on either side of the refracting surface. From the above example we can see that

$$\sin(\alpha) = \sin(4.79°) = \mu \sin(\alpha') = 1.5\sin(3.19°) \qquad (8.27)$$

When the angles refer to the angle between the extreme bounding ray of the beam and the central ray of the beam, we call the product given by equation (8.27) the **numerical aperture** of the beam. For a beam being refracted by a series of flat surfaces, the value of this quantity remains constant.

8.4.3 *Beam path length*

If a block of any material is inserted into a beam, the focus of the beam is shifted by an amount that depends upon the refractive index and thickness of

Fig. 8.26: Maximum
beam width of a cone of
rays totally internally
reflected by a prism.

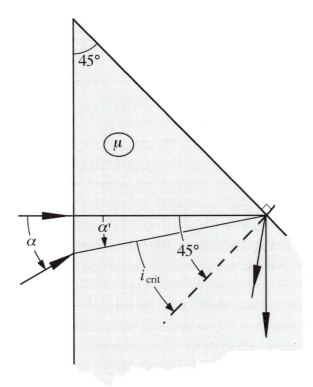

Fig. 8.27: The effect of a
block of glass on the image
position.

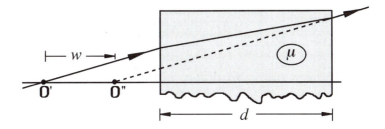

the material. For example, let us look at Figure 8.27, where an intermediate
image is formed at o'. If we now place a plane block of material in the beam
beyond o', the focus appears to be at the point o''. The shift w in focus can be
shown to be

$$w = d(\mu - 1)/\mu \qquad (8.28)$$

The length of the path of a beam can also appear to be altered by the geometry
of the beam. For example in some prism systems, the beam is reflected several
times and may spend part of the time travelling backwards. This backward path
is a type of folding of the beam and shortens the distance between adjacent
components, thus allowing the instrument length to be shortened.

Exercises and problems

8.1   Calculate the prismatic power of a prism whose index is 1.7 and apex angle is
      10°.

      ANSWER: $F_p = 12.21\ \Delta$

8.2   Find the apex angle of a thin prism whose prismatic power is 10 Δ and which is
      made from a glass of index 1.532.

      ANSWER: 10.77°

8.3   Calculate the apex angle of a 5 Δ prism made from a material with a refractive
      index of 1.7.

      ANSWER: apex angle $(\beta) = 4.09°$

8.4   Calculate the angle of incidence on the reflecting face of a 45°–90°–45° Dove
      prism with a refractive index of 1.5. What is the critical angle?

      ANSWER: 73.13°, critical angle = 41.8°

Summary of main symbols and equations

z	distance from object or image to prism
y	transverse displacement of an object or image
β	apex angle of a prism
β_{unf}	apex angle of an "unfolded" NAP prism
i_{min}	angle of incidence for the minimum angle of deviation
θ	angle of deviation of a ray passing through a prism
θ_{min}	minimum angle of the above deviation
F_p	prismatic power of a prism (in prism dioptres)
Δ	symbol for prism dioptre

Section 8.1.2: Paraxial approximation or thin prism

$$\theta = \beta(\mu - 1) \tag{8.5}$$

Section 8.1.2.1: The power of a prism and prism dioptre

$$F_p = 100\beta(\mu - 1)\ \Delta \tag{8.7}$$

References and bibliography

* Born, M. and Wolf, E. (1989). *Principles of Optics*. 6th ed. Pergamon Press, Oxford.
* MIL-HNDBK-141. (1962). *Military Standardization Handbook: Optical Design*.
  Defence Supply Agency, Washington D.C.
* Schott. (September 1992). *Optical Glass catalog*. Schott Glaswerke, Mainz, Germany.
* Smith, G., Johnston, A.W., and Maddocks, J.D. (1990). The NAP prism. *Optom. Vis.
  Sci.* 67 (20), 133–137.

9

Aperture stops and pupils, field lenses and stops

9.0   Introduction

Every optical system contains a surface or surfaces which limit the width of the beam passing through the system from each object point. These surfaces may be a lens surface, a face of a prism, a mirror or simply a plate containing an opening of suitable size. Since the amount of light in the beam depends upon the beam width, they control the image brightness. They also affect image quality and to some extent the size of the field-of-view.

The surface that controls the width of the beam from the axial object point is called the **aperture stop** and because a beam cannot be infinitely wide, every system must have an aperture stop. Typical examples are the iris of the eye and the diaphragm of a camera lens. For off-axis object points, the beam width may be controlled by other surfaces.

Figure 9.1 shows a system with two components, a simple surface that acts as the aperture stop and a simple lens. For the on-axis object point and for object points some distance off-axis, the aperture stop limits the width of the beam. As one moves farther off-axis, the lens mount begins to limit the beam. The obstruction of the rays by a surface other than the aperture stop is called **vignetting**. As one moves even farther off-axis, the lens mount finally blocks the entire beam passing through the aperture stop and the vignetting has become complete. In this example, the width of the field-of-view is limited by vignetting. In complex optical systems, more than one surface may cause vignetting.

The brightness of any point in the image depends on the amount of light passing through the aperture stop, that is the width of the beam. Therefore image brightness must decrease if vignetting reduces the beam width. If vignetting alone controls the field-of-view, a portion of the central field is free of vignetting because for object points some distance off-axis, the aperture stop alone limits the width of the beam. At the edge of this central field, vignetting begins and increases in magnitude as one proceeds away from the centre, until it finally becomes total. In these cases, once vignetting begins, the image brightness decreases gradually and becomes zero at the edge of the field. In such situations, the extent of the field-of-view may also be defined by the **field of half illumination**, whose limit is defined as the points where the image brightness is half the central or axial value.

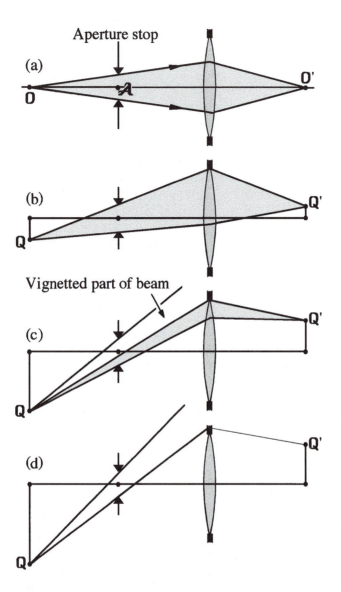

Fig. 9.1: Aperture stops and vignetting.

If the off-axis aberrations of a system are significant, vignetting is often de-signed into an optical system in order to reduce these aberrations by blocking the peripheral badly aberrated rays. While such a technique improves image quality, it does so at some cost to peripheral image brightness. Therefore high image brightness and high image quality cannot often be satisfied simultane-ously. However, if the aberrations are negligible, the wider the aperture stop, the brighter the image and the better the image quality, since the other factor af-fecting image quality, diffraction, decreases with increased size of the aperture.

Sometimes an aperture is placed at or very near the final or some intermediate image plane. This surface is called a **field stop** and at the field edge, the decrease in image brightness is sudden, leading to a well defined edge to the field-of-view. However, vignetting may also occur in optical systems containing field stops. A typical example of this case is a camera lens, in which the peripheral beam is vignetted in order to reduce the effects of the off-axis aberrations. But

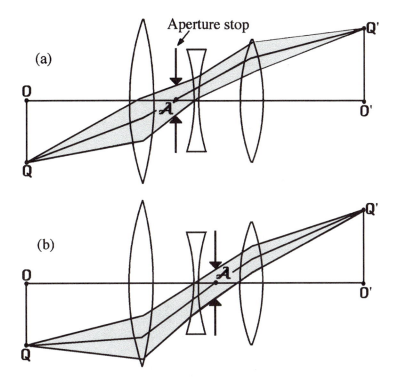

Fig. 9.2: The aperture stop in a more complex system and the effect of its position on the path of a beam traversing the system.

a camera also contains a field stop, which is a plate in the film plane designed to hold the film flat and set the shape and dimensions of the image frame.

From this discussion, it is obvious that the beam limiting apertures and surfaces are very important components of optical systems, controlling the level of aberrations, image quality, image brightness and the diameter of the field-of-view. The aperture stop also affects the depth-of-field and this aspect is discussed in the next chapter.

9.1 Aperture stops and pupils

9.1.1 The aperture stop

Aperture stops are inevitable components of optical systems, since no system can be infinitely wide. We have defined the aperture stop of a system as the surface or component that limits the width of the beam from the axial object point and, as shown in Figure 9.1, limits the beam width for some distance off the axis. Thus in this diagram, it is clear that for off-axis points, other surfaces or components may limit the beam width. However, the example shown in this diagram is a simple system and in more complex optical systems, the aperture stop is usually inside the system. Figure 9.2 shows this schematically for a simple type of three component camera lens. Another common example is the eye.

In the most common man-made systems, the aperture stop is circular or near circular. The aperture stop of the human eye is also circular although in many other vertebrate eyes, the aperture is non-circular; for example the cat has a slit-like aperture stop. In the following discussion, we will assume the aperture stop is circular unless otherwise stated.

9.1.1.1   Aperture stop diameter

The diameter of the aperture stop plays a very important role in the image quality characteristics of an optical system. The diameter of the aperture stop affects the depth-of-field (Chapter 10), image brightness level (Chapter 12), aberration level (Chapter 33) and levels of diffraction (Chapter 26) and image quality (Chapter 34). In many systems such as the eye (Chapter 13) and camera lenses (Chapter 21), the diameter is variable. In others, such as most telescopes (Chapter 17), it is fixed.

9.1.1.2   Position of the aperture stop

For a single lens system, the aperture stop may be placed at the lens itself or there may be no separate aperture stop and the lens itself acts as the aperture stop. For more complex systems, it is often a separate surface internal to the system or one of the optical surfaces. Figure 9.2 shows an example of a three lens optical system with an internal aperture stop, but with the aperture stop in two different positions. This diagram shows that the beam cannot pass centrally through all the lenses in the system and therefore must pass peripherally through some of them. We have seen in Chapter 5 that the aberrations of a beam are affected by the particular path traversed by the beam through a lens, particularly if the beam passes through the lens peripherally. Therefore the position of the aperture stop affects the level and types of aberrations present. This aspect is discussed further in Chapter 33.

9.1.2   Entrance and exit pupils

The **entrance pupil** is defined as the image of the aperture stop as seen from object space. That is, it is the image of the aperture stop formed in object space by that part of the optical system on the object space side of the aperture stop. The **exit pupil** is defined as the image of the aperture stop as seen from image space. That is, it is the image of the aperture stop formed in image space by that part of the optical system on the image space side of the aperture stop. Therefore the entrance pupil, aperture stop and exit pupil are conjugates.

While we have defined the aperture stop as the surface or component limiting the width of the beam from the axial object point, we have seen from Figure 9.1 that the aperture stop also limits the beam for a certain distance off-axis. Since the pupils are conjugates of the aperture stop, it follows that providing there is no vignetting, any ray from an off-axis object point that passes through the entrance and exit pupils must also pass through the aperture stop. Therefore, they can be further regarded as pseudo or imaginary apertures, in object and image space, through which all the light must pass, if it is to pass through the system. We have shown this schematically in Figure 9.3, for a system with external (real) pupils.

9.1.2.1   Notional position of the entrance and exit pupils

Entrance and exit pupils may be real or virtual. In some optical systems, such as Keplerian telescopes (Chapter 17), the first component is the aperture stop. In these cases, this first component is simultaneously the entrance pupil, and in the

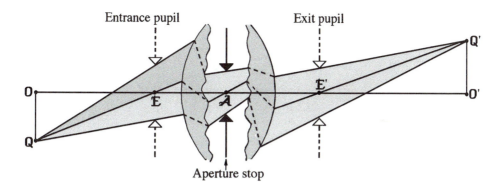

Keplerian telescope, the exit pupil is real. If the last component is the aperture stop, then it is simultaneously the exit pupil. In camera lenses, the aperture stop is inside the system and both the entrance and exit pupils are virtual.

We can visualize the location of the entrance and exit pupils as the positions where the central ray of the beam appears to cross the axis in object or image space, respectively. For example, in Figure 9.2, if we extended the object space central ray of the beam to where it would cross the axis, it would cross the axis for (a) inside the system, and for (b) to the right of the system. In both cases the entrance pupil would be virtual. Similarly, if we traced the image space central ray of the beam backwards to where it crosses the axis, we would find that the exit pupil is also formed as a virtual image on the left of the last component in both cases. Thus, in this example, both entrance and exit pupils are virtual images of the aperture stop.

However, in order to minimize cluttering of the diagrams, the pupils will often be drawn as real, as in Figure 9.3, and therefore external to the system, but we should keep in mind that they are often virtual.

9.1.3 The paraxial marginal and pupil rays

It has been shown in Chapter 3 that only two paraxial rays need to be traced in order to determine most of the paraxial properties (particularly the positions of the six cardinal points and the vertex and equivalent powers) of an optical system. These two rays are arbitrary and depend somewhat on the situation. In certain circumstances, a very useful pair of rays are the **paraxial marginal** ray and the **paraxial pupil** ray. Apart from allowing one to determine the above paraxial properties, these two rays also allow one to estimate the desired aperture radius of each component of the optical system and the Seidel aberrations (Chapter 33).

The paraxial marginal ray (PMR) is shown in Figure 9.4a and is defined as the ray from the axial object point o, touching the edge of the aperture stop (and hence touching the edges of the entrance and exit pupils) and finally intersecting the optical axis at the paraxial image o' of o. The paraxial angles and heights of this ray are denoted here by the symbols u and h.

The paraxial pupil ray (PPR) is shown in Figure 9.4b and is the ray from the point ϱ at the edge of object field, passing through the centre of the aperture stop (and hence the centres of the entrance and exit pupils) and finally intersecting the paraxial image plane at the paraxial image ϱ' of ϱ. The ray angle and height are denoted by the symbols and $\bar{u}$ and $\bar{h}$. This ray is also known as the **principal**

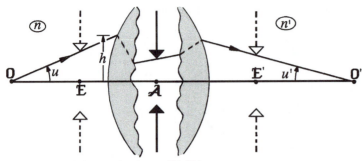

(a) The paraxial marginal ray (PMR)

Fig. 9.4: The definition of (a) the paraxial marginal ray (PMR), (b) the paraxial pupil ray (PPR) and (c) the rays in the space containing the aperture and symbol notation.

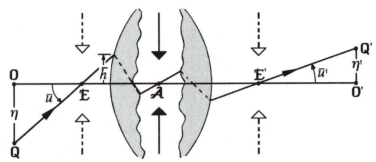

(b) The paraxial pupil ray (PPR)

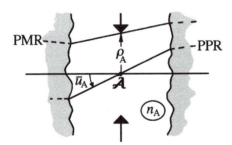

(c) The aperture stop space

or **chief** ray. The pupil ray may be regarded as the central ray of the unvignetted beam from Q, referred to in Chapter 5 and in the preceding sections of this chapter.

9.1.3.1   The optical invariant

In Appendix 1, it is shown that for any two distinct paraxial rays, the quantity

$$n_j(u_{1j}h_{2j} - u_{2j}h_{1j}) \qquad (9.1)$$

evaluated at any plane or surface (say the jth) in the system, and on either side of this plane or surface, is the same for all values of j. The quantities u_{1j} and h_{1j} refer to one ray and u_{2j} and h_{2j} refer to the other ray. The constant or invariant above is denoted by the symbol H and is called the **optical invariant**. In some other text books it is called the **Lagrange** invariant or the **Helmholtz–Smith** relation. If the two rays are chosen to be the paraxial marginal and paraxial pupil rays (dropping the j subscript),

$$H = n(\bar{u}h - u\bar{h}) \qquad (9.2)$$

where u and h are the angle and height, respectively, for the marginal ray and $\bar{u}$ and $\bar{h}$ are the angle and height for the pupil ray. While the above quantity has the same value at any plane in the system, its value does depend upon the aperture stop diameter, position of the conjugate planes and nominated size of the field-of-view, but once these are fixed, it is constant throughout the system.

Referring again to Figure 9.4, there are a number of special cases of the optical invariant. Firstly, at the object plane, the marginal ray angle is u, the marginal ray height h is zero, the pupil angle is $\bar{u}$ and the pupil ray height $\bar{h}$ is η. Thus from equation (9.2), at the object plane

$$H = -nu\eta \qquad (9.3a)$$

and similarly at the image plane

$$H = -n'u'\eta' \qquad (9.3b)$$

where here n and n' are the refractive indices of object and image space, respectively. Now the optical invariant is constant throughout the system, and therefore we must have

$$nu\eta = n'u'\eta' \qquad (9.4)$$

and since the quantity η'/η is the transverse magnification M, it follows that

$$M = \frac{nu}{n'u'} \qquad (9.5)$$

which is an alternative proof to equation (2.23b). Now consider the optical invariant at the aperture stop. At the aperture stop, the height of the marginal ray must be equal to the aperture stop radius ρ_A and the height of the pupil ray must be zero, as shown in Figure 9.4c. If the pupil ray angle at the aperture stop is $\bar{u}_A$ and the refractive index is n_A, then we have from equation (9.2)

$$H = n_A \bar{u}_A \rho_A \qquad (9.6)$$

where the subscript A refers to the aperture stop.

9.1.4 Location of entrance and exit pupils by ray tracing

We have tried to stress above that the entrance and exit pupils are not always real images (or external) to the system as shown in Figures 9.3 and 9.4. In many systems, one or both pupils are virtual. For example, in the case of camera lenses, both pupils are inside the lens, that is virtual as in the system shown in

Figure 9.2. In the Galilean telescope, to be discussed in Chapter 17, the entrance pupil is virtual and formed on the eye side of the telescope, that is in image space. In most visual optical instruments the exit pupil is real and about 12 mm away from the last surface. The reason for this will be explained in greater depth in Section 9.2 and Chapter 36.

The positions and diameters of the entrance and exit pupils of an optical system can be found by laboratory techniques or, if the constructional parameters are known, by numerical ray tracing. To find the position of the entrance pupil by ray tracing, we begin by tracing a paraxial ray from the centre of the aperture stop backwards into object space, using the ray procedures described in Chapter 2. The position and size of the entrance pupil from the results of this ray trace can be found using the equations given in Chapter 2. A similar ray traced from the aperture stop to the image space can be used to locate the position and size of the exit pupil.

9.1.5 *Pupil diameters, numerical aperture and F-number*

Because the size of the aperture stop and hence the sizes of the pupils have such an important effect on the performance of an optical system, it is often necessary to specify the pupil sizes. Sometimes pupil size is expressed simply as a linear radius or diameter. For example in the case of the eye, the entrance pupil diameter is usually expressed in millimetres. In some other cases, it is expressed as an angular radius or diameter measured from the axial object or image point. For example, the pupil size of microscope objectives is defined in terms of the sine of the angular radius α of the entrance pupil subtended from the axial object point, multiplied by the refractive index of the object space medium. This may be made a little clearer from Figure 9.5a. This product is called the **numerical aperture** (*NA*) and thus

$$NA = n \sin(\alpha) \qquad \text{(in object space)} \qquad (9.7a)$$

For some instruments, it may also be defined in image space and in this case would be written in the form

$$NA' = n' \sin(\alpha') \qquad \text{(in image space)} \qquad (9.7b)$$

where n and n' are the refractive indices of the respective space and α and α' are the angular radii as shown in the diagram. Numerical apertures are always regarded as positive and hence α and α', n and n' are always taken as positive.

9.1.5.1 *Numerical aperture and size of the aperture stop*

Sometimes we need to relate the numerical aperture to the actual aperture stop radius. This is not always easy, because the numerical aperture is defined in terms of the angle of the real and not the paraxial marginal ray. This angle may be large and well outside the paraxial region. However, if the angle is small, we may use the paraxial approximation and equate the sine of the angle with the paraxial angle. Thus when the angle is small, we can write

$$n \sin(\alpha) = nu \qquad (9.8a)$$

Fig. 9.5: Different ways
of specifying aperture stop
size. (a) The definition of
numerical aperture $n \sin(\alpha)$,
(b) numerical aperture and
aperture radius of a single
lens and
(c) *F-number* (f/D) and
its relation to numerical
aperture.

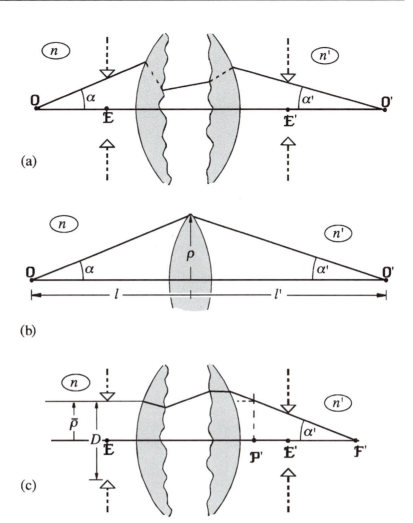

(a)

(b)

(c)

and the accuracy of this equation or the actual value of the angle α can be
checked by tracing a real marginal ray.

However, for a single thin lens the problem is simple. Let us take the example of a simple thin lens alone, as shown in Figure 9.5b. This lens must be
simultaneously the aperture stop. Let the lens aperture radius be ρ. In this case,
the angles α and α' are given by the tangents in the form

$$\tan(\alpha) = |\rho/l| \quad \text{and} \quad \tan(\alpha') = |\rho/l'| \tag{9.8b}$$

If we substitute these in equations (9.7a and b) which define numerical aperture,
we have

$$NA = n \sin[\arctan(|\rho/l|)] \quad \text{and}$$

$$NA = n' \sin[\arctan(|\rho/l'|)] \tag{9.9}$$

9.1.5.2 Specific case of the object at infinity

If the object is at infinity as shown in Figure 9.5c, the image space numerical aperture can be simply related to the diameter of the entrance pupil.

If the system is free from spherical aberration and coma, the sine condition is satisfied (see Section 5.6.1) and we can use equation (5.32), but here expressed in the form

$$\sin(\alpha') = \bar{\rho}/\mathscr{P}'\mathscr{F}' \tag{9.10}$$

where the negative sign has been dropped because the numerical aperture is always regarded as positive. Using equation (3.33), that is

$$F = n'/\mathscr{P}'\mathscr{F}' \tag{9.11}$$

where F is the equivalent power, we can write

$$NA' = n'\sin(\alpha') = \bar{\rho}F \tag{9.12a}$$

or

$$NA' = n'\sin(\alpha') = \frac{DF}{2} \tag{9.12b}$$

where D is the diameter of the entrance pupil.

For many systems which usually have the object at infinity, the entrance pupil size is often specified in terms of **F-number**. This is defined by the ratio

$$F\text{-}number = \frac{\text{equivalent focal length}}{\text{diameter of the entrance pupil}} = \frac{f}{D} = \frac{f}{2\bar{\rho}} \tag{9.13}$$

It is clear from this definition that *F-number* is only defined for air (i.e. $n = n' = 1$), because only in air is the equivalent focal length a meaningful quantity. Combining equations (9.12b) and (9.13), we can express the *F-number* directly in terms of the $\sin(\alpha')$ or the numerical aperture, thus

$$F\text{-}number = \frac{1}{2\sin(\alpha')} = \frac{1}{(2NA')} \text{ (in air)} \tag{9.14a}$$

or

$$NA' = \sin(\alpha') = \frac{1}{(2F\text{-}number)} \text{ (in air)} \tag{9.14b}$$

Note: The *F-number* (or F/no) and the above equations relating *F-number* to numerical aperture are only used and valid when the object is at infinity (see Figure 9.5c) and the system is in air.

Minimum value of F-number

In air, the maximum value of $\sin(\alpha')$ is 1.0 and therefore, from equation (9.14a), the minimum value of the *F-number* is 0.5.

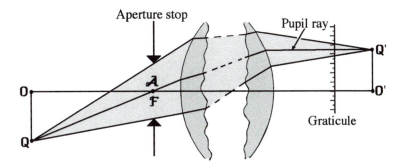

Fig. 9.6: The telecentric pupil.

9.1.6 Telecentric pupils

Occasionally, the aperture stop is placed at the front focal point of the lens or optical system as shown in Figure 9.6. In this case, the exit pupil is formed at infinity and the pupil is known as a **telecentric** pupil. The pupil ray in image space must now be parallel to the axis as shown in the diagram. We should note that since the exit pupil cannot be in the same plane as the image, the exit pupil cannot be telecentric if the image is at infinity. Since the image of visual optical systems is usually at infinity, telecentric pupils cannot normally be used with these systems.

Telecentric arrangements are useful in the measurement of the sizes of objects as telecentric pupils decrease or eliminate measurement errors caused by some types of focus errors. Figure 9.6 shows a schematic arrangement for a telecentric aperture stop. The exit pupil is formed at infinity and the image must be formed at a finite distance. Now suppose we wish to measure the dimensions of part of the image by reading a graticule placed in the image plane. However, suppose the graticule is not placed exactly in the image plane at O', but a little farther away as shown in the diagram. In this case since the pupil ray is parallel to the axis, it intersects the graticule at the same height, irrespective of the position error. Therefore if the stop is telecentric, the misplacement of the graticule has no effect on the measurement.

The effects of defocus on image size are discussed in much greater depth in Chapter 10.

9.1.7 Shape of off-axis pupil

Once vignetting begins, more than one surface blocks the rays. One way of determining the shape of the effective pupil (i.e. the cross-sectional shape of a beam) for an off-axis object point is as follows. Imagine we place our eye at the object or image point and look into the system. We would see the image of each surface and if we were off-axis, these images would be transversely displaced relative to each other. The effective pupil shape is defined by the region common to these images. As an illustration, let us take the example of a single lens and an aperture stop given in Figure 9.1. The views from O or Q are shown in Figure 9.7. In all these examples, the common region is dotted and this shows the shape of the effective pupil in each case.

Once vignetting occurs, the paraxial pupil ray (as defined in Section 9.1.3) is no longer the central ray of the beam. In this case, if we wish to retain the idea of the pupil ray as the central ray of the beam, we must redefine it but differentiate

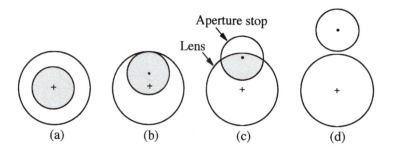

Fig. 9.7: The shape of the effective pupils when vignetting occurs. Letters (a) to (d) refer to the views from the respective object positions in Fig. 9.1.

this new pupil ray from that already defined. We will call the new pupil ray the **effective pupil ray** and define this as the central ray of the vignetted beam and define the positions of the effective entrance and exit pupils as the crossing points of this ray in the object and image space, respectively.

9.1.8   Systems with no intrinsic aperture stop

Some systems have no intrinsic aperture stop. However, in these cases, these systems are always used in conjunction with another system that provides the aperture stop for the combined system. Examples are ophthalmic lenses, the simple magnifier and the Galilean telescope. In all these cases, these systems are used with the eye, and the aperture stop of the eye (the iris) becomes the aperture stop of the combined system.

9.1.9   Effects of aberrations on the formation of the pupils

So far we have implicitly assumed that the entrance and exit pupils are defined only in terms of paraxial quantities but extended beyond the paraxial region. Therefore these pupils are strictly Gaussian pupils. In the presence of aberrations, the real entrance and exit pupils are not identical to the Gaussian pupils, in size, shape or position.

9.2   Cascading of systems and pupil matching

Often two or more optical systems are cascaded or used in tandem. The most frequent example is the eye viewing through a visual optical instrument such as a microscope. In this example, the microscope is the first system and the eye is the second. There are also examples in non-visual optical systems, for example condenser systems which provide light sources for a wide variety of optical systems such as projection lenses, microscopes and interferometers.

A very common example of cascaded systems in visual optics is where the image of the first system becomes the object for the second system. This intermediate image may be real or virtual. The relative positions and diameters of the exit pupil of the first system and the entrance pupil of the second system affect the brightness and the diameter of the field-of-view of the final image.

Consider a typical situation as shown in Figure 9.8, where the first system produces a real image in the plane at o_1', on which is an off-axis image point Q_1'. This image plane becomes the object plane for the following system and so the points o_2 and Q_2 are the same as o_1' and Q_1' for this second system. If there is no

Fig. 9.8: The cascading of two optical systems and the vignetting effect when the pupils are not superimposed.

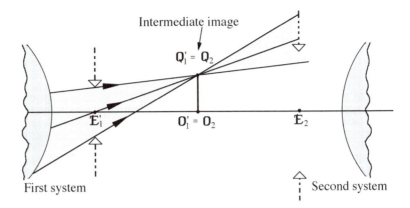

Intermediate image

$Q_1' = Q_2$

E_1' $O_1' = O_2$ E_2

First system   Second system

vignetting in the first system for this image point, the rays forming the image of Q_1' completely fill the exit pupil ε_1' of the first system. Some of these rays cannot pass through the entrance pupil at ε_2 of the second system and, therefore, while they may enter the second system, will not pass through its aperture stop. For this discussion we will say that in the situation shown in the diagram, the aperture stop of the second system vignets the beam. The amount of vignetting by the aperture stop of the second system increases as Q_1' goes farther away from the optical axis, until vignetting becomes total. Thus, when the two pupils are separated, the size of the final field-of-view is reduced by vignetting and it is clear that the greater the separation of the pupils, the greater will be the vignetting effect, and hence the smaller will be the final field-of-view.

If now the pupils are superimposed, that is ε_1' and ε_2 coincide, and the exit pupil diameter of the first system is equal to or smaller than the entrance pupil diameter of the second, all the rays that pass through the exit pupil of the first system can enter the entrance pupil of the second system. Now there will no longer be any vignetting by the aperture stop of the second system. If the exit pupil diameter of the first system is larger than the entrance pupil of the second, the aperture stop of the second system "blocks" the excess rays and the final image has a reduced brightness on axis.

The above observations and discussions lead to the conclusion that when two optical systems are cascaded, it is desirable that they should be placed so that the position of the exit pupil of the first system coincides with the entrance pupil of the second system and that these pupils should have similar diameters. In this book, this is called **pupil matching**.

Sometimes, however, pupil matching cannot be achieved by simply positioning the two systems; for example when the pupils are virtual. In these cases and when the intermediate image is real and it is still very desirable to match the pupils, a positive power lens placed at or near the intermediate image can be used to "match" the pupils, at least in position. The equivalent power of this lens is chosen to image the exit pupil of the first system onto the entrance pupil of the second. The use of such lenses is discussed in greater detail in Section 9.3.

9.2.1 Pupil matching and the eye and eye relief

A good example of position pupil matching is viewing through visual instruments such as microscopes and telescopes. In these cases, the entrance pupil of the eye should be superimposed on the exit pupil of the instrument being looked through. In a visual instrument that has an external exit pupil, the distance between the last surface of the instrument and the exit pupil is called the **eye relief** and should have a minimum value of about 12 mm. The details of the calculation of this value are given in Chapter 36.

In general the diameter of the instrument exit pupil should closely match the entrance pupil of the eye. However, since the entrance pupil of the eye varies with light level, this poses a small problem. One solution is to design various versions of the instrument for different light levels, that is with different size aperture stops and hence different diameter exit pupils. An example is the design of binoculars and this is discussed in Chapter 17.

There is one exception to the above general rule of pupil matching for diameter. Later, in Chapter 36, it will be shown that if the eye has to rotate while viewing through an instrument, the exit pupil of the instrument should be wider than the entrance pupil of the eye.

9.3   Field lenses

It has just been shown that when optical systems are cascaded without pupil matching, there will be both some loss in field-of-view and image brightness. In many situations, the loss in field-of-view is more critical than a loss of image brightness. Sometimes, the pupils can be matched in position by suitably altering the separation of the two optical systems. However, this is not always possible, for example with two instruments both of which have internal or virtual pupils. In these cases, pupil matching at least in position can be achieved by adding a third optical system with a specific power, usually a simple positive lens between the other two and placed preferably at the intermediate image plane. The purpose of this lens is to image the exit pupil of the first system onto the entrance pupil of the second, although usually with some unavoidable difference in sizes. Such a simple lens is called a **field lens**. Field lenses may also be used within a single system to control the pupil imagery within that system. An example in this category will be given later in this section.

However, it is not always possible to place the field lens in the intermediate image plane position. We will give an example of this situation shortly, but first we will consider the case where the field lens is placed at the intermediate position and the lens is required to pupil match in position.

9.3.1   *Field lens placed at an intermediate image position*

9.3.1.1   *Two cascaded systems*

Consider the example shown in Figure 9.8, where the first system produces a real image at o'_1 and this acts as the object o_2 for the second system. A lens (a field lens) placed between the two pupils can be used to image the exit pupil of the first system onto the entrance pupil plane of the second system. If this lens is placed at the intermediate image, as shown in Figure 9.9, it will have no effect on the magnification of the final image. The required power F of the

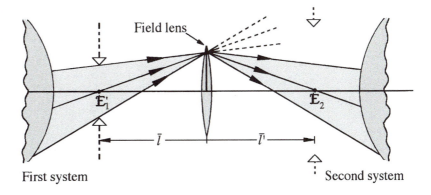

Fig. 9.9:   The use of a field
lens to match the pupils in
the cascaded system shown
in Fig. 9.8.

First system   Second system

field lens is given by the lens equation (3.14), here written as

$$\frac{1}{\bar{l'}} - \frac{1}{\bar{l}} = F \tag{9.15}$$

The size of the image of this exit pupil image can be found by applying the transverse magnification equation (3.49).

9.3.1.2   Within a single system

Within some systems, the pupil imagery can be controlled and the vignetting reduced by using field lenses placed at convenient positions in the system; usually at intermediate image positions. Typical examples are telescopes and microscopes and this aspect of their design will be discussed in the chapters dedicated to them.

9.3.2   Field lens not at the intermediate image but pupil matching in position

Ophthalmoscopy is the visual examination of the internal structure of the eye, particularly the retina. The optics of **direct ophthalmoscopy** is shown in Figure 9.10a, but without the necessary optics for illuminating the patient's retina or a lens for compensating for refractive error of either of the two eyes. Ideally, the image of the patient's retina is formed at infinity by the optics of that eye. This is formed finally on the retina of the clinician as shown in this diagram. The field-of-view is now very dependent upon the separation between the entrance pupils of the two eyes. It is a simple problem to determine the angular field-of-view as a function of this separation and this calculation is given as a problem at the end of this chapter.

Because of the small field-of-view, direct ophthalmoscopy has its limitations. However, the field-of-view or the amount of the retina seen simultaneously can be increased by the addition of a field lens that is designed to match the pupils, that is, image the patient's pupil onto the clinician's pupil. The optics of this technique is shown in Figure 9.10b. In this example, the field lens cannot be put at or near an intermediate image because the image of the patient's retina is not usually formed between the patient's eye and the clinician's eye. For example,

Patient   Clinician

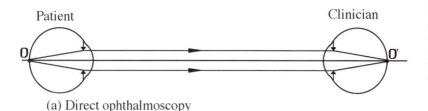

(a) Direct ophthalmoscopy

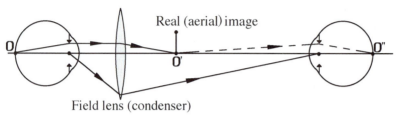

Real (aerial) image

Field lens (condenser)

(b) Indirect ophthalmoscopy

Fig. 9.10: Direct ophthalmoscopy and the use of a field lens to pupil match and hence increase the field-of-view (indirect ophthalmoscopy).

if the patient's eye is emmetropic, the image of the patient's retina is formed at infinity. The field lens, commonly known as a **condenser**, is usually placed just in front of the patient's eye with a power chosen to image the patient's entrance pupil onto the clinician's entrance pupil. The field or condenser lens also forms an image of the patient's retina at or near its back focal point. The use of the field lens increases the amount of retina seen simultaneously, but reduces the image magnification. This technique is known as **indirect ophthalmoscopy**.

9.3.3   Aberrations

Field lenses usually have a positive power and therefore increase the Petzval sum (i.e. the field curvature) of the system, thereby in general increasing field curvature. The corollary of this is that a negative lens placed at an intermediate image will decrease the field curvature, though with some loss of field size. A negative power lens used in this manner is called a **field flattener**.

9.4   Field stops

All optical systems have a finite or limited field-of-view and it is clear from the preceding discussions that this is often due to vignetting. Vignetting limits the field-of-view by a progressive blocking of rays and hence image brightness as one moves away from the optical axis, until finally the vignetting becomes complete. In these cases, ultimately the edge of the field is where the vignetting is just complete. In some other optical systems the field-of-view is limited by a physical aperture placed at or near an intermediate or the final image plane. An example of the latter is the rectangular aperture placed at the back of a camera and just in front but in contact with the photographic film. Such an aperture is called a **field stop**. Occasionally the field lens is the field stop.

Some text books refer to **entrance** and **exit ports** or **windows**. These are analogous to entrance and exit pupils and are defined as the images of the field stop in object and image space, respectively, but because many systems do not contain a specific field stop, entrance and exit ports are of much more limited use. They have no validity in systems where the field-of-view is finally limited by vignetting since more than one surface often contributes to this vignetting and therefore there is no unique "field stop" surface.

Exercises and problems

9.1   Calculate the positions and magnifications of the entrance and exit pupils and of the Le Grand full theoretical schematic eye given in Appendix 3, assuming the aperture stop of the eye is the iris, which lies in the vertex plane of the front surface of the lens.

ANSWERS:

$v\varepsilon$ = 3.04 mm and 13% larger than the real pupil

$v'\varepsilon'$ = −3.90 mm and 4% larger than the real pupil

9.2   The following data are the results of a paraxial marginal and paraxial pupil ray trace through an optical system *in air*, specified in terms of curvatures:

	Marginal ray		Pupil ray	
Component	u	h	$\bar{u}$	$\bar{h}$
	0.000		0.087489	
1		22.857		−2.49968
	−0.38095		0.099987	
2		20.952		−1.99974
	−1.09524		0.199974	
3		10.000		−0.00000
	−0.66349		0.133316	
4		8.010		0.39995
	−0.99524		0.199974	

(a)   Which surface is the aperture stop and what is its diameter? Determine the positions and diameters of the entrance and exit pupils.

(b)   Calculate the position of the object and image planes.

(c)   Calculate the distance between the third and fourth surfaces.

ANSWERS:

(a)   surface number is 3 and has a diameter 20.00; entrance pupil position $v\varepsilon$ = 28.57 and diameter = 45.71; exit pupil position $v'\varepsilon'$ = −2.00 and diameter = 20.00

(b) vo = infinity; $v'o'$ = 8.05

(c)   distance = 3

9.3   Consider a cascaded optical system. The image formed by the first system becomes the object for the second. An object is placed in the front focal plane of the first system. The exit pupil of the first system and the entrance pupil of the second system are separated by 300 mm. Draw the system. Calculate the expected radius of the field-of-view in object space if the equivalent focal length of the first lens is 50 mm. Take the diameter of all pupils as 4 mm.

ANSWER: η = (−)0.667 mm

9.4   Find the optimum field lens power for the following situation. An optical system forms an image 50 mm beyond its exit pupil. This image becomes the object for a second system. The separation between the exit pupil of the first system and the entrance pupil of the second system is 350 mm.

(a)   Sketch the arrangement and determine the power of a suitable field lens.
(b)   What is the relative size of the image of the exit pupil of system 1, formed in the plane of the entrance pupil of system 2, to the size of this entrance pupil.

ANSWERS: (a) $F = 23.3\text{m}^{-1}$, (b) six

9.5   Derive the following equation:

$$\theta = (D_p + D_c)/(2d)$$

for the angular radius θ of the field-of-view in direct ophthalmoscopy, if the distance between the two eyes is d, the patient's pupil diameter is D_p and the clinician's pupil diameter is D_c.

Summary of main symbols and equations

n_A refractive index in the aperture stop space
ρ_A aperture radius of the aperture stop
u, h paraxial marginal ray angle and height
$\bar{u}, \bar{h}$ paraxial pupil ray angle and height
$\bar{u}_A$ pupil ray angle in aperture stop space
H optical invariant
NA numerical aperture

Section 9.1.3.1: The optical invariant

$$H = n(\bar{u}h - u\bar{h}) \tag{9.2}$$

$$H = -nu\eta \quad H = -n'u'\eta' \tag{9.3a\&b}$$

$$nu\eta = n'u'\eta' \tag{9.4}$$

$$H = n_A\bar{u}_A\rho_A \tag{9.6}$$

Section 9.1.5: Numerical aperture and F-number

$$NA = n\sin(\alpha) \text{ (object space)} \tag{9.7a}$$

$$NA' = n'\sin(\alpha') \text{ (image space)} \tag{9.7b}$$

$$NA' = n' \sin(\alpha') = \bar{\rho}F \quad \text{(if object at infinity)} \tag{9.12a}$$

$$NA' = n' \sin(\alpha') = \frac{DF}{2} \quad \text{(if object at infinity)} \tag{9.12b}$$

$$F\text{-}number = \frac{\text{equivalent focal length}}{\text{diameter of the entrance pupil}} = \frac{f}{D} = \frac{f}{2\bar{\rho}} \tag{9.13}$$

$$F\text{-}number = \frac{1}{2\sin(\alpha')} = \frac{1}{(2NA')} \quad \text{(in air)} \tag{9.14a}$$

or

$$NA' = \sin(\alpha') = \frac{1}{(2F\text{-}number)} \quad \text{(in air)} \tag{9.14b}$$

10

Defocus, depth-of-field and focussing techniques

10.0   Introduction

In the discussion of image formation so far, we have always assumed that the image is correctly focussed, with only one exception – in the reference to the telecentric pupil in the preceding chapter. However, in this chapter we will show that focussing is not usually exact and therefore in any imaging system, there is some residual defocus. One reason for this focus error is that there is always a range of positions of the "image" plane within which the image appears to be correctly focussed. This range is called the **depth-of-field** or **depth-of-focus**. A defocus outside this range will, by definition, be noticeable and hence must have some effect on image quality, but within this range, there will be some loss of image quality that may be detectable under different circumstances.

Sometimes we defocus a system intentionally. For example, we may wish to understand the effect of a defocus on some particular target structure or image quality criteria. In these cases, we may need to control the level of defocus accurately and quantify it.

Therefore, whatever the cause of defocus, it is important to understand the different ways in which a defocus can occur and how to quantify the level of defocus. To meet these needs, in this chapter, we will investigate the defocussing of optical systems, the effect of a defocus on the image, sources of defocussing, and depth-of-field and finally discuss a variety of methods that can be used to focus an optical system accurately.

10.0.1   *Quantification of defocus*

Any study of defocus must look at the ways of quantifying the level of defocus. The conceptually simplest measure is the size of the point spread function, which is defined as the light distribution in the image of a point. In Gaussian optics, the focussed image of a point is a point and the defocussed point spread function is a projection of the exit pupil, through the paraxial image point Q', onto the defocussed image or observation plane, as shown in Figure 10.1. If the aperture is circular and there is no vignetting, this point spread function is a uniformly illuminated circular disc (the **defocussed blur disc**).

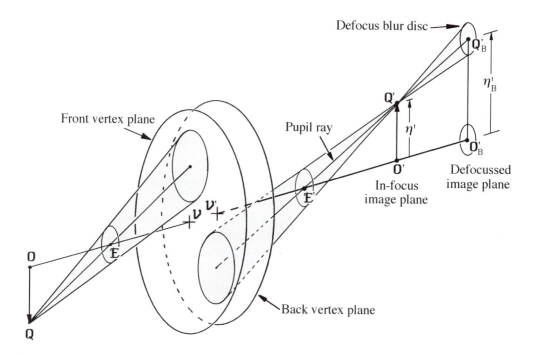

Fig. 10.1: Defocussed image and defocus blur disc (due to an error in the position of the image plane).

Providing there is no vignetting, the centre of the defocus blur disc on any observation plane is the point of intersection of the pupil ray with that plane. This is shown in Figure 10.1. The pupil ray from Q intersects the defocussed image plane at the point Q'_B. The defocussed blur disc centred on the point Q'_B may be regarded as the blurred image of the object point Q.

This model of the defocus blur disc shows very clearly that (a) the shape of the defocus blur disc is that of the aperture stop and (b) its diameter is proportional to (i) the diameter of the pupil or aperture stop and (ii) the distance between the in-focus and out-of-focus planes. Thus the effective level of a defocus will depend upon the size of the aperture stop and the image plane position error.

10.0.2 Effect of defocus on image quality

Intuitively, we know that a defocus decreases image quality. Defocus blurs out fine detail, blurs sharp edges and reduces the image contrast. If we wish to calculate the effect, we must use some measure of the level of defocus and an image quality criterion.

The actual effect of this defocus on the quality of the image depends upon the image quality criteria used. There are a number of criteria that one can use to assess general image quality and that can also be used to study the effect of defocus. Useful criteria are the point spread function and the **optical transfer function**, which are discussed in Chapter 34.

10.0.3 Effect of a defocus on the image size or magnification

Apart from reducing image quality, a defocus affects the image size and hence magnification. The factors affecting this change in magnification are the source of the defocus and the position of the aperture stop. This may be very important when image sizes have to be accurately measured. Any error in locating the position of the image plane may lead to an error in measuring the image size. However, we will show that in special cases, defocus has no effect on magnification.

Studying the effect of defocus on the size of the image presents an initial problem, the problem of how to measure the size of blurred images. Consider the image of the line OQ shown in Figure 10.1. This is in focus on the plane at O' but out of focus in the plane at O'_B. Each point on this line is imaged as a defocus blur disc (assumed to be a circular aperture) but only the discs for the points O and Q are shown in the diagram. We could define the size of the blurred image size of OQ as being measured from the bottom of the blur disc centred on O'_B to the top of the blur disc centred on Q'_B. However, the blur disc sizes depend upon the aperture stop size and thus if the above definition is adopted, the above defocussed image size will depend upon the size of the aperture stop at the time. On the other hand, it would be preferable to have a definition that would be independent of aperture stop size. A suitable definition specifies the image size as measured from the centres of the defocussed blur discs at O'_B and Q'_B, that is, where the pupil ray meets the out-of-focus image plane or the distance η'_B shown in the diagram.

10.1 Types of defocus

There are several causes of what is usually unintentional defocus. These are as follows:

(a) *Image plane shift*: Here, the object plane remains fixed but the observation plane is incorrectly positioned. A typical example in ophthalmic optics is the extreme **myopic** eye. Myopia is explained in Chapter 13. The length of this eye is often longer than normal and hence the retinal position can be regarded as being in error.

(b) *Object plane shift*: In this case, the image or observation plane is correctly positioned but the object plane position is incorrect.

(c) *Shift of the lens*: In some situations, the positions of the object and image planes are set and the system is focussed by adjusting the position of the lens. Typical examples are focussing of camera lenses and projectors. If these lenses are not in the correct position, the image will be defocussed.

(d) *Incorrect power of the imaging lens*: In this case the object, image and imaging lens positions are fixed but the equivalent power of the imaging lens is incorrect. A typical example in ophthalmic optics is the effect of the incorrect power of a refractive correction such as a spectacle lens, a contact lens or an intra-ocular lens.

(e) *Use of an auxiliary lens*: Sometimes we desire the image to be defocussed. For example, we may wish to simulate a defocus in order to investigate its effect on a given image. One convenient method of simulating defocus is to place an auxiliary (defocussing) lens in the beam at some suitable position.

In the remainder of this section, we will look at the optics of these different sources of defocus. The following discussion will be mostly qualitative and use only an understanding of Gaussian optics to investigate the effect of defocus on image formation. Equations and numerical calculations will be covered later in Section 10.2.

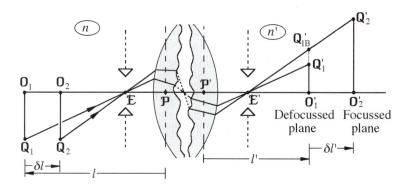

Fig. 10.2: Defocus due to change in position of the object plane.

10.1.1 Defocus due to an error in the position of the image plane

Here, the object plane remains fixed and the observation plane is not at the correct image plane position, as shown in Figure 10.1. This diagram shows an optical system with an in-focus image plane at o' and the observation plane at o'_B. The beam that forms an off-axis image point Q' intersects the observation plane as a blur disc centred on the point Q'_B.

Figure 10.1 shows that the observed image size defined as η'_B is different from the focussed image size η'. Thus this type of defocus produces some change in image size or magnification. However, if the aperture stop or entrance pupil is at the front focal point of the optical system as shown in Figure 9.6, the exit pupil will be at infinity and the pupil ray will be travelling parallel to the optical axis in image space. Therefore there will be no change in image size. A pupil placed at the front focal point is called a **telecentric** pupil and we have discussed this type of pupil imagery already in Chapter 9.

10.1.2 Defocus due to an error in the position of the object plane

Figure 10.2 shows objects at o_1 and Q_1 being imaged at o'_1 and Q'_1. If we move the object plane to o_2 but keep the observation plane fixed at o'_1, the new image plane is now at o'_2 and Q_1 is moved to Q_2, which is imaged at Q'_2. The beam from Q_2 intersects the old image plane at Q'_{1B}.

We need to know how this type of defocus affects the image size. Let us look at the pupil ray. If the object approaches the entrance pupil, as shown in Figure 10.2, the pupil ray enters the pupil at a larger angle. According to the properties of paraxial rays discussed in Chapter 2, the ouput angle is proportional to the input angle. Therefore as the object space pupil ray angle increases, so must the image space pupil ray angle. Therefore the pupil ray must pass through the exit pupil at a larger angle and must therefore intersect the old image plane at a greater height, leading to an increase in image size. If the object moves away from the entrance pupil, the image size will decrease but in both cases, this source of defocus also leads to a change in image size.

A telecentric pupil can be used to maintain a constant image size in this case but not if the entrance pupil coincides with the front focal point, as in the previous case and as shown in Figure 9.6. Instead, the exit pupil must coincide with the back focal point. The reader can confirm this by regarding the entrance pupil in Figure 10.2 as at the front focal point and then following the path of

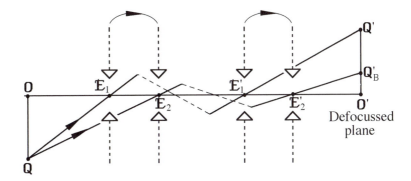

Fig. 10.3: Defocus due to error in position of the imaging lens. Only the pupils are shown in this diagram.

the pupil ray, and then repeating the exercise with the exit pupil coinciding with the back focal point.

10.1.3   Defocus due to an error in the position of the lens

In some situations, the object and image planes are fixed and the system is focussed by altering the position of the imaging lens. Figure 10.3 shows this situation but only with entrance and exit pupils shown. The image must be defocussed because it cannot remain in focus as the imaging lens is brought closer to the image plane.

If we look at the path of the pupil ray we will see that in this case the image becomes smaller. The reason is as follows. As the imaging lens is moved farther towards the image plane and hence farther from the object, the pupil ray from the point Q must make a progressively smaller angle to the axis. According to the properties of paraxial rays discussed in Chapter 2, the output angle is proportional to the input angle. Therefore as the object space pupil ray angle decreases, so must the image space pupil ray angle. The progressive decrease in pupil ray angle in image space, combined with the smaller distance between the exit pupil and the image plane, means that the pupil ray intersects the image plane closer and closer to the axis, thus decreasing the image size.

10.1.4   Defocus by addition of a defocussing lens

A system may be easily intentionally defocussed by placing an auxiliary lens somewhere in the beam. This is a very useful method of focussing for reasons that will become clear later.

Figure 10.4 shows an optical system with an auxiliary lens in the beam. If this lens were not present, the image plane would be at o'_1. With the lens in place, the beam is focussed to the plane at o'_2 but we observe the defocussed image in the old plane at o'_1. The pupil ray must be deviated by this auxiliary lens and therefore must cross the old image plane at Q'_{1B}, which is at a different height. Therefore this lens, while defocussing the image, causes a change in image size.

Let us look at the effect of placing the auxiliary lens in the plane of the exit pupil. In this case the pupil ray would pass through the lens at its centre and therefore would not be deviated and would therefore intersect the image plane at the same height. As a result, there would be no change in image size in this case.

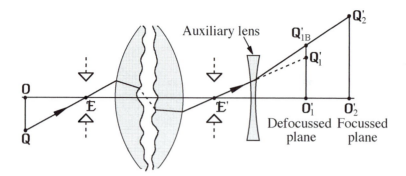

Fig. 10.4: Defocus due to an auxiliary lens.

Placing the lens at the entrance pupil or aperture stop will have the same effect since these are conjugate to the exit pupil and therefore the lens at these positions will also not deviate the pupil ray. We should note that, while the auxiliary lens can optically be placed at either pupil plane or at the aperture stop, if the pupils are virtual the auxiliary lens cannot be placed at these planes.

Finally, we should also note that the above discussion assumes that the auxiliary lens is thin. Since lenses are in reality thick, this thickness will have some effect on the image size, even if it is placed at the aperture stop or one of the pupil planes.

10.1.5 Overview

We have seen that any defocus, intentional or unintentional, usually produces a change in image size. However, in three cases there is no change in size of the image. These are as follows.

Case (1)   *Incorrect position of the image plane.* The aperture stop or entrance pupil is in the front focal plane. When the pupil is placed in this position, the pupil is said to be telecentric. This requires the exit pupil to be at infinity, but this is not practical with a visual instrument.

Case (2)   *Incorrect position of the object plane.* The aperture stop or exit pupil is in the back focal plane. This is also a telecentric pupil. This requires the entrance pupil to be at infinity.

Case (3)   *Use of a defocussing auxiliary lens.* This lens is placed in the plane of one of the pupils or the aperture stop.

It may now be clear that the pupils play a very important role in the effect of a defocus on the blurred image. The diameter of the pupils affects the diameter of the defocus blur disc and the positions of the pupils affect the change in image size.

10.2 Defocus calculations

In this section, we will develop equations that will be useful in defocus calculations and by using examples, show how the diameter of the defocus blur disc can be calculated.

10.2.1 Object and image plane shifts

Exact calculations

We will start with the lens equation [equation (3.43)], that is

$$\frac{n'}{l'} - \frac{n}{l} = F \tag{10.1}$$

Now, if the object moves by an amount δl from its initial position, the corresponding movement in the image plane will be $\delta l'$ as shown in Figure 10.2 and these two quantities are exactly related by the equation

$$\frac{n'}{(l' + \delta l')} - \frac{n}{(l + \delta l)} = F \tag{10.2}$$

These two equations can be used to derive the following relations between δl and $\delta l'$:

$$\delta l = \frac{n(l' + \delta l')}{[n' - F(l' + \delta l')]} - l \tag{10.3a}$$

$$\delta l' = \frac{n'(l + \delta l)}{[n + F(l + \delta l)]} - l' \tag{10.3b}$$

These equations show that, in general, any shift in object plane δl does not have the same magnitude as the corresponding shift $\delta l'$ in the image plane and the magnitude of $\delta l'$ also depends upon the sign of δl.

We can also express the object and image plane movements in terms of the changes δL and $\delta L'$ in the vergences. From the lens equation expressed in vergence form, equation (3.63)

$$L' - L = F \tag{10.4}$$

A change in object vergence δL leads to a change in image vergence of $\delta L'$, where

$$(L' + \delta L') - (L + \delta L) = F$$

It immediately follows that

$$\delta L' = \delta L \tag{10.5}$$

It should be noted that while δl and $\delta l'$ are not numerically equal, the vergences δL and $\delta L'$ are.

For visual optical systems, we will also find it useful to mix vergences with distances. For example, let us use the lens equation in the form

$$L' - \frac{n}{l} = F$$

A change $\delta L'$ in L' is related to a change δl in l, where

$$L' + \delta L' - \frac{n}{l + \delta l} = F$$

We can use these two equations to get

$$\delta L' = \frac{n\delta l}{l(l + \delta l)} \tag{10.6a}$$

Similarly if we start with the lens equation in the form

$$\frac{n'}{l'} - L = F$$

we can derive the following relation between $\delta l'$ and δL:

$$\delta L = -\frac{n\delta l'}{l'(l + \delta l')} \tag{10.6b}$$

Approximate calculations

For small values of δl and $\delta l'$, that is $\delta l \ll l$ and $\delta l' \ll l'$, we can differentiate the lens equation (10.1), and use small differentials to get

$$\frac{n'\delta l'}{l'^2} = \frac{n\delta l}{l^2} \tag{10.7}$$

We can also make some simplifications to the above equations. For small values of δl and $\delta l'$, equations (10.3a and b) can be approximated to

$$\delta l = \frac{nn'\delta l'}{(n' - Fl')^2} \tag{10.8a}$$

$$\delta l' = \frac{nn'\delta l}{(n + Fl)^2} \tag{10.8b}$$

and equations (10.6a and b) can be reduced to

$$\delta L' = -\frac{n\delta l}{l^2} \quad \text{and} \quad \delta L = -\frac{n'\delta l'}{l'^2} \tag{10.9a,b}$$

Since $\delta L' = \delta L$ from equation (10.5), we can use these latest equations to confirm equation (10.7) above.

> **Example 10.1:** Suppose we need to place a target in the front focal plane of a lens with a focal length of 500 mm, so that the image vergence is in the range ± 0.75 m^{-1}. To what precision must the target be placed at the focal point?
>
> **Solution:** From the above information, we have immediately $f = 500$ mm $= 0.5$ m and $\delta L' = \pm 0.75$ m^{-1}. Let us use the approximate

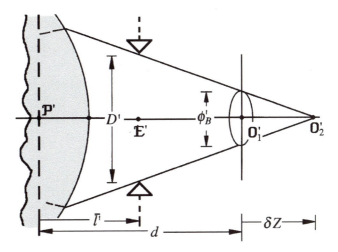

Fig. 10.5: Example of calculating diameter of the defocus blur disc.

equation (10.9a). However, before we can use this equation, we need to know the value of l. This is the nominal object or target distance and here will be the focal length f of the lens. Therefore we have

$$\delta l = -\delta L' f^2 = \pm 0.75 \times 0.5^2 = \pm 0.1875 \, \text{m} = \pm 187.5 \, \text{mm}$$

Therefore the target must be placed within 187.5 mm of the focal point to give an image vergence within the vergence range of $-0.75 \, \text{m}^{-1}$ to $+0.75 \, \text{m}^{-1}$.

10.2.2   The diameter of the defocus blur disc

When a system is defocussed, we may need to know the diameter of the defocus blur disc. Equations can be derived to give the diameter of this disc in any situation using simple trigonometry. Let us examine an example.

Figure 10.5 shows an optical system, with an exit pupil of diameter D', imaging a point on axis to o'_2, but the image (the defocus blur disc) is being observed in the plane at o'_1. The defocus blur disc diameter in the plane at o'_1 is ϕ'_B where from similar triangles

$$\frac{\phi'_B}{\delta Z} = \frac{D'}{(d - \bar{l}' + \delta Z)}$$

Solving for ϕ'_B, we have

$$\phi'_B = \frac{D' \delta Z}{(d - \bar{l}' + \delta Z)} \tag{10.10}$$

An alternative approach is to calculate the diameter in any given situation using basic geometry. We will show how this is done in the next section, by working through a numerical example.

Example 10.2: Given the following simple eye consisting of a cornea with a corneal radius of curvature of 5.5 mm, a refractive index of ocular medium of 1.3333, and an aperture stop of diameter 4 mm placed in the corneal vertex plane, calculate

(a) the diameter of the defocus blur disc for a defocussed object plane 1 m in front of the cornea and
(b) its angular diameter subtended at the nodal point.

Solution: This eye is shown in Figure 10.6a. It consists of a single refracting surface with a power

$$F = (1/5.5)(1.3333 - 1) = 60.6 \text{ m}^{-1}$$

The retina of this eye is at the back focal point $\mathcal{F}'$. The back principal point $\mathcal{P}'$ is at the corneal vertex and hence the length of the eye is

$$\mathcal{P}'\mathcal{F}' = n'/F = 1.3333/60.6 = 0.022 \text{ m or } 22.0 \text{ mm}$$

Now if an object is 1.0 m in front of the corneal vertex, the image distance l' is found from the lens equation

$$\frac{1.3333}{l'} - \frac{1}{l} = 0.0606$$

where $l = -1000$ mm. Solving for l' gives

$$l' = 22.371 \text{ mm}$$

Therefore the focussing error $\delta l'$ is

$$\delta l' = 0.371 \text{ mm}$$

We can find the diameter of the defocus blur disc using simple trigonometry. From similar triangles in Figure 10.6b, we have

$$\frac{\phi'_\text{B}}{0.371} = \frac{4}{22.371}$$

Thus

$$\phi'_\text{B} = 0.066 \text{ mm}$$

The nodal points $\mathcal{N}$ and $\mathcal{N}'$ are at the centre of curvature of the cornea as shown in the diagram and therefore the angular subtense ω of the defocus blur disc at the nodal points is

$$\omega = \frac{0.066}{16.5} = 13.7 \text{ minutes of arc}$$

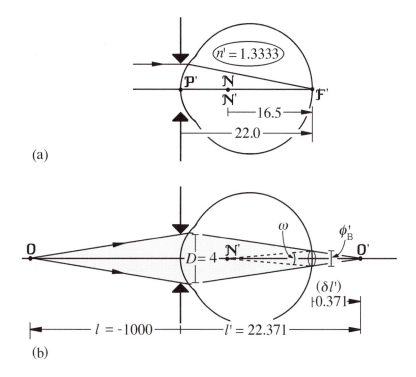

Fig. 10.6: Defocussed schematic eye for Example 10.2. Distances are in millimetres.

10.3 Depth-of-field

For a number of reasons, no optical system can be perfectly focussed. This can be the result of properties of the detector, for example the sensitivity of the detector to respond to small levels of defocus. In other words, the detector has a threshold level of defocus, below which the image appears equally focussed. Or it can result from properties of the imaging system, and one of the most important of these is the aperture diameter. The remaining factors are the structure of the image itself. For example a completely empty field would have no structure to allow focussing.

The range of defocus within which the image appears to be focussed is called the **depth-of-field**. From the definition of the depth-of-field, while any defocus of an image reduces image quality within the depth-of-field, the effects will not be observable under those conditions by that detector. Outside this region, there will be some detectable reduction in image quality.

Some textbooks draw a distinction between depth-of-field and what is called depth-of-focus. They define depth-of-field as the range in object space over which a defocus of the image is not detectable and depth-of-focus as the corresponding range in image space. The justification for this distinction is that if the values are expressed in terms of distances, they are not numerically equal, as can be seen from equation (10.3a or b). However, if the depths-of-field are expressed in vergences, equation (10.5) shows that the depth-of-fields in both object and image space are equal. Thus if depths-of-field are expressed in terms of vergences, there is nc need to make the distinction between object and image space values. In this book, we will only use the term depth-of-field, and when a distinction between object and image depths-of-field is necessary, it will be made clear.

We should note that for large depths-of-field, the distances from the conjugate plane to the extreme edges of the "in-focus" zone is not the same on either side of the in-focus plane. Thus the depth-of-field zone could be specified by the notation

$$l - \delta l_- \text{ to } l + \delta l_+ \quad \text{(object space)} \tag{10.11a}$$

and

$$l' - \delta l'_- \text{ to } l' + \delta l'_+ \quad \text{(image space)} \tag{10.11b}$$

Only if the distances are very small can we make the approximation

$$\delta l_- = \delta l_+ = \delta l \quad \text{and} \quad \delta l'_- = \delta l'_+ = \delta l' \tag{10.12}$$

We will show these differences with a numerical example.

10.3.1 Geometrical approximation to the depth-of-field

In the geometrical approximation to the depth-of-field phenomenon, one can begin with the defocus blur disc and then make assumptions about the threshold size at which the blur disc is resolvable as a disc, that is differentiable from a point. Usually, this threshold value depends upon the diameter of the detecting elements of the detector. For example, it would depend upon the size of the retinal cells in the case of the eye or on the size of the silver grains in the photographic emulsion in photography. If we take the example of the eye, the smallest retinal point spread function has an angular diameter of several minutes of arc.

> **Example 10.3:** Using the eye given in Example 10.2, determine the depth-of-field limits if this eye is correctly focussed at 1 m and the defocus blur disc threshold size is 0.05 mm. This situation is shown in Figure 10.7.

> **Solution:** Firstly, we need to find the axial length l' of the eye if it is focussed at 1 m. In Example 10.2, this distance was found to be 22.371 mm. The threshold blur disc diameter ϕ'_B is 0.05 mm.
> For the closer edge of the depth-of-field zone shown in Figure 10.7b and similar triangles, we have

> $$\frac{\delta l'}{0.05} = \frac{(22.371 + \delta l')}{4}$$

> Solving for $\delta l'$ gives

> $$\delta l' = 0.2832 \text{ mm}$$

> We can find the corresponding value of δl from equation (10.3a).

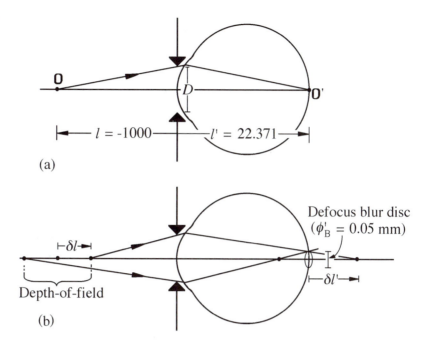

Fig. 10.7: Defocussed schematic eye for Example 10.3. Distances are in millimetres.

Substituting

$$n' = 1.3333, \quad \delta l' = +0.2832, \quad l' = 22.371, \quad n = 1,$$

$$F = 0.0606 \quad \text{and} \quad l = -1000$$

into this equation gives

$$\delta l = 427 \text{ mm}$$

Similarly, for the more distant edge of the depth-of-field zone shown in the diagram,

$$\delta l' = -0.276 \text{ mm} \quad \text{and} \quad \delta l = -2914 \text{ mm}$$

Thus the depth-of-field zone has the limits $l = -3914$ mm and $l = -573$ mm, with the centre at -1000 mm.

This example confirms that the two distances from the in-focus plane to the edges of the depth-of-field zone are not equal.

The weakness of the geometrical optics approach is that for small levels of defocus, the point spread function is more limited by diffraction and aberrations than the level of defocus. Therefore aberrations and diffraction may have a significant effect in modifying any geometrical optics predictions, but the study of these effects is beyond the scope of this chapter.

10.4   Focussing techniques

There are a number of techniques for focussing an optical system. Some of these are as follows:

(1)   Simple perception of blur, such as contrast of a simple periodic pattern
(2)   Split image and vernier acuity
(3)   Knife-edge
(4)   Scheiner principle
(5)   Laser speckle patterns
(6)   Rangefinding

This list is not exhaustive and other techniques are possible. For example, Cohen et al. (1984) described a method that uses the changing shape of the defocus blur disc as the observation plane is moved through focus, in the presence of astigmatism. In that method, astigmatic lenses are placed in the beam. In practice, an ideal focussing method has the following characteristics:

(1)   is independent of the form of the object
(2)   detects the parity of the defocus
(3)   can be used to quantify the current level of defocus and
(4)   can be automated.

However, none of the methods described here satisfies all of these requirements. Most require special forms of the object, for example a point or line, and therefore have limited applications. We will now look in more detail at some of the above methods.

10.4.1   *Simple perception of blur*

Many visual optical systems are subjectively focussed by simply maximizing the image sharpness. The accuracy of this method depends critically on the perception of defocus blur threshold of the observer, observer's visual acuity, observation distance and detail in the target.

    If point sources are present in the image, the size of the defocussed blur discs can be used and minimized. The blur disc threshold of detection is several minutes of arc for visual focussing. The correct focus is where the image of the point source has a minimum diameter.

    If a sinusoidal pattern is used as the object, the system can be focussed by maximizing the contrast of the sine wave patterns. However, with these periodic targets, **spurious resolution** (see Chapter 34) may occur, leading to a possibility of false focus. The accuracy of this method depends upon the sensitivity of the detector to a change in contrast of sinusoidal patterns and this will also depend upon the angular spatial frequency. The effect of a defocus on the image of a sinusoidal pattern is best analysed in terms of the optical transfer function, which is discussed further in Chapter 34, along with some examples.

    If the eye is the detector, it should be noted that the eye has a maximum sensitivity to absolute threshold contrasts for patterns in the range of about 2 to 10 c/deg. However, this may not be the frequency range where the eye is most sensitive to changes in contrast.

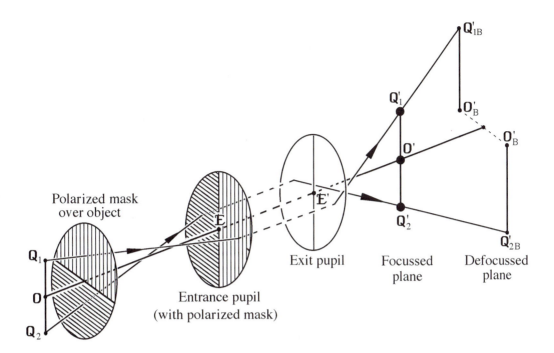

10.4.2 *Split image and vernier acuity*

The split image principle requires that the target contain at least one straight
edge which can be split into two, say at its mid-point. The defocus moves one
part of the edge sideways relative to the other. The eye then sees the straight edge
split into two and displaced sideways. When the system is correctly focussed,
there is no displacement and hence the two parts of the image are aligned.
This method makes use of the eye's **vernier acuity**, which is a measure of the
threshold of displacement in the break in an edge. Typically the eye can detect
an edge mis-alignment of about 20 seconds of arc.

Such a method can be realized using polarizing filters placed over the object
or target as shown in Figure 10.8, and another pair of polarized filters placed
over the pupil of the eye. This second pair is rotated through 90° relative to the
first pair. Part of the beam enters only one half of the pupil and the other part
of the beam enters the other half of the pupil. If the image is defocussed on
the observation plane, a vertical line in the image is split and the two parts are
displaced sideways away from each other, as shown in the diagram.

A second method is based upon the principle that if an image is viewed
through a thin prism but the image is not in the plane of the prism: the image
is displaced in the direction of the apex of the prism. This phenomenon was
explained in Chapter 8 and depicted in Figure 8.8. This method is used in split-
image rangefinders commonly found in single lens reflex cameras. In practice,
it is common to use two prisms as shown in Figure 10.9. With this arrangement,
each prism displaces the image in the opposite direction to the other; thus the
displacement is doubled for any level of defocus, doubling the sensitivity of the
method.

The accuracy of this method is affected by the sharpness of the edge. It
would be expected that factors which reduce edge sharpness, such as diffraction

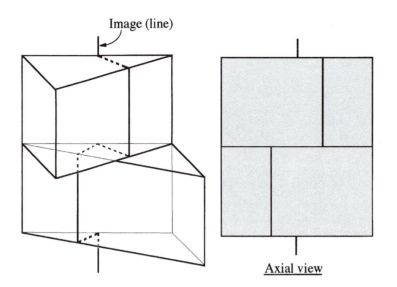

Image (line)

Axial view

Fig. 10.9: Prism based split image focussing using two prisms.

and aberrations, will decrease the accuracy of vernier alignment. The relative position of the split image lines can be used to give the parity of the defocus.

10.4.3 Knife-edge (Foucault) method (point or near point objects)

The knife-edge method requires a point source or light. Consider the situation shown in Figure 10.10. A point (very small in practice) source of light Q in the object plane at O is imaged at Q'. We assume that the image is free of aberration. If we place an eye at the position shown in the diagram, the eye will see the exit pupil of the optical system fully and uniformly illuminated. If a knife-edge is placed close to, but beyond the image, as shown in this diagram, the knife-edge blocks rays coming from the top part of the pupil, making this part of the pupil appear dark. If the knife-edge is moved farther upwards, it increasingly vignets the beam and a dark shadow moves downwards across the pupil. In this case, the shadow moves in the opposite direction to the knife movement. If now the knife-edge is placed between the lens and the image, rays from the bottom part of the pupil will be vignetted and hence the bottom part of the pupil will be dark and if the same procedure were repeated, the dark shadow would move up the pupil in the same direction as the knife movement. If now the knife-edge was placed exactly in the paraxial image plane at O', the whole pupil would go instantaneously dark with no obvious shadow edge and movement. When a focus error is present, the direction of movement of the light/dark boundary can be used to give the parity of the defocus.

This method is essentially the basis of **retinoscopy**, a method for determining the refractive state of the eye, which we discuss in greater depth in Chapter 31.

10.4.4 The Scheiner principle

In the Scheiner method, a mask containing a number of holes, usually two, is placed over or near the entrance or exit pupils as shown in Figure 10.11. If the

View of exit pupil

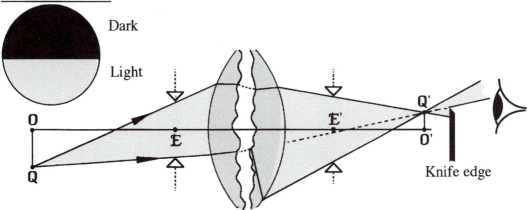

Fig. 10.10:   Knife-edge
method of focussing.

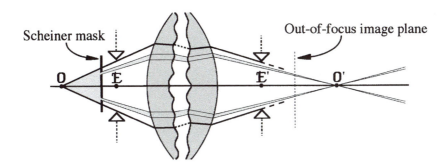

Fig. 10.11:   Scheiner
method of focussing.
object is a very small source of light, the rays passing though the two holes in
the mask intersect the image plane at different points. The observer then sees
two images of the object on the image plane. The separation of the two points
is a measure of the level of defocus.

The relative position of the hole images in the image plane depends upon
the parity of the defocus. In Figure 10.11, the image is focussed beyond the
observation plane and the upper spot in the observation plane corresponds to the
upper hole in the aperture. If the image is now formed in front of the observation
plane, the lower spot corresponds to the upper hole in the mask. Therefore the
relative position of the image holes is a clue to the parity of the defocus.

10.4.5   Laser speckle

Laser speckle is a diffraction phenomenon requiring **coherent** light. We have
not discussed coherence yet and do not do so until Chapter 25. Laser speckle
can be used to measure the presence and parity of a defocus and in Chapter 31,
we describe how this can be done when applied to the eye.

10.4.6 Rangefinding

So far, we have described methods of focussing the image by directly observing the image plane and using some technique to detect the level of defocus. An alternative approach is to measure the distance of the object and, using a knowledge of the Gaussian properties of the system, position the image plane accordingly. For example, a camera lens is marked with different distances and the camera can be focussed using the distance scale on the lens and therefore without the need to look in the viewfinder. There are a number of optical methods that can be used for measuring the distance of an object or target and these are discussed in Chapter 20.

10.4.7 Effect of aberrations

The presence of aberrations decreases the ability to detect a defocus, and therefore the accuracy of all of the above methods, in some way, is affected by aberrations. As an exercise, the reader could examine the knife-edge and Scheiner disc methods and try to predict the effect of spherical aberration.

Exercises and problems

10.1   For a single lens of power 5 m^{-1} and diameter 5 cm, calculate the depth-of-field (in object space) for the lens if it is used to image an object 25 cm from the lens and the resolvable defocus blur disc diameter in the image is 0.1 mm.

ANSWER: ± 0.125 mm

10.2   For a lens with a power of 60 m^{-1}, calculate the precision in positioning the object if the image vergence at the lens has to be in the range -0.25 m^{-1} to $+0.25 \text{ m}^{-1}$.

ANSWER: ± 0.0694 mm

10.3   For a thin lens of power 60 m^{-1}, with the object at infinity and an image space index of 1.336, what is the defocus image distance $\delta l'$ corresponding to an error in object vergence of -1.0 m^{-1} measured at the lens?

ANSWER: $+0.371$ mm

10.4   For the following simplified schematic eye:

corneal radius of curvature $= 5.0$ mm
aqueous/vitreous refractive index $= 4/3$
eye length $= 21$ mm
pupil coincident with the cornea and having diameter 5 mm

calculate

(a)   the defocus error in terms of the error in the image plane if the eye is focussed at infinity
(b)   the diameter of the defocus blur disc formed on the retina for an object 1 m in front of the cornea

ANSWERS: (a) $\delta l' = +1.0$ mm; (b) diameter $= 0.17$ mm

10.5   Examine the knife-edge method of focussing in the presence of positive spherical aberration. Describe the shadow formation as the knife is moved through the beam at the paraxial focus.

10.6   Examine the Scheiner disc method of focussing in the presence of positive spherical aberration. Will the aberration cause the system to be focussed closer to or farther away from the system? Explain your answer.

Summary of main symbols

ϕ'_B	diameter of the defocus blur disc
O', O'_1,..	axial image positions
Q'_B	position of the defocussed image of Q'
δl, $\delta l'$	shifts in object and image
δL, $\delta L'$	vergence changes corresponding to δl and $\delta l'$
η'_B	size of defocussed image

References and bibliography

Bai Han Xiang and Indebetouw G. (1983). Focussing and alignment of visual optical instruments by laser speckle. *Appl. Opt.* 22(11), 1609–1611.

Charman W.N. (1974). On the position of the plane of stationarity in laser speckle refraction. *Am. J. Optom. Physiol. Opt.* 51, 832–838.

*Cohen D.K., Gee W.H., Ludeke M., and Lewkowicz J. (1984). Automatic focus control: the astigmatic lens approach. *Appl. Opt.* 23(4), 565–570.

Kocher D.G. (1983). Automated Foucault test for focus sensing. *Appl. Opt.* 22(12), 1887–1892.

Smith G. (1982). The angular diameter of defocus blur discs. *Am. J. Optom. Physiol. Opt.* 59(11), 885–889.

11

Basic optical metrology

11.0 Introduction

It is sometimes necessary to determine the structural properties of a single lens or lens system. This information may be needed to understand the optical properties which can be further investigated by ray tracing. For example, we can use the knowledge of the refractive indices, surface curvatures and surface separations to determine the Gaussian properties such as the powers and positions of the cardinal points by paraxial ray tracing, using techniques described in Chapter 3. The ray trace results can also be used to calculate the primary or Seidel aberrations using the equations given in Chapter 33. The powers and cardinal points can also be measured directly by laboratory techniques without the need to take the system apart.

Thus this chapter is concerned with the analysis of optical systems that have been constructed and for which we do not have the constructional details. That is we do not know the refractive indices of the materials, the surface separations or the surface curvatures. If we wish to examine the optical properties of such a system, we can solve the problem in two ways.

(1) We can take the system apart, measure the refractive indices, surface separations and surface curvatures. Using paraxial ray tracing, we can determine the Gaussian properties such as equivalent power and positions of the cardinal points.

(2) We can measure the Gaussian properties directly. For example, in this chapter we will describe several methods for measuring the equivalent power of an optical system in the laboratory.

The powers and cardinal point positions are not the only Gaussian parameters that occasionally have to be verified. In some situations, image position and size, and field-of-view need to be measured. However the procedures for these measurements depend somewhat on the particular instrument, so that these discussions will be put aside until the various instruments are described in subsequent chapters.

11.1   Refractive index (solid samples)

There is a wide range of methods for measuring the refractive index of solid optical materials, varying in optical principle, sophistication and accuracy. One factor limiting the accuracy of any refractive index determination is the dispersion of the material. Therefore accurate methods require the use of monochromatic light. If a range of monochromatic sources are available, the dispersion and V-value of the material can be found.

Some methods require the material to have at least one flat surface, some require two flat parallel sides and some require the sample to be worked into the form of a prism. Therefore these methods are not readily applicable to lenses unless the lens can be destroyed by optically polishing one flat surface on it or working it into the required shape.

There are a number of commercial instruments that measure refractive index. The Abbe refractometer is an example. Some of these instruments, including the Abbe, only measure the refractive index at one particular wavelength. However in this chapter, we will not discuss these instruments and concentrate instead on measurement principles. A commercial instrument will usually be based upon one of these principles.

Some of the principles require the accurate measurement of the directions of a beam of parallel rays. The sample is mounted on a **spectrometer** (or **goniometer**) table. A spectrometer is made up of a rotatable table on which is mounted the sample. The table is connected to a circular scale so that rotation angles can be measured. A collimator and a telescope are mounted so that their optical axes are perpendicular to the axis of rotation of the table and also intersect this axis. The **collimator** and **telescope** can be independently rotated about the table axis and their angular position measured using the angle scale attached to the table. We will describe several methods of measuring refractive index using the spectrometer, but first we will describe a very simple method based upon apparent thickness.

11.1.1   From the apparent thickness

If the sample is a slab with parallel sides, perhaps the simplest method of measuring refractive index is to calculate it from a measurement of the apparent thickness. From the lens equation (2.11), it easily follows that the apparent thickness d' of a slab of material, whose refractive index is μ and actual thickness is d, is given by the equation

$$d' = d/\mu \tag{11.1}$$

The actual thickness d can be measured with a micrometer and the apparent thickness d' is measured optically and the refractive index μ found using equation (11.1), that is

$$\mu = d/d' \tag{11.1a}$$

The apparent thickness d' can be found using a microscope. A moderate to high power microscope is focussed on one face of the slab, then moved forwards or backwards until the other face is focussed. The difference in positions is the apparent thickness d'. Dust and scratches on the two surfaces usually allow clear location of the surfaces. The accuracy of this method is limited by the accuracy of the thickness measurements. For example, the fraction error $\delta\mu/\mu$ in μ is given approximately by the equation

$$\delta\mu/\mu \approx |\ \delta d/d\ | + |\ \delta d'/d'\ | \approx 2\ |\ \delta d/d\ | \qquad (11.1b)$$

We can use this equation to estimate the required accuracy of the thickness measurement. For example, if we wished to measure the refractive index to an accuracy of ±0.001, this equation shows us that, for a refractive index of 1.5, we would need to measure the thickness to a precision of

$$|\ \delta d/d\ | < 0.0003$$

This required high precision makes the application of this method to thin samples impractical.

Suppose we now have a thick single lens. The apparent thickness is modified by the refractive effect of the curvature of the surface facing the microscope. If C is the curvature of this surface, the lens equation (2.11) applied to refraction at this surface gives us the equation

$$\mu = \frac{d(d'C - 1)}{d'(dC - 1)} \qquad (11.1c)$$

This equation uses the normal sign convention and assumes that the microscope is on the right of the lens and thus d and d' will usually be negative if left to right ray tracing is assumed. The accuracy in this measurement will be less than for the parallel slab above because of the extra uncertainty arising from the errors in the measured curvature.

11.1.2 Critical angle principles

There are a number of methods based upon the critical angle effect (Section 1.5.2). The critical angle principle is schematically shown in Figure 11.1a. The diagram shows the boundary between two media, the upper one having the lower refractive index. Diffuse light in the upper medium is incident on the boundary at a range of angles but only the grazing incidence ray is shown. Some of this light crosses the interface but no ray can have an angle of refraction greater than the critical angle i'_{max}, where from Snell's law

$$\sin(i'_{max}) = n/n' \qquad (11.2)$$

and the corresponding grazing ray is shown in the diagram. One way of ensuring that the upper sample is diffusely illuminated is to pass the illuminating source through a diffuser placed adjacent to the sample.

To put this method into practice we need to be able to (a) measure the angle of refraction i'_{max} and (b) diffusely illuminate the upper medium. Figure 11.1b shows one implementation using a spectrometer table. The sample is mounted on a prism and diffusely illuminated. The telescope is used to measure the direction of the emerging refracted beam. A light/dark boundary marks the direction

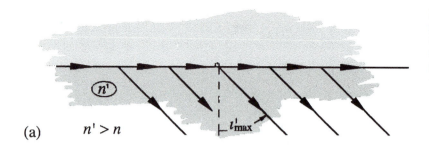

(a) $n' > n$

Fig. 11.1: (a) The principle of the critical angle method of measuring refractive index. (b) One particular implementation of the method.

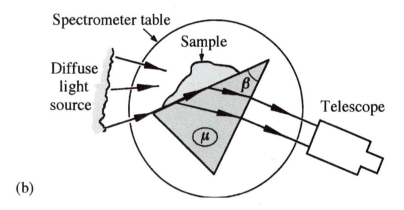

(b)

of the corresponding emerging critical angle rays. The position of the telescope does not immediately give the angle i'_{max}, which is the angle inside the prism. However there are several ways of finding the refractive index of the sample.

Assuming we know the refractive index μ and apex angle β of the prism and the direction of the normal to the output face of the prism, we can trace a ray backwards to find the angles inside the prism and thus the angle i'_{max}. Alternatively, with one fixed set-up, we could calibrate the system, using a number of samples of known refractive index.

This method requires a good optical contact between the sample and the base (a prism in this example). This requires that the contact surfaces are flat to a high accuracy and to improve optical contact, a liquid film is placed between the two surfaces. The index of the liquid must be higher than that of the sample in order to prevent total internal reflection.

One advantage of this method is that it only requires one flat surface to be ground on the sample. Other methods, such as the one to be described in the following section, require the sample to be worked into the shape of a prism.

There are several disadvantages with this type of critical angle measurement. The method cannot be applied to a sample with a refractive index greater than the base material upon which it is placed. Secondly, to measure the sample index at other wavelengths, the dispersion of the base or second material has to be known. Fortunately, there are critical angle methods that do not depend upon refraction into a second material. A very simple one to implement is one in which the sample is in the form of a prism. This is explained in the next section, under the heading "The grazing incidence method".

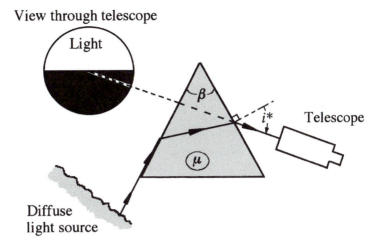

Fig. 11.2: The grazing incidence (a critical angle method) for a prism.

11.1.3 Methods using the sample in the shape of a prism

There are a number of very useful methods that are based upon refraction of light by a prism. The refractive index of the prism material is deduced from the knowledge of the refraction in particular situations. Thus in these cases, the sample has to be in the form of a prism and all the methods require a knowledge of the apex angle β of the prism. While this extra shaping of the sample may be seen as a disadvantage, the methods do not require a knowledge of the index or indices of some other material, which is the case in the critical angle method described in the preceding section.

The grazing incidence method

The grazing incidence method is a method based on the critical angle and is one method of implementing the critical angle principle described in Section 11.1.2.

Monochromatic light, from an effectively extended source, is allowed to fall on one surface of a prism as shown in Figure 11.2, such that some of the rays will meet the surface of the prism at grazing incidence. There will now be a certain limiting direction (i^*) of emergence at the second face corresponding to rays at grazing incidence on the first. Using the prism ray trace procedure given in Section 8.1.1, we can show that the final angle i^*, refractive index μ of the prism and its apex angle β are related by the equation

$$\mu^2 = 1 + \left[\frac{\sin(i^*) + \cos(\beta)}{\sin(\beta)} \right]^2 \tag{11.3}$$

Thus given the apex angle β of the prism, we only have to measure the angle i^* and we can then use this equation to calculate the refractive index of the prism. Since the angle i^* is relative to the normal to the exit surface, the measurement of this angle requires the measurement of the direction of the normal. This in turn can be found from the directions of incidence and reflection of a collimated beam reflected off this surface.

Angle of minimum deviation

In Section 8.1.1.1, we showed that when light is refracted by a prism, the angle of deviation has a minimum value θ_{min} and this angle is related to the prism index μ and apex angle β by the simple equation (8.3a),

$$\sin[(\theta_{min} + \beta)/2] = \mu \sin(\beta/2) \qquad (11.4)$$

Thus if we can measure the minimum angle of deviation θ_{min}, we can calculate the refractive index. However, this angle cannot be found by one simple measurement and instead requires some trial and error or searching for this angle. Therefore this method may not be as fast at the one above, based on grazing incidence.

The optical set-up for this measurement also uses the spectrometer, that is the set-up is similar to that for the grazing incidence method described above and shown in Figure 11.2, but with the diffuse source replaced by a collimator.

11.1.4   Liquid immersion methods

Liquid immersion methods are suitable for those samples which cannot be reshaped. Thus they are ideally suitable for samples that have an irregular shape or a lens that cannot be destroyed. These methods are based on the principle that if an object is immersed in a liquid with the same refractive index, it will be optically invisible.

This method requires a stable liquid whose refractive index can be varied. Such a requirement can be achieved by using a mixture of two different and miscible liquids; one of high index and one of low index. A typical pair of useful liquids is 1-monobromonaphthalene ($n = 1.6588$) and amyl acetate ($n = 1.400$). A wide range of suitable liquids, taken from the *Handbook of Chemistry and Physics* (1975), is listed in Table 11.1.

The sample is placed in a cell which is filled with the liquid mixture and the proportions of the two liquids are varied until the sample is no longer visible. At this point, the indices are equal and the refractive index of the liquid can be measured by a number of methods, some of which are described in Section 11.2.

While liquid immersion methods are very versatile, because they are the only methods that can be used in many cases, they do have a number of drawbacks. The first is that they can be long and tedious. The correct liquid index sometimes can only be found by much trial and error. The search for the correct index is improved if there is some test for the relative difference in the mismatch between the sample and liquid, that is whether the sample index is higher or lower than the liquid index. However, even if this is known, it is not easy to predict the magnitude of the difference.

There are a number of variations on the liquid immersion methods, such as (a) the collimated target method, (b) the Becke line method and (c) the method developed by Smith (1982) for simple lenses. The Smith method has the advantage that the procedure predicts the correct index at each stage in the search and therefore speeds up the process. These methods will now be described.

Fig. 11.3: Measurement of refractive index of irregular samples using liquid immersion.

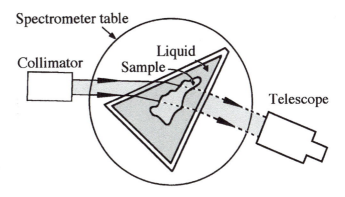

Table 11.1. *Liquids suitable for use in the liquid immersion methods of measuring refractive index*

Liquid	Refractive index n_D at 24°C
Amyl acetate	1.400
Trimethylene chloride	1.446
Cineole	1.456
Hexahydrophenol	1.466
Decahydronaphthalene	1.477
Isoamylphthalate	1.486
Tetrachloroethane	1.492
Pentachloroethane	1.501
Trimethylene Bromide	1.513
Chlorobenzene	1.523
O-Nitrotoluene	1.544
Xylidine	1.557
O-Toluidine	1.570
Aniline	1.584
Bromoform	1.595
Quinoline	1.622
α-Chloronaphthalene	1.633
1-Monobromonaphthalene	1.654
Methylene iodide	1.738

Note: The symbol n_D is the refractive index at the sodium D line (589.3 nm). A value for 1-monobromonaphthalene was not given in the above book and the value listed was measured by the authors on an Abbe refractometer.

Source: Handbook of Chemistry and Physics 1975–1976.

The collimated target method

A typical arrangement is shown in Figure 11.3. The collimator produces an image at infinity of a suitable target, such as a narrow slit and the telesope is focussed on infinity. If the indices do not match, the sample scatters or defocusses the light and the collimator target is not seen clearly through the

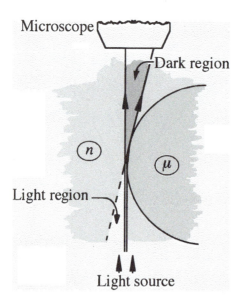

Fig. 11.4:   The principle of
the Becke line method of
measuring refractive index
of spheres. In this diagram
$n < \mu$.

telescope. The sample optically disappears when there is an index match and then the collimator target is seen clearly and in-focus through the telescope.

The liquid refractive index is varied until the target is clearly imaged in the telescope. The liquid and sample could be contained in a cell with flat parallel sides but the hollow prism is more versatile. The advantage of the prism over the parallel sided cell is that the prism set-up allows the direct measurement of the liquid index once the index match is found, using the grazing incidence method for solid prisms (see Section 11.1.3). However, if the prism is used, as the liquid index is varied, the beam deviation angle will change so that the telescope has to be continually re-aligned.

The Becke line method

The Becke line method was initially developed to determine the refractive index of small crystals (Wahlstrom 1969). It is a bright or dark line that is formed above or below a sample due to refractions and reflections as the beam passes through the sample. We can see how this line is formed in the special case of a sphere shown in Figure 11.4. In this diagram, the liquid refractive index is less than the sample index. If the microscope is focussed firstly below the sample and then moved upwards, the bright band is firstly observed in the liquid and then moves across the sample boundary until it is inside the sample when the microscope is focussed above the sample. Thus the direction of movement of the bright line gives an indication of whether the sample index is greater or less than the liquid index. If the sample index is less than the liquid index, the opposite effect is observed. Usually there is a corresponding dark band and the movement of this band also indicates the sign of the refractive index difference. When the liquid and sample index match, no bright or dark line is observed.

Once the Becke line disappears, the next step is to measure the refractive index of the liquid. This can be done by one of the methods described in Section 11.2.

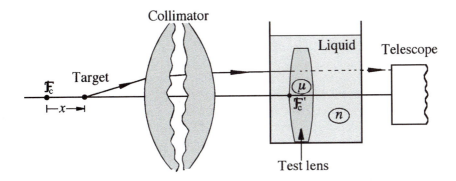

Collimator

Liquid   Telescope

Target

Test lens

Fig. 11.5: The Smith
modification of the liquid
immersion method for
measuring refractive index
of a simple lens.

However, there are better methods for large spheres and lenses, which we will now describe.

The Smith method for a single lens

If the sample is a simple lens, whether thin or thick, immersed in a medium of index n, the refractive index μ of the lens could be found in principle from measured values of the vertex power (say the front vertex power F_v), surface curvatures C_1 and C_2 and lens thickness d. If these are known, the refractive index μ could be found from the equation

$$F_v = \frac{(\mu - n)[(C_1 - C_2) + (\mu - n)C_1C_2d/\mu]}{[1 + C_2(\mu - n)d/\mu]} \tag{11.5}$$

which can be derived from the equations given in Chapter 3 for the vertex and surface powers of a thick lens. If the simple lens is in air, then the value of n is unity. Solving for the refractive index μ leads to a quadratic equation in μ. However, this method does not seem to work satisfactorily in practice, due to an accumulation of errors and the difficulty in accurately measuring the vertex power. For many simple lenses, residual aberrations such as spherical aberration make it impossible to measure the vertex power to a sufficient accuracy. If the diameter of the aperture stop is reduced in order to decrease the aberration effect, depth-of-field increases and thus decreases precision.

On the other hand, the above principle can be combined with a liquid immersion method to produce a very fast and accurate method, developed by Smith (1982). The technique is based upon the principle of the focimeter, which is described in detail in Section 11.4.2.3. The lens is immersed in a liquid and placed so that its front vertex coincides with the back focal point $\mathcal{F}_c'$ of the collimator lens, as shown in Figure 11.5. The test lens defocusses the target which has to be moved a distance x to be refocussed. It can be shown that if the collimator has an equivalent power F_c', the refractive index μ can be found by solving the quadratic equation

$$\mu^2(C_1 - C_2 + C_1C_2d) - \mu[n(C_1 - C_2) + 2nC_1C_2d$$

$$+ xF_c^2(dC_2 + 1)] + dC_2n(nC_1 + xF_c^2) = 0 \tag{11.6}$$

This equation is derived from equation (11.5) and the focimeter equation (11.21). Because equation (11.6) is quadratic, there will be two solutions. Only one of these solutions will be meaningful; the other being unrealistic.

This method is very rapid in that only the starting value of the liquid index n has to be known. If the set-up is perfect and there are no aberrations or imprecisions due to depth-of-field, the above equation will give the correct value for μ. However, errors in position of the test lens, in the measurement of collimator lens power F_c, test lens curvatures C_1 and C_2, lens thickness d and residual aberrations (mainly spherical) lead to the above solution being only approximate. So in practice, equation (11.6) only gives an estimate of μ, which is then used to find a more accurate value of liquid index n which is closer to the lens index μ. This new liquid index is used to find a better value of μ, and hence the method is iterative with a rate of convergence that depends upon the accuracy of initial set-up, accuracy in the above quantities and lens aberrations. The estimated value of μ becomes more accurate as the above quantities are known to a greater precision, because the significance of the errors and aberrations decreases as the match of the indices improves. Smith. (1982) found that with careful set-up, the procedure converges to the end-point with only three or four iterations. Finally, the refractive index of the lens is that of the liquid which can be determined by methods described in Section 11.2.

11.1.5   *The rainbow method for spheres or cylinders*

The rainbow method is applicable to spherical samples (but could be used for cylindrical samples). Even though very few optical components are spherical, this method is discussed here because small glass or plastic spheres are used to make retro-reflectors. The spheres are used either (a) alone as large single spheres or (b) in large numbers as small spheres (diameter $\approx$ 1 mm). The small spheres used in large numbers are set in a sometimes coloured binder to form coloured retro-reflecting sheets. Retro-reflectors and retro-reflecting sheets are used to make such things as traffic signs and **front projection screens**. Road marking paint is often made retro-reflecting by mixing the small spheres with the paint.

The method is based upon the rainbow phenomenon. When light enters a sphere, some of the light reflects off the back surface and is refracted once more as it passes out of the sphere, as shown in Figure 11.6a. The angle of deviation θ, shown in the diagram, is a function of the angle of incidence i or ray height h. Using this diagram and simple trigonometry, it can be easily seen that the angle of deviation θ is given by the equation

$$\theta = 4r - 2i \qquad\qquad\qquad (11.7a)$$

where

$$\mu \sin(r) = \sin(i) \qquad\qquad\qquad (11.7b)$$

from Snell's law. If we evaluate the deviation angle θ for a range of ray heights h, we will find that this angle has a maximum value for a certain value of h (or i), which corresponds to the maximum ray density. By differentiating θ in equation (11.7a) with respect to i, we have

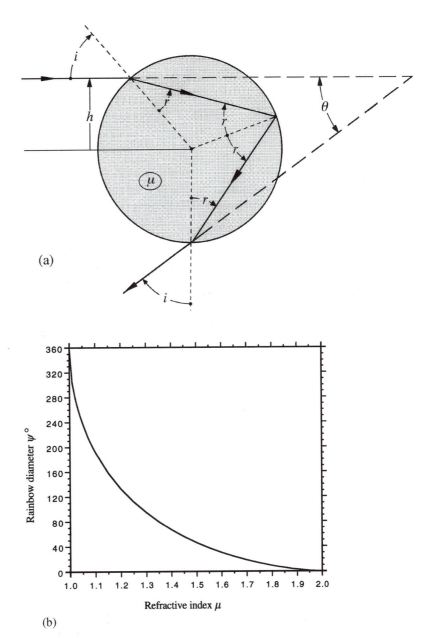

Fig. 11.6: (a) The ray path for the formation of the rainbow. (b) The angular diameter ψ of the rainbow as a function of refractive index μ. For a water index of $\mu = 1.3333$, $\psi = 84.08°$.

(a)

(b)

$$d\theta/di = 4dr/di - 2$$

The turning point and hence direction of maximum ray density occurs when $d\theta/di$ is zero, that is

$$dr/di = 0.5$$

By differentiating equation (11.7b) we can finally show that the turning point occurs at the incidence angle i given by the equation

$$\sin(i) = \sqrt{[(4 - \mu^2)/3]} \tag{11.7c}$$

At this angle, the ray density is very high leading to a high localized light level. In white light, the position of the maximum varies with wavelength, due to the dispersion of the sphere material and this produces the well known rainbow. The angular diameter ψ of the rainbow subtended at an observer is given by the equation

$$\psi = 2\theta_{max} \tag{11.8}$$

Figure 11.6b is a graph of ψ plotted against μ.

Because the angle θ_{max} defining the rainbow is a maximum angle, all the reflected rays have angles less than this value and therefore these rays fall on the inside of the rainbow. Thus, the inside of the rainbow is brighter than the outside.

A practical realization of this rainbow method is to place a sample of the spheres on the table of a spectrometer and upon a dark background such as a piece of black felt and to illuminate the sphere or spheres with collimated white light from a small source as shown in Figure 11.7. The rainbow is observed in the spectrometer telescope and its angular diameter measured. This is the method described by Smith (1977a). More precise values can be found using monochromatic light. In this case a bright edge is observed instead of a rainbow.

Since no simple equation exists giving μ explicitly in terms of ψ, the corresponding value of index μ can be found by the following two methods:

(a)   Graphically from the measured value of ψ. For example one could use a graph of the form shown in Figure 11.6b, but drawn on a finer scale if necessary.
(b)   By numerically varying the index in the above equations and finding the value that gives the correct value of ψ.

The above theory is based upon geometric optics and assumes that the radius of the spheres are considerably greater than the wavelength of light and in fact the geometrical optics results are independent of the radius. If the spheres are small (less than about 0.1 mm), a correction should be made for the physical optical effects (Yamaguchi 1975; Smith 1977b).

11.2   Refractive index (liquid samples)

The refractive index of liquids can be quickly and accurately measured on commercial refractometers such as the Abbe refractometer. However some of these instruments only measure the index at one particular wavelength. Because of this limitation or the lack of availability of a commercial instrument, there is a need for more versatile methods that can be used to give the refractive index at any wavelength. Two very useful methods are those based upon the use of prisms, as described in Section 11.1.3.

The prism methods can be adapted to the measurement of refractive index of liquids by using a hollow prism and filling it with the liquid. Providing the surfaces of the walls of the prism are parallel and of good optical quality, the walls have no effect on the measured refractive index. If one has a hollow prism and spectrometer, the measurement of the refractive index using one of the prism methods is relatively simple to perform. However, the hollow prism methods have the disadvantage in that a large volume of the liquid may be required,

Fig. 11.7: Practical arrangement for the rainbow method of measuring refractive index of spheres.

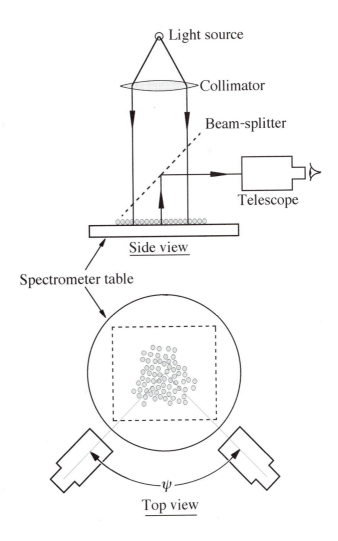

depending upon the size of the prism. Sometimes only a small volume of the liquid is available and therefore other methods more suitable to small volumes are required.

There are several methods of measuring the refractive index of small samples that are based upon the critical angle methods. The critical angle method shown in Figure 11.1 can be readily adapted for small amounts of liquids by replacing the extensive upper medium by a thin film of the liquid held between the base medium and an upper plate.

The refractive index of liquids is often significantly temperature sensitive and therefore a meaningful measurement may require the control of the temperature.

11.3 Surface curvature

There are a number of methods for measuring the radius of curvature of a spherical surface. We will divide these methods into two groups: methods that rely on some physical contact with the surface and those that do not. We will call

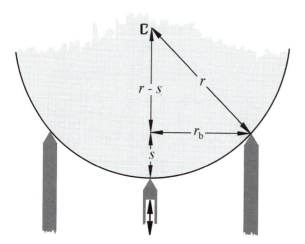

these (a) contact or mechanical methods and (b) non-contact or optical methods, respectively. The non-contact methods have been developed for surfaces which would be distorted by the physical contact with the instrument.

11.3.1 Contact or mechanical methods

In contact methods, a device is placed in contact with the surface, and some sort of distortion of the device gives a measure of the curvature.

11.3.1.1 Spherometer

The spherometer principle is shown in Figure 11.8 with a cross-section through the instrument only being illustrated. The spherometer is placed in contact with the surface and the middle pin is moved up or down, until it just touches the surface. The distance s this pin has been moved (the sag) from the contact plane is a measure of the surface radius of curvature. Applying Pythagoras' theorem and using the symbols in the diagram, the radius of curvature r of the test surface is given by the equation

$$r = (r_b^2 + s^2)/(2s) \qquad (11.9)$$

In some simple spherometers, the base is formed from three fixed pins at the apex of an equilateral triangle. The fourth moveable pin is at the centre of the triangle. More accurate instruments are more robust, have accurately machined pins and a finer scale on the central pin movement. More accurate physical constructions will be described below.

Equation (11.9) assumes that the ends of the pins have zero thickness, and the finite thickness of real pins or bases can lead to errors in the calculated measures of the radius of curvature. For example, if the end of the pins or base is flat as shown in Figure 11.9, a convex surface will make contact with the inside edge of the pins and a concave surface will make contact with the outer edge, as shown in this diagram. Thus in this case, the effective radius r_b of the spherometer base will depend upon whether the test surface is convex or concave.

Fig. 11.9: Spherometer with flat base or feet.

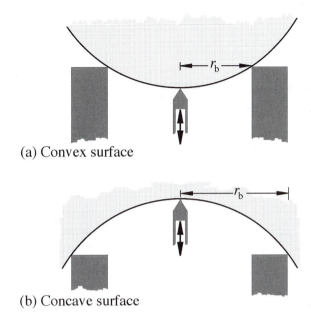

(a) Convex surface

(b) Concave surface

Fig. 11.10: Spherometer with small spheres on the base.

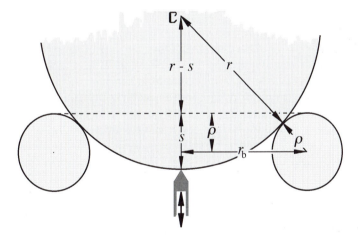

While machining an accurate flat edge on the spherometer base can increase its accuracy, the 90° edge may wear and induce errors in the final calculated value of the radius of the test surface. This problem can be overcome by using spheres to form the base of the spherometer, as shown in Figure 11.10. Now the effective radius of the base depends upon the radius of these spheres. In the diagram, we have the example of a convex surface and from Pythagoras' theorem

$$(r + \rho)^2 = r_b^2 + (r - s + \rho)^2$$

Expanding and simplifying give

$$r = \frac{r_b^2 + s^2 - 2s\rho}{2s} \quad \text{(convex surface)} \qquad (11.10a)$$

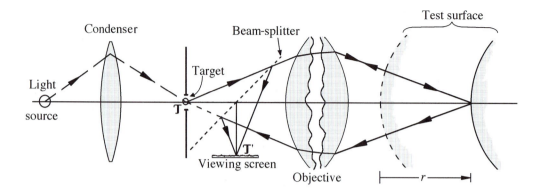

where ρ is the radius of the small spheres and r_b is the spherometer base radius measured to the centre of these spheres. Similarly for a concave surface, it is easily shown that

$$r = \frac{r_b^2 + s^2 + 2s\rho}{2s} \quad \text{(concave surface)} \qquad (11.10b)$$

Fig. 11.11: The Drysdale method of measuring the radius of curvature of a spherical surface.

Geneva lens measure

The spherometers described above can only be used to measure the curvature of spherical surfaces. They are of little use for measuring toric or aspheric surfaces. Since many ophthalmic lens surfaces are toric, a special version of the spherometer is available to measure toric curvatures. This instrument is called a **geneva lens measure**. This instrument is essentially a two dimensional version of the spherometer. It is specifically designed for ophthalmic optics and the scale is calibrated for surface power and not surface radius of curvature. Therefore there must be some assumed refractive index in the construction of the power scale.

11.3.2   Non-contact or optical methods

These methods have been devised for situations where it is impossible or undesirable to make physical contact with the surface, for example when physical contact would distort the surface and thus lead to inaccuracies. These non-contact methods are usually based upon a variety of optical principles.

11.3.2.1   Drysdale method

In the Drysdale method, an object or target is imaged towards the surface to be measured and the reflected beam observed. The method is based upon the principle that if the target is either imaged on the surface vertex or at the centre of curvature as shown in Figure 11.11, the rays are retro-reflected, and the image of the object or target is formed back on itself. This diagram shows a typical arrangement and the main imaging lens should be free of spherical aberration. The surface to be measured is moved backwards or forwards until the two retro-reflecting positions have been located. It can be seen from the

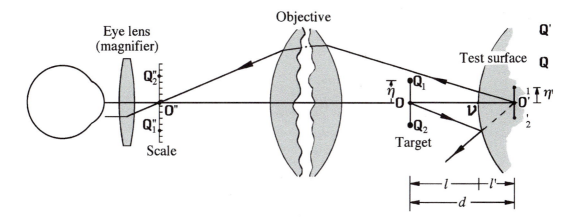

Fig. 11.12: Practical application of the keratometer principle.

above diagram that the displacement between these two positions is simply the radius of curvature r of the surface.

The Drysdale principle is usually used to measure the radius of curvature of contact lenses.

11.3.2.2   The keratometer method

The Drysdale method is very simple but cannot be readily used unless the surface position is very carefully controlled. This can be done with ophthalmic lenses but cannot with the living eye. Measurement of the radius of curvature of the cornea of the eye requires a different method, one that is not sensitive to uncontrolled movements of the surface.

The keratometer principle is shown in Figure 11.12. A target of known size η is imaged by reflection into the surface being measured. In the example shown in the diagram, the surface is convex with a radius of curvature r and a virtual image is formed inside the surface. If one applies the lens equation to this problem, the size η' of the image can be found as follows. Firstly, we recall the mirror equation (7.3),

$$\frac{1}{l'} + \frac{1}{l} = \frac{2}{r}$$

with the restriction

$$l' - l = d$$

The transverse magnification is given by equation (7.8a), that is

$$M = \eta'/\eta$$

and for a reflecting surface is given in terms of the object and image distances by equation (7.9), that is

$$M = -l'/l$$

Using these equations and solving for the radius of curvature r lead to the equation

$$r = \frac{2dM}{(1 - M^2)} \tag{11.11}$$

If the object distance l is large compared to the radius r then M is small and r can be approximated by the equation

$$r = 2dM \tag{11.11a}$$

Now since the image formed inside the convex surface is virtual, its size cannot be measured directly. To get around this problem, the image is relayed onto a scale by another lens as shown in the diagram, and this secondary image and the scale may be viewed through a magnifying lens as shown in Figure 11.12. Providing the magnification of this relay system is known and all distances are kept constant, the scale can be calibrated to give a direct reading of surface curvature or radius of curvature.

11.4 Powers, cardinal points and focal lengths

The determination of powers and focal lengths on one hand and the positions of cardinal points on the other do not always require different methods. Very often determination of a power or focal length automatically leads to the location of the cardinal points, and vice versa. There are only a few exceptions to this rule. There are a wide range of methods that are available to find the equivalent power and locate the cardinal points, and too many to cover here. Therefore only the most common and easiest to set up will be discussed.

It will become clear, as we explain different methods of measuring the powers and cardinal point positions, that most of the methods are only immediately suitable for positive power lenses. If they are to be used for negative power lenses, some modification has to be applied to allow the measurement of the sizes of virtual images. This can be done by using a positive power auxiliary lens to project any virtual image onto a real image surface.

11.4.1 Equivalent power or equivalent focal length

11.4.1.1 Two conjugate methods (in air)

These methods rely on the measurement of the transverse magnification M of the optical system, for two positions of the conjugate planes, as shown in Figure 11.13. Let these magnifications be M_1 and M_2, respectively, where

$$M_1 = \eta_1'/\eta \quad \text{and} \quad M_2 = \eta_2'/\eta \tag{11.12}$$

If we measure the shift in the conjugate planes (Alternative A), we can find the equivalent power only. On the other hand, if we measure the distance between the object and image planes (Alternative B) we can find the equivalent power and the separation of the principal planes.

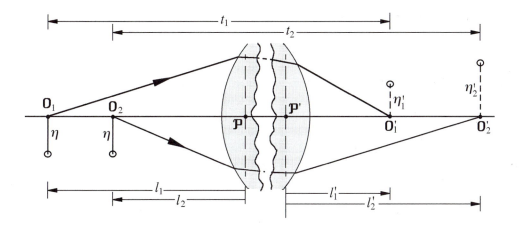

Fig. 11.13: The two-conjugate method for determining equivalent power of an optical system.

Alternative A: Equivalent power only

If we measure the shifts of the object and image planes, that is the displacements

$$l_1 - l_2, \quad l_2' - l_1'$$

we can use equations (3.51a and b) to express the equivalent power F in terms of these shifts and the above magnifications by the equations

$$F = \frac{(M_1 - M_2)}{(l_2' - l_1')} \tag{11.13a}$$

$$F = \frac{(1/M_1 - 1/M_2)}{(l_1 - l_2)} \tag{11.13b}$$

Since this method requires both a real object and a real image, it is only applicable to positive power lenses and therefore the above equations should give a positive value. It is clear that although the above equations involve distances from the principal planes, the positions of these planes do not have to be known, since the equations only involve shifts in object or image planes.

Alternative B: Equivalent power and separation of the principal planes

In method A above, only the shift or displacement of the conjugate planes is measured. However if we measure the distance (t) between the object and image planes, a distance we are calling the **throw**, we can also find the separation of the principal planes, that is the distance $\mathcal{PP}'$. Equations for these quantities can be found as follows.

In Figure 11.13, for either pair of conjugate planes, we have

$$l' - l + \mathcal{PP}' = t \tag{11.14}$$

From equations (3.50a and b) with $n = n' = 1$, we have

$$1 - M = l'F \quad \text{and} \quad 1/M - 1 = lF$$

If we now use these to eliminate l and l' from equation (11.14), we have

$$[(2 - M - 1/M)/F] + \mathcal{PP}' = t \tag{11.15}$$

As before, if we know the magnifications (M_1 and M_2) for two different pairs of conjugate planes and measure the separations t_1 and t_2 of the conjugate planes, we have

$$[(2 - M_1 - 1/M_1)/F] + \mathcal{PP}' = t_1 \tag{11.16a}$$

$$[(2 - M_2 - 1/M_2)/F] + \mathcal{PP}' = t_2 \tag{11.16b}$$

If we now eliminate the distance $\mathcal{PP}'$ from these two equations and solve for the equivalent power F, we have

$$F = [(M_1 - M_2) - (1/M_2 - 1/M_1)]/(t_1 - t_2) \tag{11.17}$$

The separation $\mathcal{PP}'$ of the principal planes can now be found by substituting the now known value of F in either equation (11.16a or b). For example, from equation (11.16a)

$$\mathcal{PP}' = t_1 - [(2 - M_1 - 1/M_1)/F] \tag{11.18}$$

As a check, we should use more than two sets of conjugate positions. These will give a number of estimates of the measured quantities and the spread of these values will give an estimate of the uncertainty in the mean value. Thus these methods can be very time consuming. There is an alternative method that is much quicker, but relies on the availability of a second lens of known equivalent power or focal length. This is known as the foco-collimator method.

11.4.1.2 The foco-collimator method

The arrangement of the foco-collimator is shown in Figure 11.14. A target of known size η is collimated by a lens of known equivalent power F_c. The lens of unknown power F is placed after the collimator as shown, close to it but the distance apart is not critical. The target is imaged in the back focal plane at $\mathcal{F}'$ of the unknown lens as shown. The image height or target size η', which can be easily measured, is related to other quantities as follows:

$$\theta = \eta/f_c = \eta'/f$$

Therefore

$$\frac{\eta'}{\eta} = -\frac{F_c}{F}$$

Fig. 11.14: The foco-collimator method for measuring equivalent power.

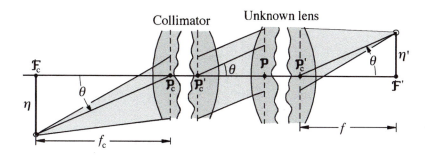

that is

$$F = -\eta F_c/\eta' \tag{11.19}$$

Therefore for a fixed set-up, η and F_c will be known and therefore ηF_c is a constant and we only need to measure the image size η' in order to find the value of F.

11.4.2 Vertex powers and vertex focal lengths

In ophthalmic optics, it is more common to specify a lens in terms of its vertex powers, rather than its equivalent power. However it should not be forgotten that if a lens is specified in terms of vertex powers, other information has to be added to make the specification complete. Usually, it is sufficient to add one surface power, the refractive index and the lens thickness to the specification. Let us look at methods for measuring the vertex powers or focal lengths.

11.4.2.1 Given a collimated beam (or a distant bright object or target)

The measurement of vertex power is much simpler than the measurement of equivalent power. For positive power lenses, the simplest method requires a collimated or distant source or target. If the test lens is placed in the beam, the target is imaged in the back focal plane $\mathcal{F}'$ of the lens. It is very easy to measure the distance from the surface vertex to the back focal plane, that is, the back vertex focal length f'_v.

11.4.2.2 The auto-collimation method

The vertex powers, focal lengths or positions of the focal points can be found to a higher accuracy using the auto-collimation arrangement shown schematically in Figure 11.15. Once again, this method is only suitable for the measurement of positive power lenses. The real image of the source via the condenser is on the object side of the lens. When it is at the front focal point, the beam in image space is collimated. A mirror placed on the image side reflects the beam backwards towards the lens and an image of the source is formed back on itself. Various techniques can be used to improve the visibility of the return image. For example, an illuminated aperture or target can be used as the source as shown in the diagram and the returning image is displaced sideways by tilting the mirror

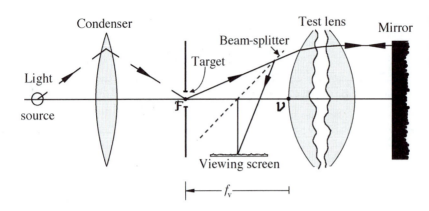

Fig. 11.15:
Auto-collimation method
for determining vertex
powers and focal lengths.

and using a white diffuse receiving surface on the screen containing the small
hole source. Alternatively, the returning beam can be reflected sideways using
a beam-splitter, as shown in the diagram. The viewing screen must be in a plane
conjugate to the target. The vertex focal distance f_v can then be measured.

11.4.2.3   The focimeter principle

The focimeter method is one of the few that are equally suitable for negative
as well as positive vertex power lenses. The schematic arrangement of the
focimeter is shown in Figure 11.16. A target at T is initially placed at the front
focal point $\mathcal{F}_c$ of the collimator lens. If there is no test lens in place, the image of
this target is seen in focus through the telescope. If a lens is now placed in front
of the telescope, it will de-collimate the beam and the target will now be out
of focus in the telescope. The target can be refocussed by moving it along the
optical axis of the collimator. If the target has to be moved a distance x towards
or away from the collimator to refocus it, the image will be a distance x' from
the back focal point of the collimating lens. Now recalling Newton's equation
[Equation (3.66)], in the notation used here, the distances x and x' are related
by the equation

$$xx' = -1/F_c^2 \tag{11.20}$$

If the lens is placed so that its front vertex point coincides with the back focal
point $\mathcal{F}'_c$ of the collimator, the distance x' is equal to the front focal length of
the test lens, that is

$$x' = f_v$$

If we now use equation (3.61a) with $n = 1$, we can write

$$x' = -1/F_v$$

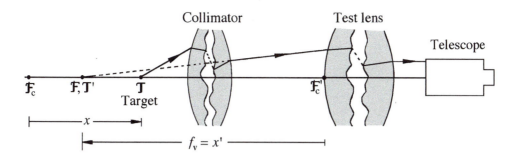

Fig. 11.16:   The focimeter.

Combining this equation with equation (11.20) gives the equation

$$F_v = xF_c^2 \tag{11.21}$$

This equation shows that the vertex power is related to the target displacement x on a linear scale, which because of its linearity can be easily calibrated.

11.4.3   Cardinal points

11.4.3.1   Focal points

The back and front focal points can be located using the same method as for the measurement of vertex powers. In fact the determination of vertex powers or focal lengths automatically determines the positions of the focal points. For example, if the vertex powers F_v and F_v' are found, the distances $\mathcal{VF}$ and $\mathcal{V}'\mathcal{F}'$ are found from equations given in Section 3.6.8, that is

$$\mathcal{VF} = -1/F_v \quad \text{and} \quad \mathcal{V}'\mathcal{F}' = 1/F_v' \quad \text{(in air)} \tag{11.22}$$

11.4.3.2   Principal points

The principal points can be located if the equivalent and vertex powers or focal lengths are known. For example referring to Figure 3.8

$$\mathcal{P}'\mathcal{V}' = \mathcal{P}'\mathcal{F}' - \mathcal{V}'\mathcal{F}' = f - f_v' \tag{11.23a}$$

and

$$\mathcal{VP} = \mathcal{FP} - \mathcal{FV} = f + f_v \tag{11.23b}$$

Alternatively, they can be located indirectly by first finding the positions of the nodal points by a method described in Section 11.4.3.3. The distance between the principal and nodal points is related by equation (3.58), that is

$$\mathcal{P}'\mathcal{N}' = \mathcal{PN} = \frac{(n' - n)}{F} \tag{11.24}$$

If the refractive indices in object space (n) and image space (n') are equal, then the principal and nodal points coincide.

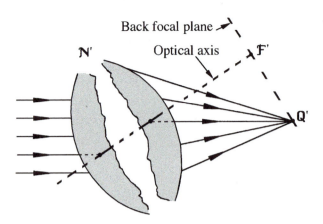

Fig. 11.17: Principle of the nodal slide.

11.4.3.3 Nodal points

The nodal points can be located using the **nodal slide**, which is a special type of lens mount that allows the system to be rotated about a vertical axis passing through any point along the optical axis.

If the lens on the nodal slide is placed in a collimated beam, as shown in Figure 11.17, the beam will be focussed in the back focal plane of the lens. If the lens is rotated about a vertical axis, the image Q' will move across the back focal plane. However, there is one position of the axis of rotation where the image will be stationary and that is when this axis passes through the back nodal point of the lens. If the lens rotates about any other point, the point Q' will move. In practice, the lens is moved backwards and forwards along the nodal slide and rotated in both directions by a small amount at each position. By trial and error the back nodal point can be located.

It should be remembered that the nodal points are paraxial quantities and thus only exist in an aberration free region. Therefore if the nodal slide is rotated too far, the collimated beam entering the lens will be sufficiently far enough off-axis to develop aberrations. Thus the nodal points should be found by using only small rotations.

Exercises and problems

11.1   Calculate the radius of curvature of a surface from the following two pin sphero-meter measurements: sag = 1 mm, pin separation = 15 mm.

ANSWER: radius = 38 mm

11.2   In a certain optical set-up, if the transverse magnifications are −2.2 and −3.0 for two positions of the conjugates and the distance between image planes for these two positions is 20 mm, calculate the equivalent focal length of the optical system.

ANSWER: $f = 25$ mm

Summary of main symbols

Sections 11.1 and 11.2: Refractive index

r angle of refraction
d lens thickness

i	angle of incidence
i^*	final angle of refraction for grazing incidence at a prism
ψ	angular diameter of the rainbow
F_c	equivalent power of a collimator

Section 11.3: Surface curvature

d	distance between object and image
r_b	radius of spherometer base
ρ	radius of small spheres on base of spherometer
s	spherometer sag

References and bibliography

Bhattacharya J.C. (1989). Measurement of the refractive index using the Talbot effect and a moire technique. *Appl. Opt.* 28(13), 2600–2604.

Diaz-Uribe R., Pedraza-Contreras J., Cardonna-Nunez O., Cordero-Davila A., and Cornejo-Rodriguez A. (1986). Cylindrical lenses: testing and radius of curvature measurements. *Appl. Opt.* 25(10), 1707–1709.

Handbook of Chemistry and Physics. 56 ed. (1975–1976). CRC Press, Boca Raton, Fla., p. E219.

Longhurst R.S. (1967). *Geometrical and Physical Optics*. 2nd ed. Longman, London.

*Smith G. (1977a). Measurement of the refractive index of glass beads for traffic marking. *Aust. Road Res.* 7(1), 26–31.

*Smith G. (1977b). The rainbow method for the measurement of the refractive index of glass beads. *Traffic Engineering and Control* 18(10), 1–3.

*Smith G. (1982). Liquid immersion method for the measurement of the refractive index of a simple lens. *Appl. Opt.* 21, 755–757.

*Wahlstrom E.E. (1969). *Optical Crystallography*, 4th ed. Wiley, New York.

Wyant J.C. and Smith F.D. (1975). Interferometer for measuring power distribution of ophthalmic lenses. *Appl. Opt.* 14(7), 1607–1612.

*Yamaguchi T. (1975). Refractive index measurement of high refractive index glass beads. *Appl. Opt.* 14(5), 1111–1115.

12

Photometry of optical systems

12.0   Introduction

Ray tracing, with either paraxial or real rays, is insufficient to assess the efficiency of optical systems or indicate the quality of the final image. Apart from the effect of aberrations, it is also necessary to understand basic photometric principles, photometric quantities such as source luminance and surface illuminance, and a number of other factors such as vignetting which affect the image plane illuminance in optical systems.

Before beginning a study of the photometry of optical systems, it is necessary to understand the four basic photometric quantities, namely luminous flux, luminous intensity, luminance and illuminance. For a long time, photometry was regarded as an independent field of study with its own fundamental units and international standards. For example, luminous intensity has been a fundamental basic physical quantity for many years with a physical standard using a sample of thorium oxide held at the melting point of platinum (2042 K) as the light source (Sanders and Jones 1962). This light source had a defined luminance of 60×10^4 cd/m². More recently, however, there has been a trend to regard photometry as a branch of radiometry and derive all photometric quantities from radiometric quantities.

Radiometry may be defined as the measurement of the energy or power in an electromagnetic beam, measured over the entire spectrum. The division of the entire electromagnetic spectrum is shown in Figure 1.2, Chapter 1. Light is only a very small part of this spectrum.

12.0.1   *The nature of light*

Light is that part of the electromagnetic spectrum that elicits a visual response in the eye and is in the range of approximately 400–700 nm. The eye is not equally responsive to electromagnetic energy in this range. The relative visual response is almost zero at the limiting values, and the spectral response curve is "bell" shaped. The spectral response function depends upon light level, with the two extremes of light level being called the **relative photopic luminous efficiency function** for high light levels and denoted by $V(\lambda)$ and the **relative scotopic luminous efficiency function** for low light level and denoted by $V'(\lambda)$.

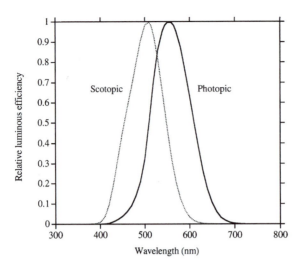

Fig. 12.1: Photopic
$[V(\lambda)]$ and scotopic $[V'(\lambda)]$
relative luminous efficiency
functions.

These functions are shown in Figure 12.1. In this book, unless otherwise specified, the photopic case will always be assumed. The values of these luminous
efficiency functions are set by the Commission Internationale de L'Eclairage
(CIE), Publication 18.2 (1983).

Radiometry and photometry

If a beam of electromagnetic energy has a spectral power density denoted by
$P(\lambda)$ then the amount of **radiant flux** or power F_R in the beam is given by the
integral

$$F_R = \int_0^\infty P(\lambda)\, d\lambda \quad \text{watt} \tag{12.1}$$

where the quantity $P(\lambda)$ has units of watt/unit of wavelength. The measurement
of radiant flux is very difficult in practice because no real detector has a constant
response over the entire spectrum. For any single real detector with a spectral
sensitivity $S(\lambda)$, the recorded radiant flux F_R will be

$$F_R = \int_0^\infty P(\lambda)S(\lambda)\, d\lambda \tag{12.2}$$

So in practice a number of separate detectors usually have to be used to cover
the spectral range $P(\lambda)$ of the beam, with calibration factors used to account
for the non-uniform spectral sensitivity $S(\lambda)$ of each detector.

Photometry, being the study and measurement of light, can be regarded as
a special case of radiometry, but using a special detector which has the same
spectral sensitivity as the eye, that is $S(\lambda)$ is proportional to the $V(\lambda)$ curve.

12.1   Photometry: Quantities, units and levels

There are four fundamental photometric quantities. We will introduce them in
this section along with their units and give some typical photometric levels.

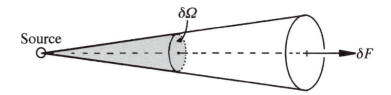

Fig. 12.2: Definition of luminous intensity.

12.1.1 Luminous flux (F)

A detector that measures light must have the same spectral response as the eye, that is its spectral sensitivity should be proportional to the relative photopic spectral luminous efficiency function $V(\lambda)$ of the eye. The luminous flux F in a beam whose spectral power density is $P(\lambda)$ can be defined mathematically as

$$F = K_\mathrm{m} \int_0^\infty P(\lambda) V(\lambda) \, \mathrm{d}\lambda \quad \text{lumen (lm)} \tag{12.3}$$

where the constant K_m is known as the photopic luminous efficiency constant or the maximum spectral luminous efficiency of radiation for photopic vision. It has a value of 683.002 lm/W [CIE Publication 18.2 (1983)]. Since the $V(\lambda)$ function is effectively zero outside the range 380 to 780 nm, we can write the above integral as

$$F = K_\mathrm{m} \int_{380}^{780} P(\lambda) V(\lambda) \, \mathrm{d}\lambda \quad \text{lumen (lm)} \tag{12.3a}$$

These integrals convert radiant flux in watts to luminous flux in lumens.

Example levels

A 60 W light bulb that is commonly used for lighting in the home will emit about 600 lumens.

12.1.2 Luminous intensity (I)

Few sources radiate isotropically and instead radiate more luminous flux in some directions and less in others. Two sources may emit the same luminous flux but have very different spatial distributions of this flux.

For any light source, it may be more important to know the luminous flux density emitted in a given direction rather than the total luminous flux emitted. For example a car headlight is designed to direct more luminous flux straight ahead than to the side and therefore is a very directional light source. Luminous intensity is a measure of this flux density and is defined as the ratio of the luminous flux (δF) contained in an infinitesimally narrow cone of solid angle ($\delta \Omega$) in a given direction, as shown in Figure 12.2. Mathematically one can define luminous intensity in the form

$$I = \delta F / \delta \Omega \quad \text{lumen/steradian or candela (cd)} \tag{12.4}$$

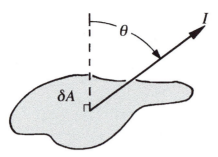

Fig. 12.3: Relationship between luminous intensity and luminance.

In practice luminous intensity is only used for sources of small angular subtense (see Section 12.2.1).

Example levels

The luminous intensity of a common candle is about 1 cd and this is the origin of the word "candela" as the candle was used as a crude standard source of light for many years. Traffic signal lights have luminous intensities on axis of about 200 to about 600 cd, but may be higher. A car headlight on full beam can have a maximum luminous intensity of about 20,000 cd.

12.1.3   Luminance (L)

Luminance is the objective measure of the subjective "brightness" of an extended source. For a small element of an extended source, it can be related to the luminous intensity of the element and a direction. Referring to Figure 12.3, if the source is a small plane element of area δA of luminance $L(\theta)$ in a direction θ, the luminous intensity $I(\theta)$ in that direction is given by the equation

$$I(\theta) = L(\theta)\, \delta A\, \cos(\theta)$$

that is

$$L(\theta) = \frac{I(\theta)}{\delta A \cos(\theta)} \quad \text{candela/metre}^2 (\text{cd/m}^2) \tag{12.5}$$

Some sources have the same "brightness" or luminance in different directions. These sources are called **Lambertian**.

Many sources of light such as the sky are effective sources because they reflect or scatter incident light. In the case of a reflecting source, if the surface is a **perfect diffuser**, it will act as a Lambertian source. Materials that come close to being perfect diffusers are magnesium oxide, soot, and barium sulphate. Roughened chalk (calcium carbonate) is also close to being a perfect diffuser. A perfectly diffusing surface is one which scatters incident light equally in all directions such that the surface luminance is the same for all directions. Thus for a Lambertian source or surface

$$L(\theta) = \text{constant} = L$$

and the luminous intensity is given by the equation

$$I(\theta) = L \, \delta A \cos(\theta) \tag{12.6a}$$

For normal viewing $\theta = 0°$ and hence

$$I = L \, \delta A (\text{normal viewing}) \tag{12.6b}$$

The opposite of a perfect diffuser is a **perfect specular** surface. In this case all the incident light is reflected according to Snell's law; the surface is smooth and is a perfect mirror. In practice, all surfaces have some specular and diffusing properties and hence lie somewhere between these two extremes.

Example levels

The sun is the main source of light and the luminance of the disc as seen from the earth's surface depends upon the elevation of the sun above the horizon and the amount of materials such as water vapour and dust in the atmosphere that scatter and absorb sunlight. A clean atmosphere also scatters and absorbs sunlight with the scatter strongly wavelength dependent and this is the cause of the blue sky and reddish sun at sunset. In a clean atmosphere, the luminance of the sun, with an elevation of greater than about $30°$, is about 2×10^9 cd/m^2. Under similar conditions, the luminance of the moon is about 2000 cd/m^2.

12.1.4 Illuminance (E)

Illuminance is a measure of the luminous flux density incident on a surface. Since it can vary over a surface it is best defined in terms of small elements of area δA. Thus

$$E = \delta F / \delta A \quad \text{lumens/metre}^2 (\text{lm/m}^2 \text{ or lux}) \tag{12.7}$$

Example levels

The solar illuminance at the earth's surface also depends upon solar elevation and the state of the atmosphere. On a bright day with a clean atmosphere, the illuminance can be about 50,000 lux. The illuminance in an office or library at a work table or desk will be in the region of about 200 to about 1000 lux.

12.1.5 SI and non-SI photometric units

In the previous section, the units given with each photometric quantity are the units specified by Système International d'Unités or SI units. A variety of other units exist and while they should be discouraged, they appear in older literature and therefore we need to be able to convert from those units to the SI units. These older units are listed in Table 12.1 along with the conversion to SI units.

Table 12.1. SI and non-SI units and their conversions

SI units

Quantity		Unit
Luminous flux	lumen	lm
Luminous intensity	candela	cd lm/st
Luminance	cd/m^2	$lm/st/m^2$
Illuminance	lm/m^2	lux

Non-SI units

Luminous intensity
  candlepower = the candela

Unit

Luminance		cd/m^2
1 stilb	$= 1\ cd/cm^2$	$= 10{,}000.0$
1 apostilb	$= (1/\pi)\ cd/m^2$	$= 0.3183$
1 lambert	$= (10{,}000/\pi)\ cd/m^2$	$= 3{,}183.0$
1 millilambert	$= 0.001$ lambert	$= 3.183$
1 footlambert	$= (1/\pi)\ cd/ft^2$	$= 3.426$
1 candela/ft^2	$=$	$= 10.76$
1 candela/in^2	$=$	$= 1{,}550.0$

Illuminance	
1 footcandle	$= 1\ lm/ft^2 = 10.76\ lm/m^2$ or lux

Note: st = steradian (angular measure of solid angle).

12.2   Some relations between photometric quantities

12.2.1   *Luminous intensity and illuminance: The inverse square law*

For a point source with a luminous intensity I, the luminous flux δF contained in a cone of solid angle $\delta\Omega$ is related to I by the definition of I embodied in equation (12.4), here written in the form

$$\delta F = I\,\delta\Omega$$

From the definition of solid angle Ω, the area δA of the curved base of this cone at a distance d from the cone vertex is

$$\delta A = d^2\,\delta\Omega$$

Therefore the illuminance on this surface is, from equation (12.7),

$$E = \frac{\delta F}{\delta A}$$

By making appropriate substitutions for δF and δA from above, we have

Fig. 12.4: Illuminance at
$\mathcal{G}$ due to a circular
Lambertian source.

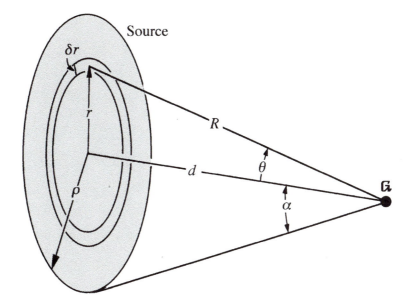

$$E = \frac{I\,\delta\Omega}{\delta A} = \frac{I\,\delta\Omega}{d^2\,\delta\Omega}$$

which reduces to

$$E = I/d^2 \tag{12.8}$$

This equation shows that for a surface illuminated by a point source and per-pendicular to the source direction, the illuminance decreases as the inverse of the square of the distance from the source, an inverse square law. If the surface is tilted by an angle θ to this source direction, then

$$E(\theta) = I\cos(\theta)/d^2 \tag{12.8a}$$

However no source is actually a point source. All have some surface area and in practice it is often very necessary to know how the surface illuminance depends upon the distance from an extended source. An idea of the accuracy of the inverse square law when applied to sources of finite area can be found by examining the illuminance along the axis as a function of distance from some simple sources, e.g. a circular Lambertian source.

12.2.2 *Luminance and illuminance: Circular Lambertian source*

Figure 12.4 shows a circular source and a point $\mathcal{G}$ on the axis. Let us find an equation for the illuminance at $\mathcal{G}$ assuming the source is Lambertian. We divide up the source into a series of annular zones. For a representative zone as shown, the illuminance δE at $\mathcal{G}$ in a plane normal to the axis is the sum of the illuminance contributions from all points on this annulus. Since they are all equal

$$\delta E = I\cos(\theta)/R^2$$

Now from equation (12.6a),

$$I = L \, \delta A \cos(\theta)$$

Therefore

$$I = \frac{L \, \delta A \cos^2(\theta)}{R^2}$$

Since $\delta A = 2\pi r \, \delta r$, we have

$$E = 2\pi L \int_0^\rho \frac{\cos^2(\theta)}{R^2} r \, dr$$

Also $R = d / \cos(\theta)$ and $r = d \tan(\theta)$ and after these substitutions, the above integral can be evaluated and reduced to

$$E = \pi L \sin^2(\alpha) \tag{12.9}$$

12.2.3 Accuracy of the inverse square law

If we now assume that the inverse square law applies at the point g in Figure 12.4, it follows that

$$E = \frac{I}{d^2} = \frac{LA}{d^2} = \frac{L\pi\rho^2}{d^2}$$

Now, since $\tan(\alpha) = \rho/d$, we have

$$E = \pi L \tan^2(\alpha) \tag{12.10}$$

if the inverse square law applies.

If we define the fractional error ϵ in applying the inverse square law as

$$\epsilon = \frac{\text{value predicted by inverse square law} - \text{exact value}}{\text{exact value}}$$

using equation (12.9) for the exact value and equation (12.10) for the inverse square law prediction, it follows that

$$\epsilon = \frac{\tan^2(\alpha) - \sin^2(\alpha)}{\sin^2(\alpha)}$$

This equation reduces to

$$\epsilon = \tan^2(\alpha) \tag{12.11}$$

Thus if the inverse square law is to be applied with a required accuracy of less than 1% error, it can be applied when

$$\epsilon = \tan^2(\alpha) < 0.01$$

That is

$$\epsilon < 5.7°$$

for a circular Lambertian source. In contrast, it can be shown that the illuminance due to a spherical Lambertian source obeys the inverse square law for all distances.

In the region of practical validity of the inverse square law, the source can be described by its luminous intensity. Outside this region, luminous intensity is no longer a valid descriptor. Instead the source must be described by its luminance and size. If the source is plane, circular and Lambertian, equation (12.9) must be used to predict illuminance along the axis. Other shaped sources and values at points off the source axis would require different equations.

12.3 Luminous flux collected by an optical system

We will now derive equations that give the luminous flux accepted by an optical system and will see that the pupils play an important role in this process. To keep the discussion as simple as possible, we will assume Gaussian optics operate and this means that there will be no pupil aberrations, that is the actual pupil will always be the Gaussian pupil. In real optical systems, pupil aberrations distort the size and shape of the pupil which in turn affect the luminous flux accepted particularly from off-axis areas of the object.

12.3.1 Small area Lambertian axial sources

12.3.1.1 Sources at a finite distance

Referring to Figure 12.5 a small plane source of area δA is on the optical axis and normal to the axis. Let its luminance be L. The luminous flux accepted by the system is equal to that which enters the entrance pupil. To find an expression for this flux, the entrance pupil plane is divided into a series of annular zones as shown in the diagram. The luminous flux δF accepted by one of these zones is given as

$$\delta F = I(\theta)\delta\Omega$$

where

$$I(\theta) = L\,\delta A \cos(\theta)$$

and

$$\delta\Omega = 2\pi r\,\delta r \cos(\theta)/R^2$$

Thus

$$F = 2\pi L\,\delta A \int_0^{\bar{\rho}} \frac{\cos^2(\theta)}{R^2} r\,\mathrm{d}r \qquad (12.12)$$

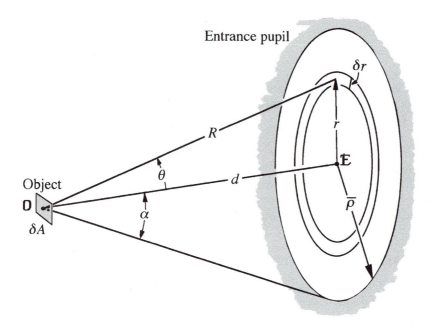

Fig. 12.5: Luminous flux
accepted by an optical
system from small area
source on-axis.

If we make the substitutions

$$r = d \tan(\theta) \quad \text{with} \quad \mathrm{d}r/\mathrm{d}\theta = d \sec^2(\theta)$$

and

$$R = d/\cos(\theta)$$

the integral equation (12.12) reduces to

$$F = \pi L \, \delta A \sin^2(\alpha) \tag{12.13a}$$

If the source is very small, it may be more convenient to express this equation in terms of the luminous intensity I of the source rather than the luminance L and area δA. Since $I = L \, \delta A$ from equation (12.6b), an alternative form to equation (12.13a) is

$$F = \pi I \sin^2(\alpha) \tag{12.13b}$$

As a further alternative, we can express this equation in terms of the source distance d and diameter D of the entrance pupil. For small values of the angle α, from Figure 12.5, it follows that

$$\sin(\alpha) \approx \tan(\alpha) = \bar{\rho}/d = D/(2d)$$

where $\bar{\rho}$ is the radius of the pupil. Within this approximation, we can now write equation (12.13b) in the form

$$F = \pi I [D/(2d)]^2 \tag{12.13c}$$

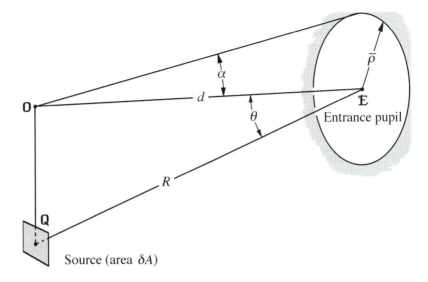

Fig. 12.6: Luminous flux accepted by an optical system from small area source off-axis.

12.3.2 *Small area Lambertian off-axis source and small pupil*

If the source is now off-axis as shown in Figure 12.6, the pupil is small and the luminous intensity in the direction of the pupil is I, then the luminous flux F accepted by the pupil can be written as

$$F = I\Omega$$

In this case

$$I = L\, \delta A \cos(\theta)$$

where Ω is the solid angle the pupil subtended at the off-axis point and is given by the equation

$$\Omega = \pi \bar{\rho}^2 \cos(\theta)/R^2$$

Thus

$$F = \pi L\, \delta A \bar{\rho}^2 \cos^2(\theta)/R^2$$

Now since $d/R = \cos(\theta)$, the flux F accepted can be written as

$$F = \pi L\, \delta A \bar{\rho}^2 \cos^4(\theta)/d^2$$

If $\bar{\rho}/d$ is replaced by $\tan(\alpha)$ then

$$F = \pi L\, \delta A \tan^2(\alpha) \cos^4(\theta) \qquad (12.14)$$

However, for small pupils or pupils of small angular subtense to the source, the angle α is small and $\sin(\alpha) \approx \tan(\alpha)$. Hence the above equation can be written

$$F \approx \pi L \, \delta A \sin^2(\alpha) \cos^4(\theta) \tag{12.15}$$

which is similar to the axial case equation (12.13a) with the addition of the extra $\cos^4(\theta)$ factor.

12.4   Transmittance of a system

Not all of the luminous flux accepted by or entering the system reaches the image plane, even if there is no vignetting. When a beam of light passes through an optical system, light is lost from the direct beam by reflection, scatter or absorption. The losses due to reflections are easily predicted and we will give equations for their calculation. In contrast, the losses due to scatter and absorption are not so readily calculated so we will only briefly discuss their origins and not give any equations for their calculation.

12.4.1   *Reflectance and transmittance at a surface*

At any boundary between media of different refractive indices some energy is reflected but at an air/glass interface most is transmitted. The fraction that is reflected and transmitted at a single interface can be found by applying the Fresnel equations, which for normal incidence are

$$R = \frac{(n' - n)^2}{(n' + n)^2} \quad \text{reflectance} \tag{12.16a}$$

and

$$T = \frac{4n'n}{(n' + n)^2} \quad \text{transmittance} \tag{12.16b}$$

where n' and n are the refractive indices of the two media. Thus, assuming the only losses are by reflection, the transmittance $T(k)$, for a system of k identical surfaces, is given by the equation

$$T(k) = T^k \tag{12.17}$$

Example 12.1: Calculate the surface reflection at an air/glass interface where the glass is Schott glass BK 7, for a single lens and a system of 10 identical surfaces.

Solution: From the Schott (1992) glass catalog, the refractive index of BK 7 glass in the middle of the visible spectrum is very close to 1.5187. Substituting $n = 1$ and $n' = 1.5187$ into equations (12.16a and b) we have

$$R = 0.042 \quad \text{and} \quad T = 0.958$$

For a lens, there are two air/glass surfaces and so the fraction of light transmitted is, from equation (12.17),

$$T(2) = 0.958^2 = 0.918$$

For an optical system with 10 air/glass surfaces, the fraction of light transmitted is

$$T(10) = 0.958^{10} = 0.651$$

Thus in this case, about 35% of the light is lost.

Some of this reflected light is not completely lost from the system. A fraction is multiply reflected and the fraction that is reflected an even number of times finally reaches the image plane. This reflected light can sometimes produce a ghost image or an image of the aperture stop or a spread out veiling glare illuminance. If these effects are severe, the image quality is reduced. **Veiling glare** illuminance acts like a fog and reduces the contrast of the image. This is discussed further in Chapter 34.

These troublesome reflections can be reduced by vacuum depositing an anti-reflection thin film coating on the glass surfaces. This usually consists of a stack of very thin layers of special materials such as magnesium fluoride (MgF_2). The optical thicknesses are usually one quarter or one half of a wavelength depending upon the stack design. With a multi-layered stack more than one material is used.

12.4.2 Absorption

Optical glass is of such quality, that in most optical systems path lengths are too short to lead to any significant absorption. Perhaps the notable exceptions are fibre optical systems. In these systems, the path lengths (in the order of metres to kilometres) can be very long and a significant amount of light is lost due to scatter and absorption.

The absorptive properties of optical glass can be quantified by its internal transmittance and this is a function of wavelength and thickness. The Schott (1992) catalog quotes a value of 0.999 at 500 nm for a 5 mm thick sample of BK 7 glass, which is one of the most common optical glasses. Compare this with a transmittance of 0.918 from Example 12.1, due to surface reflection losses alone for a single lens. This example shows that for high quality optical glass, surface reflection losses are more important than losses due to internal absorption

In complex optical systems, other source of loss are absorption by filters and metal layer coated beam splitting mirrors.

12.4.3 Scatter

Again because of its high quality, optical glass contains few defects such as bubbles and other impurities that give rise to scatter within the bulk material. However a poorly polished surface may also give rise to some scatter. The most common cause of scattered light is due to dirty surfaces, such as dust, badly

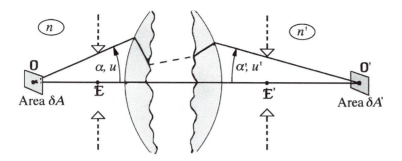

Fig. 12.7: Illuminance in the image plane.

scratched surfaces and poor quality anti-reflection and mirror coatings. As in the case of reflections, some of the light invariably reaches the image plane as a veiling glare and degrades the image.

12.5 Luminous flux leaving a system

The luminous flux F' in a beam leaving the system depends upon the luminous flux F entering and the (luminous) transmittance τ of the system. Since we define the transmittance as the fraction of light transmitted, it follows that

$$F' = \tau F \qquad (12.18)$$

The transmittance of the system is less than unity as a result of surface reflections, absorption and scatter discussed in the previous section. If we only include the effect of reflections then the value of τ for a system with uncoated surfaces is the value of $T(k)$ given by equation (12.17).

12.6 Image illuminance

In all optical systems, the illuminance of the image should be regarded as a very important aspect of image quality, particularly for visual systems, because the darker the image the more difficult it is to see the image detail. Therefore it is important to understand the factors that affect and control the image illuminance. At this stage, we must distinguish between the case of sources of finite size and those that are effectively point sources. We will begin with sources of finite but small area.

12.6.1 Sources of small finite area

12.6.1.1 Axial imagery – general

For a Gaussian system, all of the image forming light entering a system from a source of area δA must be incident on the Gaussian image, which will have an area $\delta A'$ as shown in Figure 12.7. The illuminance E' of this image is given by the equation

$$E' = \frac{F'}{\delta A'}$$

If we now use equations (12.18) and (12.13a), we have

$$E' = \tau \pi L \, \delta A \sin^2(\alpha)/\delta A' \tag{12.19}$$

We can eliminate the areas δA and $\delta A'$ from this equation and express it in terms of the angular radius α' of the exit pupil instead of the angular radius α of the entrance pupil. To derive the corresponding equation, we proceed as follows. Firstly, we recall the equation for transverse magnification, equation (3.47),

$$M = \frac{nu}{n'u'}$$

We can also write

$$M^2 = \delta A'/\delta A$$

If we now assume that the system is free of both spherical aberration and coma, then the sine condition is satisfied. Once the sine condition holds, we can use equation (5.31), that is

$$\sin(\alpha)/u = \sin(\alpha')/u'$$

where we have replaced the symbol U by α. These last three equations give

$$\sin^2(\alpha) = [n'^2 \sin^2(\alpha')M^2]/n^2 = (n'^2/n^2) \sin^2(\alpha')\delta A'/\delta A \tag{12.20}$$

If we now replace $\sin(\alpha)$ in equation (12.19), we finally have

$$E' = \tau \pi L [n' \sin(\alpha')]^2/n^2 \tag{12.21a}$$

Object plane at a great distance or infinity: If the object plane is at a great distance or infinity, we can alternatively express equation (12.21a) in terms of the diameter D of the entrance pupil. From equation (9.12b), when the object is at infinity, we have

$$\sin(\alpha') = \frac{D F_{ep}}{2n'}$$

where F_{ep} has replaced the symbol F in equation (9.12b) for equivalent power, because in this chapter F is used for luminous flux. So equation (12.21a) can now be rewritten in the form

$$E' = \frac{\tau \pi L D^2 F_{ep}^2}{4n^2} \tag{12.21b}$$

If the system is in air, we can express the image illuminance in terms of the *F-number*, which is defined by equation (9.13), as follows.

$$E' = \frac{\tau \pi L}{4(F\text{-}number)^2} \quad \text{(in air and object at infinity)} \tag{12.21c}$$

which shows that the illuminance is independent of the focal length or aperture stop diameter separately, but is dependent upon the ratio of the two, that is the *F-number*.

12.6.1.2   *Off-axis imagery with small pupils; the cos⁴ law*

For off-axis small area sources and an optical system with small pupils and no vignetting, the flux accepted by the system is given by equation (12.15). If we proceed as above, to determine the illuminance of the image, we would finally get

$$E' = \tau \pi L [n' \sin(\alpha')/n]^2 \cos^4(\theta) \tag{12.22}$$

Thus the illuminance in the image plane decreases as the fourth power of the cosine of the off-axis angle. This effect is known as the **cos⁴ law** and is particularly important in wide angle systems.

12.6.2   *Point sources*

The previous section covers the photometry of images of sources of finite though small areas. The equations are not valid for point sources for the following reasons:

(a)   Point sources cannot be assigned an area δA.
(b)   For very small sources, the size of the image $\delta A'$ is not defined by Gaussian optics but instead is limited by aberrations and diffraction effects.

Suitable equations for point sources can be found as follows.

For a nominal point source, the source can only be described by its luminous intensity I instead of its luminance L and surface area δA. In this case, the flux accepted by the optical system is given by the equation (12.13c). Combining this equation and equation (12.18), the luminous flux F' leaving the system is

$$F' = \tau \pi I D^2/(4d^2) \tag{12.23}$$

If the system is free of aberrations, all of this flux is distributed over the diffraction limited point spread function. We discuss this particular function in Chapter 26 and the reader is directed to Section 26.4.1.1, where we explain how this flux is distributed over that function.

Exercises and problems

12.1   If a light source isotropically emits 50 lm, calculate its luminous intensity.

   ANSWER: 3.97 cd

12.2   If the above light source emits the same luminous flux only in a cone with an apex angle of only 10°, calculate the new luminous intensity.

   ANSWER: 2091 cd

12.3 If an isotropic light source emits 60 cd, what is the illuminance at a distance of 3.5 m?

ANSWER: 4.90 lux

12.4 Given that a car headlight gives out light of intensity I cd in the forward direction and illuminates retro-reflective material on the road ahead and normal to the beam which reflects all the light back uniformly into a cone of apex angle $5°$, calculate the luminance of the retro-reflective material at a distance of 200 m.

ANSWER: $4.1 \times 10^{-3} \times I$ cd/m^2

12.5 Assume a quartz halogen lamp with a filament area of 2.5×5 mm has a filament (front on) luminance of 10^8 cd/m^2. Calculate the illuminance on axis in a plane 20 mm away, assuming the inverse square law.

ANSWER: 3.125×10^6 cd/m^2

12.6 Calculate the maximum illuminance at the earth's surface from a full moon (assume angular subtense of moon $= 32$ minutes of arc and luminance of moon $= 3 \times 10^3$ cd/m^2).

ANSWER: 0.204 lux

12.7 Find the illuminance at the earth's surface and parallel to the earth's surface, at (a) the equator and (b) $60°$ latitude due to the sun, assuming there is no absorption or reflectance by the atmosphere (assume angular subtense of sun $= 32$ minutes of arc and luminance of sun $= 2 \times 10^9$ cd/m^2).

ANSWERS: (a) 136,100 lux, (b) 68,051 lux

12.8 Calculate the approximate reflectance of the moon's surface if it is perfectly diffusing.

ANSWER: 0.069

Summary of main symbols and equations

F, F'	luminous flux accepted or transmitted, lumens (lm)
I	luminous intensity, candela (cd)
L	luminance, cd/m^2
E	illuminance, lm/m^2 (lux)
E'	illuminance of image, lux
θ	off-axis angle
Ω	solid angle (steradian)
K_m	photopic luminous efficiency constant $= 683$ lm/W
$V(\lambda)$	photopic relative luminous efficiency function
$P(\lambda)$	spectral power of a source
ϵ	fractional error in inverse square law
d	distance from a source
R	reflectance at a surface
T	transmittance at a surface
τ	total transmittance of a system
ρ	radius of a circular source
A	area of a source
F_{ep}	equivalent power of a system

References and bibliography

*CIE. (1983). *The Basis of Physical Photometry.* Publication CIE 18.2 (TC-1.2). Commission Internationale de L'Eclairage, Paris.

*Sanders C.L. and Jones O.C. (1962). Problem of realizing the primary standard of light. *J. Opt. Soc. Am.* 52(7), 731–746.

*Schott. (1992). *Schott Optical Glass Catalog.* Schott Glaswerke, Mainz, Germany.

Part II

Geometrical optical instruments or systems

13

The eye

13.0   Introduction

The performance of visual optical instruments cannot be fully assessed without some knowledge of the anatomy and functions of the eye, working either monocularly or binocularly. This chapter describes the optics of the eye, but its interaction with visual instruments is covered later in Chapters 36 and 37.

A cross-section of the human eye is shown in Figure 13.1, giving only the most relevant optical components. A more detailed anatomical description can be found in a number of textbooks, for example Davson (1990). Image forming light enters the eye through and is refracted by the **cornea**. It is further refracted by the **lens**, bringing it to a focus on the retina. Of the two refracting elements, the cornea has the greater refractive power. However, whereas the power of the cornea is constant, the power of the lens depends upon the level of accommodation, which is the process by which the refractive power of the eye changes to allow closer or more distant objects to be sharply imaged on the retina. The diameter of the incoming beam of light is controlled by the **iris**, which is the aperture stop of the eye.

The dimensions of the eye and its optical components vary greatly from person to person and some further depend upon accommodation level, age and certain pathological conditions. In spite of these variations, average values have been used to construct representative or schematic eyes. These are discussed further in Section 13.6.

13.0.1   The refractive components

The relaxed eye has an equivalent power of about 60 m^{-1}. The corneal power is about 40 m^{-1}, which is two-thirds of the total power. The anterior corneal surface power is a little greater than this with the posterior surface having a small negative power. In this relaxed eye, the lens has a power of about 20 m^{-1}. The actual component powers of a number of model eyes are given in Appendix 3.

The refractive index of the cornea is usually taken as 1.376. However, the anterior corneal surface is not a smooth optical surface due to its cellular structure,

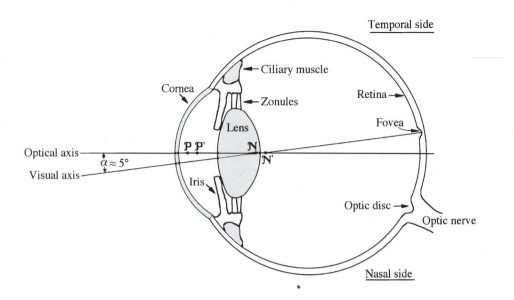

although an optically smooth surface is provided by the very thin tear film which
covers it. While this has an index less than 1.376, for most optical calculations
this tear film can be regarded as a very thin optical element consisting of two
concentric surfaces of almost equal radii of curvature and therefore has negligible power.

The lens does not have a uniform refractive index. Instead it has a gradient
index which is highest in the centre and lowest at the edge. Thus its refractive
power is due to both the front and back surface powers and the variation of
index within the lens.

13.0.2 *The point of fixation and the fovea*

When the eye fixates on an object of interest, the image is formed on the **fovea**,
which has an angular subtense of about $2°$.

13.0.3 *Retinal structure*

The retina contains two types of light sensitive cells, namely **rods** and **cones**.

Cones The cones are far less sensitive to light than rods and only function at high light levels (greater than about 1 lux). On the other hand, they are
capable of providing high spatial resolution and colour vision. They predomi-
nate in the fovea, which is rod free. Cones are thinly scattered throughout the
peripheral retina.

Rods The rods are very sensitive to light and thus are suited to and only
function at low light levels. They reach their maximum density at about $20°$
from the fovea. The neural network of rods is such that the output of about 100
rods can combine on the way to the brain, giving rods a very high sensitivity to
light but as a result, poor spatial resolution.

At high light levels, best resolution is attained by the cones in the fovea.
At low light levels the cones at the fovea do not operate and thus the fovea is
"night blind" and one must look eccentrically to see objects using rod vision.

Consequently in poor lighting conditions, maximum visual acuity is reached about 20° away from the fovea.

The neural network and vascular supply to the retina enter or leave the eye at the optic disc. Thus at the optic disc there are no cones or rods and this region is blind, hence the name the "blind spot". The optic disc is about 15° on the nasal side of the retina and so the blind spot is about 15° towards the temporal side in the visual field and approximately horizontal from the fovea. The reader can confirm the presence of the blind spot as follows. Draw two crosses on a sheet of clean paper approximately 8 cm apart. Look at the lefthand cross with the right eye (left eye closed) from a distance of about 20 cm. By varying this distance slightly, the righthand cross should disappear.

13.0.4   Optical and visual axes

The optical system of the eye is not a rotationally symmetric system due to a combination of the toroidal nature of some of the surfaces, the fact that these surfaces are sometimes tilted with respect to each other and that they may also have no symmetry. Thus the eye does not have a true optical axis. This can be confirmed by observing the images of a small light source formed by reflections from each optical surface (the **Purkinje** images). It is usually impossible to align these images. All that one can do is construct a best fit optical axis which could be defined in terms of the direction of the incident beam, from this small light source, which gives the smallest spread of Purkinje images.

The visual axis, that is the line passing through the fovea, back nodal point, front nodal point and the point of fixation, does not coincide with the best fit optical axis. An idea of the position of the nodal points of the eye can be gauged from those of a number of model eyes shown in the diagrams in Appendix 3. From those diagrams, it can be seen that the nodal points are close to the back surface of the lens. The principal and nodal points are also plotted on Figure 13.1 along with the visual and optical axes. These axes are tilted to each other by about 5° and this angle is sometimes denoted by the symbol α.

13.1   Accommodation

Within limits, the lens of the eye can change its equivalent power, while remaining essentially stationary, to ensure that objects of interest at different distances can be imaged sharply on the retina. This is in contrast to a camera, in which the lens has a fixed power and has to be moved backwards or forwards to focus on objects at different distances.

The accommodation process is shown in Figure 13.2. When the eye needs to focus on closer objects, the ciliary muscle contracts, causing the zonules or suspensory ligaments supporting the lens to relax. This allows the lens to take a more spherical shape due to the inherent elasticity of the lens capsule. As the lens takes on a more spherical shape, it thickens at the centre and the surface curvatures increase, the front surface moves slightly forward and the equivalent power increases. When the eye has to focus on more distant objects the reverse process applies.

There are physical limits to how far the ciliary muscle can relax and contract and how far the lens can be stretched and contracted. Thus there are upper and lower limits to the equivalent power of the lens and hence eye, and in turn

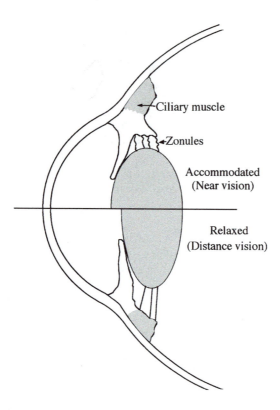

Fig. 13.2: Changes in the lens shape and position with accommodation.

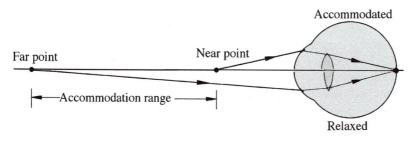

Fig. 13.3: The far and near points of the eye.

farthest and closest distances of clear vision. These extreme distances are called the **far point** and **near point**, respectively, and are shown in Figure 13.3. Thus when the ciliary muscle is completely relaxed, the eye is focussed on the far point. When the ciliary muscle is maximally contracted, that is the lens capsule maximally relaxed, the eye has its greatest equivalent power and is focussed on the near point.

The difference between the vergences of the near and far points of the eye is the **amplitude of accommodation**. There are different methods used for measuring the amplitude of accommodation. One is called the push-up method and this involves a subjective measure of the first noticeable onset of blur at the near and far points and therefore includes the effect of depth-of-field. Another is the method of stigmatoscopy, which is claimed to be independent of depth-of-field.

The amplitude of accommodation is affected by age, always decreasing with increasing age. Figure 13.4 shows the general variation of amplitude of accommodation with age, for the push-up and stigmatoscopy methods, taken from Sun

Fig. 13.4: Variation of amplitude of accommodation with age, from Sun et al. (1988).

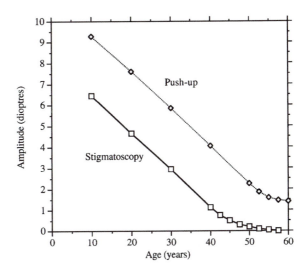

et al. (1988). At each age, there is a spread of amplitudes of accommodation, so that the curves shown in this diagram are only the sample averages. The severe reduction in amplitude of accommodation in older age is called **presbyopia**. Note that both curves shown in the diagram flatten out after about the age of 40 years. However Hofstetter (1965), by examining the reduction of amplitude of accommodation of two subjects over a few years, found that the individual curves did not flatten out but went linearly to zero. The flattening out only occurs with population averaged data and Charman (1989) has argued that the effect is a statistical artifact due to taking the means of progressively increasing truncated normal distributions.

13.2 Refractive errors

Because of large variations in the dimensions of the components of real eyes, it is not easy to define a "normal" eye. However it is usually assumed that a normal eye should be focussed at infinity when the accommodation is relaxed; that is the normal eye has a far point at infinity. This eye is termed **emmetropic**. The desired accommodation range of this eye is not defined. All other eyes are termed **ametropic** and are regarded as having some kind of **refractive error**. The magnitude of the refractive error is defined as the vergence of the far point at the eye. Refractive errors can be categorized as either (a) spherical or (b) cylindrical (or astigmatic). These will now be discussed in turn.

13.2.1 Spherical refractive errors

These are refractive errors which are independent of the azimuth in the pupil. They can be further sub-classified as myopic, hypermetropic or presbyopic.

13.2.1.1 Myopia (or short sightedness)

In the myopic eye, the far point is at a finite distance in front of the eye, as shown in Figure 13.5a. Thus in the relaxed eye, an object at infinity is focussed in front

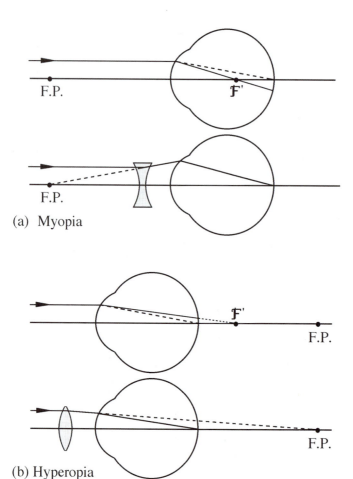

Fig. 13.5: Spherical refractive errors (ametropias) and their correction. F.P. = far point.

(a) Myopia

(b) Hyperopia

of the retina. This refractive error arises because the equivalent power of the eye is too great for its axial length. The refractive error is corrected by a spectacle or contact lens of negative power as shown in the diagram. This negative lens power counteracts the "excess" positive power of the eye. Alternatively, we can say that the lens images the object at infinity onto the back focal plane of the lens, which is coincident with the far point plane of the eye.

13.2.1.2 Hypermetropia (or hyperopia or long sightedness)

In the case of hypermetropia, the far point is "beyond infinity", that is behind the eye, as shown in Figure 13.5b. Thus in the relaxed eye, an object at infinity is focussed beyond the retina. Hyperopia is due to the power of the eye being too low for its axial length. However, this eye can focus on distant objects by accommodating. When the required accommodation is excessive, the eye is corrected with a suitable positive spectacle or contact lens as shown in the diagram. Apart from the need to reduce accommodation when viewing distant objects, hyperopic eyes are often corrected in order to improve near viewing. For the same amplitude of accommodation, the hyperope will have a more distant near point than the emmetrope, thus making it more difficult to focus on close objects. The positive power of the ophthalmic lens prescribed for the hyperope can be thought of as either compensating for the insufficient positive power of the eye or instead imaging distant scenes onto the far point plane.

13.2.1.3 Presbyopia

This is the refractive error condition of an eye with zero or little amplitude of accommodation. It is usually due to advancing age, as discussed in Section 13.1. The presbyopic eye cannot be corrected by a single power lens. Each object distance will require a different correcting power which will of course also depend upon the fixed focussing distance of the presbyopic eye. In practice, the presbyope is either given several pairs of single power lenses, each for a different working distance, or given multifocal lenses, for example bifocal or trifocal, in which different segments of the same lens have different powers. Lenses with a continuously varying power from top to bottom are also available and these were discussed in Chapter 6.

13.2.2 Cylindrical or astigmatic refractive errors

So far it has been assumed that all the refractive surfaces of the eye and the eye as a whole are rotationally symmetric. If one or more surfaces are either (a) decentred or (b) toroidal in shape, the eye has axial astigmatism and thus has a "cylindrical" component to the refractive error. Cylindrical errors are compensated by a suitable toric ophthalmic lens. This type of lens was introduced and examined in Chapter 6.

13.2.3 Population distributions of refractive errors

The population distribution of refractive errors is strongly age dependent, being considerably sensitive to age in the younger age group. However, in this text, we are only interested in the adult population as these are the main users of optical instruments. Determining the distribution of refractive errors in the population has been the aim of a number of investigations. A review of these have been given by Borish (1970). Probably the two largest surveys are by Scheerer (1928) and Betsch (1929), covering 25,000 adults. However, the details of these investigations are not readily available. The results of two smaller but more readily available studies (Stenström 1948; Sorsby et al. 1957) are shown in Figure 13.6, which are data for essentially European populations. These results show that the mean level of refractive error for the sample is between 0 and 1 m^{-1} of hyperopia. The distribution is also not a normal distribution. Pure spherical ametropia is rare as the normal refractive error usually involves some astigmatism. There are racial differences in the incidence of myopia and Borish (1970) presents some data on these differences.

13.2.4 Powers of the correcting lens

The power of a lens required to correct a refractive error can be determined from the level of refractive error and the distance of the lens from the eye. These calculations and other aspects of these corrective lenses are discussed in the next chapter.

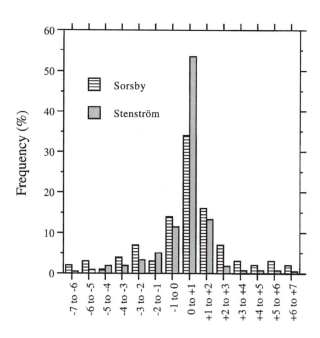

Fig. 13.6: Adult population distribution of spherical refractive errors from Stenström (1948) and Sorsby et al. (1957).

Refractive error (dioptres)

13.2.5 *Refractive errors and ocular parameters*

In the past, it has been common to divide refractive errors into what were known as **axial** or **refractive** errors. "Axial" refractive errors were supposed to be due to errors in axial length, with the refractive power of the eye assumed to be normal. "Refractive" refractive errors were supposed to be due to errors in refractive component powers and the axial length was supposed to be normal. However, the relationship between refractive error and ocular parameter values is complex and such a division is misleading. For example, ocular metrology studies have shown that emmetropic eyes may have wide variations in component powers and axial length, making it difficult to define a normal power or normal length. However, with these cautionary comments made, there are situations when it is important to know the relationships between changes in some ocular components and refractive error and sometimes the refractive error can be identified as axial or refractive. An example of each will now be given.

13.2.5.1 *Refractive error and axial length*

Let us suppose we have an emmetropic eye that for some reason experiences a change in axial length without a change in equivalent power. This eye will become ametropic and the far point will no longer be at infinity. The vergence of the new far point will be the refractive error. Let us derive an equation for this refractive error or vergence in terms of the change in axial length.

By applying the lens equation to the eye, one can derive a simple equation relating a change in axial length to a corresponding change in refractive error. Equation (10.9b), based on the lens equation, relates a change in image distance

$\delta l'$ to a change in object vergence δL, that is

$$\delta L \approx -n' \delta l' / l'^2$$

Now if the eye is initially emmetropic, we have $L = 0$ and from the lens equation, we have

$$l' = n'/F$$

and therefore

$$\delta L \approx -\delta l' F^2 / n' \tag{13.1}$$

with the approximation increasing in accuracy as the magnitude of the change $\delta l'$ decreases. This change δL in object vergence is equivalent to a refractive error. This equation indicates that for a decrease in axial length, the refractive error is positive and if the axial length increases the refractive error is negative. The signs of these refractive errors are also the signs of the powers of the correcting lens.

> **Example 13.1:** Estimate the change in axial length required to produce a change of $+1.0 \text{ m}^{-1}$ in refractive error. A vergence of $+1 \text{ m}^{-1}$ means that the far point is behind the eye and hence the eye is now hyperopic.
>
> **Solution:** In equation (13.1), let us use $F = 60.0 \text{ m}^{-1}, n' = 1.336$ and $\delta L = +1.0 \text{ m}^{-1}$. Then we have
>
> $$\delta l' = -0.37 \text{ mm}$$
>
> Thus if the axial length reduces by 0.37 mm, the eye becomes $+1 \text{ m}^{-1}$ hyperopic.

13.2.5.2   Change in corneal radius

Now let us look at the change in refractive error due to a change in the radius of curvature of the cornea of an emmetropic eye. In this case, this will produce a change in power of the eye.

The power F_c of the anterior corneal surface is given by the equation

$$F_c = C(n-1) = (n-1)/r \tag{13.2}$$

Therefore a small change of δr in surface radius of curvature r leads to a change in corneal power δF_c where

$$\delta F_c \approx -(n-1)\delta r / r^2$$

Replacing r in this equation, using equation (13.2), we have the alternative equation

$$\delta F_c \approx -\delta r \, F_c^2 / (n-1)^2 \tag{13.3}$$

which is an estimate of the change in power of the eye and is more accurate for smaller changes. This equation shows that if the radius of curvature decreases, the change in power is positive and the eye has become myopic. Thus in this case, the refractive error will have the opposite sign to that given by equation (13.3).

> **Example 13.2:** For the Gullstrand number 1 eye, given in Appendix 3, estimate the change in power if the anterior corneal radius of curvature changes by -0.1 mm.
>
> **Solution:** For the Gullstrand eye, $r = 7.7$ mm, $n = 1.376$ and $F_c = 48.83$ m^{-1}. Substituting these values into equation (13.3), we have
>
> $$\delta F_c = +1.69 \text{ m}^{-1}$$
>
> Therefore if the radius of curvature of the cornea reduces by -0.1 mm, the power of the eye increases by about 1.7 m^{-1}; that is the eye becomes myopic.

13.3   The pupil

The iris is the aperture stop of the eye. The image of the opening of the iris is called the **entrance pupil** of the eye and is usually simply called the pupil. Its diameter depends upon a number of stimuli: light level, some drugs, psychological factors and the level of accommodation. In the latter case, the pupil diameter decreases with increasing accommodation which may be to compensate for the narrower depth-of-field that occurs with closer viewing. Depth-of-field is discussed in Section 13.5. The **exit pupil** is the image of the iris formed on the retinal side of the lens.

13.3.1   *Entrance and exit pupil positions*

The positions of the entrance and exit pupils depend somewhat on the power of the refractive components of the eye, the positions of the refracting surfaces and the refractive index distribution in the lens. Taking a typical schematic eye (Gullstrand's number 1 relaxed eye, Appendix 3), the positions and magnifications of the pupils are shown in Figure 13.7. Of the two, the entrance pupil is by far the more important as there is little application for information on the exit pupil. It can be seen from the data given in Appendix 3 and the diagram, that the entrance pupil of the Gullstrand schematic eye is about 15% closer to the cornea and about 13% larger than the real pupil.

13.3.2   *Entrance pupil diameter and ambient light level*

The diameter of the pupil increases with decrease in light level. De Groot and Gebhard (1952), using all available data, proposed the equation

$$\log_{10}(D) = 0.8558 - 4.01 \times 10^{-4}[\log_{10}(L_{\text{lum}}) + 8.6]^3 \qquad (13.4)$$

to describe the relationship between the average pupil diameter D (mm) and the average field luminance denoted by L_{lum} (cd/m^2). There are large inter-subject

Fig. 13.7: Entrance and
exit pupils of the Gullstrand
number 1 relaxed eye.
Distances are in
millimetres.

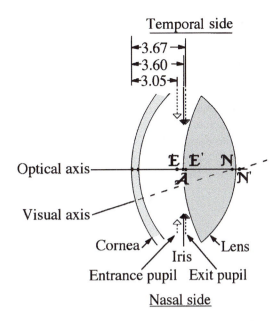

Fig. 13.7: Entrance and exit pupils of the Gullstrand number 1 relaxed eye. Distances are in millimetres.

variations. Standard deviations from data from Spring and Stiles (1948) are approximately ±1 mm for large pupil diameters and ±0.5 mm for the smallest pupil diameters.

Moon light levels are approximately in the order of 0.01 cd/m^2 and full-sun light levels are in the order of 1000 cd/m^2. These are only approximations because these ambient light levels depend upon the elevation and phase of the moon, the elevation of the sun, atmospheric conditions and the reflectance of the ambient surround. These variables can affect the ambient luminance levels by a factor of about 100.

The variation of pupil size with light level is not sufficient to ensure a constant retinal illuminance because the amount of light entering the pupil only changes by a factor of 16, when the pupil size changes from 2 to 8 mm in diameter. However, the light level difference between bright daylight and moon light varies by up to 10^7 : 1. According to Campbell and Gregory (1960) and Woodhouse (1975) the pupil size alters with light level in order to produce optimum visual acuity, rather than constant retinal illuminance.

13.3.3 Age and entrance pupil diameter

Pupil size is also affected by age. Kumnick (1954) showed that both the maximum and minimum pupil sizes decreased with age. Kadlecova et al. (1958) showed that there is a progressive decrease in pupil diameter in the dark with age. Mean values ranged from about 7.5 mm at 10 years of age to about 5 mm at 80 years of age, with standard deviations of about 0.75 mm. Weale (1961) reviewed the data of Kadlecova et al. and others and showed that the dark adapted eye diameter reduced more rapidly with age than the light adapted eye. The light adapted eye reduced from about 3 mm diameter at about 15 years of age to about 2.5 mm at 65 years of age.

13.3.4   Pupil centration

The pupil is not necessarily centred on the optical axis. According to Westheimer (1970), the pupil is usually decentred by about 0.5 mm nasally. If it were centred on the optical axis as shown in Figure 13.7, it would be decentred with respect to the visual axis. However, if it were centred on the visual axis, it would be decentred nasally from the optical axis, as can be deduced from the diagram. Taking the angle α of 5° and typical entrance pupil and nodal point distances given for the Gullstrand number 1 relaxed schematic eye in Appendix 3, the pupil would appear to be decentred nasally by 0.35 mm. Thus if Westheimer is correct, the pupil is closer to being centred on the visual axis than the optical axis.

13.4   Visual acuity

Visual acuity is a measure of the ability to resolve fine detail and is quantified in terms of the size of the finest detail that can be resolved. The eye has greatest visual acuity at the fovea and decreases rapidly with distance from the fovea. The cones provide the greater acuity and the rods the lesser. Acuity also depends upon light level and increases with light level, reaching a maximum with ambient luminances of about 1000 cd/m^2. Optimum acuity values are for the correctly focussed eye. Any defocus will reduce the acuity and we could examine this effect using the defocus blur disc model described in Chapter 10.

In practice there are a number of different ways that visual acuity may be defined and measured. Multiple definitions are necessary in order to cover the wide range visual tasks that we encounter in our daily lives. Figure 13.8 shows some of these definitions and acuity limits and details of these different measures of acuity are given below.

13.4.1   Grating acuity

Periodic gratings of equally spaced black and white bars are very useful in analysing the performance of any optical system, including the eye. These may have square or sinusoidal forms. Grating acuity is the highest grating frequency that can be resolved by the eye and a typical acuity limit is about 30–60 c/deg at optimum luminances. That is the eye cannot resolve a grating with a spatial frequency higher than about 30–60 c/deg. Some of the factors that set this limit are explained in Chapter 34, where we explain the theory of the imaging of sinusoidal gratings. Some simple grating charts are shown in Appendix 5.

13.4.2   Letter or Snellen acuity

Clinical or optometric visual acuity is usually defined as a letter acuity, using a standard letter chart observed at a set distance. The size of the smallest letters that can be recognized is taken as the clinical visual acuity of the subject.

For distance vision, the visual acuity is expressed as a fraction (the Snellen fraction) M/N, where M is the viewing distance in metres from which the chart is read and N is the distance at which the height, of the smallest letters can be recognized, is 5 minutes of arc. The reading distance is usually 6 metres and

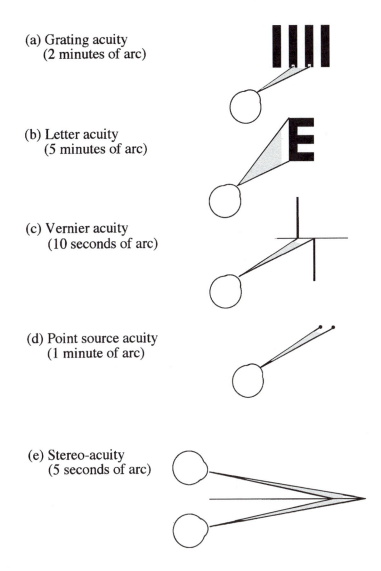

Fig. 13.8: Acuity targets and typical acuity limits.

(a) Grating acuity
(2 minutes of arc)

(b) Letter acuity
(5 minutes of arc)

(c) Vernier acuity
(10 seconds of arc)

(d) Point source acuity
(1 minute of arc)

(e) Stereo-acuity
(5 seconds of arc)

thus the letter size corresponding to 6/6 vision is 5 minutes of arc in height. An acuity of 6/120 means that the smallest letter than can be recognized by this subject would subtend 5 minutes of arc at 120 m. In fact the smallest letter that this subject can read at 6 m subtends $120 \times 5/6 = 100$ minutes of arc.

The typical acuity limit of 5 minutes of arc of letter height is equivalent to about 30 c/deg grating acuity. If one considers the 6/6 level letter E, in the vertical direction it may be regarded as being composed of three black lines and two white lines. If these are of equal width, this letter can be regarded as having a corresponding equivalent grating period of 2 minutes of arc or a spatial frequency of 30 c/deg. However, assigning equivalent periodic frequencies to non-periodic targets should be done with some caution, because in this E there are only two and a half cycles and the lines are very short. Normal grating acuity is measured using a grating field consisting of many cycles and line lengths many times longer than their widths.

13.4.3   Vernier acuity

Letter acuity is a good guide to reading performance but should not be used to predict visual performance for all other tasks, for example when reading scales or verniers. In these cases, a good judgement of the alignment of two lines or edges is required. The angular threshold of just perceptible mis-alignment of two lines or edges is called **vernier acuity** and is about 10 seconds of arc (Bennett and Rabbetts 1989, p. 31).

13.4.4   Point source acuity

Point source acuity is a measure of the ability to resolve two very close point sources. The resolving power in this case depends upon whether the sources are white on black or black on white. For white sources against a dark background the resolution is about the same as the grating acuity. That is, two sources can be distinguished if their separation is greater than about 1 minute of arc (Bennett and Rabbetts 1989, 27).

13.4.5   Stereoscopic acuity

Stereoscopic acuity is a measure of the ability to resolve the depth separation of two objects close together in space but at different distances from the observer. This type of acuity requires binocular vision. It is measured as an angular value, which may be as small as 5 seconds of arc (Bennett and Rabbetts 1989, 232). Stereoscopic acuity is discussed further in Chapter 37.

13.5   Depth-of-field

The concept of depth-of-field was introduced in Chapter 10 and is the range of object or image plane positions over which the image appears to be in focus. The depth-of-field of the eye ultimately sets the precision to which the refractive state of the eye may be measured, the precision required in focussing optical instruments and the accuracy of the accommodation system of the eye. The first of these aspects is of great importance to optometrists and ophthalmologists and the second aspect is important to designers of visual optical systems.

An estimate of the expected depth-of-field of the eye can be calculated from basic optical principles and known optical properties of the eye. Depth-of-field values have also been measured experimentally.

13.5.1   Geometric optical predictions

The geometrical optical predictions are based upon the assumption that if there is a focus error, the image of a point is a blur disc which can be assumed to be a uniformly illuminated circular disc. These assumptions neglect the effects of diffraction, regular and irregular aberrations and the Stiles-Crawford effect (see Section 13.7). According to this simple model, the eye will not detect a focus error if the blur disc is smaller than the diameter of one foveal cone.

Smith (1982) has shown that the angular diameter ω of a defocus blur disc can be accurately given by the equation

$$\omega = D\delta L \text{ rad} \tag{13.5}$$

for any level of accommodation, where D is the diameter of the entrance pupil and δL is the focus or refractive error. The diameter of a foveal cone has been estimated by Polyak (1957) to be about 0.003 mm. The angular diameter ω_{cone} of this cone is then given as

$$\omega_{\text{cone}} = 0.003/\mathcal{N}'\mathcal{F}' \text{ rad}$$

where $\mathcal{N}'\mathcal{F}'$ is the distance between the back nodal point ($\mathcal{N}'$) of the eye and the retina (at $\mathcal{F}'$). In the Gullstrand number 1 relaxed schematic eye (Appendix 3) this distance is 17.05 mm. Therefore for this eye

$$\omega_{\text{cone}} = 0.003/17.05 = 0.00018 \text{ rad} \tag{13.6}$$

If the angle ω in equation (13.5) is now the threshold diameter given by equation (13.6), the threshold of detectable defocus is thus

$$\delta L = 0.00018/D \text{ m}^{-1} \tag{13.7}$$

where D is in metres and δL is in dioptres.

> **Example 13.3:** Using the above model of depth-of-field, estimate the depth-of-field of an eye with a 4 mm pupil.
>
> **Solution:** In this example, we substitute $D = 0.004$ m in equation (13.7) and this gives
>
> $$\delta L = 0.00018/0.004 = 0.045 \text{ m}^{-1}$$

This dioptric or vergence depth-of-field can be converted to an equivalent nearest point of clear focus δl for an emmetropic eye focussed on infinity. To convert from the vergence δL depth-of-field to a distance δl, we can use the relation between distance and vergence, equation (3.62a), with $n = 1$, to give the equation

$$l \pm \delta l \approx 1/L \pm \delta L/(L^2) \tag{13.8}$$

Using this equation, a dioptric value of $\delta L = \pm 0.045$ dioptres leads to the distances in Table 13.1. The results in this table show that the depth-of-field, expressed as a distance, decreases with increase in accommodation.

13.5.2 Experimental findings

The actual depth-of-field has been evaluated by a number of investigators. There are differences in findings, which are most likely due to the differences between individuals in the population and the different criteria used. For example Ogle and Schwartz (1959) discussed four criteria for defining depth-of-field:

(a) defocus required to reduce acuity by a set amount,
(b) threshold of perceptible blur,
(c) loss of visibility or detectability through loss of contrast, and
(d) minimum defocus required to produce an accommodation response.

Table 13.1. Predicted depth-of-field distance limits for a geometrical optical estimated depth-of-field of $\pm 0.045\ m^{-1}$

Accommodation distance		Depth-of-field range		
L (m^{-1})	l (m)	from	to	range (cm)
0	∞	∞	22.7 m	
1	1.0	1.044 m	0.956 m	88.0
2	0.5	51.1 cm	48.90 cm	22.0
5	0.2	20.76 cm	19.84 cm	9.2
10	0.1	10.04 cm	9.96 cm	0.9

Table 13.2. Measured depth-of-field values

Pupil diameter (mm)

3   Campbell (1957): ± 0.3 m^{-1}

Campbell and Westheimer (1958):

	Accommodation level (m^{-1})					
	zero			1.0		
	from (m)	to (m)	range (m^{-1})	from (m)	to (cm)	range (m^{-1})
1	0	1.25	0.80	5.0	56	1.59
2	0	2.33	0.43	1.8	70	0.87
3	0	2.94	0.34	1.5	75	0.67
4	0	3.57	0.28	1.4	78	0.57

Campbell (1957) has shown that the depth-of-field also depends upon luminance, contrast and pupil diameter and found that the minimum depth-of-field, under optimum conditions and a 3 mm diameter pupil, was ± 0.3 m^{-1}. This and other values from Campbell and Westheimer (1958) are given in Table 13.2. Estimates by Charman and Whitefoot (1977), who used laser speckle (633 nm), are shown in Figure 13.9. The depth-of-field values are defined as a "total depth-of-field" and are four times the standard deviation of the subjects' error score. For a 2 mm pupil diameter, the mean depth-of-field was about 0.6 m^{-1} and dropped to about 0.3 m^{-1} at 7 mm.

A comparison of the geometrical optical prediction in Example 13.3 and these experimental findings shows that the geometrical optical model underestimates the depth-of-field of real eyes. Some causes of the difference between the geometric optics predictions and actually measured depth-of-field values are most likely due to the presence of aberrations and diffraction effects in the eye. For example, the geometric optics model assumes that the retinal image of a point is a point, while the minimum point spread function diameter is several minutes of arc.

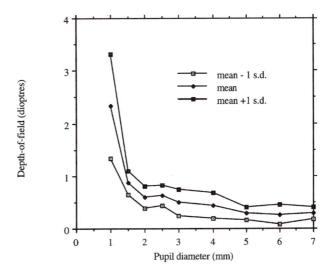

Fig. 13.9: The
depth-of-field of the eye as
a function of pupil
diameter, from Charman
and Whitefoot (1977).

13.6   Schematic eyes

So far in this chapter, we have made several references to the Gullstrand schematic eye. This is only one of a number of schematic eyes that have been developed and found useful for many optical calculations. Schematic eyes are simplified optical models of real eyes, with the constructional parameters being averages of population distributions. Most schematic eyes are **paraxial** schematic eyes. These are models which are very much oversimplified and have been found to be only useful in the paraxial region, that is, for only small pupils and small fields, and they do not accurately predict the aberration levels of real eyes at larger pupils. More accurate schematic eyes are termed **finite** or **wide angle**, but to date there has not been a universally accepted finite schematic eye. These two types are discussed in a little more detail below.

13.6.1   Paraxial schematic eyes

Paraxial schematic eyes are oversimplified versions of real eyes, in three main respects. Firstly, the refracting surfaces of schematic eyes are spherical and rotationally symmetric, while in real eyes this is not so. In real eyes, the surfaces are aspheric, often decentred and rarely rotationally symmetric, typically being toric. Secondly, in the schematic eye, the visual axes are made to coincide with the optical axis; while in real eyes, these axes differ by about 5°. Thirdly, the refractive index of the lens is assumed to be constant. In reality, this refractive index varies throughout the lens, being highest in the centre and decreasing towards the periphery.

One of the most commonly used schematic eyes is the Gullstrand schematic eye number 1, and the simplest schematic eye is the reduced eye type, which consists of only one refractive component, the cornea. The constructional and Gaussian data of these eyes and a number of other schematic eyes are given in Appendix 3.

13.6.2 Finite schematic eyes

A number of finite schematic eyes have been developed to try to improve
the aberration predictions of real eyes at large apertures and fields of view, for
example Lotmar (1971), Drasdo and Fowler (1974), Blaker (1980) and Navarro
et al. (1985). These types of schematic eyes usually use aspheric surfaces. The
Blaker model is the only one with a gradient refractive index lens.

13.7 Photometry of the eye

Equations (12.21a and b) relate the luminance L_{lum} of an extended object to the
illuminance E' of the image. Applying these to the eye, we set the object space
refractive index n equal to one. So for the eye, the illuminance E' is the retinal
illuminance and we have

$$E' = \tau \pi L_{\text{lum}}[n' \sin(\alpha')]^2 \tag{13.9a}$$

or

$$E' = \tau \pi L_{\text{lum}} D^2 F^2 / 4 \tag{13.9b}$$

where τ is the transmittance of the eye, n' is the refractive index of the vitreous
(1.336), α' is the angular radius of the exit pupil, D is the entrance pupil diameter
and F is the equivalent power of the eye (approximately $60\,\text{m}^{-1}$). However, this
is not the true perceived retinal light level because of the Stiles–Crawford effect.

13.7.1 The Stiles–Crawford effect

In 1933, Stiles and Crawford (1933) discovered that the luminous efficiency of a
thin beam of light entering the eye and incident on the fovea depended upon the
entry point in the pupil. This effect is of major importance in photometry, aberration and image quality analysis of the eye. It is a photopic and hence a cone phenomenon. This effect is thought to be due to the "waveguide" nature and directionality of the cones. It has been suggested that the cones point towards the exit
pupil and this is designed to reduce the effects of intra-ocular scattered light and
this theory has been supported by Enoch (1972) and Enoch and Bedell (1979).

Various mathematical models have been used to quantify the effect. Here we
will express the luminous efficiency for a ray entering the pupil at a distance r
from the pupil centre as $L_e(r)$. Stiles (1937) used the function

$$L_e(r) = \exp[-p(r - r_0)^2] \tag{13.10}$$

to fit his data, where r_0 is the position of the peak response in the pupil. We can
simplify this by ignoring the pupil decentration and write

$$L_e(r) = \exp(-pr^2) \tag{13.11a}$$

The value of p depends upon the individual. Stiles (1937), using one subject,
found that p depended on wavelength and varied between 0.166 at 440 nm, to
a low of 0.122 at 520 nm and up to a value of 0.142 at 720 nm, with a mean of
0.141 over the spectrum. Crawford (1937) used the same function for his own

Table 13.3. Published values of the p parameter, which is used to describe the Stiles–Crawford function given in Section 13.7.

p	Source
0.141–0.166	Stiles (1937)
0.105–0.108	Crawford (1937)
0.115	Krakau (1974)
0.108	van Meeteren (1974)
0.116 ± 0.029[a]	Applegate and Lakshminarayanan (1993)

[a] This value is the mean of the horizontal and vertical values.

eye at two different times (separated by several years) and found p had a value of 0.105 and 0.108. These and other published values of p are shown in Table 13.3.

Other mathematical forms have been suggested. Moon and Spencer (1944) recommended using the functions

$$L_e(r) = 1 - 0.0850r^2 + 0.0020r^4 \qquad (13.11b)$$

or

$$L_e(r) = 0.379 + 0.621 \cos(0.515r) \qquad (13.11c)$$

Enoch (1958) expressed the Stiles–Crawford luminous efficiency function as

$$L_e(r) = 0.25[1.0 + \cos(9.5\theta)]^2 \qquad (13.11d)$$

where

$$\tan(\theta) = 0.045r$$

and the angle θ is the angle of inclination of the ray with the optical axis inside the eye. This has a maximum value of about $10°$ for the ray passing through the edge of an 8 mm diameter pupil.

13.7.2 Photometric efficiency

We can use the Stiles–Crawford effect to relate the effective retinal illuminance to the actual illuminance. If a source of light produces a uniform illuminance E at the entrance pupil, the effective weighted total luminous flux accepted by the eye is

$$flux = E \int_0^{2\pi} \int_0^{\bar{\rho}} L_e(r) r \, dr \, d\phi$$

where $\bar{\rho}$ is the radius of the entrance pupil. If we use equation (13.11a) for $L_e(r)$, we have

$$flux = E \int_0^{2\pi} \int_0^{\bar{\rho}} \exp(-pr^2) r \, dr \, d\phi$$

This integral reduces to

$$flux = E\pi[1 - \exp(-p\bar{\rho}^2)]/p \tag{13.12}$$

If there were no Stiles–Crawford effect, the same luminous flux could be collected by a smaller pupil. Let the radius of this pupil be $\bar{\rho}^*$. Then we have

$$E\pi\bar{\rho}^{*2} = E\pi[1 - \exp(-p\bar{\rho}^2)]/p$$

that is

$$\bar{\rho}^{*2} = [1 - \exp(-p\bar{\rho}^2)]/p \tag{13.13}$$

We will call $\bar{\rho}^*$ the equivalent reduced pupil radius.

Alternatively one may use the concept of the total luminous efficiency $S(\bar{\rho})$ of Stiles and Crawford (1933). This is defined in terms of the uniformly illuminated pupil as

$$S(\bar{\rho}) = \frac{\text{effective luminous flux entering eye}}{\text{actual luminous flux entering eye}}$$

The effective luminous flux is given by equation (13.12) and the actual luminous flux entering the eye is

$$E\pi\bar{\rho}^2$$

Therefore we have

$$S(\bar{\rho}) = \frac{1 - \exp(-p\bar{\rho}^2)}{p\bar{\rho}^2} \tag{13.14}$$

and this function is identical to the ratio $(\bar{\rho}^*/\bar{\rho})^2$, where $\bar{\rho}^*$ is given by equation (13.13). Therefore

$$S(\bar{\rho}) = (\bar{\rho}^*/\bar{\rho})^2 \tag{13.15}$$

Example 13.4: Calculate the reduced pupil radius $\bar{\rho}^*$ and the "total luminous efficiency" $S(\bar{\rho})$ for a pupil radius of 4 mm, using $p = 0.126$.

Solution: In equation (13.13), we substitute $\bar{\rho} = 4$ mm and $p = 0.126$ and have

$$\bar{\rho}^* = 2.62 \text{ mm}$$

If we now substitute $\bar{\rho}^* = 2.62$ and $\bar{\rho} = 4.0$ into equation (13.15) we get

$$S(4) = 0.429$$

As a result of the Stiles–Crawford effect, the effective illuminance of a retinal image is not given by equation (13.9a or b). Instead, the effective illuminance levels should be the values given by these equations but modified by the total luminous efficiency factor $S(\bar{\rho})$; that is the effective retinal illuminance E'_{eff} should be better predicted by the equations

$$E'_{\text{eff}} = \tau \pi S(\bar{\rho}) L_{\text{lum}} [n' \sin(\alpha')]^2 \qquad (13.16a)$$

or

$$E'_{\text{eff}} = \tau \pi S(\bar{\rho}) L_{\text{lum}} D^2 F^2 / 4 \qquad (13.16b)$$

13.8   Binocular vision

When we look at an object of interest, the visual axes of the two eyes intersect at the object of interest. For distant vision, the visual axes are parallel and the optical axes turn outwards with an angle of inclination of twice the angle α. For closer objects, the visual axes must converge to the object of interest. The direction of gaze is controlled by three pairs of muscles; one pair lying in the horizontal plane for rotation about a vertical axis, one pair lying in the vertical plane for rotation about a horizontal axis and a pair of torsional muscles for rotation of the eye about the visual axis. For many situations we assume that these three axes intersect and the point of interesection is called the **centre of rotation of the eye**. A study by Fry and Hill (1962) places this at close to 15 mm from the corneal pole.

13.8.1   Binocular separation or inter-pupillary distance

The separation of the two eyes is usually quantified in terms of the distance between the centres of the two pupils of the eyes, when the visual axes are parallel. This distance is known as the **inter-pupillary distance** or often shortened to simply "PD". Population distributions of inter-pupillary distances, from Hofstetter (1972), Dudley and Brown (1978) and a number of other sources compiled by Harvey (1982) are given in Table 13.4. A further very comprehensive list of inter-pupillary distance means and standard deviations drawn from a large number of sources has been compiled by NASA (1978). The large differences in means and standard deviations between the surveys may be a result of the probable large errors in measuring the inter-pupillary distances and natural variations within the population.

The inter-pupillary distance will depend upon the angle of convergence of the two eyes and as the eyes converge, the pupils move closer together and this effect has some relevance to the design of binocular instruments that have converging eyepiece axes. We will discuss this particular aspect a little further in Chapter 37, where we discuss various aspects of binocular instruments.

13.8.2   Accommodation and convergence

As the eyes converge for viewing closer objects, each eye accommodates. Since the required amounts of convergence and accommodation are numerically correlated, it is not surprising that they are linked and associated at high levels in the

Table 13.4. Inter-pupillary distances drawn from various sources

	Sample size	Mean (mm)	Standard deviation
Males			
Hofstetter (1972)			
raw data	247	65.7	3.3
normal distribution		65.6	3.3
Dudley and Brown (1978)			
raw data	430	62.3	3.0
normal distribution		62.4	2.9
Harvey (1982)[a]			
Greece (all services)	1084	62.7	2.9
Iran Air Force	790	61.4	3.3
Iran Army	7884	60.8	3.2
Italy (all services)	1358	64.0	3.0
Turkey (all services)	595	63.2	3.0
U.K. Air Force	501	64.8	3.3
U.S. Air Force	4063	63.3	[b]
Females			
Hoffstetter (1972)			
raw data	101	62.0	2.9
normal distribution		61.9	3.1
Harvey (1982)[a]			
U.S. Army	8100	61.0	[b]

[a] Drawn from original sources.   [b] Not stated.

brain. Thus accommodation stimulates convergence and vice versa. The degree to which they can be dissociated depends very much upon the individual and age.

With unaided vision, the accommodation convergence relationship is not usually a problem. However with binocular instruments this is very important. Bad design can cause a conflict between accommodation and convergence demands. For example, if binocular tubes converge too much and the image is placed at infinity, there will be a conflict between convergence and accommodation demands. Therefore there will be some difficulty in seeing a clear single image. This subject is discussed in greater detail in Chapter 37.

Exercises and problems

13.1   Using the data for the Gullstrand number 1 relaxed eye, given in Appendix 3,

- (a)   Find the power of the cornea and the lens separately, and of the complete eye. Why is the power of the complete eye not equal to the sum of the corneal and lens powers?
- (b)   Find the positions of the six cardinal points. Accurately sketch the eye and mark on the sketch the positions of the cardinal points.
- (c)   For the emmetropic eye focussed at infinity, calculate the distance of the retina from the back surface of the lens.
- (d)   If the sun has an angular diameter of 32 min. of arc, find the size of the retinal image of the sun.

ANSWERS:

(a)   43.05 m^{-1}, 19.11 m^{-1}, 58.64 m^{-1}

(b) $v'\mathcal{F}' = 17.19$ mm, $\mathcal{P}'\mathcal{F}' = 22.78$ mm, $v'\mathcal{N}' = +0.13$ mm
      $v\mathcal{F} = -15.71$ mm, $\mathcal{P}\mathcal{F} = -17.05$ mm, $v\mathcal{N} = 7.08$ mm

(c)   17.19 mm

(d)   0.159 mm

13.2   Calculate the far point position of the Gullstrand number 2 eye, if the crystalline lens is removed.

ANSWER: 7.65 cm behind the cornea

13.3   Calculate the level of ametropia of a 60 m^{-1} reduced eye (index = 1.336)

(a)   that is 0.7 mm axially too long

(b)   for which the corneal radius is 5% too small

ANSWERS: (a) -1.89 m^{-1}, (b) -3.0 m^{-1}

13.4   For the following single surface cornea schematic eye, calculate the position and diameter of the entrance pupil.

corneal radius = 7.7 mm
aqueous refractive index = 1.336
distance of the iris to cornea = 3.5 mm
radius of iris aperture = 3.5 mm

ANSWERS: distance from cornea = 2.96 mm and diameter = 7.90 mm

13.5   For an ametropic eye with a refractive error of -1.5 m^{-1} and a pupil diameter of 3 mm, calculate the expected angular diameter of the defocus blur disc for distant vision.

ANSWER: 15.45 minutes of arc

13.6   Calculate the retinal illuminance of an object with a luminance of 1000 cd/m^2, if the eye has a pupil diameter of 4 mm.

ANSWER: $E = 42.93$ lux

Summary of main symbols

n' refractive index of vitreous (1.336)

Section 13.2: Refractive errors

n refractive index of cornea (1.376)

F equivalent power of the eye

$\delta l'$ change in axial length inducing a change in refractive error

δL refractive error induced by a change in axial length

r radius of curvature of the anterior corneal surface

δr change in corneal radius of curvature inducing a change in refractive error

δF_c change in power of the eye due to change in corneal radius

Section 13.4: The pupil

L_{lum} luminance of the object field (cd/m^2)

Section 13.5: Depth-of-field

$\pm\delta l$ depth-of-field in object space
$\pm\delta L$ corresponding vergence depth-of-field

Section 13.7: Photometry of the eye

E'_{eff} retinal illuminance (lux)
$L_e(r)$ Stiles–Crawford function
r ray height (usually in mm) in the pupil
p Stiles–Crawford function attenuation factor
$\bar{\rho}^*$ equivalent pupil radius if no Stiles–Crawford effect
$S(\bar{\rho})$ total luminous efficiency factor

References and bibliography

*Applegate R.A. and Lakshminarayanan V. (1993). Parametric representation of Stiles-Crawford functions: normal variation of peak location and directionality. *J. Opt. Soc. Am. (A)* 10(7), 1611–1623.

*Bedell H.E. and Enoch J.M. (1979). A study of the Stiles–Crawford (S-C) function at 35° in the temporal field and the stability of the S-C function peak over time. *J. Opt. Soc. Am.* 69(3), 435–442.

*Bennett A.G. and Rabbetts R.B. (1989). *Clinical Visual Optics*, 2nd ed. Butterworth-Heinemann, Oxford.

Berck R.R. (1983). An examination of the refractive error through computer simulation. *Am. J. Optom. Physiol. Opt.* 60(1), 67–73.

*Betsch A. (1929). Über die Menschliche Refractionskurve. *Klin. Monatsbl. F. Augensheilk.* 82, 365–379 (cited by Borish).

Birren J.E., Casperson R.C., and Botwinick J. (1950). Age changes in pupil size. *J. Gerontol.* 5, 216–224.

*Blaker J.W. (1980). Toward an adaptive model of the human eye. *J. Opt. Soc. Am.* 70(2), 220–223.

*Borish I.M. (1970). *Clinical Refraction*, 3rd ed. The Professional Press, Chicago, Chapter 1.

*Campbell F.W. (1957). The depth-of-field of the human eye. *Opt. Acta* 4, 157–164.

*Campbell F.W. and Gregory A.H. (1960). Effect of pupil size of visual acuity. *Nature Lond.* 187, 1121–1123.

Campbell F.W. and Robson J.G. (1968). Application of Fourier analysis to the visibility of gratings. *J. Physiol.* 197, 551–566.

*Campbell F.W. and Westheimer G. (1958). Sensitivity of the eye to differences in focus. *J. Physiol.* 143, 18.

Carroll J.P. (1980). Apodization model of the Stiles–Crawford effect. *J. Opt. Soc. Am.* 70(9), 1155–1156.

*Charman W.N. (1989). The path to presbyopia: straight or crooked? *Ophthal. Physiol. Opt.* 9, 424–430.

*Charman N.W. and Whitefoot H. (1977). Pupil diameter and the depth-of-field of the human eye as measured by laser speckle. *Opt. Acta* 24(12), 1211–1216.

*Crawford B.H. (1937). The luminous efficiency of light rays entering the eye pupil at different points and its relation to brightness threshold measurements. *Proc. Roy. Soc. (B)* 124, 81–96.

*Davson H. (1990). Physiology of the Eye, 5th ed. Pergamon Press, New York.

*De Groot S.G. and Gebhard J.W. (1952). Pupil size as determined by adapting luminance. *J. Opt. Soc. Am.* 42, 492–495.

*Drasdo N. and Fowler C.W. (1974). Non-linear projection of the retinal image in a wide angle schematic eye. *Br. J. Ophthal.* 58, 709–714.

*Dudley R.C. and Brown V.A. (1978). The distribution of eye interpupillary distance in a survey of 430 soldiers. Army Personnel Research Establishment Memorandum No. 7/78 (U.K.)

*Enoch J.M. (1958). Summated response on the retina to light entering different parts of the pupil. *J. Opt. Soc. Am.* 48, 392–405.

*Enoch J.M. (1972). Retinal receptor orientation and the role of fibre optics in vision. *Am. J. Optom. Arch. Am. Acad. Optom.* 49(6), 455–471.

Enoch J.M. and Bedwell H.E. (1979). Specification of the directionality of the Stiles-Crawford function. *Am. J. Optom. Physiol. Opt.* 56, 341–344.

*Fry G.A. and Hill W.W. (1962). The center of rotation of the eye. *Am. J. Optom.* 39, 581–595.

*Harvey R.S. (1982). Some statistics of interpupillary distance. *Optician* November 12, Number 4766, Vol. 184, p. 29.

*Hofstetter H.W. (1965). A longitudinal study of amplitude changes in presbyopia. *Am. J. Optom. Arch. Am. Acad. Optom.* 42(1), 3–8.

*Hofstetter H.W. (1972). Interpupillary distances in adult populations. *J. Am. Opt. Assoc.* 43(11), 1151–1155.

*Kadlecova V., Peleska M., and Vasko A. (1958). Dependence on age of the diameter of the pupil in the dark. *Nature* 182, 1520–1521.

*Krakau C.E.T. (1974). On the Stiles-Crawford phenomenon and resolution power. *Acta Ophthalmologica* 52, 581–583.

*Kumnick L.S. (1954). Pupillary psychosensory restitution and aging. *J. Opt. Soc. Am.* 44(9), 735–741.

*Lotmar W. (1971). Theoretical eye model with aspherics. *J. Opt. Soc. Am.* 61, 1522–1529.

*Moon P. and Spencer D.E.S. (1944). On the Stiles-Crawford effect. *J. Opt. Soc. Am.* 34, 319–329.

*NASA (1978). *Anthropometric Source Book*, Volume II. Editors – Staff of Anthropology Research Project, Webb Associates (Washington). National Aeronautics Space Administration, Scientific and Technical Information Office, Springfield, VA.

*Navarro R., Santamaría J., and Bescós J. (1985). Accommodation dependent model of the human eye with aspherics. *J. Opt. Soc. Am.* A 2(8), 1273–1281.

O'Brien B. (1946). A theory of the Stiles and Crawford effect. *J. Opt. Soc. Am.* 37, 506–509.

O'Brien B. (1947). The Stiles and Crawford effect in polarized light. *J. Opt. Soc. Am.* 37, 275–278.

*Ogle K.N. and Schwartz. J.T. (1959). Depth of focus of the human eye. *J. Opt. Soc. Am.* 49(3), 273–280.

*Polyak S. (1957). *The Vertebrate Visual System.* University of Chicago Press, Chicago.

Safir A. and Hyams L. (1969). Distribution of cone orientations as an explanation of the Stiles-Crawford effect. *J. Opt. Soc. Am.* 59, 757–765.

Safir A, Hyams L., and Philpot J. (1971). The retinal directional effect: a model based on the Gaussian distribution of cone orientations. *Vis. Res.* 11, 819–831.

Said F.S. and Sawires W.S. (1972). Age dependence of changes in pupil diameter in the dark. *Opt. Acta* 19(5), 359–362.

*Scheerer R. (1928). *Deutsche Ophthalmologische Gesellschaft Heidelberg* 47, 118 (cited by Borish).

Slataper F.J. (1950). Age norms of refraction and vision. *Arch. Ophthal.* 43, 466–481.

*Smith G. (1982). Angular diameter of defocus blur discs. *Am. J. Optom. Physiol. Opt.* 59(11), 885–889.

Snyder A.W. and Pask C. (1973). The Stiles-Crawford effect – explanation and consequences. *Vis. Res.* 13, 1115–1137.

*Sorsby A., Benjamin B., Davey J.B., Sheridan M., and Tanner J.M. (1957). Emmetropia and its aberrations. *Spec. Rep. Ser. Med. Res. Counc. Lond.* Number 293, HM Stationery Office.

*Spring K.H. and Stiles W.S. (1948). Variation of pupil size with change in angle at which the light stimulus strikes the retina. *Br. J. Ophthal.* 32, 340–345.

*Stenström S. (1948). Investigation of the variation and the correlation of the optical elements of the human eye, Parts III and IV, Chapter III. *Am. J. Optom. Arch. Am. Acad. Optom.* 25, 340–350, 438–449.

*Stiles W.S. (1937). The luminous efficiency of monochromatic rays entering the eye pupil at different points and a new colour effect. *Proc. Roy. Soc. (Lond.)* B123, 90–118.

*Stiles W.S. and Crawford B.H. (1933). The luminous efficiency of rays entering the eye pupil at different points. *Proc. Roy. Soc.* B112, 428–450.

*Sun F., Stark L., Nguyen A., Wong J., Lakshminarayanan V., and Mueller E. (1988). Changes in accommodation with age: static and dynamic. *Am. J. Optom. Physiol. Opt.* 65(6), 492–498.

*Van Meeteren A. (1974). Calculations on the optical modulation transfer function of the human eye for white light. *Opt. Acta* 21, 395–412.

*Weale R.A. (1961). Retinal illumination and age. *Trans. Illum. Eng. Soc.* 26(2), 95–100.

*Westheimer G. (1970). Image quality in the human eye. *Opt. Acta* 17(9), 641–658.

Wixson R. (1950). Frequency distribution and range of refractive errors in 500 ametropic eyes. *Pennsylvania State College of Optometry Alumni Bulletin* 3(12).

*Woodhouse J.M. (1975). The effect of pupil size on grating detection at various contrast levels. *Vis. Res.* 15, 645–648.

14

Ophthalmic lenses

14.0   Introduction

The primary role of an ophthalmic lens is to correct a refractive error of the eye, thus allowing the eye to clearly see objects at a chosen distance. The refractive error may be due to myopia, hyperopia, presbyopia or astigmatic errors and these have been explained in Chapter 13. However, optical refractive corrections have some side effects such as altering the effective positions of the near and far points, altering retinal image sizes, making it more difficult to satisfactorily use visual instruments and finally their aberrations may lead to reduced visual performance. These different aspects of ophthalmic lenses will now be discussed in detail.

14.0.1   *Spectacle lenses, contact lenses or intra-ocular lenses*

Either spectacle or contact lenses may be used to correct refractive errors. Both have their advantages and disadvantages. For example, spectacle lenses have little or no biological interaction with the tissues of the eye and therefore cause less or no biological reaction. However, they have a more restricted visual field and affect the size of the retinal image. This change in retinal image size is called **spectacle magnification** and the magnitude increases with lens power and distance of the lens from the eye or more strictly, from the entrance pupil. Contact lenses also have spectacle magnification but since these are much closer to the pupil, their spectacle magnification is much less than that of spectacle lenses. Intra-ocular lenses are artificial lenses inserted in the eye to replace the original lens after it has been removed, usually because of a cataract. Because these lenses are placed close to or in the same position as the original lens, that is close to the pupil, their spectacle magnification is almost zero.

14.1   Spectacle lenses

14.1.1   *Power of the spectacle lens*

The term "power" will be used here to mean the equivalent power of the lens. It should be remembered that there is more than one type of refractive power: namely surface power, vertex power and equivalent power. However equivalent power is rarely used to specify ophthalmic lenses. Instead an ophthalmic lens is usually specified by its back vertex power. In this chapter, we are restricting

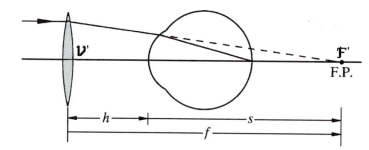

Fig. 14.1:   The power
$F(=1/f)$ of a correcting
spectacle lens. F.P. = far
point of the eye.

the discussion to thin lenses and in this case, vertex and equivalent power will be equal.

The actual power required to correct a refractive error depends upon the position of the near and far points and the desired object or working distance. With younger people, it is usual to correct for errors in distance vision. Only with older age and the onset of presbyopia and the receding near point is it also usual to correct for near vision. Therefore, because correction of distance vision is the more common case, the following discussion will only deal with ophthalmic lenses prescribed for distance vision.

The required power of the spectacle lens will depend upon the distance **(vertex distance)** of the lens from the eye and the far point distance. We pointed out in the preceding chapter that the eye can be corrected by a lens whose back focal point coincides with the far point of the eye. Figure 14.1 shows an example of hyperopia. If the far point is at a distance s from the corneal vertex, then the required focal length of the spectacle lens is

$$f = s + h$$

where h is the vertex distance to the corneal vertex and always regarded as positive. The power F is the reciprocal of the focal length f and therefore

$$F = 1/f$$

Since in practice the spectacle lens can be placed at any distance from the eye, within limits, one should write F as a function of h as follows

$$F(h) = 1/(s + h) \tag{14.1}$$

Example 14.1: Calculate the required power of a spectacle lens placed 15 mm in front of a myopic eye with a far point at a distance 55 cm from the corneal vertex.

Solution: The power of the spectacle lens is given by equation (14.1) with $s = -0.55$ m and $h = 0.015$ m, that is

$$F = 1/(-0.55 + 0.015) = -1.87 \text{ m}^{-1}$$

In clinical practice, s is never measured directly and alternative techniques are used to measure the refractive error. One method uses a number of lenses of different power placed at some vertex distance h_1, then the highest positive (or lowest negative) power $F(h_1)$ of the lens is found, which just gives clear viewing for a target at infinity. When this condition is satisfied, the far point distance s would be given by the equation

$$s = 1/F(h_1) - h_1 \qquad (14.2)$$

Sometimes, the spectacle lens is finally placed at a different vertex distance, say h_2. The required power $F(h_2)$ of the lens in this new position is given by the equation

$$F(h_2) = 1/(s + h_2)$$

If we eliminate s from the above two equations, we can express the value of $F(h_2)$ in terms of $F(h_1)$, h_1 and h_2 by the equation

$$F(h_2) = \frac{F(h_1)}{(h_2 - h_1)F(h_1) + 1} \qquad (14.3)$$

Example 14.2: If the spectacle lens power at a vertex distance of 12 mm were -3.50 m^{-1}, what would be the power as a contact lens?

Solution: For a contact lens, the vertex distance is zero and therefore $h_2 = 0$. If we substitute this value, $h_1 = 12$ mm and $F(h_1) = -3.5$ m^{-1} into equation (14.3) we get

$$F(h_2 = 0) = -3.36 \text{ m}^{-1}$$

Apart from giving the power of the spectacle lens at vertex distances other than that used to measure the refractive error, this equation is useful for determining the fitting tolerances for high power lenses.

14.1.2   Effect on the working distance

A spectacle lens changes the apparent positions of objects seen through the lens. Figure 14.2a shows a positive power lens with an object point o at a distance w. Because this is a positive power lens, the image o' must be farther from the lens. Let this distance be w'. For a negative power lens shown in Figure 14.2b, the object is imaged closer.

These two distances w and w' are connected by the lens equation (3.15) with $n = n' = 1$, here written as

$$\frac{1}{(w' + h)} - \frac{1}{(w + h)} = F$$

for the thin lens case, where w and w' are negative in Figure 14.2. Solving for w we have

$$w = \frac{w' + h}{[1 - (w' + h)F]} - h \qquad (14.4a)$$

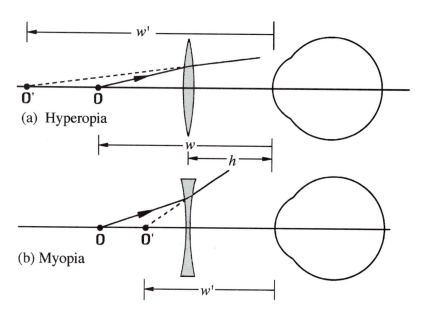

Fig. 14.2: Effect of a
spectacle lens on the
working distance of the eye.

(a) Hyperopia

(b) Myopia

or solving for w' we have instead

$$w' = \frac{w + h}{[1 - (w + h)F]} - h \qquad (14.4b)$$

14.1.2.1 Special cases: the far and near points

The purpose of the spectacle lens is to move the far point to infinity. Let us look at the effect on the near point. Naturally, if the far point moves away from the eye as in the case of the corrected myopic eye, the near point must also recede. The opposite occurs for the corrected hyperopic eye.

Suppose that the near point distance for the unaided eye is d_{np} as shown in Figure 14.3. This must also correspond to the near point distance of the image viewed through the spectacle lens and hence is the distance w' in equation (14.4a). The distance w corresponds to the distance $(d_{np})_{new}$ which is the conjugate of the near point distance as seen through the spectacle lens. Thus on substituting $w = (d_{np})_{new}$ and $w' = d_{np}$ in equation (14.4a), we have

$$(d_{np})_{new} = \frac{d_{np} + h}{[1 - (d_{np} + h)F]} - h \qquad (14.5)$$

Example 14.3: Consider a myopic eye corrected by a spectacle lens of power -2 m^{-1} placed 15 mm from the eye. Calculate the new effective near point position for the corrected eye if the unaided near point distance is 20 cm away.

Solution: Here $d_{np} = -0.20$ m, $h = 0.015$ m and $F = -2$ m^{-1}. Substitution of these values into equation (14.5) gives

Fig. 14.3: Effect of a
spectacle lens on the near
point of the eye.

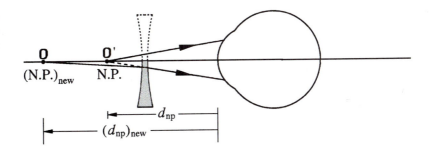

$$(d_{np})_{new} = \frac{-0.20 + 0.015}{[1 - (-0.20 + 0.015)(-2.0)]} - 0.015 = -30.9 \text{ cm}$$

Thus the effective near point has moved about 10 cm farther away from the eye, increasing the effective closest reading distance.

14.1.3 Equivalent power of the lens/eye system

If the back principal point of the spectacle lens coincides with the front focal point of the eye, the results of the discussion in Section 3.7 in Chapter 3 show that the power of the spectacle/eye system is that of the eye only. A scan of the schematic eyes in Appendix 3 shows that the front focal point of the eye is about 15 mm in front of the cornea of the relaxed eye. Typical spectacle lens vertex distances vary from about 12 to 15 mm, so that the spectacle lens is usually placed close to the front focal point of the eye. However this does not imply that there is little or no change in retinal image size as will be seen in the next section.

14.1.4 Spectacle magnification

It will now be shown that spectacle lenses change the size of retinal images; that is the size of the retinal image before and after correction is different. It will be shown that spectacle lenses produce a magnification for positive power lenses and a minification for negative power lenses. It will be seen that the magnitude of this effect depends on the spectacle lens power and its vertex distance. This magnification effect is called **spectacle magnification**.

While spectacle magnification can be defined and calculated for any distance, it is most commonly used for distance vision (i.e. the object at infinity). For near vision, the spectacle lens magnification is identical to the magnification of a simple magnifier to be discussed in the next chapter. Spectacle magnification, here denoted as SM, is defined as

$$SM = \frac{\text{retinal image size after correction (sharp)}}{\text{retinal image size before correction (blurred)}} \tag{14.6}$$

This ratio of the retinal image sizes is equal to the ratio of the angular "object" sizes, provided that the angles are measured from the entrance pupil. Therefore we can write the above equation as

$$SM = \theta'/\theta \tag{14.7}$$

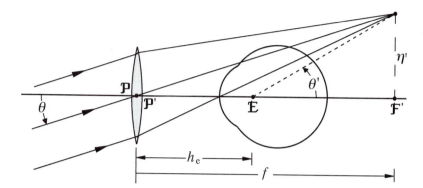

Fig. 14.4: Effect of correcting spectacle lens on the retinal image size (spectacle magnification or *SM*).

where θ is the angular size of the blurred object seen with the uncorrected eye and θ' is the angular size of the image seen through the spectacle lens, with both angles measured at the entrance pupil. However, if the object is at infinity, the angle θ can be measured from any point.

Figure 14.4 shows an example of hyperopic correction. The eye is viewing an object at infinity of angular size θ. If the eye were viewing this object directly without a spectacle correction, the apparent angular size would be θ but the retinal image would be blurred. The object is imaged in the back focal plane (at $\mathcal{F}'$) of the spectacle lens. The image size η' is given by equation (3.52) but with $n = 1$,

$$\eta' = \theta f = \theta/F$$

that is

$$\theta = \eta' F$$

where F is the equivalent power of the lens. The eye will see this object in focus on the retina and the angular size θ' of the image at the entrance pupil ε of the eye is given by the equation

$$\theta' = \eta'/(f - h_e)$$

where h_e is the vertex distance from the eye's entrance pupil at ε. Therefore equation (14.7) can now be re-expressed as

$$SM = \frac{\eta'}{(f - h_e)\eta' F}$$

which reduces to

$$SM = \frac{1}{(1 - h_e F)} \qquad (14.8)$$

This equation shows that for myopic corrections or negative power lenses there is a reduction in image size, and for hyperopic corrections or positive power lenses there is an increase in image size. In both cases, if $h_e = 0$, there is no change in image size. This result agrees with the conclusion in Chapter 10 that an auxiliary lens placed at a pupil does not change the size of the image.

> **Example 14.4:** Calculate the spectacle magnification of a spectacle lens of power -3.5 m^{-1} placed 15 mm in front of the cornea.

> **Solution:** To solve this problem, we use equation (14.8), but we need the value of the vertex distance from the entrance pupil of the eye and not the cornea. If we use Gullstrand's number 1 relaxed schematic eye, the entrance pupil is 3.05 mm inside the eye and therefore in this case

> $$h_e = 15 + 3.05 = 18.05 \text{ mm}$$

> Substituting this value and $F = -3.5$ m^{-1} into equation (14.8) gives

> $$SM = 0.941$$

> which is equivalent to a reduction of 6% in image size.

14.1.5   Effect on pupil position

Spectacles are often worn when using visual optical instruments. The presence of the spectacle lens restricts the closest distance the eye can be placed to the instrument. For those instruments with short **eye reliefs**, say 8 mm, a spectacle lens placed at a typical vertex distance of 12 mm will prevent adequate pupil matching, with a resultant loss in field-of-view. The ophthalmic lens not only prevents the eye coming close to the eye lens but also modifies the position and size of the eye entrance pupil. Let us examine how it does this.

One can obtain an idea of the effect by considering a thin lens placed a distance h in front of an eye with an entrance pupil at a distance of 3.05 mm inside the eye, the value from the Gullstrand number 1 relaxed eye given in Appendix 3. This situation is shown in Figure 14.5. A thick lens only complicates the mathematics and adds little to the understanding of the effect and the final general trends. The entrance pupil of the eye is at a distance $\bar{l}$ from the lens, where

$$\bar{l} = -(h + 3.05)\text{mm}$$

The distance $\bar{l}'$ of the new entrance pupil, measured from the ophthalmic lens, can be found using the lens equation (3.15) and the pupil magnification can be calculated using the magnification equation (3.49). For a typical value of vertex distance $h = 12$ mm, the pupil image distance $\bar{l}'$ and magnification $\bar{M}$ are given for different values of F in Table 14.1. It can be clearly seen from this table that positive lenses increase the distance and size of the entrance pupil from the lens and negative lenses decrease the distance and size.

14.1.6   Notes on thick lenses

The preceding discussion has been restricted solely to thin lenses. If one extends the discussion to thick lenses, the vertex distances h and h_e should be measured from the back principal plane of the lens. Furthermore, since ophthalmic lenses

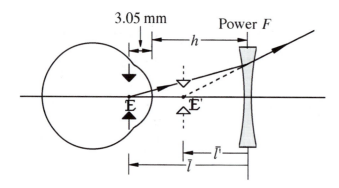

Fig. 14.5: Effect of correcting spectacle lens on the position of the entrance pupil of the eye.

Table 14.1. Distance $\bar{l}'$ of the effective eye entrance pupil from the spectacle lens and its magnification $\overline{M}$ for a thin lens as a function of power F, using a vertex distance h of 12 mm and the entrance pupil of the eye taken as being 3.05 mm inside the eye

F (m^{-1})	$\bar{l}'$ (mm)	$\|\overline{M}\|$	F (m^{-1})	$\bar{l}'$ (mm)	$\|\overline{M}\|$
1	15.3	1.02	−1	14.8	0.99
2	15.5	1.03	−2	14.6	0.97
5	16.3	1.08	−5	14.0	0.93
7	16.8	1.12	−7	14.6	0.90
10	17.7	1.18	−10	14.1	0.87

Note: The negative signs on $\bar{l}'$ have been dropped. The situation is shown in Figure 14.5.

are specified by their back vertex power instead of the equivalent power, spectacle magnification is not usually expressed in terms of equivalent power as in equation (14.8). This equation can be transformed into the form

$$SM = \frac{1}{[1 - (t F_1/\mu)](1 - h_{ev} F_v')} \tag{14.9}$$

where t is the lens thickness, F_1 is the front surface power, μ is the refractive index, h_{ev} is the vertex distance now measured from the back vertex of the lens to the entrance pupil and F_v' is the back vertex power of the lens.

14.2 Contact lenses

Contact lenses can be treated in a similar manner and hence the same equations apply, with the vertex distances $h = 0$ and h_e being about 3.0 mm. The non-zero vertex distance h_e means that contact lenses have some spectacle magnification, although much less than the corresponding spectacle lens.

14.3 Intra-ocular lenses

Intra-ocular lenses are small artificial lenses placed inside the eye to replace the crystalline lens of the eye which has been removed, usually because of disease (e.g. cataract). They are placed in either of two positions. They may be placed in front of the iris (i.e. in the anterior chamber) or behind the iris (i.e. in the posterior chamber). Since these lenses are placed almost in the same position of the natural lens, they have almost the same power.

14.4 Aberrations

The aberrations of a single lens were briefly discussed in Section 6.1.1.4 and a specific reference was made to the aberrations of spectacle lenses. Detailed aberration theory is given in Chapter 33.

14.4.1 *Spectacle lenses*

Spectacle lens/eye systems can be regarded as small aperture, wide angle systems and therefore the most important aberrations are astigmatism, field curvature and distortion. The level of field curvature cannot be changed with a single lens by any means. Astigmatism and distortion depend upon the shape of the lens but these two aberrations cannot be eliminated simultaneously. For the most common lens powers, primary astigmatism can be eliminated or reduced to acceptable levels, with an appropriate and fortunately acceptable lens shape. In contrast, distortion cannot be made zero using manufacturable lens shapes. The sign of the primary distortion is the same as the sign of the power of the spectacle lens.

14.4.2 *Contact lenses and intra-ocular lenses*

For both contact and intra-ocular lenses, the most important aberration is spherical aberration. While the spherical aberration of an ordinary lens is a function of lens shape and therefore can be altered by bending, the spherical aberration of contact lenses cannot be altered by bending since the shape is fixed by the need for it to conform to the shape of the cornea. The only way to change the level of spherical aberration in contact lenses is to aspherize the front surface. In contrast, intra-ocular lenses can be bent, but the primary spherical aberration of a single lens can only be minimized by bending and not eliminated. However, the minimum spherical aberration of an intra-ocular lens is small compared to the contribution from the cornea.

Exercises and problems

14.1   For a myopic eye with a far point at 25 cm, calculate the power of a corrective (a) spectacle lens placed at 15 mm from the eye and (b) a contact lens.

ANSWERS: (a) -4.26 m^{-1}, (b) -4.0 m^{-1}

14.2   Calculate the spectacle magnification for a thin lens of power -5 m^{-1} placed at a vertex distance of 15 mm.

ANSWER: $SM = 0.93$

14.3   For an eye corrected with a $+12 \text{ m}^{-1}$ thin lens placed 15 mm in front of the cornea, calculate the position of the effective entrance pupil of the spectacle lens/eye system.

ANSWER: distance from lens $= 23.1$ mm

Summary of main symbols

d_{np}	near point distance
$(d_{np})_{new}$	near point distance for the spectacle/eye system
F	equivalent power of a spectacle lens
f	corresponding equivalent focal length
ε	centre of the entrance pupil of the eye
s	far point distance from the corneal vertex
w, w'	general working distances from the eye (usual sign convention operates)
h, h_1, h_2	spectacle lens vertex distances measured from the lens to the corneal vertex (always positive)
h_e	vertex distance of a lens specified as the distance from the lens to the entrance pupil of the eye (always positive)
θ	angular size of a distance object seen by the uncorrected eye and measured from the entrance pupil
θ'	angular size of the image seen through the spectacle lens and measured from the entrance pupil
SM	spectacle magnification

15

Simple magnifiers and eyepieces

15.0   Introduction

Simple magnifiers are high positive power lenses that, under appropriate conditions, provide some magnification for near objects. They have a wide range of uses; for example they are used

(a)   by those with normal eyesight to examine fine detail
(b)   as magnifiers by those with low or poor vision
(c)   as the basis of cheap 35 mm photographic slide viewers and
(d)   as the basis of eyepieces (see Section 15.4).

In all these examples, they are used to enlarge detail in the object which is near or below the resolution threshold of the eye. The resolution limit of the eye in object space is best described as an angular resolution and not a linear resolution. Figure 15.1a helps to show how these are related. For small angles, an object of size η at a distance w subtends an angle θ at theeye, where

$$\theta = -\eta/w \tag{15.1}$$

The negative sign is used to be consistent with the established sign convention. The angle θ is measured from the front nodal point $\mathcal{N}$ of the eye, but usually the distance w is much larger than the distances between any of the points – the corneal vertex, front principal point, entrance pupil or front nodal point. Therefore, in practice any of these points can be taken as the reference point, without any significant error. We can use this equation to relate the spatial frequency of periodic patterns.

It follows from equation (15.1) that if η is the linear period of a periodic pattern and θ is the angular period, then the corresponding linear spatial frequency ($\sigma_L = 1/\eta$) and angular spatial frequency ($\sigma_A = 1/\theta$) are related by the equation

$$\sigma_A = w\sigma_L \tag{15.2}$$

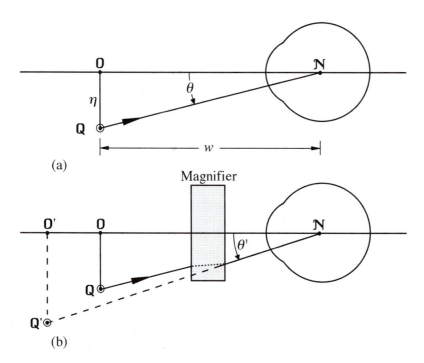

Fig. 15.1: The angular size of objects as seen by (a) the unaided eye and (b) a schematic magnifier.

with the negative sign dropped because spatial frequencies do not have a positive or negative sign. For the normal eye under optimum conditions, the resolution limit of the eye is about 30–60 c/deg. However for people with some ocular pathologies, the resolution is much lower. This equation shows that with unaided vision, the finest detail that can be seen depends upon the angular resolving power of the eye and the closest viewing distance with distinct vision, that is at the near point. Unfortunately, this recedes with age. For young people, the near point is about 7–10 cm but for people in their middle years it may have receded to 30–40 cm away. Taking 30 c/deg as a reference angular resolution limit, equation (15.2) shows that a young person focussed at 10 cm can resolve about 17 c/mm. In contrast, an older person, with the same visual acuity but with a near point at 40 cm, would only be able to resolve 4.3 c/mm at 40 cm. For advanced presbyopia, the near point may be much farther away and may be at many metres' distance or even at infinity.

If a person wishes to resolve or see more detail than is possible by simply viewing the object at the near point, more detail can be resolved with the help of optical magnifying devices. Let us now see how this is done and look at a suitable definition of magnification.

15.0.1 *Magnification*

Let us suppose now we can view the object through a magnifying system as shown in Figure 15.1b. The object plane at o is image to o' and the angular size of the image is θ'. The magnification of a magnifier is defined as an angular magnification (M) as follows

$$M = \frac{\text{angular size of image formed by magnifier } (\theta')}{\begin{array}{c}\text{angular size of object seen unaided at}\\ \text{a specified distance } (\theta)\end{array}} \qquad (15.3)$$

The position of the object seen unaided as shown in Figure 15.1a may be at any distance, but will most commonly be the near point of vision. In this chapter we will explore the conditions under which this definition leads to a magnification value that is greater than unity.

The simplest optical magnifying device is a simple positive power lens (a simple magnifier) and we will spend the rest of this chapter examing the properties of this simple type of magnifier. Other and more complex types of magnifiers such as the microscope and telescope will be discussed in subsequent chapters.

15.0.2 Pupil imagery

Simple magnifiers do not have an intrinsic pupil. When used with the eye, the iris of the eye acts as the aperture stop for the combined system. Therefore if we wish to examine certain performance criteria of a magnifier on its own, such as aberration levels, we should place a pseudo aperture stop where the entrance pupil of the eye would be. However, this will not always be very well defined since there is no fixed position of the eye when viewing through simple magnifiers.

15.0.3 Eyepieces

The foregoing discussion assumes that the object is an actual object and not an optical image. There are circumstances where the object to be magnified is an image formed by some other optical system or component. There are significant differences in ray or beam paths between these two situations so that a simple magnifier used to magnify the aerial images needs to be a little more complicated in design. This aspect will be discussed in Section 15.4.

15.0.4 Binocular viewing

Simple magnifiers are mainly designed for single eye vision, but under some circumstances can be used with the two eyes and hence provide some binocular vision. Specially designed magnifiers for this task are called **bioculars**. However, we will see that these devices have an inherently low magnification and if binocular vision is required for high magnification, then there must be one lens per eye. This type of binocular magnifier is also used in stereoscopic viewers. These aspects will be discussed in Section 15.6.

15.1 Simple magnifiers: Special case $L' = 0$

In this section, we will consider the special case in which the object is placed in the front focal plane of the magnifier and thus the image will be at infinity giving an image vergence of zero, that is $L' = 0$. This situation is shown in Figure 15.2. The more general case will covered in the next section.

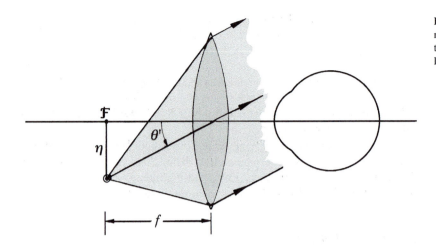

Fig. 15.2: The simple magnifier with the object at the front focal point of the lens.

15.1.1 Magnification

The object of linear size η shown in Figure 15.2 has an image of angular size θ', where

$$\theta' = -\eta/f$$

or in terms of power

$$\theta' = -\eta F \tag{15.4}$$

Because the image is at infinity, this angular size is independent of the eye position. If we compare this size with the angle θ subtended by the object at the distance w, then θ is given by equation (15.1). Substituting for this value of θ and the above value of θ' in equation (15.3) leads to

$$M_{L'=0} = wF \tag{15.5}$$

where $M_{L'=0}$ is the magnification for the image formed at infinity (i.e. zero vergence). In many texts, it is assumed that the distance w is a typical near viewing distance and has a value of about 25 cm. In this case, the above equation reduces to

$$M_{L'=0, w=25\,\text{cm}} = F/4 \tag{15.6}$$

with the equivalent power F expressed in dioptres. We will call this magnification the **nominal magnification** of a simple magnifier.

This equation indicates that a positive power lens can only provide some magnification (i.e. greater than unity) if the equivalent power is greater than $4\,\text{m}^{-1}$.

> **Example 15.1:** What is the magnification of a lens with an equivalent power of $36\,\text{m}^{-1}$?
>
> **Solution:** From equation (15.6), the magnification is $36/4 = 9$.

15.1.2   Periodic patterns

The purpose of a simple magnifier is essentially to increase the angular size of the detail. For detail of size η placed in the front focal plane of a lens of equivalent power F, the angular size θ' of the image is given by equation (15.4). It will also be useful to look at the specific case of a periodic grating. Just as equation (15.1) was converted to equation (15.2), we can write equation (15.4) in terms of the image angular spatial frequencies σ_A' as follows

$$\sigma_A' = \sigma_L/F \quad \text{c/rad} \tag{15.7a}$$

or using equation (15.5), we can express this in terms of magnification

$$\sigma_A' = w\sigma_L/M_{L'=0} \quad \text{c/rad} \tag{15.7b}$$

These equations give the angular spatial frequency σ_A' of the image of a linear pattern with a spatial frequency σ_L, seen through a simple magnifier of power F or magnification $M_{L'=0}$ defined for a reference distance w. We can write equation (15.7b) as

$$M_{L'=0} = w\sigma_L/\sigma_A' \tag{15.8}$$

which specifies the magnification required to convert a linear spatial frequency σ_L at a distance w, to an angular spatial frequency σ_A'.

> **Example 15.2:** Calculate the power and magnification of a simple magnifier required to resolve a periodic pattern of 100 c/mm at 25 cm, using 30 c/deg as the resolution threshold of the eye.
>
> **Solution:** Firstly we take
>
> $$\sigma_A' = 30 \,\text{c/deg} = 1719 \,\text{c/rad}$$
>
> and
>
> $$\sigma_L = 100 \,\text{c/mm}$$
>
> Therefore, from equation (15.7a),
>
> $$F = \sigma_L/\sigma_A' = 100/1719 = 0.0582 \,\text{mm}^{-1} = 58.2 \,\text{m}^{-1}$$
>
> Putting this value and $w = 0.25$ m into equation (15.5) or from equation (15.8) directly, we have
>
> $$M_{L'=0, w=25\,\text{cm}} = 14.5$$

Note: The above equations have been derived assuming that the angles are in radians and therefore when the above equations are used for any numerical calculations, spatial frequencies should be in cycles/radian and not cycles/degree.

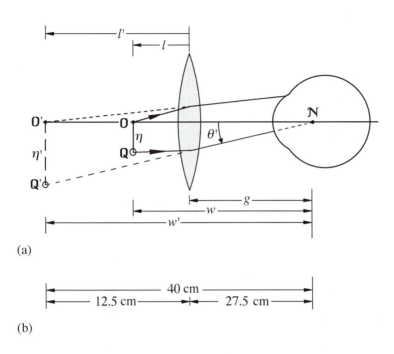

Fig. 15.3: (a) The general case of the simple magnifier. (b) A numerical example (Example 15.3) of the use of a simple magnifier with the object not at the front focal point.

(a)

(b)

15.1.3 Equivalent viewing distance

The equivalent viewing distance is the distance at which the object would have to be placed in order to have the same angular subtense as its image when viewed through the lens. This distance is the value of w in equation (15.5) that gives a corresponding magnification of 1. Therefore this distance is the focal length of the magnifier.

15.1.4 Limitations of the above equations

The two assumptions made in deriving equation (15.6) are not always valid and sometimes can lead to significant errors in predicting gains in visual acuity or performance. It is not valid for those who have near viewing distances or near points very different from 25 cm or who are myopic and cannot clearly view images at infinity. For myopes, the image must be placed at a finite distance in front of the eye; that is the object would have to be placed between the front focal point and the lens. To cope with this situation, the completely general situation will now be investigated and suitable equations derived.

15.2 Simple magnifiers: General case $L' \neq 0$

In the general case, the object will be at a distance l from the lens as shown in Figure 15.3a and not at the front focal point. Thus the image will be formed at some finite distance l' and its vergence L' at the lens will not be zero, that is $L' \neq 0$.

15.2.1 Magnification

In the general case, the angular subtense θ' of the image as seen by the eye is now

$$\theta' = -\eta'/(g - l')$$

where g is the distance from the magnifier to the eye. Using the transverse magnification defined by equation (3.46) but in the form given by equation (3.50a), with $n' = 1$, we can write this angle in the form

$$\theta' = -\eta \frac{(1 - l'F)}{(g - l')}$$

Replacing l' by its vergence L' gives

$$\theta' = -\eta \frac{(F - L')}{(1 - gL')} \tag{15.9}$$

If we define a quantity F^* by the equation

$$F^* = \frac{(F - L')}{(1 - gL')} \tag{15.10}$$

the angular size of the image can be finally written in the form

$$\theta' = -\eta F^* \tag{15.11}$$

This equation has the same form as equation (15.4) for the special case of the image at infinity, and we will call the quantity F^* the effective power of the magnifier. If we now substitute this value of θ' into equation (15.3) and use the value of θ given by equation (15.1), we have the following equation for the magnification M

$$M = wF^* \tag{15.12}$$

which has the same form as equation (15.5), except for the difference in lens power F.

One can interpret this equation as showing that a non-zero image vergence changes the ability of a positive power lens to magnify, that is changes its effective power. From equation (15.10), if the image vergence L' is negative, as it usually is, we can see that this effective power decreases with increase in the distance g of the eye from the lens. This decrease in effective power leads to a decrease in magnification. We can demonstrate this effect with the next example.

> **Example 15.3:** Consider the case of a stand magnifier with a power of 20 m^{-1} and having an image vergence of -8 m^{-1}. Let us calculate the magnification for a user with a near point of 40 cm. This value is typical for many presbyopes, the people often using simple magnifiers.

Solution: In this case, since the image has a vergence of $-8\,\text{m}^{-1}$ at the magnifier, the image is formed 12.5 cm away from the magnifier as shown in Figure 15.3b. Hence if the image is to be seen in sharp focus, the user must move his or her eye back at least a distance of 27.5 cm, that is

$$g = 0.275\,\text{m}$$

Using the following data, $F = 20\,\text{m}^{-1}$, $L' = -8\,\text{m}^{-1}$ and $g = 0.275\,\text{m}$, into equation (15.10), we have

$$F^* = 8.75\,\text{m}^{-1}$$

which is much less than the equivalent power value of $20\,\text{m}^{-1}$. Substituting this value for F^* and $w = 0.40\,\text{m}$ into equation (15.12) gives

$$M = 3.5$$

For comparison, equation (15.6) gives a corresponding value of

$$M_{L'=0,w=25\,\text{cm}} = 5$$

which, being larger, predicts a greater improvement in visual performance than is actually achieved.

Equation (15.12) is the most general form of the effective magnification of a simple magnifier and equation (15.6) is only a special case. Other special cases are discussed in Section 15.2.4.

15.2.2 Periodic patterns

Just as equation (15.4) leads to equation (15.7a), equation (15.11) leads to the equation

$$\sigma'_A = \sigma_L / F^* \tag{15.13}$$

which relates an object linear spatial frequency σ_L to an image angular spatial frequency σ'_A for the general case. This equation is identical in form to equation (15.7a) except that the equivalent power F has been replaced by the effective power F^*. In terms of the magnification, combining this equation and equation (15.12) leads to

$$M = w\sigma_L / \sigma'_A \tag{15.14}$$

which is identical to equation (15.8) for the special case of the image at infinity.

15.2.3 Equivalent viewing distance

The equivalent viewing distance is the distance at which the object would have to be placed in order to have the same angular subtense as its image when viewed through the lens. This distance is the value of w in equation (15.12) that gives a corresponding magnification of 1. Denoting this distance w_{evd} it follows that

$$w_{\text{evd}} = 1/F^* \tag{15.15}$$

15.2.4 Special cases of the magnification equation

(1) *The image vergence is zero*; that is the eye is relaxed and $L' = 0$. In this case equation (15.10) reduces to

$$F^* = F$$

and hence equation (15.12) becomes

$$M = wF \tag{15.16}$$

which is identical to equation (15.5).

(2) *The image is placed at the same distance as the object when initially viewed unaided, that is at the distance* w. In this case, the observer does not have to change the level of accommodation. Now with reference to Figure 15.3a

$$w = g - l'$$

where the negative sign occurs in front of l' because it is negative in the diagram. That is

$$L' = 1/(g - w)$$

Substituting this expression for L' into equation (15.10) leads to the equation

$$F^* = \frac{(w - g)F + 1}{w} \tag{15.17a}$$

and now substituting this expression for F^* into equation (15.12) gives

$$M = wF + 1 - gF$$

That is

$$M = M_{L'=0} + 1 - gF \tag{15.17b}$$

where $M_{L'=0}$ is given by equation (15.5). Bennett (1985) has called this special case **iso-magnification**, because the user has the same level of accommodation when viewing the object unaided as when viewing the image. Let us analyse the different components of this equation.

$M_{L'=0}$ is the magnification if the image vergence was zero

1   is due to the image being placed in the same position as the object

$-gF$ is a factor that depends upon the eye position

The sum of the first two components shows that placing the image in the original position of the object increases the magnification by one unit of magnification, compared to that if the image vergence is zero. However, the presence of the third component and the fact that it has a negative value means that this one unit gain can be lost if the eye is placed too far back from the lens. This gain is lost when

$$1 - gF < 0$$

That is

$$g > f$$

or when the eye is placed beyond the back focal point of the magnifier.

Example 15.4: Let us examine the situation given in Example 15.3, using equation (15.17b).

Solution: In that example, $F = 20\,\text{m}^{-1}$, $w = 0.4\,\text{m}$ and $g = 0.275\,\text{m}$. Substituting these in equation (15.17b), we have

$$M = (0.4 \times 20) + 1 - (0.275 \times 20)$$

$$= 8 + 1 - 5.5 = 3.5$$

as before.

The calculation in this form shows that when the eye is 27.5 cm back from the magnifier an appreciable loss in magnification results.

(3) *The eye is placed at the back focal point of the lens.* With this condition

$$g = 1/F$$

and equation (15.10) reduces to

$$F^* = F$$

and equation (15.12) becomes

$$M = wF \tag{15.18}$$

which is independent now of the image vergence L'. This is the **Badal principle**, which we will discuss further in Chapter 30. We should note that this equation is identical to equations (15.5) and (15.16) but for a different situation.

An analysis of these cases reveals the following trends.

(a)   The closer the observer's near point, the less effective is the magnification of a simple magnifier.

(b)   When the image is formed at a finite distance, the farther the eye is from the magnifier the lower the magnification.

(c)   A person who places the image nearer than infinity and accommodates may gain an increase in magnification. However, the maximum gain is 1 unit in magnification, which is less significant with higher magnifications.

It is now clear that the effective magnification or the extra amount of detail available to the eye depends not only on the equivalent power of the magnifier as implied by equation (15.5) but also on how the magnifier is used and the accommodative abilities of the user. The image vergence at the magnifier and the distance of the eye from the magnifier are of particular importance. These factors are embodied in equations (15.10) and (15.12).

The magnification is not the only important property of a simple magnifier. Other factors such as the image vergence, field-of-view and the limitations in using binocular vision are equally important, but before we look at them, we will look at an interesting application of the above general equation for magnification.

Spectacle magnification

It can be shown that the general equation for magnification given by equation (15.12) reduces to the special case of spectacle magnification given in Chapter 14. To do this we must let the object go to infinity and the image go towards the back focal point of the lens. Firstly we substitute

$$L' = F + L$$

into equation (15.10), then substitute

$$w = g - l = g - 1/L$$

into equation (15.12), combine the two equations and take the limit that L goes to zero. These operations lead to the equation

$$M = 1/(1 - gF) \quad \text{(spectacle magnification)} \tag{15.19}$$

which is the same as the equation given in Section 14.1.4. We will now proceed to examine some interesting ergonomical properties of simple magnifiers.

15.2.5   Image vergence at the eye

Referring to Figure 15.3a, the vergence W' of the image measured at the eye is

$$W' = 1/w' = 1/(g - l')$$

That is

$$W' = \frac{L'}{(gL' - 1)} \tag{15.20}$$

and this image vergence must be within the accommodation range of the user. For those with decreased accommodation ranges, limits must be placed on the position of the object being viewed and hence the image vergence of the magnifiers, particularly stand magnifiers. This limit usually means that the object must be in the vicinity of the front focal point of the lens.

> **Example 15.5:** If a presbyopic observer is wearing reading spectacles for a working distance of 33.3 cm and using a stand magnifier 20 cm from the spectacle plane, what is the desired image vergence of the magnifier? Neglect the vertex distance of the spectacles.
>
> **Solution:** We use equation (15.20) to find the value of L', with $W' = 3\,\mathrm{m}^{-1}$ and $g = 0.2\,\mathrm{m}$. Solving for L' gives
>
> $$L' = -7.5\,\mathrm{m}^{-1}$$

15.2.5.1 *Image vergence variation and object stability*

The object may not always be held steady or put in the correct position and errors in object distance cause changes in the image vergence at the eye. Variations in the position of the object plane may lead to a defocus and hence a need to change the level of accommodation. If the object to lens distance changes by an amount δl, an estimate of the change in image vergence $\delta W'$ at the eye, and hence change in accommodation demand, can be found by the differential approximation

$$\delta W' \approx \frac{\mathrm{d}W}{\mathrm{d}l}\delta l$$

On differentiating equation (15.20), we have firstly

$$\frac{\mathrm{d}W'}{\mathrm{d}l} = \frac{-1}{(gL'-1)^2}\frac{\mathrm{d}L'}{\mathrm{d}l}$$

and by differentiating the lens equation in the form

$$L' - 1/l = F$$

we have

$$\frac{\mathrm{d}L'}{\mathrm{d}l} = -(L' - F)^2$$

Thus we finally have

$$\delta W' \approx \frac{(F - L')^2}{(1 - gL')^2}\delta l = F^{*2}\delta l \tag{15.21}$$

Example 15.6: Let us consider a magnifier with a power of $20\,\text{m}^{-1}$ and an image vergence of $-3\,\text{m}^{-1}$, held 20 cm from the eye. Let us calculate the tolerance on the variation δl in the lens to object distance if the corresponding variation in image vergence $\delta W'$ has to be no greater than $\pm 0.25\,\text{m}^{-1}$, which is the approximate depth-of-field of the eye for medium pupil diameters.

Solution: In equation (15.21), we put $F = 20\,\text{m}^{-1}$, $L' = -3\,\text{m}^{-1}$ and $g = 0.2\,\text{m}$ and we have

$$\delta W' = \pm 206.6\delta l\,\text{m}^{-1}$$

Therefore if $\delta W'$ has a value of $\pm 0.25\,\text{m}^{-1}$, the corresponding value of δl is $\pm 1.2\,\text{mm}$.

It can be seen from equation (15.21) that the change in image vergence $\delta W'$ depends upon the square of the effective power F^* of the lens. Thus the accommodation demand increases approximately as the square of the equivalent power. Therefore the positioning of high power lenses is much more critical than for low power lenses. For users with reduced accommodation, this implies that the object or lens must be held very steady for lenses of high powers. This is the reason that high power simple magnifying lenses are constructed on stands and are known as stand magnifiers.

15.2.6 Field-of-view

Visual performance in many tasks depends on the size of the field-of-view. As a general rule, the larger the field-of-view the better, particularly for reading and search tasks. Unfortunately as a general rule, the higher the magnification, the smaller the object field-of-view. With simple magnifiers, the field-of-view depends greatly on the aperture diameter of the magnifier, on the distance of the eye from the magnifier and to a small extent on the diameter of the pupil of the eye.

The field-of-view can be divided into two parts, a central zone free of vignetting and an outer annular zone in which vignetting increases progressively until it becomes complete at the outer edge of this zone. Therefore for field-of-view calculations, we will consider two field sizes, one free of vignetting and the other corresponding to the boundary at which vignetting just becomes complete. The boundary between these two zones is the point where the vignetting just begins.

Figure 15.4 shows a beam of rays from an off-axis point Q that fills the entrance pupil at ε, with the magnifier lens vignetting some of the rays. The level of vignetting increases as the point Q moves farther from the optical axis.

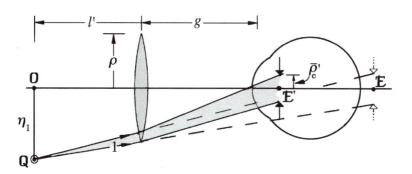

(a) Vignetting just beginning

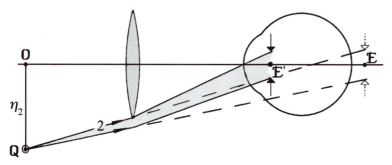

(b) Vignetting just becoming complete

Fig. 15.4:   The field-of-view of a simple magnifier.

Vignetting just begins

Vignetting just begins when ray 1 in Figure 15.4a just touches the bottom edges of the lens and entrance pupil of the eye. The corresponding object field-of-view radius η_1 can be found by tracing this ray backwards from the eye, to the object plane. This process leads to the equation

$$\eta_1 = \left[\frac{(\rho - \bar{\rho}_e)}{g} - \rho L' \right] \frac{1}{(L' - F)} \tag{15.22a}$$

According to the usual sign convention, the numerical value of η_1 will be negative from this equation.

Vignetting just becomes complete

Vignetting just becomes complete at a field radius η_2, for which ray 2 shown in Figure 15.4b just touches the bottom edge of the lens and the top edge of the entrance pupil of the eye. In deriving an expression for η_2, the only difference between this case and the preceding one is that the value of $\bar{\rho}_e$ changes sign. Therefore we can short cut any derivation by replacing $\bar{\rho}_e$ by $-\bar{\rho}_e$ in equation (15.22a) and hence

$$\eta_2 = \left[\frac{(\rho + \bar{\rho}_e)}{g} - \rho L' \right] \frac{1}{(L' - F)} \tag{15.22b}$$

Example 15.7: Calculate the field-of-view of a simple magnifier of power $40\,\text{m}^{-1}$ with an aperture radius of 8 mm and image vergence of $-10\,\text{m}^{-1}$ placed 10 mm from an eye which has a pupil radius of 2 mm.

Solution: Substituting

$$F = 40\,\text{m}^{-1} = 0.040\,\text{mm}^{-1}, \quad \rho = 8\,\text{mm}, \quad L' = -10\,\text{m}^{-1}$$

$$= -0.01\,\text{mm}^{-1}, \quad g = 10\,\text{mm} \quad \text{and} \quad \bar{\rho}_e = 2\,\text{mm}$$

into the above equations gives

$$\eta_1 = (-)13.6\,\text{mm for vignetting just beginning}$$

and

$$\eta_2 = (-)21.6\,\text{mm for vignetting just becoming complete}$$

The negative sign is not important and only occurs because the image point at ϱ in Figure 15.4 is below the axis.

Special case of the eye pupil diameter being neglected

For many calculations, the diameter of the pupil of the eye can be neglected when the aperture diameter of the magnifier is much larger than the diameter of the pupil of the eye. In this case, the zone of partial vignetting is negligible, that is has zero width, and the point of vignetting just beginning and just becoming complete is the same. Thus there is only one value of field radius in this case. An equation for this radius η can be found by setting the pupil radius $\bar{\rho}_e = 0$ in either of equations (15.22a or b), that is

$$\eta = \frac{\rho(1 - gL')}{g(L' - F)} = \frac{\rho}{gF^*} \tag{15.23}$$

Note: These equations were derived assuming the magnifier lens had no significant edge thickness, its mount does not obstruct the view and there are no aberrations.

Equation (15.23) does not show completely how the field-of-view depends upon power, because the lens aperture radius ρ must also depend upon power. As the equivalent power increases, the aperture diameter must in general decrease approximately in proportion to the focal length, that is

$$\rho \approx kf$$

or

$$\rho \approx k/F \tag{15.24}$$

where k is some unknown constant. Therefore substituting for this value of ρ

in equation (15.23) gives

$$\eta \approx \frac{k}{gFF^*} \qquad\qquad (15.25)$$

which indicates that the field-of-view of a simple magnifier is approximately inversely proportional to the square of power of the magnifier.

The field-of-view can be increased by increasing the aperture radius ρ. For a lens with spherical surfaces, this can only be done by making the lens thicker, which in turn increases the weight, and the extra field gained may be so badly aberrated to be of little use. However, the use of aspheric surfaces partly overcomes these problems. Fortuitously, the asphericity required to reduce aberrations leads to a decrease in surface power or curvature towards the periphery, which in turn leads to thinner edges and hence thinner lenses. Simple magnifiers are often constructed with aspheric surfaces.

An experimental study of the fields-of-view of a number of magnifiers has been reported by Blommaert and Neve (1987). Such a study on real magnifiers naturally includes the effects of aberrations which will reduce the actual fields-of-view below the simple theoretical values predicted by the above equations. Blommaert and Neve found that the effective field sizes were less than the theoretical values, confirming the effects of aberrations.

15.3   Simple magnifiers: Specification and verification

The equation (15.6) is often used to define the magnification of simple magnifiers which are often only specified in terms of the magnification. Thus a lens described as a four-times magnifier would be expected to have an equivalent power of $16 \, \text{m}^{-1}$. Since it has been shown that the effective magnification of a positive power simple lens depends upon the way it is used, the magnification is not a unique property of the lens, unlike equivalent power. Thus it would seem preferable to specify magnifiers in terms of equivalent power instead of magnification. If the magnifier is only described by a magnification value, users may wish to measure the equivalent power themselves. It is equally important to know the value of and know how to measure the image vergence of stand magnifiers.

In this section, we will describe methods that can be used to measure the equivalent power of a simple magnifier and also measure the image vergence of stand magnifiers.

15.3.1   *Equivalent power*

Accurate methods for the measurement of equivalent power are described in Chapter 11, but if one does not have the sophisticated equipment or does not need high accuracy, other more readily applied techniques are available, although with some loss of accuracy.

15.3.1.1   *Via vertex power*

A measurement of back focal length or back vertex power will give an estimate of the equivalent power of a simple magnifier. The focimeter can be used to measure the vertex power for any lens providing the power does not exceed the upper limit of the instrument, which is usually approximately 20 to 25 m^{-1}. The error in this method will increase with lens thickness and surface powers. From Chapter 3, the equivalent power F can be found from measured vertex powers using equations (3.24b) and (3.25b), by the equations

$$F = [1 - (dF_1/\mu)]F_v'$$
(15.26a)

and

$$F = [1 - (dF_2/\mu)]F_v$$
(15.26b)

If the vertex power is too high for the focimeter, we can measure the vertex focal length using a distant source. We can then measure the thickness and measure the surface power with a Geneva lens measure. We can use these values and an estimate of the refractive index to calculate the equivalent power from the above equations.

15.3.1.2   Direct measure of equivalent power

A simple method for measuring or estimating the equivalent focal length of any positive power lens is to image a distant object of known angular size onto or near the back (or front) focal plane of the lens, and measure the size of the image. A window several metres away, providing it is uniformly illuminated, is ideal for this purpose. If the size of the object (the window) η, its distance l and the image size η' are known, we can use the transverse magnification defined by equation (3.46) in the form given by equation (3.50b) to find the equivalent power F from the equation

$$F = (\eta/\eta' - 1)L$$
(15.27)

where $L (= 1/l)$ is the vergence of the object and η'/η and L must be negative to maintain the sign convention.

The distance l should be measured from the front principal plane of the lens but if the value of l is large, the distance can be taken to the front surface vertex of the lens, with little error.

A suitable method for measuring the size of the image is to take a piece of translucent material such as tracing paper or translucent sticky tape and place it on the scale of a transparent ruler. The ruler is then placed so that the image is sharply formed on the translucent material. The image size can then be read directly from the ruler scale.

15.3.2   Image vergence of stand magnifiers

Neutralization

(a) The image vergence of stand magnifiers can easily be measured using an array of ophthalmic lenses, for example conveniently mounted in a rack. With relaxed accommodation, an emmetropic or corrected ametropic observer views the image formed by the magnifier through the lens array held close to the magnifier. The highest positive power is used first and the power progessively decreased until the image is clear for the first time. The image vergence at the top face of the magnifier is then simply the negative value of the power of this lens.

(b) An alternative method is as follows. A ground glass or other suitable rear viewing screen material is placed in contact with the base of the magnifier and the magnifier pointed at a distant bright object. Lenses from the rack described above are placed immediately in front of the magnifier and varied until the distant object is sharply imaged on the viewing screen. The image vergence is the negative value of the power of the final lens.

Retinoscopy

An alternative method is to use retinoscopy, but the high level of aberrations normally present in a simple magnifier may prevent a satisfactory reflex from being formed. If this occurs, the aberrations can be reduced by stopping down the aperture of the magnifying lens with a mask that only uses the central few millimetres of the lens.

15.4   Eyepieces

Often the scene being viewed is not a physical object or a hard copy of some material. Instead it may be a real or aerial image formed by another lens or optical system. Such images are formed by the objective lenses of telescopes or microscopes. These images are small, usually no larger than 1 to 2 centimetres in diameter, and therefore have to be viewed with some magnification. The optical system used for viewing these images is called an **eyepiece**. The situation is analogous to viewing a hard copy or a real object with a simple magnifier. However, there are several essential differences between the two situations leading to eyepieces being constructed as modified and more complex simple magnifiers.

The first essential difference is due to the nature of the incident beam. Figure 15.5 shows in (a) a real object and in (b) an aerial image. In the case of the real object, each point in the object emits rays in all directions, some of which will pass through the magnifier. In the second case, the aerial image is formed by a preceding optical system, of which only the exit pupil is shown. Here the beam of rays leaving a point on the aerial image is limited in size and direction by the position and size of the exit pupil, and it is clear that none of the rays in the beam will pass through the magnifier. Thus because of the nature of the aerial image, the rays are all travelling obliquely away from the optical axis in a narrow cone. An eyepiece is a modified simple magnifier designed to capture all of these rays. In order to ensure a full field-of-view, eyepieces must contain a **field lens** to deviate these peripheral rays back towards the axis, as shown in Figure 15.5c, and through the simple magnifier of the eyepiece. Without a field lens in position, the field-of-view is significantly reduced. When used in an eyepiece, a simple magnifier is called an **eye lens**.

15.4.1   Power of the field lens

In general, the role of the field lens is essentially one of pupil matching, but in this case, the eye lens, like a simple magnifier, does not have an aperture stop. It is the entrance pupil of the eye which has to be matched. Therefore the power of the field lens must be chosen to match the exit pupil of the preceding optical

Fig. 15.5: Magnifying "hard copy" and aerial images using a simple magnifier.

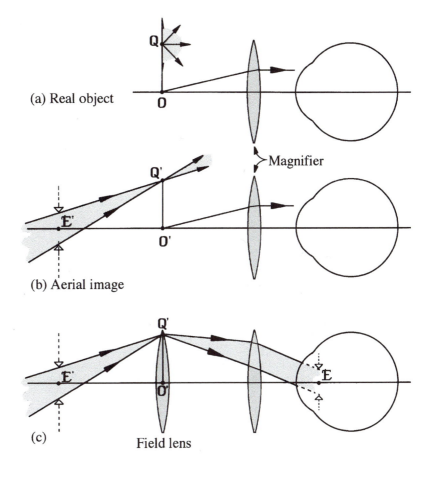

(a) Real object

Magnifier

(b) Aerial image

(c) Field lens

system to the entrance pupil of the eye which is on the opposite side of the eye lens as shown in Figure 15.5c.

15.4.2 Practical constructions of eyepieces

The schematic construction of an eyepiece is shown in Figure 15.6. In its simplest form, the eyepiece consists of two lenses: an eye lens which plays the role of the simple magnifier and a field lens. Conventionally, the image is placed in the front focal plane of the eye lens, so that the final image vergence is zero.

The field lens is usually not placed exactly in the plane of the aerial image for several reasons. Firstly, there is a need to prevent dust and scratches that may be on the field lens from being seen superimposed on the image. Secondly, graticules or reticles are sometimes placed in the aerial image plane so that measurements can be made, or sometimes cross-hairs are placed in this position to control accommodation of the observer or act as a simple reference marker.

The optical design of eyepieces is usually more complex than the schematic construction shown in Figure 15.6. The eyepiece is often designed to have aberrations that compensate for the aberrations in the remainder of the system. Aberration control is much easier with eyepieces than with simple magnifiers because while the latter have no intrinsic aperture stop, eyepieces are always

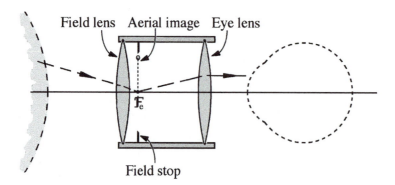

part of a system and the other part of the system contains the aperture stop. The position of the aperture stop significantly affects the aberrations, and it is easier for the optical designer to control and minimize the aberrations given a fixed position of the aperture stop.

There are a wide variety of eyepiece designs, differing in complexity and hence image quality. The degree of complexity depends upon the purpose of the eyepiece, the magnification, field-of- view, aberration control and eye relief. The two simplest and probably most common are the Huygens and Ramsden eyepieces, both of which use plano-convex lenses for both eye lenses and field lenses. These are shown in Figure 15.7. This diagram also shows the Kellner eyepiece, which is a modified Ramsden eyepiece. Further details of the Huygens and Ramsden eyepieces are as follows.

Huygens

In the Huygens eyepiece, the aerial image at o' is formed between the field and eye lenses, that is the field lens is placed farther away from the eye lens than the front focal point of the eye lens. This configuration reduces the transverse chromatic aberration and therefore gives a better image quality than the Ramsden below. The eye lens has a focal length about one half that of the field lens. This eyepiece is commonly used in microscopes.

Ramsden

Here the intermediate aerial image at o' is placed in front of the field lens, that is outside the two lenses. The eye lens and field lenses have equal focal lengths. The Ramsden design is preferred to the Huygens when a graticule is being used because the graticule and aerial image are equally affected by any aberrations in the eyepiece. On the other hand, in the Huygens eyepiece, the aerial image is affected by both the field and the eye lenses but the graticule would only be affected by the eye lens.

15.4.3 Eyepiece specification

Since the main purpose of an eyepiece is to give a magnified view of an aerial image, an eyepiece has to be specified first of all by a magnification value. The magnification of eyepieces is usually determined according to the "$F/4$" rule

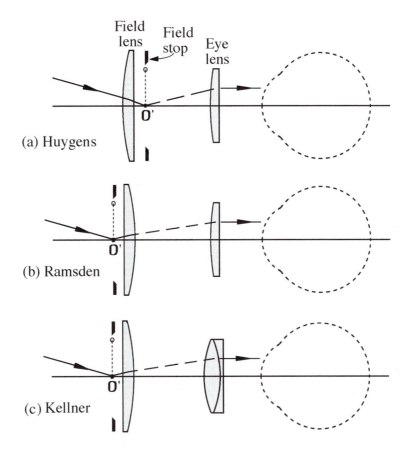

Fig. 15.7: Some simpler
eyepiece designs.

given by equation (15.6) for simple magnifiers. For example, a $15 \times$ eyepiece will have an equivalent power of $60\,\mathrm{m}^{-1}$.

15.4.4 *Exit pupil position*

Eyepieces have no intrinsic aperture stop. However, they are never used alone and use the aperture stop of the preceding part of the system. In effect, the exit pupil of this preceding part of the system acts as the entrance pupil of the eyepiece. In many instruments, as will be seen when we come to investigate microscopes and telescopes, this pupil is remote from the eyepiece, and usually much farther away than the focal length of the eyepiece. Thus the exit pupil of the eyepiece is formed near the back focal plane of the eyepiece. The higher the power of the eyepiece, the closer is the eyepiece exit pupil to its back focal point, which in turn is closer to the eye lens; thus the higher the power of the eyepiece the shorter its eye relief.

15.4.5 *Micrometer eyepieces*

Sometimes we need to make a measurement of some part of the image. This can be done in a number of ways. One method utilizes a graticule placed in the field stop or aerial image plane of the eyepiece. Size and distance measurements can then be made with the graticule. However, the accuracy of this method is limited by the interval distances on the graticule and the ability to interpolate between them. An alternative is to attach a cross-hair graticule to a micrometer gauge which moves the cross-hair transversely across the image. The accuracy of measurement is about ±0.01 mm.

15.4.6   Negative power eyepieces

An aerial image can be magnified by a negative power eye lens and the previous magnification equations, derived for a positive power simple magnifier, equally apply. However, negative power eye lenses are not routinely used because they would not be able to provide a real exit pupil or allow a graticule to be used.

15.5   Magnification limits

The upper limit of the magnification of simple magnifiers and eyepieces is limited by inherent aberrations. For a fixed pupil size (in this case the eye pupil), the aberrations of a lens increase very rapidly with equivalent power and hence magnification. This factor limits the upper magnification to about $20\times$, that is powers of $80\,\mathrm{m^{-1}}$. For higher magnifications, a microscope must be used and this has a magnification in the range 20 to about 2000. Microscopes are discussed in the next chapter.

15.6   Binocular viewing and systems

For most visual tasks, especially for prolonged viewing, binocular vision is often preferable to monocular vision. The two main advantages of binocular viewing over monocular viewing are (a) the use of two eyes to perform a visual task is less tiring than using one eye alone and (b) stereoscopic vision is possible. The simple magnifiers and eyepieces described so far are designed for monocular use. In this section, we will look at the design of magnifying systems suitable for binocular vision. These systems fall into two types. They can either consist of (a) a single lens and both eyes looking through that lens or (b) two lenses, one for each eye.

15.6.1   Single lens systems

Figure 15.8 shows a simple magnifier being used with binocular vision. In this case binocular vision is possible providing the fields-of-view of each eye overlap. So let us look at the minimum conditions for the formation of some overlap.

15.6.1.1   Conditions for binocular viewing

Figure 15.8 shows two eyes looking through the same lens. Let us derive an equation for the radius of the field of binocular overlap. Let this field be denoted by the symbol η_{bo}. From the diagram and similar triangles

$$\frac{[(p/2) - \eta'_{bo}]}{(g - l')} = \frac{[(p/2) - \rho]}{g}$$

Fig. 15.8: Binocular vision and binocular overlap with a simple magnifier.

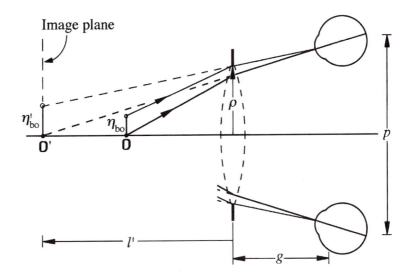

where p is the inter-pupillary distance and η'_{bo} is the image of the object field size η_{bo}. Replacing η'_{bo} by η_{bo} using the various associated transverse magnification equations given in Chapter 3, solving for η_{bo} and replacing the magnifier aperture radius ρ by its diameter ϕ, we have

$$\eta_{bo} = \frac{(\phi - p - g\phi L')}{2g(F - L')} \tag{15.28}$$

For some binocular vision, η_{bo} must be greater than zero. Since the denominator is positive, we have the condition

$$\phi - p - g\phi L' > 0 \tag{15.29}$$

Let us examine some particular cases.

Case 1: Image vergence is zero ($L' = 0$). In this case, the condition that there is some binocular vision reduces to

$$\phi > p$$

That is, providing the diameter of the magnifier is greater than the inter-pupillary distance, there will always be some binocular vision and the radius of the binocular object field will be, from equation (15.28),

$$\eta_{bo} = \frac{(\phi - p)}{2gF} \tag{15.30}$$

Case 2: Image vergence is not zero. Equation (15.28) shows that if the image vergence is not zero, there may be some binocular vision even if the inter-pupillary distance is greater than the magnifier diameter, but there is some restriction on the distance between the magnifier and eyes. There is a minimum distance, g_{min}, for binocular viewing. From equation (15.29), the minimum distance g_{min} for some binocular overlap is the value of g when η_{bo} is just zero.

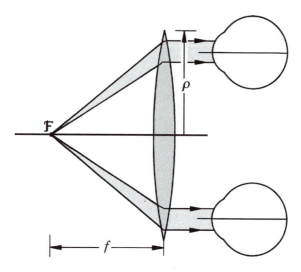

Fig. 15.9:   Biocular simple magnifier.

That is g_{min} is given by the equation

$$g_{min} = \frac{(\phi - p)}{\phi L'} \qquad (15.31)$$

If this equation gives a negative value of g_{min}, there will always be some binocular vision and if it is positive, there is a minimum viewing distance for binocular vision. Let us look at some alternative situations arising out of this equation.

Assuming the image vergence L' is negative, we have two conditions depending upon the relative values of p and ϕ.

(a)   If $p > \phi$, that is if the inter-pupillary distance is greater than the aperture diameter of the magnifier, g_{min} is positive. Therefore there is minimum viewing distance for binocular vision.

(b)   If $p < \phi$, that is the inter-pupillary distance is less than the aperture diameter, g_{min} will be negative and therefore there will always be some binocular overlap.

15.6.1.2   Bioculars

The above discussion has shown that there is always some binocular vision if the image is at infinity or providing the inter-pupillary distance is less than the diameter of the magnifier lens. In some optical magnifying systems, the image is placed nominally at infinity, in order to reduce accommodation demand, especially when there is prolonged viewing. Therefore these magnifiers must have a diameter greater than the expected inter-pupillary distance of the user. These magnifiers are called **bioculars**. A schematic biocular is shown in Figure 15.9. By allowing both eyes to be used, they cause less fatigue than monocular simple magnifiers with extended viewing. Biocular magnifiers are often used to view small cathode ray or image intensifier tubes.

One important application of bioculars is in the form of **head-up displays**. A head-up display, often referred to by its acronym **HUD**, is used in aircraft cockpits to project a cathode ray tube image onto the straight ahead field-of-view of the pilot. The display is placed below the aircraft windscreen and an arrangement of mirrors and lenses is used to project the image to the pilot's eyes. One of the mirrors is a beam-splitter that must be placed in front of the windscreen. Because of the large distances involved the pilot's eyes may be over half a metre from this beam-splitter and the biocular system. Thus the biocular must have a wide aperture to provide an adequate field-of-view. The wide aperture and aberration requirements (discussed in Chapter 37) mean that the head-up display is complex and may consist of six or more separate lenses.

Aperture diameters and field-of-view

As stated above, bioculars must have an aperture diameter greater than the inter-pupillary distance of the user and since we have assumed that the image vergence is zero, the field-of-view is given by equation (15.30). The actual field-of-view will be larger than the value given by this equation, but this remaining field will be seen with one or the other eye only and therefore will be seen only in monocular vision. A wider field-of-view can be seen by moving the head.

From Chapter 13, the population mean inter-pupillary distance is about 64 mm with a standard deviation of about 3 mm. Thus if one takes into account two standard deviations, the minimum biocular aperture diameter needs to be at least 70 mm. We can use equation (15.30) and these data to determine the required aperture diameter for a required field-of-view. Let us look at a particular numerical example.

> **Example 15.8:** Determine the minimum aperture diameter of a biocular with a magnification of four and designed to provide a binocular field of radius 25 mm for an expected inter-pupillary distance of 70 mm with the eyes placed 5 cm from the biocular.
>
> **Solution:** To find the diameter, we solve equation (15.30) for ϕ, with $\eta_{bo} = 25$ mm, $F = 16 \, \text{m}^{-1}$, $p = 0.07$ m and $g = 0.05$ m, that is
>
> $$\phi = 0.11 \, \text{m}$$

Such a large aperture diameter limits the power and hence the maximum possible magnification. The reasons are discussed below.

Magnification limits

For a lens of power F and aperture diameter ϕ, the *F-number*, as defined by equation (9.13), of the magnifier is here expressed as

$$F\text{-}number = 1/(\phi F)$$

If we now take the $F/4$ rule for magnification, then the above equation for *F-number* can be written

$$F\text{-}number = 1/(4\phi M)$$

where M is the magnification given by the $F/4$ rule. Alternatively we can write

$$M = 1/(4\phi F\text{-}number) \tag{15.32}$$

Now as explained in Chapter 9, the minimum possible F-*number* is 0.5; thus the maximum possible magnification M_{max} is given by the equation

$$M_{max} = 1/(2\phi) \quad \text{where } \phi \text{ must be in metres} \tag{15.33}$$

Example 15.9: Calculate the maximum magnification of a biocular with an aperture diameter of 110 mm (the value from Example 15.8).

Solution: Here $\phi = 0.11$ m and therefore from equation (15.33), $M_{max} = 4.5$.

Rogers (1985) quotes a maximum value of five as the upper limit of magnification of a biocular, but we have seen that it depends upon the inter-pupillary distance of the user and the desired size of the binocular field-of-view.

15.6.2   Binocular simple magnifiers

A binocular simple magnifier system consists of two simple magnifiers, one for each eye. Figure 15.10 shows two possible arrangements. In (a), the optical axes of the two magnifiers intersect at the object point of interest at o. In (b), the optical axes are parallel, and the object of interest Q now has to be off-axis. As we will see, both of these systems have problems and therefore are not ideal.

In the configuration shown in Figure 15.10a the eyes must converge to the point o, but the image may be at or near infinity, as it would be in many cases. This would require relaxed accommodation and would be in conflict with the ocular convergence. Accommodation and convergence are linked; converging eyes are usually accommodating eyes and this disparity will produce an undesirable conflict. This aspect of binocular vision is discussed further in Chapter 37.

Now let us look further at the configuration shown in Figure 15.10b. Figure 15.11 shows the same situation in more detail. The eyes are examining an object at Q and the two lenses form images at Q'_L and Q'_R in the plane through o'_L and o'_R. Now the eyes must accommodate to the points Q'_L and Q'_R but converge to the point Q where the two visual axes intersect. Because the image points Q'_L or Q'_R and the convergent point Q are at different distances, there will be a conflict between accommodation and convergence requirements. Let us look as some of the solutions to this problem.

By use of prisms

Let us take the special case of the object placed in the front focal plane of the lenses, as shown in Figure 15.12. Now the images of Q will be formed at infinity, but off-axis by an angle θ given by the equation

$$\theta = pF/2 \tag{15.34a}$$

Fig. 15.10: Two possible configurations of binocular simple magnifiers.

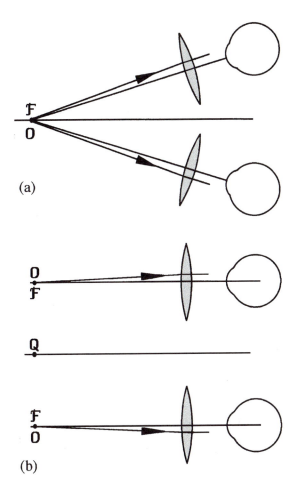

(a)

(b)

Since the images are at infinity, the visual axes of the eyes should have zero convergence; that is the images of Q should now lie on the optical axis. This requirement can be achieved by placing suitable "base-in" prisms in front of the magnifiers, as shown in the diagram. The angle of deviation θ of the prism must satisfy the above equation. Using the paraxial approximation to a thin prism discussed in Section 8.1.2.1 and equation (15.34a), the prism apex angle β must be

$$\beta = \frac{pF}{2(\mu - 1)} \tag{15.34b}$$

and the power F_p of the prism is

$$F_p = 50pF \quad \text{prism dioptres} \tag{15.34c}$$

This equation shows that the higher the power and hence magnification of the magnifying lenses, the higher the power of the prism. For high power lenses, the high prism power values may lead to undesirable transverse chromatic and other aberrations. Thus high power simple magnifiers are not suitable for binocular

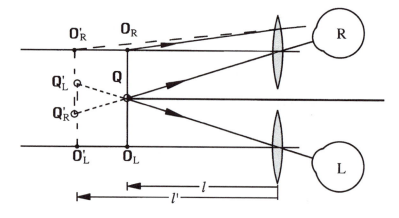

Fig. 15.11: The binocular simple magnifier shown in Figure 15.10b and effect on convergence and conflict with accommodation.

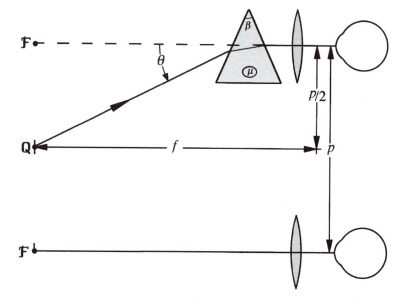

Fig. 15.12: Use of a prism to overcome the convergence and accommodation conflict when using the binocular simple magnifier shown in Figures 15.10b and 15.11.

use, unless some other more suitable arrangement is used for controlling the convergence of the binocular optical axes.

By use of decentred lenses

If instead of placing "base-in" prisms in front of or behind the magnifying lenses, one may try to achieve the same effect by using the prismatic effect of transversely displaced lenses (Section 8.1.6). In this case, if the lenses were displaced in towards the centre, they would give the same effect as a "base-in" prism. However, this possibility would not work because the displacement required is too great. From equation (8.13) a lens of power F, transversely displaced by an amount h, deviates the beam by an amount θ, given by the equation

$$\theta = (-)hF$$

Comparison of this equation with equation (15.34a) shows that the displacement h must be half the inter-pupillary distance p. A displacement of this magnitude is not practical.

15.6.3   Stereoscopic magnifiers

Stereoscopes are a binocular form of the simple magnifier, in which a pair of images are viewed through identical lenses. The optical arrangement is shown in Figure 15.13. The images are known as stereopairs. For example they may be photographs of a three dimensional scene taken from two different positions. Thus they replicate the images that would have been formed on the retina if the observer viewed the same scene. A simple stereopair is shown in the diagram. The required power of the magnifying lenses depends upon the size of the images and the required field-of-view. In typical stereoscopic magnifiers, the power of the lenses is about $10\,\mathrm{m}^{-1}$, that is have a focal length of about 100 mm.

These types of viewers are becoming increasingly used in dynamic three dimensional simulation, where images are produced on small television screens under computer control. A typical application is in **virtual reality** simulators. In these simulators, the two small television screens are placed in the front focal planes of the stereo system shown in Figure 15.13. In such applications, there is a need for as wide a field-of-view as possible. This requires a short focal length or high power. We can relate the focal length, screen size and field-of-view diameter by equation (15.4), but with the angle θ replaced by the tangent of this angle to allow extension beyond the paraxial region. For example, if the screens are 25 mm wide and the desired field-of-view is $60°$, modified equation (15.4) gives a power of $46\,\mathrm{m}^{-1}$ or a focal length of 21.7 mm. Such short focal lengths may put some strain on alignment tolerances of these devices and these aspects will be discussed further in Chapter 37. Since such short focal lengths and wide fields-of-view are likely to have large aberrations, these types of systems may be as complex as an eyepiece, in order to keep the aberrations to acceptable levels.

15.7   Aberrations

15.7.1   Simple magnifiers and eyepieces

The aberrations of simple magnifiers are difficult to control, partly because of the lack of a fixed pupil position and partly because of the simplicity of the design. The most common aberrations are astigmatism, field curvature and transverse chromatic aberration.

In contrast, eyepieces are one component in a more complex optical system and the aperture stop position is fixed by the system. Therefore the aberrations are more predictable and the aberrations inherent in the eyepiece can partly be compensated for in the remaining part of the system.

15.7.2   Bioculars

The peculiar viewing geometry of bioculars shown in Figure 15.9, that is the two visual axes that pass obliquely through the magnifier, presents difficult problems to the optical designer. For example, because the pupil ray for each eye must pass obliquely through the lens, the image will particularly suffer

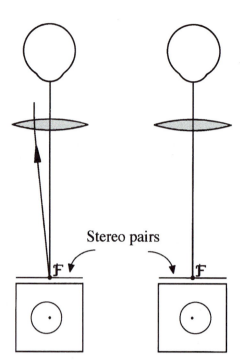

Fig. 15.13: The stereoscopic magnifier.

Stereo pairs

off-axis aberrations such as astigmatism and transverse chromatic aberration. As a result, the construction of bioculars is far more complex than that of conventional simple magnifiers, usually being multi-element design. A number of designs have been reviewed by Rogers (1985). However, the transverse chromatic aberration is not always a problem because the image is often a cathode ray tube image which may only emit narrow band light. The aberrations of these systems are discussed a little further in Chapter 37.

Exercises and problems

15.1   Given a lens with a power of $24 \, \text{m}^{-1}$ used as a simple magnifier, calculate

(a)   its nominal magnification
(b)   the angular image size of an object of size of 0.1 mm.

ANSWERS: (a) $M = 6$, (b) 8.3 minutes of arc

15.2   If the resolving power limit of the eye is 30 c/deg, calculate the minimum required magnification of a simple magnifier needed to see a grating of 100 lines/mm.

ANSWER: magnification = 14.5

15.3   A stand magnifier with a power of $10 \, \text{m}^{-1}$ is set at a height of 8 cm above the page. Calculate the image vergence at the magnifier and effective magnification for an observer who places the image at their near point, which is at 50 cm from the eye, using a reference distance w also of 50 cm. Compare this with the standard expected magnification from the $F/4$ rule.

ANSWERS: $L' = -2.5 \, \text{m}^{-1}$, $M = 5.0$ and $M(F/4) = 2.5$

15.4   In the above problem, what is the object field diameter of the field-of-view if the aperture diameter is 2.5 cm? Find the closest distance for binocular viewing if the inter-pupillary distance is 65 mm.

ANSWERS: $\eta = 12.5$ mm and $g_{min} = 64$ cm

15.5 If the field lens of an eyepiece is placed in the front focal point of the eye lens, how are the powers of the eye lens and eyepiece related and where is the front focal point of the eyepiece?

Summary of main symbols

μ	refractive index of the magnifier
d	thickness of the magnifier
g	distance between magnifier and eye
g_{min}	minimum distance between magnifier and eye for some binocular vision
ρ	aperture radius of magnifier
ϕ	corresponding aperture diameter of magnifier
p	inter-pupillary distance
F^*	"effective" power of a magnifier
F_1, F_2	front and back surface powers
F_p	prismatic power of a prism
w, w'	object and image (working) distance from the eye (always positive)
W, W'	corresponding vergences
η_1, η_2	field-of-view radii for just vignetting and complete vignetting
η_{bo}	radius of field of binocular overlap (> 0)
$\bar{\rho}_e$	entrance pupil radius of the eye
σ_L	linear spatial frequency (e.g. in c/mm)
σ_A	angular spatial frequency of an object (in c/rad)
σ'_A	angular spatial frequency of an image (in c/rad)
θ, θ'	angular size of object and image

References and bibliography

*Bennett A.G. (1985). Magnification and equivalent viewing power. *Optician* February 15, 15–17.

*Blommaert F.J.J. and Neve J.J. (1987). Reading fields of magnifying loupes. *J. Opt. Soc. Am. (A)* 4(9), 1820–1830.

Buchroeder R.A. (1988). Distortionless eyepiece. *Appl. Opt.* 27(16), 3327–3328.

Chung S.T.L. and Johnston A.W. (1990) New stand magnifiers do not meet rated levels of performance. *Clin. Exp. Opt.* 73(6), 194–199.

Ellerbrock V.J. (1946). Report on survey of optical aids for subnormal vision. *J. Opt. Soc. Am.* 36(12), 679–695.

Mehr E.B. and Freid A.N. (1975). *Low Vision Care*. The Professional Press, Chicago.

Neve J.J. (1989). On the use of hand-held magnifiers during reading. *Optom. Vis. Sci.* 66(7), 440–449.

Powell I. (1986). Dual magnification viewfinder. *Opt. Eng.* 25(1), 184–188.

*Rogers P.J. (1985). Biocular magnifiers – a review. *1985 International Lens Design Conference*, ed. W.H. Taylor and D.T. Moore. SPIE Vol. 554, Society of Photo-Instrumentation Engineers, Bellingham, Wash., pp. 362–370.

Sasieni L.S. (1975). *The Principles and Practice of Optical Dispensing and Fitting*. 3rd ed. Butterworths, London.

16

Microscopes

16.0   Introduction

The simple magnifier has an upper limit of magnification of about 20. Above this value, the lens becomes too small and the aberrations become too high to form a useful image. When higher magnifications are required, they must be achieved by a two stage process. Two stage magnification is possible by using two lenses as shown in Figure 16.1 and the extra complexity allows more freedom to control the aberrations. The first stage magnification is done by the objective and magnifications of between 10 and 100 are achieved depending upon the equivalent power of the objective. The objective forms a real, inverted and magnified image of the object. This image is further magnified by the eye lens. The eye lens is effectively a simple magnifier and therefore the upper limit of magnification is that of a simple magnifier, that is about 20. Therefore the upper limit of the magnification of the microscope as a whole is about 2000. Thus the extra magnification gained by a two component microscope over the simple magnifier is just that gained by the magnification due to the objective.

16.1   Construction and image formation

A microscope basically consists of two positive power lenses: the objective and the eye lens, as shown in Figure 16.1. The objective carries out the first stage of magnification and produces a real image of the object. The second lens (the eye lens) further magnifies the image. The objective is the aperture stop. Usually a field lens and field stop are used at or near the intermediate image plane in order to reduce vignetting and hence provide a wider field-of-view. The combination of the field lens and the eye lens forms an eyepiece, as discussed in Chapter 15. The back focal point $\mathcal{F}'_0$ of the objective and front focal point $\mathcal{F}_e$ of the eye lens are separated by a distance called the **tube length**, which is denoted here by the symbol t. We will see that this tube length affects the magnification of the objective.

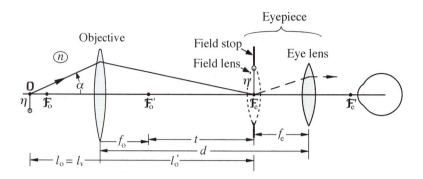

In order to produce a real image, the object must be placed beyond the front focal point $\mathcal{F}_o$ of the objective as shown in Figure 16.1. This image is conventionally formed at the front focal point $\mathcal{F}_e$ of the eye lens and thus the final image is formed at infinity. A typical image forming ray is shown in the diagram. Since the objective inverts the image and the eye lens does not, the image formed by this type of microscope construction is inverted.

Forming the final image at infinity is only nominal because in practice, users will individually focus the instrument to suit their particular refractive error. For example, myopes will need to move the image to within their far point, which is at a finite distance in front of the eye. Refocussing can be done by moving the position of the object relative to the microscope or by moving the eyepiece relative to the intermediate image.

The objective shown in Figure 16.1 is depicted as a lens. However, some microscope objectives have been designed as reflecting components. The simplest construction is a co-axial combination of concave and convex mirrors. The beam from an object point first strikes the larger concave mirror, which is behind the convex mirror. The beam reflects off the concave mirror onto the convex mirror, and after reflection from this mirror, the beam passes through a central opening in the concave mirror and forms an image behind it.

The eye lens is depicted as a positive power lens but could instead have a negative power and this would have the advantage of giving an erect image. However, apart from historical interest, this type of microscope eye lens is not used today because there would be no intermediate real image accessible for a field lens or a graticule. Without a field lens, the system would have a smaller field-of-view.

16.1.1   Cardinal points

Assuming the image is formed at infinity, the ray from the object point o in Figure 16.2a emerges from the system in image space parallel to the axis. Therefore the object point o is also the front focal point $\mathcal{F}$ of the system. The position of the front principal point $\mathcal{P}$ can be found by locating the plane of intersection of the extension of the object and image space rays as shown in the diagram. It is to the left of the front focal point.

We now trace a ray from infinity on the left, as shown in Figure 16.2b. From this ray trace we can see that the back focal and principal points are located to the right of the eye lens. The back focal point $\mathcal{F}'$ of the microscope is just to

Fig. 16.2: Cardinal points of the microscope.

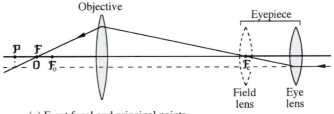

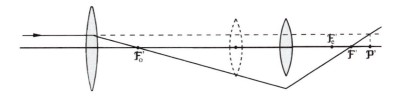

(a) Front focal and principal points

(b) Back focal and principal points

the right of the back focal point $\mathcal{F}'_e$ of the eye lens. The back principal point $\mathcal{P}'$ is to the right of the back focal point $\mathcal{F}'$.

The order of the positions of the focal and principal points in Figure 16.2 is the same as that of a negative power thick lens as shown in Figure 6.1b; that is $\mathcal{P}$ is to the left of $\mathcal{F}$ and $\mathcal{P}'$ is to the right of $\mathcal{F}'$. This implies that the microscope, as a whole, has a negative equivalent power. This observation will be confirmed by another method later in this chapter. However, $\mathcal{F}$ is on the left and $\mathcal{F}'$ is on the right of the system, which is the positioning for a positive power lens.

If the object and image space have the same refractive indices, as is the case for lower magnifications, the nodal points will coincide with the principal points. However, the highest power objectives (those with a magnification of about 100) are used with the object immersed in oil and are called oil immersion objectives. Thus the object space is oil and not air. The difference in index will displace the principal and nodal points according to equation (3.58).

The above discussion neglects the use of a field lens in the eyepiece. Suppose the eyepiece now contains a thin field lens, placed at the intermediate image. The ray from right to left in Figure 16.2a is not deviated by this ray. Therefore the field lens has no effect on the positions of the front focal and principal points. However, the ray traced from left to right, in Figure 16.2b, is deviated by the field lens and crosses the axis closer to the eye lens. Therefore a field lens would move the back focal and principal points closer to the eye lens.

16.2 Magnification

Since the microscope is essentially a two stage magnifier, we can investigate the magnification of the microscope, as a whole, by looking at the magnification of each of the two stages in turn.

16.2.1 The objective magnification

The magnification M_o of the objective or primary image is the **transverse magnification** defined in Section 3.6.5.1. From Figure 16.1 equation (3.50a), with $n' = 1$, this transverse magnification is

$$M_{\mathrm{o}} = \eta'/\eta = 1 - l_{\mathrm{o}}' F_{\mathrm{o}} \tag{16.1}$$

but

$$l_{\mathrm{o}}' = t + f_{\mathrm{o}} \tag{16.2}$$

Therefore

$$M_{\mathrm{o}} = 1 - (t + f_{\mathrm{o}}) F_{\mathrm{o}}$$

where F_{o} is the equivalent power of the objective lens, f_{o} is the corresponding equivalent focal length and t (the tube length) is the distance between the back focal point of the objective and the front focal point of the eye lens. Thus

$$M_{\mathrm{o}} = -t F_{\mathrm{o}} \tag{16.3}$$

Alternatively

$$F_{\mathrm{o}} = -M_{\mathrm{o}}/t \tag{16.4}$$

This equation shows that the magnification of a microscope objective lens depends upon its equivalent power F_{o} and the tube length t. Most microscope objectives are in fact specified by their magnification value, rather than by their equivalent power or focal length. When this is done, the most common value of the tube length (t) is 160 mm.

16.2.2   The eye lens magnification

The magnification M_{e} of the eye lens is the angular magnification of a simple magnifier and is usually expressed in the standard form of the simple magnifier, that is equation (15.6). Thus

$$M_{\mathrm{e}} = F_{\mathrm{e}}/4 \tag{16.5}$$

where F_{e} is the equivalent power of the eye lens. The field lens in the eyepiece does not affect this magnification provided that it is coincident with the intermediate image.

16.2.3   Total magnification

The total magnification M of a microscope is the product of the magnifications of the objective and eye lens, that is

$$M = M_{\mathrm{o}} M_{\mathrm{e}} \tag{16.6}$$

or

$$M = -t F_{\mathrm{o}} F_{\mathrm{e}}/4 \tag{16.7}$$

At this stage, let us pause and consider the significance of this magnification. This magnification is a product of a transverse and an angular magnification. We will see shortly that the above magnification of a microscope, as a whole, is exactly the same as the $F/4$ rule of the simple magnifier; that is it is the ratio of the angular size of the image seen at infinity through the microscope and the angular size of the object seen at a distance of 25 cm.

16.2.4   Magnification limits

As mentioned in the introduction, the upper limit of the magnification of objectives is about 100 and that of the eyepiece is about 20. Therefore the upper limit of the magnification of the microscope as a whole is about 2000. This limit is set by two factors: (i) the wavelength of the light used and (ii) the visual acuity of the observer.

Diffraction theory predicts that detail smaller than about the wavelength of the imaging radiation cannot be resolved. Therefore the upper limit of useful magnification is that which magnifies detail of the size of the wavelength to just above the threshold resolution limit of the eye. Higher magnifications would not reveal any more detail. This aspect of the microscope is discussed in greater detail in Chapter 36.

16.3   The equivalent power

The equivalent power F of the microscope can be calculated from the two lens equation (3.20) with $\mu = 1$, here written in the form

$$F = F_o + F_e - d F_o F_e \tag{16.8}$$

For two thick lenses, the distance d is the separation of the respective principal points of the pair of lenses. From Figure 16.1,

$$d = f_o + t + f_e \tag{16.9}$$

Substituting this expresssion for d in equation (16.8) leads to

$$F = -F_o F_e t \tag{16.10}$$

It should be noted that since F_o and F_e are usually positive, the equivalent power F of the microscope is negative, confirming the observations in Section 16.1.1.

We can use equation (16.7) to replace the term $-F_o F_e t$ in the above equation by the magnification M, to give

$$F = 4M \quad \text{(dioptres)} \tag{16.11}$$

where both M and F are negative. Alternatively we can write this equation as

$$M = F/4 \tag{16.12}$$

which is the same form as equation (15.6) for the nominal magnification of a simple magnifier, showing that the magnification of a microscope is the same as that of a simple magnifier with equivalent power F.

> **Example 16.1:** Calculate the equivalent power of a microscope in which both the objective and the eye lens have a magnification of 10.
>
> **Solution:** In equation (16.6), we put $M_o = -10$ and $M_e = +10$, giving
>
> $$M = (-10) \times (+10) = -100$$
>
> Now from equation (16.11), the equivalent power is
>
> $$F = 4 \times (-100) = -400 \text{ m}^{-1}$$

16.4   The equivalent simple magnifier

The forgoing discussion indicates that a microscope can be represented by an equivalent simple magnifier with an equivalent power given by equation (16.10). Since the objective and eye lens powers of a conventional microscope are positive, the power of this equivalent magnifier would be negative, as confirmed in Example 16.1. With this negative power, equation (16.12) indicates the magnification would also be negative, thus giving an inverted image.

Using the concept of the equivalent simple magnifier, we can apply some of the equations developed for simple magnifiers to microscopes. In particular, equations (15.7a and b) may be useful. Equation (15.7a) is

$$\sigma_A' = \sigma_L/F \quad \text{c/rad} \tag{16.13a}$$

which expresses the image angular spatial frequency σ_A' in terms of the object linear spatial frequency σ_L. Using the standard 25 cm reference distance, equation (15.7b) becomes

$$\sigma_A' = 0.25\sigma_L/M \quad \text{c/rad} \tag{16.13b}$$

where in this latter equation, the linear spatial frequency σ_L must be in c/m.

16.5   Working distance

One of the very important properties of a microscope is the distance between the object plane at o and the front surface of the objective. We will call this the working distance. For a thin objective lens, this distance will be to the lens itself, but for a thick real objective it will be to the front vertex of the objective. For a thin lens, we will denote this distance by the symbol l_o. Let us derive an equation for this distance in terms of the objective magnification M_o and the optical tube length t.

From the lens equation (3.15) and Figure 16.1

$$1/l_o' - n/l_o = F_o$$

If we use equation (16.2) and equation (16.3) for objective magnification M_o, it follows that

$$l_o = \frac{nt(M_o - 1)}{M_o^2} \tag{16.14}$$

Note that this working distance only depends on the objective magnification or power and not on those of the eye lens.

> **Example 16.2:** Compare the working distances for two objective magnifications (5 and 50), using an optical tube length of 160 mm and an object space of air.
>
> **Solution:** In equation (16.14), we put $t = 0.16$ m and $M_o = -5$ and $M_o = -50$ in turn. These substitutions give
>
> $$l_o = -38.4 \text{ mm } (M_o = -5) \text{ and } - 3.3 \text{ mm } (M_o = -50)$$
>
> This example shows that the higher magnifications have working distances of only a few millimetres.

Special case $M_o >> 1$:

If the objective magnification is significantly greater than unity, we can make an approximation to equation (16.14). When $M_o >> 1$, then $M_o - 1 \approx M_o$ and equation (16.14) reduces to

$$l_o \approx nt/M_o \tag{16.15}$$

As the objective magnification increases (i.e. the focal length decreases), the object point o must move towards the front focal point $\mathcal{F}_o$ of the objective. However under all conditions, because the image must be real, the point o must always be beyond this front focal point.

The working distance given by equation (16.14) is the maximum one can expect. In real cases, the objective will usually consist of a number of separate thick lenses. This complexity will usually place the front principal point inside the lens and thus reduce the distance between the object and front surface of the objective. However it is possible to design a high magnification microscope objective with a long working distance. The simplest design that satisfies this requirement is a two lens system based upon the **inverse telephoto** principle, which is discussed in detail in Chapter 21.

16.6   Pupil imagery

The aperture stop of a microscope is the objective or one sub-component of a multi-lens objective. Thus for a single thin lens objective, the objective is also the entrance pupil. The exit pupil is the image of the objective formed by the eyepiece. The distance of the exit pupil from the eye lens is called the eye relief. Figure 16.3 shows a schematic microscope, the positions of the aperture stop, pupils and the paths of the paraxial marginal and pupil rays.

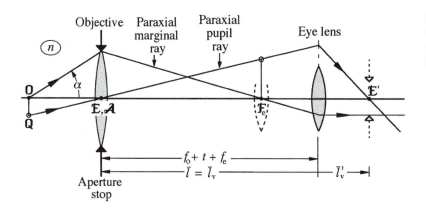

Fig. 16.3: The pupils of the microscope, the paraxial marginal ray and the paraxial pupil ray.

16.6.1 Aperture stop (or entrance pupil) size of a microscope

The size of the aperture stop is not specified in terms of its actual diameter. Instead it is defined in terms of a quantity called the numerical aperture (*NA*). With the help of Figure 16.3, this is defined as

$$NA = n \sin(\alpha) \tag{16.16}$$

where α is the angular radius of the entrance pupil as measured from the axial object point O. For low power microscopes, $n = 1$ because the objective is used in air, but for high power objectives, the object is immersed in oil so that the value of n is then the refractive index of the oil.

16.6.2 Position of the exit pupil and the eye relief

If we neglect the field lens, the distance of the exit pupil from the eye lens can be found by applying the lens equation (3.15)

$$\frac{1}{\bar{l}'} - \frac{1}{\bar{l}} = F_e$$

across the eye lens where

$$\bar{l} = -(f_o + t + f_e) \tag{16.17}$$

The distance $\bar{l}'$ is the eye relief, which we will denote by the symbol $\bar{l}'_v$, that is $\bar{l}' = \bar{l}'_v$. It then follows that

$$\bar{l}'_v = \frac{F_o + tF_oF_e + F_e}{F_e^2(1 + tF_o)} \tag{16.18a}$$

We can express the eye relief in terms of the magnifications rather than the powers, that is

$$\bar{l}'_v = \frac{4M_e - 4M_eM_o - M_o/t}{16M_e^2(1 - M_o)} \tag{16.18b}$$

However, the distance from the objective to the eye lens is usually much longer than the focal length of the eye lens and thus the exit pupil is just beyond the back focal point of the eye lens; that is the eye relief is a little longer than the focal length of the eye lens.

> **Example 16.3:** Calculate the eye relief of a microscope with an objective magnification of 100, a tube length of 160 mm and an eyepiece magnification of 20. Neglect the presence of a field lens in the eyepiece.
>
> **Solution:** Firstly $M_o = -100$, $M_e = +20$ and $t = 160$ mm. Substituting these values in equation (16.18b), we get
>
> $$\bar{l}'_v = 13.5 \text{ mm}$$
>
> compared with a value of 12.5 mm for the equivalent focal length of the eye lens.

If the microscope eyepiece also contains a field lens as it usually does, the exit pupil will be brought closer to the eye lens and be slightly smaller.

16.6.3 *Exit pupil diameter*

In Chapter 9, we showed that for a well corrected system (i.e. one that has little aberration) with the object at infinity, the entrance pupil diameter (D) and the numerical aperture in image space [$n' \sin(\alpha')$] are connected by equation (9.12b), that is

$$DF/2 = n' \sin(\alpha')$$

Microscopes are well corrected systems and therefore we can use this equation here. Since the microscope image is at infinity, we can turn this equation around and use it to express the exit pupil diameter (D') in terms of the object space numerical aperture [$n \sin(\alpha)$], by the equation

$$D' = \frac{2n \sin(\alpha)}{F} \tag{16.19a}$$

or replacing the equivalent power F by the magnification M from equation (16.11), we have

$$D' = \frac{n \sin(\alpha)}{2M} \tag{16.19b}$$

> **Example 16.4:** Calculate the exit pupil diameter for a microscope with objective and eye lens magnifications of 10, if the objective has a numerical aperture of 0.25.
>
> **Solution:** Firstly for equation (16.6), we have $M_o = -10$, $M_e = +10$ and $n \sin(\alpha) = 0.25$ and so we have
>
> $$M = M_o M_e = (-10) \times (+10) = (-)100$$

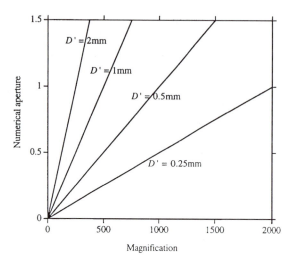

Fig. 16.4: The relationship between magnification M of the microscope and the numerical aperture of the objective for various exit pupil diameters (D') in millimetres.

Next from equation (16.19b)

$$D' = 0.25/(2 \times 100) = 1.25 \text{ mm}$$

16.6.4 Pupil matching

For optimum viewing, the eye entrance pupil should be placed in the same position as the exit pupil of the microscope and the exit pupil diameter should be of comparable size or larger. Otherwise, the image size and brightness will be reduced. However, we will see now that the exit pupil of a microscope may be much less than that of the eye and this could lead to a reduced brightness of the image.

If we wished to maintain the same exit pupil diameter for all magnifications, the numerical aperture of the objective must depend upon the magnification according to equation (16.19b), that is

$$n \sin(\alpha) = 2MD' \qquad (16.20)$$

We have plotted this equation in Figure 16.4 for a number of pupil sizes. The maximum numerical aperture in air is 1.0 and therefore this diagram shows us the approximate maximum exit pupil diameter that can be achieved with any microscope magnification. High power objectives are often "oil immersion" objectives and the immersion of the object in an oil allows the value of the numerical aperture to be a little greater than 1.0, so that with these objectives, slightly larger exit pupils may be obtained.

> **Example 16.5:** Calculate the maximum possible exit pupil diameter for a microscope with a magnification of 1500 and an objective working in air.
>
> **Solution:** Assume that the objective has the maximum possible numerical aperture, that is 1.0. If we substitute this value and $M = 1500$

in equation (16.19b), we have the answer

$$D' = 0.33 \text{ mm}$$

When the exit pupil is much smaller than the entrance pupil of the eye, the photometric efficiency of the system is not high and the image brightness will be reduced. However, in microscopy this can be overcome by providing extra illumination and this aspect will be discussed in Section 16.8. The extra illumination is also provided because some transmitting samples may have significant absorption.

16.7   Field-of-view

The size of the field-of-view of any optical instrument is an important property, so let us look at how this is affected by the design of the microscope.

The aperture stop of the microscope is the objective. In the absence of a field lens the field-of-view is therefore limited by vignetting by the eye lens. Since in practice the microscope has an eyepiece and not simply an eye lens, the field-of-view will usually be limited by the field lens or the field stop in the eyepiece. Their maximum diameter is the diameter of the tube holding them. Let us derive an equation giving an estimation of the size of the object field-of-view in terms of the diameters of these components

The objective field-of-view can be estimated as follows. The diameter of the main microscope tube is about 20 mm, but the diameter of the eyepiece field stop varies with eyepiece magnification. Lower magnification eyepieces have field stops as wide as 15 mm, but an eyepiece with a magnification of 15 may have a field stop diameter as small as 8 mm. The diameter of the field stop will limit the maximum possible usable diameter of the intermediate image. To keep the discussion as general as possible, let us assume the maximum possible radius of this intermediate image is $\eta'_{\max}$. From the definition of transverse magnification of the objective, the corresponding object radius $\eta_{\max}$ is given by the equation (16.1), that is

$$\eta_{\max} = \eta'_{\max}/M_o \qquad (16.21)$$

Example 16.6: Calculate the object field radius of a microscope with an objective with a magnification of 10 and a field stop of radius 5 mm.

Solution: Substituting $M_o = -20$ and $\eta'_{\max} = 5$ mm in equation (16.21) leads to an object field radius of

$$\eta_{\max} = (-)5/20 = (-)0.25 \text{ mm}$$

16.8   Illumination

It is often necessary to provide some auxiliary illumination of the object being examined. The most important reason for this is that very high magnifications lead to lower image luminances and smaller effective pupils of the microscope/eye system. In the case of transparent samples, auxiliary illumination also may be necessary when the sample absorbs or scatters an appreciable amount

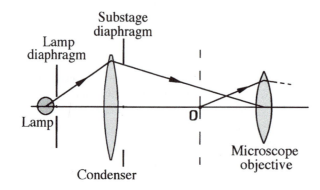

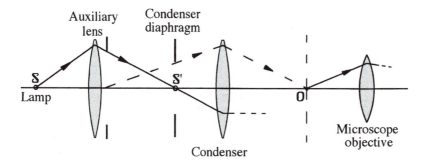

of the incident light. Reflecting samples may also need auxiliary illumination because the presence of the microscope and observer produces shadows which often reduce the ambient light level incident on the sample. The illumination systems are condenser systems.

16.8.1   Transparent Samples

When the source is fairly large, a typical illumination system is as shown in Figure 16.5. In this arrangement, the lamp source is imaged at the objective. The image of the filament or light source becomes the effective aperture stop or entrance pupil of the microscope objective and this may be smaller than the actual aperture stop.

The illuminance of the object can be varied by altering the lamp diaphragm diameter. However, since this action changes the size of the effective lamp image formed at the objective, it also may change the effective numerical aperture of the system and hence affects image quality. The substage or condenser diaphragm diameter only affects the size of the sample illuminated and hence the field-of-view.

The above illumination system is not adequate if the light source is small as this would produce a very small entrance pupil at the objective, which would in turn lead to poor image quality due to excess diffraction. In this case, the alternative arrangement, as shown in Figure 16.6, is preferable. This is known as **Köhler illumination**.

In the Köhler illumination system, the source s is imaged onto the front focal plane of the condenser lens at s' and the auxiliary lens diaphragm is imaged onto the object plane at o. The light illuminating the object is then collimated. The auxiliary lens diaphragm controls the amount of the object illuminated and the condenser diaphragm controls the illumination level of the object.

16.8.2 Reflecting samples

Reflecting samples are best observed with the incident illumination set at 45° to the optical axis of the microscope. This minimizes any specular reflections being observed.

16.9 Construction and specification of components

16.9.1 Microscope objectives

The optical construction of a microscope objective is often quite complex because of the need to reduce the aberrations to almost a negligible level. Since in general the aberrations increase with equivalent power and numerical aperture, the complexity of design increases also with power and numerical aperture.

Microscope objectives are usually specified in terms of their magnification and numerical aperture as follows.

16.9.1.1 Magnification

The magnification of a microscope objective is related to its equivalent power by equation (16.3), but this equation shows that the magnification is also dependent upon the optical tube length. Therefore if the objective is used at a different tube length, the magnification will be different.

16.9.1.2 Numerical aperture

The aperture stop of a microscope is usually contained in the objective lens. If the objective is a single lens then the aperture stop will be the lens itself. The diameter of the aperture stop affects the image brightness, image quality and depth-of-field and thus is an important part of the specification of the objective. The aperture diameter is usually specified in terms of the angular subtense of the entrance pupil measured at the object plane. In Figure 16.1, the half angular subtense is denoted by α and the aperture diameter is specified by the numerical aperture defined by equation (16.16). Table 16.1 gives the magnifications and typical numerical apertures of common objectives. A relationship between the numerical aperture of the objective, magnification of the microscope as a whole and the exit pupil diameter is given by equation (16.19b).

16.9.2 Eyepieces

The construction of some common eyepieces is discussed in Section 15.4.

Eyepieces are usually only specified by their magnification, defined by the conventional magnification equation (16.5). Occasionally, the field-of-view may be also specified.

Table 16.1. Typical magnifications of objectives and their equivalent powers and focal lengths assuming a tube length t of 160 mm

Magnification	Equivalent power (m^{-1})	Equivalent focal length (mm)	Numerical aperture (range)
Air			
4	25	40	0.1–0.16
10	62.5	16	0.25–0.4
20	125	8	0.4–0.65
40	250	4	0.55–0.95
60	375	2.666	0.8–0.9
Oil immersion			
100	625	1.6	0.9–1.35

16.10   Aberrations

Microscopes are regarded as one of the most highly corrected of all optical instruments, being almost diffraction limited. This rule particularly applies to the objectives. In general, the aberration of a lens increases rapidly with power and numerical aperture and since the objective usually has the higher power and the numerical aperture increases with power, it has potentially the most aberration. Therefore the design of objectives is often more complex and increases in complexity with power. Achromatic objectives are corrected for chromatic aberration, spherical aberration and coma. Apochromatic objectives are corrected at a third wavelength (sometimes the G-violet line). The objectives may be corrected for the spherical aberration inherent in the cover glass which is laid over the specimen. Sometimes residual aberrations in the objective are corrected in the eyepiece.

Exercises and problems

16.1 (a)   For a microscope with an optical tube length of 160 mm, calculate the equivalent focal lengths of the following objectives:

	Magnification	Answers (mm)
(i)	4	40
(ii)	10	16
(iii)	20	8
(iv)	100	1.6

(b)   Using the above optical tube length and objectives, calculate the exit pupil positions and the objective numerical apertures if the eye lens has a magnification of 10 and the exit pupil has a diameter of 2 mm. Comment on the results.

ANSWERS: (i) 28.1 mm, 0.16, (ii) 28.6 mm, 0.40, (iii) 28.7 mm, 0.80, (iv) 28.9 mm, 4.0. The numerical aperture of 4.0 is far too high and therefore it is not possible for a 1000 × microscope to have an exit pupil with a diameter of 2 mm.

16.2   Calculate (a) the magnification and (b) the eye relief of a microscope specified as follows:

(i)   objective power $= 20\ m^{-1}$

(ii)   eye lens power $= 50 \text{ m}^{-1}$

(iii)   optical tube length $= 160$ mm.

ANSWERS: (a) $M = (-)40$ and (b) eye relief $= 21.9$ mm

16.3   (a) For a microscope with an objective diameter of 4 mm, a power of 50 m^{-1} and an eye lens with a magnification of 10, calculate the diameter of the exit pupil. Take an optical tube length of 160 mm.

ANSWER: diameter $= 0.56$ mm

16.4   For a microscope fitted with an objective power of 60 m^{-1} and an eye lens power of 20 m^{-1}, what would be the equivalent power of a simple magnifier giving the same magnification?

ANSWER: $F = -192 \text{ m}^{-1}$

Summary of main symbols

d	distance between objective and eye lens
F	equivalent power of a microscope
f	corresponding equivalent focal length
F_o	equivalent power of an objective lens
f_o	corresponding equivalent focal length of the objective
F_e	equivalent power of an eye lens
f_e	corresponding equivalent focal length of the eye lens
$\mathcal{F}, \mathcal{F}'$	front and back focal points of a microscope
$\mathcal{F}_o, \mathcal{F}'_o$	front and back focal points of the objective
$\mathcal{F}_e, \mathcal{F}'_e$	front and back focal points of the eye lens
$\mathcal{P}, \mathcal{P}'$	front and back principal points of the microscope
$\varepsilon, \varepsilon'$	centres of entrance and exit pupils of the microscope
l_o, l'_o	object and image distances from the objective
$\bar{l}, \bar{l}'_v$	pupil distances from the eye lens ($\bar{l}'_v$ is the eye relief)
η_{max}	maximum object size set by maximum image size η'_{max}
η'_{max}	maximum image field size inside in microscope tube
α	angular radius of objective (from o)
t	optical tube length (usually 160 mm)
M_o	transverse magnification of objective (numerically negative)
M_e	angular magnification of eye lens (numerically positive)
M	magnification of the complete microscope
σ_L	linear spatial frequency in object
σ'_A	angular spatial frequency in image

References and bibliography

Benford J.R. and Rosenberger H.E. (1978). Microscope objectives and eyepieces. In *Handbook of Optics*. ed. W.G. Driscoll and W. Vaughan. McGraw-Hill, New York, p. 6.3.

Laikin M. (1991). *Lens Design*. Marcel Dekker, New York.

Longhurst R.S. (1973). *Geometrical and Physical Optics*, 3rd ed. Longman, London.

Martin L.C. (1966). *The Theory of the Microscope*. Blackie, London.

Oates C.W. and Young M. (1987). Microscope objectives, cover slips and spherical aberration. *Appl. Opt.* 26(11), 2043.

17
Telescopes

17.0   Introduction

As magnifying devices, simple magnifying lenses are restricted to short working distances. The magnification and working distance are closely related and therefore single lens magnifiers are only useful for magnifying objects close to the lens and the higher the magnification, the closer this distance has to be. However, if one goes to two lens systems, magnification and working distance can be made independent and in principle, one can design a two lens system to give any magnification for any working distance. The microscope may be regarded as an exception to this rule in the sense that the higher magnifications lead to shorter working distances. Thus microscopes act more like simple magnifiers and do not make full use of the potential independence of working distance and magnification that are possible with a system consisting of two lenses.

In this chapter, another special case will be considered, one in which the working distance is infinite or very long. The completely general case will be left until the next chapter. The class of instruments used for providing magnification for very distant objects are called **telescopes**. These are usually regarded as **afocal**; that is their equivalent (refractive) power is zero. Telescopes are not only used as magnifying devices. Their unique properties make them ideal as aligning or focussing devices in a range of optical instruments. Binoculars are made from two identical telescopes with parallel optical axes.

Telescopes may be classified as either refracting or reflecting. The distinction is the type of leading component called the objective. In a refracting telescope, the objective is a lens or lens system. In the reflecting telescope, the objective is a mirror or mirror system.

17.0.1   *Refracting telescopes*

There are two fundamentally different designs of refracting telescopes (namely Keplerian and Galilean), and these are shown in Figure 17.1. They will be

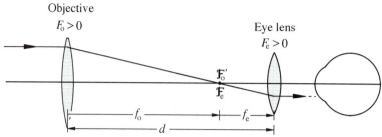

Objective
$F_o > 0$

Eye lens
$F_e > 0$

(a) Keplerian or astronomical telescope

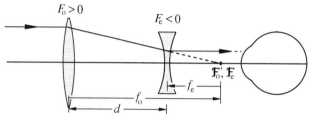

$F_o > 0$

$F_e < 0$

(b) Galilean telescope

Fig. 17.1: The Keplerian and Galilean types of refracting telescopes.

described in greater detail later but for the moment a brief description as follows will suffice. The principle of magnification will be explained in Section 17.2.

17.0.1.1 *Keplerian or astronomical*

The schematic construction of the Keplerian telescope is shown in Figure 17.1a. It has a positive power objective and positive power eye lens. The objective forms a real image of an object at infinity, in its back focal plane at $\mathcal{F}'_o$, which is also the front focal point $\mathcal{F}_e$ of the eye lens. This image is then viewed with some magnification by the eye lens, which acts as a simple magnifier. The final image is inverted and formed at infinity.

17.0.1.2 *Galilean*

The schematic construction of the Galilean telescope is shown in Figure 17.1b. This telescope has a positive power objective and a negative power eye lens. As in the above case of the Keplerian telescope, the image of an object at infinity is formed in the back focal plane of the objective and this image is viewed by the eye lens. However, in this case the eye lens has a negative power and the image is erect.

A comparison of properties of these two telescope types is given in Table 17.1 (Section 17.7.2).

17.0.2 *Reflecting telescopes*

In a number of applications such as astronomy, large diameter objective lenses are required (for example 5 m) and it would be difficult to make these from glass because of the mass of the glass lens and the difficulty in making such

Fig. 17.2: The Newtonian
and Cassegrain reflecting
telescopes.

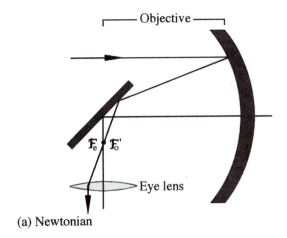

(a) Newtonian

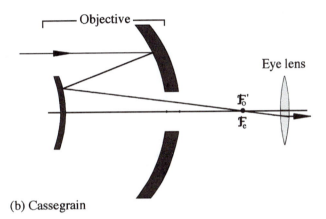

(b) Cassegrain

a large piece of glass with sufficient uniformity in refractive index. In these cases, it is better to make the objective as a reflecting component, that is a mirror. Such telescopes are called reflecting telescopes, but only the objective is reflecting. The eyepiece is a conventional (refracting) eyepiece. Reflecting telescopes are usually used only for high magnifications (approximately > 20). Apart from the smaller mass, reflecting objectives have other advantages. Most importantly, chromatic aberration of the telescope, which usually increases with magnification, is significantly reduced because the reflecting objective has no chromatic aberration. The only chromatic aberration is then due to the eye lens or eyepiece.

There are two main types of reflecting telescopes, namely the Newtonian and the Cassegrain. They are described briefly below.

17.0.2.1 Newtonian

The Newtonian reflecting telescope is shown in Figure 17.2a. The objective is a concave reflector and a plane mirror is used to deviate the beam through 90° to the eyepiece or eye lens.

17.0.2.2   Cassegrain

The Cassegrain reflecting telescope is shown in Figure 17.2b. In this design, the objective consists of two curved reflectors. The primary mirror is concave as in the Newtonian telescope, but it has a central hole cut away. The secondary mirror is convex and images the beam through this hole and into the eye lens or eyepiece. The advantage of this design is that the beam is not deviated through 90° as in the Newtonian design.

In both designs, the beam is centrally obscured by the second mirror.

17.1   The afocal condition

Any optical system that takes an object at infinity and images it also at infinity must have zero "equivalent" power. This result follows from the paraxial refraction equation (3.32) applied to a general system. Therefore telescopes have zero power and such systems are termed **afocal**.

Telescopes are in their simplest form two lens systems and if F_o and F_e are the equivalent powers of the objective and eye lens respectively and the lenses are separated by a distance d as shown in Figure 17.1, then the equivalent power F given by equation (3.20) is zero, that is

$$F = F_o + F_e - dF_oF_e = 0 \qquad (17.1)$$

For thick components, the distance d must be measured from the back principal point of the objective to the front principal point of the eye lens. The condition of zero power leads to two interesting and important conclusions.

(1) The lens separation d is given by the equation

$$d = 1/F_o + 1/F_e \qquad (17.2)$$

that is the separation d is the sum of the (equivalent) focal lengths of the objective f_o and the eye lens f_e, that is

$$d = f_o + f_e \qquad (17.2a)$$

(2) The telescope has no cardinal points and it is therefore not possible to analyse the imagery of the telescope as a single unit, using the conventional paraxial transfer and refraction equations. However, image formation can be studied by applying these equations progressively through the system, lens by lens or surface by surface.

17.2   Image formation: Infinite conjugates

The telescope is designed to view distant objects which for many purposes can be regarded as at infinity. The image is also placed at infinity so that the emmetropic user can view the image with relaxed accommodation. Thus the conventional conjugates of a telescope are at infinity and as a result the system has zero power. However, while having no power, the telescope provides some magnification.

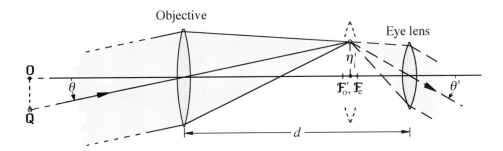

17.2.1 (Angular) magnification

The fact that telescopes have the ability to magnify objects without having any equivalent power is interesting in itself. The lack of an equivalent power limits the use of the transverse magnification equations given in Chapters 2 and 3 and we have to be wary of the use of the term "transverse magnification" here because the object and images, being at infinity, would have infinite linear sizes η and η', respectively.

The alternative is to consider angular magnification M which is defined as

$$M = \frac{\text{angular size of image } (\theta')}{\text{angular size of object } (\theta)} \tag{17.3}$$

but only for small angles, so that we stay within the paraxial approximation. Since the object and image are both at infinity, it does not matter at what point these angles are measured from, for example objective or eye lens. This definition is essentially the same as that used previously for simple magnifiers and microscopes, except that if a conjugate is at a finite distance, the angular size depends upon the position from where one measures the angle.

Consider a Keplerian telescope as shown in Figure 17.3 with an object off-axis at infinity and with an angular size θ. The image has an angular size θ' and there is some magnification if $\theta' > \theta$, as is the case in the diagram. We can find an equation for this magnification in terms of the component powers as follows.

The intermediate image shown in Figure 17.3 is formed in the back focal plane of the objective at $\mathcal{F}'_o$, and has a height η'_i, where from equation (3.52)

$$\eta'_i = \theta/F_o$$

This image is the object for the eye lens and because it is in the front focal plane of the eye lens, it is finally imaged at infinity, with an angular size θ' where

$$\theta' = -\eta'_i F_e$$

If we substitute for θ and θ', using the expressions from these last two equations, into equation (17.3), we have

$$M = -\frac{F_e}{F_o} \tag{17.4}$$

It will be left as an exercise for the reader to confirm that this equation equally applies to Galilean telescopes.

Since the Keplerian is made of two positive lenses, equation (17.4) shows that its magnification is negative and hence it gives an inverted image. In contrast, the Galilean has a positive power objective lens and a negative power eye lens and therefore has a positive magnification and hence an erect image. Having an erect image makes the Galilean immediately useful for terrestrial use and where a small mass is essential. For example, the Galilean telescope is the design used in "opera glasses" and many medical applications. The image of a Keplerian telescope can be erected by adding erecting lenses or prisms but these can significantly increase its mass, and if lenses are used, the length is increased.

Given a telescope of length d and angular magnification M, one can find the objective and eye lens powers by solving for F_o and F_e using equations (17.1) and (17.4), that is

$$F_o = \frac{(M-1)}{Md} \quad \text{and} \quad F_e = -\frac{(M-1)}{d} \tag{17.5}$$

and these equations can be used to design a schematic telescope for any magnification and length. While any telescope with a certain magnification M can be designed to have any length, it will be seen later in this chapter that the choice of d affects the eye relief in the Keplerian designs.

17.2.2 *Magnification limits*

Just as there are limits to the magnification of simple magnifiers and microscopes, there are also limits to the magnification of a telescope. This upper limit is usually set by inherent aberrations and since Keplerian and Galilean telescopes suffer different types and amounts of aberrations, their upper limits of magnifications are different. The upper limit of a Keplerian telescope is about 20 and that of a Galilean about 4.

17.3   Image formation: Finite conjugates

Conventionally, we regard the telescope as working with infinite conjugates. However, they are used frequently to view objects at finite distances. Therefore let us explore the performance of a telescope if the object is at some finite distance instead of at infinity.

17.3.1 *Image position*

When the telescope is in the afocal mode and the object is not at infinity, neither is the image but the image position cannot be found by application of the lens equation to the telescope as a whole since it has no cardinal points and no equivalent power. One alternative is to apply the paraxial refraction and transfer equations or lens equation, lens by lens through the system. The difficulty with this method is that it requires a knowledge of the individual lens powers and separations which may not be readily known. Fortunately, by algebraic ray tracing, it is possible to derive a simple equation relating object vergence L_v at the objective to the image vergence L_v' at the eye lens, in terms

Fig. 17.4: Formation of the image of an object at a finite distance – object and image distance notation.

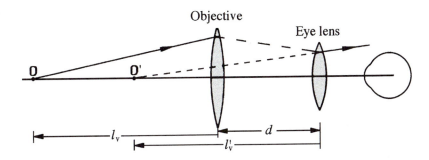

of the telescope magnification M and lens separation d. For a two lens afocal system, Fried (1977) and Smith (1979) showed that these quantities are related by the equation

$$L'_v = \frac{M^2 L_v}{(1 - dM L_v)} \tag{17.6}$$

Alternatively one can express this equation in terms of distances instead of vergences, as follows:

$$l'_v = \frac{l_v - dM}{M^2} \tag{17.6a}$$

This latter equation shows that the object and image distances are linearly related. These distances are shown in Figure 17.4 for a Keplerian telescope but once again they equally apply to a Galilean telescope.

If we transpose equations (17.6) and (17.6a) to give the object position in terms of the image position, we have

$$L_v = \frac{L'_v}{M(M + dL'_v)} \tag{17.7}$$

and

$$l_v = M(M l'_v + d) \tag{17.7a}$$

These equations are only valid for an afocal two lens system and thus do not apply to a Keplerian telescope which contains a field lens. However, the equations apply to thick lens telescopes providing the distances are measured from the respective principal planes of the objective and eye lens.

For conjugates at large distances, L_v will be small and for sufficiently large distances, the term $dM L_v$ is small relative to unity. Equation (17.6) then reduces to

$$L'_v \approx M^2 L_v \tag{17.8}$$

or in terms of distances

$$l'_v \approx l_v / M^2 \tag{17.8a}$$

Equation (17.8) shows that the image vergence increases more rapidly than the object vergence and approximately as the square of the magnification. Thus, close objects may place a heavy demand on the accommodation of the user. This accommodation demand can be reduced by refocussing the telescope, that is reduce the image vergence. This can be achieved by altering the telescope length. The properties of refocussed telescopes are discussed in Section 17.7.

17.3.2 *Transverse magnification*

In Section 3.6.5.1, we showed that if the equivalent power of a system is zero, the transverse magnification is independent of conjugate position. Therefore the transverse magnification would be the same for all conjugate positions. This result was established by a different approach in Appendix 1, Section A1.7.1. It is also shown in that appendix that this transverse magnification M_T of an afocal system is given by the equation

$$M_T = 1/M \tag{17.9}$$

where M is the angular magnification. This equation states that the transverse magnification M_T of an object at a finite distance is the inverse of the angular magnification M of the telescope.

This result may seem unexpected. If the angular magnification for infinite conjugates is M, it may seem a contradiction that the transverse magnification for the same object at a finite distance is the inverse of this value. However, there is no contradiction and the explanation can be found in Figure 17.5. In this diagram, we have assumed that the object is at a great but finite distance and in this situation we can neglect the telescope length d. The object is at a distance l_v and, from equation (17.8a), the image is at a distance l_v/M^2. If the object has a size η, then from equation (17.9), the image size η' is η/M. These dimensions are drawn to scale on the diagram with the value of M at 2. It can be seen in this diagram that the angular subtense θ of the image is greater than the angular subtense of the object and, within the paraxial approximation, these are related by the equation

$$\theta' \approx M\theta$$

showing that while the image is smaller than the object, it is much closer to the telescope, giving it a large angular size. The approximations made here become increasingly accurate as the object moves towards infinity.

17.3.3 *Longitudinal magnification*

Longitudinal magnification M_L was introduced and defined in Chapter 3. It is defined as

$$M_L = \delta l'/\delta l$$

where $\delta l'$ and δl are small shifts in the object and image positions. Thus we can

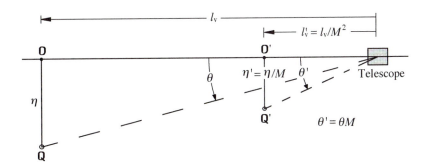

Fig. 17.5: The magnification of an object at a finite distance.

also write this as

$$M_{\mathrm{L}} = \delta l_{\mathrm{v}}'/\delta l_{\mathrm{v}} \tag{17.10}$$

in terms of image shifts from the telescope vertices. In the case of telescopes, we can use equation (17.6a) to show that this longitudinal magnification is given by the equation

$$M_{\mathrm{L}} = 1/M^2 \tag{17.11}$$

which is independent of conjugate position. Thus for objects at finite distances, both the transverse and longitudinal magnifications are independent of the object position.

17.4 Keplerian or astronomical telescopes

The schematic design of the Keplerian telescope is shown in detail in Figure 17.6a along with the path of the paraxial marginal and pupil rays. In this design, the objective forms a real inverted image in its back focal plane at $\mathcal{F}_{\mathrm{o}}'$ which coincides with the front focal plane at $\mathcal{F}_{\mathrm{e}}$ of the eye lens. The formation of this real image allows a field lens to be used to increase the field-of-view and a graticule or reticle can be used for alignment or measurement. The physical length of this telescope is given by equation (17.2a) and note that the focal lengths are both positive.

17.4.1 Pupils

The objective acts as the aperture stop of the telescope and hence is also the entrance pupil and therefore the exit pupil is the image of the objective as formed by the eye lens. Because the objective is beyond the front focal point of the eye lens, the exit pupil lies beyond the back focal point of the eye lens. Its distance from the eye lens (eye relief) can be found from applying the lens equation (3.15) to the situation shown in Figure 17.6b. Here the equation applied to the pupil imagery is

$$\frac{1}{\overline{l}'} - \frac{1}{\overline{l}} = F_{\mathrm{e}}$$

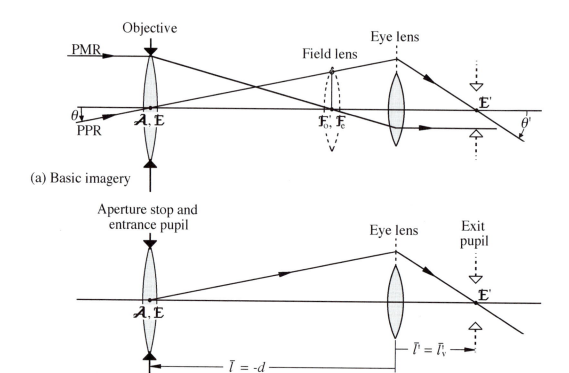

(a) Basic imagery

(b) Pupil imagery

where

Fig. 17.6: The optics of
the Keplerian telescope.
PMR = paraxial marginal
ray, PPR = paraxial
pupil ray.

$$\bar{l} = -d \qquad (17.12)$$

Solving for $\bar{l}'$, which we will now denote as the eye relief $\bar{l}'_v$, gives

$$\bar{l}'_v = \frac{d}{(dF_e - 1)}$$

If we now replace F_e by the expression given in equation (17.5), we have

$$\bar{l}'_v = -d/M \qquad (17.13)$$

Alternatively, we can take equation (17.6a) and put $l_v = 0$ and we will get the same result. The diameter D' of the exit pupil is related to the aperture diameter D of the objective (which is also the entrance pupil) by the equation

$$\frac{D'}{D} = \overline{M} = \frac{\bar{l}'(= \bar{l}'_v)}{\bar{l}}$$

Using equation (17.13) for $\bar{l}_v$, and equation (17.12) for $\bar{l}_v$, the above equations reduce to

$$D' = D/M \quad \text{or} \quad = D\overline{M} \qquad (17.14)$$

and

$$\overline{M} = 1/M \qquad (17.15)$$

Equation (17.15) is a special case of equation (17.9), here applied to the pupil planes, which can be regarded as one special pair of conjugates which are at a finite distance. An alternative way of expressing this result is as follows.

$$M = \frac{\text{diameter of entrance pupil (objective)}}{\text{diameter of exit pupil}} \qquad (17.16)$$

These equations provide a very useful means for measuring the angular magnification of telescopes, indirectly from the diameters of the entrance and exit pupils. It is relatively easy to measure accurately the diameters of the objective and exit pupils. In fact any pair of conjugate planes can be used, not necessarily pupil planes.

If a field lens is placed in or near the intermediate image position to increase the field-of-view (see Section 17.4.2), it has little or no effect on the angular magnification of the telescope. However field lenses do have some effect on the position of the exit pupil or eye relief but they do not affect the diameter of the exit pupil.

For viewing extended objects with maximum image luminance, the exit pupil of the telescope should have a larger diameter than the entrance pupil of the eye. Thus telescopes designed for night or low light level work should have exit pupils at least 7 mm in diameter.

Those designed for day or high light level usually have exit pupil diameters in the region of 2–4 mm. These requirements do not apply when viewing point sources, such as viewing stars as in astronomy. This aspect is discussed further in Chapter 36.

17.4.2 Field lenses

The optics of field lenses were introduced in Chapter 9. Essentially the purpose of a field lens is to decrease vignetting and thus increase the field-of-view. Here, its purpose is to reduce vignetting by the eye lens and if it is placed exactly at the intermediate image position it does not change the position or magnification of the image. It changes the eye relief but does not change the diameter of the exit pupil. The exit pupil diameter does not change if the path of the paraxial marginal ray is unaltered by the presence of the field lens but the lens does change the path of the pupil ray. Let us look at the effect on eye relief.

Since the pupil ray does not pass through the centre of the field lens, it must be refracted and deviated by it, as shown in Figure 17.7. As a result, the field lens shifts the exit pupil and moves it towards the eye lens. The amount of the shift depends upon the power of the field lens. Referring to the diagram, the new eye relief can be found by locating the position of the image of the objective, now imaged by a field lens of power F_{fl} and the eye lens of power F_{e}. The conventional ray tracing equations can be used to show that the new eye relief $(\bar{l}'_{\text{v}})_{\text{new}}$ is given by the equation

$$(\bar{l}'_{\text{v}})_{\text{new}} = (\bar{l}'_{\text{v}})_{\text{old}} - F_{\text{fl}}/F_{\text{e}}^2 \qquad (17.17)$$

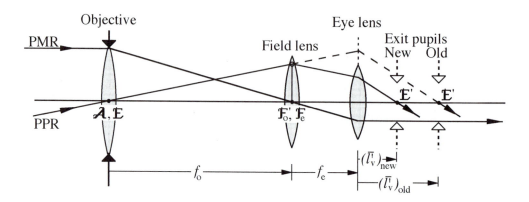

Fig. 17.7:   The effect of a field lens on exit pupil position or eye relief of a Keplerian telescope. PMR = paraxial marginal ray, PPR = paraxial pupil ray.

where $(\bar{l}'_v)_{old}$ is given by equation (17.13). We will examine the effect of the field lens on field-of-view in the next sub-section.

17.4.3 Field-of-view

If the telescope does not contain a field lens, an arrangement which is rare since most eyepieces automatically contain them, the field-of-view is limited by the diameter of the eye lens. In this case, the larger the eye lens, the wider the field-of-view. When a field lens is used, the situation is more complex and the field-of-view then depends upon the power of the field lens, eye lens diameter and perhaps occasionally the field lens diameter.

Often the field-of-view is limited by a field stop placed at the intermediate image position. When this is used, it is usually present to reduce the field-of-view to one of good image quality and eliminate peripheral images which may be badly aberrated.

We will now derive equations for the field-of-view in image space, in terms of the angle θ' shown in Figure 17.8. For small angles, this angle is connected to the corresponding object space angle θ by equation (17.3) but for field-of-view calculations an equation more suitable for larger angles is required. A more meaningful alternative is

$$\tan(\theta') = M \tan(\theta) \tag{17.18}$$

Now the field-of-view may be defined as where vignetting just begins, just becomes complete or the average of these two values. We will now derive equations for the two extreme field-of-view sizes but ignore the effects of aberrations.

Vignetting just beginning

We can examine the size of the central field free of vignetting by using Figure 17.8b. In this diagram, the vignetting just begins when the top ray of the emerging beam just touches the top edge of the eye lens as shown. At this stage, the beam subtends an angle θ'_v to the axis. From this diagram, we can readily

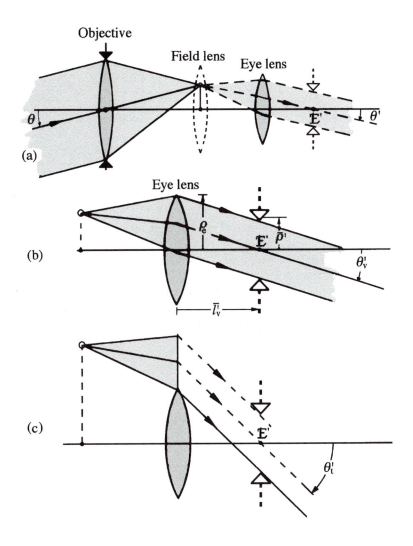

deduce that

$$\tan(\theta_v') = (-)\frac{(\rho_e - \bar{\rho}')}{\bar{l}_v'} \tag{17.19a}$$

The corresponding object space angle can then be found from equation (17.18).

Vignetting just becomes complete

The point in the field where vignetting just becomes complete can be found with the help of Figure 17.8c. The vignetting just becomes complete when the bottom ray of the emerging beam just touches the top edge of the eye lens as shown. At this stage, the beam subtends an angle θ_t' to the axis. From this diagram, we can readily deduce that

$$\tan(\theta_t') = (-)\frac{(\rho_e + \bar{\rho}')}{\bar{l}_v'} \tag{17.19b}$$

and the corresponding object space angle can then be found from equation (17.18).

Example 17.1: Given the following telescope:

objective: $F_o = 10 \text{ m}^{-1}$,   aperture diameter $= 20$ mm
eye lens: $F_e = 50 \text{ m}^{-1}$,   aperture diameter $= 10$ mm (i.e. $\rho_e = 5$ mm)

calculate

(a)   the field-of-view with no field lens being used and
(b)   the field-of-view if a field lens of power 20 m^{-1} is used.

Assume that the field lens does not vignet the beam.

Solution: From equation (17.2a), this telescope must have a length (d) of 120 mm. From equation (17.4), it has a magnification (M) of -5. From the data, the entrance pupil diameter $D = 20$ mm and therefore from equation (17.14), the exit pupil diameter (D') is 4 mm.

Condition (a): Without a field lens. From equation (17.13), the eye relief is

$$\bar{l}_v' = -120/(-5) = 24 \text{ mm}$$

Substituting into equations (17.19a) and (17.19b) along with equation (17.18), we have the image and object space fields-of-view as follows:

$\tan(\theta_v')$	$\tan(\theta_v)$	θ_v°	$\tan(\theta_t')$	$\tan(\theta_t)$	θ_t°
0.125	0.025	1.4	0.292	0.058	3.3

Condition (b): With a field lens. From equation (17.17), the new eye relief is

$$\bar{l}_v' = 16 \text{ mm}$$

and the new field-of-view values are

$\tan(\theta_v')$	$\tan(\theta_v)$	θ_v°	$\tan(\theta_t')$	$\tan(\theta_t)$	θ_t°
0.188	0.038	2.1	0.438	0.087	5.0

17.4.4   *Image erection*

For terrestrial use, the image is usually erected by using lenses or prisms. The term "erection" is rather misleading. The image has to be rotated through 180° rather than simply turned upside down. While lenses rotate the image in one single step, prisms using reflections can only flip the image over about one axis with each reflection. Thus prism systems must use more than one reflection to achieve the 180° rotation.

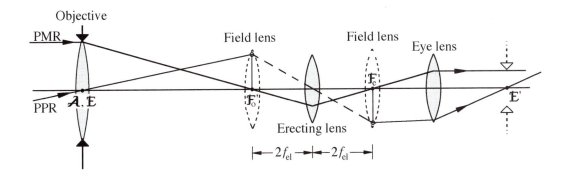

17.4.4.1 Erecting lenses

In some telescopes, image erection is performed by an auxiliary lens system. This has the disadvantage that it increases the length of the telescope and the extra lenses may increase the aberrations, particularly field curvature. However, the presence of an erecting lens increases the eye relief and therefore can be used to control its value. The erecting lens will not change the afocal nature of the telescope. As it does not change the angular magnification, it cannot change the diameter of the exit pupil.

Effect on exit pupil position

Figure 17.9 shows an example of the use of an erecting lens that has a transverse magnification of -1. If the power of this lens is F_{el}, it can be shown that the new eye relief $(\bar{l}'_v)_{new}$ is given by the equation

$$(\bar{l}'_v)_{new} = (\bar{l}'_v)_{old} + F_{el}/F_e^2 \tag{17.20}$$

and $(\bar{l}'_v)_{old}$ is given by equation (17.13). Equation (17.20) shows that the higher the power of the erecting lens, the longer the new eye relief.

17.4.4.2 Erecting prisms

The optical principles of a range of prisms are given in Chapter 8. Two of these are suitable for image erecting in Keplerian telescopes: the Porro and Pechan prisms.

Porro prism

A typical construction is shown in Figure 17.10a. Notice that the optical axis is transversely displaced. The main advantage of this system is the cheapness of production of the Porro prism, which is essentially composed of two simple $45° - 90° - 45°$ prisms. On the other hand this system is less robust. A drop or a knock can displace or rotate the prisms relative to each other. If this happens, the optical axis is deviated or the image is not exactly rotated through $180°$.

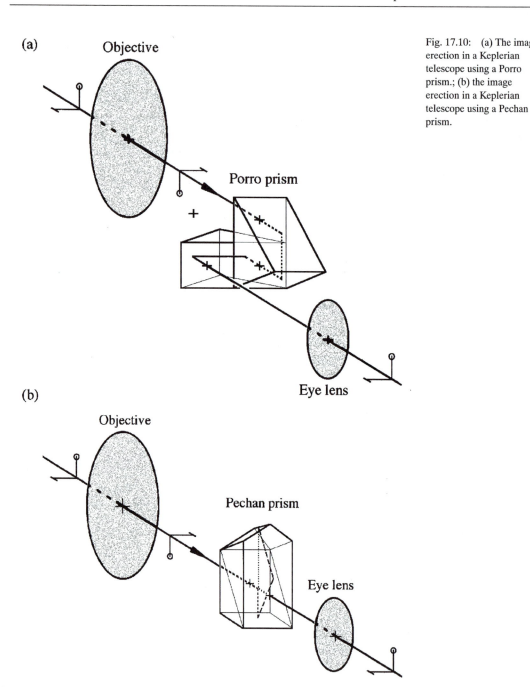

Fig. 17.10: (a) The image erection in a Keplerian telescope using a Porro prism.; (b) the image erection in a Keplerian telescope using a Pechan prism.

Pechan prism

Figure 17.10b shows a typical construction using a Pechan prism. This prism is less common than the Porro prism, because the cost of production of the Pechan prism is much greater. On the other hand it has the advantage that there is no displacement of the optical axis, as there is with the Porro prism.

Fig. 17.11: The optics of
a Galilean telescope. PMR
= paraxial marginal ray,
PPR = paraxial pupil ray.

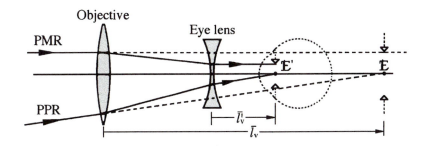

17.5 Galilean telescopes

The construction of the Galilean telescope is shown in Figure 17.11. In the Galilean construction, the back focal point of the objective is beyond the eye lens, though co-incident with the front focal point of the negative power eye lens. Thus for distant objects, a real intermediate image is not formed, and hence a graticule cannot be used for any measurement, a field lens cannot be used to increase the field-of-view and a field stop cannot be inserted to give a sharp edge to the field.

17.5.1 Pupils

Usually the Galilean telescope has no intrinsic aperture stop, unlike the Keplerian telescope, in which the objective acts as the aperture stop and entrance pupil. When it is used with the eye, the iris of the eye becomes the aperture stop of the complete system and the entrance pupil of the eye can be regarded as the exit pupil of the telescope. When this is the case, optical calculations show that the entrance pupil is virtual and appears to be in image space beyond the eye lens and exit pupil.

Position of the entrance pupil

Figure 17.11 shows the eye placed near the eye lens such that the entrance pupil of the eye is at ε', which becomes the nominal position of the exit pupil of the telescope. If $\bar{l}'_v$ is the distance of the eye from the eye lens, the entrance pupil distance $\bar{l}_v$ from the objective is given by equation (17.7a), that is

$$\bar{l}_v = M(M\bar{l}'_v + d) \tag{17.21}$$

where the conjugates are now the entrance and exit pupils. Since M, d and $\bar{l}'_v$ are all positive, the entrance pupil distance $\bar{l}_v$ must always be positive and therefore the entrance pupil appears to be on the eye lens side of the telescope.

Pupil magnification

The transverse magnification of the pupils is related to the angular magnification of the telescope as it was for the Keplerian telescope, that is equation (17.14).

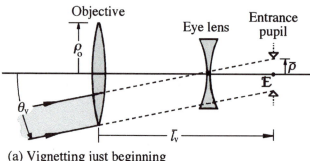

(a) Vignetting just beginning

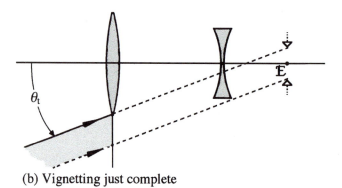

(b) Vignetting just complete

Fig. 17.12: The field-of-view of a Galilean telescope.

We will write this in terms of pupil radii and in the form

$$\bar{\rho} = M\bar{\rho}' \qquad (17.22)$$

where $\bar{\rho}$ is the radius of the entrance pupil and $\bar{\rho}'$ is the radius of the exit pupil.

Paraxial marginal and pupil rays

Figure 17.11 shows the paths of the paraxial marginal and pupil rays. We should note that the pupil ray passes through the objective lens at a distance from the optical axis.

17.5.2 Field-of-view

In the Galilean telescope, the eye forms the exit pupil and the field-of-view is limited by vignetting due to the objective. Thus the larger the objective, the larger the field-of-view. Once again the field-of-view may be defined as where vignetting just begins, where it just becomes complete or the mean of these two values.

Figure 17.12a shows the entrance pupil of the telescope at a distance $\bar{l}_v$ from the objective. Its value is given by equation (17.21) in terms of the distance $\bar{l}'_v$ of the eye from the eye lens. Let the radius of the entrance pupil be $\bar{\rho}$. The value of this radius is related to the radius of the exit pupil $\bar{\rho}'$ by the magnification equation (17.22). We will now use this diagram to derive equations for the field-of-view.

Vignetting just beginning

Figure 17.12a shows a beam from an off-axis point entering the entrance pupil. As the point moves away from the axis, vignetting first begins when the bottom ray of the beam just touches the bottom edge of the objective, as shown. This ray makes an angle θ_v to the axis where, from the diagram

$$\tan(\theta_v) = \frac{(\rho_o - \bar{\rho})}{\bar{l}_v} \tag{17.23a}$$

Vignetting just complete

As the point moves farther off-axis, vignetting finally becomes complete when the top ray of the beam just touches the bottom edge of the objective lens, as shown in Figure 17.12b. This ray makes an angle θ_t to the axis, where from the diagram

$$\tan(\theta_t) = \frac{(\rho_o + \bar{\rho})}{\bar{l}_v} \tag{17.23b}$$

Example 17.2: Let us compare the field-of-view of the Keplerian telescope given in Example 17.1 with an equivalent Galilean design, that is

objective:	$F_o = +10 \text{ m}^{-1}$,	aperture diameter $= 20$ mm
eye lens:	$F_e = -50 \text{ m}^{-1}$,	aperture diameter $= 10$ mm

Solution: The Galilean does not have an intrinsic aperture stop and hence does not have an exit pupil. To allow a comparison, let us place the eye pupil at the same distance from the eye lens as the eye relief in the Keplerian design, that is $\bar{l}'_v = 24$ mm, which gives $\bar{l}_v = 1000$ mm from equation (17.21), and with the same entrance pupil diameter of 20 mm.

Using the above value of $\bar{l}_v$, $\rho_o = 10$ mm and $\bar{\rho} = 10$ mm in equations (17.23a) and (17.23b), we have the image space fields-of-view as follows.

$\tan(\theta'_v)$	$\tan(\theta_v)$	θ_v°	$\tan(\theta'_t)$	$\tan(\theta_t)$	θ_t°
0.00	0.00	0.0	0.100	0.020	1.1

In this example, there is no central field free of vignetting.

A comparison of these results with those for the equivalent Keplerian values from Example 17.1 suggests that Galilean telescopes have a smaller field-of-view than Keplerian telescopes.

If one neglects the diameter of the pupil of the eye, the field-of-view is given by the equation

$$\tan(\theta) = \frac{\rho_o}{\bar{l}_v} \tag{17.24}$$

Since the diameter of the pupils has been neglected here, there is no distinction between the field-of-view free of vignetting and that up to vignetting just becoming complete.

17.6   The thick lens telescope

A single thick lens can be made into a telescope, with a magnification that increases with lens thickness. However, physical limitations restrict this magnification to only a few percent difference from unity and these lenses are sometimes used to simulate **aniseikonia**. We can find equations describing the properties of this type of lens beginning with the standard equation for the power of a thick lens.

The equivalent power F of a thick lens is given by equation (3.20), that is

$$F = F_1 + F_2 - (dF_1F_2/\mu)$$

where F_1 and F_2 are the front and back surface powers, respectively, d is the thickness and μ is the refractive index. For an afocal thick lens, the equivalent power F must be zero and thus

$$F_1 + F_2 - (dF_1F_2/\mu) = 0 \qquad (17.25)$$

By a similar argument used to derive the angular magnification equation (17.4) for a conventional telescope, we can show that the angular magnification M of this thick lens telescope is given by exactly the same equation but written here as

$$M = -F_2/F_1 \qquad (17.26)$$

If we now replace F_1 and F_2 by their surface powers

$$F_1 = C_1(\mu - 1) \quad \text{and} \quad F_2 = C_2(1 - \mu) \qquad (17.27)$$

the above equation for magnification reduces to

$$M = C_2/C_1 \qquad (17.28)$$

Useful equations for designing a thick lens telescope can be derived by solving equations (17.25) and (17.26) for the two powers. Thus

$$F_1 = \frac{(M-1)\mu}{Md}, \quad F_2 = -\frac{(M-1)\mu}{d} \qquad (17.29)$$

Replacing the surface powers by curvatures, we have

$$C_1 = \frac{(M-1)}{Md}\frac{\mu}{(\mu-1)}, \quad C_2 = \frac{(M-1)}{d}\frac{\mu}{(\mu-1)} \qquad (17.30)$$

The lens thickness d can be expressed in terms of the other parameters by

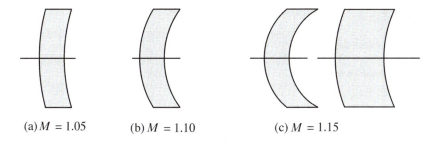

(a) $M = 1.05$ (b) $M = 1.10$ (c) $M = 1.15$

Fig. 17.13: Some thick lens telescopes with magnifications of 1.05, 1.10 and 1.15, drawn to scale (see Example 17.3). These are types of Galilean telescopes.

transposing the above equations, that is

$$d = \frac{(M-1)}{MC_1} \frac{\mu}{(\mu-1)} \tag{17.31a}$$

and

$$d = \frac{(M-1)}{C_2} \frac{\mu}{(\mu-1)} \tag{17.31b}$$

Example 17.3: Design a thick lens telescope with a desired magnification (M) of 1.05, made from a material with a refractive index (μ) of 1.5 and with a thickness of 10 mm.

Solution: Using equations (17.30)

$$C_1 = \frac{(1.05-1) \times 1.5}{1.05 \times 10 \times (1.5-1)} = 0.0143 \text{ mm}^{-1}$$

and

$$C_2 = \frac{(1.05-1) \times 1.5}{10 \times (1.5-1)} = 0.0150 \text{ mm}^{-1}$$

This lens is drawn to scale in Figure 17.13a.

A few worked examples will show that thick lens telescopes with large magnifications become very thick or the surface powers become very high for only a few percent in magnification. These effects are demonstrated by the examples shown in Figure 17.13. In all of these cases, the refractive index is kept constant at 1.5. Comparison of the lens in (a) with those in (b) and (c) shows that for the same thickness, the curvatures increase with magnification. The design next to (c) shows that high curvatures can be reduced by increasing the thickness.

The above example shows that the front and back surface powers are of opposite sign and since the magnification is positive, this thick lens telescope has the typical properties of a Galilean telescope. If the magnification were less than 1.0 in the above problem, the front surface power would have been negative and the back surface power would have been positive.

17.7   Refocussing and its effects

Telescopes are nominally afocal and therefore have no refractive power. They
are designed for objects at infinity and since the power is zero, the image is also
at infinity. However, telescopes are not often used exactly in this manner, for
the following two reasons.

(1)   If the object is not at infinity, the image is closer than the object and the
      resulting image vergence may cause excessive accommodation demand.
(2)   The user may have a refractive error that makes it difficult to focus on an
      image formed at infinity.

To cope with these cases, telescopes are usually designed with an adjustable
length to allow the user to refocus the image according to the need. This refocussing changes the effective magnification. We will now explore the effects
of the refocussing but restrict ourselves to looking only at two lens telescopes;
that is we will assume that Keplerian telescopes have no field lenses.

17.7.1   Finite working distance

Length adjustment

When the object is at a finite distance, the image formed by the objective is
formed beyond its back focal point $\mathcal{F}_0'$ as shown in Figure 17.14a for a Keplerian
telescope and Figure 17.14c for a Galilean telescope. If the final image is to be
formed at infinity, the eye lens must be moved to place the intermediate image at
the front focal point $\mathcal{F}_e$ of the eye lens, as shown in the diagram. Figures 17.14b
and d show that for both types of telescopes, the eye lens must be moved back
and hence the telescope must be lengthened.

   If the object or working distance vergence is W and positive as shown in
Figure 17.14, it can be shown that the image vergence can be made zero by
changing the length by an amount

$$\Delta d = \frac{W}{F_0(F_0 - W)} \tag{17.32}$$

which is independent of the power of the eye lens. The sign convention is such
that a positive value of Δd means an increase in length.

Effect on magnification

Sloan (1977, 29) argued that when the afocal telescope is used to view a near
object and has to be refocussed to prevent excessive accommodation, some of
the power of the objective is used to overcome the divergence of the incident
light, that is re-collimate it. The remaining lower power of the objective and
the eye lens forms a new afocal telescope with a new length. According to
equation (17.4), the magnification of the new afocal telescope increases since
the effective power of the objective has been decreased.

   If the object or working distance has a vergence W measured at the objective
with the magnitude being positive, the new effective power F_0' of the objective

Objective

Eye lens

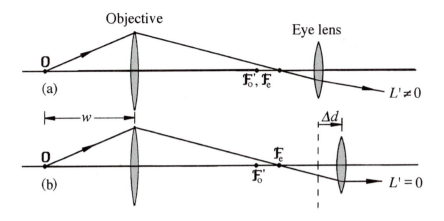

(a) $L' \neq 0$

(b) $L' = 0$

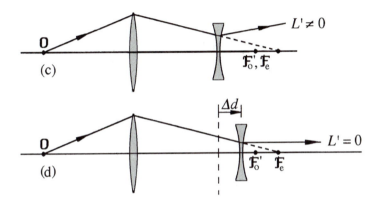

(c) $L' \neq 0$

(d) $L' = 0$

Fig. 17.14: Refocussing a telescope for a finite working distance, placing the image at infinity.

becomes, according to Sloan (1977),

$$F_o' = F_o - W \tag{17.33}$$

as shown in Figure 17.15. From equation (17.4), the new effective magnification M' is

$$M' = -F_e/F_o' = -F_e/(F_o - W)$$

If we use equation (17.4) to eliminate F_e, we can write this equation as

$$M' = \frac{M}{(1 - W/F_o)} \tag{17.34}$$

where M is the angular magnification of the original afocal telescope. For both types of telescopes, the objective power is positive and since the working distance vergence is always positive, the denominator in this equation must always be less than unity. As a result the effective magnification M' increases for both types of telescopes.

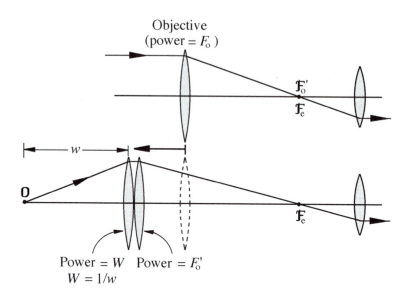

Fig. 17.15: Adjustment of a telescope for a finite working distance (Sloan model).

The above model assumes that the length of the telescope is negligible, relative to the working distance. Therefore, the above equation will not be exact and will have an error that will increase as the working distance decreases.

Example 17.4: Given the following Galilean telescope

objective: $F_{\mathrm{o}} = +10 \text{ m}^{-1}$ eye lens: $F_{\mathrm{e}} = -50 \text{ m}^{-1}$

calculate the change in length and new effective magnification if this telescope is refocussed for a working distance of 1 m.

Solution: Here from equation (17.4), we have

$$M = +5$$

and from equation (17.32),

$$\Delta d = \frac{1}{10(10-1)} \text{ m} = 11.1 \text{ mm}$$

and from equation (17.34)

$$M' = \frac{5}{(1 - 1/10)} = 5.56$$

17.7.2   *Refractive error*

Length adjustment

If the user has a refractive error and is viewing an object at infinity, he or she may not be able to focus on the image that is also at infinity. While a hyperope

Fig. 17.16: Refocussing a
telescope for a user
refractive error, for an
object at infinity.

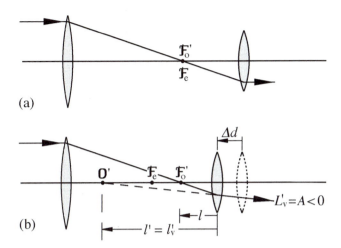

(a)

(b)

will be able to accommodate on the image, the myope will have to refocus the telescope to bring the image at least up to the far point. The telescope can be refocussed by changing its length. If an image vergence A, as shown in Figure 17.16, is desired, the telescope length has to be adjusted by an amount

$$\Delta d = \frac{A}{F_e(F_e - A)} \tag{17.35}$$

where in the diagram

$$l' = l'_v = 1/A$$

which is independent of the power of the objective. For a myopic user, A is negative and thus the telescope length is decreased for both Keplerian and Galilean types. For a hyperopic user, A is positive and thus the telescope length is increased for both types.

Effect on magnification

Sloan (1977, 29) states that when a myope focuses an afocal telescope, the magnification is decreased and explains this effect as follows. The refractive error of the myope has to be taken up by some of the negative power of the eye lens. This effectively reduces the power of the eye lens and accordingly, from equation (17.4), the magnification must decrease. According to this argument, the original afocal telescope is split into a corrective refractive lens next to the eye and another afocal telescope of lower magnification and different length.

In this case, let the refractive error be represented by A, where the magnitude is negative for myopia and positive for hyperopia. Therefore, according to the argument of Sloan (1977), we split the eye lens into two components of powers F'_e and A, where

$$F_e = F'_e + A$$

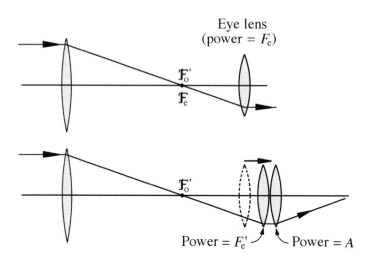

and where F'_e is the power of the eye lens of the new afocal telescope. Thus

$$F'_e = F_e - A \qquad (17.36)$$

as shown in Figure 17.17. From equation (17.4), the new effective magnification M' can be written as

$$M' = -F'_e/F_o$$

that is

$$= -(F_e - A)/F_o$$

Therefore

$$M' = M + A/F_o \qquad (17.37)$$

where M is the angular magnification of the original afocal telescope. Since the power F_o of the objective is positive for both types of telescopes, therefore the sign of the term

$$A/F_o$$

is negative for myopes and positive for hyperopes. If we now analyse equation (17.37), we reach the following conclusions.

(1) The magnification M' increases for
       Galilean telescopes refocussed by hyperopes and
       Keplerian telescopes refocussed by myopes.
(2) The magnification M' decreases for
       Galilean telescopes refocussed by myopes and
       Keplerian telescopes refocussed by hyperopes.

   The above model assumes that the angular magnification of the image, which is now at a finite distance, is measured from the eye lens rather than the exit

Table 17.1. *A comparison of Keplerian and Galilean telescopes*

	Keplerian	Galilean
General		
Magnification	up to 20–30	usually less than 4
Image	inverted	erect
Field-of-view	large	small
System length	longer	shorter
Pupils	exit pupil	no exit pupil
Telescope length changed for (Section 17.7)		
Myopia	decreases	decreases
Hyperopia	increases	increases
Finite working distance	increases	increases
Effective magnification (Section 17.7)		
Myopia	decreases	increases
Hyperopia	increases	decreases
Finite working distance	increases	increases

pupil. This will lead to actual observed changes in magnification being different from those predicted by the above equations and these errors will increase with level of refractive error.

Example 17.5: For the Galilean telescope given in Example 17.4, find the change in telescope length and new magnification if it is refocussed to give an image vergence (myopia) A of -3 m^{-1}.

Solution: From equation (17.35), we have

$$\Delta d = \frac{-3}{(-50)[(-50) - (-3)]} = -0.00127 \text{ m} = -1.3 \text{ mm}$$

From equation (17.37), we have

$$M' = 5 + (-3)/10 = 4.7$$

This latter result shows that on refocussing the telescope, the magnification reduces.

The results of this section are summarized in Table 17.1.

17.8 Telescope specification and verification

It is apparent from the previous discussions that to specify fully the performance of a telescope, the magnification, objective diameter, length, eye relief and field-of-view should be given, known or determined. Sometimes not all these quantities are known and if they are required, need to be measured in some manner. Sometimes even if they are known, they may need to be checked or verified. This section concentrates on simple methods for the verification of the above parameters.

17.8.1   Specification of parameters

17.8.1.1   Magnification

The magnification of a telescope is an angular magnification and is sometimes referred to as the **power** of the telescope. The use of the word "power" is not advisable as it can be confused with refractive power. In the case of Keplerian telescopes, the magnification is usually written on the telescope body as one of a pair of numbers, for example

> 7, 50

where the first number is the magnification and the second is the diameter of the objective in millimetres. In contrast, the magnification of Galilean telescopes is rarely written anywhere on the telescope.

17.8.1.2   Objective diameter and pupil size

In the case of Keplerian telescopes, the diameter of the objective is normally specified and its value in millimetres is written on the telescope body as in the above example. This value is important because, being the aperture stop, its diameter sets the diameter of the exit pupil, which can be found from equation (17.14). The diameter of the exit pupil in turn indicates whether the telescope is intended for day or night use. For instance, in the above example, equation (17.14) gives an exit pupil diameter of 7 and 1/7 mm indicating the telescope is designed for night use.

   Rarely is the objective diameter specified for a Galilean telescope or written on the telescope body; the reason being that the objective is not the aperture stop, as it is for the Keplerian telescope and therefore it has no effect on the size of the exit pupil. However, it is important because it affects the field-of-view.

17.8.1.3   Eye relief

The eye relief of a telescope must have a minimum value in order to provide comfortable viewing. We show in Chapter 36 that for those who do not wear spectacles, the minimum eye relief should be about 12 mm, but for spectacle wearers it should be about 20 mm. Gun sight telescopes may need even longer eye reliefs.

   The eye relief cannot be specified for a Galilean telescope because a Galilean telescope does not have an intrinsic exit pupil.

17.8.1.4   Field-of-view

The image field-of-view of many visual optical instruments is about $50°$. Thus in the case of telescopes, the object field-of-view diameter is approximately this value divided by the magnification. Strictly, the fields-of-view are related by the tangents of the angles, that is by equation (17.18) rather than the paraxial equation (17.3), that is

$$\tan(\theta') = M \tan(\theta) \tag{17.38}$$

Based upon the 50° value, the object field-of-view diameters of two common binoculars are as given below.

for a magnification of 7:   object field-of-view = 7.6°
for a magnification of 8:   object field-of-view = 6.7°

On many telescopes the field-of-view is specified as an angular measure in degrees, and in others as a ratio, for example in the form 1 : 100. This ratio means that the field is say 1 m wide at a distance of 100 m or 10 m at 1000 m and so on.

17.8.1.5 Closest working distance

By having an adjustable length, a telescope can be refocussed to allow comfortable viewing of near objects. However, for any length adjustment, the shortest working distance that can be seen in focus will depend upon the maximum level of accommodation that the user can exert. Therefore the closest working distance will depend both upon the amount of length adjustment possible and the maximum accommodation possible by the user.

The closest working distance is not normally specified and if a near working distance were to be specified, an accommodation level would have to be assumed. This would be similar to the setting of 25 cm as a near viewing distance used to specify the magnification of simple magnifiers.

17.8.2 Verification of parameters

17.8.2.1 Magnification

Visual comparison method

In the visual comparison method, an observer views a distant regular pattern, such as brickwork, with both eyes open. One eye views through the telescope and the other views the pattern directly. A comparison of sizes, by counting the number of small bricks plus a fraction which equals one big brick, will give a good estimate of the magnification.

One disadvantage of this method is that it is sometimes difficult, if not impossible, to view two very disparate images simultaneously. Occasionally and depending upon the individual, one image fades out. This is known as suppression and often occurs when the brain tries to cope with two disparate images. If the disparity is too great, the brain may turn off or suppress the image from one of the eyes.

Entrance and exit pupil magnification (Keplerian telescopes only)

The entrance and exit pupil magnification method makes use of the relation between angular magnification and pupil magnification discussed in Section 17.4.1. In principle, the objective (the entrance pupil) and exit pupil diameters are measured and the magnification determined from equation (17.16). The objective diameter is easily and accurately measured but the exit pupil being much smaller cannot be measured to the same relative accuracy with the same ease. A quick and simple method of measuring the exit pupil diameter is to take a piece of diffusing material such as translucent sticky tape and stick it onto the scale side of a transparent ruler. If the telescope is aimed at a large area light source, the exit pupil can then be easily located and measured. Typically it is unlikely that this quick method will give an accuracy of better than ±0.2 mm. A more accurate method would require a micrometer eyepiece or a simple magnifier with a scale attached.

An alternative method, with good accuracy, is as follows. Finely ruled parallel lines are drawn on a thin sheet of glass or solid plastic. The separation of these lines must be measured accurately though only once. This is used as a test target with a diffusing material behind it and then placed in the exit pupil plane, or closer to the eye lens and the image of the target located on the objective side of the telescope. Being much enlarged, the separation of the lines on the image of the target can be measured much more accurately than an equivalent image near the exit pupil. Since the transverse magnification is independent of distance, the position of the target is not critical, but it must be between the exit pupil and the eye lens in order to give a real image beyond the objective.

Image vergence method (two lens telescopes)

If an object is placed at a finite distance with a vergence L_v from the objective and the image vergence L'_v is measured from the eye lens, the magnification M of the telescope can be estimated using equation (17.6). In this case, we need to solve for M, leading to the equation

$$M^2 L_v + M d L_v L'_v - L'_v = 0 \tag{17.39}$$

which is quadratic in M. This method has been described by Bailey (1978a, 1978b).

Collimation/de-collimation method

A target of known size is collimated by a good quality collimating lens. A second, not necessarily identical, collimating lens is used to de-collimate the image and thus form the image of the target in the back focal plane of this second lens. The transverse magnification of this arrangement is measured. The telescope is then placed between the two collimators. If the telescope is in the afocal mode, the target should still be in focus on the viewing screen but magnified. The ratio of the two image sizes is the magnification of the telescope.

17.8.3   Accuracy of afocal adjustment

Telescopes are nominally afocal, but there will always be some residual power due to uncertainties or errors in the setting of the lens separation. The residual power can be measured using a **dioptric telescope**, which is described in Section 17.9.2.

The telescope to be tested is aimed at a very distant object. If the telescope is afocal, and the object is at infinity, the image will also be formed at infinity. If the telescope is not afocal, the image will be formed at a finite distance; in the back focal plane of what is effectively a focal telesope. The image is viewed through the dioptric telescope which can measure the vergence of the image.

This image vergence is the back vertex power of the telescope. This is read directly from the dioptric telescope scale.

17.9 Special applications and special telescopes

Telescopes are probably more commonly regarded as magnifying devices. However, their special properties make them suitable as alignment and focussing devices, where the magnification is of secondary importance. We will look at some of these applications in this section, both as magnifiers and focussing or as alignment tools.

17.9.1 As alignment devices

Telescopes are excellent tools for sighting or alignment. Telescopes with high magnification can be used to sight on very distant objects and if the telescope is placed on a rotatable graduated scale, the direction of the object can be accurately measured. Typical uses are as gun sights and in surveying instruments.

The image vergence magnification property of telescopes, exemplified by equation (17.6), makes the telescope a very useful device for checking the focus level of many other instruments that produce an image at or near infinity, for example for checking the level of collimation of the collimators used in a wide range of instruments such as spectrometers, goniometers and focimeters and setting the image vergence of eyepiece graticules.

17.9.2 Special telescopes

The auto-collimating telescope

The auto-collimating telescope is a telescope accurately focussed on infinity, but the special property of this telescope is that it has an internally illuminated graticule as shown in Figure 17.18. The telescope is usually used with a mirror placed in front of the objective, as shown in the diagram. The mirror images the graticule back into the telescope and onto a second, sometimes micrometer moveable graticule. Any angular tilt of the mirror leads to a transverse displacement of the image of the first graticule. One use of this arrangement is to measure the flatness of a surface. The mirror is moved backwards and forwards over the surface along a rail. Any departure from flatness causes the mirror to tilt and the tilt causes a deviation in the returning beam and hence a displacement between the graticule and its image.

Dioptric telescopes

Dioptric telescopes are used to check the residual power in afocal telescopes or set the vergence of graticules within eyepieces. The eyepiece graticule is viewed through the dioptric telescope and the focus of the telescope is adjusted until the graticule is in focus on the cross-hairs of the dioptric telescope. A focus adjustment and scale on the dioptric telescope allows the residual vergence of the graticule in the eyepiece to be measured.

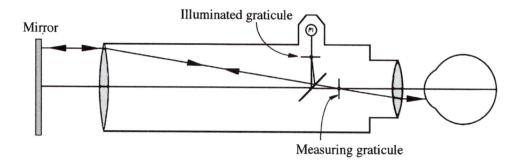

Mirror

Illuminated graticule

Measuring graticule

Security door viewers

Fig. 17.18: The optical
construction of an
auto-collimating telescope.

Security door viewers are small Galilean telescopes inserted into doors to allow those inside to have a wide field-of-view outside the door. These are normally oriented so that the image seen from the inside is reduced and wide angle. In this orientation, the telescope must have a magnification less than unity and hence the system is a reverse Galilean telescope.

17.9.3   As vision aids

Spectacle/contact lens telescopes

Spectacle/contact lens telescopes have been suggested and designed, in order to control the magnification differences between the two eyes when they have significant aniseikonia. In these cases, the objective is a spectacle lens and the eye lens is a contact lens. Byer (1986) has described the use of this type of telescope for magnifications up to 5. Because they must be short and give an erect image, they must be of the Galilean type.

Low vision aids

Telescopes are also used extensively as optical aids by people suffering from a number of ocular diseases in which the visual acuity is either very low because of a diseased retina, lenticular cataract or because large areas of the retina are effectively non-functioning (scotomas). The visual acuity can be improved if the material they wish to view is magnified. For close material, simple magnifiers are ideal. Telescopes are used to magnify distant material such as street names and bus numbers. In the case of scotomas, small detail may be imaged on the scotoma and not seen. This problem can sometimes be alleviated or overcome by using eye and head movement. An alternative is to magnify the detail so that it is much larger than the scotoma, and hence more likely to be seen. Conventional telescopes are often used for this task.

However, in contrast, if the scotoma is entirely and extensively peripheral, objects in the periphery are not observed and this can occasionally make mobility difficult and sometimes dangerous. This problem can be partly reduced by using a telescope in reverse. Conventional telescopes take a narrow object field and expand it into a wider image field with every object in the field correspondingly magnified. On the other hand, if a telescope is used in reverse, it takes a wide object field and contracts it into a narrow image field, with each object in the field reduced in size. Objects in the peripheral field are thus brought much closer to the central visual field or the visual axis, though with reduced size. Because of the requirements for erect images and a convenient short length, Galilean telescopes are usually used for this application and when used in this mode are often called **field expanders**.

17.10 Binoculars

Binoculars or a pair of binoculars are two identical telescopes joined together and hinged so that their separation can be varied to suit the inter-pupillary distance of the user. For reasons that will be discussed in Chapter 37, the image of binoculars is usually erect; therefore binoculars must be made either from Keplerian telescopes containing image erecting optics or from Galilean telescopes. "Opera glasses" are Galilean binoculars.

Figure 17.19 shows two possible forms of binoculars based upon the Keplerian telescope, one (a) using Porro prisms and another (b) using the roof type of erecting prism (for example the Pechan prism). The presence of the Porro prisms causes the optical axis to be displaced and therefore the centres of the objectives can be placed farther apart than the centres of the eyepieces. This increase in separation increases the stereoscopic effect and this is discussed further in Chapter 37. If a Pechan or a similar roof prism system is used, the optical axis is not displaced and hence the centres of the objectives must be the same distance apart as the inter-pupillary distance of the user. This places limitations on the maximum diameter of the objectives and hence limits on the exit pupil size.

Since no two telescopes can be perfectly identical, differences in optical properties between the two parts must be accepted but kept to a minimum to prevent binocular vision problems. These differences, their visual effects and tolerances are discussed in Chapter 37.

17.11 Aberrations

Telescopes usually have small object angular field sizes and therefore do not suffer greatly from aberrations such as distortion and transverse chromatic aberration. However, the exact amount of each aberration present depends to some extent on the type of telescope, for example Keplerian or Galilean, but in both cases, spherical aberration, coma and longitudinal aberrations are usually the most important aberrations and these arise mostly in the objective. These aberrations can be greatly reduced by using an achromatic doublet as the objective. The aberration theory of these doublets is discussed in Chapter 33. With a suitable choice of glasses and shape, all three aberrations can be almost eliminated.

The aberrations in the eye lens or eyepiece can be reduced by choosing an appropriate design. Some common designs are shown in Chapter 15. For critical work, the objective and the eyepiece should be designed as a single system, balancing the aberrations between the two sub-components.

Let us look at the specific aberrations of the two types of telescopes.

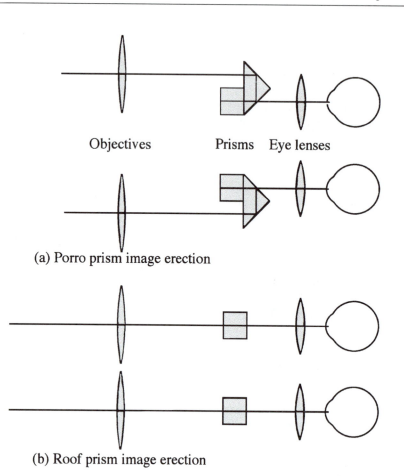

Fig. 17.19: The layout of a pair of binoculars with image erection using (a) Porro prisms and (b) a roof prism such as a Pechan prism.

Objectives　　　　　Prisms　Eye lenses

(a) Porro prism image erection

(b) Roof prism image erection

17.11.1　Keplerian telescopes

Schematic Keplerian telescopes are constructed from positive power lenses and therefore will suffer from spherical aberration, field curvature and longitudinal chromatic aberration as these aberrations have the same sign as the power of the lens and therefore accumulate. As stated above, spherical aberration and longitudinal chromatic aberration can be reduced by using an achromatic doublet as the objective. The theory of achromatic doublets is discussed in Chapter 33. However, doublets have little effect on Petzval curvature. This can only be reduced by using more complex designs containing negative power components.

In very high magnification telescopes, longitudinal chromatic aberration in the objective is a particularly serious problem and cannot be entirely overcome by using achromatic doublets. It can only be eliminated entirely by using a reflecting objective, for example the Newtonian or Cassegrain objective.

17.11.2　Galilean telescopes

Spherical aberration, Petzval curvature and longitudinal chromatic aberration tend not to be serious in Galilean telescopes because these aberrations have the same sign as the lens powers and tend to cancel out. However, in this telescope, off-axis aberrations are particularly troublesome because of the position of the entrance pupil. Because it is on the eye lens side of the telescope, the pupil ray, and hence the beams from off-axis points, intersect the objective at some distance from the optical axis. This has the effect of introducing considerable off-axis aberrations, particularly astigmatism.

Exercises and problems

17.1 Given the following Keplerian telescope

   (i) objective power $= +20\,\text{m}^{-1}$ and diameter $= 20\,\text{mm}$
   (ii) eye lens power $= +80\,\text{m}^{-1}$

   Calculate

   (a) the magnification
   (b) the telescope length
   (c) the eye relief
   (d) the exit pupil diameter
   (e) the position of a field lens, if used

   > ANSWERS: (a) $M = -4$, (b) length $= 62.5$ mm, (c) eye relief $= 15.6$ mm, (d) exit pupil diameter $= 5$ mm, (e) distance from objective $= 50$ mm

   If the above telescope were a Galilean instead (that is the eye lens had a negative power of $-80\,\text{m}^{-1}$), what would be the lens separation?

   > ANSWER: lens separation $= 37.5$ mm

17.2 For a Galilean telescope with the following construction:

   > objective lens power $= 0.1\,\text{cm}^{-1}$ and aperture radius $= 2.5$ cm
   > eye lens power $= -0.2\,\text{cm}^{-1}$ and aperture radius $= 0.5$ cm
   > separation $= 5$ cm

   calculate

   (a) the angular magnification for infinite conjugates
   (b) the entrance pupil position for an eye placed 15 mm back from the eye lens
   (c) the angular diameter of the object field-of-view neglecting the diameter of the eye pupil
   (d) the closest working distance for a user whose near point is at 25 cm

   > ANSWERS: (a) $M = 2$, (b) $\bar{l}_v = 16$ cm, (c) diameter $= 17.76°$, (d) 90 cm

17.3 Compare the field-of-view of the following Keplerian and Galilean afocal telescopes:

   > objective power $= +10\,\text{m}^{-1}$
   > eye lens power $= \pm40\,\text{m}^{-1}$
   > objective aperture radius $= 2.5$ cm
   > eye lens aperture radius $= 1.0$ cm

   Assume the Galilean telescope uses the same eye relief and exit pupil diameter as the Keplerian telescope.

ANSWERS:

	Object field radius (°)	
	Vignetting free	Total
Keplerian	1.7	7.4
Galilean	0.0	3.6

17.4   A reverse Galilean telescope is to be used to overcome the image size difference
in monocular aphakia already corrected by a contact lens. The objective lens is
in the spectacle plane and the eye lens is a further correction to the contact lens.
Assume the desired magnification is $+0.91$ and the spectacle vertex distance is
15 mm. Calculate the powers of the spectacle lens (F_{sp}) and the overcorrection
of the contact lens (ΔF_{cl}).

ANSWERS: $F_{sp} = -6.59 \text{ m}^{-1}$ and $\Delta F_{cl} = +6.00 \text{ m}^{-1}$

17.5   Calculate the effective magnification when a Galilean afocal telescope, with
a magnification of 3 and length of 3 cm, is refocussed to view an object at a
distance of 2 m and the image remains at infinity.

ANSWER: 3.07

Summary of main symbols

d	telescope length when afocal
ρ_o	aperture radius of objective lens (always positive)
ρ_e	aperture radius of eye lens (always positive)
F	equivalent power of a telescope
F_o	equivalent power of an objective lens
f_o	corresponding equivalent focal length
F_e	equivalent power of an eye lens
f_e	corresponding equivalent focal length
f_{el}	focal length of an erecting lens
$\mathcal{F}_o, \mathcal{F}'_o$	front and back focal points of objective
$\mathcal{F}_e, \mathcal{F}'_e$	front and back focal points of eye lens
F_1, F_2	front and back surface powers of the thick lens telescope
l_v, l'_v	object and image distances from objective and eye lens
L_v, L'_v	corresponding (reduced) vergences
$\bar{l}_v, \bar{l}'_v$	distance of entrance and exit pupils from eye lens; $\bar{l}'_v$ is the eye relief
M	angular magnification of an afocal telescope (object at infinity)
M_T	transverse magnification for an afocal telescope (finite conjugates)
M_L	longitudinal magnification for an afocal telescope (finite conjugates)
θ, θ'	angular sizes of object and image
θ_v	angular radius of field-of-view free of vignetting
θ_t	angular radius of total or full field-of-view

Section 17.7: Refocussing and its effects

Δd adjustment of telescope length
A refractive error level measured at the eye lens (negative for myopes)
w working distance measured from the objective (always positive)
W corresponding working distance vergence (always positive)
M' corresponding effective magnification when not used afocally

References and bibliography

*Bailey I.L. (1978a). New method for determining the magnifying power of telescopes. *Am. J. Optom. Physiol. Opt.* 55(3), 203–207.

*Bailey I.L. (1978b). Measuring the magnifying power of Keplerian telescopes. *Appl. Opt.* 17, 3520–3521.

*Byer A. (1986). Magnification limitations of a contact lens telescope. *Am. J. Optom. Physiol. Opt.* 63(1), 71–75.

Drasdo N. (1976). Visual field expanders. *Am. J. Optom. Physiol. Opt.* 53(9), 464–467.

Drasdo N. and Murray I.J. (1978). A pilot study on the use of visual field expanders. *Br. J. Physiol. Opt.* 32, 22–29.

Faye E.E. (1976). *Clinical Low Vision*. Little, Brown, Boston.

*Fried A.N. (1977). Telescopes, light vergence and accommodation. *Am. J. Optom. Physiol. Opt.* 54(6), 365–373.

Krefman R.A. (1981). Reversed telescopes on visual efficiency scores in field-restricted patients. *Am. J. Optom. Physiol. Opt.* 58(2), 159–162.

Longhurst R.S. (1973). *Geometrical and Physical Optics*. 3rd ed. Longman, London.

Mehr E.B. and Fried A.N. (1975). *Low Vision Care*. The Professional Press, Chicago.

*Sloan L.L. (1977). *Reading Aids for the Partially Sighted*. Williams & Wilkins, Baltimore.

*Smith G. (1979). Variation in image vergence with change in object distance for telescopes: The general case. *Am. J. Optom. Physiol. Opt.* 56(11), 696–703.

18

Macroscopes

18.0   Introduction

Telescopes are conventionally designed for viewing distant objects, but are often used for viewing closer objects. Equation (17.6a) shows that for telescopes, as an object comes closer, the image also comes closer, thus requiring the user to accommodate but to a greater level than the vergence of the object. In practice, this accommodation demand can be eased, because most telescopes can be refocussed by adjusting their length. These aspects have been discussed in Chapter 17. The closest working distance will then depend upon the length of adjustment and the maximum possible accommodation that the user can exert.

On the other hand, some specialized "telescopes" are designed for working distances that are relatively short, for example 20 cm to several metres. These have sometimes been called "near point telescopes", "intermediate distance telescopes" or "tele-microscopes". Perhaps a better term would be "**macroscope**," and this term would cover what are basically two lens magnifiers with working distances intermediate between microscopes and telescopes. However instruments of this type are often not readily available and when working distances are short, it is often more convenient to adapt a conventional (afocal) telescope (see Section 18.5).

The use of terms containing the words "telescope" and "microscope" to describe these instruments leads to confusion in the interpretation of magnification. Telescopes and microscopes use different definitions of magnifications and these in turn will be different from that defined for macroscopes. These points are discussed in greater detail in Section 18.3 and a numerical example will be given showing how the differences in definition of magnification can lead to very different outcomes.

Macroscopes have similar constructions to telescopes; for example they may be of the Keplerian type or Galilean type as shown in Figure 18.1. The Keplerian macroscope is also similar to the conventional microscope, which consists of two positive lenses, and the power of both the macroscope and microscope is negative. However, the working distance of a macroscope is much longer. In contrast, Galilean macroscopes have a positive equivalent power.

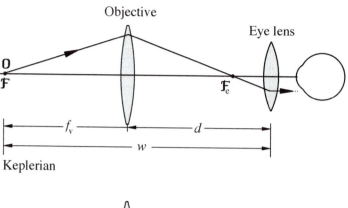

Fig. 18.1: Keplerian and
Galilean type macroscopes.

Keplerian

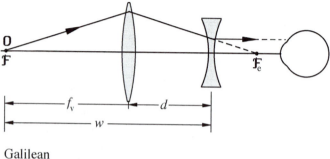

Galilean

In this chapter, we will look at some of the special properties of macroscopes, but we will not analyse these systems as extensively we did in the previous chapter. We will not discuss the image erection in Keplerian macroscopes and effects of refractive error as these will be similar to those for telescopes.

18.1   Image formation and equivalent power

Macroscopes are designed for near viewing and thus the object will be at some finite distance which may be as small as a few tens of centimetres. The image may be regarded as at infinity, in order to allow viewing with relaxed accommodation. The assumption that the image is at infinity does however neglect the needs of those users with refractive errors and also neglects instrument accommodation (see Chapter 36).

If the powers of the objective and eye lens are F_o and F_e, respectively, then the equivalent power F of the macroscope is given by equation (3.20) for two lenses, that is

$$F = F_o + F_e - dF_oF_e \qquad (18.1)$$

and this must be non-zero. We will see shortly that this power is negative for a Keplerian design and positive for a Galilean design. Since the image is at infinity, the object position o coincides with the front focal point $\mathcal{F}$, as shown in the diagram. The front focal length f_v of a two lens system is given by equation

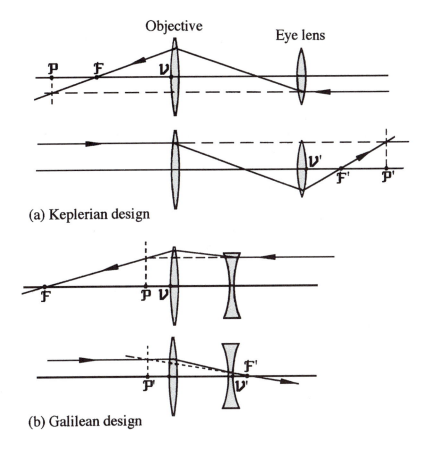

Fig. 18.2:   The cardinal
points of macroscopes.

(a) Keplerian design

(b) Galilean design

(3.23), but with n and μ set to 1. In terms of the notation used here,

$$f_v = (dF_e - 1)/F \qquad (18.2)$$

The working distance w, which is shown in the diagram, is defined here as the distance from the front focal point at $\mathcal{F}$ to the eye lens and always taken as positive. It is given by the equation

$$w = d - f_v \qquad (18.3a)$$

that is

$$w = d + [(1 - dF_e)/F] \qquad (18.3b)$$

We will shortly use these equations to derive an equation for the magnification.

18.2   Cardinal points

The Keplerian and Galilean type macroscopes are shown in Figure 18.2 with the paths of the two rays used to define the positions of the cardinal points, $\mathcal{F}$, $\mathcal{F}'$, $\mathcal{P}$ and $\mathcal{P}'$. For any particular system, the cardinal points can be found by tracing these particular rays or, if the macroscope is a two lens system (that is it is either a Galilean system or a Keplerian without a field lens), we can use equations given in Sections 3.3/3.4.1. In Example 18.1, we will use this latter method.

18.2.1   The Keplerian type

The Keplerian macroscope is similar in construction to the microscope described in Chapter 16, but with a longer working distance. It was shown in Chapter 16 that a microscope has negative power; therefore, because of the similarity in construction, the equivalent power of the Keplerian macroscope will also be negative. The negative power also follows from relative positions of the principal and focal points shown in Figure 18.2a. In this diagram, the front focal point $\mathcal{F}$ is to the right of the front principal point $\mathcal{P}$, indicating a negative equivalent power. However, this diagram is only schematic and not to scale and the actual positions will depend upon the specific design.

18.2.2   The Galilean type

The relative positions of cardinal points of the Galilean type of macroscope shown in Figure 18.2b are those of a positive power lens, but with the principal points outside and on the object side of the system. This is demonstrated in Example 18.1.

18.3   Magnification

The use of the term "telescope" to describe magnifiers working at intermediate distances is misleading. In particular it may lead to an incorrect specification and estimate of effective magnification. For a telescope, the (afocal) magnification is given by equation (17.4), that is

$$M = -F_\mathrm{e}/F_\mathrm{o} \quad \text{(telescope)} \tag{18.4}$$

and it has been shown in Chapter 17 that when afocal telescopes are adjusted for near viewing by changing their lengths, the magnification is increased. On the other hand some people regard macroscopes as microscopes or as simple magnifiers and specify the magnification by the usual simple magnifier form, equation (16.12), that is

$$M = F/4 \quad \text{(microscope)} \tag{18.5}$$

where the equivalent power F in this equation must be in dioptres. We will see, in Example 18.1, that these equations can give the wrong measure of magnification.

In Chapter 17, we found that refocussing a telescope for a finite working distance changes the magnification. We should expect the same rule to apply to the macroscope. Therefore, the specification of the magnification of a macroscope should include a reference working distance. If macroscopes are to be used for other working distances, the effective magnification may be very different. Referring to Figure 18.3, the most appropriate definition of the magnification of a macroscope that includes a working distance is as follows.

Fig. 18.3: Object and image formation and definition of magnification. The image is formed at infinity.

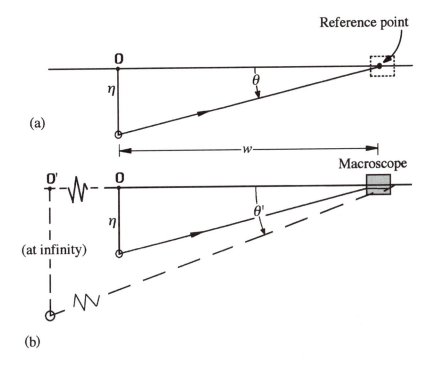

(a)

(b)

$$M = \frac{\text{angular size } (\theta') \text{ of image (at infinity) seen through the instrument}}{\text{angular size } (\theta) \text{ of object at the working distance } w} \quad (18.6)$$

If the object has a size η, then from Figure 18.3a its angular size θ seen from the reference point will be given by the equation

$$\theta = \eta/w \quad (18.7)$$

The angular size of the image of an object placed in the front focal plane of an optical system with an equivalent power F can be found from equation (3.52). Applying this equation to this situation gives an image size of

$$\theta' = \eta F \quad (18.8)$$

Combining equations (18.7) and (18.8) and substituting into equation (18.6) gives

$$M = wF \quad (18.9)$$

which expresses the magnification in terms of the working distance and equivalent power of the macroscope. This is also the same as equation (18.5) for the magnification of a conventional microscope.

At this point there may be some ambiguity about the actual definition of w. In equations (18.3a and b) we took the distance from the eye lens as shown in Figure 18.1, but in the derivation of equation (18.9), we assumed that the distance was measured from the eye. However, because we cannot be sure of the eye position, especially with a Galilean macroscope, we will assume that w is measured from the eye lens as shown in the diagrams, because it is the closest fixed and known point to the eye. Providing the eye is not too far from the eye lens and the working distance w is much greater than this distance, this assumption will not lead to a serious error.

We can show that equation (18.9) is a more general definition for magnification than the afocal case given by equation (18.4). If we replace w in equation (18.9) by the expression from equation (18.3b), we have

$$M = d(F - F_e) + 1$$

If we further replace F by zero and take d as the sum of the focal lengths (the afocal length), we see that this equation reduces to the afocal equation (18.4).

18.3.1   *Powers of the objective and eye lens*

Assuming the image is at infinity, the object will be at the front focal point, as shown in Figure 18.1. To design a macroscope, one must find the powers of the objective and eye lens. We will assume that we are given the desired values of the magnification M, lens separation d and the working distance w and use these to find the powers of the objective and eye lens.

To find equations for these lens powers in terms of the above quantities, it is easier to first find an equation for the eye lens power. Starting with equation (18.3b) to solve for F_e and then using equation (18.9) to replace F by M/w, we have

$$F_e = \frac{[w - M(w - d)]}{wd} \tag{18.10a}$$

To find an equation for F_o, we use equation (18.9) to replace F in equation (18.1) by M/w and solve for F_o to give

$$F_o = \frac{M - wF_e}{w(1 - dF_e)}$$

If we replace F_e by the expression given by equation (18.10a), we have finally

$$F_o = \frac{w(M - 1)}{dM(w - d)} \tag{18.10b}$$

Example 18.1: Find the thin lens powers and positions of the cardinal points of a macroscope designed to have a magnification of 3.2, length of 4.0 cm and working distance of 32 cm measured from the eye lens.

Solution: *Powers.* Here $w = 0.32\,\mathrm{m}$, $d = 0.04\,\mathrm{m}$ and $M = 3.2$. From equation (18.10a)

$$F_e = -45.000\,\mathrm{m}^{-1}$$

and from equation (18.10b)

$$F_{o} = 19.643 \, \text{m}^{-1}$$

We can check these values by firstly finding the equivalent power from equation (18.1), that is

$$F = 19.643 - 45.000 + (19.643 \times 45.000 \times 0.04) = 10.0 \, \text{m}^{-1}$$

and then calculating the magnification using equation (18.9), that is

$$M = 0.32 \times 10.0 = 3.2$$

which agrees with the desired value.

In contrast, the telescope magnification equation (18.4) gives

$$M = -(-45.000)/19.643 = 2.29$$

and the microscope rule for magnification (18.5) gives

$$M = 10.0/4 = 2.50$$

Note that both of these values differ from the macroscope value and both underestimate the actual effective magnification of the macroscope.

Cardinal points. We can find the position of the cardinal points using equations given in Section 3.3, with the following substitutions

$$n = n' = \mu = 1, \quad d = 0.04 \, \text{m}, \quad F_1 = F_o = 19.643 \, \text{m}^{-1}$$

$$\text{and} \quad F_2 = F_e = -45.000 \, \text{m}^{-1}$$

Now from equation (3.22),

$$\mathcal{V}'\mathcal{F}' = \frac{1 - (0.04 \times 19.643)}{10} = 0.0214 \, \text{m}$$

from equation (3.23)

$$\mathcal{V}\mathcal{F} = \frac{[0.04 \times (-45.0)] - 1}{10} = -0.280 \, \text{m}$$

from equation (3.27a)

$$\mathcal{V}'\mathcal{P}' = \frac{0.04 \times 19.643}{10} = -0.0786 \, \text{m}$$

and from equation (3.27b)

$$\mathcal{V}\mathcal{P} = \frac{0.04 \times (-45.0)}{10} = -0.180 \, \text{m}$$

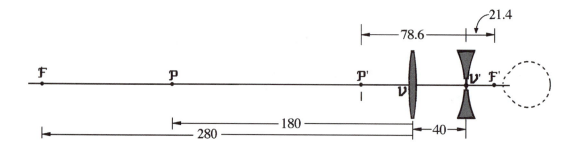

Figure 18.4 shows the schematic macroscope with the positions of its cardinal points. Since the system is in air, the nodal points will coincide with the principal points.

Fig. 18.4: A thin lens design of a Galilean macroscope (see Example 18.1 for details). Units are in millimetres.

18.3.2   Linear size of the object and angular size of the image

Given an object of a certain size, say η, we may need to know the angular size θ' of the image in terms of the angular magnification M and working distance w. We can find a suitable equation; by combining equations (18.8) and (18.9) we have

$$\theta' = \eta M / w \qquad\qquad (18.11)$$

18.4   Pupils

The pupil properties of macroscopes are similar to those of the telescopes. Equations can be derived for finding the positions and pupil magnification of the pupils of macroscopes. We begin by reviewing the definition of the pupil magnification. For both Keplerian and Galilean type macroscopes, pupil magnification $\bar{M}$ is defined by the equation

$$\bar{M} = \bar{\rho}'/\bar{\rho} = D'/D \qquad\qquad (18.12)$$

where $\bar{\rho}$ and $\bar{\rho}'$ are the radii and D and D' are diameters of the entrance and exit pupils, respectively. This equation can also be expressed in terms of the magnification M, working distance w, lens separation d and lens powers. The resulting equation will depend upon whether the instrument is a Keplerian or Galilean type.

18.4.1   Keplerian type macroscopes

As with Keplerian telescopes, the objective of the Keplerian macroscope is the aperture stop and the exit pupil is real and is the image of the objective, formed by the eye lens (or eyepiece if a field lens is present). For the two lens macroscope (i.e., no field lens) shown in Figure 18.5a, the exit pupil is shown at a distance $\bar{l}'(= \bar{l}'_v)$ from the eye lens. An equation for this distance (the eye relief) can be found by applying the lens equation (3.15), with $\bar{l}_v = -d$ the lens separation. After some manipulation we can get

Fig. 18.5: Pupil formation
in macroscopes.

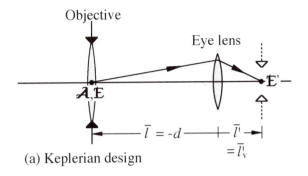

(a) Keplerian design

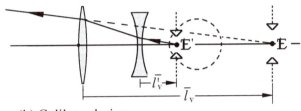

(b) Galilean design

$$\bar{l}'_v = \frac{dw}{M(d-w)} \qquad (18.13a)$$

The pupil magnification $\bar{M}$, defined by equation (18.12), is also equal to $\bar{l}'_v/\bar{l}$ and we can use the above results to express it in terms of M, w and d, as follows:

$$\bar{M} = \frac{w}{M(w-d)} \qquad (18.13b)$$

Note that these last two equations are not the same as those for an afocal system which are given by equations (17.13) and (17.15), respectively. However, if the working distance w becomes infinite, these equations reduce to the afocal values.

Example 18.2: Find the position and magnification of the exit pupil of a Keplerian macroscope with a magnification of $(-)3.2$, a length of 4 cm, a working distance of 32 cm and an objective diameter of 3 cm.

Solution: From equation (18.13a), we have

$$\bar{l}'_v = \frac{4 \times 32}{(-3.2)(4-32)} = +1.429\,\text{cm}$$

and from equation (18.13b)

$$\bar{M} = \frac{32}{(-3.2)(32-4)} = -0.357$$

If we had regarded the macroscope as a telescope and used the afocal equations (17.13) and (17.15) we would have the values

$$\bar{l}'_v = -4/(-3.2) = +1.250 \, \text{cm} \quad \text{(telescope)}$$

and

$$\bar{M} = 1/(-3.2) = -0.3125 \quad \text{(telescope)}$$

which are noticeably different from the correct or macroscope values.

18.4.2   Galilean type macroscopes

As in the Galilean telescope, the Galilean type macroscope has no intrinsic aperture stop and therefore the entrance pupil of the eye acts as the exit pupil of the macroscope. Figure 18.5b shows the exit pupil at ε' a distance $\bar{l}'_v$ from the eye lens. For field-of-view calculations and other purposes, it may be necessary to locate the position and size of the corresponding entrance pupil. We can find an equation for its distance $\bar{l}_v$ from the objective in terms of the component powers, their separation d, system magnification M, working distance w and the distance $\bar{l}'_v$ of the exit pupil from the eye lens. By applying the lens equation (3.15) at each lens in turn we can show that

$$\bar{l}_v = \frac{\bar{l}'_v(1 - dF_e) + d}{1 - (\bar{l}'_v M/w) - dF_o} \tag{18.14a}$$

and at the same time show that the pupil magnification $\bar{M}$ is given by the equation

$$\bar{M} = 1 - (\bar{l}'_v M/w) - dF_o \tag{18.14b}$$

Note that this is different from the afocal equation given by equation (17.15), but it is left to the reader, as an exercise, to show that this equation reduces to equation (17.15), in the limit that w goes to infinity.

> **Example 18.3:** Find the position and magnification of the entrance pupil of a Galilean macroscope with a magnification M of 3.2, a length d of 4 cm and a working distance w of 32 cm, if the eye is placed 15 mm beyond the eye lens. This is the same macroscope specified in Example 18.1.
>
> **Solution:** For this type of macroscope, we first need to find the values of the component powers using equations (18.10a) and (18.10b). These equations give
>
> $$F_o = +19.643 \, \text{m}^{-1} \quad \text{and} \quad F_e = -45.00 \, \text{m}^{-1}$$
>
> These values were also calculated in Example 18.1.

Substituting these powers and the above values of M, w and d into equation (18.14a) gives a value of

$$\bar{l}_v = +127.6\,\text{cm}$$

which indicates that the entrance pupil is on the image side of the system as it always is for a Galilean telescope (see Section 17.5.1). However it is not necessarily always so for a Galilean macroscope. It depends upon the position of the exit pupil relative to the back focal point of the macroscope. In this example (see Example 18.1 and Figure 18.4), the back focal point is 2.14 cm beyond the eye lens. If the exit pupil is between this focal point and the eye lens, then the entrance pupil will be on the image space side of the system. However, if the exit pupil is beyond the back focal point, the entrance pupil will be on the object space side. If the exit pupil coincides exactly with the focal point, the entrance pupil will be formed at infinity.

Substituting the appropriate values into equation (18.14b) gives the magnification

$$\bar{M} = 0.0643$$

If we had assumed that the equivalent telescope equations applied, equation (17.21) would give

$$\bar{l}_v = +28.16\,\text{cm} \quad \text{(telescope)}$$

Equation (17.15) for the pupil magnification of the Keplerian telescope also applies to the Galilean telescope and this equation gives

$$\bar{M} = 1/3.2 = 0.3125 \quad \text{(telescope)}$$

These are different from the correct macroscope values.

18.5   Adapting telescopes

Telescopes can be readily converted to macroscopes by either (a) altering the length of the telescope, that is refocussing it for the finite working distance or (b) placing a suitable auxiliary lens over the objective or eye lens (near point caps). In both cases, there is some change in the magnification.

18.5.1   Adjusting length

The effect of refocussing a telescope for a finite working distance has been discussed in Chapter 17, where it was shown that refocussing changes the effective magnification. However, for very short working distances, it is not possible to refocus the telescope if the object is within the front focal point of the objective as the image formed by the objective would be virtual. In these cases, near point caps are the only alternatives.

18.5.2   Using near point caps

Suppose a telescope is used to view an object at finite distance l_v from the objective, that is at a vergence $L_v (= 1/l_v)$. If the telescope is a two lens system, the corresponding image vergence L'_v at the eye lens is given by equation (17.6). Thus the image will no longer be at infinity, but it can be placed there if the beam entering the telescope is collimated. This can be achieved by placing a positive power lens with a power F'_A immediately in front of the objective, where

$$F_A = -L_v \qquad (18.15)$$

This auxiliary lens function is known as a **near point cap**.

Instead of placing an auxiliary lens over the objective, we can place one over the eye lens, but the power of the lens must be different. Because the image vergence is L'_v, the image can be put at infinity by placing a lens of equivalent power F'_A over the eye lens, where

$$F'_A = -L'_A \qquad (18.16)$$

Assuming that equation (17.6) is applicable, we can combine it and equation (18.16) to give

$$F'_A = \frac{M_T^2 F_A}{(1 + d M_T F_A)} \qquad (18.17)$$

where M_T is the magnification of the telescope and d is the distance between the two components. If $d M_T F_A \ll 1$ then we can write this as

$$F'_A \approx M_T^2 F_A \qquad (18.17a)$$

If the telescope is more complex than a two lens system, e.g. it contains a field lens, equation (18.17) will only be approximate.

18.5.2.1   Required aperture diameter of the near point cap

Keplerian telescopes

The minimum aperture diameter of the near point cap must be that of the objective or eye lens, depending on whether it is placed over the objective or eye lens, respectively. Thus, if the diameter of the telescope objective is D, then this is the minimum aperture diameter of the near point cap placed over the objective. In the Keplerian telescope, the objective is the aperture stop and hence D is also the diameter of the entrance pupil. The diameter of the eye lens must be greater than the diameter D' of the exit pupil, which is related to the entrance pupil diameter by equation (17.14), but written here as

$$D' = D/M_T \qquad (18.18)$$

and this is the minimum diameter of the near point cap placed over the eye lens. In Chapter 6, we gave equation (6.8), which sets a rule of thumb for checking whether a particular lens could be manufactured given its power and aperture radius. In terms of the aperture diameter, this equation is

$$|FD| < 0.5 \qquad (18.19)$$

and the greater this product the thicker will be the lens and the greater will be the aberrations. Let us look at the product $|F'_A D'|$ for the near point cap placed over the eye lens. Using equations (18.17a) and (18.18), we can write

$$|F'_A D'| \approx |(M_T^2 F_A)(D/M_T)| = M_T|F_A D| \qquad (18.20)$$

This equation shows that if the magnification is greater than 1.0, as it usually is, the product $|F'_A D'|$ of a near point cap placed over the eye lens is M_T times greater than the value $|F_A D|$ for the near point cap placed over the objective. Thus a near point cap placed over the eye lens will be relatively thicker and have greater aberrations with a potential reduction in image quality than if the corresponding near point cap were placed over the objective. Therefore it is preferable to place the near point cap over the objective.

Galilean telescopes

The objective of the Galilean telescope is not the aperture stop and therefore the above argument does not strictly apply in this case and there is no clear relationship between the diameters of the objective and eye lens. Therefore in this case, before one decides which is preferable, a near point cap over the objective or over the eye lens, one would have to look at the actual diameters of the two components, determine the required powers and then look at the product of the power and aperture diameter. The one with the lower value would be the preferable alternative.

18.5.2.2 *Effect on magnification*

We may expect the use of an auxiliary lens or near point cap to change the effective magnification of the telescope. We can investigate this magnification effect as follows. From equation (17.1), the power F of the telescope is given by the equation

$$F = F_o + F_e - dF_oF_e = 0$$

which is the afocal condition. If we add a lens of power F_A, in contact, to the objective then this power becomes

$$F = (F_A + F_o) + F_e - d(F_A + F_o)F_e \neq 0$$

Because of the above afocal condition, this equation reduces to

$$F = F_A(1 - dF_e)$$

If we now recall equation (17.2), which sets the length of the telescope, that is

$$d = 1/F_o + 1/F_e$$

the new power F is now

$$F = F_A(-F_e/F_o)$$

By recalling equation (17.4), for the afocal magnification of the telescope, this equation can be written

$$F = F_A M_T \qquad\qquad (18.21)$$

Substituting this value of F in equation (18.9) gives the magnification of the macroscope

$$M = w F_A M_T \qquad\qquad (18.22)$$

Since

$$w = d + 1/F_A$$

we can write equation (18.22) in the form

$$M = M_T + d F_A M_T \qquad\qquad (18.23)$$

The product $d F_A$ is always positive and therefore the resulting magnification is always greater than the telescopic value.

18.6   Conversion to a telescope

In the previous section, we have shown that a telescope can be converted to a macroscope by altering its length or adding near point caps. We will now look at the conversion of a macroscope to a telescope by adjusting its length.

To become a telescope, equation (17.2a) states the lens separation must be the sum of the focal lengths. We will denote this lens separation by the symbol d_{afocal} and thus

$$d_{\text{afocal}} = f_o + f_e$$

or in terms of powers, equation (17.2) gives

$$d_{\text{afocal}} = 1/F_o + 1/F_e$$

If we replace the component powers by expressions containing the design magnification M, lens separation d and working distance w, using equations (18.10a) and (18.10b), then we have

$$d_{\text{afocal}} = \frac{dM(w - d)}{w(M - 1)} + \frac{wd}{w - M(w - d)} \qquad\qquad (18.24)$$

In terms of the component powers F_o and F_e, the magnification of a telescope is given by equation (17.4). Referring to this magnification as M_{afocal}, then we have

$$M_{afocal} = -F_e/F_o$$

Once again, if we replace the component powers by expressions containing the design magnification M, lens separation d and working distance w, using equations (18.10a) and (18.10b), then we have

$$M_{afocal} = \frac{M[M(w-d) - w](w-d)}{w^2(M-1)} \qquad (18.25)$$

Example 18.4: Given the Galilean macroscope specified in Example 18.1, calculate the length and magnification if that macroscope is refocussed for an infinite working distance.

Solution: Here $M = +3.2$, $w = 0.32$ m and $d = 0.04$ m. Substituting these values in equation (18.24) gives

$$d_{afocal} = 0.02869 \, m = 28.69 \, mm$$

which is shorter than the design length. If we now substitute the values in equation (18.25) we have

$$M_{afocal} = +2.29$$

which is less than the macroscope value.

18.7 Aberrations

Since macroscopes have similar constructions to telescopes and microscopes, they would be expected to have similar aberrations. However, they are not as highly corrected as microscopes and their aberration levels are more similar to those in telescopes. See Section 17.11 for a discussion of the aberrations of telescopes.

Exercises and problems

18.1   Find the thin lens powers of a macroscope to be used as a low vision magnifier to watch television, if the television is at 3 m and the desired magnification is $+3$. Take the lens separation as 3 cm.

ANSWERS: $F_o = +22.45 \, m^{-1}$, $F_e = -66.00 \, m^{-1}$

18.2   Find the position and diameter of the entrance pupil of the following macroscope, with an eye relief of 10 mm and an eye pupil diameter of 4 mm.

$$M = 8, w = 0.5 \, m, d = 0.05 \, m$$

ANSWERS: $\bar{l}_v = 0.386$ and entrance pupil diameter $= 10.03$ mm

18.3   Calculate the near point cap power for adapting a Keplerian telescope (magnification of 6 and length of 6 cm) to a working distance of 1 m.

ANSWERS: Objective cap $F_A = +1\,\mathrm{m}^{-1}$ and eye lens cap $F'_A = +56\,\mathrm{m}^{-1}$

Summary of main symbols

d	length of macroscope
M	magnification of macroscope
w	working distance (positive) measured from eye lens
F_o	equivalent power of objective
f_o	corresponding equivalent focal length
F_e	equivalent power of eye lens
f_e	corresponding equivalent focal length
F	equivalent power of a macroscope
f	corresponding equivalent focal length
f_v	corresponding front vertex focal length
F_A, F'_A	powers of auxiliary lenses placed at the objective or eye lens
θ, θ'	angular sizes of object and images
M_T	magnification of a telescope
d_{afocal}	length of the macroscope when converted to a telescope
M_{afocal}	magnification of the macroscope when converted to a telescope

References and bibliography

Jackson J. and Silver J. (1983). Visual disability, part 7, Telescopic systems for near vision (2). *Ophthal. Optician* September 24, 597–604.

Smith G. (1979). Variation of image vergence with change in object distance for telescopes: The general case. *Am. J. Optom. Physiol. Opt.* 56(11), 696–703.

19

Relay systems

19.0   Introduction

So far we have looked at optical systems that we could call "direct vision" systems; that is the optical system is pointed directly at the object and, apart from the effect of image erecting prisms, the optical axis joining the object and image is a straight line. In contrast, in a number of situations, there is a need to bend an image forming beam around a "corner" or several "corners". Usually this also requires the beam to be restricted to lie within a tube of a certain size and be transmitted over a distance that is long relative to the diameter of the beam. If the beam width is to be so constrained, it is often necessary to form intermediate images along the beam path, which in turn requires the use of what are known as **relay lenses** or **relay systems**. The main use of relay systems is to view normally inaccessible or hazardous places or simply to transmit luminous flux (light) to places remote from the source.

The earliest type of relay system was the **periscope**, which has been developed to a high degree for use in submarines. Much smaller systems have been developed to examine the inside of bodily organs or machines. The most recently developed type of relay system is based upon **optical fibres**. Optical fibres are usually circular in cross-section, much longer than their diameter and usually made from glass or plastic. They operate on the principle that light entering one end is constrained in the fibre by successive total internal reflections at the internal walls. Optical fibre bundles may be also used to transmit luminous flux (incoherent bundles) as well as images (coherent bundles).

19.1   Periscopes

The simplest periscope consists only of mirrors as shown in Figure 19.1. This example uses two mirrors to provide the indirect view and an erect image. The field-of-view of this construction is simply limited by the size of the mirrors. Often they have the same size and in this case, the field-of-view is limited by the objective mirror. If the periscope has a length d, that is the distance between the two mirrors is d, and if a dimension of the objective mirror is ϕ, the size of the field-of-view is simply the angular size of the objective mirror. Thus the field-of-view diameter θ, as seen at the second mirror, is given by the equation

$$\tan(\theta/2) = \phi/(2d) \tag{19.1}$$

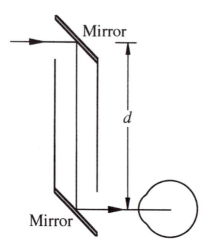

For a fixed mirror size, the longer the periscope, the smaller the field. This factor limits the practical length of such simple periscopes.

The field-of-view can be increased, without increasing the diameter of the mirrors, by the addition of an optical system. Such an optical system can also provide some magnification. A common example of such a periscope is the submarine periscope, which may be up to 10 m in length. A simplified construction of a submarine periscope is shown in Figure 19.2. In this system, right-angle prisms, described in Section 8.3, are used instead of mirrors to deflect the beam through 90°. An objective lens forms a primary image on the first field lens at o', where a graticule may be placed. A relay lens then transfers this image farther along to the point o''. Since the beam is reflected twice in the same plane by the prisms, the image would be erect if no lenses were used. The addition of an objective lens and the formation of an intermediate image would result in an inverted image unless some image erection system is used. In the example shown in the diagram, the relay system acts as an erecting lens system which is composed of two lenses between which the beam is collimated. The relay or erecting system forms the final image at the field lens of the eyepiece. Very long periscopes may incorporate more than one relay system. However, to maintain an erect image, there must be either (a) an odd number of relay systems, that is an even number of real images formed within the periscope, or (b) an even number of relay lenses plus an erecting prism system. The focal lengths of the objective, relay system and eyepiece can be chosen to provide the desired magnification.

19.2   The general lens relay system

The periscope may be regarded as a special case of a more general class of lens relay systems. The principle of the general lens relay system is that the system can be broken down into almost identical units placed in tandem. Each unit is an optical system which relays the image farther along with no change in image size. Mirrors or prisms may be used to deflect the beam through any desired angle.

The schematic simplest lens relay system consists of a sequence of single lenses as shown in Figure 19.3. The sequence alternates between a field lens and relay lens as shown in this diagram. Each relay lens takes the image from the

Fig. 19.2: Schematic arrangement of a general periscope.

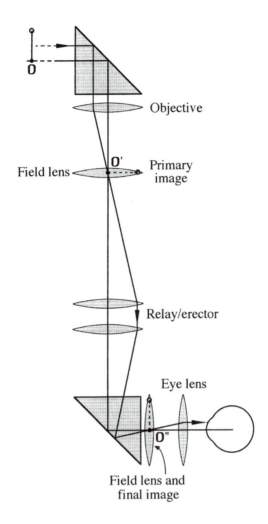

preceding relay lens and using it as its object, forms a new image farther down the line, usually with a transverse magnification of -1. This new image is then used as the object for the following relay lens. Field lenses are placed at each intermediate image position; otherwise vignetting would cause a progressive loss of field-of-view.

The initial image is formed by a suitable objective lens and the final image is usually viewed with an eyepiece at some suitable magnification. This system has two distinct drawbacks.

(i)   The first is the rigidity of the system. It is not flexible, although mirrors or prisms are sometimes placed at the object or eyepiece ends to allow greater flexibility of direction or field-of-view. This is done in the periscope.

(ii)   The other disadvantage is the tendency to have large amounts of field curvature, which is due to the large number of positive power lenses in the system.

In Figure 19.3, the relay lens is shown as a single lens. In actual designs, it may be more complex. For example in the submarine periscope shown in

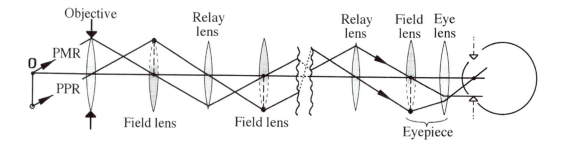

Figure 19.2, the relay lens unit is a system of two lenses designed to have
collimated light between them. There are other possibilities. For example a
telescope may be used as a relay unit and this has some useful properties that
are worth further discussion.

19.2.1   The telescope as a relay unit

A telescope has some special properties which make it potentially useful as a
relay unit. Some of these properties have already been discussed in Chapter 17.
One of these is that the transverse magnification is independent of position.
Figure 19.4 shows a telescope being used as a relay system. It images an object
at o to o' and the image size η' will be independent of object position. The
transverse magnification M_T is given by equation (17.9) and thus we have

$$\eta'/\eta = M_T = 1/M \tag{19.2}$$

where M is the angular magnification of the telescope. It is clear from this
equation that the transverse magnification and hence image size are independent
of object position.

However, for use as a relay element in a relay array of the type such as
shown in Figure 19.3, it is necessary that the image be real. Now using equation
(17.6a) the distances l_v' and l_v are related by the equation

$$l_v' = (l_v - dM)/M^2 \tag{19.3}$$

The image is real if the distance l_v' is positive, that is

$$l_v' > 0$$

From equation (19.3), it follows that this condition is satisfied if

$$l_v > dM \tag{19.4}$$

This condition requires that the distance l_v be more positive (that is less negative)
than the value dM. Now for a Keplerian telescope, the magnification is negative
and since l_v is usually negative, it follows that the object should be closer to the
objective than the distance dM.

Fig. 19.4: A telescopic lens relay unit. Here the transverse magnification $M_T = \eta'/\eta$ is independent of the position of the conjugates.

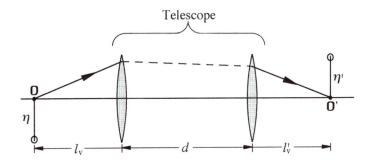

Let us now look at the throw oo'. From Figure 19.4, we have

$$oo' = l'_v + d - l_v$$

From equation (19.3) this distance can be expressed in terms of l_v as follows.

$$oo' = [(l_v - dM)/M^2] + d - l_v$$

which can be transformed to

$$oo' = l_v[(1/M^2) - 1] + d[(1 - (1/M)] \tag{19.5}$$

This equation shows that if $M = \pm 1$, the throw oo' is independent of l_v. If the transverse magnification is -1 (i.e. Keplerian), we have the result that

$$oo' = 2d \tag{19.6}$$

In contrast, if the magnification is $+1$, equation (19.5) shows that the throw would be zero, which is not practical. Furthermore, such a telescope would be of the Galilean type and it is not possible to construct a Galilean telescope with a magnification of $+1$.

19.2.2 *Different types of relay lens systems*

The detailed construction of a lens relay system depends upon its purpose. The periscope is one example and this usually has the construction as shown in Figures 19.1 and 19.2. It is distinguished by the $90°$ beam deviations at either end, which are achieved by the use of simple mirrors or prisms. Other common types are the **endoscope** and **boroscope**.

19.2.2.1 *Endoscopes and boroscopes*

Endoscopes are medical optical instruments designed for insertion into a body cavity through natural or surgically created body openings, for the purposes of examination, diagnosis and therapy. Endoscopes are usually rigid but the object end may have a mirror or a prism attached to it which can be remotely controlled to alter the direction of view. In the endoscope literature, the terms "distal" and "proximal" are frequently used to describe the object and image ends of the

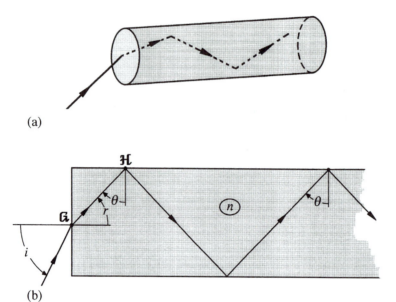

Fig. 19.5:   Optical fibre and a typical ray path.

(a)

(b)

endoscope respectively. The name "endoscope" is the family name for this class of instrument. There are many sub-classifications, such as laproscopes and cystoscopes.

Boroscopes are essentially the same as endoscopes but are designed for use in engineering to examine the internal cavities of machines.

These systems are effectively rigid and their rigidity restricts their use in many situations, where flexibility is essential. For example to examine the colon and large intestine, the instrument must follow a curved path and to examine the inside of some machines, the instrument must turn a few sharp corners. In these cases, the relay system must be flexible and a rigid lens relay system is not suitable. Optical relay systems using optical fibres as the relay element meet this requirement of flexibility.

19.3   Optical fibre systems

Optical fibre systems are essentially an ensemble of thin glass or plastic circular fibres. Light can be transmitted along and constrained within individual fibres by the principle of total internal reflection. If a general (skew) ray of light enters the end face of a fibre as shown in Figure 19.5a, the ray will be constrained inside this fibre, providing total internal reflection takes place at the walls. Referring to Figure 19.5b, the ray will be totally internally reflected at each reflection if the angle of incidence θ on the side wall of the rod is greater than the critical angle θ_{crit}, which is given by equation (1.13), that is

$$n \sin(\theta_{\text{crit}}) = \sin(90°) = 1$$

This condition places limitations on the maximum incidence angle at the input face.

Figure 19.5a shows a general skew ray but we will now simplify the discussion by restricting our analysis to the paths of meridional rays. Including skew rays significantly complicates the mathematics but adds little to understanding the properties of optical fibre systems as optical image relay devices. Figure 19.5b shows a meridional ray entering a fibre with an incidence angle i. It can easily be shown using the diagram and Snell's law applied at points g and $\mathcal{H}$ that

$$\sin^2(i) \leq n^2 - 1 \qquad (19.7)$$

In this simple explanation of the principle of the optical fibre, we have assumed that the principles of geometrical optics apply. This approach is valid provided the fibre diameter is much larger than the wavelength of the radiation. If the diameter is close to or less than the wavelength, then propagation of light in the fibre must be explained by physical optics.

19.3.1 Numerical aperture

The maximum limiting value i_{max} of i at the input face is the angle at which total internal reflection just occurs on the inside walls of the fibre. We now denote this angle i_{max} by the symbol α, thus from equation (19.7), it follows that

$$\sin(\alpha) = \sqrt{(n^2 - 1)} \qquad (19.8)$$

which is the numerical aperture of the fibre.

> **Example 19.1:** Calculate the numerical aperture of a fibre made with a glass of index 1.5.
>
> **Solution:** From equation (19.8), $\sin(\alpha) = \sqrt{(1.5^2 - 1)} = 1.12$. Since $\sin(\alpha)$ cannot be greater than unity, the actual numerical aperture is limited to 1.0, and thus rays striking the end face at an incidence angle of 90° and entering the fibre would be constrained in the fibre by total internal reflection.

19.3.2 Fibre diameters

Optical fibres can be of any diameter but the smaller the diameter, the more flexible they are (see Section 19.3.3). A typical diameter is usually about 25 μm. However, for long fibres with diameters less than about 10 μm, physical optics become significant and under these circumstances, geometrical optics cannot be used adequately to describe propagation of light along the fibre. Very short fibres (those used as field curvature correcting face plates) may have diameters down to about 5 μm.

19.3.3 Bending of fibres

The above theory assumes that the fibre is straight and does not bend. However one of the supreme advantages of fibres is their flexibility, but bending leads to some light loss. As a ray moves around a bend, its angle of incidence at the walls changes. The angle of incidence at the outside wall becomes less and the

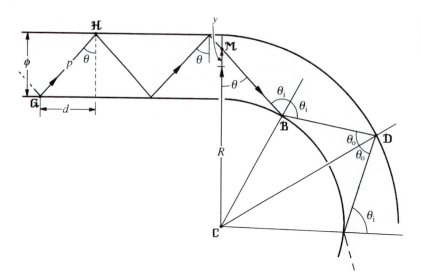

Fig. 19.6:   Ray path
around a bend in a simple
fibre.

angle of incidence at the inside wall becomes greater as shown in Figure 19.6.
If the angle of the ray on the outside wall is now less than the critical angle, a
part of the energy in the ray will pass out of the fibre.

Figure 19.6 shows a ray being propagated along a fibre of diameter ϕ and
bent in a circular arc of radius R. This ray has an angle of incidence θ at the
surface of the straight part of the fibre. The angles of incidence θ_o at the outside
wall and θ_i at the inside wall of the curved section depend upon the ray height
y where the ray enters the curved section and can be found as follows.

To find the incidence angle θ_i on the inside wall, we proceed as follows. In
triangle MBC, the sine rule gives

$$\sin(\theta_i)/(R + y) = \sin(\theta)/[R - (\phi/2)]$$

and hence

$$\sin(\theta_i) = \sin(\theta)(R + y)/[R - (\phi/2)] \tag{19.9}$$

Now since $(R + y)/[R - (\phi/2)] > 1, \sin(\theta_i) > \sin(\theta)$ and therefore $\theta_i > \theta$,
proving that the angle on the inside wall has a larger value than that for the
straight section.

To find the incidence angle θ_o on the outside wall, we proceed in a similar
manner. In triangle BDC, and using the sine rule

$$\sin(\theta_o)/[R - (\phi/2)] = \sin(\theta_i)/[R + (\phi/2)]$$

Therefore

$$\sin(\theta_o) = \sin(\theta_i)[R - (\phi/2)]/[R + (\phi/2)]$$

Now using equation (19.8)

$$\sin(\theta_o) = \sin(\theta)(R + y)/[R + (\phi/2)] \tag{19.10}$$

Now since $(R + y)/[R + (\phi/2)] < 1$, $\sin(\theta_0) < \sin(\theta)$ and therefore $\theta_0 < \theta$, proving that the angle on the outside wall is smaller than that for the straight section.

Equations (19.9) and (19.10) show that the effect of bending decreases as the ratio ϕ/R decreases and therefore light losses due to bending will also decrease as this ratio decreases. As a rule of thumb, fibres can be bent to a radius of 20 times their own diameter without any appreciable loss.

19.3.4 Path lengths

In the geometrical analysis of light transmission along fibres, a typical ray path is longer than the length of the fibre and will depend on the angle of incidence of that ray at the front face. A typical ray is shown in Figure 19.6. The ratio of the total path travelled p (path $\mathcal{GH}$), to the distance d moved along the fibre, is given by the equation

$$p = d/\sin(\theta) \tag{19.11}$$

that is

$$p/d = 1/\sin(\theta) \tag{19.11a}$$

This difference will affect travel times but this aspect is beyond the scope of this book as it is not relevant to visual optical instrumentation. However, it also effects transmission losses through absorption.

19.3.5 Transmission losses

Light passing into, through and out of optical materials suffers some losses or attenuation, no matter how transparent the material appears to be. There are three sources of attenuation: surface reflections, absorption and scatter. For thin materials, the absorption and scatter are usually insignificant. Thus for lens systems, absorption and scatter losses are usually negligible and the greatest loss is due to reflections at surfaces and these have been discussed in Chapter 12 and are discussed again in Chapter 34. On the other hand, an optical fibre may be as long as several metres. With these lengths, even the most minute absorption or scatter becomes significant. For a pure material, scatter is ultimately set by diffraction which closely follows Rayleigh's fourth power scatter law, which states that the amount of scatter is inversely proportional to the fourth power of the wavelength.

The internal transmittance $T(p)$ as a function of ray path length p follows Lambert's law and is thus given by the equation

$$T(p) = e^{-bp} \tag{19.12}$$

where b is the attenuation coefficient which includes attenuation losses due to both scatter and absorption. The value of b is a function of wavelength. In the vicinity of the visible spectrum, it increases with a decrease in wavelength. That is, the transmittance decreases with decrease in wavelength. For optical quality glass, the value of b in the middle of the visible spectrum is about 0.2 m^{-1}.

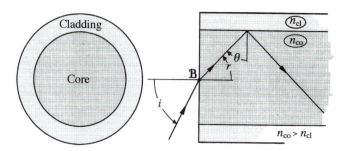

We will now look at the effect on the rays striking the wall at different angles of incidence. Using equation (19.11), we can replace p by the angle of incidence θ and the longitudinal distance d travelled along the fibre. Thus

$$T(\theta, p) = T(\theta) = e^{-bd/\sin(\theta)} \tag{19.13}$$

This equation shows that the smaller the angle of incidence (that is the smaller the angle θ), the greater the attenuation.

19.3.6   Transmission of polarized light

Conventional fibres with a circular cross-section do not maintain the polarization state of an injected polarized beam. Therefore a polarized beam directed into the fibre loses its state of polarization and the beam leaving the fibre at the output end is unpolarized.

19.3.7   Fibre insulation

Single optical fibres must be insulated to ensure that light does not leak out of the fibre when it is incident on a part of a wall that is in contact with another fibre or some other material. One method of doing this is to clad each fibre with a sheath of material that has a lower index than the core as shown in Figure 19.7. These types of fibres are known as **cladded** fibres. The cladding usually reduces the effective numerical aperture of the fibre.

Assume the core fibre has a refractive index of n_{co} and the cladding material has an index of n_{cl} as shown in Figure 19.7. The numerical aperture of this fibre is now given by the equation

$$\sin(\alpha) = \sqrt{(n_{co}^2 - n_{cl}^2)} \tag{19.14}$$

instead of equation (19.8). These cladded fibres are also sometimes known as **stepped index** fibres because the refractive index distribution has the form of a step.

> **Example 19.2:** Calculate the numerical aperture of a cladded fibre which has a core index of 1.70 and a cladding index of 1.52.
>
> **Solution:** From equation (19.14),
>
> $$\sin(\alpha) = \sqrt{(1.70^2 - 1.52^2)} = 0.80$$

Fig. 19.8: Typical ray path
in a gradient index fibre.

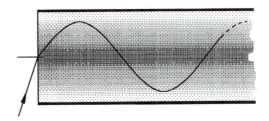

Fig. 19.9: Imaging light
into a fibre and numerical
aperture matching.

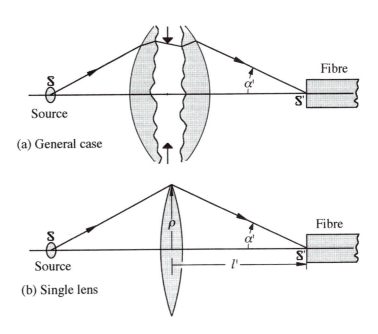

An alternative to the cladded fibre is the gradient index fibre. In this type of fibre, the refractive index decreases continuously from the centre of the fibre to the wall. In gradient index materials, a ray is continuously refracted and turns towards the region of higher index. This has the effect of constraining the rays in the fibre as shown in Figure 19.8.

19.3.8 Imaging light into a fibre and numerical aperture matching

Sometimes a beam of light is imaged into a fibre with a lens as shown in Figure 19.9a. If the amount of luminous flux being transmitted by the fibre is to be maximized, the image space numerical aperture of the lens should be the same as the numerical aperture of the fibre given by equation (19.14). If the numerical aperture of the lens is greater, the extreme rays will not be held in the fibres by total internal reflection and will be lost from the beam. If the numerical aperture of the fibres is greater, the fibres will be operating at less than their maximum capacity. If a single lens is used to image light into a fibre as shown in Figure 19.9b, its aperture radius sets its working numerical aperture. Suppose that the lens has an aperture radius ρ and the distance from the lens to the fibre is l', then we have

$$\tan(\alpha') = \rho/l'$$

and thus the numerical aperture $[n' \sin(\alpha')]$ of the lens is

$$\sin(\alpha') = \sin[\arctan(\rho/l')] \tag{19.15}$$

should be the same as that of the fibre given by equation (19.14).

Light entering an optical fibre is completely mixed so that the light is evenly distributed over the output face. This is due to the fact that rays entering the fibre at different angles follow different paths through the fibre. Thus any spatial detail that may be present on the input face is lost in transmission.

19.3.9   *Fibre bundles*

While a thinner fibre is more flexible, the amount of light it can transmit decreases as the square of its diameter. Thus in many applications, fibres are used in bundles. Such fibre optical systems are flexible while having a large effective diameter and therefore good for transmitting luminous flux or an image. However, because adjacent fibres are in contact, each fibre must be insulated from its neighbour in some manner.

There are two types of fibre bundles, known as **incoherent** and **coherent** bundles. They are distinguished by the relative order of fibres at the input and output faces of the bundle. In the incoherent bundle, the fibres orientation at the output face is a jumbled arrangement at the input face. In contrast, in the coherent bundles, the individual fibre arrangement is the same at each end.

19.3.9.1   *Transmission of luminous flux*

A single fibre or a bundle of fibres can be used to transmit light or luminous flux over long distances. For example this technique is often used when the actual light source such as a tungsten lamp and the heat it usually produces have to be remote from the object being illuminated. Typical applications occur in medicine and biology where the tissue could be damaged by the excess heat from the light source. These types of light sources are often called **cold light sources**.

While incoherent and coherent bundles may be used for transmission of luminous flux, coherent bundles are more expensive and therefore incoherent bundles are usually used for this task.

19.3.9.2   *Transmission of images*

A fibre bundle can be used to transmit images as well as light over relatively long distances. In principle, if an object is put into contact with the input face or imaged onto the face of a bundle as shown in Figure 19.10, the image will appear on the end face, where it can be viewed by an eye lens which acts as a magnifier and puts the image close to infinity for relaxed accommodation viewing. Thus in general, the fibre optical relay system consists of three basic components: the objective, the optical fibre bundle and the eye lens.

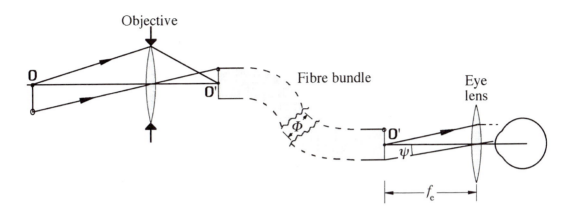

Fig. 19.10: Transmission of an image by a coherent fibre bundle.

The objective

The role of the objective is to image the scene of interest onto the input face of the bundle. Its focal length, aperture and the conjugate distances must be chosen to satisfy two requirements: (a) the image of the scene should fill the bundle input face, in order to have maximum resolution, and (b) the numerical aperture of the lens should match that of the fibre.

The choice of focal length of the objective depends upon the distance of the conjugates, the size of object and the diameter of the fibre bundle. The equations for determining the appropriate focal length can be found in Chapter 3.

The fibre bundle

Each fibre only transmits a portion of the image; the part of the image that is formed on the input face of that fibre. However, the image will only be correctly formed at the output face, if the individual fibres have the same relative positions; that is the bundle must be coherent.

The eye lens

Because of the small diameter of the fibre bundle, an eye lens or simple magnifier must be used to observe and magnify the image formed on the exit face of the bundle. The magnification should be adequate to pick up the detail in the image, but need be no larger than required to just resolve the individual fibres. The optics of eye lens and eyepieces are discussed in Chapter 15.

If we view an optical fibre bundle of diameter Φ with an eye lens or simple magnifier of focal length f_e as shown in Figure 19.10, then the angular radius ψ of the image field will be given by the equation

$$\tan(\psi) = (-)0.5\Phi/f_e \tag{19.16}$$

Resolving power of the fibre bundle

When a beam of light is imaged onto one end of a single fibre, the light is completely spatially mixed across the face of a fibre. Thus if a single point source is imaged on one end of a fibre, the other end of the fibre will be uniformly illuminated. If two point sources are to be resolved, they must illuminate separate fibres with at least one unilluminated fibre between them. It follows that the resolving power is approximately equal to twice the individual fibre diameter.

The illumination system

In many optical fibre imaging systems, it is common to have a central coherent core carrying the image and, surrounding this, an incoherent annular bundle to carry the illuminating light.

> **Example 19.3:** An object 50 mm in diameter is to be imaged into an optical fibre system which has a diameter of 2.5 mm, using a lens of focal length 100 mm. Calculate
>
> (a)   the object and image distances for this set-up,
> (b)   the aperture diameter of the lens if the fibres have a numerical aperture of 0.25, and
> (c)   the eyepiece focal length if the image is to subtend an angle of 30° at the eye.
>
> **Solution:**
>
> (a)   The image must be reduced by a factor of 2.5/50; that is the imaging lens will work at a transverse magnification M of $(-)0.05$. If l and l' are the object and image distances, then from equation (3.49), in air we have
>
> $$M = l'/l = -0.05$$
>
> This magnification is related to the distance l' by equation (3.50a), and so solving for l' gives $l' = 105$ mm and so $l = -2100$ mm.
> (b)   To find the numerical value of the aperture radius ρ of the lens, we can use equation (19.15), with the value of 0.25 for $\sin(\alpha')$, the image distance l' as 105 mm and solve for ρ. Thus we have
>
> $$0.25 = \sin[\arctan(\rho/105)]$$
>
> Solving for ρ gives
>
> $$\rho = 27.1 \text{ mm}$$
>
> giving a lens aperture diameter of 54 mm.
> (c)   To find the focal length of the eyepiece lens, we use equation (19.16) with $\psi = 15°$ and $\Phi = 2.5$ mm. Thus
>
> $$\tan(15°) = 0.5 \times 2.5/f_e$$
>
> giving
>
> $$f_e = 4.67 \text{ mm}$$

19.3.10   Applications of coherent fibre bundles

Flexible coherent fibre bundles have a wide range of applications where an image has to be relayed along a complex curved path or around a number of corners. A number of commercial relay instruments are available both in medicine and engineering that are based upon optical fibres. Those instruments that are designed to look into enclosed spaces with no internal illumination require light to be piped in from an external light source, usually along a second set of incoherent fibres.

Coherent optical fibres also have some special applications. For example, in some image recording applications, the field curvature causes unacceptable loss of image quality towards the edge of the field. One can compensate for this aberration by using a fibre optical correcting plate. This is a short coherent bundle the same cross-sectional size as the image. The front face is curved to match the field curvature of the lens and thus the image falling on this face is in focus over the entire field. The back face is flat and in contact with the recording surface. Thus the bundle transmits and transforms a curved focussed image onto a flat recording surface.

Exercises and problems

19.1   For a cladded optical fibre (core index $= 1.65$ and cladding index $= 1.50$), calculate the maximum angle of incidence for a ray at the input face if it is to be propagated down the fibre by total internal reflection.

ANSWER: $43.4°$

19.2   If the desired numerical aperture of a cladded optical fibre is 0.5 and the core refractive index is 1.8, what is the index of the cladding?

ANSWER: refractive index $= 1.73$

19.3   An object 50 mm in diameter is to be imaged into an optical fibre system which has a diameter of 5 mm, using a lens of focal length 50 mm. Calculate object and image distances for this set-up.

ANSWERS: $l' = +55$ mm and $l = -550$ mm

Summary of main symbols and equations

n_{co}	refractive index of fibre core
n_{cl}	refractive index of fibre cladding
ρ	aperture radius of the lens
ϕ	diameter of an individual fibre
Φ	diameter of a fibre bundle
p	path distance of an oblique ray
f_e	equivalent focal length of the eye lens
ψ	angular radius of the final image
$\sin(\alpha)$	numerical aperture $[n \sin(\alpha), n = 1]$ of a fibre
$\sin(\alpha')$	image space numerical aperture $[n' \sin(\alpha'), n' = 1]$ of a lens

Section 19.3.7: Fibre insulation

Operating numerical aperture

$$\sin(\alpha) = \sqrt{(n_{\text{co}}^2 - n_{\text{cl}}^2)} \tag{19.14}$$

References and bibliography

Fuki Y. Takeda T. and Iida T. (1990). Systematic generation method of relay optical systems. *Appl. Opt.* 29(13), 1947–1959.

Hecht E. and Zajac A. (1974). *Optics*, Sect. 5.6, p. 135. Addison Wesley, Sydney.

Kingslake R. (1969). *Applied Optics and Optical Engineering*, Volumes I and V. Academic Press, New York.

20

Angle and distance measuring instruments

20.0   Introduction

A number of visual optical instruments are designed to measure the angles between two distant objects or the distance of an object. The determination of the angle between two distant objects is frequently done in surveying. Using simple rules of trigonometry, the location of any point can be determined if the direction of two other points and the distances between any two pairs of points are known. The determination of the elevation of celestial objects combined with astronomical tables and an accurate clock can be used to determine a position on the earth's surface, for surveying and navigation.

In many applications, these instruments have been superseded by other quicker or more accurate means. For example distance measurement can now be done with laser rangefinders and satellites are used for routine navigation. However, visual optical instruments are still used in some applications and are still worthy of some attention.

20.1   Angle measuring instruments

20.1.1   The theodolite

The theodolite is in essence a telescope of medium magnification with an eye-piece containing an alignment graticule. It is mounted such that it can rotate through horizontal and vertical circles, so that its horizontal and vertical direction of pointing can be measured. The theodolite is fitted with a levelling bubble so that the horizontal scale or table can be accurately made horizontal. If the vertical scale is zero for this horizontal scale then the elevation of an object can be absolutely read from the vertical scale.

The theodolite is an accurate instrument for measuring angles, but has the drawback that, at least in the horizontal plane, the two angles have to be measured at separate times. If the objects or the observer are moving, then this method fails to give a meaningful result. In these cases, the only alternative is to use a method in which both objects are seen simultaneously. We will call these methods co-incidence methods since the objects are imaged through a system that superimposes two images.

20.1.2   Co-incidence methods

The principle of these methods is shown in Figure 20.1. An observer, using a telescope, views object at $\mathcal{G}$ directly, but object $\mathcal{H}$ is viewed by reflection off

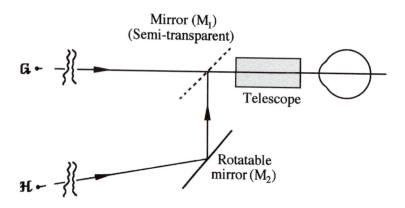

Mirror (M_1)
(Semi-transparent)

Telescope

Rotatable
mirror (M_2)

Fig. 20.1:   The principle of
measuring the angle
between two distant objects.

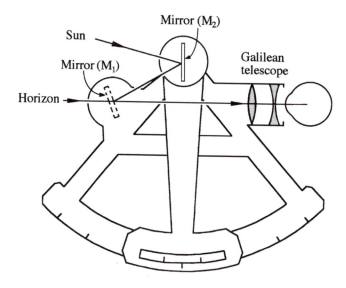

Sun

Mirror (M_2)

Mirror (M_1)

Galilean
telescope

Horizon

Fig. 20.2:   Schematic
construction of a sextant.

the mirror M_2 and both are viewed simultaneously through the beam splitting
mirror M_1. Mirror M_2 is rotated to bring the two objects into co-incidence and
the mirror rotation is a measure of the angular separation of the two objects.
Historically one of the most interesting co-incidence methods is the sextant.

20.1.2.1   *Sextants*

According to Smith (1947), the sextant was developed by Hadley in 1731 as
an aid in determining latitude at sea. It was designed to measure the angular
elevation of the sun or other astronomical bodies above the horizon. This elevation at a particular (Greenwich Mean) time could then be used to calculate
latitude from published tables. In principle, the sextant can be used to measure the angular distance between any two objects or measure the angular size
of an object.

The construction of a typical sextant is shown in Figure 20.2. One of the
objects is seen on transmittance through the semi-transparent mirror M_1. The

other object is seen after reflection from mirror M_2 and M_1. The mirror M_2 is attached to a rotatable arm, which rotates about the centre of this mirror. The mirror and arm are rotated until the two objects being viewed are superimposed. The angular separation is then read directly from the scale at the bottom of the sextant. The accuracy of alignment then depends upon the vernier acuity of the user. Normal vernier acuity is about 10 seconds of arc [Bennett and Rabbetts 1989, 31).

When the sun or some other very bright object is being observed, high density filters must be placed in the appropriate beam path to prevent damage to the eye.

The objects are viewed through a Galilean telescope as shown attached to the sextant in Figure 20.2. Galilean telescopes are used here because they are lighter and shorter than the equivalent Keplerian telescope and give an erect image. Since there is no requirement for a wide field-of-view, the small field-of-view of the Galilean is no disadvantage.

When used to measure the elevation of astronomical bodies, one of the objects will be the horizon, but it must be the true horizon. At sea, this would be the junction between sea and sky, but on land, because of the presence of hills and valleys, the true horizon is rarely seen. However, the presence of bad weather on land or sea can prevent the true horizon being seen. To cope with these situations, an artificial horizon or accurate horizontal surface must be used. Typically, a small reservoir of liquid such as mercury is used, which is ideal because of its high reflectance and low evaporation. Taking as an example solar elevation, one solar image is formed directly and the other is formed by reflection from the liquid surface. If the angular separation of these two images is ϕ, the solar elevation angle θ is

$$\theta = \phi/2 \tag{20.1}$$

20.2 Distance measuring instruments (rangefinders)

To help explain the different types of rangefinders, we will discuss them in two classifications: those designed to be used monocularly and those requiring binocular vision.

20.2.1 Monocular rangefinders

There are a number of different optical methods that one can use to estimate or measure the distance of an object. Three that we will describe here are as follows:

(1) Measuring the angular size of the object if its physical size is known.
(2) Focussing methods similar to that used in cameras.
(3) Image doubling using an extended base line.

The first of these three methods requires an object of known size or a target of known size to be placed at the object distance. This is possible if the distance is not too far or inaccessible. However, in many situations, the object distance is inaccessible and other methods must be chosen. The latter two of the above methods are applicable in this situation.

20.2.1.1   Measurement of the angular size of objects of known physical size

The measurement of the distance of an object of known size h can be calculated from a knowledge of that size if the angular subtense θ of the object is known. Thus the problem reduces to one of determining the angular subtense and we have discussed this measurement in Section 20.1. Once we know the value of θ, the distance s is then given by the simple equation

$$s = h\theta \tag{20.2}$$

where the angle must be in radians and is assumed to be small.

20.2.1.2   Focussing methods

The focussing of a camera is a simple type of ranging measurement. The image formed by the camera lens is observed on the viewing screen. By moving the camera lens in and out, the photographer finds the camera lens position which gives the sharpest image on the screen. The distance is then read from the camera lens focussing scale. The accuracy of this method is highest for close object distances and longer focal length lenses. In a similar manner, the focussing of a telescope can be used to estimate distances.

The telescope method works on the same principle as the dioptric telescope. The principle of operation is as follows. If the telescope is initially afocal, the image of an object at a finite distance will not be formed at infinity. For a two component telescope, equation (17.6) relates the image vergence L'_v at the eye lens to the object vergence L_v at the objective. This equation is

$$L'_\text{v} = \frac{M^2 L_\text{v}}{(1 - dML_\text{v})} \tag{20.3}$$

where M is the telescope magnification and d is the distance between the respective principal planes of the two components. While this equation only strictly applies to a two lens telescope (i.e. not one with a field lens), it indicates that on viewing an object at a finite distance with an afocal telescope, the image is formed closer than the object and this places an accommodation demand on the viewer. In order to reduce the accommodation demand and put the image back at infinity, the viewer will tend to refocus the telescope by increasing its length. Thus the telescope length has to be increased for closer objects. The telescope refocussing mechanism can be calibrated to read the range directly. Equation (20.3) predicts that the accommodation demand increases as the square of the telescope magnification and therefore the higher the magnification the more accurate the estimate of range. Such telescopes are called **ranging telescopes**. If the telescope length is calibrated against distance, then it does not matter whether the telescope is a two lens or multi-lens system.

20.2.1.3   Image doubling using an extended base line

These systems work on a similar system to the one described in Section 20.1 for angle measurement. The differences in construction are that (a) the distance

Fig. 20.3: The principle of the monocular rangefinder.

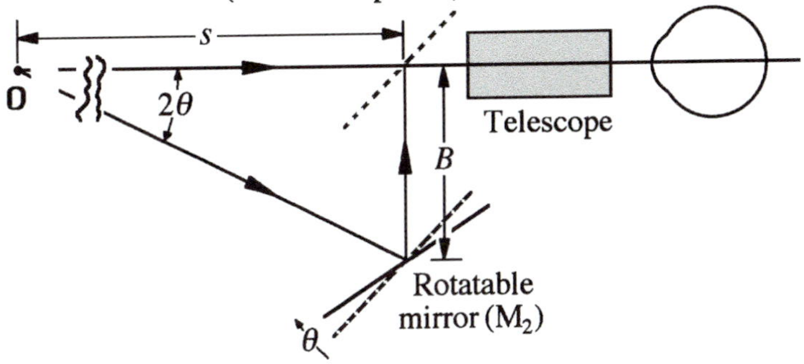

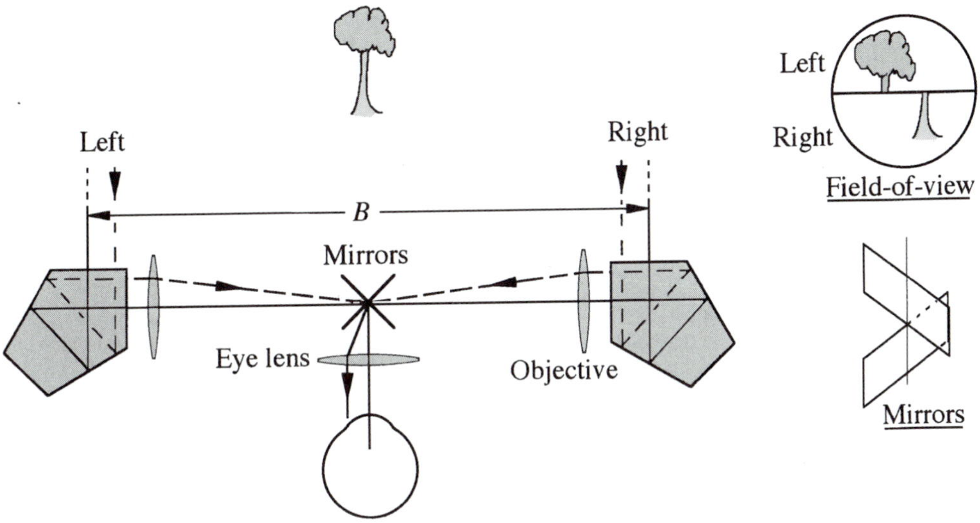

Fig. 20.4: Practical monocular rangefinder.

between the two mirrors is critical and should be as large as possible and (b) the rotating mirror reflects the single object of interest into the telescope such that two images of the object are seen in co-incidence. The required amount of rotation θ of the mirror depends upon the distance of the object and the distance between the two mirrors. This arrangement is shown in Figure 20.3 and if the angles are small, the distance s of the object is given by the equation

$$s = 2B\theta \qquad (20.4)$$

where the angle θ must be in radians.

A more complex schematic system is shown in Figure 20.4. In this design, the mirrors form a combined image which consists of the top part of one and the bottom part of the other as shown in the diagram. If the object was at infinity, the two half images would be in line. However, for targets at a finite distance, the images are laterally displaced as shown in the diagram inset. The instrument is provided with some means of aligning the two images. There are a number of ways the images can be aligned. These are as follows.

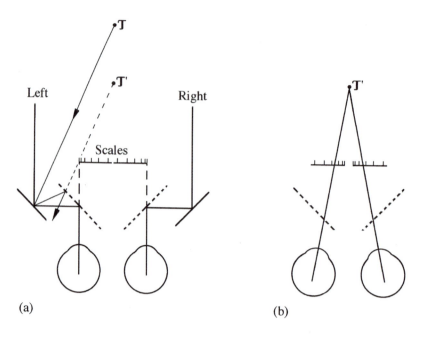

(a)   (b)

(1)   One of the lenses in one of the arms can be transversely shifted. This is
      often used in camera rangefinders.
(2)   One of the mirrors can be rotated.
(3)   A thin prism can be placed in one arm and the beam can be displaced
      by moving the prism backwards and forwards in the beam. The use of
      prisms for this purpose is described in Chapter 8.

The lenses shown in the diagram form a Keplerian telescope for the magnification of the object although this telescope is not always necessary. If a Keplerian
telescope is used as shown in this diagram, it will invert the image. In the system
shown here, the penta prisms and mirrors provide three reflections in the horizontal plane and therefore give a correct left/right image orientation but do not
vertically invert the image. Thus the image has the correct left/right orientation
but is inverted vertically. The penta prisms deviate the beam through 90° and
are superior to a simple mirror or a 45°−90°−45° prism because they are less
sensitive to mis-alignment and do not invert the image in any manner.
    The accuracy of these types of rangefinders depends upon the length of
the base line B and increases with base line length. Lengths of hand held
rangefinders are about 1 m.

20.2.2   Binocular rangefinders

In binocular rangefinders, both eyes are used to view the image and can make use
of stereoscopic vision. One type with no moving parts is shown in Figure 20.5a.
The two eyes view the image T' of the target T through a binocular telescopic
system with a long base, but the lenses are not shown in the diagram. If the eyes
are parallel, a target at a finite distance would be seen as a double image. The
two eyes would then have to converge to fixate the target images and thus give

Fig. 20.6: Rangefinder
scales of a binocular
stereoscopic rangefinder.

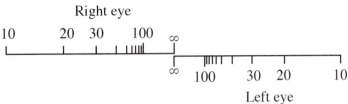

(a) Scales for the target at infinity

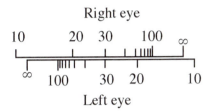

(b) Scales for target at 30 m

a single image as shown in Figure 20.5b. The amount of convergence is then a measure of the target distance.

The amount of convergence can be used to give a direct reading of the distance. Using beam-splitters the distance measuring scales, as shown in Figure 20.5, are presented to each eye as shown in Figures 20.6. When the eyes are parallel, the two scales do not overlap, but just touch as shown in this diagram. As the eyes converge, the two scales slide past each other, and only at one point do the same distance marks coincide. This is the distance of the target. Because there are no moving parts to adjust, this type of rangefinder can give a distance reading in a shorter time than the monocular rangefinders discussed in the previous subsection.

As in the monocular rangefinders, the accuracy of these types of stereoscopic rangefinders depends upon the base length of the optical system. The longer the base length, the greater the accuracy.

References and bibliography

*Bennett A.G. and Rabbetts R.B. (1989). *Clinical Visual Optics*, 2nd ed. Butterworths, London.

Kingslake R. (1969). *Applied Optics and Optical Engineering*, Vol. 5, Part II, Chapter 7(III). Academic Press, New York.

*Smith C.J. (1947). *Intermediate Physics*, 3d ed. Edward Arnold, London.

21

Cameras and camera lenses

21.0   Introduction

Camera lenses are optical systems designed to give a real image of distant objects. Therefore they must have a positive power with their aberrations corrected to produce reasonably good image quality for an object plane at infinity, that is for the image formed in the back focal plane of the lens. Some camera lenses known as "macro-lenses" are designed to give optimum image quality at intermediate or close object distances rather than for the object at infinity. Examples of macro-type lenses are those used in photocopiers and for photographing documents.

Camera lenses usually have a variable aperture stop, known as the diaphragm, whose diameter is varied within limits, for control of the image illuminance and hence film exposure (Section 21.4) and the depth-of-field (Section 21.6).

While a camera lens may consist of a single positive power lens, such a lens is not suitable when the focal length is very long or very short. Also a single lens would give less than adequate image quality for many applications, particularly those requiring large apertures or wide fields-of-view. Thus most camera lenses are complex optical systems, often consisting of two or more separate lenses. Designs suitable for very long and very short focal lengths are discussed in Section 21.2 and aberration considerations for wider apertures and field angles are discussed in Section 21.10.

21.1   Field-of-view, focal length and image sizes

The field-of-view of a camera system is usually limited by a field stop placed in the image or film plane. In 35 mm cameras, the field stop is 35×24 mm; thus the image or frame size on the film has these dimensions. The angular diameter of the field-of-view will then depend upon the particular dimension on the film, for example horizontally, vertically or diagonally. In this section, we will look at how the focal length of the camera lens, field-of-view and image size are related, and in the following discussion, we will restrict the object plane to be at infinity.

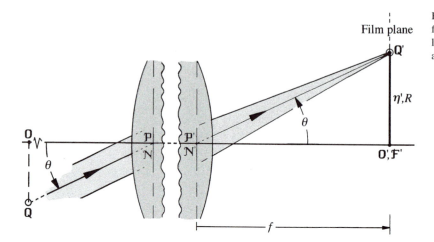

Fig. 21.1: The field-of-view of a camera lens, assuming the object is at infinity.

21.1.1 Field-of-view and focal length

In Figure 21.1, an object at infinity, with an angular subtense of θ, is imaged in the back focal plane at $\mathcal{F}'$ of a lens of equivalent focal length f. In air, the principal and nodal points coincide as shown in this diagram, so the angular subtense of the object measured from the front principal point is the same as the angular subtense of the image measured from the back principal point. For any lens of equivalent focal length f, the angular radius θ of the object field-of-view, for an object at infinity, is given by the equation

$$\tan(\theta) = R/f \tag{21.1}$$

where R is the radius of the image field.

> **Example 21.1:** Calculate the angular diameter of the field-of-view of a camera lens with an equivalent focal length of 50 mm, used with a film format of 35×24 mm.
>
> **Solution:** From equation (21.1), we have,
>
> for the 35 mm dimension: diameter $= 2 \arctan(17.5/50) = 38.6°$
> for the 24 mm dimension: diameter $= 2 \arctan(12.0/50) = 27.0°$
>
> A 35×24 mm format has a diagonal of 42.4 mm; therefore along the diagonal, the field-of-view diameter would be $46.0°$.

21.1.2 Image size and focal length

An alternative form of equation (21.1) is to express the image size η' of a particular object in terms of its angular size θ and the equivalent focal length f of the lens, that is

$$\eta' = f \tan(\theta) \tag{21.2a}$$

Fig. 21.2: (a) Focal length
and image size. (b) Focal
length and angular
field-of-view when the
image plane width is fixed.

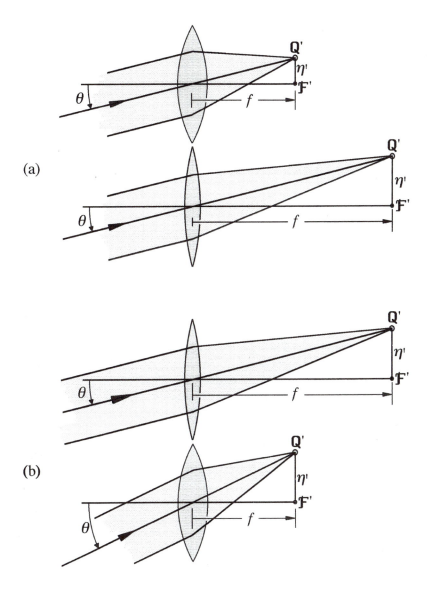

This equation shows that the image size on the film is proportional to the focal length and this is demonstrated further in Figure 21.2a. Therefore, if we want a larger image we must either get close to the object and hence increase the angle θ or use a longer focal length lens. Often getting closer is impossible, especially if the object is very distant. If we use a longer focal length lens and we desire an object of angular size θ to have an image of a certain size η', then the required focal length is given by the equation

$$f = \eta' / \tan(\theta) \qquad (21.2b)$$

Example 21.2: Calculate the equivalent focal length of a lens required to photograph the moon if the image is to just fill the frame of a 35 mm camera. The angular diameter of the moon is 0.5°.

Solution: Now the shortest dimension of a 35 mm film frame is 24 mm, so let us take an image size a little less than this, say 22 mm. Therefore we take η' as 11 mm.

We substitute 11 mm for η' and 0.25° for θ in equation (21.2b) and get

$$f = 11/\tan(0.25°) = 2521 \text{ mm}$$

Compare this with the standard focal length of about 50 mm.

In contrast, we may need to photograph an object that has a very wide angular subtense. If this is to be completely imaged in the frame, then its image size η' has to be less than the frame dimensions. Equation (21.2b) indicates that for a given image size η', the larger the angular subtense θ of the object, the shorter must be the focal length f, as demonstrated in Figure 21.2b.

Example 21.3: Calculate the focal length of a lens required to photograph the front of a building that is 35 m wide from across the street, which is only 12 m wide, with the image just filling the frame.

Solution: The above lengths give an angular diameter of 111°. Let us suppose this has to fit into the 35 mm length of the 35 × 24 mm frame, but we should choose an image size slightly smaller, say 33 mm.

Thus we substitute $\eta' = 16.5$ mm and $\theta = 55°$ into equation (21.2b); we get a focal length value of $f = 11.6$ mm. Once again, compare this with the standard value of about 50 mm.

This discussion and the examples have shown that there may be a need for much longer and much shorter focal length lenses than the normal focal length, which is about 50 mm for a 35 mm camera. Such extreme focal lengths pose severe problems if they only consist of a single lens. These problems can be overcome with the use of special types of lens systems. One such type of lens with a long equivalent focal length is called a telephoto lens and that designed with a short equivalent focal length is called an inverse telephoto lens. We will look at the properties of these lens systems in the next section.

21.2   Telephoto and inverse telephoto lenses

We have just seen that there is sometimes a need for lenses with very long or very short equivalent focal lengths. A simple lens with a very long focal length would be impractical in that it would need to be placed a long way from the camera body. For example a lens with a 2 m focal length would be placed 2 m out in front of the camera body. A single lens with a very short focal length is also impractical because it has to be placed close to the film plane and this would then interfere with the viewing system (Section 21.8). Both of these restrictions can be removed if we use a special system that, in its simplest form, consists of two single lenses and places the principal points well away from the lens system. Let us start with the telephoto lens to see how this is done.

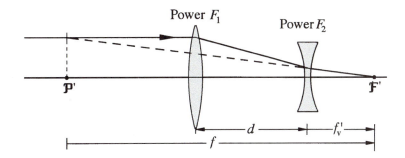

Fig. 21.3: Telephoto camera lens: long equivalent focal length f and a short back focal length f'_v and a short system length.

21.2.1 The telephoto lens

Telephoto lenses have a long equivalent focal length but have both a short back focal length and a short system length. The schematic construction of a telephoto lens is shown in Figure 21.3 with a ray showing the position of the back principle point $\mathcal{P}'$, which is well out in front of the lens. This construction replaces a single lens placed at $\mathcal{P}'$. For any given equivalent focal length f, back focal length f'_v, and lens separation d, the component powers F_1 and F_2 can be found by simultaneously solving equations (3.20) and (3.22), which for two lenses in air are

$$F = F_1 + F_2 - F_1 F_2 d \tag{21.3a}$$

and

$$f'_v = (1 - dF_1)/F \tag{21.3b}$$

where $f = 1/F$. The required equations are, firstly from equation (21.3b),

$$F_1 = (1 - f'_v F)/d \tag{21.4a}$$

and then after substituting for F_1 in equation (21.3a)

$$F_2 = (F - F_1)/(1 - dF_1) \tag{21.4b}$$

Note that the system is only effectively a telephoto construction if the sum of the distances d and f'_v is less than f, that is

$$d + f'_v \ll f \tag{21.5}$$

The ratio

$$\frac{d + f'_v}{f} \tag{21.6}$$

may be defined as the **telephoto ratio** and must be less than 1.0. In the design of these systems, the distance d can often be chosen to have any value provided the above condition is met.

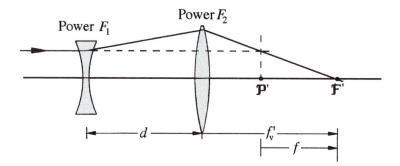

Fig. 21.4:   Inverse or
reverse telephoto lens.

Example 21.4: Find the powers of a telephoto lens with the following specification: $f = 1000$ mm, $f'_v = 100$ mm and $d = 100$ mm.

Solution: Now since $f = 1000$ mm, $F = 0.001$ mm^{-1} and from equation (21.4a)

$$F_1 = [1 - (100 \times 0.001)]/100 = 0.009 \, \text{mm}^{-1}$$

and from equation (21.4b)

$$F_2 = (0.001 - 0.009)/[1 - (100 \times 0.009)] = -0.08 \, \text{mm}^{-1}$$

Then in terms of equivalent focal lengths, the solution is

$$f_1 = 111.1 \, \text{mm} \quad \text{and} \quad f_2 = -12.5 \, \text{mm}$$

From equation (21.6), this design has a telephoto ratio of 0.2.

It is clear from these results that the telephoto lens has a positive power leading lens and a negative power following lens.

21.2.1.1   Reflecting forms

A telephoto "lens" can also be constructed from mirrors. A suitable form is similar to the Cassegrain telescope objective shown in Figure 17.2b. Such a form clearly pushes the back principal plane well out in front of the system.

21.2.2   Inverse telephoto lens

The inverse telephoto lens is designed to have a short equivalent focal length but a long back focal length and has the construction shown in Figure 21.4, which shows the position of the back principal plane. The equations used to find the two component powers are the same as those for the telephoto case; that is equations (21.4a) and (21.4b), and for this type of lens system, the telephoto ratio, defined by equation (21.6), must be greater than 1.0.

Example 21.5: Find the powers of an inverse telephoto lens with the following specification: $f = 20$ mm, $f'_v = 50$ mm and $d = 25$ mm.

Solution: Since $f = 20$ mm, $F = 0.05$ mm^{-1} and from equation (21.4a)

$$F_1 = [1 - (50 \times 0.05)]/25 = -0.06 \, \text{mm}^{-1}$$

and from equation (21.4b)

$$F_2 = [0.05 - (-0.06)]/[1 - 25 \times (-0.06)] = 0.044 \, \text{mm}^{-1}$$

Then in terms of equivalent focal lengths, the solution is

$$f_1 = -16.7 \, \text{mm} \quad \text{and} \quad f_2 = +22.73 \, \text{mm}$$

From equation (21.6), this design has a telephoto ratio of 3.75.

Actual telephoto and inverse telephoto lenses are usually more complex than shown in Figures 21.3 and 21.4, because of the need to minimize aberrations.

21.2.3 Tele-converters

The equivalent focal length of a camera lens or any other positive power lens can be increased by placing a suitably designed negative lens between the lens and its back focal point. In the case of a camera lens, this is between the lens and the camera body. This has the effect of converting the lens to a type of telephoto lens. Such a lens is known as a **tele-converter** or a **Barlow lens** and is also used in telescopes to increase the focal length of the objective.

The power of the Barlow lens can be calculated by using equation (21.4b). We would start with an initial lens equivalent focal length or power and the desired equivalent focal length or power. In equation (21.4b), these powers would be F_1 and F, respectively. The power F_2 would be the power of the Barlow lens and d would be the distance between the back principal point of the original lens and the Barlow lens. It is clear from equation (21.4b) that the distance d is a free variable. Let us look at an example.

Example 21.6: Calculate the focal length of a Barlow lens required to increase the equivalent focal length of a 50 mm lens to 100 mm.

Solution: In this problem,

$$F_1 = 1/50 = 0.02 \, \text{mm}^{-1}$$

$$F = 1/100 = 0.01 \, \text{mm}^{-1}$$

and we need to solve for F_2, which is the power of the Barlow lens. Therefore from equation (21.4b)

$$F_2 = (0.01 - 0.02)/(1 - d \times 0.02) = -0.01/(1 - d \times 0.02)$$

Since d is not defined, we are free to choose its value. We should note that the smaller the value of d, the lower the power of the Barlow lens.

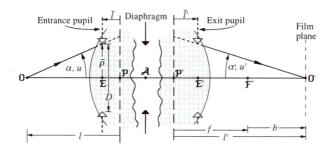

Fig. 21.5: The schematic camera lens, showing the internal virtual pupils of a camera lens. This diagram is also used to examine the effect of finite object distance on the image plane illuminance.

Let us choose a value of 20 mm and with this value, the power and focal length of the Barlow lens are

$$F_2 = -0.0166\,\text{mm}^{-1} \quad \text{and thus} \quad f_2 = -60\,\text{mm}$$

21.3   Apertures and pupils

The camera lens usually contains an aperture stop of variable diameter, which is known as the diaphragm. It is not usually circular since it is constructed from a number of movable blades. Thus a diaphragm made of six blades will have approximately a hexagonal shape. Except for the single lens type camera lens, the diaphragm is placed inside the system, as indicated in Figure 21.5, and close to the centre of the lens as this helps to minimize some of the aberrations.

21.3.1   Positions of the entrance and exit pupils

The entrance and exit pupils of a camera lens are virtual images of the aperture stop or diaphragm and are internal to the lens system as shown in Figure 21.5. The fact that the pupils are internal is not a problem, because camera lens systems are not used directly with the eye and therefore there is no need for an external exit pupil or pupil matching.

The diameter of the diaphragm can be varied by rotating the diaphragm ring on the lens body. This diameter governs the image plane illuminance and depth-of-field. It also affects image quality since its diameter affects the level of both aberrations and diffraction.

21.3.2   Aperture stop size and F-number

The diameter of the aperture stop of a camera lens is not specified in terms of its actual diameter but indirectly in terms of a quantity called the *F-number*. This is defined as follows.

$$F\text{-}number = \frac{\text{equivalent focal length } (f)}{\text{diameter of the entrance pupil}(D)} = \frac{f}{D} \qquad (21.7)$$

where D is shown in Figure 21.5.

Example 21.7: Calculate the entrance pupil diameter of a camera lens, with an equivalent focal length of 100 mm and operating with an *F-number* of 2(*F/*2).

Solution: From equation (21.7)

$$D = 100/2 = 50 \, \text{mm}$$

Note: With this definition, the *F-number* is inversely proportional to entrance pupil diameter and therefore as this diameter increases, the *F-number* decreases. This means that a smaller *F-number* gives a larger entrance pupil.

The *F-number* scale on the diaphragm ring is expressed in what are called **F-stops**. These are *F-number*, which generally use the following sequence

$$1.4, 2, 2.8, 4, 5.6, 8, 11, 16, 22 \tag{21.8}$$

in which the consecutive values are in the ratio of approximately $\sqrt{2}$. The reason for this type of scale is explained in the next section.

The *F-number* is used as an indicator of the image plane illuminance or film exposure (see next section) and depth-of-field (Section 21.6). In Section 21.4, we will see that for objects at infinity, the image plane illuminance is inversely proportional to the square of the *F-number*. We will also see that the relationship is more complex if the object is at a finite distance.

21.4   Image plane illuminance and film exposure

The amount of light falling on the film during an exposure time t is best quantified as the product of the film plane illuminance E' and time t. This quantity is called the exposure and will be denoted by the symbol E_{exp}. Thus

$$E_{\text{exp}} = E't \quad \text{with units of lux·s} \tag{21.9}$$

For the general case, the image illuminance E' close to the axis is given by equation (12.21a). Here with the system in air (that is the object and image space indices are unity) we have

$$E' = \tau \pi L_{\text{um}} \sin^2(\alpha') \tag{21.10}$$

where τ is the average luminous transmittance of the camera lens, L_{um} is the luminance of the object and α' is the angular radius of the exit pupil as seen from the axial image point and as shown in Figure 21.5.

For wide angle lenses, the $\cos^4$ law may apply, and in this case, the illuminance for points towards the edge follows equation (12.22). However, this equation ignores pupil aberration and vignetting and is not applicable if these are present. Pupil aberration may be designed into the system to reduce the $\cos^4$ effect and vignetting reduces the effective pupil size at a different rate.

21.4.1   Object at infinity

If the object plane is at a great distance or infinity, we can alternatively express
the illuminance E' in terms of the *F-number* of the lens, that is equation (12.21c),
here written as

$$E' = \frac{\tau \pi L_{um}}{4(F\text{-}number)^2} \qquad\qquad (21.10a)$$

since the object and image space refractive index n and n' are unity. Thus for a
distant object

$$E_{exp} = \frac{\tau \pi L_{um}}{4} \, \frac{t}{(F\text{-}number)^2} \qquad\qquad (21.11)$$

For a fixed exposure level E_{exp}, the exposure time t and *F-number* can be
varied as long as the quantity

$$t/(F\text{-}number)^2 = \text{constant} \qquad\qquad (21.12)$$

When a photograph is being taken, the actual values of the exposure time t and *F-
number* used depend upon a number of factors such as the ambient or available
light, depth-of-field, any expected object movement and image quality. For
example, fast moving objects require short exposure times and hence smaller
F-numbers. On the other hand, if a large depth-of-field is required, the *F-
number* must be large and hence the exposure time must be long. Depth-of-
field is discussed in detail in Section 21.6. The image quality depends upon
the *F-number* because of the effect of entrance pupil diameter on aberrations
and diffraction.

In cameras with manual exposure controls, the camera shutter speeds have
discrete values which vary by a factor of approximately 2. For example the
available shutter speeds may be in the sequence

$$1, 1/2, 1/4, 1/8, 1/15, 1/30, 1/60, \ldots s$$

If in a particular situation, the shutter speed is increased by a factor of 2 but the
film exposure is to be kept constant, equation (21.12) shows that the square of
the *F-number* must also be increased by a factor of 2. Thus the *F-number* can
be changed in steps (**F-stops**) which are in the ratio of $\sqrt{2}$ and this is the reason
for the form of the *F-number* sequence given by equation (21.8).

21.4.2   Object at a finite distance

If the object is at a finite distance as shown in Figure 21.5, the image plane
illuminance is not given accurately by equation (21.11), because this equation
only applies if the object is at infinity. Instead one must use equation (21.10),
which is accurate for any object distance. As the object distance decreases, the
image moves farther from the lens. As a consequence, the angle α' decreases and
this in turn decreases the image illuminance as indicated by equation (21.10).

However, equation (21.10) is not the most appropriate equation in this case as the angle α' is not readily known. A better equation is one that expresses the film plane exposure E_{exp} in terms of equation (21.11), assuming the object was at infinity, and a **correction factor,** which depends upon either the object distance or the change in camera lens to film distance. We will begin the derivation of an appropriate equation by firstly defining the **(exposure) correction factor** as

$$\text{(exposure) correction factor} = \frac{\substack{\text{illuminance predicted for a distance}\\ \text{object by equation (21.10a)}}}{\substack{\text{actual film plane illuminance}\\ \text{predicted by equation (21.10)}}}$$

(21.13)

and then for close objects, the exposure has to be increased by this factor. Substituting expressions from equations (21.10) and (21.10a) into (21.13), we have firstly

$$\text{correction factor} = \frac{1}{4\sin^2(\alpha')F\text{-}number^2}$$

(21.14)

The next step is to replace $\sin(\alpha')$ by an expression containing the object distance or one containing the change in image distance. Let us firstly look at the object distance.

In terms of object distance

We begin with the transverse magnification equation [from equations (3.47) and (3.49) in air] which gives

$$u/u' = l'/l$$

Let us now refer to Figure 21.5, and assume the system is free of spherical aberration and coma. We can use the sine condition [equation (5.31)], with α replacing U, that is

$$u/u' = \sin(\alpha)/\sin(\alpha')$$

Combining the above two equations gives us

$$\sin(\alpha') = -l\sin(\alpha)/l'$$

(21.15)

The negative sign is used because according to our standard sign convention, l is negative but α and α' are always positive. From the diagram, we can show that

$$\sin(\alpha) = -\bar{\rho}/(l - \bar{l})$$

where in this equation, the usual sign convention holds; that is l and $\bar{l}$ will be negative if to the left of the front principal plane, which is the case shown in the diagram. We can now use this equation to eliminate $\sin(\alpha)$ from equation

(21.15) and eliminate l' using equation (3.45b). We finally have

$$\sin(\alpha') = \frac{f+l}{2F\text{-}number\,(l-\bar{l})} \tag{21.16}$$

where l is the object distance from the front principal plane and $\bar{l}$ is the distance of the entrance pupil from the front principal plane. Using this equation, equation (21.14) for the correction factor becomes

$$\text{correction factor} = \frac{(l-\bar{l})^2}{(f+l)^2} \tag{21.17}$$

Now in general, the distance $\bar{l}$ between the entrance pupil and the front principal plane will not be readily known and will depend upon the design of the particular camera lens, but if it is negligible compared to l, then it can be omitted from the above equation and we can write

$$\text{correction factor} \approx \frac{l^2}{(f+l)^2} \tag{21.17a}$$

Note: This correction factor is always greater than or equal to 1.0.

In terms of change of image distance

As the camera lens is focussed for closer objects, the camera lens is moved farther from the camera body. For extremely close objects, special extension tubes or a bellows unit can be inserted between the camera lens and camera body. Let us derive an equation for the correction factor in terms of the distance b the camera lens is moved away from the camera body in order to focus on the close object. This length b may be the length of an extension tube or bellows assuming the camera lens is focussed on infinity.

All we have to do in the derivation is to replace the object space distances l and $\bar{l}$ in equation (21.17) by their corresponding image space distances l' and $\bar{l}'$. Using the lens equation (3.45a) and after some simplification and replacing l' by $f+b$, we have

$$\text{correction factor} = \frac{(f+b-\bar{l}')^2}{(f-\bar{l}')^2} \tag{21.18}$$

Once again, the pupil position $\bar{l}'$ will depend upon the particular camera lens and therefore its value will be generally unknown and for convenience all we can do is make the assumption that it is negligible with respect to $f+b$ and b. With this assumption

$$\text{correction factor} \approx \frac{(f+b)^2}{f^2} \tag{21.18a}$$

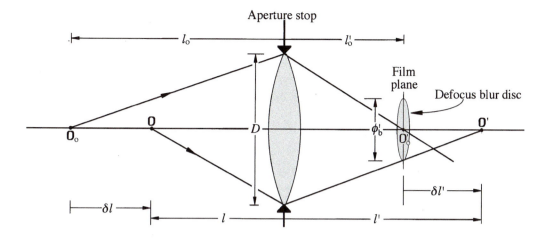

Aperture stop

Fig. 21.6: The defocus
blur disc and depth-of-field.

21.5   Lens specification

A camera lens is usually specified in terms of its equivalent focal length f and
its smallest *F-number*, that is the widest aperture or entrance pupil diameter.
A typical specification may be

$$(f =) \ 50\,\text{mm} F/4 \text{ or } 1:4$$

For most common camera lenses, the highest *F-number* is about 16 or 22.

21.6   Depth-of-field

The general concept of depth-of-field has already been discussed in Section 10.3
and the evaluation of depth-of-field in terms of defocus blur discs was discussed
extensively in that chapter. We will now apply those basic principles to the
special case of photography. In angular terms, the visual threshold diameter of
the defocus blur disc is about 1 minute of arc and this value will be taken in any
numerical calculation.

21.6.1   *Depth-of-field (object space)*

For simplicity, we will consider the camera lens as a single lens, in air, with
the stop at the lens. In Figure 21.6, the conjugate planes o_o and o'_o are in focus
on the film plane and the conjugate planes o and o' are out of focus on the film
plane. From the lens equation (3.45b)

$$l' = \frac{lf}{(l+f)} \quad \text{and} \quad l'_o = \frac{l_o f}{(l_o + f)}$$

but with the power F replaced by the focal length f. The longitudinal focus
error $\delta l'$ is

$$\delta l' = l' - l'_o = \frac{lf}{(l+f)} - \frac{l_o f}{(l_o + f)}$$

which reduces to

$$\delta l' = \frac{f^2(l - l_\mathrm{o})}{(l + f)(l_\mathrm{o} + f)}$$

Now from similar triangles, the diameter ϕ_b' of the defocus blur disc on the film plane, as shown in the diagram, is

$$\frac{\phi_\mathrm{b}'}{\delta l'} = \frac{D}{l'}$$

where D is the diameter of the aperture stop at the lens. Therefore

$$\phi_\mathrm{b}' = \frac{D f^2(l - l_\mathrm{o})}{l'(l + f)(l_\mathrm{o} + f)} \tag{21.19}$$

If we replace l' by the expression from above, we have

$$l' = \frac{lf}{(l + f)}$$

The equation (21.19) for ϕ_b' now reduces to

$$\phi_\mathrm{b}' = \frac{D f(l - l_\mathrm{o})}{l(l_\mathrm{o} + f)}$$

If we replace the (entrance) pupil diameter D by the *F-number* and equivalent focal length, using equation (21.7), then

$$\phi_\mathrm{b}' = \frac{f^2(l - l_\mathrm{o})}{l(l_\mathrm{o} + f)(F\text{-}number)} \tag{21.20a}$$

If the in-focus object plane is at infinity, l_o is infinite and the above expression becomes indeterminate. However, if we replace l and l_o by their corresponding vergences ($L = 1/l$ and $L_\mathrm{o} = 1/l_\mathrm{o}$, in air), equation (21.20a) can be written in the form

$$\phi_\mathrm{b}' = \frac{f^2(L_\mathrm{o} - L)}{(1 + f L_\mathrm{o})(F\text{-}number)} \tag{21.20b}$$

which is now determinate for all object distances.

Now let us assume that the image on the film is enlarged by a magnification factor M; the blur disc diameter ϕ_b'' on the new image will be

$$\phi_\mathrm{b}'' = M\phi_\mathrm{b}' \tag{21.21}$$

From a viewing distance s, the observed angular diameter ω (in radians) of the blur disc in this final image will be

$$\omega = \phi''/s = M\phi_\mathrm{b}'/s \tag{21.22}$$

Typically, s will be the normal viewing distance, which has a value of about 250 mm.

Finally let us assume that this angular size is the threshold of detection for a blurred point, then by substitution in equation (21.20b) and solving for L, we have

$$L = L_0 - \frac{\omega s (1 + f L_0)(F\text{-}number)}{M f^2} \qquad (21.23)$$

Since L_0 corresponds to the in-focus object plane, there is a limiting distance either side of this value which defines the edges of the apparent region of sharp focus. The above equation only gives one of the limiting values, the nearer one. The other, the farther value, can be found be replacing the angular blur disc diameter ω by $-\omega$. Therefore, we could re-express equation (21.23) in the form

$$L = L_0 \pm \frac{\omega s (1 + f L_0)(F\text{-}number)}{M f^2} \qquad (21.24)$$

where the two values of L correspond to the extreme edges of the apparent zone of sharp focus.

If the viewed image has the same angular subtense as the field-of-view of the camera, the magnification is related to viewing distance s and focal length f by the equation

$$M = s/f \qquad (21.25)$$

A magnification of this value ensures that the image has the same apparent size as the object viewed from the position of the camera lens.

Equation (21.24) shows that the depth-of-field increases as the focal length decreases and as the $F\text{-}number$ increases. This equation also shows that it depends upon the choice of threshold blur disc diameter ω, the viewing distance s and the magnification M of the print or final image. Therefore, camera lens manufacturers must make some assumptions about viewing conditions when they state depth-of-field values with their lenses. The threshold value of ω will be in the region of several minutes of arc, the viewing distance s is often assumed to be 250 mm and the magnification M can be taken as that given by equation (21.25).

> **Example 21.8:** Let us calculate the depth-of-field for a photograph, enlarged from a 35 × 24 mm negative to a 175 × 120 mm print and the print is viewed from 250 mm.

> **Solution:** For the above enlargement, the magnification would be $M = 5$, which is that given by equation (21.25), so that the image will have the same apparent size as the object. If we choose a threshold defocus blur disc diameter of 2 minutes of arc, we then only have to choose a focal length, $F\text{-}number$ and distance. If we choose an

F-number of 10, a focal length of 50 mm and a focussing distance of 2 m, then the depth-of-field range is from 1.63 m to 2.59 m.

21.6.2   *Hyperfocal distance*

If a camera is focussed at infinity, the depth-of-field extends from a finite distance to infinity and beyond. The apparently clear field beyond infinity is wasted. For the same aperture, the usable depth-of-field can be increased by focussing the camera to a point such that the farther limit of the depth-of-field is at infinity. The distance corresponding to the correct in-focus object point is called the **hyperfocal distance**. We can find an equation for this quantity using the above equations. The hyperfocal distance vergence L_{hf} is the value of L_o in equation (21.21) with $L = 0$ and taking the "+" sign in the "±" alternative. That is

$$L_{hf} = -\frac{\omega s(F\text{-}number)}{Mf^2 + \omega s f(F\text{-}number)} \qquad (21.26)$$

with L_{hf} being negative according to our sign convention. Once the camera lens is focussed on this distance, the far edge of the in-focus zone will be at infinity. The near edge has a vergence

$$L(\text{near edge}) = -\frac{2\omega s F\text{-}number}{(Mf^2 + \omega s f F\text{-}number)} \qquad (21.27)$$

Example 21.9: Calculate the near edge distance of the in-focus zone of a 50 mm focal length lens set at $F/2$ and focussed on infinity, using a 35 mm image enlarged by a magnification of 5, a viewing distance of 250 mm and a threshold blur disc diameter of 2 minutes of arc. Also calculate the hyperfocal distance.

Solution: From equation (21.24), the near edge is at 49.8 m and from equation (21.26), the hyperfocal distance is 49.8 m.

    If the camera is now focussed at this distance, equation (21.27) gives the new near edge at 24.9 m.

Choice of threshold blur disc size ω

In the above calculations, we have assumed that the threshold blur disc size is 2 minutes of arc. While this seems to be the value assumed in photography, it may be too small and better estimates are discussed in Chapter 36.

Effect of aberrations and diffraction

The defocus blur disc model assumes that the light distribution in the image of a point is adequately described by geometric optics of an aberration free beam. Both aberrations and diffraction will spread out the beam in the region of focus, and therefore make it more difficult to detect when the image of a point has changed its size as a result of defocus alone. Therefore the above modelling should only be regarded as a guide.

21.7 Focussing range

Cameras are usually designed to photograph distant objects and therefore can readily be focussed on infinity. They can also be used to photograph objects at finite distances but to do this, the camera lens to film plane distance must be increased. The lens equation (3.45b) relates object to image distance, and, using focal length rather than power, this equation is

$$l' = \frac{fl}{(f+l)}$$

The distance b a lens has to be moved away from the camera body, in order to focus to a distance l, is

$$l' - f = b$$

Solving for b gives

$$b = -\frac{f^2}{(f+l)} \tag{21.28a}$$

If we wish to express l in terms of b, we have

$$l = -\frac{f(f+b)}{b} \tag{21.28b}$$

Example 21.10: Calculate the nearest focussing distance of an 80 mm focal length lens, if the lens can be moved through a distance of 30 mm.

Solution: In equation (21.28b), we put $b = 30$ mm, $f = 80$ mm and solve for l to get

$$l = -293 \, \text{mm}$$

which is the closest focussing distance of that camera lens.

Thus the closest working distance depends upon the maximum distance the camera lens can be moved away from the film plane and this is usually purely a physical limitation set by the mechanical design of the lens. Closer working distances can be achieved by using extension tubes, bellows and close-up lenses.

Sometimes lenses are designed to give a much closer minimum focussing distance than the standard lenses and are also designed to give optimum image quality at these closer distances. These lenses are called **macro** lenses.

21.8 Viewing system

Cameras have a viewing system which has a visual image field size identical or close to that falling on the film. Cameras fall into two basic types: (a) those with two separate lens systems and (b) single lens reflex cameras. In the first

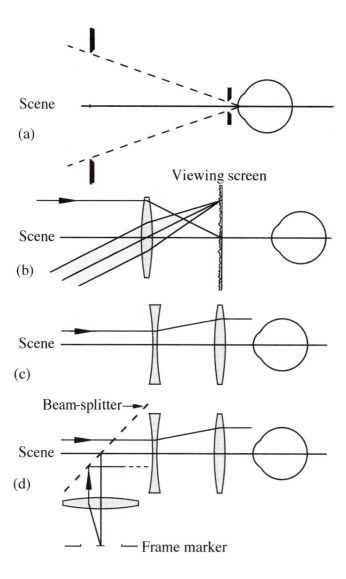

Fig. 21.7:   Various forms
of viewfinders used in some
cameras.

of these, one lens system is the main lens and only used to form the image
on the film and the other lens system is the viewing system, often called a
viewfinder. In single lens reflex cameras, the field-of-view is observed through
the same lens that forms the image on the film. These two systems are described
in more detail below.

21.8.1   *Those with a separate viewfinder*

The simplest type of viewfinder for this type of camera is a simple lens-less
aperture as shown in Figure 21.7a. An alternative is to project the image onto
a viewing screen as shown in Figure 21.7b, but in this case the image will be
inverted on the viewing screen. A common but more complex viewfinder is
the reversed Galilean telescope, shown in Figure 21.7c. This system provides
an erect but a reduced image size, because it must give a wide field and the

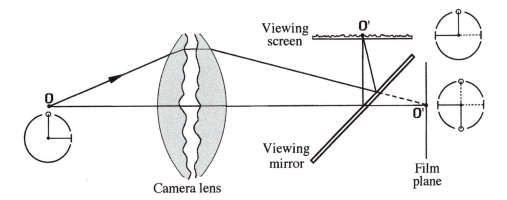

Viewing screen

O'

Camera lens

Viewing mirror

Film plane

O

O'

best way to achieve this is to use a magnification that is less than unity. Some viewfinders are based upon the Keplerian telescope, but since these give an inverted image, an inverting system may have to be incorporated in the design. Because of the physical separation of the optical axes of the two systems, these types of viewfinders suffer from parallax with near photography.

Frame marker

Some viewfinders contain a frame marker, which should be imaged at the object plane of focus of the main camera lens, but for simplicity is usually imaged at infinity. This can be easily incorporated in the Keplerian design by placing it at the intermediate image position, that is in the front focal plane of the eye lens.

In the Galilean viewfinder, this is not accessible to a real object, so a beam splitting mirror would have to be used to project a collimated image of the frame marker onto the back focal plane of the eye lens. A schematic arrangement is shown in Figure 21.7d. The same projection system can be used to produce a frame marker in the lens-less simple viewfinders as shown in Figure 21.7a.

21.8.2 Single lens reflex camera

In the single lens reflex construction, the photographer observes the image through the camera lens via a complex optical arrangement. The leading part of this system is shown in Figure 21.8. The image formed by the camera lens and falling on the film is rotated through 180°, that is inverted top to bottom and left to right. The mirror reflects the image onto a viewing screen as shown in the diagram. This mirror moves out of the way when the photograph is being taken.The viewing screen is conjugate to the film plane, and the circles shown in this diagram and patterns within these circles indicate the orientation of the image at the various positions in the system. Note that the image on the viewing screen is erect but inverted left to right.

There are several possible geometries for viewing this image. For example, if the image on the viewing screen is viewed from above, the image will be erect but inverted left to right. However, if the viewing screen is viewed horizontally by the use of a mirror as shown in Figure 21.9a, the image will be inverted top to bottom by this mirror, so the final image will be upside down and inverted

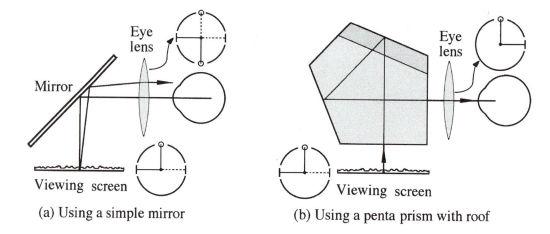

(a) Using a simple mirror   (b) Using a penta prism with roof

left to right, as shown in the diagram. Both of these viewing arrangements are therefore unsatisfactory.

The ideal viewing system should deviate the beam through 90°, keep the correct image orientation in the vertical direction, but swap the image left for right in the horizontal direction. A suitable system that satisfies these requirements is the penta prism with a roof. This prism, which is shown in Figure 8.23, has two reflections in the vertical plane so there is no inversion in this plane. The roof then turns the image left to right. Thus the image is finally seen with the correct orientation. This viewing system is shown in Figure 21.9b.

As shown in Figure 21.9, an eye lens, acting as a simple magnifier, is used to view and magnify the image formed on the viewing screen. The screen is usually fitted with a double prism system to aid in focussing, similar to that shown in Figure 10.9 and described in Section 10.4.2. Because the viewing lens is a simple magnifier, there is no exit pupil, and as in a simple magnifier, the closer the eye to the lens, the greater the eye's field-of-view.

Fig. 21.9: Two geometries for observing the image formed on the viewing screen. Use this diagram in conjunction with Figure 21.8.

21.8.2.1 *Image vergence of the viewing screen*

If the viewing screen is placed at the front focal point of the viewing eye lens, the image will be formed at infinity; that is it will have an image vergence of zero. An image vergence of zero makes it difficult for myopes to see the viewing screen clearly unless they use their ophthalmic corrections, that is their contact or spectacle lenses. However, if one uses spectacle lenses, the eye must be placed farther away from the viewfinder with a resulting decrease in field-of-view. This problem could be overcome by either placing the image closer to the eye, that is giving it a vergence of say -1 to $-2\,\mathrm{m}^{-1}$, or preferably by providing a variable focussing viewfinder. The negative image vergence can be seen clearly by the emmetrope or hyperope if they accommodate.

21.9 The Scheimpflug principle

In Chapter 2, we showed that according to paraxial optics, if the object plane is perpendicular to the optical axis, the image is formed also on a plane perpendicular to the optical axis. However, occasionally, either the object plane or the

Fig. 21.10:   The
Scheimpflug principle.

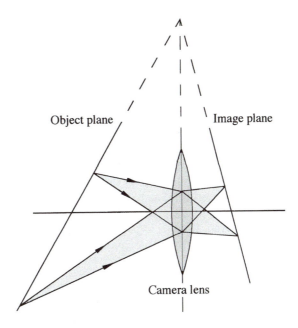

image plane is tilted with respect to the optical axis. For example if the object
plane was tilted and the recording image plane remains perpendicular to the
optical axis, the image will be out of focus with the level of defocus changing
with position along the image plane. This blur can be overcome by tilting the
image plane. The rule for setting the amount of tilt is known as the Scheimpflug
principle, which is explained below.

If the plane of the lens, the object plane and the image plane are co-incident
as shown in Figure 21.10, the image is in focus all along the image plane. This
result is known as the Scheimpflug principle and is used frequently in certain
types of photography. In cameras with the Scheimpflug option, the lens and
film plane must be separately tiltable. However, while the image is in focus
all along the plane, the magnification is not constant along the plane. This is
because different parts of the plane are at different distances from the lens. A
proof of the Scheimpflug principle, which uses the transverse and longitudinal
magnification equations, can be found in Kingslake (1965, 211).

21.10   Aberrations

In comparison with microscopes and telescopes, camera lenses, in general, are
not regarded as highly corrected optical systems. This is partly due to the fact that
the recording medium, for example film, has a relatively low resolving power
and there is little point in designing a camera lens that has an image quality
much in excess of the resolving power of the film. There are exceptions to this
rule, for example the camera systems used for printing integrated electronic
circuits, which are not limited by standard photographic emulsions and require
much higher resolving powers.

The type and amount of the dominant aberrations in a particular lens depend upon the operational requirements of the lens, for example the aperture diameter or *F-number* and field-of-view. Since the aperture diameter is

variable, the aberrations will usually increase with diameter. For example spherical aberration may increase by at least the fourth power of diameter. For wide angle lenses the off-axis aberrations, astigmatism, field curvature and distortion, become important, with distortion becoming particularly a problem with very wide angle lenses. The presence of distortion can be detected by the bowing inwards or outwards of vertical or horizontal straight lines. All the aberrations except distortion can be reduced by reducing the aperture diameter.

At the design stage, the aberrations of a camera lens can be decreased by making the system more complex than a single positive power lens and the greater the reduction in aberrations, the more complex the system usually has to be. The use of more components gives the optical designer flexibility to balance aberrations between them. Some aberrations (coma and distortion) can be greatly reduced by making the system as symmetric as possible about the aperture stop. Field curvature can be a problem in camera lenses, but this is partly reduced by using a negative power component in the system.

Exercises and problems

21.1 (a) For a 80 mm focal length camera lens, on a camera with a 35×24 mm film format, calculate the object field-of-view diameter in the longer direction.

   ANSWER: Object field-of-view diameter $= 24.7°$

21.2 Given a $f = 40$ mm, $F/2$ camera lens, calculate

   (a) the entrance pupil diameter
   (b) the numerical aperture in image space
   (c) the minimum diameter of the first surface in the camera lens

   ANSWERS: (a) 20 mm, (b) 0.25, (c) 20 mm

21.3 What will be the powers of a telephoto lens which has an equivalent focal length of 1000 mm, a back focal length of 200 mm and a first component to back focal point distance of 400 mm?

   ANSWERS: $F_1 = 0.004\,\text{mm}^{-1}$ and $F_2 = -0.015\,\text{m}^{-1}$

21.4 Calculate the power of a Barlow lens designed to increase the equivalent focal length of a $f = 50$ mm camera lens to 100 mm, if the Barlow lens is to be placed 30 mm from the back focal point of the camera lens.

   ANSWER: power $= -16.7\,\text{m}^{-1}$

21.5 Given a $f = 100$ mm camera lens set at $F/5.6$, calculate the change in film plane illuminance in the following cases.

   (a) the *F-number* is changed to 16
   (b) the*F-number* is changed to 2

   ANSWERS: (a) 0.123, (b) 8 or 7.84

Summary of main symbols

Note: Object and image distances and vergences have the usual sign convention.

   F equivalent power of a camera lens
   f corresponding equivalent focal length
   $\mathcal{A}$ aperture stop or diaphragm.

Section 21.1: Field-of-view, focal length and image sizes

R radius of film field
η' image size
θ angular radius of an object or object field

Section 21.2: Telephoto and inverse telephoto lenses

F_1, F_2 equivalent power of the lenses in telephoto and inverse telephoto lenses
f_1, f_2 corresponding equivalent focal lengths
F'_v back vertex power
f'_v back focal length
d lens separation in telephoto and inverse telephoto lenses

Section 21.4: Image plane illuminance and film exposure

E' film plane illuminance
E_{exp} film plane exposure (= illuminance × exposure time)
L_{um} luminance of an object
t exposure time
τ lens transmittance
b distance of lens moved from camera body or "extension tube" length

Section 21.6: Depth-of-field

ω visual threshold angular (radians) diameter of defocus blur disc
M magnification of image (e.g. on enlargement)
s viewing distance of enlarged image
l_o, l'_o object and image distances from respective vertex planes for in-focus object
L_o, L'_o corresponding vergences
L_{hf} hyperfocal distance (as a vergence)
ϕ'_b diameter of defocus blur disc in the image plane

Section 21.7: Focussing range

b maximum allowed movement of lens from body
l closest focussing distance

References and bibliography

*Kingslake R. (1965). *Applied Optics and Optical Engineering*, Vol. 1. Academic Press, New York.

Ohzu H. and Shimojima T. (1972). Optimum diopter value for a view-finder of photographic camera. *Opt. Acta* 19(5), 343–345.

22

Projectors

22.0 Introduction

Projection systems are optical systems designed to project images of solid objects and photographic objects such as transparencies, usually with some magnification, onto an observation screen. Typical uses are as profile projectors in engineering, 35 mm photographic slide projectors, motion picture projectors, microfilm and microfiche readers and photographic enlargers. They consist of a projection lens, an illumination system and a screen on which a real image is observed. A typical projection system is shown in Figure 22.1. Some projection systems, though very few, have no screen because the image is virtual and this is projected directly into the eye using an eyepiece. Figure 22.1 shows the object being trans-illuminated, but some objects are opaque and the reflected light is used to form the projected image. In the common photographic projector, the object is commonly a piece of photographic film either in positive or negative form.

22.1 The projection lens

The projection lens is a positive equivalent power lens, usually well corrected for aberrations. It usually does not contain an aperture stop. Instead, the aperture stop is provided by the illuminating system with the image of the light source being imaged into the projection lens and acting as the effective entrance pupil of the projection lens. We will discuss the pupils of a projection system in Section 22.3.

The optics of the projection system are shown in Figure 22.1, where the projection lens is depicted as a thin lens. In reality, the projection lens will be more complex mainly because of the need to give good image quality for a wide aperture and over a wide field. This lens images the object onto the screen as shown. The object distance l, image distance l' and equivalent focal length f of the projection lens are related by the lens equation (3.15), which can be expressed in terms of focal length instead of power, as

$$1/l' - 1/l = 1/f \qquad (22.1)$$

The choice of the focal length f of the projection lens will be set by such requirements as the transverse magnification, the throw (distance of object to image) or the distance between the projector and screen.

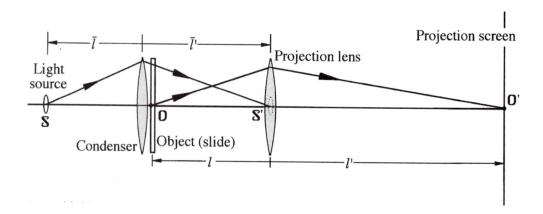

If the magnification M and distance l' from the projection lens to the projection screen are fixed, we can recall equation (3.50a), that is

Fig. 22.1: The optics of a projection system.

$$M = 1 - (l'/f) \qquad (22.2)$$

where we have replaced the power F in equation (3.50a) by the focal length f. This equation allows us to find the focal length f from the other two quantities.

> **Example 22.1:** Calculate the focal length of a projection lens needed to project a 35 mm photographic slide onto a screen which has a width of 1.5 m and is 5 m from the projection lens.
>
> **Solution:** In this case, we can assume that the image distance l' is 5 m. From the data, we can determine the required magnification. The 35 mm slide has dimensions 35×24 mm. The longer dimension (35 mm) must fit on the screen, therefore the magnification M is
>
> $$M = \text{image size/object size} = (-)1500/35 = -42.86$$
>
> If we now substitute $M = -42.86$, $l' = 5000$ mm in equation (22.2), we solve for f to get
>
> $$f = 5000/[1 - (-42.86)] = 114.0 \, \text{mm}$$

22.1.1 Viewing geometry

If the observer wishes to view the image so that it appears to be the same size (that is same angular subtense) as the original object seen from the camera lens, then the observer must be placed at a distance from the viewing screen given by the equation

$$\text{distance} = \text{distance from projector to screen}$$

$$\times \frac{\text{focal length of camera}}{\text{focal length of projection lens}} \qquad (22.3)$$

Fig. 22.2: Two types of illumination used in a projection system: (a) simple, (b) diffusing plate. The condenser form shown in Figure 22.1 is a third type.

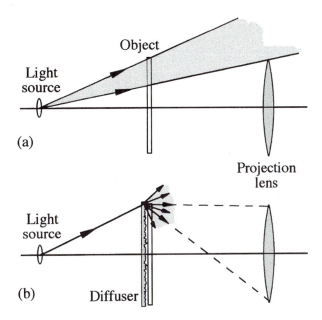

22.2 The illumination system

In Figure 22.1, we have shown the illumination system as a condenser system. We could ask, "Why is there a need for an illumination system and in particular the condenser lens?" We can find the answer to this question by constructing the simplest projection system possible and analysing its peformance.

Let us begin by examining why we need an illumination system. Firstly, the projection lens collects only part of the light coming from the object. This light is imaged onto the projection screen but is spread out over an area that is larger than the object by a factor that is equal to the square of the magnification. The magnification can be high. For example in home photography, a 35 mm photographic slide could be projected onto a screen that is about 1 m in width. The corresponding magnification would be about 30, leading to an illuminance on the screen less than 1/900 that of the object illuminance. The resulting image would be too dull to see. Thus in projection systems, the object has to be illuminated to a very high level, well above that of the ambient level on the projection screen. This high light level is achieved by placing a very bright light source close to the object as shown in Figure 22.2a.

Figure 22.2a shows a very simple projection system with a light source illuminating the object. We can see from this diagram that, while the peripheral parts of the object are illuminated by the source, the illuminating rays do not pass through the projection lens and therefore the peripheral parts of the object are not imaged on the screen. We can overcome this problem by three methods, but only two are practical. The first is that we could make the projection lens wider. While it may be possible to do this in some situations, it is not a preferred option, because the extra aperture width or diameter increases the aberrations. The larger lens diameter makes the lens thicker and more difficult to make (see Section 22.2.2.2). Therefore this option is not feasible.

Two practical options are (i) using a diffuse scatterer in front of the object or (ii) using a condenser. We will look at these two options now, starting with the scattering plate.

22.2.1   The use of a diffusing plate

Figure 22.2b shows a scattering plate placed just in front of the object. The plate scatters an incident ray into a wide range of angles. Some of the scattered light will be deflected into the projection lens, but most is not and represents wasted light. Therefore while this option will provide a full image of the object, it wastes a lot of light and hence the image will not be as bright as it could be.

22.2.2   The condenser system

If we place a lens between the light source and the object as shown in Figure 22.1, and of appropriate focal length or power such that the lens images the light source into the projection lens and the image of the light source is smaller than the aperture diameter of the projection lens, then all the rays that illuminate the peripheral parts of the object will pass through the lens and thus form the image. Since the projection lens captures all of the rays that illuminate the object, the light collection has maximum efficiency and the image will have maximum brightness.

22.2.2.1   Illuminance in the object plane and on the projection screen

If the light source has an axial luminous intensity I and the distance from the light source to object is $\bar{l}$, as shown in Figure 22.1, and we neglect any light losses as the light passes through the condenser, the axial illuminance E on the object can be estimated from the inverse square law [equation (12.8)], that is

$$E = I/\bar{l}^2 \tag{22.4}$$

If we now consider an element of the object of area A and assume the above illuminance is uniform over the object, then the luminous flux (*Flux*) falling on the element of area A is

$$Flux = AE \tag{22.5}$$

If the object does not absorb or scatter any of this light and the condenser images the light source into the projection lens with the lens being larger than the source image, all of this flux is collected by the projection lens and imaged onto the projection screen. If the transverse magnification is M, then the object element of area A is imaged to an element of area A', where

$$A' = AM^2 \tag{22.6}$$

The illuminance E' on the projection screen is then given as

$$E' = Flux/A' \tag{22.7}$$

If we combine equations (22.4) to (22.7), we can write the projection screen illuminance as

$$E' = \frac{I}{\bar{l}^2 M^2} \tag{22.8}$$

Thus if we have a high magnification and also require a high screen illuminance, we must have either a source with a high luminous intensity or have the source close to the object. Since we have seen in Chapter 12 that the luminous intensity of a small source is a product of the source area and luminance, the first of these conditions requires a source of high luminance or large area or a combination of both. The source area can be effectively doubled by placing a reflector behind the light source and slightly displacing the light source transversely so that the reflector images the source back in the same plane as the source and just to the side, so that the source and its image appear side by side thus appearing to be a source of twice the actual size. However, there is a limitation in source size, in that the source image should not be greater than the aperture of the projection lens, otherwise not all the light collected by the condenser is collected by the projection lens.

The higher the object plane illuminance the greater the amount of radiant and conducted heat reaching the object. Protection from excessive heat is usually provided by (a) placing a heat absorbing filter between the source and the condenser lens and/or (b) using fan forced air to cool the object and carry away the heat.

22.2.2.2   *The power and aperture diameter of the condenser*

For the condenser to operate efficiently, it must image the light source into the projection lens as shown in Figure 22.1. Normally the condenser is placed as close to the object as possible as this leads to the lowest power and smallest aperture diameter. Here we will assume it is in contact with the object. If the distance from the condenser to the light source is $\bar{l}$ and that to the projection lens $\bar{l}'$ as shown in the diagram, then the power $F_{condenser}$ is given by the lens equation (3.15), that is

$$F_{condenser} = 1/\bar{l}' - 1/\bar{l} \tag{22.9}$$

We have seen in the preceding subsection that for high screen illuminance, the distance $\bar{l}$ must be as small as possible. From the above equation, it follows that the condenser lens may have to have a high power, which in turn may lead to unwanted aberrations and difficulties in its manufacture.

The reason for this latter problem is that the aperture radius of the condenser must be the same as or slightly greater than that of the object being imaged. If the object has a radius ρ, then this must also be the minimum aperture radius of the condenser. We saw in Chapter 6 that for a lens to be constructable, there was an upper limit to the product of the power F and aperture radius ρ. In Section 6.1.1.5, we showed that as a rule of thumb a lens could not be made if

$$|\rho F| > 1$$

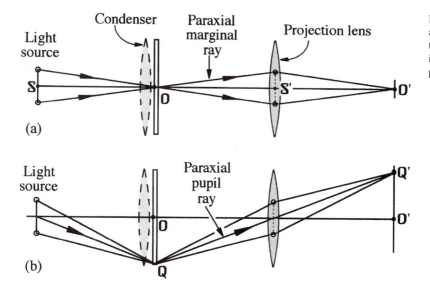

Fig. 22.3: The light source as the aperture stop of the total system and whose image forms the entrance pupil of the projection lens.

In some projection systems, the product is often very high, but there are several ways this condition can be overcome. These are as follows:

(1)   The single condenser lens can be split into two or more separate components.
(2)   One or both surfaces can be aspherized, such that the local radius of curvature reduces towards the edge of the lens. This option has the added advantage that spherical aberration is reduced.
(3)   The conventional condenser lens is replaced by a Fresnel lens (which is described in Section 6.1.3). This is the option used in overhead projectors.

A condenser lens can be regarded as a type of field lens since its function is to provide a full field-of-view by bending peripheral rays back towards the axis and into the following imaging lens.

22.3   Aperture stop and pupils

Every optical system must have an aperture stop and hence entrance and exit pupils. In a projection system, the light source is the effective aperture stop of the system as a whole and the image of the light source formed in the projection lens acts as the entrance pupil for that lens. We can explain this phenomenon by looking at Figure 22.3a. This diagram shows the axial point on the object being illuminated by rays from all parts of the light source. These rays must pass through the projection lens but through corresponding points on the image of the light source. Therefore this image sets the size of the cone of rays leaving the axial object point and therefore is a "pseudo" aperture stop. Image forming rays cannot pass through the projection lens outside the source image. The extreme marginal ray of the beam is the paraxial marginal ray. Similarly Figure 22.3b shows rays illuminating the edge of the object and once again, the set of rays must pass through the projection lens but only through the source image.

Fig. 22.4:   The geometry
for a projection system
using (a) front projection
screen and (b) rear
projection screen.

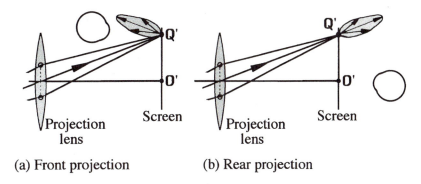

(a) Front projection   (b) Rear projection

The paraxial pupil ray is the central ray of this beam. The exit pupil is the image
of the source seen from image space.

Thus the effective aperture stop of a projection lens is the image of the
light source and thus its position depends upon the choice of focal length of
the condenser. For a single thin lens projection lens, the source image should
be coincident with the lens. However, for a more complex projection system,
which may consist of several and separated components, there is no readily
defined best position for the source image plane. This is of some importance
since we have seen in Chapter 5 that the aberrations of a lens can depend upon
the position of the aperture stop. The aberrations of the projection lens are
discussed further in Section 22.5.

22.4   The projection screen

Projection screens are surfaces coated with some type of diffusing material.
They may be either front projection type as shown in Figure 22.4a or the rear
projection type shown in Figure 22.4b. With front projection systems, the image
is seen on reflection and with rear projection screens, the image is seen on
transmission. The screen material should be highly reflective (front screens)
or highly transmissive (rear screens) but in both cases highly diffusing. The
diffusing properties of the screen materials are critical in affecting the uniformity
of image luminance. As a general rule, materials used for coating a front surface
screen are more diffusing than those used for rear projection screens.

22.4.1   *Front projection screens*

Front projection screens are usually coated with a highly reflective material
and the image is seen on reflection. While this gives a bright image, the high
reflectance of front projection screens makes them very vunerable to ambient
lights and thus they have to be shielded from them.

Suitable materials for front screens are magnesium oxide (MgO) and barium
sulphate ($BaSO_4$). These are almost perfectly Lambertian and the screen would
appear to have equal brightness from all directions of view. This is ideal if the
screen is observed from a wide range of directions. However, very often the
screen is only observed over a narrow range of angles centred on the axis of the illuminating beam. In this case the light reflected off to the side is wasted. If the reflected light could be concentrated back in the direction of incidence, that is back

towards the observers, then the image would be brighter in that direction and hence less susceptible to ambient lights. Concentrating the reflected light back into a narrow cone centred on the incident beam is called **retro-reflection** and can be achieved by mixing small transparent (usually glass) spheres in the diffusing material. These spheres act as retro-reflectors and were first encountered as a problem (Problem 2.6) in Chapter 2. Alternatively circular grooves can be cut into the screen such that the face of each groove is perpendicular to the incident beam. The surface acts as a "Fresnel" reflector which is analogous to the Fresnel lens discussed in Chapter 6 and thus acts as a field reflector instead of a field lens.

22.4.2   Rear projection screens

With rear projection screens, the image is seen on transmission. The screen material should be highly transmitting and highly diffusing but have very low reflectance to reduce the effects of ambient lights. The material could be described as **translucent**. Rear projection screens are usually much more directional than front projection screens and therefore the diffusing properties are a very important performance criterion. We will now look at this and other properties of rear projection screens.

22.4.2.1   Gain $G(\theta)$

When a narrow beam of light falls upon a screen, whether front or rear, the light is scattered but not equally in all directions, as suggested in Figure 22.4, and this leads to a variation in luminance across the screen. Thus one of the important properties of any screen material is the angular variation of screen luminance. The relative variation with viewing angle θ is called the **gain** of the screen. Writing this as $G(\theta)$, it is defined as

$$G(\theta) = \frac{\text{screen luminance } L(\theta)}{\text{screen illuminance } E} \qquad (22.10)$$

Examination of some materials used for rear projection screens, such as ground glass, shows that the gain can be approximated by an equation of the form

$$G(\theta) = G(0)e^{-b\theta^m} \qquad (22.11)$$

Values of $G(0)$, b and m for two samples examined by the authors are as follows.

A piece of ground glass diffuser: $G(0) = 5.6$, $b = 0.022$ and $m = 1.6$ and for a piece of commercial rear projection screen material: $G(0) = 0.83$, $b = 0.0036$ and $m = 1.8$.

We can note that the smaller the value of b, the more uniform the screen luminance. We could also note that for a perfectly uniformly scattering material (that is Lambertian), the gain would be constant and have a value of $(1/\pi) \approx 0.32$.

22.4.2.2   Scintillation

Scintillation is the small scale visible variations in screen luminance, which are due to microscopic reflections and refractions within the screen material.

22.4.2.3   Resolving power

A screen material scatters light usually because it has a particulate surface or bulk nature. The size of the scattering particles limits the size of the detail that can be imaged on the screen. In point spread function terms, a point image is scattered into a finite light distribution within the screen material.

The resolving power of a material can be measured by using a standard resolving power chart. The authors carried out a survey of three commercially but unlabelled screen materials and measured their resolving powers using the USAAF resolving power chart (Appendix 5). The resolving powers ranged from 19 to 25 c/mm. Alternatively, the resolving power could be measured by means of the modulation transfer function (Kuttner 1972).

22.4.2.4   Polarization effects

Occasionally it is suggested that three dimensional images can be formed by using two projectors to project stereo pairs of photographs on to the screen. The suggestion is that the two beams are polarized at $90°$ to each other, by placing polarizers perpendicular to each other over each projection lens. According to the suggestion, the light reflected from the screen retains its polarization and the observer views the images through a pair of polarizing goggles. With suitable orientation of the polarizers in front of each eye, each eye only sees one of the images and this leads to a stereoscopic image. However, this does not occur with conventional projection screens because the scattering properties of the screen material depolarize the incident light.

22.5   Aberrations

22.5.1   The projector lens

The projector lens usually works with a large aperture in order to maximize image brightness and wide angle. Therefore all the aberrations are important. However, the control of these aberrations is complicated by the fact that projection lenses do not have an inherent aperture stop since its effective aperture stop is the image of the light source. The aperture stop position affects the level and type of aberrations and it is difficult to predict where this source will be imaged.

22.5.2   The condenser lens

In condenser lenses, the most important aberration is spherical aberration and this can be reduced by splitting the lens into two or more components or by aspherizing one or more surfaces. Spherical aberration is the most important aberration because the light source is usually small and therefore most of the rays arise from points close to the axis. If spherical aberration were present, the real marginal rays would be excessively deviated and would not be imaged into the entrance pupil of the imaging lens. For example in Figure 22.1, if the marginal ray from the light source to the projection lens as shown had too much uncorrected spherical aberration, it would intersect the projection lens below its bottom rim. Thus the part of the object illuminated by this ray would not be imaged by the system, and therefore the field-of-view would be reduced.

Exercises and problems

22.1   Calculate the focal length of a projection lens required to image a 35 mm object
       onto a screen with dimensions 3.5 m × 2.4 m if the distance from the projector
       to screen is 7 m.

      ANSWER: $f = 69.3$ mm

22.2   For an $f = 100$ mm projection lens, used to project an image of a 35 mm object
       at a distance of 10 m, estimate the power of a suitable condenser lens if the light
       source is 50 mm from the object. If the length of the object diagonal is 40 mm,
       can the condenser be made from a single component?

      ANSWERS: $30 \, \text{m}^{-1}$, yes

Summary of main symbols

f	equivalent focal length of the projection lens
F	corresponding power
$F_{\text{condenser}}$	power of the condenser lens
E'	image illuminance
I	luminous intensity of the light source
$L(\theta)$	screen luminance as a function of angle of view θ
$G(\theta)$	screen gain as a function of angle of view θ
$\bar{l}, \bar{l}'$	distances of condenser lens from light source and projection lens
M	transverse magnification

References and bibliography

*Kuttner P. (1972). Modulation transfer function of ground glass screens. *Appl. Opt.*
11(9), 2024–2027.

23

Collimators

23.0   Introduction

Collimators are optical systems designed to produce a reasonable quality image of a target (or light source or some other object) at optical infinity. The angular size of the image is usually small; therefore the field-of-view is small and thus the system is relatively simple. Since the target has to be imaged at optical infinity, it must be placed at the front focal point of the collimator lens.

Collimated light is often referred to incorrectly as parallel light. No doubt, the term arises because paraxial or unaberrated real rays from a single point in the object or target are all parallel to each other in image space. This term often leads to the misunderstanding that a collimated beam has parallel sides. If this were true, a collimated beam would have zero divergence. In reality, a collimated beam diverges and there are three causes of this divergence: (1) the finite size of the source or target, (2) aberrations and (3) diffraction. Diffraction usually only dominates the divergence if the beam has a small diameter, say several millimetres or less. The diameter of collimators used in visual optics is usually much wider than this and therefore source size and aberrations are the dominant causes of beam divergence. Let us look at these in turn.

23.0.1   Effect of source size

In Gaussian optics, the beam must diverge and the amount of divergence is proportional to the size of the source or target. This can be easily demonstrated using Figure 23.1, which shows a source or target of radius η at the front focal point $\mathcal{F}$ of the collimating lens. Representative rays are shown leaving the object and being refracted by the lens. All of the rays, from the extreme ends of the source, subtend the same angle θ to the axis in image space, where

$$\theta = \eta/f \tag{23.1}$$

and f is the (equivalent) focal length of the collimator. Therefore the beam must diverge by an amount 2θ. Thus a knowledge of the source or target size will give an indication of the beam width at any distance. From the diagram, if θ is small, the beam width $\phi(d)$ at a distance d must be given by the equation

$$\phi(d) = 2(\rho + d\theta) \tag{23.2}$$

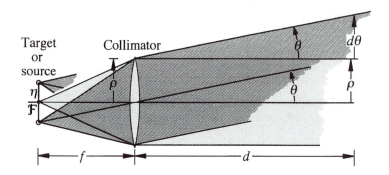

where ρ is the aperture radius of the collimating lens. Combining equations (23.1) and (23.2), we have

$$\phi(d) = 2[\rho + (d\eta/f)] \tag{23.3}$$

which confirms that the beam divergence is proportional to the source radius η.

> **Example 23.1:** If a collimator of focal length of 50 cm and aperture diameter of 6 cm is collimating a source that is 5 mm in diameter, calculate
>
> (a) the beam divergence angle θ and
> (b) the beam width at a distance of 1.0 m away from a collimating lens.
>
> **Solution:**
>
> (a) From the data we have, $\eta = 2.5$ mm, $f = 500$ mm; therefore from equation (23.1)
>
> $$\theta = 2.5/500 = 0.005 \text{ rad} = 0.286°$$
>
> Therefore the beam divergence angle $\theta = 0.286°$.
> (b) Now $\rho = 30$ mm and $d = 1000$ mm; therefore from equation (23.2),
>
> $$\phi = 2[30 + (1000 \times 0.005)] = 70 \text{ mm}$$
>
> That is the beam width increases by 10 mm to 70 mm over the first 1 m.

While Figure 23.1 shows that a collimated beam diverges, we could ask, "Is there a source position that produces a beam that does not converge or diverge?" The answer is yes, but for only one source position and only over a limited region beyond the lens. That source position corresponds to a position giving an image the same size as the diameter of the collimator. This situation is shown in Figure 23.2a, in which the source at point o must be farther from the lens than the front focal point $\mathcal{F}$. A beam leaves the edge of the source as shown in the diagram and the upper ray of the refracted beam is the ray bounding the beam in the region between the lens and the image. A little time spent sketching

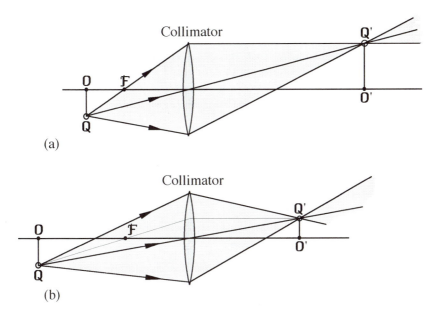

(a)

(b)

Fig. 23.2: (a) Example of an uncollimated beam that is "parallel" over a certain distance from the collimating lens. (b) Effect of placing the source farther from the collimating lens than shown in (a).

rays will show that all rays from the bottom half of the source must stay below the upper horizontal ray, but only in the region between the lens and the image at O'. Therefore, we can see from this diagram that if the collimating lens forms a real image of the source with the same size as the diameter of the lens, then the beam is "parallel" from the lens to the image, but then diverges. We should note that the corresponding source position for this situation to occur depends upon the source size.

If the source is farther away from the collimator, as shown in Figure 23.2b, the beam will initially converge after leaving the lens, but begin to diverge after the beam has passed through the image.

23.0.2 Effect of aberrations

Apart from the beam divergence due to the finite size of the source, beam spread also results from the effects of aberrations. For a well aligned collimator made from a single lens, spherical and longitudinal chromatic aberrations will be the main aberrations that cause the beam to converge or diverge. For example, the effect of the positive spherical aberration is to initially converge the beam but then eventually diverge it. Longitudinal chromatic aberration is equivalent to a shift of focus or defocus with wavelength. This longitudinal chromatic aberration induced defocus will cause the beam to converge for shorter wavelengths and diverge for longer wavelengths as it leaves the lens. These aberrations are discussed further in Section 23.5.

23.1 Applications

Collimators have a number of uses. They may be used simply to provide the eye with an image at infinity, say to simulate distance vision and sometimes to stimulate the relaxation of the accommodation. They are also used to check the settings of afocal systems such as telescopes or the infinity setting of cameras, the alignment of binoculars, and sights for many aiming devices. They are an integral part of many optical instruments such as focimeters, foco-collimators and spectrometers. Some aspects of collimators used for visual applications are discussed further in Section 23.4.

23.2   Methods of collimation

There are a number of methods for the collimation of a target, depending upon the nature of the target to be collimated. Two useful methods are (a) visual focussing using a telescope and (b) auto-collimation.

23.2.1   *Visual collimation using a telescope*

Visual collimation using a telescope is a simple method that initially requires a telescope to be focussed on a distant object. The telescope is then used to view the target through the collimator. The distance between the target and the collimator is adjusted until the target is in clear focus through the telescope, and at this point, the image should be as distant as the target used to initially focus the telescope. Errors can occur due to variations and fluctuations in the accommodation of the observer. These can be minimized by using one of the following two procedures.

(a) *By using a high magnification telescope.* We have shown in Chapter 17 that the object vergence L_v at the objective of a two-lens telescope is related to the image vergence L'_v at the eye lens by equation (17.6), but if the vergences are small, then the relationship can be reduced to the approximation given by equation (17.8), that is

$$L'_v \approx M^2 L_v \tag{23.4}$$

We can use this equation to show that errors due to fluctuations in the accommodation of the observer are reduced if the observer looks through a telescope, and the higher the magnification, the smaller the error.

Suppose that on viewing through the telescope, instead of using relaxed accommodation, the observer accommodates by an amount $\delta A'$ and sets the target position so that it is clearly in focus for this level of accommodation. This is the image vergence at the eye lens. Equation (23.4) shows that the corresponding image vergence δA at the objective of the telescope is

$$\delta A \approx \delta A'/M^2 \tag{23.5}$$

Thus providing M is greater than unity, the magnitude of δA is less than $\delta A'$ and is reduced by the square of the magnification of the telescope. The maximum possible value of $\delta A'$ is the amplitude of accommodation of the observer, which varies with the individual and with such personal factors as age. For a young observer, the amplitude of accommodation may be as large as $10\,\text{m}^{-1}$. The effect of age on amplitude of accommodation has been discussed in Chapter 13.

Example 23.2: Suppose an observer with an amplitude of accommodation of $5\ \text{m}^{-1}$ is using a telescope with a magnification of four to collimate a target. Calculate the expected possible error in image vergence.

Solution: From equation (23.5), take $\delta A' = 5\ \text{m}^{-1}$ and since $M = 4$, we have

$$\delta A' = 5/4^2 = 5/16 = 0.31\text{m}^{-1} = \text{error}$$

(b) *Using a telescope with cross-hairs in the eyepiece.* The cross-hairs aid in maintaining a fixed level of accommodation and thus reduce fluctuations in accommodation.

23.2.2   Auto-collimation

Auto-collimation is an accurate method and does not depend upon the maintenance of accurate or steady accommodation of the eye. A typical arrangement is shown in Figure 11.15. That figure was used to show how to measure the vertex focal length of a lens. At the same time, it collimates the target. The target is brightly illuminated. If the target is perfectly collimated, an image of the target will be formed in the plane of itself and also on the viewing screen, which must be conjugate to the target. If the reflected image is not in focus, the lens is moved backwards or forwards until the return image is focussed. While this system is not completely objective, as it does require a subjective judgement of best focus, errors are not induced by accommodation fluctuations of the observer. The accuracy is high since any effect of a focussing error position of the target is doubled because of the reflection from the mirror. However, it should be noted that errors in positioning the viewing screen will lead to errors in collimation and therefore it must be set accurately.

The inherent accuracy depends upon the depth-of-field, which can be reduced by using a wider aperture beam. However, as the beam width is increased, residual aberrations in the collimator may offset this gain if the beam width becomes too great.

23.3   Positioning accuracy of source or target

In this section, we will discuss the relationship between an error in positioning the target and the vergence of the collimated beam. Consider the lens equation (3.43) in the form

$$L' - 1/l = 1/f \tag{23.6}$$

where we have neglected n and n' since we are working in air and we have replaced the image distance l' by its vergence L' and the equivalent power F by the equivalent focal length f. Let us suppose that the target has a small position error of δl. The resulting image will not be collimated but will have a residual vergence $\delta L'$, which is then given by the equation

$$\delta L' + \delta l/l^2 = 0$$

that is

$$\delta L' = -\delta l/l^2$$

But at the correct setting, $L' = 0$ and from equation (23.6)

$$l = -f$$

Thus

$$\delta L' = -\delta l/f^2 \tag{23.7}$$

Example 23.3: Suppose we have precisely collimated a diaphragm using the auto-collimation technique described in the previous section, using a lens with a focal length of 20 cm. If we now replace this diaphragm with an alternative target, calculate the allowed error in positioning the target if the image vergence error must not be greater than ±0.5 m^{-1}.

Solution: From the above data, $\delta L' = \pm0.5$ m^{-1} and $f = 200$ mm, therefore from equation (23.7),

$$\pm0.5 = -\delta l/0.2^2$$

that is

$$\delta l = \pm20 \text{ mm}$$

23.4   Visual collimators

For the purpose of this book, we will call visual collimators those collimators that are specifically designed to present an image at infinity for the eye. Typical situations are head-up displays and simple magnifiers. Ideally the image should be at infinity, that is have zero image vergence. However, we have seen in the previous section that errors in setting the position of the target will lead to a non-zero image vergence. We need to look at tolerances to these errors keeping in mind the defocus tolerance or depth-of-field of the eye. We also need to take a brief look at the effect of eye position. In this section, we will look at both of these issues, starting with the focus tolerances.

23.4.1   Focussing tolerance on visual collimators

When collimators are being used for visual work, the error in collimation should be less than the depth-of-field of the eye or the smallest measurable refractive error. Both of these values are about ±0.12 m^{-1}. Therefore visual collimators should be collimated to better than ±0.12 m^{-1}.

The corresponding tolerance on target position can be found using equation (23.7). The tolerance δl on the target position is found by substituting $\delta L' = \pm 0.12$ m^{-1} into equation (23.7) and solving for δl, that is

$$\delta l = \pm 0.12 f^2 \qquad (23.8)$$

where the quantities δl and f must be in metres.

> **Example 23.4:** With a 50 cm focal length collimator, calculate the required precision in positioning the target if the image vergence is to be within the range ± 0.12 m^{-1}.
>
> **Solution:** From the above, $f = 0.5$ m. Therefore from equation (23.8),
>
> $$\delta l = \pm 0.12 \times 0.5^2 = \pm 0.03 \text{ m} = \pm 30 \text{ mm}$$

23.4.2   Eye position

We could begin by asking the question, "When viewing collimated images, is the eye position important?" This question can be answered in several parts. Firstly, collimators are similar in principle to simple magnifiers and, as a consequence, do not have an intrinsic aperture stop. Therefore there is no exit pupil to fix the position of the eye. With no fixed eye position, where is the best eye position? To answer this we need to take two factors into account: (a) the image vergence and (b) the field-of-view.

Ideally, the image vergence at the collimator is zero and therefore the distance of the eye from the collimator does not affect the image vergence at the eye. It would only be a problem if the image vergence were substantially non-zero, and if this were the case, the collimator would not be functioning correctly as a collimator.

When viewing through a system that does not have an intrinsic exit pupil, the closer the eye is to the instrument, the larger the field-of-view and vice versa. However, this does not mean that the eye must be placed as close to the collimator as possible. If the target being viewed is small, it may be fully seen over a wide range of viewing positions. Figure 23.3 shows such a situation. The image of a target of radius η collimated by a collimator with a focal length f has an angular radius θ, given by equation (23.1), and if the viewing distance is d as shown in the diagram, the lens subtends an angle

$$\psi = \rho/d \qquad (23.9)$$

In both of these equations it is assumed that the angles are small. Otherwise the angles must be replaced by their tangents.

Now θ is independent of d and therefore always subtends the same angle provided it is not vignetted by the collimator. In contrast, the angular size of the collimator depends upon eye position. These two angular sizes will be equal at a certain eye position d, where from Figure 23.3,

$$\frac{\eta}{f} = \frac{\rho}{d}$$

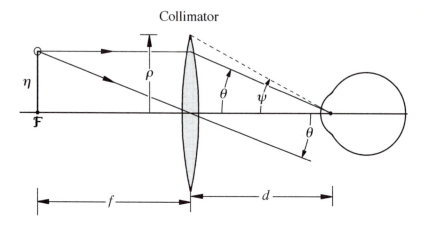

Collimator

Fig. 23.3: Eye position
and field-of-view.

that is

$$d = \rho f / \eta \qquad (23.10)$$

For larger distances, the target will be vignetted by the collimator, and therefore, provided d is less than the value given by equation (23.10), then the field-of-view of the target will be independent of eye position provided the eye is positioned along the axis.

23.5   Aberrations

When using small targets, the most important aberrations in a collimator are spherical aberration and longitudinal chromatic aberration. For larger targets, coma may have to be considered. The other off-axis aberrations are not usually significant because collimators usually work with small fields-of-view.

While single simple lenses are not ideal for use as collimators, because they cannot be corrected for spherical aberration and longitudinal chromatic aberration, they may be adequate if the *F-number* is sufficiently high. The amount of spherical aberration in such a lens depends upon its *F-number* and its shape (that is plano-convex, bi-convex or meniscus). Seidel aberration theory, presented in Chapter 33, predicts that Seidel spherical aberration of a simple lens cannot be made zero by any lens shape but can be minimized and the ideal lens shape for a collimator, with minimum Seidel spherical aberration, is an almost plano-convex but bi-convex lens. The actual shape does not depend upon the focal length but does depend upon the refractive index. For an index of 1.5, the ideal shape has front and back radii of curvatures in the ratio of exactly 6:1 with the surface of larger radius of curvature facing the target being collimated.

Fortunately spherical aberration and longitudinal chromatic aberrations can be substantially reduced by using an achromatic doublet, which has far less spherical aberration than the best shaped single lens. Achromatic doublets can also be designed to minimize coma. These are the reasons that in practice, collimating lenses are usually achromatic doublets. The design of an achromatic doublet suitable as a collimator is given in Chapter 33.

Exercises and problems

23.1 If a circular light source has a diameter of 2 mm and is collimated by a 100 mm focal length lens with a diameter of 6 cm, what will be (a) the angular divergence of the beam and (b) the diameter of the beam 1.0 m from the lens?

ANSWERS: (a) $1.146°$ and (b) 80 mm

23.2 In the telescopic method of collimation, discuss the significance of the distance between the collimator and the telescope.

23.3 Calculate the required minimum magnification of a telescope to be used for visual accommodation, if the observer has an amplitude of accommodation of 5 m^{-1} and the required error of collimation is less than 0.1 m^{-1}.

ANSWER: magnification $= 7.1$

23.4 Given that the allowed uncertainty in image vergence is $\pm 0.12 \text{ m}^{-1}$, calculate the required accuracy in positioning the target, using an $F = 1 \text{ m}^{-1}$ collimator.

ANSWER: $\delta l = \pm 120$ mm

23.5 Comment on the statement "A beam can be collimated by observing the beam width as it leaves the collimator. The source position is adjusted until the beam does not diverge".

Summary of main symbols

η radius of source or target
θ angular radius of the image of the target $=$ half the beam divergence angle
ϕ collimated beam diameter at a distance d from the collimator
d distance from the collimator for calculating beam width ϕ
ρ aperture radius of the collimator
f equivalent focal length of the collimator
A' amplitude of accommodation of the observer
$\delta A'$ fluctuation or error in accommodation of the observer
$\delta L'$ error in collimation (difference from zero)
δl error in positioning the target

24

Photometers and colorimeters

24.0   Introduction

A number of photometric and colorimetric instruments are visual optical systems as these devices have a viewing system to observe a scene either for alignment or to make a subjective judgement of brightness (luminance) or colour.

24.1   Photometers

Photometers are either subjective or objective instruments and are designed to measure the absolute or comparative brightness of sources or scenes. In the subjective type, the observer has to make a match between the brightness of a standard scene and that of a second scene. In the objective instruments, an observer uses the viewing system only for alignment.

24.1.1   Subjective photometers

24.1.1.1   The Lummer–Brodhun photometer

The Lummer–Brodhun photometer is designed to compare the luminous intensities of two light sources. Often one is a secondary standard source of known luminous intensity, and in this case, the luminous intensity of the second or unknown source can be found.

The instrument is shown in Figure 24.1. The light from the two sources illuminates both sides of a white diffuse block as shown in the diagram. Both sides of the block are viewed simultaneously through an optical system which contains a prism assembly and an eyepiece. One side of the block (the left side in the diagram) is seen through the central part of the prism and the other side of the white block is seen through the peripheral part. The two beams are brought together but separated by a sharp boundary by the prisms. The eyepiece is focussed on the sharp boundary. The field-of-view seen through the eyepiece is shown in the diagram. The relative luminances of both fields are adjusted by moving the photometer backwards and forwards along the line joining the two sources, until the luminances match. Providing the two sides of the block are identical, if the luminances are equal, then the illuminances are equal. Once the illuminances are equal, the inverse square law, equation (12.8), can be used to calculate the luminous intensity of the unknown source from the luminous intensity of the known source.

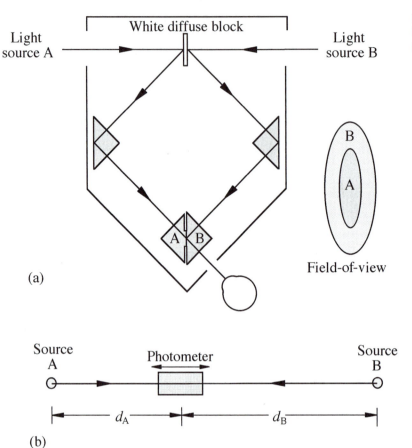

Fig. 24.1:   The Lummer–Brodhun photometer.

24.1.2 *Objective photometers*

Objective photometers are essentially meters designed to measure the luminance of a source and require an optical system to image the source onto an illuminance cell and another sub-assembly to align the instrument. The output of the illuminance cell is then processed electronically to give an output in luminance units.

An optical arrangement is shown in Figure 24.2. The 45° beam splitting mirror deflects part of the beam onto the photo-cell, which is at the focus of the beam. The remainder of the beam passes through into the eye lens. A circular mark corresponding to the edge of the photo-cell is marked on a transparent plate placed in the viewfinder image plane. The observer is then able to see which part of the scene is being imaged on the photo-cell. With this arrangement, the image of the object being measured must be larger than the photo-cell.

24.2 Colorimeters

Colorimeters are designed to allow comparison of two colours each made from a mixture of primaries. They have a number of important uses; perhaps the most important are

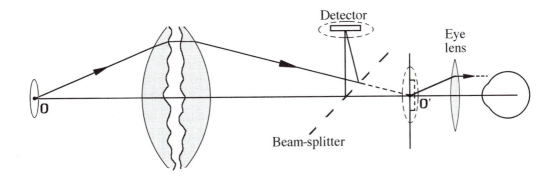

Fig. 24.2: Typical construction of a luminance meter.

Fig. 24.3: Colorimeter for measuring the chromaticity of object colours. The Donaldson colorimeter is based upon this principle.

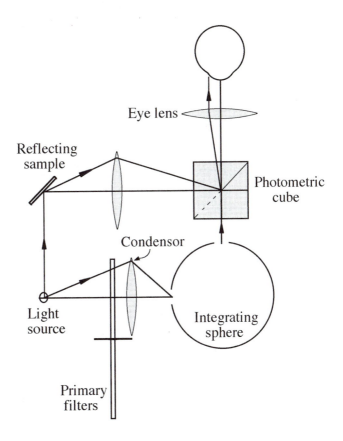

(1) the subjective but quantitative measurement of the colour of an unknown colour sample,

(2) fundamental experiments in colour vision and the assessment of colour deficiencies.

All colorimeters consist of a set of primary light sources, with the capability to mix the primaries in carefully controlled amounts. The mixing is usually

Light source

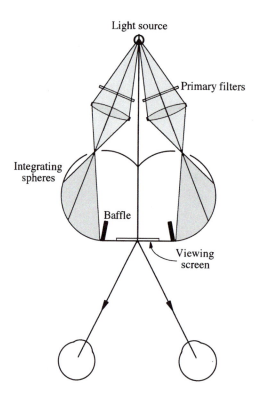

Primary filters

Integrating
spheres

Baffle

Viewing
screen

Fig. 24.4:
Monocular/binocular
colorimeter for measuring
just noticeable colour
differences. The MacAdam
colorimeter is based upon
this principle.

done in an integrating sphere, which is a hollow sphere painted on the inside
with a diffuse white material. Suitable paints use magnesium oxide or barium
sulphate as the base material. The colorimeter also requires the projection of a
patch of the wall of the sphere side by side with the unknown colour. The colour
match is then made by altering the primary components until the observer is
happy with the colour match of the sample under investigation. There are a
number of colorimeters that have been designed over the years. Two common
constructions are shown in Figures 24.3 and 24.4.

Figure 24.3 shows a schematic form of an additive colorimeter. This instrument allows the measurement of the chromaticity co-ordinates of a reflecting
sample to be measured using an additive mixture of primaries. The primaries
are generated using light from the light source which has passed through the
coloured filters. The condensor projects the transmitted coloured beams into an
integrating sphere where they are mixed. Some of this light leaves the sphere
through the exit aperture and is relayed through a photometric beam splitting
cube into the viewing eyepiece. The light from the reflecting sample is also relayed through this cube into the eyepiece. The observer sees a circular bipartite
field with a sharp edge delineating the two colours. To obtain a sharp edge, the
eye lens must be focussed on the beam splitting cube. By trial and error, the
colour of the sample is then matched with a suitable combination of the primaries. The Donaldson additive colorimeter is based on this principle. This colorimeter is only used to determine the chromaticity of sample or object colours
and is essentially a monocular instrument unless a binocular eyepiece is used.

For research into colour differences, a different type of colorimeter is required, one which generates two very similar colours. Such an instrument is

shown in Figure 24.4. The system uses two identical integrating spheres which contain baffles to prevent any directly reflected light falling on the viewing screen. The colours produced by varying the ratio of primaries in each sphere can be varied independently. In each sphere, the light from each colour is mixed thoroughly in the sphere. The mixture then falls evenly on the viewing screen. This system has the advantage that it allows either monocular or binocular viewing. The MacAdam binocular colorimeter is based upon this schematic design.

References and bibliography

ASTM E 259-93. Standard practice for preparation of pressed white reflectance factor transfer standards for hemispherical geometry. American Society for Testing Materials. Vol. 06.01. Philadelphia, 1994.

British Standard Specification BS354-1961. (1985). *Recommendations for Photometric Integrators*. British Standards Institution, London.

Grum F. and Luckey G.W. (1968). Optical sphere paint and a working standard of reflectance. *Appl. Opt.* 7(11), 2289–2294.

Grum F. and Wightman T.E. (1977). Absolute reflectance of Eastman White reflectance standard. *Appl. Opt.* 16(11), 2775–2776.

Longhurst R.S. (1973). *Geometrical and Physical Optics*. 3rd ed. Longman, London, Chapter 18.

Nonaka M. (1974). Improved paint coating for photometric integrators. *Lighting Res. Tech.* 6(1), 30–31.

Part III

Physical optics and physical optical instruments

25

Interferometry and interferometers

25.0 Introduction

So far, all the visual optical instruments that have been discussed have been essentially describable in terms of geometrical or ray optics. In this and the following chapter, we will describe some important visual optical instruments and devices based upon physical or wave optics. The essential difference between geometrical and physical optics is the difference between the geometrical description of the propagation of light by rays and the physical optical description using waves. The wave nature of light gives rise to two very important phenomena, **interference** and **diffraction**. This chapter is concerned with interference and the next will look at diffraction.

The phenomenon of interference allows us to produce visual stimuli in which the light level varies sinusoidally across the pattern. Since early work by Campbell and Green (1965), the visual system is now regularly analysed in terms of its response to sinusoidally varying luminance profiles. This is often done by measuring the threshold contrast of sinusoidal patterns at a range of spatial frequencies. The resulting function is known as the **contrast sensitivity function** (see Chapter 35). Sinusoidal patterns can be produced in several ways. They may be produced in printed form, for example as photographs, or can be produced on television screens or by optical interference techniques. This chapter is concerned with the last mode of production.

There are two fundamental differences between the production of sinusoidal patterns by interference techniques and those in printed or television screen forms. These are as follows:

(1) The interference techniques essentially by-pass the optics of the eye and therefore are not as susceptible to the effects of ocular aberrations as are the alternative techniques.

(2) In general, interference techniques use monochromatic light, in contrast to the alternatives, which use white light or broad band light. However, in a later section, we will briefly discuss an interference technique using white light.

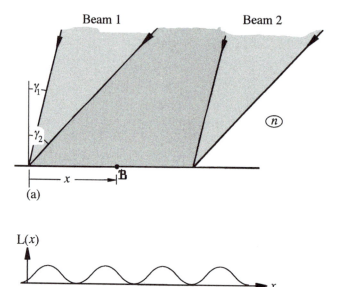

Fig. 25.1: (a) Interference between two collimated and tilted beams. (b) Sinusoidal form of the interference fringes.

Techniques for producing sinusoidal light profiles directly on the retina use the principles of interferometry, which we will briefly introduce here before we discuss two very useful interferometers, namely the Young's slit interferometer and the Twyman–Green interferometer. Interferometry is the study of the effect of the superposition of one or more sources of wave disturbances, which in our case is light. A discussion of interferometry is best begun by stating the **principle of superposition**, followed by the necessary conditions for the formation of interference fringes.

25.1 Principle of superposition

Interferometers are instruments in which two or more mutually **coherent** beams are combined to produce an interference pattern. We will define and discuss the nature and significance of coherence in the next major section, Section 25.2. The resulting amplitude of the disturbance is the sum of the amplitudes of all the interfering beams. This is known as the **principle of superposition**. The form of the light level profile depends upon the number of beams and their relative intensities and phases. Let us use the principle of superposition to develop an understanding of the formation of interference fringes and their light level profile, using the simple case of two beams.

25.1.1 Addition of two plane waves of the same frequency

Suppose two plane waves of the same vacuum wavelength λ and frequency υ but of different amplitudes and phases illuminate a flat surface in a medium of refractive index n as shown in Figure 25.1a. Along this plane, at a distance x from the arbitrary origin and at a time t, the amplitudes are

$$E_1(x, t) = E_{01} \cos(k_1 x + \omega t + \alpha_1) \tag{25.1a}$$

and

$$E_2(x, t) = E_{02} \cos(k_2 x + \omega t + \alpha_2) \tag{25.1b}$$

where

E_{01} and E_{02} are taken as always positive,

$$k_1 = (2\pi n/\lambda) \sin(\gamma_1) \text{ and } k_2 = (2\pi n/\lambda) \sin(\gamma_2) \tag{25.2a}$$

γ_1, γ_2 are the angles of inclination of the beams to the normal

$$\omega = 2\pi \upsilon \tag{25.2b}$$

and

α_1 and α_2 are arbitrary phase factors.

If these two disturbances interfere at the point $\mathcal{B}(x)$, then according to the principle of superposition, the resulting disturbance is

$$E(x, t) = E_1(x, t) + E_2(x, t)$$

To perform this sum, it is easier to use the complex notation introduced in Section 1.1. In the complex notation we would write

$$E_1(x, t) = E_{01} e^{i(k_1 x + \omega t + \alpha_1)} \tag{25.3a}$$

and

$$E_2(x, t) = E_{02} e^{i(k_2 x + \omega t + \alpha_2)} \tag{25.3b}$$

In this notation, the two amplitudes $E_1(x, t)$ and $E_2(x, t)$ are vectors in complex space with the real components being

$$E_{01} \cos(k_1 x + \omega t + \alpha_1) \quad \text{and} \quad E_{02} \cos(k_2 x + \omega t + \alpha_2) \tag{25.4a}$$

and the imaginary components being

$$E_{01} \sin(k_1 x + \omega t + \alpha_1) \quad \text{and} \quad E_{02} sin(k_2 x + \omega t + \alpha_2) \tag{25.4b}$$

We can draw these components on a vector diagram as shown in Figure 25.2a and add them using the rules of vector addition. In this diagram, E_{01} and E_{02} are the lengths of the vectors and the real (horizontal) components are given by equations (25.4a) and the imaginary (vertical) components are given by equations (25.4b). Using the rules of vector addition, their sum is shown in Figure 25.2b and we can write the sum as

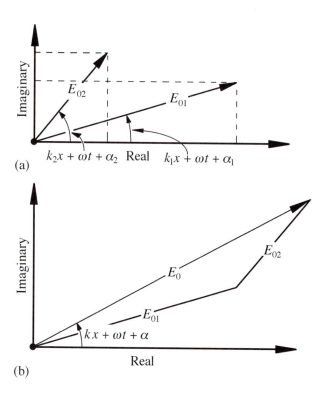

$$E(x, t) = E_{01}e^{i(k_1x+\omega t+\alpha_1)} + E_{02}e^{i(k_2x+\omega t+\alpha_2)} \tag{25.5}$$

That is

$$E(x, t) = e^{(i\omega t)}[E_{01}e^{i(k_1x+\alpha_1)} + E_{02}e^{i(k_2x+\alpha_2)}] \tag{25.6}$$

Light level profile of the fringes

When two plane wave beams interfere, the complex amplitude of the interference pattern is sinusoidal and this has negative and positive values. However, we only observe the light level (illuminance or luminance), which must always be positive. Therefore the light level distribution in these sinusoids is always positive. We have shown in Section 1.1 that the observed intensity, as opposed to the amplitude, is the product of the complex disturbance $E(x, t)$ and its complex conjugate $E^*(x, t)$. That is, the observed intensity $L(x)$ (that is say a luminance) is

$$L(x) = E(x, t)E^*(x, t)$$

Thus

$$L(x) = \{e^{(i\omega t)}[E_{01}e^{i(k_1x+\alpha_1)} + E_{02}e^{i(k_2x+\alpha_2)}]\}\{e^{(-i\omega t)}[E_{01}e^{-i(k_1x+\alpha_1)}$$

$$+ E_{02}e^{-i(k_2x+\alpha_2)}]\}$$

which can be reduced to

$$L(x) = E_{01}^2 + E_{02}^2 + E_{01}E_{02}[e^{i(k_1x+\alpha_1)-i(k_2x+\alpha_2)}]$$

$$+ E_{01}E_{02}[e^{-i(k_1x+\alpha_1)+i(k_2x+\alpha_2)}]$$

We can simplify this equation further. Noting that the righthand side is no longer a function of t, we can write

$$L(x) = E_{01}^2 + E_{02}^2 + 2E_{01}E_{02}\cos[(k_1 - k_2)x + (\alpha_1 - \alpha_2)]$$

which can be written in the form

$$L(x) = E_{01}^2 + E_{02}^2 + 2E_{01}E_{02}\{2\cos^2[(k_1 - k_2)x/2$$

$$+ (\alpha_1 - \alpha_2)/2] - 1\} \tag{25.7}$$

which shows that the light level $L(x)$ is always positive and expresses the light level as a function of the distance x and is sinusoidal (or cosinusoidal) for a two beam interferometer, as indicated in Figure 25.1b.

25.1.2 *Multiple beam interferometers*

In multiple beam interferometers, the interference pattern is more complex and the illuminance or luminance profile depends upon the number of beams, their relative amplitude and phase differences. Perhaps the most important example of this type is the Fabry–Perot interferometer (Born and Wolf 1989). However, to date, multiple beam interferometers have little application in vision and visual instruments, except for the potential use of diffraction gratings in white light interferometers.

25.2 Coherence

The most important requirement for the formation of interference fringes is that the interfering beams be **mutually coherent**. Before we define coherence we need to give some background to the random production of photons and its effect on the stability of interference fringes. A wavetrain of light is not a continuous and perfect sine wave as indicated by equation (25.1a or b). Because light is emitted in discrete energy packets or photons which must have a finite emission time and hence length, and because successive photons are usually emitted randomly in time, a wavetrain consists of a stream of photons random in phase. That is, if we use equations of the type (25.1a or b), to describe the wave motion, the numerical values of α_1 and α_2 in the preceding equations undergo random changes. If we look at equation (25.7), we see that the position of the interference fringes depends upon the phase difference

$$\alpha_1 - \alpha_2$$

of the two beams. If this phase difference undergoes a change, the fringes move sideways by an amount that depends upon the magnitude of the phase change.

Now the temporal length of photons is about 10^{-8} s and the minimum temporal resolution of the eye is of the order of about 10^{-1}s. Therefore the fringes will chance position about 10^7 times over this period. Therefore they will appear to be blurred and hence have reduced contrast. As the level of the phase changes increases, the fringes move more with a resulting decrease in fringe contrast. The **degree of coherence** is a measure of constancy of the phase difference between the two beams and since the amount of the changes in the phase difference correlates with the fringe contrast, we could use the fringe contrast, which can be measured, to determine the degree of coherence. Fringe contrast can be defined as

$$\text{contrast} = \frac{L_{max} - L_{min}}{L_{max} + L_{min}} \tag{25.8}$$

where L_{max} and L_{min} are maximum and minimum light levels, respectively, in the fringe pattern and thus contrast has values on a scale from 0 to 1. For two disturbances of equal amplitude and a constant phase difference, the fringes have a maximum contrast of unity. This degree of coherence would also be unity. At the other extreme, if changes in the phase difference made the fringes move by many periods, the observed fringe contrast would be zero, the two disturbances would be totally incoherent and therefore the degree of coherence would be zero.

Now given two disturbances or two beams that we wish to use to form interference patterns, we need to know how we can produce two mutually coherent beams. If the two beams were derived from separate sources, there would be no way of ensuring that the phase difference was stable, but if the two beams were derived from the same source and identical in all respects, the above phase difference would be constant and zero. However, while the two disturbances must be derived from the same source, they may not be exactly identical. For example one beam may have been emitted from the source at a different time or may have arisen from a different part of the source. In these cases, the phase difference may not be constant with time, but may not vary by a large amount.

Thus the beams that we use to form interference fringes must be coherent and we say they must be mutually coherent. Apart from the requirement of zero or minimum change in phase difference, the two beams must have the same wavelength.

25.2.1 *Production of two mutually coherent beams*

The simplest method of producing two mutually coherent beams is to take light from a single source and divide it into two beams of roughly equal amplitude and having almost the same optical path length to the plane of interference and mutually inclined to produce a continuous phase difference across the plane of observation. Beams from two different sources are not sufficiently coherent to form observable fringes.

There are two principal methods of producing two mutually coherent beams from a single source:

(a)   division of wavefront and
(b)   division of amplitude

Fig. 25.3: Two examples
of division of wavefront
interferometers. A third is
Young's interferometer
shown in Figure 25.6.

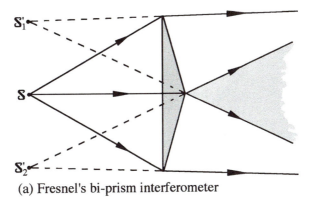

(a) Fresnel's bi-prism interferometer

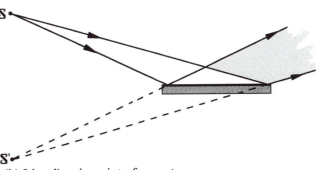

(b) Lloyd's mirror interferometer

These can be explained by the use of diagrams and examples.

Two examples of division of wavefront are shown in Figure 25.3. In the Fresnel biprism, the prism forms two separate but identical images, s_1' and s_2', of the source s. In the Lloyds mirror, the source s appears to interfere with a mirror image s' of itself. A third is Young's double slit interferometer, shown in Figure 25.6. In this interferometer, light from a slit source illuminates two slits s_1 and s_2. Light passes through the slits and diffracts. In these cases, wherever the two beams overlap (that is the shaded region), there will be an interference pattern, but not necessarily with the same fringe contrast.

Two examples of division of amplitude interferometers are shown in Figure 25.4. In these cases, a beam is split into two beams of more or less equal amplitude and hence the term "division of amplitude". In the interference plane, these two beams interfere with each other. In these examples, there may be a path difference between the two halves of the divided beam. A third configuration is the Twyman–Green interferometer shown in Figure 25.7.

However, while the beams must be coherent and therefore have to be taken from the same source, coherence is a complex phenomenon. In the general sense, no two beams are perfectly coherent; instead the coherence between them is only partial and the degree of coherence depends upon which parts of the two beams are interfering with each other. This complexity leads to the identification of two types of coherence – spatial and temporal.

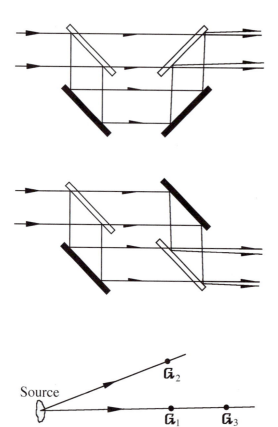

Fig. 25.4: Two examples of division of amplitude interferometers. A third is the Twyman–Green interferometer, shown in Figure 25.7.

Fig. 25.5: Spatial and temporal coherence.

Spatial and temporal coherence

Let us consider the situation shown in Figure 25.5. Two general rays emitted by the source are shown. In the division of wavefront interferometers, light at point G_1 on one ray will be made to interfere with light at point G_2 on the second ray. Now the degree of coherence between these two points depends upon the distance between the two points, their distance from the source and the size of the source. This type of coherence is called **spatial coherence**. If the source size were zero, it would have perfect spatial coherence, irrespective of the separation and distance of G_1 and G_2.

If we now consider interferometry by division of amplitude, we are usually interfering disturbances at different points on the same ray in the beam, for example the points G_1 and G_3 in Figure 25.5. In this case, the degree of coherence depends upon the path difference between G_1 and G_3 and the **spectral band-width** of the source. No source is perfectly monochromatic; it has a finite range of wavelengths. This range of wavelengths is known as the spectral band-width and the finite band-width is due to the fact that photons have a finite length; the longer a photon, the narrower the band-width.

This type of coherence is called **temporal coherence**. The maximum optical path difference at which fringes are observable is called the **coherence length**, which is inversely proportional to the band-width and may be thought of as a measure of the length of the photons in the beam. Instead of using coherence

Table 25.1. Coherence lengths for some spectral sources

Source	Line wavelength (nm)	Coherence length
Argon	488.0	1.5 cm
Krypton	605.8	21 cm
Mercury	546.1(high pressure)	8 μm
Neon	632.8	6.4 cm

Note: The above values are only approximate because they depend upon a number of factors such as temperature, which affects the Doppler width and pressure.

length to describe the coherence properties of a source, we can use **coherence time**, which is the coherence length divided by the speed of light in the medium. The wider the band-width, the lower the degree of coherence. The coherence length is of the order of millimetres for spectral line sources such as sodium and mercury discharge lamps. Stabilized lasers have very narrow band-widths and therefore can have long coherence lengths, of the order of metres.

25.2.2 Coherent and incoherent sources

We have argued that for interference to take place we must have two mutually coherent and monochromatic beams and have argued that the degree of monochromacy or spectral band-width is related to the temporal coherence of the source. In practice no light source is perfectly monochromatic or perfectly coherent, but a few are sufficiently monochromatic, that is have a sufficiently narrow range of wavelengths or band-width. These sources fall into two distinct classes: (a) those we will call spectral light sources and (b) lasers.

25.2.2.1 Spectral light sources

Spectral light sources such as the sodium and mercury lamps can be used as light sources in interferometers. These consist of a volume of gas emitting radiation. Since the molecules emit their photons independently of each other these are spatially incoherent across the source. Because of this spatial incoherence, the source is usually stopped down so that only a small portion is visible. However, they have some temporal coherence and hence coherence length which will depend upon the band-width of the particular line of the source. We have listed a few spectral sources and typical coherence lengths in Table 25.1.

25.2.2.2 Lasers

Compared with the above spectral light sources, lasers have very high temporal coherence and hence very long coherence lengths. This is because the individual molecules or atoms are driven to emit their photons in phase with each other by a process called stimulated emission. Because of this high temporal coherence, lasers are known as coherent sources. They are almost point sources so that

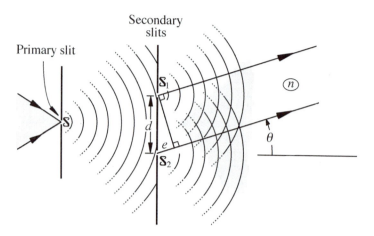

spatial coherence is also very high. Depending upon their construction, they
may have coherence lengths of kilometres.

25.3   Young's double slit interferometer

The construction of Young's double slit interferometer is shown schematically
in Figure 25.6. Light from a source falls on the primary slit s. The light passing
through the slit diffracts and some of this light illuminates the secondary slits
s_1 and s_2. The light passing through these slits also diffracts. In the region of
overlap, interference fringes are formed. The fringes may be formed on a screen
at a finite distance or formed at infinity as implied by the parallel rays shown
in the diagram. This is an interferometer based upon the division of wavefront
and therefore the fringe contrast depends on spatial coherence between the
disturbances at the secondary slits.

25.3.1   Spatial frequency of the fringe pattern (far field)

Let us assume that we observe the fringes visually by placing the eye close to
the secondary slits. If the eye uses relaxed accommodation, the fringe pattern
observed will be that formed at infinity. In this case, an equation for fringe
frequency can easily be derived. Figure 25.6 shows two slits illuminated by a
primary slit and the distances between the primary slit and each of the secondary
slits are not usually exactly equal. This difference in path length will give rise
to a phase difference between the disturbances at the secondary slits, but this
phase difference should be stable if we have sufficiently high spatial coherence.
In the following discussion of the properties of Young's interferometers, we
will assume that this phase difference is zero. Now consider two rays inclined
at an angle θ to the axis as shown in the diagram. The optical path difference
(e) between these two rays will be

$$e = nd \sin(\theta) \tag{25.9a}$$

where d is the separation of the slits. Usually, θ will be small and thus one can write

$$e = nd\theta \qquad (25.9b)$$

The maxima will occur every time this path difference varies by one wavelength λ, that is

$$nd\theta_m = m\lambda \qquad (25.10)$$

where n is the refractive index of the medium and $m = 0, \pm 1, \pm 2$, and so on, is the order of the fringe and θ_m is the direction of the fringe. The angular period of the fringe p_a will be

$$p_a = \theta_m - \theta_{m-1} = m\lambda/(nd) - (m - 1)\lambda/(nd)$$

that is

$$p_a = \lambda/(nd) \quad \text{rad} \qquad (25.11)$$

and the corresponding angular spatial frequency σ_a defined as

$$\sigma_a = 1/p_a \quad \text{c/rad} \qquad (25.12)$$

will be given by the equation

$$\sigma_a = nd/\lambda \quad \text{c/rad}(\theta \text{small}) \qquad (25.13)$$

Therefore the spatial frequency of the fringe system depends upon the refractive index of the medium, separation d of the two secondary slits and the wavelength λ.

25.3.2 Fringe contrast and size of field

While the above equations show that the fringe spatial frequency can be changed by changing the slit separation d, changing this separation will also change the fringe contrast, because this interferometer depends upon spatial coherence, which depends upon (a) the slit separation and (b) the width of the primary slit. Increasing the secondary slit separation or increasing the primary slit width decreases the degree of spatial coherence between s_1 and s_2 and hence fringe contrast. Therefore, the degree of spatial coherence between the secondary slits can be increased by decreasing the primary slit width, but this reduces the fringe brightness.

In this interferometer, the extent of the interference fringes depends upon the width w of the secondary slits, due to diffraction effects. The interference fringes due to interference between the two slits are enveloped by the spatial intensity distribution arising from the light diffracted by each slit and interfering with itself, that is the single slit diffraction pattern, which is discussed in Chapter 26.

The normalized far-field intensity pattern $I(\theta)$ due to light diffracted from a slit of width w is, from equation (26.37),

$$I(\theta) = [\sin(\zeta)/(\zeta)]^2 \qquad\qquad (25.14)$$

where

$$\zeta = \pi n \sin(\theta) w / \lambda$$

But for the small angles occurring here, we can write

$$\zeta = \pi n \theta w / \lambda \qquad\qquad (25.14a)$$

This function is plotted in Figure 26.12. The width of this envelope can be defined as the width to the first zero, which occurs at

$$\zeta = \pi n \theta w / \lambda = \pi$$

that is

$$\theta = \lambda / (nw)$$

and thus

$$\text{width} = 2\theta = 2\lambda / (nw) \quad \text{rad} \qquad\qquad (25.15)$$

Example 25.1: Calculate the slit separation if the observed fringes of a Young's interferometer are to have a spatial frequency of 20 c/deg in air for a wavelength of 555 nm. Also calculate the width of the fringe pattern if the secondary slits have a width of 0.1 mm.

Solution: Now 20 c/deg corresponds to 1146 c/rad. Using this value for σ_a in equation (25.13), we have

$$d = 1146 \times 555 \times 10^{-9} = 0.000636 \,\text{m} = 0.636 \,\text{mm}$$

From equation (25.15), the extent of the fringe pattern has a width of

$$2 \times 555 \times 10^{-6} / 0.1 = 0.0111 \,\text{rad} = 38.2 \,\text{minutes of arc}$$

Now the angular period of the fringes, given by equation (25.12), is

$$p_a = 1/1146 = 0.000873 \,\text{rad}$$

and thus the above width of the fringe pattern is equivalent to about 13 fringes.

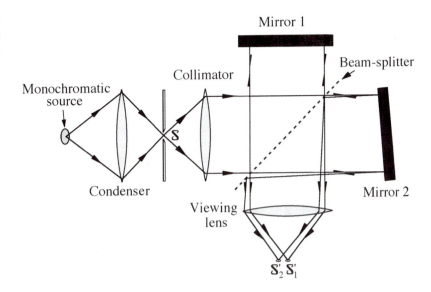

25.3.3 Production of a Young's interferometer

A very simple Young's interferometer can be made photographically. Black lines of suitable width and separation are drawn on a white background and photographically reduced onto high contrast negative film. The pieces of film can be mounted and held together several centimetres apart with the slits or lines carefully aligned.

25.4 The Twyman–Green interferometer

The Twyman–Green interferometer is one of the most common interferometers that uses division of amplitude, and its schematic construction is shown in Figure 25.7. To achieve adequate coherence between the two beams using an incoherent source, the effective source has to be small. In this case the effective source is a pinhole that has been illuminated by a spectral source such as a mercury lamp with filters to eliminate the unwanted spectral lines. The smaller the pinhole the greater the coherence but the lower the brightness of the interference fringes. Therefore a compromise has to be set between coherence and image brightness. Typical diameters are about 0.5 to 1 mm for a collimator focal length of about 25 cm. The actual source is imaged by a condenser onto the pinhole, which is at the front focal point of the collimating lens. The collimated beam is divided into two beams of equal amplitude by the beam-splitter, which reflects and transmits close to 50% of the incident light. One beam is reflected to the mirror 1, the other transmitted to the mirror 2. The two beams are reflected from the mirrors and after being reflected or transmitted by the beam-splitter again, part of the beams pass through the de-collimating lens. The de-collimator focusses the two beams onto its back focal plane as shown in the diagram, and the images formed are images s_1' and s_2' of the source pinhole.

   If the two beams have a relative tilt as shown in Figure 25.8, the two source images are imaged side by side, by a separation proportional to the amount of tilt. Now if one of the mirrors is tilted through an angle $\Delta \gamma / 2$, the two beams

Field of view

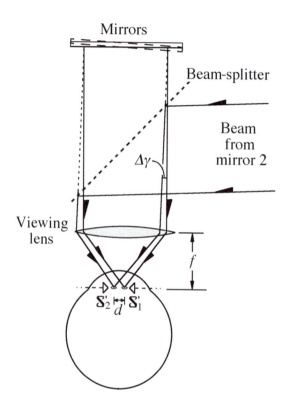

Fig. 25.8: Visual
observation of the fringes in
the Twyman–Green
interferometer.

have a relative tilt of $\Delta\gamma$ and the two source images s'_1 and s'_2 are separated by
a distance d given by the equation

$$d = f\,\Delta\gamma \qquad\qquad (25.16)$$

where f is the focal length of the de-collimating lens.

The fringes may be observed visually and to do this, observers place an eye
so that the pinhole images are co-incident with the eye pupil, as shown in Figure
25.8. This is equivalent to a **Maxwellian** view. Maxwellian view is a viewing
arrangement in which the, usually small, illuminating source is imaged in the
plane of the observer's pupil, giving a usually wide and uniformly bright field-
of-view. This particular viewing arrangement is discussed further in Chapter
36. Fringes are formed on the retina but can be thought of as initially being
formed in the vicinity of the mirrors. In quantitative terms they are formed, but
virtually, in a plane conjugate to the retina, a plane which lies in the vicinity

of the mirrors. If the mirrors lie close to the front focal plane and within the focal length of the de-collimating or viewing lens, the image of the fringes will be formed at infinity or closer and therefore easily viewed with relaxed or near relaxed accommodation.

25.4.1 Fringe formation

Fringe formation requires one of the mirrors to be tilted. If the mirrors are perfectly flat, the fringes are parallel, straight, equally spaced and formed in the region of the mirrors. A typical fringe pattern is shown in the inset of Figure 25.8. These fringes correspond to contours of equal optical path difference, with the separation of two adjacent fringes being equivalent to a difference of $\lambda/2$ in optical path difference.

25.4.2 Fringe frequency

In the Twyman–Green interferometer, the two collimated beams have a relative tilt and interfere as shown in Figure 25.8, in a plane close to the two mirrors and with one of the beams tilted relative to the other. Let us suppose that we are observing the interference fringes in a plane perpendicular to the axis of the system. In this case, the situation is as depicted in Figure 25.1, but with $\gamma_1 = \Delta\gamma$ and $\gamma_2 = 0$. The light intensity across the fringe pattern is given by equation (25.7). This equation shows that fringe maxima occur when

$$2\cos^2[(k_1 - k_2)x/2 + (\alpha_1 - \alpha_2)/2] - 1 = +1$$

where

$$k_1 = (2\pi n/\lambda)\sin(\Delta\gamma) \quad \text{and} \quad k_2 = 0$$

That is

$$\cos[(k_1 x/2) + (\alpha_1 - \alpha_2)/2] = \pm 1$$

or

$$(k_1 x/2) + (\alpha_1 - \alpha_2)/2 = 0, \pm\pi, \pm 2\pi, \cdots$$

The period p is the distance between successive values of x for these maxima and it follows that

$$p = \frac{\lambda}{n\sin(\Delta\gamma)} \tag{25.17}$$

Now in visual interferometers, the angles will be small enough to allow replacement of the sine of the angle by the angle itself and thus we can write

$$p = \frac{\lambda}{n\Delta\gamma} \tag{25.18}$$

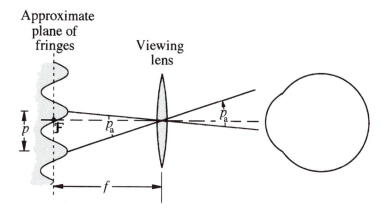

If these fringes are now viewed through a lens with the focal length f chosen so that the fringe plane is in the front focal plane at $\mathcal{F}$ of this lens, as shown in Figure 25.9, the angular period p_a is simply

$$p_a = p/f \qquad (25.19)$$

and thus the angular spatial frequency σ_a, defined as

$$\sigma_a = 1/p_a$$

is

$$\sigma_a = f/p$$

which, using equation (25.18), can be expressed in the form

$$\sigma_a = \frac{f n \Delta \gamma}{\lambda} \quad \text{c/rad} \qquad (25.20)$$

Example 25.2: If the focal length of the de-collimation (viewing) lens is 25 cm, what is the angular separation of the two beams in air, for a fringe frequency of 30 c/deg using light of wavelength 632.8 nm?

Solution: Here, $n = 1$, $\sigma_a = 30\,\text{c/deg} = 30 \times 180/\pi\,\text{c/rad} = 1719\,\text{c/rad}$, $\lambda = 632.8 \times 10^{-9}$ m and $f = 0.25$ m. Thus from equation (25.20)

$$\Delta \gamma = \frac{1719 \times 632.8 \times 10^{-9}}{0.25}$$

$$= 0.00435 \text{ rad} = 0.249°$$

This example and Exercise 25.3 at the end of this chapter confirm that the angular tilt of the two beams must be small if one is to visually observe the fringes.

If we now replace $\Delta\gamma$ in equation (25.20) by the source image separation d, using equation (25.16), we can express the spatial frequency σ_a in terms of this separation by the equation

$$\sigma_a = nd/\lambda \quad \text{c/rad} \tag{25.21}$$

which is the same as equation (25.13) for Young's interferometer.

25.4.3 Fringe localization

The fringes are formed in a volume of space defined by the region of overlap of the two beams and centered on the region of maximum overlap of the two beams. The degree of coherence will vary across this volume and therefore so will the fringe contrast. The plane of localization of the observed fringes will be in the region of maximum fringe contrast.

25.4.4 The adjustment of fringe contrast

The fringes have maximum contrast when

(a)   the optical path difference between the two beams is zero; that is the mirrors are the same optical path distance from the source and
(b)   the source is as small as possible while maintaining adequate fringe brightness.

These conditions are usually easily achieved. Sometimes variable contrast fringes may be desired. This variability can be achieved by changing the optical path distance between the two beams, which alters the degree of temporal coherence. As the optical path difference approaches the coherence length of the source, the fringe contrast will approach zero.

25.4.5 Aberrations of the interferometer lenses

The beam collimated by the collimator should be free from aberrations. Since the source is small and on the axis, the main aberration to consider is spherical aberration. This can be eliminated by the use of a suitably designed achromatic doublet.

25.5   Effect of ocular aberrations

It is commonly accepted that aberrations of the eye do not affect the contrast of the fringes formed on the retina in the Twyman–Green interferometer. This is because the light passing into the eye passes through the images of the sources which are small compared to the pupil size of the eye. The diameters are usually less than 1 millimetre and such narrow beams do not suffer significant aberration. The only effect of any ocular aberration may be to induce a phase difference or extra optical path length difference between the two beams since the two beams enter the eye through different parts of the pupil. The effect of this phase difference would be merely to cause a sideways shift of the fringes which would not normally be detectable.

25.6   White light interferometers

Interferometers can be used with white or polychromatic light and a fringe pattern may be observed. Each wavelength in the beam will produce its own set of interference fringes with its own spatial frequency or period. Having different spatial frequencies, the continuum of patterns will prevent any clear maxima or minima being formed. Therefore it is unlikely that a clear fringe pattern will be observed. However, sometimes fringes may be observed with white light, but not usually with maximum contrast.

There is one fringe maxima, which is in the same position for all wavelengths. This is the fringe in which the optical path length from the source to the fringe is the same for both beams. Thus the path difference is zero for all wavelengths and therefore there is constructive interference along the fringe. This fringe is known as the **zero order fringe** and is readily observed in Young's interferometer. If the secondary slits are equi-distant from the primary slit then the zero order fringe occurs for $m = 0$ in equation (25.10) and this white light fringe is formed in the centre of the fringe pattern. In the Twyman–Green interferometer the white light or zero order fringe is formed at the line of intersection of the mirrors shown in Figure 25.8. Whether this zero order fringe appears in the field-of-view depends upon the position of this intersection line. For example, in the situation shown in the diagram, the intersection line is in the field-of-view, but if one of the mirrors were moved backwards or forwards enough, the intersection line would be off the mirrors and therefore outside the beam. In practice, the Twyman–Green arrangement shown in Figure 25.7 needs a small modification before the white light fringe can be observed. A plate identical in material and thickness to the beam-splitter must be inserted into the beam that only passes once through the beam-splitter in order to compensate for the chromatic dispersive effect of the beam-splitter. This plate is usually known as a **compensating plate**.

There have been some attempts to design interferometers that produce extended white light fringes such that spatial frequency and position of the fringe pattern is the same for all wavelengths. We can use equation (25.21) to establish a condition for the formation of white light fringes. If the fringe frequency is to be the same for all wavelengths, then the separation d of the source images in the plane of the eye pupil must vary with wavelength and do so according to the equation

$$d(\lambda) = \lambda \sigma_a / n \tag{25.22}$$

where σ_a is the desired angular spatial frequency of the fringes. Such a situation can be achieved and one example is as shown in Figure 25.10, in which a collimated white light beam is passed through a **diffraction grating** and then through a de-collimating lens and one of the diffracted orders formed in the back focal plane of this lens becomes the light source for the Lloyds mirror shown in Figure 25.3. The principle of the diffraction grating is explained in the next chapter. It produces a transverse spectrum of the light source which is formed in the back focal plane as shown in the diagram. The diffraction grating

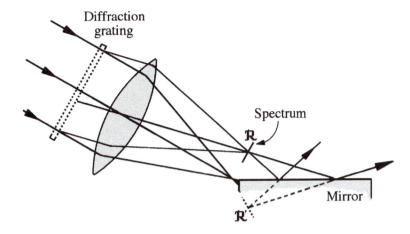

Fig. 25.10: Modified Lloyds mirror, using a diffraction grating that will give white light fringes.

has the property that for small angles of diffraction, the transverse shift for each colour is proportional to the wavelength and thus satisfies equation (25.22). However without some modification, it is not suitable as a visual instrument because the spectra cannot be made coincident with the entrance pupil of the eye. An interferometer based on the principle and designed for visual work has been suggested by Lotmar (1980).

25.6.1   *The Lotmar white light interferometer*

Lotmar (1980) described a method of producing white light or polychromatic fringes using the moire fringes generated by two identical diffraction gratings in contact with each other, with one rotated about the optical axis relative to the other. This device produces diffraction patterns that consist of an array of lines with each line containing the spectrum of the light source. These lines become sources which are mutually coherent. If the separation of the source position of each wavelength satisfies equation (25.22), then the fringe frequency will be the same for all wavelengths. If two neighbouring lines of the same order are selected by a suitable mask, they will interfere to produce a set of achromatic fringes. The fringes are due to the diffraction from the moire pattern of the tilted gratings and the fringe frequency thus depends upon the angle of tilt of the gratings.

It would appear, however, that this type of interferometer, unlike the monochromatic types, is sensitive to the aberrations of the eye. Thibos (1990) has shown that if there is lateral displacement of the two beams in the pupil, the chromatic aberration of the eye leads to a significant loss of contrast of the white light fringes.

25.7   Angular spatial frequency limit

There is a maximum limit to the angular spatial frequency that can be visually observed in these interferometers. We have seen that the angular spatial frequency is directly proportional to the physical separation d of the two interfering sources or their images at or near the entrance pupil of the eye and the actual relationship is given by equation (25.13) or (25.21). This distance d

cannot be greater than the pupil diameter and therefore the pupil diameter sets an upper limit to the angular spatial frequency for any particular wavelength. Since the maximum pupil diameter is about 8 mm and less for some people, the above equations show that the maximum angular spatial frequency that can be imaged on the retina for light of wavelength 555 nm is about 252 c/deg. We should note that this is well above the visually resolvable limit of about 60 c/deg.

Exercises and problems

25.1 For a two slit interferometer (Young's slits) in air and using light of wavelength 589 nm, calculate the slit separation if the directly visually observed fringes have a spatial frequency of 20 c/deg.

ANSWER: 0.67 mm

25.2 Calculate the maximum inclination angle of the mirrors of the Twyman–Green interferometer, if the observed fringes in air have a spatial frequency of 60 c/deg which is the upper limit of human spatial vision, under the following conditions. The fringes are formed in the plane of the mirrors, which are placed 20 cm from a de-collimating lens with a power of $5\,\mathrm{m}^{-1}$, and the light has a wavelength of 560 nm.

ANSWER: 0.0048 rad

25.3 Calculate the maximum fringe frequency possible in a Twyman–Green interferometer for a 2 mm diameter pupil. Assume a wavelength of 500 nm and an air medium.

ANSWER: 59 c/deg

Summary of main symbols

e	path difference
f	equivalent focal length of de-collimating (viewing) lens in the Twyman–Green interferometer
x	distance along observation plane
p	fringe period
p_a	angular fringe period
σ_a	angular spatial frequency of fringes
$\Delta\gamma$	angle between the two beams of the Twyman-Green interferometer (small)
m	integer number

References and bibliography

*Born M. and Wolf E. (1989). *Principles of Optics*, 6th (corrected) ed. Pergamon Press, Oxford.

*Campbell F.W. and Green D.G. (1965). Optical and retinal factors affecting visual resolution. *J. Physiol.* 181, 576–593.

Geddes L.A., Patel B.J., and Bradley A. (1990). Comparison of Snellen and interferometry visual acuity in an aging non-cataractous population. *Optom. Vis. Sci.* 67(5), 361–365.

Lotmar W. (1972). Use of moire fringes for testing visual acuity of the retina. *Appl. Opt.* 11, 1266–1268.

*Lotmar W. (1980). Apparatus for the measurement of retinal visual acuity by moire fringes. *Invest. Ophthal. Vis. Sci.* 19, 393–400.

*Thibos L.N. (1990). Optical limitations of the Maxwellian view interferometer. *Appl. Opt.* 29(10), 1411–1419.

Westheimer G. (1966). The Maxwellian view. *Vis. Res.* 6, 669–682.

26

Diffraction and diffractive devices

26.0   Introduction

Along with interference, diffraction is a manifestation of the wave nature of light and cannot be explained using geometrical optics or ray theory alone. In this chapter, we will explore the nature of diffraction and some applications to visual optical instrumentation, such as the Fresnel zone plate, speckle patterns and Fraunhofer diffraction. In order to explain the theory of these processes fully, we will need to introduce certain aspects of diffraction theory and important equations. The development of some of these equations will be beyond the scope of this book and therefore will be taken from other texts. Unless otherwise stated, the principal source of these equations and the historical development of diffraction will be Born and Wolf (1989).

26.1   The nature and cause of diffraction

In a number of situations, it can be observed that when a wave motion passes through an aperture, the waves bend around the edge of the aperture. For example, sea waves on entering an enclosed harbour will bend around the sea wall. Other examples are sound travelling around corners and a number of observable phenomena in the propagation of light. This effect is called diffraction and returning to the above example of sea waves entering a quiet harbour, as shown in Figure 26.1, let us look at what would happen if diffraction did not take place. Let us assume that the waves entering the harbour are parallel. They would remain parallel with a sharp edge at the boundary with the quiet water, producing a vertical wall of water with a sinusoidal profile at the edge as shown in the diagram. This of course is not possible. If such a wall of water did try to develop, gravity would force water to flow one way or the other to even out the edge gradient and this would cause the wave motion to move into the quiet section.

  According to Born and Wolf (1989), diffraction effects were known to Leonardo da Vinci (1452–1519) but not accurately described until the mid-seventeenth century. An explanation of diffraction took another 150 years or so to be developed. The slow development was due to the traditional belief that light was made up of particles and it took many years for the wave theory to be accepted. Without this wave theory, diffraction could not be explained.

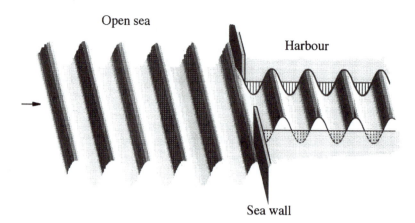

Open sea

Harbour

Sea wall

Fig. 26.1: Water waves passing through an opening in a sea wall into a quiet region without any diffraction.

Some of the earlier contributors to its development were Huygens (1629–1695), Young (1773–1829), Fresnel (1788–1827) and Kirchhoff (1832–1887). Young in 1802 tried to explain diffraction at an edge as due to an interference effect between the unobstructed wave and a wave reflected from the edge. This theory was presented in only a qualitative form and thus was not readily accepted. Young's theory became known as "the boundary diffracted wave" or the "boundary wave" theory. While Young's presentation was only qualitative it has some historical interest and can be quantitatively analysed.

26.1.1 Diffraction at an edge – Young's boundary wave

To examine the "boundary wave" explanation, let us consider a simple situation as shown in Figure 26.2a. Here a collimated beam of monochromatic light with plane wavefronts is passing an obstruction parallel to the wavefronts and illuminating a screen, also parallel to the wavefronts, some distance behind the obstruction. If one measures the light level across the screen, the relative light level has the profile shown in Figure 26.2b.

According to the boundary wave model, the fringes formed in the region illuminated by the plane waves are due to the interference between the plane waves and boundary waves originating at the edge of the obstruction and the phase difference between the two increases with distance from the shadow edge. The amplitude of the fringes decreases with distance from the shadow edge because the boundary wave weakens with distance from its source and thus the interference progressively weakens with distance from the shadow edge.

The lack of a plane wave in the geometric shadow is the reason why there are no fringes in this region. Only one wave, the boundary wave, illuminates this region and because it weakens with distance, the light level decreases with distance from the shadow edge.

Let us assume this is the correct explanation and find the position of the maxima. Let us also assume (an assumption that is not correct) that the plane waves and the boundary wave are in phase at the edge of the obstruction. The optical path difference e between the two waves at any point $\mathcal{B}$ on the observation screen at a distance y from the shadow edge, shown in Figure 26.2a, is

$$e = n'[\sqrt{(y^2 + d^2)} - d] \tag{26.1}$$

Fig. 26.2: The diffraction
at an edge and Young's
boundary wave.

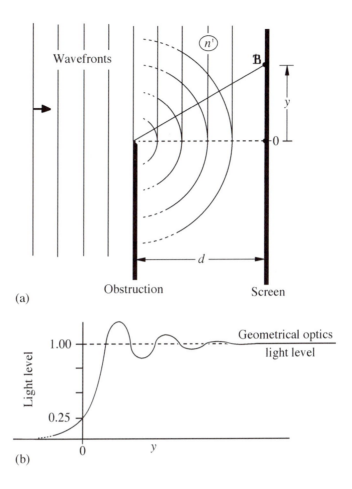

Wavefronts

(n')

B

y

0

d

Obstruction

Screen

(a)

Geometrical optics
light level

1.00

Light level

0.25

0

y

(b)

If $y \ll d$, then we can expand the square root using the binomial theorem and simplify to get

$$e = n'y^2/(2d)$$

Interference maxima only occur when this path difference is a whole number of wavelengths, that is

$$e = m\lambda \tag{26.2}$$

Thus we get maxima when

$$y = \sqrt{(2dm\lambda/n')} \qquad m = 0, 1, 2, \ldots \tag{26.3}$$

We assumed above that the two waves were in phase at the obstruction edge. This is invalid because equation (26.1) predicts that there should be a maximum at the geometric shadow edge (that is $y = 0$) when in fact the light level value is one quarter of the geometric expected value, as shown in Figure 26.2b, and the first maxima occurs just inside the shadow edge. This difference can be explained

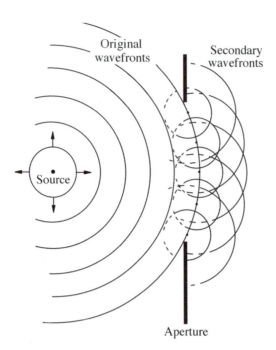

in terms of a non-zero phase difference between the two disturbances at the obstruction edge. This phase difference would shift interference pattern sideways.

Young's boundary wave had some conceptual problems, mainly the source
or mechanism of the formation of the reflected boundary wave. In opposition
to Young's theory, Fresnel explained diffraction in terms of a much earlier
suggestion by Huygens. We will look at this theory next.

26.2 Huygens' principle and the Kirchhoff integral

Huygens (1690) suggested that every point on a wave motion becomes a new
and secondary spherical source with an amplitude equal to the amplitude of the
original wave at that point. This is now known as Huygens' principle. For an
unbounded wave, the effect of these secondary waves cancels out except in the
original direction of the wave, and in that direction their cumulative effect is
identical to the original wave motion. They only become manifest near the edge
of a bounded wave, for example when the wave passes through an aperture as
shown in Figure 26.3. Fresnel explained diffraction in terms of this principle
and the interference between all of the secondary sources. This development
is known as the Huygens–Fresnel theory and has enabled the mathematical
development of the theory of diffraction and the calculation of the light level at
any point beyond the aperture due to diffraction at the aperture. The Huygens–
Fresnel theory was further developed by Kirchhoff.

26.2.1 Kirchhoff's diffraction theory

Figure 26.4 shows a point source of monochromatic radiation at s_0 emitting
spherical waves with some of the radiation passing through an aperture. Using
the Huygens–Fresnel theory, Kirchhoff showed that the amplitude A_G at the
point G is given by the following integral over the aperture

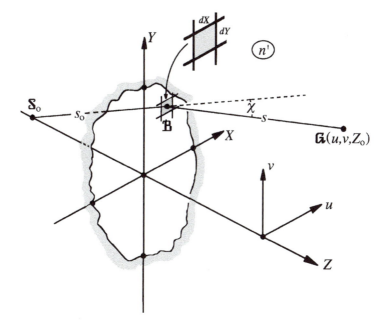

Fig. 26.4: Kirchhoff's
integral for calculating the
light level diffracted by an
aperture.

Fig. 26.4: Kirchhoff's integral for calculating the light level diffracted by an aperture.

$$A_G = -\frac{i A n' e^{(iks_0)}}{2\lambda s_0} \int \frac{e^{(iks)}}{s} [1 + \cos(\chi)] \mathrm{d}X \, \mathrm{d}Y \qquad (26.4)$$

where A is the amplitude of the disturbance a unit distance from the source and $k = 2\pi n'/\lambda$. The presence of the $[1 + \cos(\chi)]$ factor ensures that no light is diffracted directly backwards from the aperture. This integral is a simplified form of the more general Fresnel–Kirchhoff diffraction formula given by Born and Wolf (1989).

This equation allows a number of diffraction problems to be solved and we will give two useful examples shortly. However, one other interesting example is the diffraction at a straight edge. According to Born and Wolf (1989, 449), Sommerfeld in 1894 showed that diffraction of a plane wave at a straight edge was in fact identical to the combination of an uninterrupted wave and a boundary wave as suggested by Young.

26.2.2 *Fresnel and Fraunhofer diffraction*

In general, the diffraction pattern at a surface beyond an aperture or edge is known as Fresnel diffraction. For light diffracted on passing through an aperture, there is no unique Fresnel diffraction pattern because the diffraction pattern will vary with position of the observation surface. However if the beam is focussed beyond the aperture, the diffraction pattern in the plane of focus is unique and known as Fraunhofer diffraction. Fraunhofer diffraction can be defined as the diffraction effect in the focus of a beam or at infinity for a collimated beam and is therefore a special case of the more general Fresnel diffraction.

In visual optical instruments, Fraunhofer diffraction is the more important because it is relevant to the light distribution in the image surface and therefore has an important effect on image quality. We will discuss these aspects in Section 26.4. Fresnel diffraction has a minor although an important application

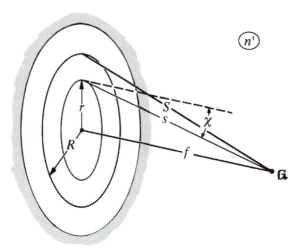

Fig. 26.5: Fresnel diffraction on the axis of a circular aperture.

in visual optical instruments. Fresnel zone plate "lenses" are now being used in vision and we will describe these in the next section. A second application of Fresnel diffraction is the speckle pattern effect seen when an expanded laser beam illuminates a rough surface.

26.3   Fresnel diffraction

Fresnel diffraction is defined as the diffraction pattern at any surface not at the focus of the beam. In any particular situation, the light distribution at a surface can be solved by evaluating Kirchhoff's integral, equation (26.4), for that case. The diffraction at an edge discussed in Section 26.1 is an example of Fresnel diffraction. Fresnel diffraction is not unique because the form of the diffraction pattern depends upon the distance of the surface from the aperture. Aspects of Fresnel diffraction of particular interest and importance are Fresnel zones and the Fresnel zone plate.

26.3.1   The Fresnel zones and zone plate

Let us consider a monochromatic beam with a plane wavefront incident on a circular aperture shown in Figure 26.5 and consider the disturbance, due to diffraction by the aperture, at the point G on the axis and which is at a distance f from the aperture. All points on the wavefront passing through the aperture contribute, by diffraction, to the disturbance at G. We can use equation (26.4) to determine the amplitude at G by all the radiation in a central zone of radius R. Since there is rotational symmetry here, we can write equation (26.4) in the polar form

$$A_G = -\frac{i A n'}{2\lambda} \int_0^R \frac{e^{(iks)}}{s}[1 + \cos(\chi)] r \, dr \, d\phi \qquad (26.5)$$

By making some assumptions, this equation can be further simplified. If the distance f is much greater than the aperture radius R, then the $\cos(\chi)$ factor does not change appreciably over the aperture. For small values of χ, we can make the approximation

$$1 + \cos(\chi) \approx 2 \tag{26.6}$$

Also while s will change, its presence in the exponential function has a much greater effect on the integral than its presence in the denominator and we can replace this occurrence by f and move it outside the integral. With these assumptions, the facts that the plane wavefront has a constant amplitude over the aperture and the integrand is circularly symmetric, we can write the above integral as

$$A_G \approx -\frac{2i\pi An'}{\lambda f} \int_0^R e^{iks} r \, dr \tag{26.7}$$

To solve this integral analytically, we need to relate r and s. From Figure 26.5, we have

$$s = \sqrt{(f^2 + r^2)} = f\sqrt{[1 + (r/f)^2]}$$

Since R is much less than f, r is also much smaller than f. Therefore we can make use of the binomial expansion to write

$$s \approx f + (0.5r^2/f)$$

If we differentiate and convert to finite differentials, we get

$$ds = r \, dr / f$$

We can use this relationship to replace the integrating variable r by s, to get

$$A_G = -\frac{2i\pi An'}{\lambda} \int_{s=f}^{s=S} e^{iks} ds \tag{26.8}$$

This integral can be solved to give

$$A_G = -A[e^{(ikS)} - e^{(ikf)}] \tag{26.9}$$

Now A_G is the amplitude of the resulting disturbance at G and, as discussed in Section 1.1, we observe the light level or intensity and not the above amplitude. The intensity is given by the product

$$\text{Intensity}_G = A_G A_G^* \tag{26.10}$$

where A_G^* is the complex conjugate of A_G. It now follows that

$$\text{Intensity}_G \approx 2A^2\{1 - \cos[k(S - f)]\} \tag{26.11}$$

Equation (26.11) shows that the intensity at G is a positive co-sinusoidal function in S. Since a cosine function varies between -1 and $+1$, the above intensity oscillates between 0 and $4A^2$. The variation in intensity is due to regular periods of alternating constructive and destructive interference. Changeovers from constructive to destructive interference, or vice versa, occur whenever the intensity reaches a maximum or a minimum, that is when

$$\cos[k(S - f)] = \pm 1$$

that is

$$k(S - f) = m\pi \quad \text{where } m = 0, 1, 2, \ldots \tag{26.12}$$

Now since $k = 2\pi n'/\lambda$, the corresponding values of S, denoted by S_m, are

$$S_m = f + [m\lambda/(2n')] \tag{26.13}$$

26.3.1.1 Fresnel zones

The boundaries defined by the above values of S_m delineate annular zones which alternate in their positive or negative contribution to the amplitude at G. So let us divide the aperture into a series of concentric circular annuli whose radii are defined by the values given by equation (26.13), as shown in Figure 26.6. The corresponding radii R_m of the zones in the aperture are given by the Pythogoras rule, that is

$$R_m = \sqrt{[S_m^2 - f^2]} \tag{26.14}$$

That is

$$R_m = \sqrt{\{\{f + [m\lambda/(2n')]\}^2 - f^2\}}$$

This reduces to

$$R_m = \sqrt{\{(fm\lambda/n') + [m\lambda/(2n')]^2\}} \tag{26.15}$$

If $[m\lambda/(2n')]^2 \ll f$, we can use the binomial theorem and some simplification to give

$$R_m \approx \sqrt{(f\lambda m/n')} \tag{26.16}$$

or

$$R_m \approx r_1\sqrt{(m)} \tag{26.17}$$

where

$$R_1 \approx \sqrt{(f\lambda/n')} \tag{26.17a}$$

Example 26.1: Calculate the radii of the boundaries between the first eight Fresnel zones for a wavelength of 589 nm and a point G

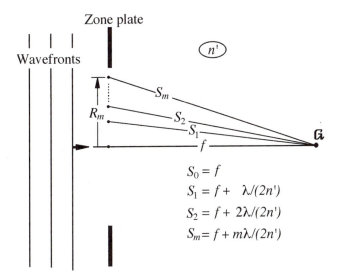

Fig. 26.6: Fresnel half-period zones of a circular aperture.

Zone plate

Wavefronts

n'

R_m

S_m

S_2

S_1

f

$\mathcal{G}$

$$S_0 = f$$
$$S_1 = f + \lambda/(2n')$$
$$S_2 = f + 2\lambda/(2n')$$
$$S_m = f + m\lambda/(2n')$$

Side view

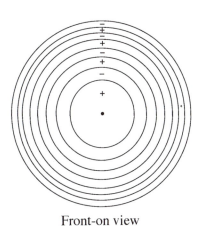

Front-on view

at a distance of 100 mm from the aperture, taking the image space medium as air.

Solution: Here we have $n' = 1.0$, $\lambda = 589$ nm and $f = 100$ mm and thus we have from equation (26.16),

$$R_m \approx \sqrt{(f\lambda m/n')} = \sqrt{(0.0589m)}$$

and we have

$$R_1 = 0.243 \text{ mm}, \quad R_2 = 0.343 \text{ mm}, \quad R_3 = 0.420 \text{ mm}$$

$$R_4 = 0.485 \text{ mm}, \quad R_5 = 0.543 \text{ mm}, \quad R_6 = 0.594 \text{ mm}$$

$$R_7 = 0.642 \text{ mm}, \quad R_8 = 0.686 \text{ mm}$$

The intensity at G due to the free wave and that due to the first zone

From the discussion following equation (26.11), the intensity at G due to each zone is

$$\text{Intensity}_G \approx 4A^2 \tag{26.18}$$

Adjacent zones have opposite phase and therefore destructively interfere. Therefore if we opened up the aperture one zone at a time, the intensity at G should oscillate between $4A^2$ and 0. In actuality, this oscillation of the intensity only strictly applies for small aperture radii. For larger radii, some of the approximations made in the derivation of equation (26.18) no longer apply. For example, as S increases, the variation in the cosine factor neglected above becomes significant, as does the s in the denominator. These two quantities progressively reduce the magnitude of the contribution from the successive zones. Born and Wolf (1989) have shown that this progressive decrease leads to the result that the amplitude of the disturbance at G due to the first zone is twice that of the unobstructed wave (that is, due to all the zones). This result explains why the intensity given by equation (26.18) is four times the expected unobstructed wave value of A^2.

26.3.1.2   The Fresnel zone plate

Now let us suppose that we block every alternate zone, those with odd values of m. In this case, all the transmitting zones are in phase at G and since there is now no cancellation or destructive interference, the intensity at G will increase. Equation (26.18) gives the amplitude at G for one zone. If there are N unobstructed zones and all constructively interfering with the same contribution, then the amplitude for N zones will be NA_{G1}, where A_{G1} is the amplitude from one zone and so from equations (26.10) and (26.18), the resulting intensity will be approximately

$$4N^2A^2 \tag{26.19}$$

which is N^2 times the intensity due to one zone. This type of aperture, like a lens, has a focussing property and is known as a Fresnel zone plate. From equation (26.17a), its focal length f can be written in terms of the first zone radius R_1 and the wavelength as

$$f = n'R_1^2/\lambda \tag{26.20}$$

A Fresnel zone plate based upon the design considered in Example 26.1 is drawn to scale in Figure 26.7, but with only the first four unobstructed zones shown.

The efficiency of the Fresnel phase can be increased if the opaque zones are made transparent but coated with a thin film that retards the phase of the wavefront by π. When this is done, all the zones now constructively interfere, and the amplitude at G will be twice that of the simple zone plate and thus the intensity now increases by a further factor of 4. This type of Fresnel zone plate is called a **phase zone plate**.

Fig. 26.7: The Fresnel
zone plate, based upon data
from Example 26.1.

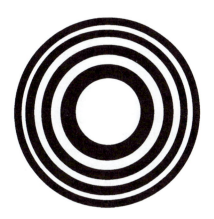

Fig. 26.8: The Fresnel
zone plate for general
conjugates.

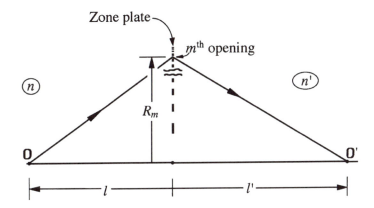

26.3.1.3   The zone plate as a lens

We have just shown that the zone plate focuses a plane wave (that is, from
infinity) to a point beyond the lens given by the above equations and thus seems
to act like a positive power lens. We will now show that the zone plate will
focus other conjugates and these conjugates are given by the conventional lens
equation. Figure 26.8 shows a zone plate with two points o and o', and a ray
passing through the mth and unobstructed zone of the zone plate. Light from o
passing through this zone will diffract to o' and constructively interfere, if the
optical path lengths satisfy the condition

$$[-n\sqrt{(l^2 + R_m^2)} + n'\sqrt{(l'^2 + R_m^2)}] - (-nl + n'l') = m\lambda/(2n')$$

where the negative signs are used to maintain the usual sign convention for
distances and where R_m is given by equation (26.16). If we write this equation
in the form

$$[-nl\sqrt{(1 + R_m^2/l^2)} + n'l'\sqrt{(1 + R_m^2/l'^2)}] - (-nl + n'l') = m\lambda/(2n')$$

and assume that R_m is much less than l and l', then we can expand the square roots using the binomial theorem and replacing R_m by the expression from equation (26.16), the above equation reduces to

$$n/l' - n'/l = 1/f \qquad (26.21)$$

where f is focal length defined by equation (26.20). Equation (26.21) is the conventional lens equation, showing that the Fresnel zone plate acts like a lens.

While the Fresnel zone plate obeys the lens equation and thus has similar imaging properties to a lens, it has several special properties not possessed by a lens. These are (a) the dual sign of the power or focal length, (b) multiple foci and (c) an opposite level of longitudinal chromatic aberration.

The dual sign of the power

We have demonstrated that the Fresnel zone plate acts as a lens with a positive focal length. Since light diffracts in all directions and symmetrically about the original direction, some light will diffract upwards after passing through the zone plate apertures. Therefore some light will appear to come from a point behind the zone plate. Thus the lens also appears to have a negative focal length as well as a positive focal length.

Multiple foci

Unlike a refracting lens, the Fresnel zone plate has other subsidiary focal lengths. Some of the light passing through the plate is not diffracted and this portion is called the zero order beam. For an incident collimated beam, this zero order beam remains collimated and thus the zone plate also has an infinite focal length. We will call the focal length defined by equation (26.20) the primary focal length.

The Fresnel zone plate also has focal lengths shorter than the primary focal length. We can discover what these are by looking for the other positions of G which satisfy the condition set by equation (26.14). Let us denote the primary focal length by the symbol f_o. Then from equation (26.14), we have

$$S_m = \sqrt{(f_o^2 + R_m^2)} \qquad (26.22)$$

For the same zone plate, let us assume that there is another set of values of S, which we will denote as $S_j (j = 0, 1, 2$ etc.), which will give a new focal length f. Then we have

$$S_j = \sqrt{(f^2 + R_m^2)} \qquad (26.23)$$

Now equation (26.13), applied to this condition, can be written

$$S_j - f = \sqrt{(f^2 + R_m^2)} - f = j\lambda/(2n')$$

where like m, j has positive integer values. We now replace R_m^2 by the expression from equation (26.16) remembering that the focal length in that equation is the primary focal length f_o. This substitution gives us

$$\sqrt{[f^2 + (f_o\lambda m/n')]} - f = j\lambda/(2n')$$

Using the binomial expansion and solving for f, we have

$$f \approx f_o m/j \qquad (26.24)$$

Now m follows the sequence 0, 1, 2, and so on, and j has a similar sequence. However, we will see that not all these integer values of j are possible. Let us begin by looking at the case of $m = 1$. If we now look at the possible values of j, the first case of $j = 1$ corresponds to the design focal length and is a trivial solution. If we now take j as an even integer, the first zone, for the new focal length f, as seen from the new focal point, contains two zones of opposite amplitude and therefore there is no constructive interference. If now j is odd, the first zone will appear to have an odd number of zones of opposite amplitude, giving a net light level at the new focal point. Therefore the zone plate will have subsidiary focal lengths defined by odd values of j. If we now include the infinite focal length and primary focal length, the zone plate has the focal lengths given by the sequence

$$f_o/p \text{ with } p = 0, \pm1, \pm3, \pm5, \dots \qquad (26.25)$$

Different amounts of light are diffracted into these various foci. The brightest focus is that with the infinite focal length, that is arising from the undiffracted beam. The light levels at the other focii have been given by Sussman (1960) and the levels relative to that of the infinite focal length are given by the equation

$$\text{light level} = \frac{4}{p^2\pi^2} \quad p = \pm1, \pm3, \pm5, \dots \qquad (26.26)$$

These relative light levels are sometimes known as **diffraction efficiencies**.

Chromatic effects

Since it is a diffractive device, a Fresnel zone plate must be designed for a given wavelength. At any other wavelength, it will have a different primary focal length. Let us find an equation for this focal length. Let the zone plate be designed to have a primary focal length $f(\lambda_o)$ for the wavelength λ_o. The radius R_1 of the first zone is given by equation (26.17a), that is

$$R_1 \approx \sqrt{[f(\lambda_o)\lambda_o/n'(\lambda_o)]}$$

This equation can be satisfied for any other pair of values of focal length $f(\lambda)$ and wavelength λ such that

$$R_1 \approx \sqrt{[f(\lambda)\lambda/n'(\lambda)]}$$

Thus we have

$$f(\lambda)\lambda/n'(\lambda) \approx f(\lambda_o)\lambda_o/n'(\lambda_o)$$

or finally

$$f(\lambda) \approx \frac{f(\lambda_o)\lambda_o n'(\lambda)}{\lambda n'(\lambda_o)} \tag{26.27}$$

For common optical materials, this focal length is dominated much more by the ratio of the wavelengths than by the ratio of the refractive indices. Thus this equation shows that the focal length decreases with increase in wavelength, which is opposite to the effect due to dispersion in a refracting lens.

26.3.1.4 Applications to vision

Fresnel zone plates have been suggested as an alternative to conventional bifocal lenses for the correction of presbyopia. A bifocal lens consists of two separate portions, of different power. One type consists of an inner central portion with one power and an annular portion with the other power. The disadvantage of this type is that it does not work when the pupil is small, as occurs in bright light, since the iris obstructs the beam that passes through the annular portion of the contact lens. A Fresnel phase zone lens can overcome this problem.

As we have seen in Section 26.3.1.3, the Fresnel zone plate has similar focussing properties to a lens, but has multiple focal lengths, that is multiple powers. To use a Fresnel zone plate as a bifocal lens, we would use two of these focal lengths or powers, one for distance viewing and one for the near viewing. However, we can see from equation (26.26) that the light is unequally distributed among the different focal lengths. Ideally, the light should only be directed into two focal lengths and equally distributed between them. That is we have to modify the diffraction efficiency of the zone plate. This can be done by modifying the physical profile of the zone plate which alters the phase of the wavefronts passing through the plate. By choice of a suitable profile, most of the diffracted light can be directed equally to only two foci.

The powers of the Fresnel zone plate lens can be further modified by combining the zone plate with a conventional lens.

26.3.2 Speckle patterns

Coherent light speckle is a diffraction phenomenon, visible when a broad and highly coherent beam of light, for example an expanded laser beam, is scattered from a diffusely reflecting surface. As shown in Figure 26.9, the reflected beam diffracts and forms a complex interference pattern (a Fresnel diffraction pattern) all along the reflected beam. Because it is usually observed using a laser source it is often called **laser speckle**. If this pattern is imaged by an optical system, the light distribution in the image appears to be "speckled". The size of the speckles depends upon the wavelength, the roughness of the reflecting surface and the aperture diameter of the imaging system. The speckle size decreases as the aperture diameter increases.

This speckle phenomenon can be used to measure focus errors of the eye. If a person looks at the speckle pattern, a speckle pattern is formed on the retina.

Fig. 26.9: Example of Fresnel diffraction – laser speckle.

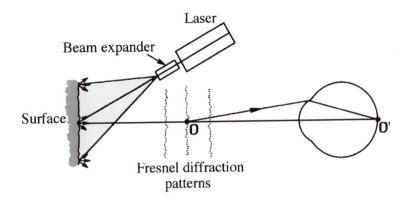

Fig. 26.10: Diffraction pattern along an aberration free focussed beam.

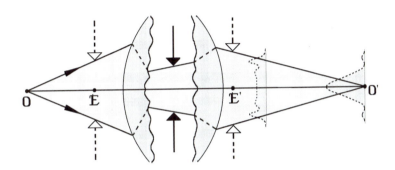

The interference pattern formed on the retina is that formed in a plane conjugate to the retina. If the observer is not perfectly focussed for the surface, the retinal conjugate surface containing the speckle pattern will not be in the same plane as the illuminated surface. Thus when the observer moves his or her head or eye a parallax movement between the interference pattern and the illuminated surface will be observed. The use of laser speckle to measure ocular refractive error is discussed further in Chapter 31.

26.4 Fraunhofer diffraction

Figure 26.10 shows a focussed monochromatic and aberration free beam, arising from an axial monochromatic point source. The continuous lines are the geometrical optics boundary and the dotted lines indicate the actual form of the light distributions across the beam at two representative positions along the beam. For the position closer to the optical system, the light distribution is similar to the light distribution at the edge of the beam, as shown already in Figure 26.2. In the centre of the beam, where there is no fringing, the light distribution is as predicted by geometrical optics. If we now move this plane towards the focus at o', the edges of the beam become closer, the flat central portion of the light distribution narrows and finally, the fringe systems at the edges merge and at the focus at o', the light level finally takes on the form shown.

The light level distribution or diffraction pattern at the focus is known as Fraunhofer diffraction and is a special case of Fresnel diffraction. Because it is the light distribution in the image of a point source for an aberration free beam, it is also known as the **diffraction limited point spread function**. If the beam is collimated, it is the far field diffraction pattern. Because it is the diffraction at the focus of a beam it is relevant to image formation because, along with aberrations, it predicts the light distribution in the image plane.

For any diffraction calculation, the starting point is the Kirchhoff diffraction integral given by equation (26.4). For the special case of the diffraction at the focus of the beam, that integral reduces to a much simpler form. Firstly, the angle χ varies very little over the diffraction pattern and therefore can be set to zero. Secondly, the distance s in the denominator, while varying, does not vary the integrand significantly. However, its presence in the exponential term is significant. From Figure 26.4, we can write

$$s = \sqrt{[(u - X)^2 + (v - Y)^2 + Z_0^2]}$$

Now Z_0^2 will be much greater than the term $(u - X)^2$ or $(v - Y)^2$ and we can write this as

$$s = Z_0 \sqrt{\{1 + [(u - X)^2/Z_0^2] + [(v - Y)^2/Z_0^2]\}}$$

If we expand the square root term using the binomial theorem up to the first set of terms, we get

$$s = Z_0 + [(u^2 + v^2)/(2Z_0)] - [(uX + vY)/Z_0] + [(X^2 + Y^2)/(2Z_0)]$$

The first two terms in this expansion do not depend upon X and Y. Therefore they can be taken outside the integral. The last term corresponds to a phase factor which corresponds to a spherical wave of radius of curvature Z_0 centred on the point $u = 0, v = 0, Z = Z_0$. Therefore if the originally collimated beam at the aperture is now made to converge to this point, the last term in the above equation disappears and we are left only with the term $-(uX + vY)/Z_0$. Substituting this expression for s in the exponential function in equation (26.4) leads to the following form for the Fraunhofer diffraction integral

$$A_G = K \int_A e^{-ik(uX + vY)/Z_0} dX \, dY \tag{26.28}$$

where K is a complex constant which incorporates the terms $Z_0 + [(u^2 + v^2)/(2Z_0)]$ above and the term in front of the integral in equation (26.4). The integration is carried out over the whole aperture.

26.4.1 Fraunhofer diffraction due to a circular aperture

Figure 26.11 shows an aberration free beam leaving the exit pupil at ε' of an optical system and being focussed to an axial point o'. The light distribution in the image plane depends upon the shape of the aperture. Since visual optical instruments mostly have circular apertures, diffraction by this shaped aperture is very important in understanding the image quality of these optical instruments. An equation for the light distribution in the image plane can be derived from the Fraunhofer diffraction integral equation (26.28). However, the derivation is beyond the scope of this book so we will simply quote the final result. Born and Wolf (1989 Section 8.5.2) have shown that for a circular aperture,

Fig. 26.11: Fraunhofer
diffraction – diffraction
pattern at the focus of a
beam passing through a
circular aperture.

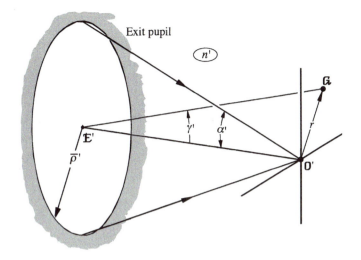

the illuminance level $E(r)$ at the point $\mathcal{G}$ in the image plane is rotationally
symmetric and equation (26.28) becomes

$$E(r) = E(0)\frac{[2J_1(\zeta)]^2}{\zeta^2} \tag{26.29}$$

where $E(0)$ is the illuminance at the centre of the diffraction pattern and referring to the diagram

$$
\begin{aligned}
J_1(\zeta) &= \text{a Bessel function which is explained further below} \\
\zeta &= k\overline{\rho}'\sin(\gamma') \text{ or } kr\sin(\alpha') &(26.29a)\\
k &= 2\pi n'/\lambda \\
n' &= \text{refractive index of image space} \\
\lambda &= \text{wavelength in vacuum} \\
\overline{\rho}' &= \text{exit pupil radius} \\
\gamma' &= \text{angular subtense of the distance } r \text{ measured at the exit pupil} \\
\alpha' &= \text{angular radius of the exit pupil measured at } o' \\
r &= \text{distance in the image plane}
\end{aligned}
$$

The Bessel function $J_1(\zeta)$ has similar properties to the sine and cosine
trigonometric functions, but unlike these trigonometric functions, it has no
fixed period and approaches zero as ζ approaches infinity. Abramowitz and
Stegun (1965, 370) have given it in the following polynomial approximation.
If $|\zeta| \leq 3$ then with $\psi = \zeta/3$

$$J_1(\zeta) = \zeta(0.5 - 0.56249985\psi^2 + 0.21093573\psi^4$$

$$- 0.03954289\psi^6 + 0.00443319\psi^8$$

$$- 0.00031761\psi^{10} + 0.00001109\psi^{12}) \tag{26.30}$$

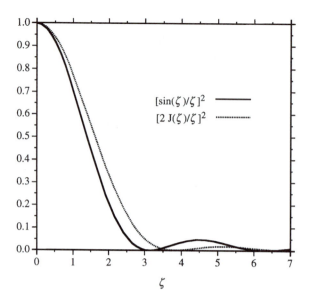

Fig. 26.12: The normalized Fraunhofer diffraction pattern for a circular aperture $[2J_1(\zeta)/\zeta]^2$ and a slit aperture $[\sin(\zeta)/\zeta]^2$.

If $|\zeta| \geq 3$ then $\psi = 3/\zeta$

$$J_1(\zeta) = g\cos(\omega)/\sqrt{(\zeta)} \qquad (26.31)$$

where

$$g = +0.79788456 + 0.00000156\psi + 0.01659667\psi^2$$

$$+ 0.00017105\psi^3 - 0.00249511\psi^4 + 0.00113653\psi^5$$

$$- 0.00020033\psi^6 \qquad (26.31a)$$

$$\omega = \zeta - 2.35619449 + 0.12499612\psi + 0.00005650\psi^2$$

$$- 0.00637879\psi^3 + 0.00074348\psi^4 + 0.00079824\psi^5$$

$$- 0.00029166\psi^6 \qquad (26.31b)$$

Plotting the function $J_1(\zeta)$ would show that it decays with increase in ζ. It can be seen from equation (26.30) that in the limit $\zeta = 0$, the function $J_1(\zeta)/\zeta$ is equal to 0.5 and hence the function $[2J_1(\zeta)/\zeta]^2$ is a normalized function, that is, is equal to 1.0 when $\zeta = 0$. This normalized function is shown in Figure 26.12.

The normalized form of $E(r)$ shown in Figure 26.12, being rotationally symmetric, has a central disc, called the **Airy disc**, surrounded by an infinite number of annular rings with ever decreasing amplitude. It has an infinite number of zeros, and these occur when $J_1(\zeta)$ is zero, except for the first zero of $J_1(\zeta)$, which occurs at $\zeta = 0$. Examination of the above polynomial approximations to $J_1(\zeta)$ shows that the next zero of $J_1(\zeta)$, and hence the first zero of $E(r)$, occurs when

$$\zeta = 3.8317$$

With this value of ζ and recalling the definition of ζ from equation (26.29a), we have

$$\sin(\gamma') = 0.61\lambda/(n'\overline{\rho}') = 1.22\lambda/(n'D')$$

where D' is the diameter of the exit pupil. However, in most cases the angle γ' is very small and we can make the approximation that $\sin(\gamma') = \gamma'$ and rewrite this equation as

$$\gamma' = 1.22\lambda/(n'D') \tag{26.32a}$$

If we use the second righthand expression for ζ in equation (26.29a), we can express the position of the first zero in terms of the physical coordinate r on the image plane, that is

$$r = 0.61\lambda/[n'\sin(\alpha')]$$

This can be written in the form

$$r = 0.61\lambda/NA' \tag{26.32b}$$

where $NA' = n'\sin(\alpha')$ is the numerical aperture of the beam in image space. If we restrict the object space medium to air and the object at infinity, we can replace the numerical aperture by the *F-number* from equation (9.12b) and have

$$r = 1.22\lambda \ F\text{-}number \tag{26.32c}$$

Example 26.2: Calculate the retinal radius of the Airy disc of a distant monochromatic point source with a wavelength of 555 nm, for the Gullstrand number 1 schematic eye given in Appendix 3, for an entrance pupil diameter of 8 mm.

Solution: The schematic eye data in Appendix 3 are given with an entrance pupil diameter of 8 mm, and from the tabulated data we have for the relaxed Gullstrand eye

$$NA' = 0.231 \text{ for an entrance pupil diameter of 8 mm}$$

Now we also have $\lambda = 555 \times 10^{-6}$ mm. Substituting these values into equation (26.32b) gives

$$r = 0.61 \times 555 \times 10^{-6}/0.231 = 0.00147 \text{ mm}$$

26.4.1.1 Absolute illuminance in the diffraction pattern

The actual illuminance distribution in the function $E(r)$ depends upon the amount of luminous flux entering the system, the pupil diameters and the wavelength. These factors are implicit in the quantity $E(0)$. Born and Wolf (1989) have derived an expression for this quantity, which in our notation is

$$E(0) =$$

$$\frac{\text{luminous flux } (flux') \text{ passing out of exit pupil} \times \text{area of exit pupil}}{(\text{wavelength} \times \text{distance } \varepsilon'o')^2}$$

(26.33)

Providing the angle α' in Figure 26.11 is small, we can use the approximation

$$\sin(\alpha') = \overline{\rho}'/\varepsilon'o'$$

to show that equation (26.33) can be expressed in the alternative form

$$E(0) = flux'\pi \left[\frac{n' \sin(\alpha')}{\lambda} \right]^2$$

(26.34)

The luminous flux $(flux')$ falling on the image plane is related to the flux $(flux)$ entering the optical system and equations for this quantity are given in Chapter 12. Since we have restricted the discussion to a point source, the relevant equation for this flux is equation (12.13c), that is

$$flux = \frac{\pi I D^2}{4d^2}$$

(26.35a)

where I is the luminous intensity of the point source, D is the diameter of the entrance pupil, and d is the distance to the point source. If the luminous transmittance of the system is τ, then the flux $(flux')$ leaving the system is

$$flux' = \tau \times flux$$

and thus

$$flux' = \frac{\tau \pi I D^2}{4d^2}$$

(26.35b)

and we can finally write

$$E(0) = \tau I \left[\frac{\pi D n' \sin(\alpha')}{2d\lambda} \right]^2$$

(26.36)

26.4.1.2 Special case of the source at a great distance

If the source is at a great distance, we can assume that the diffraction image is formed in the back focal plane of the optical system and in this case, we can use equation (9.12b) to replace $\sin(\alpha')$ above by the entrance pupil diameter D and the equivalent power F of the system. On doing this we get

$$E(0) = \tau I \left(\frac{\pi F}{4d\lambda n'} \right)^2 D^4 \quad \text{(point source at infinity)} \qquad (26.36a)$$

This equation shows that central or peak illuminance depends upon the fourth power of the entrance pupil diameter. This relationship is due to two factors, namely (a) the amount of light collected by the system increases as the square of the entrance pupil diameter (that is as the pupil area) and (b) the area of the diffraction pattern reduces by this factor. That is, as the pupil diameter increases more light is collected and is distributed over a smaller area. However, equation (26.35b) shows that the total luminous flux in this diffraction pattern only varies as the square of the pupil diameter.

26.4.2   Single slit diffraction pattern

While very few visual optical systems have slit apertures, diffraction by a slit aperture was mentioned in the discussion of Young's interferometer described in the preceding chapter. Born and Wolf (1989) have shown that the Fraunhofer diffraction pattern of a slit illuminated by an incoherent line source is given by the equation

$$E(r) = E(0)[\sin(\zeta)/\zeta]^2 \qquad (26.37)$$

where ζ has the same meaning as given by equation (26.29a), but in terms of the width w of the slit we can write

$$\zeta = \pi w n' \sin(\gamma')/\lambda \qquad (26.38)$$

This function is plotted in Figure 26.12 and has regularly spaced zeros with the first occurring at

$$\zeta = \pi$$

and it easily follows that the equivalent equations to equations (26.32a and b) are

$$\gamma' = \lambda/D' \qquad (26.39a)$$

and

$$r = 0.5\lambda/[n' \sin(\alpha')] \qquad (26.39b)$$

which give slightly smaller values than those for a circular aperture whose aperture radius is the same as the half width of the slit.

26.5   Diffraction and the aberrated beam

What we have just learned so far in this chapter can be applied to a problem that often arises – when is it safe to use geometrical optics to predict the light

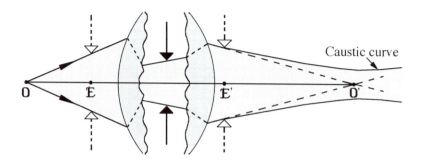

Fig. 26.13: Shape of a badly aberrated beam. The caustic is the envelope of all the refracted real and aberrated rays.

Caustic curve

level in a beam and when should we use physical optics?

The example of diffraction at an edge in Figure 26.2 shows that away from the edge, physical optics predicts the same light level as geometric optics. It is only near the edge that geometrical optics fails to accurately predict the light level. The width of this physical optics zone depends upon the width of the beam and the wavelength. The shorter the wavelength, the less diffraction affects the light level in the vicinity of the edge of the beam and thus the narrower the border zone of interference fringing. In the limit as the wavelength goes to zero, this zone goes to zero and there is no diffraction effect, leading to one definition of geometrical optics.

> *Geometrical optics is a special case of physical optics in which the wavelength is zero.*

Let us extend what we have learned to look at the light level in the aberration free focussed beam shown in Figure 26.10. We can conclude that geometric optics is accurate inside the beam but not near the edge. Now for this diffraction limited beam, the geometric focus o' is a point which is also at the edge and therefore geometric optics is invalid at this focus.

Now, let us take a very badly aberrated version of the beam as shown in Figure 26.13. In this case, there is no clearly defined focus and in the paraxial focal plane, the beam is very wide. Therefore we can use geometrical optics to predict the variation in light level over most of the light distribution.

Thus in conclusion, in a beam that has no or little aberration, we cannot safely use geometrical optics to predict the light distribution in or around the focus of the beam. On the other hand, if it is a badly aberrated beam, geometrical optics can be used and will accurately predict the light level in the vicinity of the focus.

26.6 The diffraction grating

Diffraction gratings are traditionally used to analyse the spectral content of a light or other radiation source and hence are not particularly relevant to visual optical instruments. However, they can be used to construct interferometers that produce white light fringes. Since they have been referred to in the previous chapter for this purpose, a short discussion of the principles of diffraction gratings is in place here. Diffraction gratings are either the transmission type or the reflection type and the principles are similar, so we will restrict this discussion to only one of these, the transmission grating.

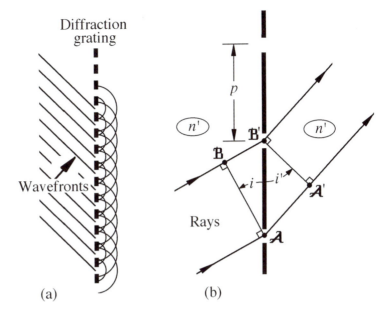

A transmission diffraction grating is a slab of high quality transparent optical
material, on which is deposited a thin layer of an opaque material. A fine stylus
is then used to rule a large number of parallel and equally spaced lines on the
surface. On each line, the material is removed and the line then acts as a slit.
Incident light diffracts after passing through the slits. In a reflection grating,
the slab is coated with a highly reflecting material such as aluminium and the
incident light reflects off the fine lines of aluminium left behind.

Diffraction gratings are usually used in collimated light and Figure 26.14a
shows a collimated beam represented by a set of parallel wavefronts, incident
on a transmission grating. The light passing through each slit diffracts and
becomes a secondary line source. Figure 26.14b shows three adjacent slits and
parallel rays from two of these that are to be used to determine the diffraction
pattern at infinity. The optical path difference e between the two rays is

$$e = n'(\mathcal{A}\mathcal{A}' - \mathcal{B}\mathcal{B}')$$

where

$$\mathcal{A}\mathcal{A}' = +p\sin(i'), \quad \mathcal{B}\mathcal{B}' = p\sin(i)$$

and p is the slit or line separation, that is the period of the grating. In the diagram,
the angles i and i' have the same sign and may be regarded as positive.

Note: The perpendicular line joining the diffracted rays shown in Figure
26.14b is not a wavefront because the two rays arise from different sources.

Thus, we have

$$e = n'p[\sin(i') - \sin(i)] \tag{26.40}$$

These two rays will constructively interfere if the path difference is a whole number of wavelengths, that is

$$n'p[\sin(i') - \sin(i)] = m\lambda \text{ with } m = 0, \pm1, \pm2, \ldots$$

That is

$$\sin(i') - \sin(i) = m\sigma\lambda/n' \qquad\qquad (26.41)$$

where

$\sigma = 1/p$ is the grating frequency or number of lines/unit distance

m is the diffraction order

If we fix the direction i of the incident beam, then the intensity of light diffracted in different directions i' depends upon the interference between the diffracted rays from each slit in the grating with the direction of the maxima or complete constructive interference given by equation (26.41). If the grating is illuminated with a beam containing different wavelengths, the maxima directions will depend upon the wavelength and therefore the diffraction grating has the ability to break up a polychromatic beam into its different wavelength components.

The different values of m relate to different diffraction orders and the positive and negative values relate to the two possible directions of the diffracted rays shown in Figure 26.14b. The diagram shows the diffracted rays rotated anti-clockwise from the incident direction and this corresponds to a positive value of m. There is another solution with the rays rotated clockwise and these correspond to negative values of m. The zero diffraction order, that is $m = 0$, corresponds to a part of the beam that passes through without diffraction.

26.6.1 Small angle approximation

If we assume that the angles i and i' in equation (26.41) are small, then we can replace the sines of the angles by the angles themselves. This gives us the approximation

$$i' \approx m\sigma\lambda/n' + i \qquad\qquad (26.41a)$$

which shows that for small angles, the diffraction angle is linear with wavelength.

26.6.1.1 Angular dispersion

Angular dispersion is the variation in diffraction angle with wavelength, for specific wavelengths and diffraction orders. For the exact equation given by equation (26.41) this can be found by differentiation but for the small angle approximation, equation (26.41a) shows that this variation is linear.

In most applications, the diffracted beam is de-collimated and a spectrum is formed in the back focal plane of the de-collimating lens (or mirror). Figure 26.15 shows the example of a de-collimating lens, with the grating normal

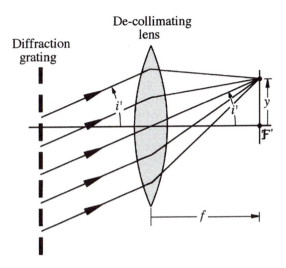

Fig. 26.15: Use of de-collimating lens to focus the diffraction pattern onto a plane at a finite distance.

to the axis of the lens. If the diffraction angle i' is small, the corresponding position on the back focal plane is given by the equation

$$y = i' f \tag{26.42}$$

where f is the equivalent focal length of the de-collimating lens. Substituting for i' from equation (26.41a) gives

$$y(m, \lambda) = [(m\sigma\lambda/n') - i] f \tag{26.43}$$

which shows that the physical separation of the maxima is proportional to the wavelength. It follows from this last equation that the physical separation $\Delta y(m, \lambda)$ of the plus and minus orders is

$$\Delta y(m, \lambda) = 2m\sigma\lambda f/n' \tag{26.44}$$

which, within the small angle approximation, is independent of the angle of incidence i.

26.6.2 Resolving power

A diffraction grating does not perfectly separate the different wavelengths in a polychromatic beam. This is because for each wavelength and maxima, all of the light is not diffracted into one single direction. Instead it is spread out over an angular region and the width of this angular spread depends upon the number of lines in the grating and the width of the slit. The theory of the light distribution for a general grating is presented in other texts such as Born and Wolf (1989).

26.6.2.1 Effect of number of lines in the grating

If we firstly consider a very simple and trivial diffraction grating consisting of only two lines, the grating is similar to a Young's double slit interferometer, described in the previous chapter. We showed in that chapter that the light distribution of two beam interferometers varied in a sinusoidal manner. Therefore the diffraction profile of a two line diffraction grating would also be sinusoidal and so the light in each maxima is considerably spread out. We will call these the primary maxima. The light level profile, in a two line grating, has only one minimum between consecutive maxima and is half way between these maxima. Diffraction grating theory shows that a grating with N lines has $N - 1$ minima equally spaced between these primary maxima and therefore has $N - 2$ secondary maxima between the primary maxima. As the number of lines increases, the light in these secondary maxima decreases and so the diffracted light is concentrated more and more in the primary maxima. The resolving power is sometimes defined as the separation of the minima in wavelength units, and this separation is inversely proportional to N.

26.6.2.2   Effect of line width

The width of the primary maxima is also affected by the finite width of lines of the grating arising from the effect of single slit diffraction, which we have already discussed in Section 26.4.2, and the maxima cannot be narrower than the equivalent single slit diffraction pattern.

> **Example 26.3:** Suppose we wish to produce two diffraction spectra of first order ($m = \pm 1$) that are 2 mm apart for light of wavelength 555 nm using a diffraction grating with 50 lines/mm. Calculate the desired focal length of the de-collimating lens and determine the length of each spectrum in the back focal plane of the de-collimating lens for light from 400 to 700 nm. Assume the medium is air.
>
> **Solution:** For this problem we use equation (26.44), firstly with $\Delta y(1, \lambda) = 2$ mm, $m = +1$, $\sigma = 50$ lines/mm and $\lambda = 555 \times 10^{-6}$ mm and solve for the focal length f. The substitutions give
>
> $$f = 2/(2 \times 50 \times 555 \times 10^{-6}) = 36.0 \text{ mm}$$
>
> The length of the spectrum from 400 to 700 nm is, from equation (26.43),
>
> $$y(1, 700) - y(1, 400)$$
>
> and therefore
>
> $$\text{length} = y(1, 700) - y(1, 400) = 36.0 \times 1 \times 50$$
>
> $$\times (700 - 400) \times 10^{-6} = 0.54 \text{ mm}$$

Exercises and problems

26.1   For a collimated beam of light partially blocked by a straight screen, use Young's boundary wave model to predict the expected distance from the geometrical edge of the shadow to the first fringe minimum if the screen is 2 m from the viewing plane, for light of wavelength of 500 nm. Assume that the boundary wave is out of phase by π to the undiffracted beam.

ANSWER: 1.414 mm

26.2 Calculate the outer radius of the central zone of a Fresnel zone plate designed for a wavelength of 633 nm and a focal length of 200 mm.

ANSWER: 0.356 mm

26.3 Calculate the focal plane diameter of the Airy disc of a diffraction limited camera lens with a focal length of 50 mm and working at $F/2$ and imaging light of wavelength 600 nm.

ANSWER: 0.00292 mm

26.4 Why does a single slit have a better diffraction limited resolving power than a circular aperture with the same diameter as the slit width?

Summary of main symbols

k $\quad 2\pi n'/\lambda$

Section 26.3.1: The Fresnel zones and zone plate

r_m $\quad$ outer radius of mth Fresnel zone
m $\quad$ positive integer

Section 26.4: Fraunhofer diffraction

γ' $\quad$ direction of a point on the image plane relative to the optical axis and centre of the exit pupil
r $\quad$ distance in the image plane
$E(0)$ $\quad$ illuminance at the centre of the diffraction pattern

Section 26.6: The diffraction grating

p $\quad$ period of a diffraction grating
σ $\quad$ line density of grating ($\sigma = 1/p$)
i $\quad$ angle of incident relative to grating normal
i' $\quad$ diffraction angle of maxima relative to grating normal
m $\quad$ order of diffraction ($m = 0, \pm 1, \pm 2, \ldots$)

References and bibliography

*Abramowitz M. and Stegun I.A. (1965). *Handbook of Mathematical Functions*. Dover, New York.
*Born M. and Wolf E. (1989). *Principles of Optics*, 6th (corrected) ed. Pergamon Press, Oxford.
*Sussman M. (1960). Elementary diffraction theory of zone plates. *Am. J. Phys.* 28, 394–398.

Part IV

Ophthalmic instruments

27

Focimeters

27.0 Introduction

There are a number of visual optical instruments used in ophthalmic optics, designed to measure such quantities as vertex powers and surface radii of curvature of both lenses and the cornea. The principles of these instruments have been discussed already in Chapter 11. In this and the following five chapters, we will examine these a little further and introduce some other useful ophthalmic instruments. In this chapter, we will look at the measurement of vertex power.

In ophthalmic optics, spectacle lenses are rarely specified by their equivalent power. Instead they are specified in terms of their back vertex power, front surface power, refractive index and thickness. Thus the measurement of vertex powers is perhaps the most common power measurement in ophthalmic optics, and fortunately it is far simpler to measure vertex power than to measure equivalent power. The focimeter is a simple instrument for measuring vertex powers and has already been described in Chapter 11.

Focimeters may be visual, projection or fully automatic. In the visual type shown in Figure 11.16, the image of a target is viewed and focussed through an optical system consisting basically of a collimator and a viewing telescope. The scale reading may also be presented in the eyepiece of the telescope. In the projection type, the telescope is replaced by an optical system which projects the image of the target onto a viewing screen. In both of these types of instruments, the operator has to make some judgement on the best focus of the target. In automatic instruments, the instrument automatically locates the optimum focus and the final powers are given directly on some type of display.

27.1 Principle of operation

The theory of this instrument has already been discussed in Section 11.4.2.3. The essential features of the instrument are shown in Figure 11.16 and the principles of operation are as follows. With no test lens in position, a target T placed at the front focal point $\mathcal{F}_c$ of the collimator will appear in focus through the telescope. If a lens is now placed in the beam, as shown in the diagram, between the collimator and the telescope, the beam will be made to converge or diverge, depending upon the power of the test lens, and the image seen through the telescope will no longer be in focus. However, the image can be brought back into focus by moving the target to some suitable new position. The amount of movement will depend upon the position of the test lens and its power.

If the lens to be measured is placed with its surface vertex coinciding with the back focal point $\mathcal{F}'_c$ of the collimator, as shown in Figure 11.16, then the vertex power of that lens is measured. The relationship between this vertex power F_v and the target displacement x from the front focal point of the collimator is given by equation (11.21), that is

$$F_v = x F_c^2 \tag{27.1}$$

where F_c is the equivalent power of the collimator. This equation shows that the vertex power is linear in target displacement. To measure the other vertex power, the lens is turned around.

27.1.1   *Limitations on the range of vertex powers*

Equation (27.1) shows that for positive power lenses, the value of target displacement x must be positive and hence the target is moved towards the collimator lens. For negative power lenses, the target is moved away from the collimator. For positive power lenses, the highest measurable power is thus limited by how far the target can be moved forward. Assuming that the collimator is a thin lens, the target is also thin and there are no other obstructions, the target can only move forward a distance equal to the focal length of the collimator, that is

$$x_{\text{max}} = f_c$$

Therefore from equation (27.1), the maximum measurable power is

$$(F_v)_{\text{max}} = +F_c \tag{27.2}$$

which is the same as the power of the collimator.

In the case of negative lenses, these restrictions do not apply, so there is no theoretical optical limit on measurable power. The only limit is the distance the target can be moved backwards within the instrument and this is set by the physical design of the instrument. In commercial instruments, the vertex powers are limited to within about ± 25 m^{-1}.

27.1.2   *Target structure and measurement of the powers of an astigmatic lens*

The focimeter must be able to measure the powers in the principal meridians of an astigmatic lens, and the meridian directions. This facility requires a suitable target design. One configuration is a set of small dots arranged around the circumference of a circle. With an astigmatic lens, as the focus of the instrument is varied, there will be two positions in which the dots will be in focus in one direction but blurred out forming long lines in the other direction. At these two positions, the dots form lines that are the thinnest but longest possible. The directions of the sets of lines at the two foci are perpendicular to each other and are the directions of the principal meridians. These directions can be read from some suitable scale.

27.1.3 Measurement of prismatic power

The focimeter is also able to measure the prismatic effect of a lens. The presence of a prism deviates the beam and for thin prisms, the deviation is proportional to the prismatic power. If there is some prism in the lens at the point of measurement, the focussed image is displaced from the centre of the scale and the amount of displacement is read from some suitable scale in prism dioptres.

27.1.4 Particular features and accuracy

The magnification of the telescope is approximately 5 to 6. The aperture diameter of the collimator/telescope system is about 5 mm but often can be varied by the use of stops usually placed at the exit face of the collimator. These particular characteristics limit the accuracy of the measured vertex power to about $\pm 1/8 \text{ m}^{-1}$. Larger aperture diameters would reduce depth-of-field and possibly decrease this uncertainty but would have the potential to increase errors because of the increase in spherical aberration from the test lens.

27.1.5 Effect of error in collimation

The above theory assumes that when the scale reading is zero, the target is collimated. Since it is impossible to collimate a collimator perfectly, we should look at the effect of a collimator error on the measured vertex power of a lens. Smith (1982) has shown that providing (a) the telescope is refocussed for the zero scale reading with no lens in position and (b) the thickness of the lens being measured can be ignored, then the collimation error has no effect on the accuracy of the measured vertex power.

27.2 Measurement of residual power in a telescope

The focimeter is designed to measure the vertex powers of ophthalmic lenses but may also be used to measure the vertex power of any lens system providing it can fit between the collimator and telescope. One particular useful application would be the measurement of the residual back vertex power of a telescope. While telescopes are nominally afocal and hence have zero back vertex power, we may wish to check an individual telescope. In principle a focimeter can be used for this task. The telescope is placed with its eyepiece facing the collimator and the usual focusing procedure carried out. The scale reading is the back vertex power of the telescope.

However, Smith (1982) has shown that, unlike a spectacle lens, collimation errors will give an erroneous result in this application. This is because of the "vergence magnification" effect of telescopes exemplified by equation (17.6). For a telescope magnification greater than unity and the eyepiece facing the collimator, any collimation error in the collimator may be magnified by the square of the magnification of the telescope under test. Thus the measurement of the back vertex power of a telescope is sensitive to errors in collimation in the focimeter. The corollary of this result is that if the telescope was perfectly afocal, it could be used to examine these collimation errors, since magnification makes them easier to detect and measure.

Summary of main symbols

F_c equivalent power of focimeter collimator
f_c corresponding equivalent focal length
F_v vertex power of test lens
x displacement of focimeter target from front focal point of colli-
            mator
$\mathcal{F}_c, \mathcal{F}'_c$ front and back focal points of focimeter collimator
$\mathcal{F}$ front focal point of test lens
$\mathcal{T}, \mathcal{T}'$ position of focimeter target

References and bibliography

Henson D.B. (1983). *Optometric Instrumentation*. Butterworths, London, Chapter 11.
*Smith G. (1982). Vertometers, their zero errors and measurement of instrument accommodation. *Aust. J. Optom.* 65(2), 69–71.

28

Radiuscopes and keratometers

28.0   Introduction

The measurement of the radii of curvature of surfaces is one of the most common optical metrological measurements. Different methods of measuring the radius of curvature of a spherical or toric surface are presented in Chapter 11. In ophthalmic optics, we are concerned with the radii of curvature of the surfaces of spectacle or contact lenses and the radii of curvature of the anterior corneal surface. Because of the potential toric nature of these surfaces any clinical instrument should be able to measure the radii of curvature in a section and hence the radii of cuvature of the principal meridians of a toric surface. There are three main types of ophthalmic instruments for this task – Geneva lens measures, radiuscopes and keratometers. The Geneva lens measure is designed to measure the surface curvature or power of a spectacle lens and is a contact method based upon the spherometer. The radiuscope is a non-contact or optical method and is based upon the Drysdale principle. The keratometer is also a non-contact or optical method and is designed to measure the surface curvature of the anterior surface of the cornea. The Geneva lens measure has been adequately described in Chapter 11 and we will not spend any more time on discussing it. We will describe the radiuscope and keratometer in the following sections.

28.1   The radiuscope

Radiuscopes are instruments designed to measure the radius of curvature, mainly of contact lenses. Because these may be soft and easily deformed, a non-contact or optical method must be used. A typical construction of a radiuscope based upon the Drysdale method is shown in Figure 11.11 and the Drysdale principle is described in Section 11.3.2.1. Briefly the principle is as follows. A target at T is imaged by a beam-splitter and an objective lens onto the surface being measured and the beam is reflected back towards a viewing screen, through the objective lens and beam-splitter as shown in the diagram. The image on the viewing screen may be viewed through an eye lens, thus giving some magnification and allowing relaxed accommodation. The surface being analysed is moved backwards and forwards along the axis of the beam and there are two positions in which the target is imaged in focus on the viewing screen. The two positions correspond to

(a)   the target being imaged on the test surface and
(b)   the rays in the beam from the axial point in the target meeting the test surface normal to the surface.

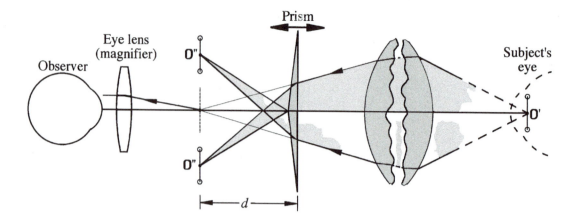

Fig. 28.1: Image doubling in the keratometer, using a movable prism.

The distance the test surface has been moved between these two positions is the radius of curvature of the test surface. Thus the largest radius that can be measured is limited by the distance from the objective lens to the focus of the beam.

Soft contact lenses are often best analysed while they are supported in a liquid medium. In this case, because of the very close refractive index match between the liquid and the test lens, the reflectivity is very low and hence the returned image is very dull. Steel and Noack (1977) and Steel and Freund (1985) have suggested modifications to the standard radiuscope design that will improve the quality of the image.

28.2 The keratometer

The keratometer is an ophthalmic instrument specifically designed for measuring the radius of curvature of the anterior surface of the cornea by an optical or non-contact method. The optical arrangement is shown in Figure 11.12, with the cornea replacing the test lens, and the theory has already been presented in Section 11.3.2.2, but a brief discription of the essential features will not be out of place here. An object or target at o is imaged onto the cornea. The size of the image at o' is related to the corneal radius of curvature by equations given in Section 11.3.2.2. However since this image is virtual, being formed inside the cornea, its size cannot be measured directly. To overcome this problem, the image at o' is reimaged by the objective lens to form a real image at o''. The size of this real image size can be measured, for example using a scale placed in the plane at o''. If all distances and powers are fixed and known, this final image size can be used to calculate the corneal radius of curvature, using equations given in Chapter 3 and Section 11.3.2.2.

However, in measurements on the living eye, when the eye moves, the image of the target on the scale would also move, making it very difficult to take a reading of the image size with such a simple optical system as described above. So a modification is required to overcome this problem. In the actual keratometer, the size of the image is determined by means of an image doubling device.

28.2.1 Image doubling

Prisms used as image doubling devices are discussed in detail in Chapter 8. Figure 28.1 shows a prism being used to double the final image seen through the eye lens. In some instruments, the amount of doubling can be varied and in the example shown in the diagram, this can be achieved by moving the doubling prism backwards or forwards. The amount of doubling can be varied until the two adjacent ends of the doubled images just touch. The amount of doubling thus required depends upon the size of the images and hence the radius of curvature of the anterior corneal surface. Therefore the prism can be attached to a scale that is marked directly in corneal radius of curvature. The adjustment can be readily done even if the images are in motion as a result of eye movements. In some instruments, the amount of doubling is fixed and the image size is adjusted by varying the size of the object or target.

References and bibliography

Henson D.B. (1983). *Optometric Instrumentation.* Butterworths, London. Chapters 5 and 11.

Sasieni L.S. (1982). A guide to keratometers. *Optician* 183, 11–12, 14, 19 (January 22), 24–25 (February 5).

*Steel W.H. and Freund C.H. (1985). A microscope for measuring soft corneal lenses. *Aust. J. Optom.* 68(3), 96–99.

*Steel W.H. and Noack D.B. (1977). Measuring the radius of curvature of soft corneal lenses. *Appl. Opt.* 16, 778–779.

29

Ophthalmoscopes

29.0 Introduction

Ophthalmoscopes are optical instruments designed for the visual inspection of the internal structure of the eye, but most commonly the retina. However, because the amount of light reflected from the subject's eye is very low, the observer will only see an image if the subject's retina is well illuminated and thus an auxiliary illuminating system is an essential component of an ophthalmoscope. Thus ophthalmoscopes consist of two main components: a viewing system and an illuminating system.

There are a number of different designs for the viewing system, and these can be divided into two groups, as follows:

(a)   Direct ophthalmoscopes. These are the simpler of the two types. They are discussed in detail in the next section, Section 29.1.

(b)   Indirect ophthalmoscopes. The viewing system is more complicated than for direct ophthalmoscopes, having extra lenses between the subject's and observer's eyes. The extra complexity allows independent control over field-of-view and magnification. Indirect ophthalmoscopes are discussed in detail in Sections 29.2 and 29.3.

In this chapter, we will concentrate on the viewing system and only make a brief mention of the illumination system of the direct ophthalmoscope.

The magnification in direct ophthalmoscopy is often quoted as 15. In contrast, the magnification of indirect ophthalmoscopy is quoted as being much lower, usually in the region of about $(-)3$. However, once we look at the construction of indirect ophthalmoscopes, we will see that there is some potential ambiguity in the way that the magnification of an indirect ophthalmoscope is defined, but for the moment we will leave that problem aside and begin by looking at the properties of the direct ophthalmoscope.

29.1 Direct ophthalmoscopes

The direct ophthalmoscope consists of two components: the viewing system and the illuminating system. Let us look first at the viewing system.

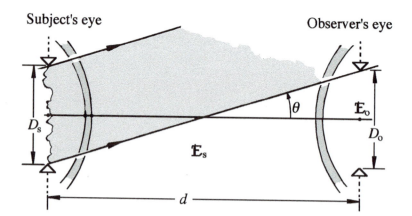

Fig. 29.1:   The angular field-of-view of the direct ophthalmoscope.

29.1.1   Viewing system

The schematic construction of the viewing system of the direct ophthalmoscope has already been shown in Figure 9.10a. The subject's and observer's eyes are aligned as shown in this diagram. Assuming the subject and observer are both emmetropic, the light leaving the subject's retina is collimated by the dioptric apparatus of that eye. Some of this light will enter the observer's eye and the image of the subject's retina will be formed on the observer's retina.

However, the amount of ambient light reflecting from the retina is far too low for an image of the subject's retina to be seen and therefore the subject's retina must be illuminated to a much higher level using an auxiliary light source and the construction of the illuminating system is discussed in Section 29.1.1.4.

29.1.1.1   Field-of-view

The amount of the subject's retina that can be seen by the observer depends upon the distance between the pupils of the subject and observer and the pupil diameters. An equation for the field-of-view can be found as follows.

Assume the subject's eye is emmetropic and firstly consider the collimated beam between the two eyes, as shown in Figure 29.1. This beam arises from a retinal point some distance from the optical axis of the subject's eye. Now the edge of the field-of-view is the point when the off-axis collimated beam, as shown in the diagram, just becomes blocked or vignetted by the observer's pupil. From the diagram and for small angles, the angular radius θ of the field-of-view is simply given by the equation

$$\theta = (D_s + D_o)/(2d) \quad \text{rad} \tag{29.1}$$

where D_s and D_o are the diameters of the subject's and observer's pupils, respectively, and d is the distance between the two pupils. This equation shows that the field-of-view is inversely proportional to the pupil separation. The observer examines other regions of the retina by either asking the subject to rotate his or her eye or moving his or her own eye in an arc around the subject's eye.

Let us now determine the corresponding size of the retinal field. The collimated beam shown in Figure 29.1 arises from some off-axis retinal point ϱ, as

Fig. 29.2: The retinal field-of-view of the direct ophthalmoscope.

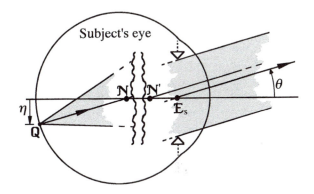

shown in Figure 29.2. We can use this value of θ to determine the actual radius η of the retinal field-of-view. Within the paraxial approximation, we can use equation (3.52) to relate these two quantities. That equation relates the physical size of an image with the angular size of an object at infinity. Here we have the image at infinity and the object in the front focal plane of the optical system and so we have

$$\eta = (-)\theta/F_s \tag{29.2}$$

where F_s is the equivalent power of the subject's eye. This is a paraxial equation but the angles involved are small enough to make this equation accurate. From the data given in Appendix 3, the equivalent power of the Gullstrand number 1 schematic relaxed eye has the value of 58.6 m^{-1}. For our example calculations we will take a rounded value as

$$F_s = 60 \text{ m}^{-1}$$

Example 29.1: If the subject's and observer's pupils have diameters of 4 mm and 3 mm respectively, and their separation is 25 cm, calculate the angular field-of-view and the actual size of the retinal field. Finally compare the angular field with that for a separation of 5 cm.

Solution: From equation (29.1), the field-of-view would have a radius of

$$\theta = (4 + 3)/(2 \times 250) = 0.0140 \text{ rad} = 0.80°$$

Now from equation (29.2), the retinal field radius would be

$$\eta = 0.0140/60 \text{ m} = 0.23 \text{ mm}$$

If we decrease the distance d to 5 cm, the radius of the field-of-view increases to about $4.0°$.

It is now obvious that the small field-of-view is limited by the distance between the pupils of the subject and observer and the diameters of the subject's and observer's pupils. If the two pupils could be made coincident, the field-of-view would be that of an eye (just greater than 180°). However since the entrance pupils of the eyes are about 3 mm inside the eye, this is not possible without some optical aid and this is the topic of discussion in Sections 29.2 and 29.3. There is little the observer can do about the pupils' sizes, short of using dilatory drugs.

The pupil size factor is complicated by the fact that the observer does not use his or her own natural pupil. Instead, the observer views through an aperture (the sight hole) in the ophthalmoscope mirror used to reflect the light into the subject's pupil and this becomes the effective pupil of the observer. It is usually smaller than the natural pupil. The choice of viewing hole size is a compromise between light flux directed to the subject's eye, depth-of-field and field-of-view.

29.1.1.2 Magnification

The magnification in direct ophthalmoscopy is often quoted as approximately 15, with the actual value depending upon the equivalent power of the subject's eye. The argument used to justify the value of 15 is as follows.

The equation for the ophthalmoscope magnification is derived from the analogy with a simple magnifier. The observer views the subject's retina through the optical system of the subject's eye, which acts as a simple magnifier. If the subject's eye has an equivalent power F_s, the magnification M_d is then assumed to be given by the $F/4$ rule [equation (15.6)], that is

$$M_d = F_s/4 \tag{29.3}$$

While this equation was derived for a lens in air, it also applies to the emmetropic eye, in which the object space medium is the vitreous humour with a refractive index of 1.336. Therefore we can use this equation to calculate the magnification of the direct ophthalmoscope and since the equivalent power F_s of the eye is close to 60 m^{-1},

$$M_d \approx 15$$

with the actual value being dependent upon the equivalent power of the subject's eye, which is usually unknown and very difficult to measure.

29.1.1.3 Adjustment for refractive errors

So far we have assumed that the subject and observer are emmetropic with relaxed accommodation. In practice, it is likely that one or both will have a refractive error. To cope with this situation, direct ophthalmoscopes contain a range of auxiliary lenses, set in a wheel which can be rotated to position individual lenses in front of the sight hole.

The magnification also varies with the subject's refractive error. This issue has been covered by Emsley (1952) and Bennett and Rabbetts (1989).

29.1.1.4 Illumination system

A schematic illuminating optical layout is shown in Figure 29.3. The beam from the light source is reflected into the eye by the mirror. The source is usually

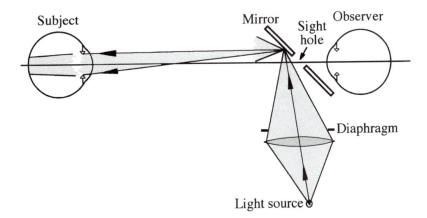

Fig. 29.3: The illumination system of the direct ophthalmoscope.

imaged onto the mirror. The diameter of the beam can be controlled by a suitable diaphragm placed near the light source focussing lens.

Some of the light will be reflecting back from the cornea and appear as an image of the light source a few millimetres behind the cornea, but while this image will be out of focus, it can still be troublesome. The reflection can be reduced by off-setting the illumination beam slightly to one side of the sight hole.

29.2 Indirect ophthalmoscopes: Single lens

It has been shown that the small field-of-view in the direct ophthalmoscope is due to the large distance between the subject's and observer's pupils, that is due to the position mis-match between these pupils. The field-of-view can be increased substantially by position matching these pupils with a type of field lens of suitable power placed between the two eyes. The use of field lenses to match pupils is discussed in Chapter 9. The power of this lens must satisfy two requirements.

(a) The image of the subject's pupil formed by the field lens should be in the same plane as the observer's entrance pupil. However, satisfying this requirement means that a real image of the retina is formed between the lens and the observer's eye.

(b) This real retinal image must be beyond the near point of the observer's eye, unless the observer uses a focussing aid.

The optics of this situation is shown in Figure 9.10b and again in Figure 29.4. In this instrument, the field lens (called a condenser) is placed between the two eyes but closer to the subject's eye. Optically, the purpose of this lens is to image the subject's pupil onto the pupil of the observer, that is position pupil match. The ideal power of the condenser depends upon the distances involved. Let the distances $\bar{l}$ and $\bar{l}'$ of the condenser lens from the two pupils be as shown in the diagram. The power F_c of the condenser lens is found from the lens equation, equation (3.43), here written as

$$F_c = 1/\bar{l}' - 1/\bar{l} \qquad (29.4)$$

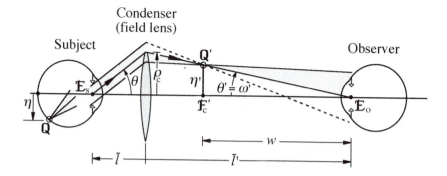

and the usual sign convention applies. The image of the subject's pupil will be magnified by an amount M, where

$$M = \bar{l}'/\bar{l} \qquad (29.5)$$

Fig. 29.4: The single lens indirect ophthalmoscope, showing its pupil matching properties and the field-of-view.

which comes from the transverse magnification equation (3.49).

> **Example 29.2:** Assume the two eyes are 30 cm apart, and the condenser lens is placed 5 cm in front of the subject's eye; find the required power of the condenser power F_c and the magnification of the subject's pupil.
>
> **Solution:** Firstly from the data, we have $\bar{l} = -5$ cm and hence $\bar{l}' = 25$ cm and therefore in the lens equation (29.4), we have $\bar{l} = -0.05$ m and $\bar{l}' = 0.25$ m, giving
>
> $$F_c = 24 \text{ m}^{-1}$$
>
> The pupil magnification is given by equation (29.5), that is
>
> $$M = 25/(-5) = (-)5$$
>
> Thus if the two pupils are the same size, the image of the subject's pupil superimposed on the observer's pupil will be five times larger than the observer's pupil.

Because the beam leaving the subject's eye is collimated or nearly so, the condenser forms a real retinal image in or near the back focal plane of the condenser as shown in Figures 29.4. The observer now has to accommodate to view this real image. By contrast, in direct ophthalmoscopy, the retinal image is nominally at infinity and therefore the emmetropic observer views with relaxed accommodation.

In the next sub-section, we will confirm that this arrangement gives a wider field-of-view than the direct ophthalmoscope. While this makes the indirect ophthalmoscope more useful when large fields-of-view are required, the observer's eye position now becomes more critical. In the indirect ophthalmoscope, the observer only sees the subject's retinal image when the two pupils are superimposed; that is, the observer's pupil falls within the image of the subject's pupil. If we look at the results of Example 29.2, we find that the subject's pupil is magnified five times by the condenser lens. If we assume a typical pupil diameter of 4 mm, the image of the subject's pupil would be 20 mm in diameter. Thus the observer must keep his or her own pupil within a circle of this diameter. If it moves outside this circle, no image at all would be seen. In contrast, in direct ophthalmoscopy, eye position is less critical in that a retinal image is always seen from any eye position.

29.2.1 Field-of-view

From the observer's eye position, there will be some light appearing to come from the edge of the condenser lens and therefore the condenser lens limits the size of the field-of-view by vignetting. In this case, there is some ambiguity about the definition of field-of-view; whether to define it as the field size free of vignetting or the edge of the field where vignetting just becomes complete or some intermediate field size. Here we will define it as at an intermediate level of vignetting. Figure 29.4 shows a collimated beam from some off-axis point Q in the subject's eye. The beam shown is being partly vignetted by the condenser lens and we define the edge of the field as where the pupil ray just touches the edge of the condenser lens as shown in the diagram. Let us now determine the angular field-of-view as seen by the observer and finally the size of the retinal field.

29.2.1.1 Observed angular field-of-view

Providing the condenser is reasonably free of spherical aberration, the angular field-of-view seen by the observer is simply the angular subtense of the condenser lens as shown in Figure 29.4. If the condenser lens has an aperture radius of ρ_c and at a distance $\bar{l}'$ from the observer's eye, then the angular radius θ' of the image field-of-view is

$$\tan(\theta') = \rho_c/\bar{l}' \qquad (29.6)$$

In this equation, we have departed from the usual paraxial approximations and used the tangent instead of the angle itself. This is because the angles may be greater than those normally assumed to be within the paraxial region.

29.2.1.2 Size of the retinal field

If the collimated beam leaving the eye subtends an angle θ to the axis, then from Figure 29.4

$$\tan(\theta) = \rho_c/\bar{l} \qquad (29.7)$$

We can use this value of θ to determine the actual radius η of the retinal field-of-view. Within the paraxial approximation, we can use equation (29.2) to relate the object angle and corresponding retinal field. Outside the paraxial approximation, the use of $\tan(\theta)$ instead of θ alone would be more accurate, but this would only be true for a flat retinal surface. However, since the retinal surface is curved with a radius of curvature of about 12 mm, the use of θ instead of $\tan(\theta)$ would be more appropriate and therefore equation (29.2) is probably a good approximation.

Example 29.3: Find the observed field-of-view and the size of the corresponding retinal field of the indirect ophthalmoscope of Example 29.2 if the aperture diameter of the condenser is 50 mm.

Solution: In Example 29.2, the distance $\bar{l}'$ is 250 mm and from the above data, the aperture radius of the condenser (ρ_c) is 25 mm. Therefore from equation (29.6), the radius of the field-of-view is

$$\tan(\theta') = 25/250$$

giving

$$\theta' = 5.7°$$

which is about 30 times larger than the angular field-of-view of the direct ophthalmoscope for the same eye separation.

Now to find the size on the corresponding retinal field, we use the value of $\bar{l} = (-)50$ mm from Example 29.2 and $\rho_c = 25$ mm from above and substitute in equation (29.7) to get

$$\tan(\theta) = 25/50$$

giving

$$\theta = 26.6° = 0.464 \text{ rad}$$

Finally using equation (29.2) and a power of the eye of 60 m^{-1}, an estimate of the radius of the retinal field is

$$\eta = 7.7 \text{ mm}$$

If we compare these values with those of Example 29.1 for the direct ophthalmoscope, we can see that this indirect ophthalmoscope gives a retinal field-of-view 30 times larger for the same eye separation. However, if we look at the observed angular field-of-view, the increase is only about seven times. The reason for the difference is the fact that this type of indirect ophthalmoscope has a smaller magnification than the direct ophthalmoscope. Let us look at this magnification.

29.2.2 Magnification

We will now show that this type of indirect ophthalmoscope, while giving a larger field-of-view of the retina, gives a smaller magnification than the direct ophthalmoscope. Let us derive an equation for this magnification.

A first step is to find an expression for the size of the image formed in the back focal plane of the condenser lens. The optical situation is shown in Figure 29.4 and we can use equation (29.2) to relate η and θ and apply this equation once more to relate θ and η' and end up with

$$\eta' = -\eta F_s/F_c \tag{29.8}$$

where F_s is the equivalent power of the subject's eye and F_c is the equivalent power of the condenser. The minus sign indicates that the image is inverted. In contrast, the image of the direct ophthalmoscope is erect.

Now we need to determine the angular size of this image seen by the observer, as shown in Figure 29.4. When viewed from the distance w, the angular size $\omega' (= \theta')$ of this image is

$$\omega' = \eta'/w = -\eta F_s/(wF_c) \qquad (29.9)$$

When compared with the retinal field size, if seen directly from the same distance w, whose angular size ω would be

$$\omega = \eta/w \qquad (29.10)$$

the effective magnification M_i of this indirect ophthalmoscope is defined as

$$M_i = \omega'/\omega \qquad (29.11)$$

Combining these last three equations gives

$$M_i = -F_s/F_c \qquad (29.12)$$

Because the two lens powers must be positive, the magnification must be negative, implying an inverted image.

We can compare this magnification with that of the direct ophthalmoscope by replacing the power F_s of the subject's eye by the magnification M_d from equation (29.3) and then we have

$$M_i = -4M_d/F_c \qquad (29.13)$$

The magnification equation in this form shows that the indirect magnification M_i is only greater than the direct ophthalmoscope magnification M_d if

$$F_c < 4 \text{ m}^{-1}$$

In practice the condenser power is about 20 m^{-1} so this condition is unlikely to be met in practice.

As is the case for direct ophthalmoscopy, the magnification is affected by refractive error (Emsley 1952).

Example 29.4: Calculate the magnification of the indirect ophthalmoscope referred to in Example 29.2 assuming an eye equivalent power of $F_s = 60 \text{ m}^{-1}$.

Solution: From Example 29.2, $F_c = 24 \text{ m}^{-1}$. Therefore from equation (29.12), the magnification is

$$M_i = -60/24 = -2.5$$

which is one sixth of the corresponding nominal value of 15 quoted for indirect ophthalmoscopy.

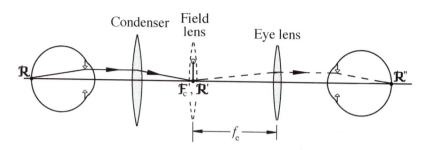

Fig. 29.5: The single lens indirect ophthalmoscope with a viewing eye lens.

We have confirmed that while giving a much wider field-of-view of the retina than the direct ophthalmoscope, this indirect ophthalmoscope provides a smaller magnification. However, we will show that this indirect ophthalmoscope can be modified to increase its magnification without any necessary loss in the field-of-view. The structure and properties of these modified ophthalmoscopes are looked at in the next section.

29.3 Indirect ophthalmoscopes: Multi-lens or telescopic systems

The image magnification of the indirect ophthalmoscope described in the preceding section can be increased by viewing the retinal image through a simple magnifier or eye lens, as shown in Figure 29.5. Apart from increasing the magnification, this would also reduce the accommodation demand.

If the eye lens has a magnification M_e, the new magnification M_i will be the product of this magnification and that given by equation (29.13); that is it will have the value

$$M_i = -4M_d M_e / F_c$$

But from equation (15.6),

$$M_e = F_e / 4 \tag{29.14}$$

where F_e is the power of the eye lens. Therefore

$$M_i = -M_d(F_e / F_c) \tag{29.15}$$

This equation shows that if the eye lens power is the same as the condenser, then the new magnification is the same as the direct ophthalmoscope, M_d. By suitable choice of eye lens power, the magnification of this modified indirect ophthalmoscope can be substantially increased above that of the simple single lens indirect ophthalmoscope.

29.3.1 Effect of eye lens on field-of-view

In principle the magnification of the eye lens can be as high as the upper limit of the magnification of simple magnifiers and eyepieces, a limit normally taken as about 20. However, as the magnification increases, the focal length of the eye lens decreases and the eye lens must be placed closer to the intermediate image. Two problems may arise.

(a)   The condenser lens and intermediate image subtend an increasing angle to the eye lens, thus increasing the angular field-of-view, with a consequent increase in aberrations.

(b)   Pupil matching may be lost with a corresponding decrease in field-of-view. To regain control over pupil matching a field lens can be placed at the intermediate image plane as shown in Figure 29.5. A lens placed in this position does not affect the final magnification.

29.3.2   Equivalent telescope

If the subject and observer are emmetropic, this type of indirect ophthalmoscope, condenser plus eye lens, is essentially a telescope, with a magnification M_t,

$$M_t = -F_e/F_c \tag{29.16}$$

where the first condenser is the objective of the telescope. Using this equation, we can replace the ratio F_e/F_c in equation (29.15) to get a new equation for magnification M_i of this indirect ophthalmoscope in the form

$$M_i = -M_d M_t \tag{29.17}$$

This type of indirect ophthalmoscope is thus in effect a telescope, with two distinct magnifications. It has an ophthalmoscope magnification given by equation (29.17) but a telescopic magnification given by equation (29.16). These two magnification values are in the ratio of the magnification M_d of the direct ophthalmoscope.

29.3.2.1   Aperture stop and pupil imagery

We have shown that this type of indirect ophthalmoscope is in effect an afocal telescope, in fact a Keplerian telescope. However, in Keplerian telescopes, the objective is the aperture stop and hence the entrance pupil. This is not the case here. In this case, the entrance and exit pupils coincide with the planes of the pupils of the subject and observer and the pupil that limits the diameter of the axial beam acts as the aperture stop of the system.

29.4   Binocular indirect ophthalmoscopes

With the construction shown in Figure 29.6, it is possible to observe the aerial image with both eyes simultaneously. If an eye lens or macroscope is placed as shown, it is also possible to magnify the aerial image. The macroscope may be either a Galilean type or a Keplerian type, but we have to consider the image orientation. There is some advantage in having an erect retinal image. With binocular systems details are seen in stereoscopic vision and this may be very useful in the diagnosis of some retinal abnormalities.

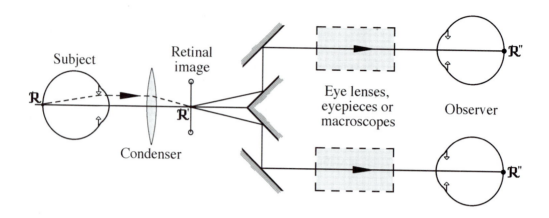

Fig. 29.6:   A binocular
indirect ophthalmoscope.

Summary of main symbols

d	distance between subject's and observer's pupils
ρ_c	aperture radius of condenser
F_s	equivalent power of the subject's eye
F_c	equivalent power of the condenser
F_e	equivalent power of the eye lens
f_e	corresponding equivalent focal length
$\mathcal{F}'_c$	back focal point of condenser
$\mathcal{F}_e$	front focal point of eye lens
$\mathcal{E}_s$	centre of the entrance pupil of the subject
$\mathcal{E}_o$	entrance pupil of the observer
$\mathcal{R}, \mathcal{R}', \mathcal{R}''$	retinal conjugate points
D_s	diameter of the pupil of the subject
D_o	diameter of the pupil of the observer
M_d	magnification of the direct ophthalmoscope
M_i	magnification of the indirect ophthalmoscope
M_t	magnification of the equivalent telescope
θ, θ'	angular radius field-of-view of object and image

References and bibliography

*Bennett A.G. and Rabbetts R.B. (1989). *Clinical Visual Optics*, 2nd ed. Butterworth-
    Heinemann, London. Chapter 16.
*Emsley H.H. (1952). *Visual Optics*, Vol. 1, 5th ed. Butterworths, London. Chapter 7.
Henson D.B. (1983). *Optometric Instrumentation*. Butterworths, London. Chapter 1.

30

The Badal optometer

30.0   Introduction

The Badal optometer (Badal 1876) is an optical instrument based upon the Badal principle, which is described in the next section. It is used to present a target of constant angular size at a range of vergences to the eye. In optical structure it has many properties similar to those of a simple magnifier, as we will see shortly.

30.1   The Badal principle

The simplest arrangement of the Badal optometer is shown in Figure 30.1. This particular arrangement shows the Badal lens as a single although thick lens, but it may be a more complex system. The Badal principle applies not only to this simple system but to more complex systems. It has many similarities to a simple magnifier and reference was made to the Badal principle in Chapter 15. The Badal principle states that if the eye is placed at the back focal point of the (Badal) lens then

(a)   the target image vergence is proportional to the distance of the target from the front focal point of the lens and

(b)   the angular size of the image is independent of target vergence.

The proofs are as follows.
   Newton's equation, equation (3.66) in Chapter 3, states that (in air)

$$xx' = -1/F^2 \tag{30.1}$$

where F is the equivalent power of the Badal lens and as shown in Figure 30.1, x and x' are the object and image distances of the target T from their respective focal points. The image vergence X' of the image distance x', which is also the distance from the eye, is thus

$$X' = -xF^2 \tag{30.2}$$

This equation shows that the image vergence X' at the back focal point of the lens is a linear function of object or target distance x from the front focal point of the Badal lens. This property of the Badal lens allows it to be easily calibrated.

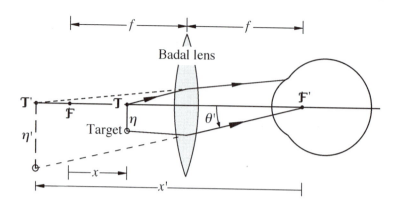

Fig. 30.1: Schematic construction of the Badal optometer.

We will now show that the angular size of the target image, seen from the back focal point of the Badal lens, is independent of its vergence. Referring to Figure 30.1, the angular size θ' of the image at the back focal point of the Badal lens is

$$\theta' = \eta'/x' \qquad (30.3)$$

where η' is the linear size of the image. If we now recall the definition of transverse magnification given by equation (3.46) and express this in the form given by equation (3.67b), noting that we are now working in air, we have

$$\eta' = -x'\eta F \qquad (30.4)$$

If we now use this equation to eliminate η' from equation (30.3), we finally have

$$\theta' = (-)\eta F \qquad (30.5)$$

The minus sign is bracketed because it is not important in this situation. This equation confirms that the angular size θ' of the image measured at the back focal point of the Badal lens is independent of the target position.

In the simple arrangement shown in Figure 30.1, the maximum negative image vergence is limited by the equivalent power of the Badal lens. If one neglects the lens thickness, the maximum negative image vergence occurs when the target has moved forwards and is at the lens, that is $x = f$ and thus from equation (30.2),

$$X'_{\max} = -F \qquad (30.6)$$

The finite lens thickness and space taken up by mounting of the lens and target will lead to a substantially lower value. On the other hand, there is no such restriction on the maximum positive image vergence.

In this arrangement of the Badal optometer, high negative image vergences require higher power lenses and hence shorter focal lengths. Shorter focal length lenses in turn imply shorter working distances which restrict the space between the Badal lens and the eye. This space is often used for beam-splitters which

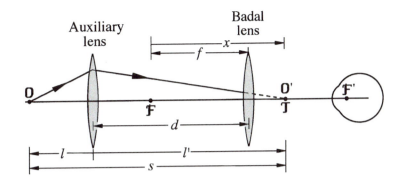

Fig. 30.2: Use of an
auxiliary lens with an aerial
image target which
increases the negative
vergence range of the Badal
optometer.

bring other visual stimuli into the field-of-view. This shorter distance may also lead to claustrophobic effects or psychological feelings of "crowding in", and may lead to unwanted instrument accommodation.

It would be very desirable if the working distance or back focal length could be increased while maintaining the high equivalent power of the Badal lens. An inverse telephoto lens satisfies this requirement, but unfortunately is not appropriate since both principal planes are on the same (image) side of the lens. This prevents the target from being brought close to the front principal point and hence the maximum value of X given by equation (30.6) cannot be achieved. However, there are at least two solutions to the problem, and these will be discussed in the next major section.

30.1.1 Field-of-view

In the Badal optometer shown in Figure 30.1, the Badal lens will vignet the off-axis beam and therefore its diameter controls the field-of-view. Since the optical arrangement is similar to that of a simple magnifier, the field-of-view equations given in Chapter 15 can be used to examine the actual field-of-view of this Badal optometer. We should note, however, that the object field-of-view will depend upon the target position.

30.2   Alternative optical constructions

30.2.1   Solution 1

In solution 1, the Badal optometer single lens can still be used but the real target is now replaced by an aerial image target. This arrangement has been used by Crane and Steele (1978). In this arrangement, shown schematically in Figure 30.2, an auxiliary lens projects the real target o to o', which is now an aerial image target that can be projected beyond the Badal lens, thus providing negative image vergences higher than that given by equation (30.6).

If the target image o' is formed a distance l' from the auxiliary lens, then this image is formed at a distance x from the front focal point $\mathcal{F}$ of the Badal lens, where

$$x = l' + f - d \tag{30.7}$$

Thus the image vergence X' at the eye is, from equation (30.2),

$$X' = -(l' + f - d)F^2 \qquad (30.8)$$

Now the highest negative image vergence occurs when the auxiliary lens is in contact with the Badal lens (that is $d = 0$) and thus from equation (30.8)

$$X'_{max} = -F^2(l' + f)$$

That is

$$X'_{max} = -F(Fl' + 1) \qquad (30.9)$$

Example 30.1: Given a Badal lens with an equivalent power of 10 m^{-1}, calculate the maximum possible negative image vergence if an auxiliary lens is used which projects a target image 20 cm beyond itself.

Solution: In this case, we have $F = 10$ m^{-1} and $l' = +0.20$ m. We substitute these values in equation (30.9), which gives

$$X'_{max} = -10(10 \times 0.2 + 1) = -30 \text{ m}^{-1}$$

Compared with the maximum image vergence of -10 m^{-1} for a real target, this is a significant improvement.

In this arrangement, the auxiliary lens and target must move together and the length of the system is increased by an amount

$$s = -l + l'$$

If we eliminate l from this equation using equation (3.45a), we have

$$s = -\frac{l'^2 F_a}{1 - l' F_a} \qquad (30.10)$$

where F_a is the power of the auxiliary lens. The system length is a minimum when $l = -2f_a$ and $l' = 2f_a$, that is

$$s_{min} = 4f_a \qquad (30.11)$$

For this minimum length condition, the maximum image vergence X'_{max} is, from equation (30.9),

$$X'_{max} = -F[(2F/F_a) + 1] \qquad (30.12)$$

Equation (30.9) or (30.12) can be used to solve the complete requirements for a particular situation. Consider the following example.

Example 30.2: Calculate the powers of the Badal and auxiliary lenses that provide a maximum vergence of -15 m^{-1} when a Badal lens to eye clearance of 20 cm is required.

Solution: Firstly, the eye clearance is the focal length of the Badal lens. Therefore

$$F = 1/0.2 = 5 \text{ m}^{-1}$$

Now substituting this value, and $X'_{\text{max}} = -15$, into equation (30.9) leads to

$$-15 = -(l'F^2 + F)$$

That is

$$l' \times 5^2 = 15 - 5$$

and finally

$$l' = 0.4 \text{ m}$$

A vergence range of 0 to -15 m^{-1} would then require a physical movement of the target equal to the value of l' plus the focal length of the above lens, that is the distance

$$0.4 \text{ m} + 0.2 \text{ m} = 0.6 \text{ m}$$

Finally the power of the auxiliary lens giving the overall minimum length can be found from equation (30.12)

$$-15 = -5[(2 \times 5/F_\text{a}) + 1]$$

That is

$$F_\text{a} = 5 \text{ m}^{-1}$$

30.2.2   Solution 2

Solution 1 keeps the target fixed relative to the auxiliary lens and the target/auxiliary lens is moved backwards or forwards as a single unit, and as we have seen, this may require a long system. There is an alternative solution that does not use an auxiliary lens but requires a more complex Badal lens. The ideal Badal lens system is shown in Figure 30.3a, in which the principal points $\mathcal{P}$ and $\mathcal{P}'$ are external to the system shown in the diagram. However any system with external principal points on opposite sides of the system must form an internal intermediate image of any object beyond the front principal focal point. An example of a simple system satisfying this requirement is shown in Figure 30.3b. However if we calculate the equivalent power of this system, we find it has a negative power. Fortunately we can alter its power, but not alter the positions of the principal planes, by adding a third lens halfway between these two lenses, as shown in Figure 30.3c. Varying the power of this middle lens will vary the equivalent power of the system as a whole. For example if we

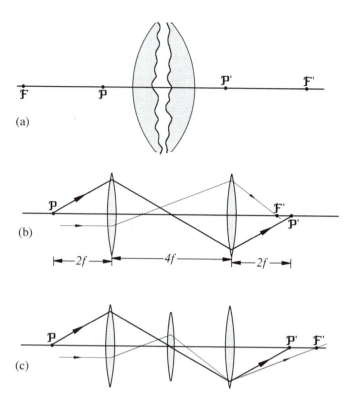

Fig. 30.3: A more general
ideal Badal lens system:
one with external principal
points. See Section 30.2.2
for details.

set the lens powers as F, $3F$ and F, equation (3.31) shows that the equivalent power of the combination will be F.

While this arrangement overcomes the limitations of using a single lens, as set by equation (30.6), it consists of a number of positive lenses and therefore, the aberration of Petzval sum or curvature may be a problem. Therefore a more complex design may be required that, while maintaining the Gaussian properties, has much better correction for Petzval curvature and any other aberration that may be troublesome.

30.3 Eye position

As yet we have not stated which point of the eye is made coincident with the back focal point of the Badal lens. There are several possibilities: the front focal point of the eye, the front principal point, the front nodal point and the entrance pupil.

Of these, providing the image is in focus, the front nodal point is the correct choice since the retinal image size is proportional to the angular size measured at the nodal points. However, in many uses of the Badal optometer, the image is defocussed or blurred. In this case, retinal image size is not proportional to the angular size measured at the nodal points. For blurred images, the correct reference point is the entrance pupil.

Summary of main symbols and equations

s	length of auxiliary system
d	distance of the auxiliary lens from the Badal lens (Section 30.2)

F	equivalent power of the Badal lens
f	corresponding equivalent focal length
F_a	power of auxiliary lens
f_a	focal length of auxiliary lens
$\mathcal{T}, \mathcal{T}'$	target conjugates
$\mathcal{F}, \mathcal{F}'$	front and back focal points of Badal lens
$\mathcal{P}, \mathcal{P}'$	front and back principal points
l, l'	object and image distances from respective principal planes
x, x'	object and image distances from respective focal points of Badal lens
X, X'	corresponding vergences
η, η'	object (or target) and image sizes
θ, θ'	angular sizes of object and image

$$X' = -xF^2 \tag{30.2}$$

$$\theta' = (-)\eta F \tag{30.5}$$

References and bibliography

*Badal J. (1876). Pour la mesure simultanée de la réfraction et d l'acuité visuelle même chez les illettrés. *Annales D'Oculistique* 75(11), March and April.

*Crane H.D. and Steele C.M. (1978). Accurate three-dimensional eye-tracker. *Appl. Opt.* 17(5), 691–705.

Wittenberg S. (1988). The Badal optometer paradox. *Am. J. Optom. Physiol. Opt.* 65(4), 285–291.

31

Optometers

31.0   Introduction

Optometers are instruments designed to determine the accommodative or refractive state of the eye. In Chapter 10, we discussed a number of methods that
could be used to measure the accuracy of focus of an optical system. However,
in all those cases, the image plane was directly accessible. In the eye, this is
not the case and therefore we have to access it indirectly. We can do this in
two ways. In one method, we can ask the subject to make some judgement of
the quality of the retinal image or focus level and we will call these subjective
methods. Alternatively, a second person or observer examines the light reflected
from the retina and makes a judgement of the focus error. We will call these
objective methods. A number of optometers that are available make the focus
judgements automatically, by electronic processing. These are based upon one
of the objective methods and are beyond the scope of this book, but have been
discussed by Bennett and Rabbetts (1989).

   The ultimate accuracy of any optometer method is limited by the depth-
of-field of the subject's eye. The depth-of-field of the eye is dependent upon
the pupil diameter and typical values have been given in Chapter 13. Errors
can also be caused by unwanted accommodation responses of the subject's
eye. Many people accommodate when viewing through optical instruments
(a phenomenon called **instrument myopia**) and we discuss this phenomenon
further in Chapter 36. This has to be overcome when measuring refractive error
and can be done by using fixation targets.

31.1   Subjective versus objective optometer methods

Below is a list of methods that could be used to measure the refractive error of
the eye. They are broken up into three groups: those that are subjective, those
that may be subjective or objective and those that can only be objective.

(1)   Subjective only
      –   simple perception of blur
      –   laser speckle

(2)   Subjective or objective
      –   longitudinal chromatic aberration of the eye

- split image/vernier alignment
- the Scheiner principle

(3) Objective only
- knife edge, for example retinoscopy
- photography

Ideally, subjective and objective methods should give the same results within the expected depth-of-field of the eye. However, differences do exist between the two measurements, partly because of the different layers of the retina where absorption and reflections occur. Subjective responses will place the best image on the retinal layer that contains the cones. In contrast objective methods may observe the light reflected from a different layer. In general, the longer the wavelength of the light source, the deeper is the reflection from the retina and choroid. This will give a more myopic refraction, but this is counteracted by the decrease in refractive index of the eye with increase in wavelength. For broad spectrum visible sources, the combined effects are about 0.20 m^{-1} (Charman 1975).

We will now look at some of the above optometer principles, but we will restrict the descriptions to only spherical ametropias or refractive errors. For astigmatic eyes, the techniques involve locating the orientation of the two principal meridians and determining the refractive error in each of them.

31.2   Subjective optometers

In subjective optometry the subject observes a target and the vergence of the target is varied by moving it directly or by using lenses (trial case lenses) placed just in front of the eye to change the vergence of the target image.

31.2.1   *Simple perception of blur*

The refractive state of the eye can be determined by asking the subject to observe a suitable target, such as acuity chart letters, and make judgements on the best focus. The judgement of best focus can be aided by the subject being given a recognition or detection task since the recognition or detection of fine or critical detail should be maximized at the best or correct focus. This is in fact the method used in routine eye testing where the patient views an acuity chart through a range of trial lenses of different power until maximum acuity is reached.

31.2.2   *Longitudinal chromatic aberration of the eye*

As we will see in Chapter 35, the eye is uncorrected for chromatic aberration and contains significant longitudinal chromatic aberration. Between the wavelengths of 400 and 700 nm it contains about 2.0 m^{-1} of longitudinal chromatic aberration. This can be used to assess the refractive state of the eye. For instance, the subject is asked to view a white point source of light through a filter that only transmits the red and blue ends of the spectrum, such as cobalt glass. Longitudinal chromatic aberration leads to the observations given in Table 31.1. While this observation is mostly qualitative, it is used in the **duochrome** test, which is part of the clinical measure of refractive error.

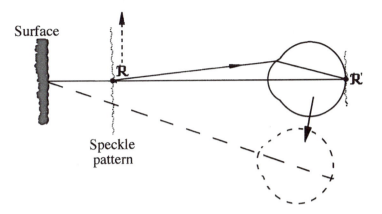

Fig. 31.1: General principle of the laser optometer.

Surface

Speckle pattern

Table 31.1. *The colour distribution of the image of a point source of white light in the presence of emmetropia and ametropia*

Refractive state	Observation
Emmetropia	purple blur disc
Myopia	red centre and blue surround
Hyperopia	blue centre and red surround

31.2.3 The laser optometer

This optometer makes use of the phenomenon of laser speckle, which is an example of Fresnel diffraction and whose origin is explained in Chapter 26 and shown in Figure 26.9. This diagram shows an expanded beam illuminating a rough surface. The reflected beam diffracts and forms an infinite number of Fresnel diffraction (speckle) patterns at different distances from the surface, and one of these will be conjugate to the retina of the eye, as shown in the diagram. If the observer moves his or her head transversely, the speckle pattern appears to move relative to the background. Figure 31.1 shows the case of a myopic eye focused at the speckle pattern at $\mathcal{R}$. A movement of the head downwards causes the speckle pattern to move upwards relative to the reflecting surface, that is in the opposite direction to the head movement. In the case of hyperopia, the movement is in the same direction. If the eye is focussed on the reflecting surface, there is no apparent speckle movement.

Rather than have the subject make repetitive head movements, the surface can be moved instead. A convenient arrangement is to illuminate a rotating drum, as shown in Figure 31.2. The optimum rotation period of the drum is about 5 minutes. However, the surface of the drum is curved and if there is no speckle movement, we need to know where the eye is focussed. This position of this plane is called the "position of the plane of stationarity".

Charman (1974; 1979) has shown that the "plane of stationarity" is not on the drum surface but inside the drum at a distance that depends upon the drum radius r and the laser beam vergence at the drum surface. Figure 31.2 shows a

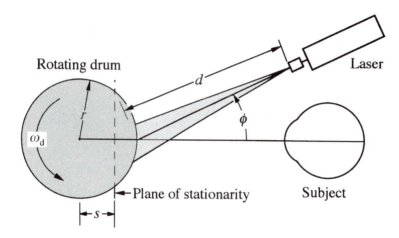

Fig. 31.2: Principle of the rotating drum laser speckle optometer.

typical set-up geometry. Charman (1979) showed that the distance of the plane of stationarity from the drum rotation axis is given by the equation

$$s = \frac{r[d\cos(\phi) + r\cos^2(\phi)]}{d[1 + \cos(\phi)] + r\cos(\phi)} \tag{31.1}$$

where s, r, d and ϕ are defined in the diagram.

As the drum rotates, if the observer is not focussing on the plane of stationarity, the speckle pattern appears to move across the face of the drum. The speed of the speckle motion depends on the drum speed and refractive error relative to the plane of stationarity. Charman (1979) showed that the angular speed ω_s of the speckle pattern is given by the equation

$$\omega_s = -\frac{r\omega_d \Delta F}{1 + dV_{ps}} \tag{31.2}$$

where ω_d is the angular speed of the drum, V_{ps} is the vergence of the plane of stationarity at the eye and ΔF is the refractive error of the eye relative to the plane of stationarity. If the eye is myopic, ΔF is positive, and if the eye is hyperopic, ΔF is negative. Thus for example, in the case of myopia, the two angular speeds have opposite signs indicating that they are in the opposite direction.

31.2.4 The Scheiner principle

In the Scheiner method, a mask containing a number of holes, usually two, is placed or imaged near or onto the entrance pupil of the subject's eye at equal distances from the pupil centre. If the object is a single point, the rays passing though the two holes in the mask intersect the out-of-focus image plane at different points and the defocussed eye sees two images of the object point. The amount of separation of the two points is a measure of the level of defocus. For more complex objects, one would observe a double overlapped image, with a transverse separation.

31.2.5   The split image/vernier alignment

The split image/vernier alignment principle requires a target with one straight edge which can be split into two. Any defocus moves one part sideways relative to the other. The subject then sees the straight edge split into two and displaced sideways. When the system is correctly focussed, the two parts of the image are aligned. The advantage of this method is that the eye is very sensitive to what is called vernier mis-alignment. Values for vernier acuity are given in Chapter 13.

Such a method can be realized using polarized filters placed over the target as shown in Figure 10.8, and another pair of polarized filters placed over the pupil of the eye (Simonelli 1980). This second pair is oriented 90° relative to the first pair. When applied to the eye, the defocussed plane coincides with the retina. Part of the beam enters only one half of the pupil and the other part of the beam enters the other half of the pupil.

31.3   Objective optometers

31.3.1   Split image/vernier acuity

The principle of the split image/vernier alignment and a subjective implementation have been explained in Section 31.2.5. One objective implementation is that used in the Fincham coincidence optometer.

The Fincham coincidence optometer

In the Fincham coincidence optometer, as shown in Figure 31.3a, a target at T, usually a thin line, is imaged through one side of the pupil of the subject's eye by the collimating/de-collimating lens. If the subject is correctly focussed on this target, the image at T' will be formed sharply on the retina. On the other hand, if the subject is ametropic, the target is focussed either in front of or behind the retina. The example in the diagram is for a myopic eye and the focussed image at T' is formed in front of the retina and retinal blurred image of the target is formed to the side of the axis at T'_B.

Some of the light forming the image is reflected back diffusely out of the eye, as shown in Figure 31.3b. Because the reflection is diffuse, the reflected beam fills the pupil. If the subject is emmetropic, this beam will emerge from the eye collimated. Thus the collimating/de-collimating lens images the retinal reflex onto its back focal plane at F' in the case of emmetropia and in the vicinity of this plane for ametropia.

For ametropic eyes, this target image is formed off-axis. In the region of the collimating/de-collimating lens's back focal plane is a Dove prism which intersects half the beam. This intersected part is refracted, reflected, refracted and finally emerges from the prism apparently from an image on the opposite side of the axis, as shown in Figure 31.3b. The observer views the two images through an eyepiece and sees two images of the target as shown in the diagram. Thus for the ametropic eye, the two images are transversely displaced and the amount of displacement is a measure of the refractive error. The observer aligns the two images by adjusting the position of the target. The new target position is also a measure of refractive error. Since the alignment is one of vernier acuity, the alignment can be made to a high degree of precision.

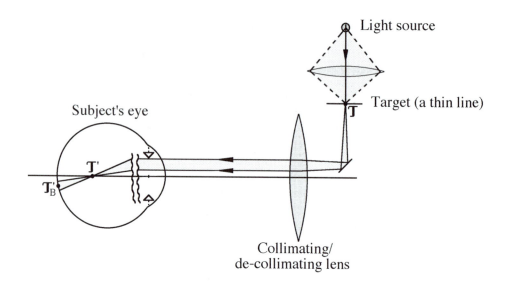

Light source

Target (a thin line)

Subject's eye

Collimating/
de-collimating lens

Fig. 31.3a: The Fincham
coincidence optometer : the
target projection system.

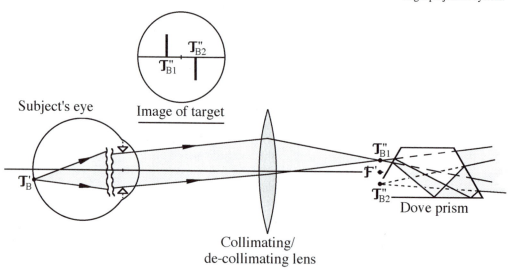

Image of target

Subject's eye

Dove prism

Collimating/
de-collimating lens

31.3.2 *Retinoscopy*

Retinoscopy is probably the most common method of measuring the refractive
state of the eye and it is a method closely related to the knife-edge test described
in Chapter 11, which is used to locate a paraxial image and to examine the
aberrations in a beam.

The schematic layout of the retinoscope is shown in Figure 31.4a. The small
source of light at s (spot or streak) is imaged by a mirror into the subject's
eye, producing an illuminated patch of light on the retina. The mirror is rotated
backwards and forwards, as shown, through a few degrees to move the patch

Fig. 31.3b: The Fincham
coincidence optometer: the
reflected beam.

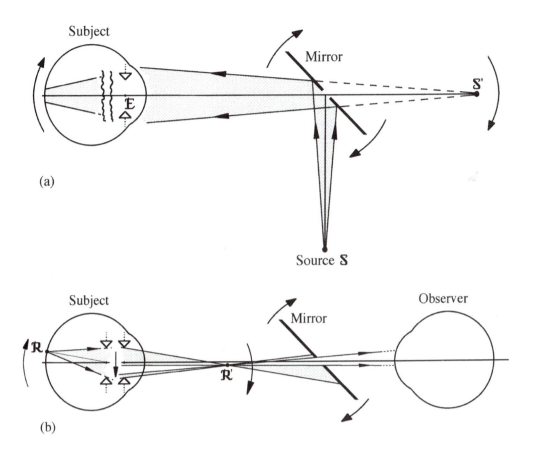

(a)

(b)

Fig. 31.4:   (a) Illumination system of the retinoscope; (b) retinal reflex movement for an eye myopic relative to the mirror.

of light over the retina, and the movement of the retinal reflex in the plane of the subject's pupil is used as an indicator of the refractive error relative to the mirror.

Let us now assume that the subject's eye is myopic relative to the mirror and the light patch on the retina is a point source at R, as shown in Figure 31.4b. The point R is now imaged by the eye to a position at R', which is in front of the retinoscope observation viewing aperture in the centre of the mirror. The observer's eye is placed behind the viewing aperture as shown in the diagram. As the mirror is rotated, the point R and hence its image R' move in arcs as shown in this diagram. The observer's eye is focussed on the pupil of the subject's eye. From the diagram, it is clear that some of the light passing through the subject's pupil is vignetted by the mirror and the corresponding areas of the subject's pupil will appear dark to the observer. Thus the observer sees only a part of the subject's pupil illuminated. As the image R' sweeps past the viewing aperture, this bright patch will move across the subject's pupil. In this example, the bright patch on the pupil moves downwards and in the opposite direction to the illuminating beam. Hence an eye, myopic relative to the mirror, produces an "against" motion. For eyes hyperopic relative to the mirror, the motion is "with".

If the point R' is now imaged in the same plane as the mirror or the viewing aperture; that is the degree of myopia is equivalent to the distance of the subject's eye to the viewing aperture, vignetting is either nil or complete. Thus as the point

$\mathcal{R}'$ is swept past the viewing aperture, either the pupil is completely illuminated or not illuminated at all and therefore the rate of movement can be regarded as infinite. This is called the point of reversal.

In the process of retinoscopy, auxiliary trial lenses are placed in front of the subject's eye to obtain the point of reversal. The correcting trial lenses give the refraction of the eye relative to the position of the retinoscope.

We have assumed in the above explanation of the process that the illuminated region of light on the retina is a point. In reality, it must have a finite size but in many retinoscopes the light source is a linear filament so that the retinal region of illumination is a long slit. The direction of mirror rotation is always perpendicular to the filament and the slit.

In the above explanation, we have also assumed that the refractive error is spherical in nature, that is myopia or hypermetropia. When astigmatism is present, the appearance of the bright patch seen by the observer is more complicated. If the mirror rotation is not aligned in one of the principal meridians of the eye, the bright patch seen in the pupil does not move in either the same or the opposite direction to the mirror, but at some angle to it. If the light source is a linear filament, the orientation of the bright patch is also different from that of the light source. In the presence of astigmatism, one of the principal meridians must be determined before reversal is attempted. The process is then similar to that described above, with reversal being obtained along each principal meridian to determine the refractive error.

A fuller account of the procedure of retinoscopy is given in texts such as Bennett and Rabbetts (1989).

31.3.3   *Photographic systems*

The use of photographic systems for determining the refraction of the eye is termed photorefraction (Howland and Howland 1974). The method has been used for infants and young children. Essentially, a flash photograph is taken of the eyes, with the flash source near the plane of the camera. The size of the pupil reflex recorded by the camera indicates the degree of refractive error.

In its simplest form, a small source of light is mounted immediately in front of the camera. If the eye is focused for the source, the light returned from the eye is occluded from the camera because it returns to the source. The pupil will appear dark. As the eye becomes more out of focus, the size of the light patch returning to the camera increases. Howland et al. (1979) took photographs with the camera focussed both in front of and behind the pupil plane in order to resolve ambiguity of the sign of the refractive error. The refractive error can be mathematically related to the film image size and the other variables. Astigmatism is apparent as an elliptical shape in the film image plane, with the axis corresponding to the astigmatic axes. Howland et al. (1979) referred to their method as isotropic refraction.

In eccentric photorefraction, the light source is decentred beyond the edge of the camera lens. An illuminated pupil reflex in the shape of a gibbous moon is imaged onto the camera film plane. In myopia, the reflex is on the same side as the source and the reverse occurs for hypemetropia. The gibbous shaped reflex size can be shown to be related to the amount of refractive error. This theory has been described and evaluated by Bobier and Braddick (1985).

Summary of main symbols

s, s'	source and its image
$\mathcal{R}, \mathcal{R}', \mathcal{R}''$	retinal conjugates of subject
$\mathcal{T}, \mathcal{T}'$	target conjugates
o, o'	object and image positions (axial case)
Q, Q', etc.	off-axis image points
$\varepsilon, \varepsilon'$	centres of subject's entrance and exit pupils
$\mathcal{F}'$	back focal point of Fincham coincidence optometer collimating/de-collimating lens

Section 31.2.2: Laser speckle optometer

ΔF	refractive error of subject
r	radius of drum
d	distance between drum and subject's eye
s	distance of plane of stationarity from drum axis
ϕ	direction of laser beam from line of sight

References and bibliography

*Bennett A.G. and Rabbetts R.B. (1989). *Clinical Visual Optics*, 2nd ed. Butterworth-Heinemann, London, Chapters 17 and 18.

*Bobier W.R. and Braddick O. (1985). Eccentric photorefraction: Optical analysis and empirical measures. *Am. J. Optom. Physiol. Opt.* 62, 614–620.

*Charman W.N. (1974). On the position of the plane of stationarity in laser refraction. *Am. J. Optom. Physiol. Opt.* 51(11), 832–837.

*Charman W.N. (1975). Some sources of discrepancy between static retinoscopy of subjective refraction. *Br. J. Physiol. Opt.* 30, 108–118.

*Charman W.N. (1979). Speckle movement in laser refraction. I. Theory. *Am. J. Optom. Physiol. Opt.* 65(4), 219–227.

*Charman W.N. and Heron G. (1975). A simple infra-red optometer for accommodation studies. *Br. J. Physiol. Opt.* 30(1), 1–12.

Charman W.N. and Whitefoot H. (1979). Speckle motion in laser refraction. II. Experimental. *Am. J. Optom. Physiol. Opt.* 56(5), 295–304.

Cornsweet T.N. and Crane H.D. (1970). Servo-controlled infra-red optometer. *J. Opt. Soc. Am.* 60(4), 548–554.

Crane H.D. and Steele C.M. (1978). Accurate three-dimensional eye-tracker. *Appl. Opt.* 17(5), 691–705.

Henson D.B. (1983). *Optometric Instrumentation*. Butterworths, London, Chapter 8.

*Howland H.C., Atkinson J., and Braddick O. (1979). A new method of photographic refraction of the eye. *J. Opt. Soc. Am.* 69, 1486.

Howland H.C., Atkinson J., Braddick O., and French J. (1978). Infant astigmatism measured by photorefraction. *Science* 202, 331–333.

*Howland H.C. and Howland B. (1974). Photorefraction: A technique for study of refractive state at a distance. *J. Opt. Soc. Am.* 64(2), 240–249.

Kruger P.B. (1979). Infra-red recording retinoscope for monitoring accommodation. *Am. J. Optom. Physiol. Opt.* 56(2), 116–123.

Lovasik J.V. (1983). A simple continuously recording infra-red optometer. *Am. J. Optom. Physiol. Opt.* 60(1), 80–87.

Polse K.A. and Kerr K.E. (1975). An automatic objective optometer. *Arch. Ophthal.* 93(3), 225–231.

*Simonelli N.M. (1980). Polarized vernier optometer. *Behav. Res. Meth. Instrument.* 12(3), 293–296.

Wesemann W. and Rassow B. (1987). Automatic infra-red refractors: A comparative study. *Am. J. Optom. Physiol. Opt.* 64(8), 627–638.

32

Binocular vision testing instruments

32.0 Introduction

There are a number of ophthalmic instruments that are designed to assess various aspects of binocular vision and functions, such as

(a)   simultaneous perception, fusion and stereopsis
(b)   accommodation/convergence relationships
(c)   presence and magnitude of phorias.

These will now be discussed in greater detail.

32.0.1 *Simultaneous perception, fusion and stereopsis*

The simplest level of binocular vision is **simultaneous perception**, the ability to see two images, one formed on each retina simultaneously. For normal binocular vision, it must be possible to see two images, one formed at the fovea of each retina, and to superimpose them. This is called **simultaneous foveal perception**.

Fusion is a more refined level of binocular vision, in which two similar images, one formed on each retina, are seen at the same time and are blended as one. That there is a blending, and not just a superposition, can be tested by placing a small prism in front of one eye. This will displace the retinal image from the fovea. If fusion is present, a corrective eye movement will be made to place the image back at the fovea.

When a solid object is viewed, the two retinal images are slightly different because the two eyes view the object from different perspectives. **Stereopsis** is the ability to see slightly different images and to blend them as one with a perception or appreciation of depth. This is the most refined level of binocular vision.

A more detailed account of these binocular vision levels can be found in texts such as Duke-Elder (1973).

32.0.2 *Accommodation/convergence relationships*

When the eyes accommodate, convergence is elicited. This is called **accommodative convergence**. The reverse is also true; that is convergence can elicit accommodation. The relation between accommodation and convergence is referred to as synkinesis. Accommodation/convergence relationships and binocular instruments are also discussed in Chapter 37.

The convergence response to a unit stimulus to accommodate is termed the **AC/A ratio**. There are different ways this can be measured. Ideally, the ratio would be the same as the ratio of convergence and accommodation required to maintain clear single binocular vision. For example, someone with an inter-pupillary distance of 60 mm would ideally have an AC/A ratio of 6 m^{-1}/m^{-1}. The AC/A ratio is lower than this in the majority of people, leaving a relative divergence of the visual axes in near vision when there is no stimulus to fusion. Fortunately, in most people there is a flexibility of the synkinesis, in which either accommodation or convergence can be increased or relaxed while the other remains relatively steady. The range within which convergence can be changed, while accommodation to a stimulus is fixed, without producing diplopia is called the amplitude of relative convergence. The range within which accommodation can be changed, while convergence is fixed, without producing blurring is called the amplitude of relative accommodation.

A more detailed discussion of the accommodation–convergence relationship can be found in texts such as Duke-Elder (1973).

32.0.3   *Heterophorias and heterotropias*

As discussed in Section 32.0.1, in normal binocular vision, the eyes must align or fixate so that only one image is perceived. In these instances, the visual axes of the two eyes may be regarded as intersecting at the point of fixation or object of interest. However in many instances, if there is no stimulus to fusion, the eyes will align in different directions. This is related to the AC/A ratio. This is easily achieved by presenting dissimilar images to each eye. This tendency to mis-alignment is called a **heterophoria**. However if mis-alignment occurs even when the images are similar or identical, that is there is a stimulus to fusion, this condition is called a **heterotropia**. Double vision (**diplopia**) is often experienced, but if there is a suppression of one eye, only one image will be seen. When the eyes misalign this may be

> ESOphoria (or tropia) – inwards
> EXOphoria (or tropia) – outwards
> HYPERphoria (or tropia) – upwards
> HYPOphoria (or tropia) – downwards
> CYCLOphoria – rotation

Many binocular instruments have been designed to diagnose these anomalies of binocular vision. The instruments work by dividing space into two areas, one area being seen by each eye. Division of space is accomplished by the use of lenses, mirrors, septums and coloured filters. A detailed account of such instruments is given by Lyle and Wybar (1970). We will now look at some of the most common instruments.

32.1   The synoptophore

The synoptophore is the most complex of all binocular vision testing instruments. It is designed to enable the presentation of similar or dissimilar images at various levels of accommodation and convergence and to vary that convergence and accommodation independently. When the images are dissimilar, any

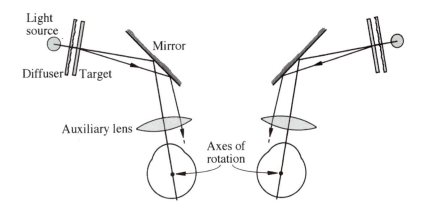

Fig. 32.1: The optical construction of the synoptophore.

Light source

Mirror

Diffuser Target

Auxiliary lens

Axes of rotation

tendency of the eyes to disassociate will be enhanced. The images may be independently transversely shifted or rotated in order to measure heterophoria or cyclophoria.

The basic design is shown in Figure 32.1. The accommodation stimulus for each eye is controlled by the auxiliary lens. The assemblies are encompassed within tubes, each of which can be rotated about a vertical axis positioned to pass approximately through the centre of rotation of the corresponding eye. This rotation changes the convergence demand of the eyes.

Different types of targets are available to test different aspects of binocular functions. The horizontal movement of the targets alters the stimulus to convergence. It is also possible to move the targets vertically. The target movements allow heterophorias and heterotropias to be measured on scales on the instrument. The targets can also be rotated, allowing cyclophorias to be investigated.

32.1.1 *Pupils*

Each target of the synoptophore is viewed directly or through a simple lens that is often a trial case spectacle lenses. Therefore the synoptophore has no intrinsic aperture stops and the entrance pupils of the eyes become the exit pupils of each arm of the synoptophore optics. Because the trial case lenses are much wider than the pupils of the eyes, accurate location of the eye is not required.

32.2 Stereoscopes

Stereoscopes are instruments for viewing stereoscopic targets or pairs of images, either to demonstrate stereopsis or to measure stereopsis or stereoscopic acuity. A very simple stereoscope is shown in Figure 15.13. The stereoscopic images are viewed through separate lenses. The optical axes are parallel and the stereoscopic pairs placed at or near the front focal points $\mathcal{F}$ of the viewing lenses. Thus the images are formed at infinity with parallel visual axes to prevent any conflict between convergence and accommodation requirements. The stereoscopic target position can often be moved sideways to stimulate different levels of convergence or divergence.

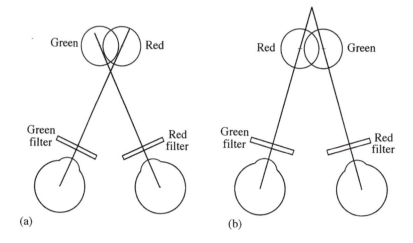

32.3   Anaglyphs

Anaglyphs consist of red and green targets on a white background. One eye looks through a green filter, so the green target cannot be seen against the background and the red target appears black. The other eye looks through a red filter, so that the red target is not seen and the green target appears black.

Different types of targets may be used to test the same binocular functions as the synoptophore targets. Figure 32.2 shows an anaglyph target used to stimulate relative divergence or convergence. In (a), the green circle is to the left of the red circle. The green circle is seen by the right eye and the red circle is seen by the left eye. To attain fusion, there must be additional convergence of the eyes relative to that usually required for the viewing distance. In (b), the target has been reversed. To attain fusion, there must be less convergence of the eyes than is usually required for the viewing distance.

References and bibliography

*Duke-Elder W.S. (ed.). (1973). *System of Ophthalmology*, Vol. VI. *Ocular Motility and Strabismus*, Henry Kimpton, London, Chapters V and VI.
*Lyle T.K. and Wybar K.C. (1970). *Practical Orthoptics in the Treatment of Squint*. 5th. ed. H.K. Lewis, London, Chapter 7.

Part V

Aberrations and image quality

33

Aberration theory

33.0 Introduction

Aberrations were introduced in Chapter 5 but only discussed qualitatively. Now they will be discussed quantitatively and in greater detail. Equations will be presented for calculating aberration levels as a function of system construction parameters, aperture stop size, conjugate plane positions and position of the object point in the field-of-view for any rotationally symmetric system. However, since the derivations of these equations are complex, space consuming and adequately covered in other texts, most of the equations will be presented here without any derivation. The equations will be mostly drawn from two texts: Hopkins (1950) and Welford (1986). Derivations of equations will only be included if the derivations are not adequately or suitably covered elsewhere.

The calculation of exact aberrations requires time consuming and tedious tracing of real rays. On the other hand, an estimate of the aberration levels can be found relatively simply from the results of two suitable paraxial ray traces. For many purposes, these estimates of aberration levels are adequate. The two paraxial rays are the marginal and pupil rays. The ray angles $\{u\}$ and ray heights $\{h\}$ along with other system constructional parameters are fed into equations for the calculations of these aberrations. One such set of equations is the **Seidel aberration** equations and the resulting estimates of aberrations are called Seidel aberrations. These equations will be introduced and discussed in the next section. While these equations are approximate, they have three very useful attributes: (a) they allow the identification and quantification of different aberration types such as spherical aberration and coma, (b) they give the aberration contribution by each surface and (c) they become more accurate the smaller the aperture and field size. Thus they usually give an accurate estimate of aberration levels for small aperture and small field systems.

33.1   Seidel aberration theory: The general lens system

Seidel aberrations are path length aberrations and can be thought of as arising at surfaces, as a result of refractions at the surfaces. Thus the aberrations can be calculated at each surface independently and the final system aberration found by adding the contributions from each surface (Seidel sums). Seidel aberrations are an approximation to the total or exact aberration of a system and therefore the surface sum rule is only valid when the aberrations are small. For large levels of aberrations finite ray tracing can be used directly to determine total or exact aberrations. However, standard finite ray tracing does not readily give the aberration contribution due to each surface.

The following equations for Seidel aberrations give the aberrations for a ray traced from the edge of the field and passing through the edge of the pupil, as defined by the marginal and pupil rays. Since, in general, aberrations increase with distance from the axis, they are usually the maximum aberrations to be expected from the system. The aberrations for a ray traced from a different point in the field and through a different point in the pupil can be found by a procedure explained in Section 33.1.3.

33.1.1   *Aberrations at a surface and Seidel sums*

Seidel aberrations are calculated at each surface and the total aberrations are the sums of the contributions from each surface.

A Seidel aberration or primary aberration analysis can be carried out using the results of two paraxial rays, namely the paraxial marginal and the paraxial pupil rays. The two rays are defined in Chapter 9 and diagrammatically in Figure 9.4 (a and b). The notation used for the ray tracing is that shown in Figure 2.8. There are seven Seidel aberrations. Five are monochromatic aberrations, that is they occur in monochromatic light, and two are chromatic aberrations, which only occur in non-monochromatic light. Hopkins (1950) and Welford (1986) gave the following equations for the Seidel aberrations at the jth surface.

Monochromatic aberrations

$$S_1 = A_j^2 h_j \Delta(u/n)_j \qquad\qquad \text{spherical} \qquad (33.1\text{a})$$

$$S_2 = A_j B_j h_j \Delta(u/n)_j \qquad\qquad \text{coma} \qquad (33.1\text{b})$$

$$S_3 = B_j^2 h_j \Delta(u/n)_j \qquad\qquad \text{astigmatism} \qquad (33.1\text{c})$$

$$S_4 = H^2 P_j \qquad\qquad\qquad \text{Petzval sum} \qquad (33.1\text{d})$$

$$S_5 = B_j[H^2 P_j + B_j^2 h_j \Delta(u/n)_j]/A_j \quad \text{distortion} \qquad (33.1\text{e})$$

Chromatic aberrations

$$C_{\text{L}} = A_j h_j \Delta(\delta n/n)_j \qquad\qquad \text{longitudinal} \qquad (33.1\text{f})$$

$$C_{\text{T}} = B_j h_j \Delta(\delta n/n)_j \qquad\qquad \text{transverse} \qquad (33.1\text{g})$$

where

u_j and h_j = paraxial marginal ray angles and heights

$\bar{u}_j$ and $\bar{h}_j$ = the paraxial pupil ray angles and heights

$$A_j = n_j(h_j C_j + u_j) = n'_j(h_j C_j + u'_j) \qquad (33.2a)$$

$$n_j(h_j C_j + u_j) = ni$$

$$n'_j(h_j C_j + u'_j) = n'i'$$

$n'i' = ni$ (paraxial version of Snell's law)

i and i' = angles of incidence and refraction, respectively

C_j = curvature of the surface

$$B_j = n_j(\bar{h}_j C_j + \bar{u}_j) = n'_j(\bar{h}_j C_j + \bar{u}'_j) \qquad (33.2b)$$

$$P_j = -C_j[(1/n'_j) - (1/n_j)] \qquad (33.2c)$$

$$\Delta(u/n)_j = (-u'_j/n'_j) + (u_j/n_j) \qquad (33.2d)$$

$$H = n_j(\bar{u}_j h_j - u_j \bar{h}_j) = n'_j(\bar{u}'_j h_j - u'_j \bar{h}_j) \text{ is the} \qquad (33.2e)$$
optical invariant

$$\delta n = n_F - n_C = (n-1)/V \qquad (33.2f)$$

$$\Delta(\delta n/n) = \delta n'_j/n'_j - \delta n_j/n_j \qquad (33.2g)$$

n_F = refractive index at the hydrogen
blue line ($\lambda_F = 486.1$ nm)

n_C = refractive index at the hydrogen
red line ($\lambda_C = 656.3$ nm)

n = refractive index at the helium yellow line
($\lambda_d = 587.6$ nm) sometimes denoted in the literature as n_d

V = Abbe V-value usually denoted by the symbol V_d
and defined by the equation

$$V = (n-1)/(n_F - n_C) \qquad (33.2h)$$

These equations give the magnitude of the aberrations for rays traced from the edge of the field and through the edge of the aperture.

Seidel sums

The Seidel aberrations for the whole system are then found by summing the individual aberrations at each surface. Thus for any Seidel aberration S_i

$$S_i = \sum_{j=1}^{k} S_{i,j} \tag{33.3}$$

where k is the number of surfaces and $S_{i,j}$ is the ith Seidel aberration at the jth surface. The above summations are sometimes known as Seidel sums.

33.1.2   *Units*

The Seidel aberrations are measures of the optical path differences, that is they are wave aberrations and therefore have units of length, for example millimetres or metres. Rather than express their numerical values in terms of these units, it is far more meaningful to express them in terms of units of a reference wavelength. For example, suppose that $S_1 = 0.0123$ mm and the reference wavelength is $\lambda = 587.6$ nm, then

$$S_1 = 0.0123/0.0005876 = 20.9\lambda$$

in wavelength units.

However, it is not common to express distortion and the transverse chromatic aberration in this form. A more useful representation is the corresponding fractional distortion, as defined by equation (5.3), and fractional transverse chromatic aberration. These are related to the above Seidel aberrations by the following equations:

$$\text{fractional distortion} = \frac{\bar{\eta}' - \eta'}{\eta'}$$

Welford (1986, 146) has shown that

$$\text{fractional distortion} = -\frac{S_5}{2H} \tag{33.4}$$

Similarly

$$\text{fractional transverse chromatic aberration} = \frac{(\eta_F - \eta_C)}{\eta_d} = -\frac{C_T}{H}$$

$$\tag{33.5}$$

where
η_F = image height for light of the wavelength λ_F

η_C = image height for light of the wavelength λ_C

η_d = image height for light of the wavelength λ_d

Table 33.1. Effect of changing the aperture and field size on the Seidel aberrations [see equation (33.6)]

		Wave		Transverse		Longitudinal	
		p	q	p	q	p	q
Monochromatic							
Spherical	S_1	4	0	3	0	2	0
Coma	S_2	3	1	2	1	1	1
Astigmatism	S_3	2	2	1	2	0	2
Petzval sum	S_4	2	2	1	2	0	2
Distortion	S_5	1	3	0	3	—	—
Distortion	%	—	—	0	2	—	—
Chromatic							
Longitudinal	C_L	2	0	1	0	0	0
Transverse	C_T	1	1	0	1	—	—
Transverse	%	—	—	0	0	—	—

Note: p is the aperture value and q is the field value.

As fractional values, these aberrations are in a transverse aberration form.

The above equations were used to calculate the Seidel aberrations of some of the schematic eyes listed in Appendix 3. This appendix contains both the constructional data for each eye and the paraxial marginal and pupil ray trace data and thus the aberration data given in the appendix can be used to check computer programs written to determine Seidel aberrations.

33.1.3   Effect of aperture diameter and field size

Wave aberration

By analysing the aperture and field size dependence of each of the above aberrations, it can be seen that each aberration is proportional to the aperture and field sizes according to the rule

$$\text{(aperture radius)}^p \times \text{(field radius)}^q$$

where p and q are given in Table 33.1. Thus, once the above Seidel aberrations are known for particular aperture and field sizes, they can be easily calculated for any other aperture and field size by multiplying the old values by the following factor

$$\text{factor} = \text{(new aperture radius/old aperture radius)}^p$$

$$\times \text{(new field radius/old field radius)}^q \tag{33.6}$$

For example, suppose the Seidel coma has a value of 100 units for a given system. If the aperture were reduced to 0.7 of its initial value and the field increased by 1.3 of its old value, then from equation (33.6)

$$\text{factor} = 0.7^p \times 1.3^q$$

From Table 33.1, for coma, $p = 3$ and $q = 1$. Therefore the new level of coma is

$$100 \times (0.7^3 \times 1.3^1) = 44.6 \text{ units}$$

This property of Seidel aberrations is very useful in optical system assessment and design.

Transverse aberration

Fractional distortion and transverse chromatic aberration are measures of the transverse aberration. The aperture dependency and field dependency also vary with aperture and field size by a factor of the form given by equation (33.6), but here p and q have different values which are given by Table 33.1.

Longitudinal aberration

Longitudinal aberrations are dependent upon aperture and field according to the same rules but with different values of p and q that are given in Table 33.1.

33.1.4 *The aplanatic surface*

Looking back to equations (33.1a, b, and c), the term $\Delta(u/n)_j$ occurs in each case. Thus if its value is zero for a particular surface, then the aberrations S_1, S_2 and S_3 are zero. Such a surface is known as an **aplanatic** surface and the above condition requires that

$$(u'_j/n'_j) - (u_j/n_j) = 0 \tag{33.7}$$

However the surface is only aplanatic for a particular pair of conjugate planes and not for any other pair. Equations for the position of these planes can easily be found and are

$$l = [1 + (n'/n)]/C \quad \text{and} \quad l' = [(n/n') + 1]/C \tag{33.8}$$

These equations show that the aplanatic points lie on the same side of the surface and on the same side as the centre of curvature.

While the above discussion only refers to the Seidel aberrations, an aplanatic surface has zero exact spherical aberration; that is all orders of spherical aberration are zero.

33.2 Monochromatic Seidel aberrations: The thin lens

In the calculation of the aberrations of a thin lens, it is conventional, convenient and instructive to firstly calculate the aberrations with the aperture stop placed at the lens. The resulting aberrations are called **central** aberrations.

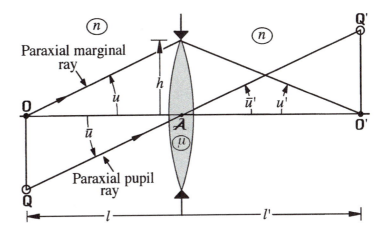

Fig. 33.1: Seidel aberrations of a thin lens with the aperture stop at the lens: definition of variables.

33.2.1 Central aberrations

The Seidel aberrations of a thin lens immersed in a medium of index n, of equivalent power F, made of a material with an index μ and with the aperture stop at the lens can be given by a set of simple equations. Welford (1986) presented a set of equations in terms of the relative index μ_r, defined as

$$\mu_r = \mu/n \tag{33.9}$$

If we express the lens index in terms of actual index μ instead of the relative index μ_r, Welford's equations become

$$S_1^* = \frac{h^4 F^3}{4n^2} \left[\frac{n^2(\mu + 2n)}{\mu(\mu - n)^2} \Gamma^2 - \frac{4n(\mu + n)}{\mu(\mu - n)} \Gamma\Omega + \frac{\mu^2}{(\mu - n)^2} \right.$$

$$\left. + \frac{(3\mu + 2n)}{\mu} \Omega^2 \right] \tag{33.10a}$$

$$S_2^* = \frac{H h^2 F^2}{2n^2} \left[\frac{n(\mu + n)}{\mu(\mu - n)} \Gamma - \frac{(2\mu + n)}{\mu} \Omega \right] \tag{33.10b}$$

$$S_3^* = \frac{H^2 F}{n^2} \tag{33.10c}$$

$$S_4^* = \frac{H^2 F}{n\mu} \tag{33.10d}$$

$$S_5^* = 0 \tag{33.10e}$$

Some of the symbols are defined and explained in Figure 33.1. The others are defined and explained below.

(i) Γ is the shape factor of the lens introduced in Chapter 6. It is defined in terms of the surface curvatures C_1 and C_2 as follows

$$\Gamma = \frac{(C_1 + C_2)}{(C_1 - C_2)} \tag{33.11}$$

The effect of the shape of a lens on the numerical value of Γ is shown in Figure 6.1. Since the thin lens power is given by the equation

$$F = (C_1 - C_2)(\mu - n) \tag{33.12}$$

it follows that

$$C_1 = \frac{F(\Gamma + 1)}{2(\mu - n)} \quad \text{and} \quad C_2 = \frac{F(\Gamma - 1)}{2(\mu - n)} \tag{33.13}$$

The back surface power F_2 can be expressed in terms of Γ as follows

$$F_2 = -F(\Gamma - 1)/2 \tag{33.14a}$$

or

$$\Gamma = 1 - (2F_2/F) \tag{33.14b}$$

and for interest

$$F_1 = F(\Gamma + 1)/2 \tag{33.14c}$$

(ii) Ω specifies the position of the conjugates and is given by the equation

$$\Omega = \frac{(u' + u)}{(u' - u)} = \frac{(l + l')}{(l - l')} = \frac{(nL' + n'L)}{(nL' - n'L)} \tag{33.15}$$

The effect of conjugate position on the value of Ω is shown in Figure 33.2.

(iii) H is the optical invariant and given by any of the equations

$$H = n(\bar{u}h - u\bar{h}) = n'(\bar{u}'h - u'\bar{h}) = -nu\eta = -n'u'\eta'$$

$$= n_A \bar{u}_A \bar{\rho}_A \tag{33.16}$$

Example 33.1: Examine the effect of the shape on the monochromatic Seidel aberrations of a spectacle lens of power $-10\,\text{m}^{-1}$ with refractive index of 1.5, when the object is at infinity (that is $\Omega = +1$). The aperture stop is at the lens, with a marginal ray height at the lens of 3.150 mm (this corresponds to a stop radius of 4 mm when the stop is moved 27 mm back from the lens), an angular field radius of $10°\,[\bar{u} = \tan(10°) = 0.176327]$ and a reference wavelength of 555 nm.

Solution: Figure 33.3 shows the results of the calculations.

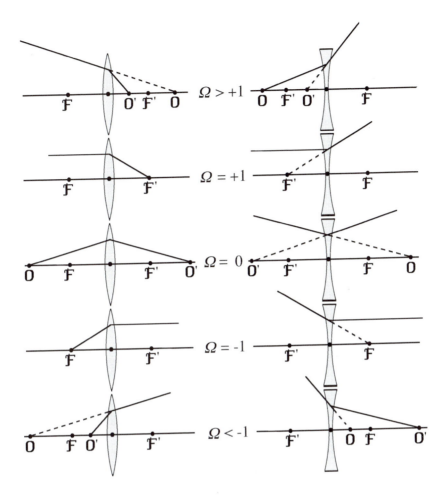

If we now return to the equation for the Seidel spherical aberration, we can see that it is quadratic in shape factor Γ, that is can be expressed in the form

$$S_1 = a_1 + b_1 \Gamma + c_1 \Gamma^2 \tag{33.17}$$

where algebraic expressions for a_1, b_1 and c_1 can be found from equation (33.10a). Now quadratic equations have two different roots, two identical roots or no roots. In general, the above quadratic for spherical aberration has no roots, indicating that spherical aberration cannot be eliminated for a single lens, only minimized. The above quadratic has a minimum when

$$\Gamma_{\min} = -b_1/(2c_1)$$

That is

$$\Gamma_{\min} = \frac{2(\mu^2 - n^2)}{n(\mu + 2n)} \Omega \tag{33.18}$$

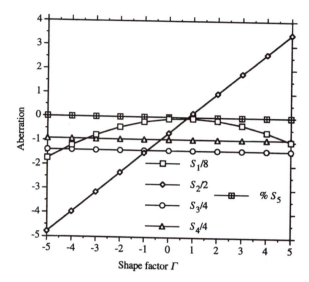

Fig. 33.3: Seidel aberrations of a thin (spectacle) lens of power $-10\ \mathrm{m}^{-1}$ with the aperture stop of radius 3.149 mm at the lens. The lens has a refractive index of 1.5.

The corresponding minimum value of S_1 is given by the equation

$$(S_1)_{\min} = \frac{h^4 F^3}{4n^2}\left[\frac{\mu^2}{(\mu-n)^2} - \frac{\mu}{(\mu+2n)}\Omega^2\right] \tag{33.19}$$

Example 33.2: For a single thin lens collimator of power 1 m^{-1} and made with a material of refractive index 1.5, find the shape and surface curvatures that give minimum spherical aberration. This design was given as an example in Section 6.1.1.4.

Solution: From the above results, we can conclude that the spherical aberration cannot be made zero, only minimized and equation (33.18) gives this minimum shape, which is independent of the power. From equation (33.15), we have $\Omega = -1$ and we also have $\mu = 1.5$ and $n = 1$. Therefore from equation (33.18), we have

$$\Gamma_{\min} = \frac{2(1.5^2 - 1)}{(1.5 + 2)}(-1) = -0.7143$$

Substituting this value in equations (33.13) with $F = 1$, the front and back surface curvatures are

$$C_1 = +0.2857\ \mathrm{m}^{-1} \quad C_2 = -1.7143\ \mathrm{m}^{-1}$$

Now the above minimum spherical aberration $(S_1)_{\min}$ is only zero when

$$\Omega = \pm\sqrt{[\mu(\mu+2n)]}/(\mu - n) \tag{33.20}$$

For example if $n = 1$ and $\mu = 1.5$, spherical aberration is only zero if $\Omega = \pm 4.58$.

Fig. 33.4: The stop shift
for the Seidel aberrations:
definition of variables.

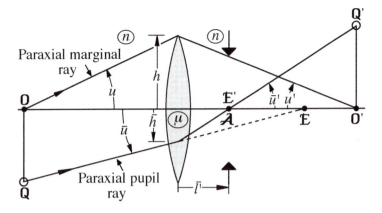

While spherical aberration is quadratic in the shape factor Γ, coma is linear in Γ and therefore there is always some shape factor which will make coma zero. This shape factor is, from equation (33.10b),

$$\Gamma = \frac{(2\mu + n)(\mu - n)}{n(\mu + n)}\Omega \quad \text{(zero coma)} \tag{33.21}$$

If we now find the value of the shape factor Γ which gives minimum spherical aberration and zero coma simultaneously, we must equate the righthand sides of equations (33.18) and (33.21). It is easily seen that this condition requires that the conjugate factor Ω be zero, that is the transverse magnification $M = -1$ from equation (33.15). It now follows from either of the above equations that the shape factor Γ must also be zero. Thus coma can only be zero simultaneously with minimum spherical aberration if the lens is used with symmetric conjugates and the lens is equiconvex or equiconcave.

33.2.2 *Non-central aberrations, effect of shift of the aperture stop*

It will now be seen that the off-axis aberrations depend upon the position of the aperture stop. Suppose that the aperture stop is now moved from being coincident with the lens to some arbitrary distance $\bar{l}'$ (see Figure 33.4) from the lens plane and on the image side. Now we introduce a stop shift parameter E defined by Welford (1986) as

$$E = \bar{h}/h \tag{33.22}$$

where h has the same value as before and $\bar{h}$ will be known from the path of the paraxial pupil ray.

The new Seidel aberrations for this new stop position are given by the stop shift formulae of Hopkins (1950), which are as follows.

$$S_1 = S_1^* \tag{33.23a}$$

$$S_2 = S_2^* + E S_1^* \tag{33.23b}$$

$$S_3 = S_3^* + 2E S_2^* + E^2 S_1^* \tag{33.23c}$$

$$S_4 = S_4^* \tag{33.23d}$$

$$S_5 = S_5^* + E(3S_3^* + S_4^*) + 3E^2 S_2^* + E^3 S_1^* \tag{33.23e}$$

It is now clear that the Seidel aberrations S_1, S_2, S_3 and S_5 are quadratic functions of the shape factor Γ and the conjugate factor Ω, since they all contain S_1^*, which is quadratic in these two quantities. However, in ophthalmic lens and system design, the shape factor is of greater interest and importance than the conjugate factor.

33.2.3   Quadratic form of the Seidel aberrations for a thin lens

Each of the above Seidel aberrations can now be expressed in a quadratic form as follows:

$$S_i = a_i + b_i \Gamma + c_i \Gamma^2 \tag{33.24}$$

where a_i, b_i and c_i are the coefficients for the ith aberration, for example $i = 2$ corresponds to coma. The coefficients a_i, b_i and c_i could be expressed in terms of the lens constructional and conjugate parameters but the development of these equations will not be pursued here.

> **Example 33.3:** Examine the effect of the shape factor Γ on the aberrations of the lens given in Example 33.1, but now with the aperture stop 27 mm to the right of the lens with a marginal ray height at the lens of 3.150 mm (this corresponds to an aperture stop radius of 4 mm), an angular field radius of $10°[\bar{u} = \tan(10°) = 0.176327]$ and a reference wavelength of 555 nm. This simulates a spectacle lens with a vertex distance of 12 mm and with the stop at the centre of rotation of the eye, which is assumed to be 15 mm behind the cornea. The object is at infinity (that is $\Omega = +1$).
>
> **Solution:** The results are shown in Figure 33.5 and the quadratic nature of the Seidel aberration dependence on shape factor can easily be seen for some of the aberrations. Compare this diagram with Figure 33.3, which shows the aberrations for the same condition but with the aperture stop at the lens. This comparison shows clearly the dependence of some aberrations on aperture stop position.

33.3   Chromatic Seidel aberrations: The thin lens

In the case of the monochromatic Seidel aberrations of a thin lens, it was very useful to begin the study of the aberrations by first examining the central aberrations, that is the aberrations with the aperture stop at the lens, and then apply the stop shift equations. We will follow this same procedure for the chromatic aberrations.

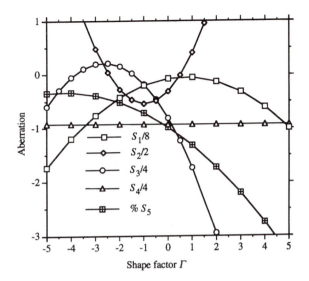

Fig. 33.5: Seidel aberrations of a thin (spectacle) lens of power -10 m^{-1} made with a refractive index of 1.5, with the stop of diameter 4 mm, placed 27 mm after the lens.

33.3.1 Central values

The chromatic Seidel aberrations of a thin lens immersed in a medium of index n, of equivalent power F, made of a material with an index μ and with the aperture stop at the lens can be calculated from the general equation given by equation (33.1f). In terms of the dispersion differences between the F and C lines the longitudinal aberration is

$$C_L^* = \left(\frac{\mu_F - \mu_C}{\mu} - \frac{n_F - n_C}{n} \right) \frac{\mu}{(\mu - n)} h^2 F \qquad (33.25\text{a})$$

and the transverse aberration is zero, that is

$$C_T^* = 0 \qquad (33.25\text{b})$$

If the lens is in air rather than some general medium, then the equation for longitudinal chromatic aberration is

$$C_L^* = \frac{h^2 F}{V} \qquad (33.26)$$

where V is the Abbe V-value given by equation (33.2h). The fact that the transverse chromatic aberration is zero if the aperture stop is at the lens is not surprising since rays that go through the centre of a thin lens are not deviated, irrespective of power or refractive index of the lens.

33.3.2 Non-central aberrations, effect of shift of the aperture stop

Using the same definition of E as for monochromatic aberrations, if the stop is not at the lens then the chromatic aberrations are

$$C_L = C_L^* = h^2 F / V \tag{33.27a}$$

$$C_T = C_T^* + E C_L^* = (\bar{h}/h) C_L = h \bar{h} F / V \tag{33.27b}$$

Note that the longitudinal aberration does not depend upon the stop position.

33.3.3 Chromatic aberration and change in the power with wavelength (in air)

The power F of a thin lens in air is given by the equation (3.12a) and using the fact that the refractive index is a function of wavelength, we can write this equation in the form

$$F(\lambda) = \Delta C [\mu(\lambda) - 1] \tag{33.28}$$

where

$$\Delta C = C_1 - C_2$$

Generally, refractive index increases (and as a result power also increases) with a decrease in wavelength. If we specify the nominal lens power as the power at the helium d line, we can write

$$F(\lambda_d) = F_d = \Delta C (\mu_d - 1) \tag{33.29}$$

where $\mu_d = \mu(\lambda_d)$. If we eliminate ΔC from equations (33.28) and (33.29), we have

$$F(\lambda) = \frac{F_d [\mu(\lambda) - 1]}{(\mu_d - 1)} \tag{33.30}$$

The change in power ΔF_{FC} between the F and C lines is thus given by the equation

$$\Delta F_{FC} = F(\lambda_F) - F(\lambda_C) \tag{33.31}$$

Using equation (33.30), the above equation can be written as

$$\Delta F_{FC} = \frac{F_d [\mu(\lambda_F) - 1]}{(\mu_d - 1)} - \frac{F_d [\mu(\lambda_C) - 1]}{(\mu_d - 1)} \tag{33.32}$$

which can be reduced to

$$\Delta F_{FC} = F_d / V \tag{33.33}$$

where V is given by equation (33.2) with n in that equation equal to μ_d and μ_d, μ_F and μ_C are the refractive indices at the d, F and C wavelengths defined in the symbol list given in the introductory notes at the beginning of the book. Now comparing equations (33.27a) and (33.33), and regarding F_d as equivalent to F, it is clear that

$$C_L = h^2 \Delta F_{FC} \qquad (33.34)$$

or

$$\Delta F_{FC} = C_L / h^2 \qquad (33.35)$$

which shows that the longitudinal chromatic aberration C_L is a measure of the difference in power between the F and C wavelengths.

33.3.4 The achromatic doublet (in air)

Since V-values are always positive, it follows from equation (33.33) that the power difference ΔF_{FC} is positive for a positive power lens and negative for a negative power lens. It will now be shown that a single lens can be replaced by a combination of a positive power lens and a negative power lens, such that the two contributions of ΔF_{FC} (one positive and one negative) cancel out. Let the single lens have a power F and let this lens be replaced by two lenses of powers F_1 and F_2. Let us suppose that these two lenses are thin and placed in contact. Then it is first necessary that

$$F_1 + F_2 = F \qquad (33.36)$$

Now if these two lenses are made from different glasses with dispersions V_1 and V_2 respectively, from equation (33.27a) the combined longitudinal chromatic aberration of the two lenses is

$$C_L = h^2 [(F_1/V_1) + (F_2/V_2)] \qquad (33.37)$$

That is the power difference between the F and C wavelengths is now, from equation (33.35),

$$\Delta F_{FC} = C_L / h^2 = (F_1/V_1) + (F_2/V_2) \qquad (33.38)$$

Thus the longitudinal chromatic aberration and resulting power difference ΔF_{FC} is zero if

$$(F_1/V_1) + (F_2/V_2) = 0 \qquad (33.39)$$

We can call this the **achromatic condition**. If we now solve equations (33.36) and (33.39) for F_1 and F_2, we have

$$F_1 = \frac{F V_1}{(V_1 - V_2)} \quad \text{and} \quad F_2 = \frac{F V_2}{(V_2 - V_1)} \qquad (33.40)$$

Since V-values are always positive, it follows from these equations that F_1 and F_2 are opposite in sign.

In practice, solving the above equations requires the choice of only the V-values of the materials and the refractive indices are irrelevant. The V-values of commercial optical glass can be found from a glass catalog. It is easy to see that the two V-values should be widely spaced; otherwise the powers of the individual components may be unacceptably large.

It should be made clear that the solution of these equations only leads to a zero power difference between the F and the C lines. There is still a variation of power with wavelength, and this residual variation gives rise to what is known as a **secondary spectrum**. An example will show the efficiency of an achromatic doublet in reducing chromatic aberration and the resulting level of the secondary spectrum.

> **Example 33.4:** Find the powers of an achromatic doublet with a power of 1.0 m^{-1} at the helium d line using Schott BK7 glass ($n_d =$ 1.51680, $V_d = 64.17$) and Schott SF9 glass ($n_d = 1.65446$, $V_d =$ 33.65)
>
> **Solution:** From equations (33.40) the thin lens powers are
>
> $$\text{BK7}: F_1 = +2.103 \text{ m}^{-1} \quad \text{and} \quad \text{SF9}: F_2 = -1.103 \text{ m}^{-1}$$
>
> We should note that these powers are independent of the refractive index of the particular materials.

Figure 33.6 shows the wavelength variation in power of a 1.0 m^{-1} thin lens made of Schott BK7 glass, using dispersion data from the Schott glass catalog and of the achromatic doublet design given in Example 33.4. Such a calculation requires a dispersion equation such as that given in Chapter 1 and dispersion data for each glass. The resulting variation of power of the combination with wavelength is also shown in the same diagram. Between 400 nm and 700 nm, the variation in power of the single lens is 0.034 m^{-1} compared to 0.005 m^{-1} for the achromatic doublet. That is, the variation in power of the doublet is about 1/7 that of the single lens, which is a considerable reduction in chromatic aberration. We can see from the diagram that an achromatic doublet has the same power at two wavelengths, thus reducing the level of chromatic aberration. This aberration can be further reduced by setting the power to be the same at three wavelengths, usually at a red, a blue and a green wavelength. Such lenses are called **apochromats**.

Solving the above equations only gives the power of the two lenses. The optical designer is then free to choose the shape of the lenses. In practice, achromatic doublets very often come in the "cemented" form; that is the two "inside" surfaces have the same radius of curvature and they are cemented together at this surface. Therefore, if the shape of one lens is chosen, the other becomes automatically fixed. Some possible shapes are shown in Figure 6.11b.

The shape does not affect the longitudinal chromatic aberration and therefore can be used to minimize or eliminate one or more of the monochromatic aberrations, commonly spherical aberration. Achromatic doublets can be used to eliminate spherical aberration, because the positive component provides

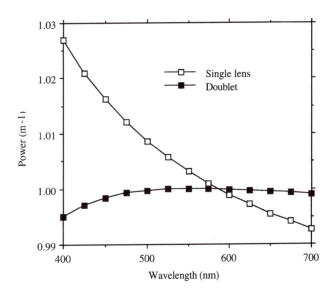

Fig. 33.6: Variation of power with wavelength of a single lens (Schott BK7) with a power of 1 m^{-1} and an equivalent achromatic doublet, made from Schott BK7 ($F = +2.103$ m^{-1}) and SF9 ($F = -1.103$ m^{-1}) glasses.

positive spherical aberration and the negative component contributes negative spherical aberration, and under some circumstances, these two contributions cancel out at some shape factor. This is a very useful property of these doublets, because while a single lens cannot be usually bent to give zero Seidel spherical aberration, a doublet often can, depending upon the choice of glass types.

Achromatic doublets are commonly used as objectives in a number of instruments such as microscopes, telescopes and collimators. Let us use the above results to design an achromatic doublet for use as a collimator.

Example 33.5: Using an achromatic doublet, design a 1 m focal length collimator which is free of Seidel spherical aberration.

Solution: In Example 33.2, we found the best shape of a single thin lens that was to be used as a collimator. However, the spherical aberration of this lens could not be made zero, only minimized. Here we will use an achromatic doublet instead. In Example 33.4, we found the powers of an achromatic doublet with a focal length of 1 m.

For small apertures we may assume that higher order aberrations will be sufficiently small and we can just aim for a lens free of Seidel spherical aberration. We can bend an achromatic doublet to vary the monochromatic aberrations, but we also have a choice of whether the positive or the negative component is the first component. For any shape, the level of monochromatic aberrations will depend upon this choice; therefore in the aberration analysis of doublets we should look at both orientations. A Seidel spherical analysis of the doublet solution given in Example 33.4 used as a collimator, that is with the object at the front focal point, will show that the Seidel spherical aberration cannot be made zero if the positive component is first. However, it can if the negative component is first.

Figure 33.7 shows how the Seidel spherical aberration of that doublet depends upon the shape of the doublet for an aperture stop at the

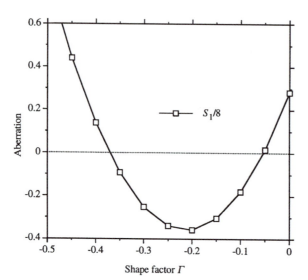

Fig. 33.7: Variation of Seidel spherical aberration with the shape Γ of the positive component of the achromatic doublet examined in Example 33.5.

lens of diameter 100 mm in wavelength units of 555 nm. From that diagram, we can see that this doublet is free of Seidel spherical aberration at two shapes, at positive component shapes of $\Gamma = -0.372$ and -0.054. Since there are two solutions, we are free to choose either one. However, there are other constraints such as the contribution to higher order spherical aberration or other Seidel aberrations. At this stage we have no way of determining the level of higher order spherical aberration and it is also preferable that collimators are free of coma. Therefore we should aim at the above solution that gives minimum coma. We can use the equations presented so far in this chapter to show that this lens is free of coma for only one shape factor and that is $\Gamma = -0.196$ for the positive power component. This is closer to the spherical aberration solution of $\Gamma = -0.054$, so we will use this solution but rounded to -0.05 for the design.

Using the positive power component shape factor of $\Gamma = -0.05$, the surface curvatures are

$$C_1 = +0.000458 \text{ mm}^{-1}, \quad C_2 = +0.002143 \text{ mm}^{-1} \quad \text{and}$$

$$C_3 = -0.001925 \text{ mm}^{-1}$$

This lens has a very similar shape to the second from the right in Figure 6.11b.

33.3.4.1   Transverse chromatic aberration

Achromatizing a lens can also control the transverse chromatic aberration. From equation (33.27b) it is clear that if the longitudinal chromatic aberration of a lens is zero so is the transverse chromatic aberration. Therefore the transverse chromatic aberration of an achromatic doublet is also zero for any position of the aperture stop.

33.4 Seidel aberrations: A thick parallel plate

The Seidel aberration equations, given in Section 33.2, imply that the aberrations of a beam passing through a plane parallel plate are zero, since the power F is zero. However, the equations only apply to a thin lens, that is, of zero thickness, and hence to a plane parallel plate of zero thickness. However, if the glass slab has a non-zero or finite thickness a beam passing through the slab does suffer from aberrations. According to Hopkins (1950) the aberrations of a thick parallel plate in air are as follows.

$$S_1 = -(\mu^2 - 1)u^4 d/\mu^3 \tag{33.41a}$$

$$S_2 = (\bar{u}/u)S_1 \tag{33.41b}$$

$$S_3 = (\bar{u}/u)^2 S_1 \tag{33.41c}$$

$$S_4 = 0 \tag{33.41d}$$

$$S_5 = (\bar{u}/u)^3 S_1 \tag{33.41e}$$

$$C_L = -(\mu - 1)u^2 d/(\mu^2 V) \tag{33.41f}$$

$$C_T = (\bar{u}/u)C_L \tag{33.41g}$$

where μ is the refractive index of the plate, d is the thickness of the plate and u and $\bar{u}$ are defined in Section 33.1.1. These equations are very useful in determining the aberration contributions of prisms, especially those prisms that have a particularly long optical path.

33.5 Seidel aberrations: Effect of surface asphericity

The above equations for the Seidel aberrations only apply to spherical surfaces. If a surface is aspheric, the asphericity leads to a change in the preceding aberrations and the change can be calculated from the value of surface asphericity. The most common aspheric surface shape is the conicoid, which was discussed in some depth in Chapter 5 but will be briefly reviewed here.

33.5.1 Conicoid surfaces

The simplest and perhaps the most common aspheric surface is the conicoid, which can be expressed in the form given by equation (5.17), but here with the radius of curvature C replacing the radius r ($C = 1/r$), as

$$C[p^2 + (1 + Q)Z^2] - 2Z = 0 \tag{33.42}$$

where

$$p^2 = X^2 + Y^2$$

and the asphericity Q sets the type of conicoid according to the rules given in Section 5.3.3.1. In that section, we showed that Z can be expressed in the two forms

$$Z = \frac{1 - \sqrt{[1 - C^2(1+Q)p^2]}}{C(1+Q)} \tag{33.43a}$$

and

$$Z = \frac{Cp^2}{1 + \sqrt{[1 - C^2(1+Q)p^2]}} \tag{33.43b}$$

The first of these is not applicable when the curvature C is zero or the asphericity Q is -1. On the other hand, the second form is always calculable and therefore is the preferred form for numerical calculations. Occasionally, it is useful to express Z as a power series in p^2, and in this case, the first equation is easier to expand using the binomial theorem to give

$$Z = \sum_{i=1}^{\infty} (-1)^{i+1} \, ^{1/2}\mathbb{C}_i C^{2i-1}(1+Q)^{i-1}p^{2i} \tag{33.44a}$$

where $^{1/2}\mathbb{C}_i$ is the combination function. For some purposes a better form is as follows:

$$Z = Cp^2/2 + (1/8)C^3(1+Q)p^4$$

$$+ \sum_{i=3}^{\infty} (-1)^{i+1} \, ^{1/2}\mathbb{C}_i C^{2i-1}(1+Q)^{i-1}p^{2i} \tag{33.44b}$$

This equation shows that the asphericity Q does not affect the power of the surface since this only depends upon the coefficient of p^2. However, since it is part of the coefficient of p^4, it has an effect on the level of the Seidel or primary and higher order aberrations.

If we neglect higher orders than those of order 2, the power series for Z reduces to

$$Z = Cp^2/2 \tag{33.44c}$$

Thus all conicoids have this same first term in common and which is quadratic in p.

Once a surface is aspherized by such a conic function, the Seidel aberrations change by amounts depending on the level of the asphericity factor Q. Hopkins (1950) and Welford (1986) defined a surface aspheric aberration contribution factor κ as

$$\kappa = C^3 Q h^4 (n' - n) \tag{33.45}$$

In terms of κ, the changes in the above Seidel aberrations at the surface, as given by Hopkins and Welford, are as follows.

$$\Delta S_1 = \kappa \tag{33.46a}$$

$$\Delta S_2 = E\kappa \tag{33.46b}$$

$$\Delta S_3 = E^2\kappa \tag{33.46c}$$

$$\Delta S_4 = 0 \tag{33.46d}$$

$$\Delta S_5 = E^3\kappa \tag{33.46e}$$

$$\Delta C_{\mathrm{L}} = 0 \tag{33.46f}$$

$$\Delta C_{\mathrm{T}} = 0 \tag{33.46g}$$

Note: S_4 is not affected by aspherizing the surface and it is also not affected by bending a lens.

It should also be noted from these equations that apart from the aperture stop position factor E, the aspheric aberration contributions do not depend upon the conjugate position Ω or shape factor Γ.

Example 33.6: Calculate the required asphericity to give zero Seidel spherical aberration of an equi-convex lens of power 20 m^{-1}, which is to be used as a condenser projecting the source with a unit (-1) magnification. Assume the lens is thin and has a refractive index of 1.6 and will be equally aspherized on both surfaces.

Solution: If we first calculate the Seidel spherical aberration for this lens, we can use equation (33.10a) with $F = 0.02\,\mathrm{mm}^{-1}$, $\Gamma = 0$, $\Omega = 0$, $\mu = 1.6$, $n = 1$ and an arbitrary ray height h at the lens of 20 mm. This equation gives

$$S_1 = 2.276\,\mathrm{mm}$$

which is distributed equally between the two surfaces and therefore there is 1.138 mm of aberration per surface. To find the required asphericity we need to find the value of Q in equation (33.45) which gives an aspheric contribution to the spherical aberration given by equation (33.46a) that is equal and opposite to the spherical contribution. If we consider the front surface, we need to find the value of Q which satisfies the following equation

$$C_1^3 Q h^4 (\mu - n) = -1.138$$

Here we have $\mu = 1.6$, $n = 1$, $h = 20$ mm and we find the value of C_1 from equation (33.13) with $F = 0.02\,\mathrm{mm}^{-1}$, $\Gamma = 0$, $\mu = 1.6$ and $n = 1$, giving $C_1 = 0.016667\,\mathrm{mm}^{-1}$. Substituting these values in the above equation gives $Q = -2.56$, and because the lens is symmetric, this is also the value for the back surface.

We can show that once a surface of a thin lens is aspherized, the Seidel aberrations that were previously quadratic in shape factor Γ now become cubic in this variable. We will demonstrate this by using the example of aspherizing the front surface of the lens. Now if the front surface is aspheric

$$\kappa = C_1^3 Q h^4 (\mu - n) \tag{33.47}$$

or by using equation (33.13) to replace C_1 by shape factor Γ

$$\kappa = \frac{[F(\Gamma + 1)]^3}{8(\mu - n)^3} Q h^4 (\mu - n) \tag{33.48}$$

which shows that κ is cubic in shape factor Γ and therefore those Seidel aberrations in equations (33.46) that contain a κ term must also be cubic in shape factor. That is the aberrations S_1, S_2, S_3 and S_5 can written in the form

$$S_i = a_i + b_i \Gamma + c_i \Gamma^2 + d_i \Gamma^3 \tag{33.49}$$

where expressions for the coefficients a_i, b_i, c_i and d_i are functions of refractive indices, lens power, conjugate positions and ray heights. Since a cubic equation always has a solution, it follows that for an aspherized thin lens, all of the monochromatic aberrations except field curvature can be set to any desired level by some appropriate choice of asphericity Q and shape factor Γ, but not simultaneously. However, not all of these solutions will be practical.

33.5.2 *Figured conicoid surfaces*

An aspheric surface of more generality than the conicoid is the "figured" conicoid. This is an extension of the conicoid equation (33.43a or b) and can be written as

$$Z = Z_c + \sum_{i=2}^{\infty} f_{2i} p^{2i} \tag{33.50}$$

where Z_c is the conicoid Z equation given by equation (33.43a or b) and f_4, f_6, f_8 and so on, are the figuring coefficients. The $f_4 p^4$ figuring term can be omitted provided the value of Q is changed and changes are also made to the other figuring coefficients. By comparing the figured expanded conicoid [equation (33.50)] with the expansion of Z_c given by equation (33.44a), with and without the f_4 figuring term, if f_4 is omitted then the new value Q' of Q is

$$Q' = Q + (8 f_4 / C^3) \tag{33.51}$$

and new figuring coefficients f'_{2n} are

$$f'_4 = 0$$

$$f'_{2i} = (-1)^{i+1} \, {}^{1/2} \mathbb{C}_i [(1 + Q)^{i-1} - (1 + Q')^{i-1}] C^{2i-1} + f_{2i} \tag{33.52}$$

for $i = 3, 4, 5 \ldots$ as far as is required. Only if $C = 0$ is a p^4 term necessary.

Hence, the Seidel aberrations can be calculated for any figured conicoid, after calculating the new asphericity value (Q'), but these equations show that only the f_4 term contributes to the Seidel aberrations.

33.6 Finite or exact aberrations

Exact aberrations must be found by tracing real or finite meridional and skew rays. The finite ray tracing is similar in principle to paraxial ray tracing in that it has two main components, a refraction algorithm and a transfer algorithm. However, the algorithms are more complex than their paraxial counterparts for several reasons. Firstly, the refraction at the surface must be exact and no approximations can be made. Secondly because of the need to trace skew rays, the equations must be in three dimensions not two. Thirdly, the procedures must also be able to cope with aspheric surfaces. Finite ray trace equations can be found in chapter 4 of Welford (1986).

In the quantification of aberrations, it is possible to use various representations as explained in Chapter 5. These are (1) longitudinal aberrations, (2) transverse aberrations and (3) wave aberrations. They will now be briefly explained in turn.

33.6.1 Longitudinal finite aberrations

The longitudinal aberration of a ray is the distance between the point where the ray intersects the optical axis in image space and the Gaussian image point o'. This distance is easily calculated from a finite ray trace. However, this distance is only meaningful if the image is at a finite distance. If the paraxial image is at infinity, the vergence of the ray is a better measure of the longitudinal aberration.

33.6.2 Transverse finite aberration

The transverse aberration is the distance of the point of intersection of the ray with the Gaussian image plane and the Gaussian image point ϱ'. For a skew ray, the aberration has two components $\delta\xi'$ and $\delta\eta'$, as shown in Figure 5.2b and here in Figure 33.8. These two values can be found very easily from the finite ray trace results. Once again, if the image plane is at infinity, these two values will be infinite and therefore meaningless. A more useful way of representing the transverse ray aberrations is to use the reduced values $\delta G'$ and $\delta H'$, which are defined in terms of $\delta\xi'$ and $\delta\eta'$ as follows:

$$\delta G' = \frac{n' \sin(\alpha')}{\lambda} \delta\xi' \quad \text{and} \quad \delta H' = \frac{n' \sin(\alpha')}{\lambda} \delta\eta' \tag{33.53}$$

These new quantities are always finite, and as defined above, are related to the width of the Airy disc of the diffraction limited point spread function. A unit value of $\delta G'$ and $\delta H'$ corresponds to a transverse aberration of 1.64 times the radius of the Airy disc, that is the radius of the Airy disc is 0.61 units.

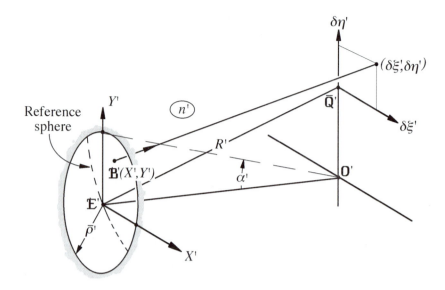

33.6.3   Wave aberrations

The wave aberration of a ray is a difference in optical path length from object
to image, between that of the ray and that of the pupil ray. We will explore this
in much greater detail in the next section.

33.7   The wave aberration function and polynomial

The concept of wave aberration was introduced in Chapter 5 and defined as a
path length aberration. We can examine this concept as follows. Consider a point
source in object space in an isotropic and homogeneous medium. The source
will emit spherical waves and thus the wavefronts will be spherical, as shown
diagramatically in Figure 2.2a. A section of each wavefront will pass into the
system and if the system is free of any aberration, these wavefronts will remain
spherical and finally leave the system with a spherical form. In image space,
they will converge to the Gaussian image point Q' as shown in the diagram.

   Now if the system has some aberration, the effect of the aberration is to distort
the wavefront as it passes through the optical system, as shown in Figure g. The
diagram shows a spherical wave entering the system through the entrance pupil
and leaving the system through the exit pupil. The wave aberration function
is a measure of this distortion. The ideal undistorted (spherical) wavefront at
the exit pupil is known as the **reference sphere**. In the diagram, the reference
sphere is shown as a dashed line, centred on the paraxial image point Q'.

33.7.1   Definition

The wave aberration W is the optical distance along a ray between the reference
sphere and the corresponding distorted wavefront. That is

$$W = [\mathcal{B}'\mathcal{B}'']$$

(33.54)

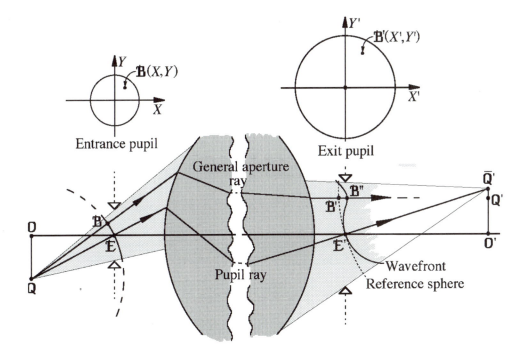

The square brackets denote optical paths. Now firstly

$$[\mathcal{BB}''] = [\mathcal{BB}'] + [\mathcal{B}'\mathcal{B}'']$$

and secondly, the segments of any two rays joining the same two wavefronts must have the same optical path lengths along the rays, that is

$$[\varepsilon\varepsilon'] = [\mathcal{BB}'']$$

It follows that

$$W = [\varepsilon\varepsilon'] - [\mathcal{BB}'] \tag{33.55}$$

Therefore the pupil ray is defined to be aberration free.

It is a function of the co-ordinates of the intersection point of the ray with the entrance or exit pupil. The wave aberration will usually be different for different rays traced from the same object point and for different object points with the same position in the exit pupil. Therefore the wave aberration is a function of position co-ordinates in the object plane (η) or image plane (η') and in the entrance pupil plane (X, Y) or exit pupil plane (X', Y') as shown in the insets of Figure 33.9. Thus we should write the wave aberration W as a function of these parameters, say in the form

$$W(\eta; X, Y)$$

However, the object or image plane or both or the entrance or exit pupils or both may be at infinity. In these cases, one cannot express the wave aberration in terms of the above co-ordinates. Fortunately this problem can be easily overcome by using relative co-ordinates for the various object/image plane and pupil plane co-ordinates. If $\eta_{\max}$ and $\eta'_{\max}$ are the radii of the object and image fields and if η and η' are actual co-ordinates in these planes, then the corresponding relative field point co-ordinate τ is defined as

$$\tau = \eta/\eta_{\max} \quad \text{and} \quad \tau' = \eta'/\eta'_{\max} \tag{33.56a}$$

Similarly, if the point $\mathcal{B}$ in the entrance pupil has co-ordinates (X, Y) and $\bar{\rho}$ is the pupil radius, then the relative co-ordinates are

$$x = X/\bar{\rho} \quad \text{and} \quad y = Y/\bar{\rho} \tag{33.56b}$$

In the exit pupil, the corresponding co-ordinates would be (x', y'), and providing the system is well corrected and the sine condition is reasonably satisfied,

$$x' = x, y' = y, \quad \text{and} \quad \tau' = \tau \tag{33.57}$$

The sine condition was defined and explained in Chapter 5. Thus we now should write

$$W(\eta; X, Y) \Rightarrow W(\tau; x, y) \quad \text{or} \quad W(\tau'; x', y') \tag{33.58}$$

33.7.2   The wave aberration polynomial

Now the wave aberration function W smoothly varies with these parameters and therefore can be expanded as a power series of these parameters. For an optical system having rotational symmetry about the optical axis, these variables only occur in a rotationally invariant form. That is the wave aberration function is a polynomial or power series in the variables

$$\tau^2, (x^2 + y^2) \quad \text{and} \quad \tau y$$

If we now use polar co-ordinates (r, ϕ) instead, where ϕ is measured from the positive y-axis and is positive for a clockwise direction as shown in Figure 33.10, we can write

$$W(\tau; x, y) \text{ as } W(\tau; r, \phi)$$

where

$$x = r \sin(\phi) \quad \text{and} \quad y = r \cos(\phi) \tag{33.59}$$

The aberration function can now be expressed as a power series in the variables

$$\tau^2, r^2 \quad \text{and} \quad \tau r \cos(\phi)$$

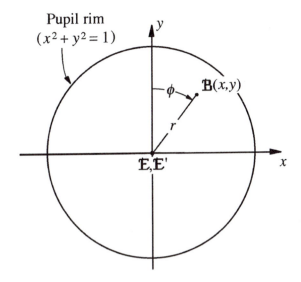

Fig. 33.10: The cartesian
and polar co-ordinate
system for rays passing
through a pupil.

Thus the expansion of $W(\tau; r, \phi)$ is of the form

$$W(\tau; r, \phi) = {}_0w_{0,0} + {}_2w_{0,0}\tau^2 + {}_0w_{2,0}r^2 + {}_1w_{1,1}\tau r \cos(\phi)$$

$$+ {}_4w_{0,0}\tau^4 + {}_0w_{4,0}r^4 + {}_1w_{3,1}\tau r^3 \cos(\phi)$$

$$+ {}_2w_{2,0}\tau^2 r^2 + {}_2w_{2,2}\tau^2 r^2 \cos^2(\phi)$$

$$+ {}_3w_{1,1}\tau^3 r \cos(\phi) + \text{higher order terms} \qquad (33.60)$$

The coefficients can be written generally as ${}_kw_{m,n}$, where k is the power of τ, m is the power of r and n is the power of $\cos(\phi)$. These coefficients have a numerical value depending on the actual situation. The terms in the extended expansion can be grouped in sets of terms, in the form

$$[\tau^2 + r^2 + \tau r \cos(\phi)]^{i+1} - \tau^{2i+2} \quad i = 1, 2, 3 \ldots$$

The number of terms in this set is $(i + 1)(i + 4)/2$ and the degree of each is $(2i + 2)$. Thus

$$k + m = 2i + 2$$

Table 33.2 gives the number of terms for the first few values of i. The terms in which $i = 1$ are called the **primary aberrations**, and those where $i > 1$ are called the **higher order aberrations**.

It can be shown that the coefficients

$${}_0w_{0,0}, \quad {}_2w_{0,0} \quad \text{and} \quad {}_4w_{0,0}$$

in the expansion given by equation (33.60) are zero, providing that the wave aberration is zero for the ray passing through the centre of the pupil, that is

Table 33.2. Number of terms in the expansion of the wave aberration polynomial

i	$(i+1)(i+4)$	$2i+2$	
1	5	4	primary
2	9	6	secondary
3	14	8	tertiary
4	24	10	quaternary

Note: The primary aberrations ($i = 1$) are often known as third order or Seidel aberrations.

through the point $r = 0$. This condition is defined to be so in the definition of the wave aberration given by equation (33.55). This equation implies that the pupil ray is free of aberration and thus $W(\tau; 0, 0) = 0$. If we take the remaining terms, then we can express the wave aberration polynomial [equation (33.60)] as

$$W(\tau; r, \phi) = {}_0w_{2,0}r^2 + {}_1w_{1,1}\tau r \cos(\phi) + {}_0w_{4,0}r^4 + {}_1w_{3,1}\tau r^3 \cos(\phi)$$

$$+ {}_2w_{2,0}\tau^2 r^2 + {}_2w_{2,2}\tau^2 r^2 \cos^2(\phi)$$

$$+ {}_3w_{1,1}\tau^3 r \cos(\phi) \qquad (33.60a)$$

The interpretation of the terms in this equation is as follows.

Term ${}_0w_{2,0}r^2$ – corresponds to a shift in the Gaussian image plane and is equivalent to a defocus of the Gaussian image plane. We can confirm this using Figure 33.11, which shows the Gaussian image plane at o' and a defocussed plane at o'_B. The wave aberration W is

$$W = [\mathcal{B}'\mathcal{B}'']$$

Now the distance $\mathcal{B}'\mathcal{B}''$ is along the ray but we can approximate this distance as the horizontal distance between the reference sphere and the wavefront. Hence we have

$$W = n'(Z' - Z'_B)$$

The Z-distances are the distances between the pupil plane and a spherical surface and hence are given by equation (33.44b), in which we only need to take the first term. Replacing the curvature by radius of curvature, in the Y–Z section we have

$$Z' = Y'^2/(2R') \quad \text{and} \quad Z'_B = Y'^2/(2R'_B) \qquad (33.61)$$

We can write the above wave aberration W as

$$W = n'(Y'^2/2)[(1/R') - (1/R'_B)] \qquad (33.62)$$

From equation (33.56b), $Y' = y\bar{\rho}'$ and for the rotationally symmetic aberration we can replace y by r, so we can write

$$W = n'(r^2\bar{\rho}'^2/2)[(1/R') - (1/R'_B)] \tag{33.63}$$

We can write this equation as

$$W = {_0w_{2,0}}r^2 \tag{33.64}$$

where

$${_0w_{2,0}} = n'(\bar{\rho}'^2/2)[(1/R') - (1/R'_B)] \tag{33.65a}$$

This can be written in the following alternative forms, which can be derived using Figure 33.11.

$${_0w_{2,0}} = -n'(\bar{\rho}'^2/2)(1/R'_B) \quad \text{(Gaussian image plane at infinity)} \tag{33.65b}$$

$${_0w_{2,0}} = n'\sin^2(\alpha')\delta l'/2 \quad \text{(Gaussian image at a finite distance)} \tag{33.65c}$$

This defocus term can also be expressed in terms of corresponding object space quantities, simply by replacing the image space quantity by the object space quantity, that is by dropping the prime superscript.

Term ${_1w_{1,1}}\tau r \cos(\phi)$ – corresponds to a transverse shift in image position or a change in scale or magnification.

The remaining five terms, in which the sums

$$k + m = 4$$

are the main five primary (or third order or Seidel) aberrations. Each of these terms corresponds to a specific aberration, as follows:

Term ${_0w_{4,0}}r^4$	spherical aberration
Term ${_1w_{3,1}}\tau r^3 \cos(\phi)$	coma
Term ${_2w_{2,0}}\tau^2 r^2$	image surface curvature
Term ${_2w_{2,2}}\tau^2 r^2 \cos^2(\phi)$	astigmatism
Term ${_3w_{1,1}}\tau^3 r \cos(\phi)$	distortion

The reader can examine the effect of these individual aberrations on the shape of the wave aberration function by sketching the above functions.

If we define an aberration as leading to a departure from sphericity of the wavefront, strictly speaking only spherical aberration, coma and astigmatism are true aberrations. The others, field curvature and distortion, only lead to a change in curvature or direction of the spherical wavefront.

As an alternative to the above power series expansion representation of the wave aberration polynomial, Zernike polynomials have been used.

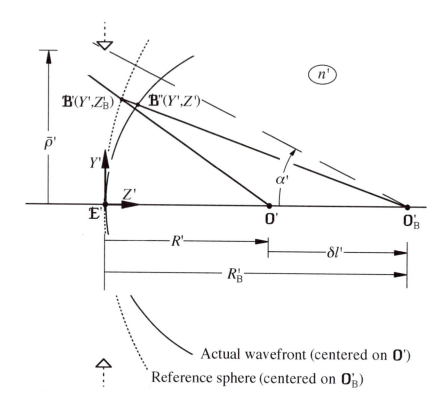

Actual wavefront (centered on **O'**)
Reference sphere (centered on **O'**$_B$)

For small fields and apertures, the higher order terms can be neglected as their values are usually very small. However, as the aperture or field sizes increase, higher order aberrations increase in size and significance and must eventually be taken into account.

Fig. 33.11: A defocus and the wave aberration.

33.7.3   *Methods of calculation*

The wave aberration function can be found by tracing finite rays from a selected field point τ and through selected points (x, y) in the entrance pupil. For each ray, the wave aberration $W(\tau; x, y)$ is defined by equation (33.54) and calculated using equation (33.55). Only a few rays have to be traced to determine the shape of the wave aberration function.

In practice, special algorithms are required to perform these calculations because of the high accuracy required. Equation (33.55) may involve the subtraction of two very similar and usually large numbers of the order of centimetres or metres to give a result of the order of the wavelength. Therefore it is very important to choose an algorithm that has high accuracy. Hopkins (1981) described a method using equally inclined chords. Given the wave aberration at a selected number of points in the pupil, the data can be fitted by some suitable algorithm, such as a least squares fit to the wave aberration polynomial given by equation (33.60), and the numerical value of the wave aberration coefficients found.

The primary terms of the wave aberration polynomial are directly related to the Seidel aberrations and therefore the Seidel aberrations can be used to approximate the exact wave aberration function. Equations for calculating the Seidel aberrations are given in Section 33.1. The primary wave aberration terms and the Seidel aberration terms are connected by the following equations (Welford 1986, 120):

$$_0w_{4,0} = S_1/8 \tag{33.66a}$$

$$_1w_{3,1} = S_2/2 \tag{33.66b}$$

$$_2w_{2,0} = (S_3 + S_4)/4 \tag{33.66c}$$

$$_2w_{2,2} = S_3/2 \tag{33.66d}$$

$$_3w_{1,1} = S_5/2 \tag{33.66e}$$

33.8 Wave and transverse aberrations

So far, aberrations have been discussed in terms of wave or path length aberrations. In many instances, transverse or longitudinal aberrations are more meaningful. However, the best representation depends upon the situation. For example, in the calculation of the optical transfer function (discussed in the next chapter), the wave aberrations are used. On the other hand in the case of visual optics, longitudinal aberrations are often more meaningful as these can be related directly to an accommodation demand, equivalent refractive error or defocus.

If we wish to evaluate the transverse and longitudinal aberrations of an optical system for a given ray traced through any part of the pupil, we can find the exact value by finite ray tracing. As an alternative, one can estimate their values from the wave aberration function and vice versa.

In this section, equations will be presented to show how one can estimate the transverse and longitudinal aberrations given the wave aberration either in exact form or from the Seidel approximation.

The wave W and transverse $\delta\xi'$, $\delta\eta'$ aberrations are related and if the aberrations are small, the mathematical relationship is simple. Welford (1986, 98) has stated that

$$\delta\xi' = -\frac{R'}{n'}\frac{\partial W(X',Y')}{\partial X'} \quad \delta\eta' = -\frac{R'}{n'}\frac{\partial W(X',Y')}{\partial Y'} \tag{33.67}$$

where R' is the radius of the reference sphere, as shown in Figure 33.8, and X' and Y' are the actual co-ordinates in the pupil and not the reduced as defined by equations (33.56b) and (33.57). In the above equations, Welford (1986) used lower case letters x and y to denote actual co-ordinate points X and Y. The above equations are only approximations, but increase in accuracy as the aberrations decrease. On the other hand, the equations are exact if the distance R' is taken as the distance along the ray between its points of intersection on the reference sphere and image plane. We can convert the actual pupil co-ordinates X' and Y' in the above equation to normalized co-ordinates x' and y' by equation (33.56b). We also need the following equation, which follows from the diagram, that

$$R' = \bar{\rho}' / \sin(\alpha') \tag{33.68}$$

This holds, providing the exit pupil radius $\bar{\rho}'$ is measured at the reference sphere in the pupil. Using these equations, we have

$$\delta\xi' = -\frac{1}{n'\sin(\alpha')}\frac{\partial W(x', y')}{\partial x'}, \quad \delta\eta' = -\frac{1}{n'\sin(\alpha')}\frac{\partial W(x', y')}{\partial y'}$$

$$\tag{33.69a}$$

If we assume there is no pupil aberration, we can express W in terms of entrance pupil normalized co-ordinates x and y and thus write

$$\delta\xi' = -\frac{1}{n'\sin(\alpha')}\frac{\partial W(x, y)}{\partial x} \quad \delta\eta' = -\frac{1}{n'\sin(\alpha')}\frac{\partial W(x, y)}{\partial y}$$

$$\tag{33.69b}$$

Combining equations (33.53) and (33.69b), we have

$$\delta G' = -\frac{1}{\lambda}\frac{\partial W(x, y)}{\partial x} \quad \text{and} \quad \delta H' = -\frac{1}{\lambda}\frac{\partial W(x, y)}{\partial y} \tag{33.70a \& b}$$

These equations have also been derived by Hopkins (1965) and can be used to determine the transverse ray distribution in the image plane and the spot diagrams for individual aberrations. Let us take the example of coma and look at its spot diagram.

Coma

If the only aberration present is coma, then

$$W = {}_1w_{3,1}\tau r^3 \cos(\phi) = {}_1w_{3,1}\tau(x^2 + y^2)y$$

Now from equation (33.70a), but omitting the λ by assuming that the coefficient ${}_1w_{3,1}$ is in terms of λ units,

$$\delta G' = -\partial W/\partial x = -{}_1w_{3,1}\tau 2xy = -{}_1w_{3,1}\tau 2r^2 \sin(\phi)\cos(\phi)$$

That is

$$\delta G' = -{}_1w_{3,1}\tau r^2 \sin(2\phi) \tag{33.71a}$$

Also from equation (33.70b)

$$\delta H' = -\partial W/\partial y = -{}_1w_{3,1}\tau(x^2 + 3y^2)$$

$$= -{}_1w_{3,1}\tau r^2[\sin^2(\phi) + 3\cos^2(\phi)]$$

After some reduction, we have

*Table 33.3. The components of the transverse
aberrations $\delta G'$ and $\delta H'$ for the five primary
aberrations*

	$\delta G'$	$\delta H'$
$_0w_{4,0}$	$-4r^3\sin(\phi)$	$-4r^3\cos(\phi)$
$_1w_{3,1}$	$-\tau r^2\sin(2\phi)$	$-\tau r^2[2+\cos(2\phi)]$
$_2w_{2,0}$	$-2\tau^2 r\sin(\phi)$	$-2\tau^2 r\cos(\phi)$
$_2w_{2,2}$	0	$-2\tau^2 r\cos(\phi)$
$_3w_{1,1}$	0	$-\tau^3$

$$\delta H' = -_1w_{3,1}\tau r^2[2+\cos(2\phi)] \qquad (33.71\text{b})$$

These two equations, (33.71a) and (33.71b), can be used to confirm the ray
intersection diagrams (spot diagrams) for coma have the form shown in Figure
5.3 and would also show that a semi-circular ray pattern in the pupil is mapped
onto a circular ray pattern on the paraxial image plane. If we now take these
two equations and eliminate ϕ, then the reduced transverse aberrations $\delta G'$ and
$\delta H'$ are related by the equation

$$(\delta G')^2 + (\delta H' + 2_1w_{3,1}\tau r^2)^2 = (_1w_{3,1}\tau r^2)^2 \qquad (33.72)$$

This is an equation of a circle with a centre at $(0, -2_1w_{3,1}\tau r^2)$ and radius of
$_1w_{3,1}\tau r^2$. Thus if we consider circles of radius r in the pupil, they form a family
of circles of radius $_1w_{3,1}\tau r^2$ in the image plane but shifted transversely by an
amount $2_1w_{3,1}\tau r^2$. The co-ordinate of the circle centre and its radius indicate
that the apex angle of the coma spot diagram is $60°$.

Equations (33.71a) and (33.71b) also give us the dimensions of the coma
flare from maximum values of the transverse aberrations $\delta G'$ and $\delta H'$. These
occur for $r = 1$ and $\phi = 45°(\delta G')$ and $\phi = 0°$ and $\phi = 90°(\delta H')$. The width
is twice the value of $\delta G'$ given by equation (33.71a) and the length is the value
of $\delta H'$ given by equation (33.71b). Thus, neglecting the minus signs

$$\text{width} = 2\tau_1 w_{3,1} \quad \text{and} \quad \text{length} = 3\tau_1 w_{3,1} \qquad (33.73)$$

where the distance units are those of $\delta G'$ and $\delta H'$, which are related to the
actual physical dimensions by equations (33.53). Equation (33.72) shows that
if the coefficient $_1w_{3,1}$ is positive, the $\delta H'$ transverse aberration of the centre
of the circle is negative and hence towards the axis, that is the coma flare is on
the axis side of the image point $\bar{Q}'$.

By the same method, the transverse aberration components $\delta G'$ and $\delta H'$ for
each of the other primary aberration can be found and the results are given in
Table 33.3.

33.9 Wave, transverse and longitudinal aberrations: The axial case

In visual optics, longitudinal aberrations, especially when expressed as an equivalent power error, are, at times, very useful. In this section, we will show how the spherical wave aberration is related to the longitudinal aberration, but before we begin to look at these relationships, we will express the wave aberration polynomial in a slightly different form to that considered previously. Up to this point, we have expressed the wave aberration polynomial in terms of normalized co-ordinates (x, y) and this had the advantage that the equations were meaningful even if the pupils were at infinity. However, in visual optical systems, the entrance and exit pupil are normally at finite distance and therefore the normalized representation is unnecessary. Therefore let us revert to representing the wave aberration polynomial in terms of the actual co-ordinates (X, Y) or (X', Y') where the ray intersects the respective pupil. Because of rotational symmetry, we can express the axial wave aberration polynomial in one co-ordinate. We will choose Y and Y'. Thus in terms of these pupil co-ordinates the axial wave aberration can be written as

$$W(Y) = \sum_{m=2,4,6...} {}_0W_{m,0}Y^m \tag{33.74a}$$

or

$$W(Y') = \sum_{m=2,4,6...} {}_0W'_{m,0}Y'^m \tag{33.74b}$$

where ${}_0W_{m,0}$ and ${}_0W'_{m,0}$ are the wave aberration polynomial coefficients evaluated at the edge of a pupil of unit radius in contrast to the ${}_0w_{m,0}$ coefficients which are evaluated at the edge of the full pupil. These sets of coefficients are related by the equation

$$_0W_{m,0} = {}_0w_{m,0}/\bar{\rho}^m \tag{33.75a}$$

$$_0W'_{m,0} = {}_0w_{m,0}/\bar{\rho}'^m \tag{33.75b}$$

Let us now look at the aberrations in the image space. We will recall equation (33.67) and write

$$\delta\eta' = -\frac{R'}{n'}\frac{\partial W(Y')}{\partial Y'} \tag{33.76}$$

We will now seek a relationship between transverse aberration $\delta\eta'$ and the longitudinal aberration $\delta l'$ shown in Figure 33.12. From this diagram we have

$$\frac{\delta\eta'}{\delta l'} = \frac{Y'}{R' + \delta l' - Z'} \tag{33.77}$$

If we assume that the aberration is small, then we can reduce this equation to

Fig. 33.12: Relation between the longitudinal $\delta l'$ and transverse aberration $\delta \eta'$.

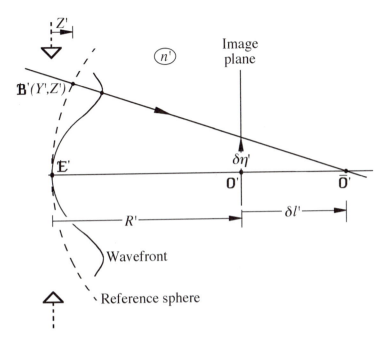

the approximation

$$\delta \eta' = \frac{Y'}{R'} \delta l' \tag{33.78}$$

This equation is used to replace $\delta \eta'$ in equation (33.76) by $\delta l'$, giving

$$\delta l' = -\frac{R'^2}{Y'n'} \frac{\partial W(Y')}{\partial Y'} \tag{33.79}$$

It will also be useful to express this equation in terms of the entrance pupil ray height Y instead of the exit pupil ray height Y'. We can make this transformation as follows. If we replace R' by the expression from equation (33.68) and replace Y' by Y from the following equation which follows from equation (33.57),

$$Y'/\bar{\rho}' = Y/\bar{\rho} \tag{33.80}$$

then we have the equation

$$\delta l' = -\frac{\bar{\rho}^2}{n' \sin^2(\alpha')Y} \frac{\partial W(Y)}{\partial Y} \tag{33.81}$$

Special case of the object at infinity

If the object is at infinity, equation (33.81) can be expressed in a different and more useful form for visual optics. When the object is at infinity, we can use equation (9.12b) to eliminate $\sin(\alpha')$ from equation (33.81), to give

$$\delta l'(Y) = -\frac{n'}{YF^2}\frac{\partial W(Y)}{\partial Y} \text{ (object at infinity)} \tag{33.82}$$

33.9.1 *The corresponding power error for the object at infinity*

The wave aberration can be related to a corresponding power error, for example a defocus which can be simulated by a $_0W_{2,0}$ term in the wave aberration polynomial and in visual optics where the eye can be defocussed by placing a defocussing lens in front of the eye. A second example is an interpretation and measurement of spherical aberration. The increase in spherical aberration with increase in ray height at a spherical surface can be interpreted as due to an unwanted increase in surface power of the surface with ray height. This concept applies to a method of the measurement of the spherical aberration of the eye. The value of the mean rotationally symmetric spherical aberration of the eye can be measured by correcting the spherical aberration in different annular zones of the pupil by a correcting lens or change in object vergence. More is given on this topic in Chapter 35. Therefore, there is a need to find equations relating the wave aberration to a corresponding power error.

The above longitudinal aberration $\delta l'$ can readily be related to an equivalent power error or aberration δF, measured at the principal planes. Let us begin with the lens equation (3.43) with the object at infinity, that is

$$\frac{n'}{l'} = F$$

After differentiating and taking small differentials, we can write this as

$$-\frac{n'\delta l'}{l'^2} = \delta F \tag{33.83}$$

and since the object is at infinity

$$l' = n'/F$$

and finally we have

$$\delta F = -F^2\delta l'/n' \tag{33.84}$$

We can express this in terms of the wave aberration using equation (33.82) to eliminate $\delta l'$ and have

$$\delta F(Y) = \frac{1}{Y}\frac{\partial W(Y)}{\partial Y} \tag{33.85}$$

We will now apply these equations to defocus and primary spherical aberration.

33.9.1.1   The defocus wave aberration term for the object at infinity

Longitudinal aberration

In many situations, it is often necessary to examine the effect of a defocus on the quality of the image and fortunately the wave aberration function allows a means for this. The term $_0W_{2,0}$ is equivalent to a defocus term and can be related to a longitudinal shift $\delta l'$ in equation (33.65c) via equation (33.75a). Let us take the wave aberration function $W(Y)$ as being purely due to defocus, that is

$$W(Y) = {}_0W_{2,0}Y^2 \tag{33.86}$$

Differentiating gives

$$\partial W(Y)/\partial Y = 2\,{}_0W_{2,0}Y \tag{33.87}$$

and therefore using equation (33.82), we have

$$\delta l' = \frac{-2n'\,{}_0W_{2,0}}{F^2} \tag{33.88}$$

which is independent of Y.

Corresponding power error

We can alternatively express $_0W_{2,0}$ in terms of a defocus δF of the object plane, assumed to be at infinity when in focus. In this case we simply substitute the differential function (33.87) in equation (33.85) and this gives

$$\delta F(Y) = 2\,{}_0W_{2,0} \tag{33.89}$$

where, since the righthand side is independent of Y so is $\delta F(Y)$, which could therefore be written simply as δF.

33.9.1.2   Seidel spherical aberration and the corresponding power error for the object at infinity

Let us look at the Seidel or primary spherical aberration. In this case, the wave aberration is

$$W(Y) = {}_0W_{4,0}Y^4$$

Therefore

$$\partial W(Y)/\partial Y = 4\,{}_0W_{4,0}Y^3 \tag{33.90}$$

After substituting this differential into equation (33.82), we have the equivalent longitudinal aberration $\delta l'(Y)$ expressed in the form

$$\delta l'(Y) = -\frac{4n'\,_0W_{4,0}Y^2}{F^2} \tag{33.91}$$

Using equations (33.66a) and (33.75a), we can express $\delta l'(Y)$ in terms of S_1 by the equation

$$\delta l'(Y) = -\frac{n'S_1Y^2}{2\bar{\rho}^4 F^2} \tag{33.91a}$$

Repeating the same procedure for the power error given by equation (33.85), we have the equations

$$\delta F(Y) = 4\,_0W_{4,0}Y^2 \tag{33.92}$$

and

$$\delta F(Y) = S_1 Y^2/(2\bar{\rho}^4) \tag{33.92a}$$

These equations show that the power error is quadratic in Y or ray height in the pupil. We will use these equations to examine the spherical aberration of schematic eyes, in Chapter 35.

33.10   Sagittal and tangential surfaces

In visual optical systems, astigmatism and the positions of the sagittal and tangential image surfaces are usually a major consideration. To determine the exact positions of these surfaces it is necessary to locate the path of the finite pupil ray and then trace the sagittal and tangential "paraxial rays" around the finite pupil ray. Welford (1986) gives equations for these procedures. However, the approximate positions of these surfaces may be found from the Seidel aberrations and if the aberrations are small, these approximations will be sufficiently accurate for many applications.

As the object point moves over the object plane, the sagittal and tangential image points move over curved surfaces. Within the Seidel approximation, these surfaces are spherical. According to Welford (1986, 126), the curvatures of the sagittal C_S and tangential C_T surfaces are

$$C_S = -\frac{n'(S_3 + S_4)}{H^2} \quad \text{sagittal surface curvature} \tag{33.93a}$$

$$C_T = -\frac{n'(3S_3 + S_4)}{H^2} \quad \text{tangential curvature} \tag{33.93b}$$

except for the negative sign, which we have added to be consistent with our sign convention.

In the absence of astigmatism, that is $S_3 = 0$, the image is formed on the Petzval surface. The curvature C_P of this surface can be found by setting $S_3 = 0$

in either of the above equations to give

$$C_P = -\frac{n'S_4}{H^2} \quad \text{Petzval curvature} \tag{33.93c}$$

Also of some interest is the curvature of the surface containing the disc of the least confusion. Its curvature C_{dlc} is the mean of the sagittal and tangential surface curvatures and hence is given by the equation

$$C_{dlc} = -\frac{n'(2S_3 + S_4)}{H^2} \quad \text{disc of least confusion} \tag{33.93d}$$

The difference between the astigmatic surface curvatures and the Petzval surface curvatures are

$$C_P - C_S = n'S_3/H^2 \quad \text{and} \quad C_P - C_T = 3n'S_3/H^2 \tag{33.94a \& b}$$

Now since the distance or sag between two centred spherical surfaces is proportional to the difference in curvatures as implied by equation (33.44c), equations (33.94a & b) show that the tangential surface is three times farther from the Petzval surface than the sagittal surface.

Example 33.7: Determine the radii of the sagittal and tangential surfaces of the (spectacle) lens given in Example 33.3.

Solution: To solve this problem we use equations (33.93a and b). The initial ray data given for Example 33.3 set an optical invariant H of 0.5554 and the image space is air so that the refractive index $n' = 1$. To find the curvatures of the sagittal and tangential surfaces we need values of S_3 and S_4. The value of S_3 depends upon the shape factor of the lens and the dependency is graphed in Figure 33.5. Let us determine the shape factors for the following values of Γ : -5, -3.855, -2.646, -1.436 and -0.5. The values of -3.855 and -1.436 correspond to $S_3 = 0$ and the value of -2.646 is the mean of these two values. The values of S_3 shown in the diagram are in units of wavelengths ($\lambda = 555$ nm) and therefore have to be converted to millimetres. The final results are shown in Figure 33.13.

If the Gaussian image is at infinity, the above surface curvatures are not meaningful and as an alternative the vergence of points on the surfaces are more useful, especially in visual optics. We can find equations for these vergences as follows. From equation (33.60a), the wave aberration due only to primary astigmatism and field curvature is

$$W(\tau; r, \phi) = {}_2w_{2,0}\tau^2 r^2 + {}_2w_{2,2}\tau^2 r^2 \cos^2(\phi) \tag{33.95}$$

We define the sagittal wave aberration $W_S(\tau)$ and tangential aberration $W_T(\tau)$ to be

$$W_S(\tau) = W(\tau; r = 1, \phi = 90°) \tag{33.96a}$$

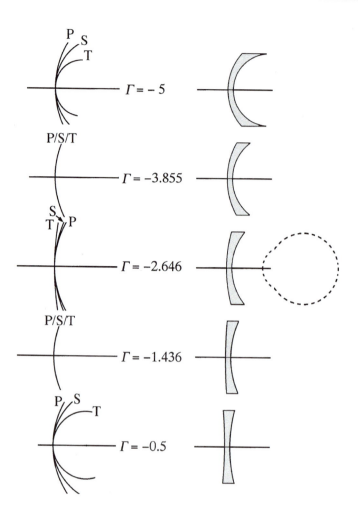

Fig. 33.13: The Petzval (P), sagittal (S) and tangential (T) image surfaces and changes of shape with lens shape of the spectacle lens examined in Example 33.7.

and

$$W_T(\tau) = W(\tau; r = 1, \phi = 0°) \tag{33.96b}$$

From the wave aberration polynomial [equation (33.95)] and equations (33.66c) and (33.66d), it follows that

$$W_S(\tau) = {}_2w_{2,0}\tau^2 = (S_3 + S_4)\tau^2/4 \tag{33.97a}$$

$$W_T(\tau) = ({}_2w_{2,0} + {}_2w_{2,2})\tau^2 = (3S_3 + S_4)\tau^2/4 \tag{33.97b}$$

Because astigmatism and Petzval curvature are effectively a defocus from the Gaussian image plane, we can represent these by a term of the same form as given in equation (33.63). If we substitute the expression for $W_S(\tau)$ given by equation (33.97a) for W in equation (33.63), but with the aberration taken at the edge of the pupil (that is $r = 1$), we have

$$1/R' - 1/R'_B = \frac{(S_3 + S_4)\tau^2}{2n'\bar{\rho}'^2} \tag{33.98}$$

In this equation, R' is the distance to the focussed image and R'_B is the distance to the defocussed image. Here we will denote the distance to the sagittal focussed image as R'_S and this is the same as R' and not R'_B. The Gaussian image is now out of focus and this distance is R'_B. If we now replace the distances R' and R'_S by corresponding vergences $L'_S(= n'/R')$ and $L'_G(= n'/R'_B)$ and after rearranging terms, we have

$$L'_S(\tau) = L'_G + \frac{(S_3 + S_4)\tau^2}{2\bar{\rho}'^2} \qquad (33.99a)$$

Similarly for the tangential case, we have

$$L'_T(\tau) = L'_G + \frac{(3S_3 + S_4)\tau^2}{2\bar{\rho}'^2} \qquad (33.99b)$$

where these vergences should be measured from the centre of the exit pupil. For the common case of the image at infinity, the vergence L'_G will be zero.

When L'_G is zero, the sign of $L'_S(\tau)$ and $L'_T(\tau)$ depend upon the signs of S_3 and S_4. If the system is free of Seidel astigmatism, that is $S_3 = 0$, then $L'_S(\tau)$ and $L'_T(\tau)$ are equal and have the same sign as S_4.

Given the vergences of the sagittal and tangential image points, we can easily deduce the level of astigmatism, which we will denote by the symbol $\Delta L'_{ST}(\tau)$. It is the difference between the sagittal and tangential vergences, that is

$$\Delta L'_{ST}(\tau) = L'_S(\tau) - L'_T(\tau) = \frac{S_3\tau^2}{\bar{\rho}'^2} \qquad (33.100)$$

We can express equations (33.99a and b) in terms of the respective surface curvatures C_S and C_T using equation (33.93a and b) as

$$L'_S(\tau) = L'_G - \frac{C_S H^2 \tau^2}{2n'\bar{\rho}'^2} \qquad (33.101a)$$

and

$$L'_T(\tau) = L'_G - \frac{C_T H^2 \tau^2}{2n'\bar{\rho}'^2} \qquad (33.101b)$$

Significance of the sign of $L'_S(\tau)$ and $L'_T(\tau)$

According to our sign and left-to-right ray tracing conventions, if the above vergences are negative, then the image surfaces are formed to the left of the exit pupil and therefore in front of the eye. If they are positive, they are to the right of the exit pupil and therefore behind the eye.

Example 33.8: Determine the vergence of the sagittal and tangential surfaces of a collimator lens with the same design as that given in Example 33.5, but with a focal length of 10 cm instead of 1 m and assuming the eye is placed 50 mm back from the collimator.

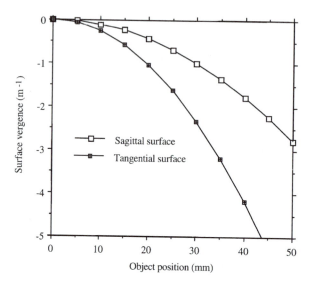

Fig. 33.14:   Vergence of
the sagittal and tangential
images of the collimator
examined in Example 33.8.

Solution: The solution given in Example 33.5 is for a 1 m focal length
lens and we can scale it to give an achromatic doublet of focal length
10 cm. This lens will then have the following thin lens construction

$$C_1 = +0.00458 \text{ mm}^{-1}, C_2 = +0.02143 \text{ mm}^{-1} \quad \text{and}$$

$$C_3 = -0.01925 \text{ mm}^{-1}$$

refractive indices: first lens $\mu = 1.65446$,

second lens $\mu = 1.5168$

With an aperture stop of diameter 8 mm placed 50 mm beyond the lens
and an object field-of-view radius 50 mm, the equations in Section
33.1.1 give the following Seidel aberrations

$$S_3 = 107.8\lambda, S_4 = 51.8\lambda \text{ with } \lambda = 555 \text{ nm}$$

Using these values and the optical invariant value of 2.0, we can now
use equations (33.101a and b) to calculate the vergences of the sagittal
and tangential image surfaces and the results of these calculations are
shown in Figure 33.14.

33.11   Analysis of aberrations by laboratory methods

Up to this point, we have shown how an optical system can be analysed for
aberrations, although we have concentrated on the Seidel or primary aberrations and only made a brief mention of the exact and higher order aberrations.
Sometimes, we have an assembled optical system and may wish to analyse its
aberrations. We can do this in two ways. We can take the system apart and measure all the constructional parameters by techniques described in Chapter 11

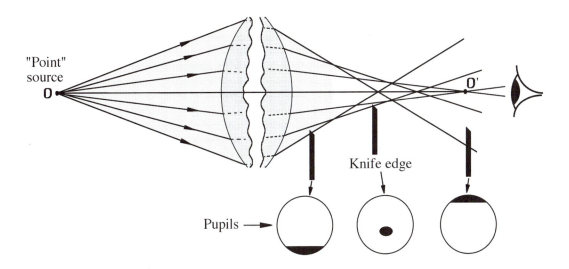

Fig. 33.15: Foucault
knife-edge test for
analysing the aberrations in
an optical system using an
example of spherical
aberration. The different
positions of the knife-edge
lead to different light
distributions in the pupil as
shown.

and then trace suitable rays to determine the aberrations. Alternatively we may
try to measure the aberrations by laboratory techniques. There are several ways
this can be done with various levels of accuracy and sophistication. We will
look at three in this section.

33.11.1   *The Foucault knife-edge test*

In the Foucault knife-edge test, a "point" or very small source is used to analyse
the aberrations in an optical system. A sharp (or knife) edge is passed across
the beam and in the vicinity of the focus. The eye is placed just beyond the
knife-edge and the light distribution across the exit pupil is observed as the
knife-edge is moved across the beam. As the knife-edge is moved across the
beam it blocks rays coming from certain parts of the pupil. These parts go dark
as the rays are blocked. The progression of the dark shadow gives some idea
of the aberrations present, but the progression depends upon the position of the
knife-edge relative to the paraxial focus.

Figure 33.15 shows the example of the shadow pattern if there is spherical
aberration. If there were no aberration and the knife-edge was placed exactly
at the paraxial focus, the whole pupil would go dark uniformly and no shadow
would be seen progressing over the pupil. Therefore this method can be used
to locate the focus as discussed in Chapter 10 and shown in Figure 10.10. This
method is not easy to interpret quantitatively.

33.11.2   *From the wave aberration function*

The measurement of the wave aberration function is quantitative and a much
more sophisticated approach to the analysis of the aberrations of an optical system. The wave aberration function can be measured using a modified Twyman-
Green interferometer. The arrangement is shown in Figures 33.16a and b, for
a general off-axis object point at infinity. This object point is represented by
plane wavefronts entering the optical system. These become distorted on passing through the system, reflect off a spherical mirror that is centred on paraxial

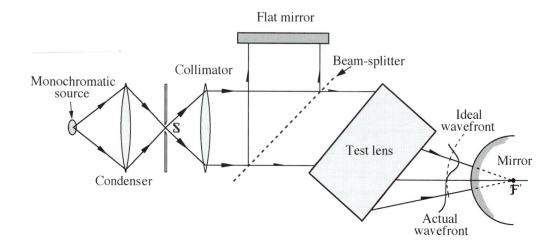

Fig. 33.16a: Use of a
modified Twyman–Green
interferometer to measure
the wave aberration
function of an optical
system: first pass of
wavefronts

focus of the converging beam, and, on passing through the system for the second time, the wavefront aberration is doubled. The resulting distorted wavefront is interfered with a plane wavefront from the other mirror and the number of interference fringes and fringe shape indicate the type type and magnitude of the aberrations present. For example, circular fringes indicate either spherical aberration or a defocus which, for off-axis points, may be due to field curvature. Elliptical fringes indicate astigmatism.

The fringe pattern can be analysed to reconstruct the wave aberration by a polynomial fitting routine and the results can be used to determine the point spread function or optical transfer function, which are described in the next chapter.

33.11.3   The star test

As we have seen from Chapter 5, the different aberrations have different effects on the image of a point and therefore if we examine the image of a point, by examining its shape, we will be able to detect and quantify the different aberrations present. This test is called the star test and is discussed in more detail in the next chapter in Section 34.3.2.2 and the reader is directed to that section for further information.

Exercises and problems

33.1   Given the following Seidel aberrations

$$S_1 = 20\lambda, S_2 = -15\lambda, S_3 = 4\lambda, S_4 = 23\lambda,$$

$$S_5 = 10\%, C_L = 7\lambda, C_T = 6\%$$

calculate the new aberration values if the aperture stop diameter is reduced by 10% and the field-of-view is increased by 20%.

ANSWERS: $13.12\lambda, -13.1\lambda, 4.7\lambda, 26.8\lambda, 14.4\%, 5.67\lambda$ and 6.0%

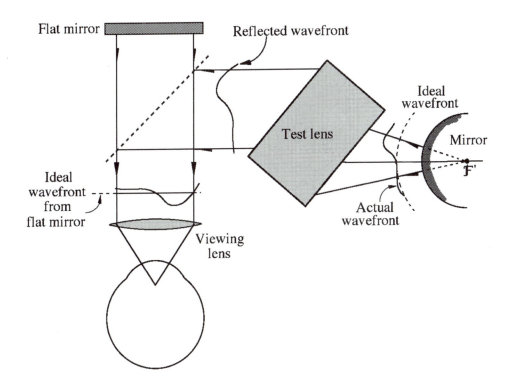

Flat mirror

Reflected wavefront

Ideal wavefront

Test lens

Mirror

F'

Ideal wavefront from flat mirror

Actual wavefront

Viewing lens

Fig. 33.16b: Use of a modified Twyman–Green interferometer to measure the wave aberration function of an optical system: reflected wavefronts and interference with plane (unaberrated) wavefront.

33.2 If the primary distortion is 50% at the edge of a 5 cm radius Gaussian image, plot the actual image shape of the square of a Gaussian image of sides 10×10 cm.

33.3 Calculate the difference between the powers at the hydrogen C and F lines for a lens of power 60 m^{-1} ($n_d = 1.5$) with a V-value of 53.

ANSWER: $\Delta F_{CF} = 1.2$ m^{-1}

33.4 Given the following wave aberration polynomial

$$W(\tau; r, \phi) = 4r^4 - 3\tau r^3 \cos(\phi) + 2\tau^2 r^2 + 4\tau^2 r^2 \cos^2(\phi)$$

$$- 3\tau^3 r \cos(\phi)$$

estimate the wave aberration function at the edge of the field in the sagittal and tangential sections for $r = 0, 0.7$ and 1.0. Sketch the difference between the wavefront and the reference sphere in these sections.

ANSWERS:

r	ϕ	$W(\tau; r, \phi)$	r	ϕ	$W(\tau; r, \phi)$	r	ϕ	$W(\tau; r, \phi)$
0.7	0	0.771	0.7	90	1.940	0.7	180	7.029
1.0	0	4.000	1.0	90	6.000	1.0	180	16.000

Summary of main symbols

| n | refractive index at the helium yellow line ($\lambda = 587.6$ nm), sometimes denoted as n_d |
| n_A | refractive index of the medium to the left of the aperture |

	stop
κ	aspheric aberration contribution
$\bar{Q}'$	real image position, in contrast to the Gaussian position Q'
$\mathcal{B}, \mathcal{B}'$	intersection of rays with wavefronts
$\delta l'$	longitudinal aberration for an on-axis conjugate
η_{max}	maximum object size
Γ	shape factor
Ω	conjugate factor
u, h	paraxial marginal ray angles and heights
$\bar{u}, \bar{h}$	paraxial pupil ray angles and heights
$\bar{u}_A$	pupil ray angle to the left of the aperture stop
H	optical invariant
$\bar{\rho}_A$	ratios of aperture stop
E	stop shift factor
$\delta\xi', \delta\eta'$	transverse aberrations
$\delta G', \delta H'$	corresponding reduced transverse aberrations
Z	distance along optical axis
X, Y	pupil co-ordinates
x, y	reduced pupil co-ordinates
r, ϕ	corresponding polar co-ordinates
$W(\tau; x, y)$	wave aberration function in cartesian co-ordinates
$W(\tau; r, \phi)$	wave aberration function in polar co-ordinates
τ	normalized field position of object
R'	radius of curvature of wavefront or reference sphere
C_L	longitudinal chromatic aberration
C_T	transverse chromatic aberration (Sections 33.1, 33.3 and 33.4); curvature of tangential surface (Section 33.10)

References and bibliography

Emsley H.H. (1956). *Aberrations of thin lenses.* Constable, London.

*Hopkins H.H. (1950). *The Wave Theory of Aberrations.* Clarendon Press, Oxford.

Hopkins H.H. (1952). Invariant focii. *Proc. Phys. Soc.* 3, LXV, 934.

*Hopkins H.H. (1965). Canonical pupil co-ordinates in geometrical and diffraction image theory. *Jpn. J. Appl. Phys.* 4 (suppl. 1), 32.

*Hopkins H.H. (1981). Calculation of the aberrations and image assessment for a general optical system. *Opt. Acta* 28(5), 667–714.

*Welford W.T. (1986). *Aberrations of Optical Systems.* Adam Hilger, Bristol.

34

Image quality criteria

34.0   Introduction

Image quality criteria are used to assess how faithfully an optical system can image an object. The aberrations discussed in the previous chapter directly do not give this information, except for the distortion aberration expressed as a fraction or percentage difference in magnification. As will be seen in this chapter, aberration values can be used to calculate some but not all of these image quality criteria. For example, aberrations by themselves do not take into account veiling glare although some of the image quality criteria, using aberrations as input, do take diffraction effects into account. In fact, the quality of the image depends upon the following three factors:

(a)   Veiling glare, which is due to unwanted reflections or scatter from surfaces
(b)   Monochromatic and chromatic aberrations, discussed in the previous chapter
(c)   Diffraction, discussed in Chapter 26

Because aberration and diffraction levels depend upon the size of the pupil and the aberrations also depend upon the position of the image point in the field, the form and magnitude of the image quality criteria depend upon pupil size and field position. Thus to assess fully the quality of a system, the image quality criteria should be evaluated for various pupil sizes and at several field positions.

Many image quality criteria are multi-dimensional functions and as a result are not always easy to use as criteria for comparing two similar systems. Therefore most image quality criteria have a reduced and sometimes approximate equivalent one dimensional form. That is, the image quality can be described by a single number on some suitable scale. Examples will be given at appropriate sections in this chapter.

As we have stated above, some image quality criteria can be predicted theoretically from the aberrations levels and as we have shown in the previous chapter, aberrations can found from the numerical results of ray tracing through a system. Therefore these image quality criteria can be predicted from the known construction and application of the system. Image quality criteria can also be determined given an assembled lens by laboratory methods and we will look at some of these in this chapter.

There are four basic criteria commonly used to assess image quality:

(1)   Veiling glare
(2)   The wave aberration function
(3)   The point spread function
(4)   The optical transfer function

As we will see, the point spread function and the optical transfer function can be calculated from the wave aberration function. We will also see that the point spread function and optical transfer functions can be derived from diffraction theory and therefore are wavelength dependent. Thus one approach is to define the wave aberration, point spread and optical transfer functions initially at a particular wavelength and later if desired, use these monochromatic functions at a number of wavelengths to derive a **white light point spread function** and **white light optical transfer function**. However, that process is beyond the scope of this book.

34.1   Veiling glare

Definition: *Veiling glare is light that has been reflected or scattered out of the image forming beam and somehow finds its way ultimately to the image plane. As it has suffered some unwanted reflections or scattering, it usually does not form an image and instead is spread over the image. It may be fairly uniformly distributed over the image plane and thus gives the appearance of looking through a fog and has the effect of reducing the contrast of the image. However, if the reflected beam is focused at or near the image plane, it is called a ghost image.*

It would be ideal if every ray entering the optical system were either wholly transmitted or completely vignetted. Unfortunately, because of various causes, some light is subtracted from the image forming beam but still falls on the image plane. The main causes are as follows:

(a)   Scattering at air/glass surfaces due to dirt, dust and scratches. This effect can be minimized by ensuring that the glass surfaces are kept as clean and as scratch free as possible.
(b)   Light falling on the barrel walls and other vignetting surfaces but not being completely absorbed. Some of this light is reflected or scattered forward onto the image plane. This source of veiling glare can be minimized by coating the barrel, the aperture stop surfaces and all other non-transmitting or reflecting surfaces with a matte black finish and by placing suitably designed baffles at appropriate positions in the barrel.
(c)   Simple reflections at air/glass surfaces. In this case, the reflections at air/glass surfaces are inherent in the physical optical nature of waves and light. At every interface between two media with different propagation velocities, a fraction of the incident energy is reflected. At normal incidence, the fraction of light that is reflected (R) and transmitted (T) can

be found from the Fresnel equations

$$R = \frac{(n' - n)^2}{(n' + n)^2} \qquad\qquad (34.1a)$$

$$T = \frac{4nn'}{(n' + n)^2} \qquad\qquad (34.1b)$$

where n and n' are the refractive indices on either side of the surface.

At an air/glass interface with the glass having a refractive index of 1.5, the amount of energy reflected is thus 0.04 or 4% and the amount transmitted is 0.96 or 96%. Since the glass is non-conducting, there is no loss of light at the interface and hence

$$R + T = 1 \qquad\qquad (34.1c)$$

We can now estimate the attenuation of light passing through an optical system of k air/glass surfaces. If T is the transmittance at a surface, then the transmittance $T(k)$ after k similar surfaces is simply

$$T(k) = T^k \qquad\qquad (34.2)$$

If we consider an example of a system of 10 air/glass surfaces with each glass having an index of 1.5, then

$$T(10) = 0.96^{10} = 0.664$$

Thus about 34% of the light is lost from the image forming beam and some of this light will suffer further reflections. The proportion that is reflected an even number of times will finally reach the image plane to form a ghost image or veiling glare, unless it is vignetted.

Since the minimum number of even reflections is two, the minimum attenuation of a veiling glare ray reaching the image is

$$(0.04)^2 \times (0.96)^2 = 0.00147$$

for a glass with an index of 1.5. This may be negligible in many instances, but if bright objects, such as the sun and street lights at night, are within the field-of-view, light from these sources even though greatly attenuated by multiple reflections and scatter can produce a veiling glare that is brighter than the background. For example consider the sun which has a luminance of 1.5×10^9 cd/m^2 and assume a background luminance of about 1000 cd/m^2. If a ghost image of the sun is formed in the image plane, the ratio of the luminances of the ghost image of the sun and the background is of the magnitude

$$1.5 \times 10^9 \times 0.00147/1000 = 2212$$

That is the ghost image of the sun is approximately 2000 times brighter than the background. Even bright sources outside the field-of-view may, by multiple reflections or scatter, may lead to veiling glare in the image plane or form ghost images in the field-of-view.

Smith (1971) has described a simple paraxial ray tracing procedure that can be used to determine whether there are any significant reflections within an optical system, identify the surfaces involved, determine the extent of the veiling glare spread and hence the brightness of the veiling glare. The reflections from these surfaces can be reduced by depositing dielectric anti-reflection coatings on the surfaces. The theory and structure of anti-reflections coatings, which are beyond the scope of this book, are covered in a number of other texts such as Macleod (1969).

34.2   The wave aberration function

Definition: *The wave aberration function is the optical path length difference along a ray between the actual wavefront and the ideal wavefront (reference sphere) at the exit pupil. It is expressed as a function of the co-ordinates of intersection of the ray with the reference sphere in the pupil.*

The wave aberration function was introduced in the preceding chapter and before readers proceed any further, they should become familiar with the relevant sections of that chapter. There, it was shown that the wave aberration function could be expressed as a polynomial in the pupil co-ordinates and the primary terms of the polynomial are related to the Seidel aberrations. We can use either cartesian co-ordinates (x, y) or polar co-ordinates (r, ϕ) for this task. If we are to use the wave aberration function as an image quality criterion in any given situation, we must be able to quantify it, either in terms of the wave aberration at different points in the pupil or in terms of the coefficients of the wave aberration polynomial.

If the system is completely free of aberrations, that is diffraction limited, then the wave aberration function is zero and each term in the wave aberration polynomial is zero. Whether aberration free or not, the wave aberration function can be used to examine the effect of defocus on the quality of the image. The defocus term is $_0w_{2,0}$ and various expressions for this term are given by equations (33.65a, b & c).

34.2.1   *Derived criteria*

Derived criteria are usually a reduced form of an image quality criterion allowing the complex multi-dimensional function to be reduced to a single number on a suitable scale. In the case of the wave aberration function, there is only one common reduction and that is its variance.

34.2.1.1   *Variance of the wave aberration*

The variance V of the wave aberration function has the same definition as the variance in statistics. Writing the wave aberration function in terms of cartesian co-ordinates $W(x, y)$, this variance can be written

$$V = \left\{ \iint_E [W(x, y) - \bar{W}]^2 \, dx \, dy \right\} / A \qquad (34.3)$$

where E indicates integration over the pupil of area A. Alternatively and more conveniently for calculations, it can be expressed in the form "mean of the squares minus the square of the mean", that is

$$V = \left[\iint_E W^2(x, y) \mathrm{d}x \ \mathrm{d}y \right] / A - \left[\iint_E W(x, y) \mathrm{d}x \ \mathrm{d}y / A \right]^2$$

(34.4a)

or in terms of the polar co-ordinates

$$V = \left[\iint_E W^2(r, \phi) r \ \mathrm{d}r \ \mathrm{d}\phi \right] / A - \left[\iint_E W(r, \phi) r \ \mathrm{d}r \ \mathrm{d}\phi / A \right]^2$$

(34.4b)

For circular pupils, the polar co-ordinate representation is the more useful in actual calculations. The value of the variance ranges from zero upwards. If the system is diffraction limited for a field point, the variance is zero. Therefore the greater the variance, the worse the image quality. If we determine the variance for a pupil radius p, using equation (34.4b), we can write this variance as

$$V = \left[\int_0^{2\pi} \int_0^p W^2(r, \phi) r \ \mathrm{d}r \ \mathrm{d}\phi \right] / (\pi p^2)$$

$$- \left[\int_0^{2\pi} \int_0^p W(r, \phi) r \ \mathrm{d}r \ \mathrm{d}\phi / (\pi p^2) \right]^2 \qquad (34.5)$$

Hopkins (1966) has discussed the use of this quantity as an image quality criterion and cites Maréchal and Françon (1960), who found that for highly corrected systems, a decrease in this variance increased the Strehl intensity ratio (see Section 34.3), but only for Strehl ratios greater than approximately 0.8. For poorly corrected systems, Hopkins stated that the optical transfer function, described in Section 34.4, was a more appropriate image quality criterion.

Born and Wolf (1989, 463–464) have shown that for small levels of aberrations, the relative illuminance E at the Gaussian image is given by the equation

$$E \approx 1 - (2\pi/\lambda)^2 V \qquad (34.6)$$

Thus the variance can be used as an image quality criterion, since it follows from equation (34.6) that the larger the value of E, the smaller must be V. The illuminance E is normalized to 1.0, so that if $V = 0$, $E = 1$.

34.2.1.2 Applications of the variance of the wave aberration function

The Maréchal criterion

It is often assumed (Born and Wolf 1989, 469) that a system is reasonably well corrected if the relative illuminance E (the Strehl intensity ratio) is greater than

0.8. If we set $E > 0.8$ in equation (34.6) above, it follows that

$$V < \lambda^2/(20\pi^2) \qquad\qquad (34.7)$$

That is

$$V < \lambda^2/197.4$$

This can be written in terms of the square root of the variance (that is $\sqrt{V}$) and in this case

$$\sqrt{V} < \lambda/14 \qquad\qquad (34.8)$$

rounding the denominator to an integer. This is known as the Maréchal criterion (Born and Wolf 1989, 469).

Tolerance to spherical aberration

We can use the Maréchal criterion to examine the maximum spherical aberration that may be tolerated in an optical system. Let us assume that the wave aberration function contains only one general spherical aberration term of the form $_0w_{m,0}r^m$, that is

$$W(r) = {}_0w_{m,0}r^m \qquad\qquad (34.9)$$

and the quantity $_0w_{m,0}$ is the aberration level at the edge of a pupil of normalized unit radius ($r = 1$), but which has an actual radius $\bar{\rho}$. Let us determine the variance for a pupil with a normalized radius p, that is for an actual pupil of radius $p\bar{\rho}$ and determine the relationship between p and $_0w_{m,0}$ for which the Marechal criterion is just satisfied. We substitute this expression for $W(r)$ in the integral equation (34.5) and get

$$V = +m^2({}_0w_{m,0})^2 p^{2m}/[(m + 1)(m + 2)^2] \qquad\qquad (34.10)$$

If the Maréchal criterion is to be satisfied, the maximum pupil radius p is found by substituting V from equation (34.10) into equation (34.7) to give the equation

$$({}_0w_{m,0})^2 p^{2m} < \frac{\lambda^2(m + 1)(m + 2)^2}{20\pi^2 m^2} \qquad\qquad (34.11)$$

Let us use this equation to find the value of the primary spherical aberration coefficient $_0w_{4,0}$ that will satisfy the Maréchal condition for the full pupil, that is for $p = 1$. If we solve for $_0w_{4,0}$ in equation (34.11) we have

$$_0w_{4,0} < 3\lambda/4\pi$$

So that the upper limit of $_0w_{4,0}$ is

$$_0w_{4,0} \approx \lambda/4 \qquad\qquad (34.12)$$

This is sometimes known as the "quarter wavelength" rule or condition.

Optimum image plane position in the presence of spherical aberration

In the presence of spherical aberration, the best image plane is not the paraxial position but depends upon the level of aberration. Let us examine this by assuming that the wave aberration polynomial consists of only one spherical aberration term, which is given by equation (34.9), but if we wish to examine the effect of a defocus, we must add the defocus term $_0 w_{2,0} r^2$ and so have

$$W(r) = {}_0 w_{2,0} r^2 + {}_0 w_{m,0} r^m \tag{34.13}$$

From equation (34.4b) above, for a pupil of radius p, the variance is

$$V = ({}_0 w_{2,0})^2 p^4 / 12 + [2m \, {}_0 w_{2,0} \, {}_0 w_{m,0} p^{(m+2)}] / [(m+2)(m+4)]$$

$$+ [m^2 ({}_0 w_{m,0})^2 p^{2m}] / [(m+1)(m+2)^2] \tag{34.14}$$

This function is quadratic in the defocus term $_0 w_{2,0}$ and the optimum level of defocus occurs when this function is minimum, that is when

$$_0 w_{2,0} = -\frac{12 m p^{(m-2)}}{(m+2)(m+4)} {}_0 w_{m,0} \tag{34.15}$$

If we consider the case of primary spherical aberration for a normalized pupil radius of $p = 1$, then the optimum image quality occurs for

$$_0 w_{2,0} = -{}_0 w_{4,0} \tag{34.16}$$

Given the numerical value of $_0 w_{2,0}$, the corresponding object or image plane shift is given by equations (33.65a, b and c).

Numerical examples involving these equations and concepts will be given in the next chapter, where we investigate the aberrations of the eye.

34.3 The point spread function

Definition: *The (intensity) point spread function (PSF) is the illuminance or luminance distribution in the image of a point. This may be in the Gaussian image plane or a nearby plane and for a well corrected optical system, this is within the realm of physical optics. Thus physical optics (predominantly diffraction) equations should be used to determine the point spread function. However, if the system is poorly corrected, geometrical optics may be used to calculate this function.*

We will denote the point spread function as $g(u', v')$ where u' and v' are the actual co-ordinates in the image plane. We should also distinguish this function from the amplitude point spread function, denoted as $g_a(u', v')$, which is the complex amplitude of the light distribution in the image of a point. What we have called the point spread function could be called the intensity point spread function, but we do not usually use the suffix "intensity". By the rule given in Chapter 1, these are related by the standard rule

$$g(u', v') = g_a(u', v') g_a^*(u', v') \tag{34.17}$$

The aberration free form of this function, known as the diffraction limited point spread function, is given by the Fraunhofer diffraction integral, equation (26.28),

$$g_a(u', v') = K \iint_E e^{-ik(u'X+v'Y)/Z_0} \, dX \, dY$$

where X and Y are actual pupil co-ordinates and Z_0 is the distance from the exit pupil to the image plane. If we normalize the pupil co-ordinates X and Y using equations (33.56b), this equation can be written as

$$g_a(u', v') = K \iint_E e^{-ik(u'x+v'y)\bar{\rho}'/Z_0} \, dX \, dY$$

We can now use equation (33.68) to replace $\bar{\rho}'/Z_0$ by $\sin(\alpha')$, where $Z_0 = R'$ in that equation and so we now have

$$g_a(u', v') = K\bar{\rho}'^2 \iint_E e^{-ik(u'x+v'y)\sin(\alpha')} \, dx \, dy \qquad (34.18)$$

The integration is carried out over the whole pupil and the result can be normalized against the equivalent diffraction limited function.

We can now take aberrations into account by assuming that the aberrations produce a variation in complex phase over the pupil. Let us also assume that the transmittance varies over the pupil. Both of these factors are incorporated in the **pupil function** $P(x, y)$, which is the complex amplitude in the pupil. It is mathematically defined as

$$P(x, y) = A(x, y)e^{\{-ikW(x,y)/n'\}} \qquad (34.19)$$

where $A(x, y)$ is the amplitude transmittance at the point (x, y) in the pupil. This function is necessary if the pupil is apodized, for example if the Stiles–Crawford effect is simulated with an equivalent pupil apodization. If the pupil is not apodized then

$$A(x, y) = 1$$

To take this pupil function into account, we modify equation (34.18) as follows:

$$g_a(u', v') = \iint_E P(x, y)e^{-ik(u'x+v'y)\sin(\alpha')} \, dx \, dy \qquad (34.20)$$

where we have dropped the quantities in front of the integral, so this integral now only gives the relative light distribution. Of more importance than the absolute light level is the light distribution relative to the equivalent diffraction limited point spread function. The point spread function $g(u', v')$ is then found from equation (34.17).

Fig. 34.1: The diffraction
limited point spread
function, an aberrated
function with 0.5λ of coma
and a defocussed point
spread function with 0.5λ
of defocus.

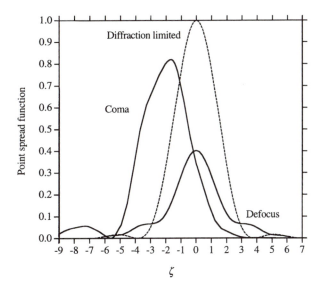

34.3.1 *The diffraction limited point spread function (circular aperture)*

In the absence of aberrations, the point spread function is the Fraunhofer diffraction pattern introduced in Chapter 26 and for any particular case, it can be found by solving equation (34.18). For a circular pupil, the point spread function $g(u', v')$ reduces to one variable say u' and it has been stated in Chapter 26 that the corresponding diffraction limited point spread function is given by equation (26.29), that is

$$g(u') = \frac{[2J_1(\zeta)]^2}{\zeta^2} \tag{34.21}$$

where from Chapter 26,

$J_1(\zeta)$	$=$	Bessel function described in Section 26.4.1	(34.22)
ζ	$=$	$ku' \sin(\alpha')$	
k	$=$	$2\pi n'/\lambda$	
n'	$=$	refractive index of image space	
λ	$=$	wavelength in vacuum	
α'	$=$	angular radius of the exit pupil measured at the axial image point	

These quantities are also explained in Figure 26.11, where r in that diagram is the same as the u' above. This function is plotted as the dashed line in Figure 34.1, along with two examples of the aberrated point spread function which will be discussed later.

As we stated in Chapter 26, the first zero of the diffraction limited point spread function occurs when

$$\zeta = 3.8317$$

and the corresponding value of u' at which this zero occurs, denoted as u'_0, is

$$u'_0 = \frac{0.610\lambda}{n'\sin(\alpha')} \tag{34.23}$$

and this is the physical radius of the Airy disc.

Units for plotting the point spread function

The point spread function can be plotted on several scales. In Chapter 26, it is plotted in units of the parameter ζ. It could be plotted in terms of the the the real distance unit u'. However, an alternative and useful scale unit is the radius of the Airy disc. Using equation (34.23), we can introduce a new scaled unit $\mathbf{u'}$, where

$$\mathbf{u'} = \frac{n'\sin(\alpha')}{0.610\lambda}u'_0 \tag{34.24}$$

and on this scale, the Airy disc has a radius of $\mathbf{u} = 1$. This is a very useful scale when plotting aberrated point spread functions as it allows a visual comparison of the width of the actual point spread function with the diffraction limited case.

34.3.2 *Methods for its determination or calculation*

34.3.2.1 *Theoretical methods*

Geometrical optics

In geometrical optics, the ray density on an image plane is an estimate of the local light level. Therefore, if the optical system is badly aberrated, we can estimate the point spread function from the density of rays. A large number of finite rays are traced from a point in the object, through the system to the chosen image plane. The ray density diagram is built up and the distribution, sometimes known as a **spot diagram**, is the geometric optical estimate of the point spread function.

Physical optics

Given the wave aberration polynomial coefficients, the point spread function can be calculated by numerically solving the integral of equation (34.20). Yzuel and Hopkins (1970) have described a routine method for the calculation of the point spread function using this method.

34.3.2.2 *Laboratory methods*

The star test

The image of a "point" source of monochromatic light, that is the point spread function, formed by the optical system, can be examined using a microscope at some suitable magnification. A typical setup is shown in Figure 34.2. A very small source ("the star") is placed at the appropriate object distance. In many situations, the star should be placed at infinity and of course in the laboratory,

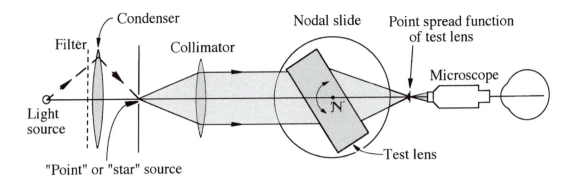

Fig. 34.2: Set-up for the star test.

this can only be done by collimating the source. If the white light point spread function is to be examined, a mirror collimator is preferable to a lens collimator in order to avoid the chromatic aberration contribution from the collimator.

If the source is collimated, one can make use of the nodal slide described in Chapter 11. If the lens is placed on a nodal slide and set in the nodal slide position, the change in shape of the point spread function with field angle can be easily observed. The shape of the point spread function gives an idea of the types of aberrations present. Observing the effect of closing or opening the aperture stop on the size and shape of the point spread function also helps to identify the types of aberration present.

The disadvantage with this method is that a very small bright source of light is needed. If one uses a conventional light source, it is necessary that the size of the source must be such that its geometrical image is smaller than the Airy disc for that lens and a field stop aperture is used to control the effective size of the source. However the smaller this aperture, the less light is available to form the point spread function and hence the duller it is. We can find an estimate of the maximum size of the source using equation (34.23). This equation gives the physical size of the Airy disc in the image plane, and therefore the radius η' of the source image should satisfy the condition

$$\eta' < \frac{0.610\lambda}{n'\sin(\alpha')} \tag{34.25}$$

but the smaller the value of η', the better. The corresponding source size can be found from the transverse magnification equation given in Chapter 3 and a knowledge of the conjugate distances.

Example 34.1: Calculate the maximum source radius placed at a distance of 4 m from an $f = 50$ mm, $F/4$ lens that is to be used to examine the point spread function via the star test, using light with a wavelength of 500 nm. Assume an air medium.

Solution: Since the source is at a distance that is much larger than the focal length, we can assume that the image of plane is the back focal plane and hence the operating numerical aperture can be determined from the *F-number* by equation (9.14b), that is

$$n'\sin(\alpha') = \sin(\alpha') = 1/(2F\text{-}number) = 1/8$$

Therefore we have

$$\eta' = 0.610 \times 500 \times 10^{-6} \times 8 = 0.00244 \, \text{mm}$$

The source size η is then given by the transverse magnification

$$\eta/\eta' = 4000/50 \quad \text{giving } \eta = 0.195 \, \text{mm}$$

$$= \text{maximum source radius}$$

34.3.3   Effect of aberrations and defocus

34.3.3.1   Low levels of aberrations

For low levels of aberrations, the shape of the point spread function is dominated by diffraction and the presence of aberrations modifies its shape. Figure 34.1 shows a typical point spread function with a small amount of coma ($_1w_{3,1} = 0.5\lambda$). Here the plot is only along the longer direction of the coma flare. On the same graph the diffraction limited point spread function is plotted (as the dashed line) for comparison. It is clear from this diagram that the effect of the aberration is to spread the light out and reduce the peak value of the point spread function.

34.3.3.2   High levels of aberrations

By definition, at high levels of aberrations, aberrations dominate the shape of the point spread function and the diffraction effects are negligible in comparison. For each aberration acting alone, the point spread function has specific shapes and some of these are uniquely affected by variations in the aperture stop diameter. Of the five monochromatic aberrations, only three (spherical, coma and astigmatism) do not produce point to point images in the geometrical approximation. The other two, field curvature and distortion, do produce point-to-point images, but these images are not in the position predicted by Gaussian optics. So let us look at the point spread functions of the other three aberrations, assuming there is only one aberration acting at a time. In the following, we will assume that the aperture stop is circular.

Spherical aberration: Spherical aberration produces a circular point spread function in which the extreme rays form the edge of the point spread function as indicated in Figure 2.6a. Therefore if we alter the diameter of the aperture stop, the point spread function contracts or expands equally in all directions, as shown in Figure 34.3.

Coma: The shape of the coma point spread function in the Gaussian image plane is shown in Figure 5.3. It is clear from this diagram that the rays at the rim of the aperture stop form a circle, which in part forms the outer edge of the blunt end of the coma flare. The pointed end is formed by the rays passing through the centre of the aperture stop. Therefore, altering the diameter of the aperture stop causes the coma flare to expand or contract from the blunt end and the pointy end is not affected. This is shown in Figure 34.3.

Astigmatism: The point spread function for astigmatism at the Gaussian image plane is an elliptical blur disc, but if we defocus the observation plane, the

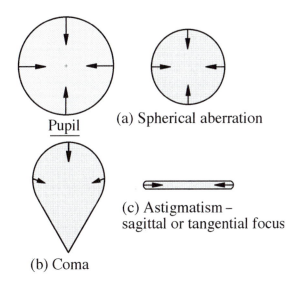

Fig. 34.3: The geometric optics point spread function base shapes for the three aberrations that do not produce point-to-point images, and with each aberration acting alone. The effect of varying the aperture stop diameter on shape and size is also shown.

Pupil

(a) Spherical aberration

(c) Astigmatism –
sagittal or tangential focus

(b) Coma

point spread function takes on the shapes shown in Figure 5.6. At two positions, the point spread functions are lines and these two lines are perpendicular to each other. The rays that form the ends of the lines pass through the system at the rim of the aperture stop. Therefore if we change the diameter of the aperture stop these lines change in length equally from both ends as shown in Figure 34.3.

34.3.3.3 Effect of defocus

We know from the study of defocus in Chapter 10 that a defocus spreads light out. In geometrical optics, the diffraction and aberration free distribution is given by the defocus blur disc. However, this is not a valid approach for small levels of defocus, where diffraction dominates and in this case, we must look at the diffraction prediction. The effect of a small amount of defocus ($_0w_{2,0} = 0.5\lambda$) on the point spread function is shown in Figure 34.1. It is clear that a small amount of defocus has a serious effect on the shape of the point spread function, reducing the height and spreading out the light.

34.3.4 Derived criteria

One disadvantage of the point spread function is that it is a two dimensional function and difficult to use as a merit function for comparing two systems. It is easier to make a comparison using a one dimensional merit function scale. There are two ways in which such a merit function can be derived from the point spread function.

34.3.4.1 Strehl intensity ratio (E)

The Strehl intensity ratio is a measure of the effect of the aberrations in reducing the maximum or peak value of the point spread function. It is defined as follows

$$E = \frac{\text{maximum light level value of } g(u', v')}{\text{maximum light level value of } g(u', v')_{W=0}} \tag{34.26}$$

where the denominator is the maximum value of the point spread function for the corresponding aberration free system (that is $W = 0$). Since the effect of aberrations is to spread the point spread function out and decrease the maximum peak height, the Strehl intensity ratio must always be less than or equal to one. Since the greater the aberrations, the lower the peak height, the greater the aberrations, the lower the value of the Strehl intensity ratio and the poorer the image quality. For the defocus example shown in Figure 34.1, the Strehl intensity ratio is about 0.4.

Strehl intensity ratio and the wave aberration function

Born and Wolf (1989, 463–464) have shown that, for small levels of aberrations, the Strehl intensity ratio (E) is related to the variance of the wave aberration by equation (34.6).

34.3.4.2 Half-width

The half-width of the point spread function is the width where the point spread function drops to half its maximum height. However since the point spread function is a two dimensional function, the half-width will depend upon the azimuth and will have maximum and minimum values in azimuths 90° apart. This will certainly be so in the presence of coma and astigmatism. However on axis, the point spread function will be circularly symmetric and hence there will be only a single half-width value.

34.3.5 *Effect of aperture stop diameter*

If one considers only the effect of diffraction, the point spread function decreases in width with increase in aperture stop diameter, as indicated by equation (34.23). On the other hand, in the presence of aberrations only, the point spread function increases in diameter with an increase in stop diameter, because the aberrations increase with diameter of the stop. In practice, both diffraction and aberrations are present. Diffraction dominates at small stop diameters and aberrations at large stop diameters. Thus there is an optimum stop diameter where the point spread function width is minimum, as shown in Figure 34.4. This is the diameter giving best image quality, but will depend upon aberration level.

34.3.6 *Line and edge spread functions*

In many practical situations, it is easier to examine the image of a line or edge instead of a point. This is simply because in lines or edges, there is much more light to work with. The light distributions are called the **line spread function** and the **edge spread function** respectively. These are related to the point spread functions by integrals. The line spread function is the sum of a line of point spread functions and the edge spread function is a sum of appropriate line spread functions.

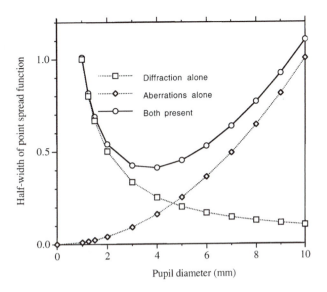

Fig. 34.4: Width of the point spread function as a function of pupil diameter. The axes values are not values of any real system.

34.3.7 Resolution limit – the Rayleigh criterion

The Rayleigh resolution criterion states that for a diffraction limited system, two point sources can just be resolved if the peak of one of the point spread functions lies on the first minimum of the other. The situation is shown in Figure 34.5. Now since the radius of the first dark ring of the diffraction limited point spread function is given by equation (34.23), it follows that this is the minimum angular resolution according to the Rayleigh criterion, that is the peaks of the two point spread functions must be separated by a distance $\Delta u'$ in the image plane, where

$$\Delta u' = \frac{0.610\lambda}{n' \sin(\alpha')} \tag{34.27}$$

Special case of the object plane (in air) at infinity

If the object plane is at infinity, the physical separation $\Delta u'$ of the images corresponds to an angular separation $\Delta\theta$ of the point sources in object space and these two quantities are related by equation (3.52), which relates the angular size of an object at infinity to the physical size of its image in the back focal plane of the lens. Using this equation, we can write

$$\Delta\theta = \frac{0.610\lambda F}{n' \sin(\alpha')} \tag{34.28}$$

where F is the equivalent power of the optical system. Now for the object at infinity, the image space numerical aperture $[n' \sin(\alpha')]$ is related to the equivalent power F and entrance pupil diameter D by equation (9.12b). We can use that equation to eliminate the numerical aperture in equation (34.28)

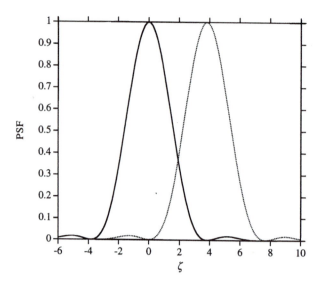

and doing so we have

$$\Delta\theta = \frac{1.22\lambda}{D} \qquad\qquad (34.28a)$$

and this equation is the mathematical form of the Rayleigh criterion.

Example 34.2: Estimate the resolution of a telescope with an objective diameter of 100 mm, assuming the Rayleigh criterion.

Solution: The Rayleigh criterion requires a wavelength, so we have to choose one and a convenient value is 555 nm, the middle of the visible spectrum. The objective of the telescope is the entrance pupil and its diameter $D = 100$ mm, so we insert these values in equation (34.28a) and get

$$\Delta\theta = 6.8 \times 10^{-6}\,\text{rad} = 0.02\,\text{minutes of arc}$$

34.4   The optical transfer function

Definition: *The optical transfer function (OTF) is a measure of the reduction in contrast and change in phase of sinusoidal patterns imaged by an optical system. It is a function of the spatial frequency and orientation of the pattern.*

To begin the mathematical treatment of the optical transfer function, let us consider a one dimensional case in which the sinusoidal grating of frequency σ and phase δ has an object light level profile (illuminance or luminance) given by the equation

$$A(u) = A_o + A\sin(2\pi u\sigma + \delta) \qquad\qquad (34.29a)$$

Over a region of the image plane, the image will also be sinusoidal and of the form

$$A'(u') = A_0 + A' \sin(2\pi u'\sigma' + \delta') \tag{34.29b}$$

The distances u and u' are connected by the transverse magnification M by the equation

$$u' = uM \tag{34.30a}$$

and the spatial frequencies σ and σ' are related as follows

$$\sigma' = \sigma/M \tag{34.30b}$$

Thus

$$u'\sigma' = u\sigma \tag{34.31}$$

The effect of aberrations and diffraction is to change the contrast of the sinusoidal pattern. We define the contrast of the object or image pattern as

$$\text{contrast} = \frac{\text{maximum light level} - \text{minimum light level}}{\text{maximum light level} + \text{minimum light level}} \tag{34.32}$$

We define the **modulation transfer function**, $G_c(\sigma)$, as the ratio of image to object contrasts, that is

$$G_c(\sigma) = \text{image contrast/object contrast} \tag{34.33}$$

Using equations (34.29a & b) and (34.32), it readily follows that

$$G_c(\sigma) = A'/A \tag{34.33a}$$

Some aberrations, distortion being the best example, move the pattern away from the expected Gaussian position. Effectively, this is the same as a change of phase, that is δ and δ' will have different values. We define the **phase transfer function**, $G_p(\sigma)$ as

$$G_p(\sigma) = e^{i(\delta'-\delta)} \tag{34.34}$$

Both the modulation and phase transfer functions are functions of spatial frequency σ.

We can create a function that combines both the modulation and phase transfer functions and call this the **optical transfer function** which we denote as $G(\sigma)$. If we express equations (34.29a and b) in complex form, define the contrast above in terms of complex amplitudes and note that $u\sigma = u'\sigma'$, it follows that the optical transfer function can be written as the product of the modulation and phase transfer functions, that is

$$G(\sigma) = G_c(\sigma)G_p(\sigma) \tag{34.35}$$

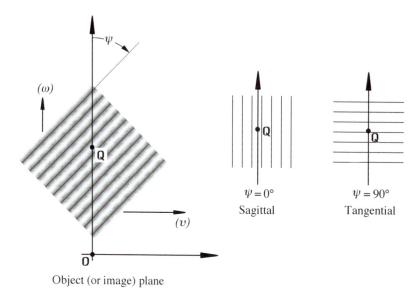

Object (or image) plane

If we now extend these ideas to two dimensions, we have to consider the orientation of the sinusoidal pattern. We define the orientation as the angle ψ as shown in Figure 34.6. Thus we can think of the sinusoidal pattern of spatial frequency σ as having a sagittal frequency υ and a tangential frequency ω, where

$$\upsilon = \sigma \cos(\psi) \quad \text{and} \quad \omega = \sigma \sin(\psi) \tag{34.36}$$

Thus $\psi = 0°$ corresponds to a pattern in the sagittal direction and $\psi = 90°$ corresponds to the pattern in the tangential direction as shown in the diagram. In terms of these parameters, we can denote the optical transfer function as $G(\sigma, \psi)$ or $G(\upsilon, \omega)$. The modulation and phase transfer functions can also be expressed as a function of these two frequencies.

It is useful to express the actual spatial frequency σ in terms of a "reduced" spatial frequency, here denoted by the symbol s, as follows:

$$s = \frac{\sigma \lambda}{n \sin(\alpha)} \quad \text{(finite conjugate)} \tag{34.37a}$$

or

$$s = \frac{\sigma \lambda}{n \bar{\rho}} \quad \text{(infinite conjugate)} \tag{34.37b}$$

where $n \sin(\alpha)$ is the numerical aperture, $\bar{\rho}$ is the radius of the pupil and σ is the spatial frequency in c/unit distance for a finite conjugate or c/radian for a conjugate at infinity. These equations apply equally to object and image spaces.

34.4.1   Methods for its determination or calculation

34.4.1.1   Theoretical methods

In the preceding main section, we showed that the point spread function and wave aberration function were related by an integral equation. Similarly the optical transfer function is related to the point spread function by an integral equation. Therefore the optical transfer function is related to the wave aberration function. In practice, it is easier to use the wave aberration function.

From the point spread function

Because u (or v) and u' (or v') are linearly related by equation (34.30a), we can express the point spread function $g(u', v')$ in terms of u and v rather than u' and v', provided we make the appropriate scaling. In terms of $g(u, v)$, the optical transfer function is given by the integral equation

$$G(\sigma, \psi) \quad \text{or} \quad G(v, \omega) = \iint g(u, v)e^{-i2\pi(uv+v\omega)} \, du \, dv \qquad (34.38)$$

This type of integral is known as a **Fourier transform**. The computation of this particular integral is complicated by two facts. One is that the point spread function is often not known and would have to be calculated first, say from the wave aberration function. The second is that, even if it were known, the point spread function has no definite limit. It extends to infinity and thus the region of integration extends from zero to infinity. Fortunately, there is a better computational method: the auto-correlation method.

Auto-correlation of the pupil function

The optical transfer function is related to the wave aberration function by the auto-correlation integral (Macdonald, 1971),

$$G(v, \omega) = (1/\pi) \iint_c e^{\{i2\pi[W(x+\underline{v}/2, y+\underline{\omega}/2)-W(x-\underline{v}/2, y-\underline{\omega}/2)]\}} \, dx \, dy$$

$$(34.39)$$

where $\underline{v}$ and $\underline{\omega}$ are reduced forms of the actual spatial frequencies v and ω. Thus they are related by equation (34.37a or b), in which $\underline{v}$ or $\underline{\omega}$ replace s and v or ω replace σ. The wave aberration function $W(x, y)$ is in the normalized form and the coefficients must be in units of wavelength. The region of integration c is the shaded common region of the sheared pupils shown in Figure 34.7. One distinct advantage of this integral over the one in equation (34.38) is the limited bounds of the functions and hence region of integration.

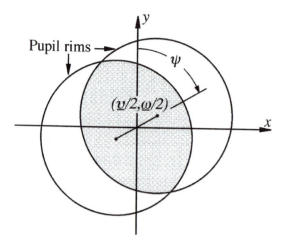

Fig. 34.7:   Sheared pupil
(auto-correlation) method
of calculating the optical
transfer function.

34.4.1.2   Laboratory methods

The optical transfer or modulation transfer function can be determined in the laboratory by imaging a set of sinusoidal wave gratings. The contrasts in the object and image are measured and their ratio is the modulation transfer function for that spatial frequency. However in practice square wave gratings are usually used because they are easier to make, but the use of square wave gratings requires some extra data processing to extract the sinusoidal components.

34.4.2   *The diffraction limited optical transfer function*

If the system is diffraction limited, then the wave aberration function is zero and the optical transfer function given by the above equation (34.39) reduces to the area common to the sheared pupils and since the optical transfer function will be now independent of ψ, we can reduce the optical transfer function $G(\sigma, \psi)$ to simply $G(\sigma)$ or in terms of the reduced spatial frequency s,

$$G(s) = [\beta - \sin(\beta)]/\pi \qquad (34.40)$$

where

$$\beta = 2 \arccos (s/2) \qquad (34.40a)$$

where s is defined by equations (34.37a and b).

It is clear that $G(s)$ goes to zero at $s = 2$. This limiting value of s corresponds to the diffraction limited resolution of the system; that is the maximum spatial frequency that can be imaged by the system. From equation (34.37a or b), these resolution limit frequencies are as follows:

$$\sigma_{\max} = \frac{2n \sin(\alpha)}{\lambda} \quad \text{c/unit distance (finite conjugate)} \qquad (34.41a)$$

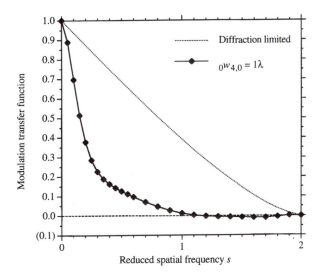

Fig. 34.8: Examples of the optical (modulation) transfer function. The diffraction limited case and one with 1 wavelength of primary spherical aberration.

or

$$\sigma_{\max} = \frac{2n\bar{\rho}}{\lambda} \quad \text{c/rad (infinite conjugate)} \tag{34.41b}$$

Example 34.3: Calculate the resolution limit in c/deg for object space, of an eye with a 2 mm diameter pupil and for light with a wavelength of 555 nm. Take the object at infinity.

Solution: Since the object is at infinity, we must use equation (34.41b). Substituting in this equation, we have

$$\sigma = 2 \times 1 \times 1/0.000555 = 62.9 \, \text{c/deg}$$

34.4.3 *Effect of aberrations*

We can examine the effect of aberrations by calculating the optical transfer function by either of the methods described in Section 34.4.1.1. Figure 34.8 shows the optical transfer function for spherical aberration of magnitude $_0w_{4,0} = 1\lambda$. It is clear from this diagram that the aberrated modulation transfer function is depressed compared with the diffraction limited function. Some aberrations change the contrast only (e.g. spherical aberration). Other aberrations change the phase only (e.g. distortion) and some change the contrast and phase (e.g. coma).

34.4.4 *Effect of defocus*

The physical optics defocussed optical transfer function can be determined by using a wave aberration function polynomial term $_0w_{2,0}$. This term is related to a physical defocus by equations (33.65a, b and c). Figure 34.9 shows the optical transfer functions for a range of values of $_0w_{2,0}$. Note that for the higher defocus levels, the optical transfer function is negative over some spatial frequency ranges.

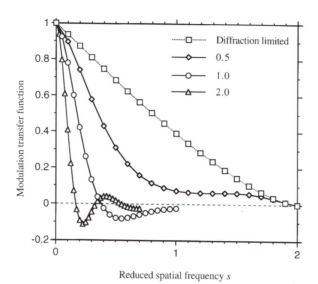

Fig. 34.9: Examples of
the defocussed optical
(modulation) transfer
function, for three values of
defocus ($_0 w_{2,0}$) in units of
wavelength.

Equation (34.38) states that the optical transfer function is a Fourier transform of the point spread function. For high levels of defocus, the point spread function is the defocus blur disc, which we have discussed in Chapter 10. We can then look at the optical transfer function in terms of geometrical optics.

The geometrical approximation using the defocus blur disc concept

For an optical system with a circular pupil, the geometric optical defocussed point spread function is rotationally symmetric and has a constant light level. Thus if it has a radius ρ'_B, the light level $g(u)$ in the spread function can be written in the form

$$g(u) = A \text{ if } |u| < \rho'_B \quad \text{and} \quad g(u) = 0 \text{ if } |u| > \rho'_B \qquad (34.42)$$

where A is the light level of the defocussed point spread function. The optical transfer function, represented here as $G(\sigma)$, is the Fourier transform of the above point spread function, that is given by equation (34.38) and in polar co-ordinates, that equation can be expressed as

$$G(\sigma) = \int_0^{2\pi} \int_0^{\rho'_B} g(u) e^{(-2\pi i \sigma u)} u \, du \, d\phi \qquad (34.43)$$

which reduces to

$$G(\sigma) = \frac{2 J_1 (2\pi \sigma \rho'_B)}{2\pi \sigma \rho'_B} \qquad (34.44)$$

This function has the same form as the diffraction limited point spread function, given by equation (34.21) except for the square on the diffraction limited point spread function. The function $G(\sigma)$ is plotted in Figure 34.10 for $\rho'_B = 1$.

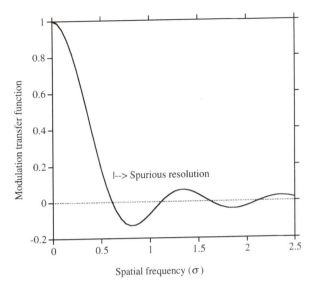

Fig. 34.10: The geometrical optics defocussed modulation optical transfer function for a defocussed blur disc of unit radius.

Like the diffraction limited point spread function, the function $G(\sigma)$ is zero and crosses the axis at a number of points, with the first at

$$2\pi\sigma\rho_B' = 3.8317$$

From this diagram, it is clear that the function $G(\sigma)$ alternately goes from positive to negative in value with ever decreasing magnitudes. The regions where the function is negative implies that a sinusoidal grating of these frequencies will be imaged but with negative contrast. The region beyond the first zero is called the region of **spurious resolution**. Spurious resolution can be readily observed using the Siemen's star, which is described in Appendix 5. Figure 34.11 shows a defocussed Siemen's star and spurious resolution. Spurious resolution can be observed in the defocussed eye and this phenomenon has been discussed by Smith (1982).

34.4.5 Derived criteria – resolving power

One could define the resolving power as the spatial frequency where the modulation transfer function first goes to zero. In the diffraction limited optical transfer function, the resolving power limit is given by equation (34.41a) or (34.41b). If aberrations are present, the resolving power limit must be less than this value and the actual value will depend upon the aberrations and their levels. However this definition is not always practical since the sinusoidal grating may not be detectable even with a non-zero contrast. This is because many detectors have a finite contrast threshold, below which the grating is not detected and this limit is partly set by inherent noise in the detector. For example, the eye has a contrast threshold which is spatial frequency dependent and we will be examining some implications of this in Chapter 36.

Fig. 34.11: Defocussed Siemen's star, showing spurious resolution.

34.5 Resolving power

Definition: Resolving power is a measure of the smallest detail that can be detected or discriminated in an image.

Resolving power is a very useful image quality criterion and we have seen that it arises naturally out of the point spread function and particularly the optical transfer function. For diffraction limited systems, we have seen that there are very simple equations that allow the calculation of the resolving power. Aberrations make the image quality worse than that predicted from diffraction theory. Since real systems contain some level of aberrations, it is possible but not simple to predict the resolving power from a real system from its system construction. Therefore in many instances, it is easier to measure the resolving power in the laboratory. This involves imaging a resolving power chart and determining the smallest detail in the chart that can be resolved. The simple resolving power charts, of the type described in Appendix 5, can be used to assess image quality.

Because both the aberrations and diffraction effects depend upon the aperture stop size, such resolving power should be measured at different pupil sizes. The effect of chromatic aberration also depends upon the spectral content of the light source used and therefore, this will also affect the result. Finally the aberrations depend upon the distance off-axis and the orientation of the chart. Therefore these should be taken into account.

Exercises and problems

34.1   For an optical system consisting of five lenses each made from the same glass with a refractive index of 1.5, calculate the light lost from the main imaging forming beam due to surface reflections.

ANSWER: 37.9%

34.2   Calculate the radius of the Airy disc for a lens with an *F-number* of 2, if the Airy disc is formed in the back focal plane of the lens, for light of wavelength of 500 nm.

   ANSWER: 0.00122 mm

34.3   For the star test, calculate the minimum diameter of a circular source if its geometrical image has to be less than 1/10 the diameter of the Airy disc in the following case:

   source distance = 3.5 m
   lens aperture = $F/2$ focal length = 75 mm
   wavelength = 633 nm

   ANSWER: Diameter = 0.014 mm

34.4   Calculate the Rayleigh diffraction limited resolving power of the eye if the pupil diameter is 6 mm. Compare this with the value corresponding to 6/6 visual acuity.

   ANSWER: 0.388 minutes of arc; 6/6 acuity corresponds to 2 minutes of arc

34.5   Calculate the optical transfer function resolution limit for an $F/4$, $f = 500$ mm diffraction limited lens, for light of wavelength 550 nm.

   ANSWER: 455 c/mm

34.6   What is the optical transfer function diffraction limited object space resolving power of a microscope objective with a magnification of 20 and a numerical aperture of 0.62, using light with a wavelength of 550 nm?

   ANSWER: 2255 c/mm

Summary of main symbols

k	$2\pi n'/\lambda$
n, n'	refractive indices of object and image spaces
$_0W_{2,0}$	defocus wave aberration coefficient in the normalized pupil co-ordinates
ρ_B'	geometrical defocus blur disc radius
X, Y	actual co-ordinates in pupils
x, y	corresponding reduced or normalized pupil co-ordinates
u, v	actual object/image plane co-ordinates relative to the object/image point
$W(x, y)$	wave aberration function in normalized pupil co-ordinates x and y
E	Strehl intensity ratio or normalized illuminance

Section 34.1: Veiling glare

R	surface reflectance
T	surface transmittance
$T(k)$	transmittance of k surfaces

Section 34.2: The wave aberration function

V variance of wave aberration function

Section 34.3: The point spread function

$P(x, y)$ pupil function
$A(x, y)$ amplitude transmittance at the point (x, y) in the pupil
$g_a(u, v)$ amplitude point spread function
$g(u, v)$ (intensity) point spread function
u_0 u value of the radius of the Airy disc
$\boldsymbol{u}$ scaled distance unit of u so that one unit of $\boldsymbol{u}$ is equivalent to the radius of the Airy disc
$\Delta\theta$ angular radius of Airy disc

Section 34.4: The optical transfer function

υ, ω spatial frequencies (cartesian representation)
σ, ψ spatial frequency σ and orientation ψ (polar representation)
σ_{max} diffraction limited value of σ
s reduced spatial frequency
$\underline{\upsilon}, \underline{\omega}$ reduced forms of υ, ω
$G(\upsilon, \omega)$ optical transfer function (cartesian representation)
$G(\sigma, \psi)$ optical transfer function (polar representation)
$G(s)$ diffraction limited optical transfer function in reduced spatial frequency form

References and bibliography

Barton N.P. (1972). Application of the optical transfer function to visual instruments. *Opt. Acta* 19(6), 473–484.

*Born M. and Wolf E. (1989). *Principles of Optics*, 6th ed. (corrected). Pergamon Press, Oxford.

Hopkins H.H. (1955). The frequency response of a defocussed optical system. *Proc. Roy. Soc. (A)* 231, 91–103.

Hopkins H.H. (1965). Canonical pupil co-ordinates in geometrical and diffraction image theory. *Jpn. J. Appl. Phys.* Suppl. 4, 31–35.

*Hopkins H.H. (1966). The use of diffraction-based criteria for the image quality in automatic optical design. *Opt. Acta* 13(4), 343–369.

*Macdonald J. (1971). The calculation of the optical transfer function. *Opt. Acta* 18, 269–290.

*Macleod H.A. (1969). *Thin Film Optical Filters*. Adam Hilger, London.

*Maréchal A. and Françon M. (1960). *Diffraction, Structure des Images*. Editions de le Rev. d'Opt. Masson SA, Paris.

*Smith G. (1971). Veiling glare due to reflections from component surfaces: the paraxial approximation. *Opt. Acta* 18(11), 815–827.

*Smith G. (1982). Ocular defocus, spurious resolution and contrast reversal. *Ophthal. Physiol. Opt.* 2(1), 5–23.

*Yzuel M. and Hopkins H.H. (1970). The computations of diffraction patterns in the presence of aberrations. *Opt. Acta* 17(3), 151–182.

35

Aberrations of the eye and retinal image quality

35.0   Introduction

The eye can be expected to suffer from all the aberrations that we find in other optical systems, but with one essential difference. Man-made systems are usually designed with some symmetry. For example, most visual optical systems have rotational symmetry. In contrast, the eye is not rotationally symmetric, which is mostly due to uneven growth patterns in the different components. The lack of symmetry means that there is a degree of irregularity in the conventional aberrations. These can be readily observed on axis by slightly defocussing the eye while viewing bright point light sources. The irregular star shaped image is due to irregular aberrations, because regular aberrations would produce a uniform circular or elliptical (if astigmatism is present) defocus blur disc.

The significance of optical aberrations, whether regular or irregular, is not at all clear. The eye generally uses only foveal viewing for the discrimination of fine detail and thus it could be argued that the eye only requires good optical image quality over the region of the fovea, which has an angular subtense of about 2°. Since the fovea is about 5° off-axis, one would expect off-axis aberrations, such as coma and transverse chromatic aberration, to be present at the fovea. The other off-axis aberrations (astigmatism, field curvature and distortion) would also be present but perhaps not in significant amounts. This is because, as shown in Table 33.1, these aberrations vary as the square of the field angle or cubic for distortion and therefore build up slowly with field angle for small field angles. In the periphery of the retina, visual acuity falls rapidly with distance from the fovea and beyond a certain distance off-axis, the perceived image quality is limited more by the neural system than by aberrations or diffraction.

The role of ocular aberrations and retinal image quality in the design and performance of visual optical instruments is also not very clear. There is little evidence that aberrations of the eye seriously affect the performance of visual optical systems or that optical designers consider the ocular aberrations in their designs. Perhaps to the designer, the only critical quantity is the ocular tolerance to residual instrumental aberrations.

However, while ocular aberrations may not be very significant in the design of the traditional viewing aids such as telescopes and microscopes, they are very important in the design of optical instruments used to examine the retina, such as the retinoscope and the fundus camera. If we look into an eye to examine the retina, we would desire good image quality. Therefore we need to know what ocular aberrations are to be expected and have some idea of their magnitude.

In the following three sections, we will explore the aberrations of the eye. Most importantly, we will examine the aberrations of real eyes, but we will also compare their values with those of schematic eyes. This comparison will show that paraxial schematic eyes predict some aberrations well and others badly. This poor prediction is partly due to the simplified nature of the surface shapes chosen and the refractive index of the lens. The surface shapes of real eyes are aspherical compared to the spherical shapes chosen for paraxial schematic eyes and the refractive index of real lenses is a complex gradient index structure compared with the constant index of the schematic models. This weakness of paraxial schematic eyes has led to a development of more accurate models known as "finite aperture" or "wide angle" models. The ideal exact model has aspherical surfaces and a varying index in the crystalline lens, either in the form of a multiple layered shell structure or as a continuously varying index. A number of these "improved" models have been proposed (Lotmar 1971; Drasdo and Fowler 1974; Kooijman 1983; Blaker 1980; Navarro et al. 1985). These have aspheric surfaces or a gradient index lens or both. However while some of the most recent data of corneal shape were incorporated in these models, the shape of the lens and its refractive index distribution, particularly the refractive index distribution has not been based upon recent accurate measurements on real eyes. Therefore, in our opinion none of these improved models has been sufficiently validated to warrant further discussion here. As a result, we will restrict our discussion to only the paraxial schematic eyes. We will see that for some aberrations these paraxial schematic eyes are sufficiently accurate for some applications.

35.1   Regular monochromatic aberrations

We can estimate the aberrations of the eye by calculating the aberrations of a standard schematic eye. We will restrict our analysis to Seidel aberrations, except for spherical aberration. One very useful property of Seidel aberration theory is that it predicts the aberration contribution of each surface and the total aberration of the system is the sum of the surface contributions. The Seidel aberration values of the Gullstrand number 1 relaxed schematic eye with an 8 mm diameter entrance pupil were calculated using the equations given in Chapter 33 and are given with the details of that eye in Appendix 3. While this is the most complex of classical paraxial models, the aberrations of the other simpler models are much the same.

How accurately these aberration values represent those of real eyes depends upon the aberration. For example, as we will see the spherical aberration of paraxial schematic eyes is much larger than in real eyes, yet the chromatic aberration is not greatly different. The reason for this is that the schematic eyes have spherical surfaces and therefore neglect the asphericity of the real surfaces of the eye. This asphericity has a profound effect on spherical aberration but no effect on chromatic aberration.

Table 35.1. Spherical aberration of the Gullstrand number 1 schematic eye.

Ray height (mm)	$S_1/8$	W	$\delta l'$ (mm) Seidel	$\delta l'$ (mm) Exact	δF (m^{-1}) Seidel	δF (m^{-1}) Exact
0.0			0.0	0.0	0.0	0.0
0.4			−0.0223	−0.0223	0.057	0.058
0.8			−0.0892	−0.0897	0.230	0.233
1.2			−0.2006	−0.2033	0.516	0.534
1.6			−0.3567	−0.3652	0.918	0.974
2.0			−0.5573	−0.5788	1.434	1.560
2.4			−0.8025	−0.8490	2.065	2.383
2.8			−1.092	−1.183	2.811	3.447
3.2			−1.427	−1.592	3.672	4.872
3.6			−1.806	−2.095	4.647	6.839
4.0	41.35	53.095	−2.229	−2.721 (−18%)	5.737	9.015 (−36%)

Note: The Seidel ($S_1/8$) and wave aberration (W) levels are expressed in units of wavelength (555 nm).

We will briefly go through each of these aberrations in turn, looking at the aberrations of both the Gullstrand number 1 relaxed schematic eyes and those of real eyes.

35.1.1 Spherical aberration

From the data in Appendix 3, the Gullstrand number 1 schematic eye has 330.78 wavelengths of the Seidel spherical aberration at the edge of an 8 mm diameter pupil. Using equations (33.91a) and (33.92a) the above aberration value was used to predict the longitudinal $\delta l'$ and power error δF aberrations for rays entering the pupil at different ray heights and the results are listed in Table 35.1.

We can examine the accuracy of this Seidel aberration and the associated values of $\delta l'$ and δF by calculating the exact values from finite ray tracing. Using the finite ray trace procedure described by Welford (1986), the longitudinal aberration $\delta l'$ was evaluated for a set of equally spaced rays traced from infinity into the eye through an 8 mm diameter entrance pupil. The wave aberration W was calculated at the edge of this pupil and the equivalent power error δF was calculated using a similar set of rays traced out of the eye from the axial point on the retina. This power error is the vergence of crossing point of the ray with the axis in object space. A lens with this power but of opposite sign placed at the corneal vertex will correct this power error by imaging the crossing point at infinity. These exact aberrations are given in Table 35.1 and the power error is also plotted, as the dashed line, in Figure 35.1.

The spherical aberration of real eyes has been the subject of a number of studies, for example Koomen et al. (1949), Ivanoff (1956), Jenkins (1963a), Schober et al. (1968) and Charman et al. (1978). These investigators determined the aberration in terms of the above power error and the results are shown in Figure 35.1. As can be seen from this diagram, all of these investigations found that the mean spherical aberration was positive for the relaxed eye. Some

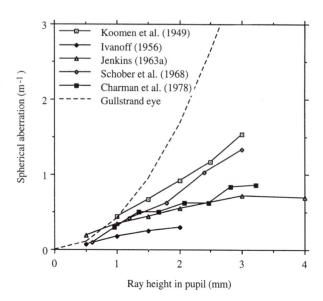

looked at the effect of accommodation and found that it reduced with increase in accommodation and for some subjects passed through zero and finally became negative. Thus in general, the eye is under-corrected for relaxed accommodation and probably over-corrected for high levels of accommodation.

According to equation (33.92), if the eye contains only primary aberration, the ray aberration as a power error is quadratic in ray height Y in the pupil. Therefore, within this approximation, we can write equation (33.92) in the form

$$\delta F(Y) = kY^2 \tag{35.1}$$

By comparison with equation (33.92a),

$$k = S_1/(2\bar{\rho}^4) \tag{35.2}$$

A number of investigators have assumed this rule can be applied to the real eye, that is they have assumed that the higher order spherical aberration is negligible. Van Meeteren (1974) averaged some experimental data, assumed the above rule and found a k value of 0.18 for Y expressed in millimetres and δF expressed in dioptres. Howland and Howland (1977) found k to have the value 0.0342, which they noted was much less than van Meeteren's. These and other values determined from the published values of spherical aberration shown in Figure 35.1 are listed in Table 35.2. These values of k show a large variation which is probably, at least partly, due to the variation in spherical aberration within the population. From Table 35.2, the mean value of k is

$$k = 0.10 \pm 0.06 \tag{35.3}$$

with a range from 0.034 to 0.190.

Given these clinically determined values of k, we can determine the expected equivalent value of the Seidel spherical aberration S_1 of real eyes by substituting

Table 35.2. Values of the coefficient "k" of Y^2 in equation (35.1) using the experimentally evaluated spherical aberration from various sources

	k
Koomen et al. (1949)[a]	0.190
Ivanoff (1956)[a]	0.091
Jenkins (1963a)[a]	0.059
Schober et al. (1968)[a]	0.161
Howland and Howland (1977)[b]	0.034
Charman et al. (1978)[a]	0.079
Millodot and Sivak (1979)[b]	0.058
Mean of above	0.096 ± 0.058
Mean of Van Meeteren (1974)[b]	0.18
Gullstrand 1 relaxed eye	0.359

Note: The value of k is evaluated for the ray height Y in millimetres and the power δF in metres^{-1}.
[a] Values were calculated from the published data shown in Figure 35.1.
[b] Values were calculated by the authors of the study.

a value of k into equation (35.2) to find the value of S_1. However in doing this, we have to be careful because of the mixed units of millimetres and metres. If we express the wavelength λ and entrance pupil radius $\bar{\rho}$ in metres, we must multiply the final value of S_1 by 10^6. Following this rule, a k value of 0.10 and an 8 mm diameter pupil, we have an expected value of S_1 of

$$S_1 = (51.2 \pm 30.7) \times 10^{-6} \text{ m} \quad \text{or}$$

$$92 \pm 55 \text{ wavelengths } (\lambda = 555 \text{ nm}) \tag{35.4}$$

for real eyes. We can convert this value to the wave aberration coefficient value as follows. From equation (33.66a),

$$_0w_{4,0} = 0.0064(\pm 0.0038) \text{ mm} \tag{35.5a}$$

and from equation (33.75b)

$$_0W_{4,0} = 0.000025(\pm 0.000015) \text{ mm}^{-3} \tag{35.5b}$$

From Table 35.1, the corresponding value of S_1 for Gullstrand's number 1 schematic eye is 331 wavelengths, which is about three and a half times greater and thus the corresponding value of k for this schematic eye calculated from equation (35.2) is 0.359, which is also shown in Table 35.2. Thus paraxial schematic eyes do not accurately predict the level of spherical aberration in real eyes. For the accommodated eye, the discrepancy is much greater. The above Gullstrand schematic eye, accommodated to a distance of 9.2 cm in front of the eye, has 2.7 times the spherical aberration of the relaxed eye, compared with generally a small negative value in real eyes.

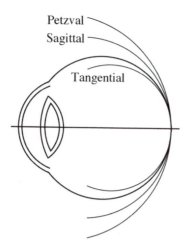

Petzval
Sagittal

Tangential

Fig. 35.2: The Seidel
estimates of the sagittal,
tangential and Petzval
surfaces of the Gullstrand
number 1 schematic eye
drawn to scale on the
cross-section of an eye with
a retinal radius of curvature
of 12 mm.

35.1.2   Coma

From Appendix 3, the Gullstrand number 1 relaxed schematic eye has $+13.21$ wavelengths of Seidel coma at $5°$ off-axis ($\lambda = 555$ nm). From equation (33.66b), the corresponding value of $_1w_{3,1}$ is 6.61λ or 0.00367 mm. From equation (33.73), the corresponding reduced transverse aberrations length of the coma flare is 19.82 units of $\delta H'$. We can convert this value to a physical value on the retina using equation (33.53), where from the appendix, the numerical aperture $[n' \sin(\alpha')]$ has a value of 0.231. Using these values in equation (33.53) gives a value of $\delta\eta'$ of 0.0479 mm.

The amount of coma in real eyes has not been studied directly, but has been found from the studies of irregular aberration, which we investigate in Section 35.2.

35.1.3   Astigmatism and field curvature

The Seidel astigmatism and field curvature values given for the Gullstrand number 1 relaxed eye are given in Appendix 3. These values are path length aberrations and since astigmatism is a defocus aberration, expressing them as a power error or vergence will be more meaningful. These values can be converted to the sagittal, tangential and Petzval surface curvatures using equations (33.93a, b and c). The radii of curvatures are given with the other aberrations in Appendix 3. These surfaces are plotted to scale on a cross-section of an eye with a retinal radius of curvature of 12 mm in Figure 35.2. We can see from this diagram that the sagittal and tangential surfaces straddle the retina. While there is some astigmatism and field curvature, the field curvature is partly ameliorated by the curved nature of the retina. These results show that the schematic eye is hyperopic in the sagittal section and myopic in the tangential section.

The sagittal and tangential power errors of real eyes can be found by determining the refractive error of nominally emmetropic eyes in the sagittal and tangential meridians as a function of field angle. Results of the studies by Ferree et al. (1931; 1932), Jenkins (1963b), Rempt et al. (1971), Millodot and Lamont (1974), Millodot et al. (1975), Rempt et al. (1976) and Smith et al. (1988) are shown in Figure 35.3. As predicted by the schematic eye, the real eye is

Fig. 35.3: Mean values of the sagittal and tangential power errors from various sources. The dashed curves are the corresponding errors predicted from the Seidel aberration values of the Gullstrand number 1 relaxed schematic eye with a retina with a 12 mm radius of curvature.

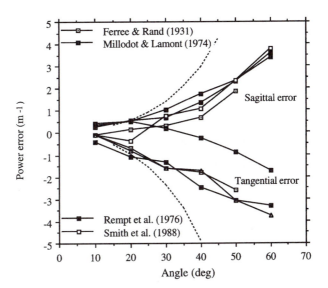

hyperopic in the sagittal section and myopic in the tangential section. However there is much variation between individuals and in different meridians with the same eye.

We can compare these values for the real eye with corresponding values from the Gullstrand number 1 schematic eye using equations (33.101a and b) to convert the sagittal and tangential curvatures of the sagittal and tangential surfaces to corresponding vergences in object space and the results are plotted on Figure 35.3 as the dashed curves. We have taken the curved nature of the retina into account by assigning to the terms L'_G, the value

$$L'_G = -\frac{C_R H^2 \tau^2}{2n' \bar{\rho}'^2} \tag{35.6}$$

where C_R is the curvature of the retina and is given a value of $1/12$ mm^{-1}. The variations in the values for real eyes depend upon the shape of the retina. It is probably clear from this approach that the actual value of the sagittal and tangential errors will depend upon the radius, or more correctly the shape, of the retina. It can be seen from the diagram that the schematic eye predictions are greater than those in the real eyes.

Astigmatism is the difference between the sagittal and tangential powers and the results shown in Figure 35.3 indicate that once again the schematic eye predicts higher levels of astigmatism than found in real eyes. Since astigmatism is the difference between the two power values, it is reasonably independent of the retinal shape.

35.1.4 Distortion

While the results shown in Appendix 3 indicate the eye has some distortion, it is doubtful that it has any practical significance for several reasons. Firstly distortion builds up only slowly with field angle, for small field angles, and does not usually become significant until the field angle is large. In the eye at large field angles, visual acuity is so poor it is unlikely that it would have sufficient resolution to detect distortion. Secondly, any image displacement due to distortion is complicated by the curved nature of the retina and could be compensated by the neural system. Thirdly since it gives point to point imagery, it does not add to the factors that degrade the perceived image. Therefore distortion is unlikely to be significant in the visual system and will not be discussed any further.

35.2   Irregular monochromatic aberration

If one measures the spherical aberration of the eye in different azimuths, it is soon apparent that the aberration is not rotationally symmetric. The presence of asymmetries in spherical aberration has led some to question the validity of the concept of conventional spherical aberration in the eye. For example, Smirnov (1961) stated that "by virtue of the asymmetry, the very concept of spherical aberration is not applicable to the eye" and referred to Helmholtz to support his view. Whether we take this view or the opposing view are matters of judgement which depend upon the purpose of the measurement or study. For example, the irregularity of the spherical aberration is severe enough to invalidate the use of isolated single azimuths of the pupil. However, annular zonal measurements are useful because they provide a mean value averaged around the pupil. Thus in real eyes, while a rotationally symmetric spherical aberration is an approximation to the true state, it is useful for some very important reasons. For example, these data are useful in the construction of rotationally symmetric finite schematic eyes. The causes of the irregularity are probably that the refractive surfaces are non-rotationally symmetric and decentred and there are local variations in the refractive index distribution in the lens.

The irregular spherical aberration has been measured by Smirnov (1961), Van den Brink (1962), Berny and Slansky (1970), Howland and Howland (1976), Howland and Howland (1977), and Walsh and Charman (1985). To quantify the irregular aberration, Howland and Howland (1977) proposed the general wave aberration function:

$$W(X, Y) = W_0 + W_1 X + W_2 Y + W_3 X^2 + W_4 XY + W_5 Y^2$$

$$+ W_6 X^3 + W_7 X^2 Y + W_8 XY^2 + W_9 Y^3$$

$$+ W_{10} X^4 + W_{11} X^3 Y + W_{12} X^2 Y^2 + W_{13} XY^3 + W_{14} Y^4$$

$$+ \text{higher order terms} \tag{35.7}$$

where X and Y are co-ordinates in the entrance pupil. Since then, there have been a number of studies (Walsh et al. 1984; Walsh and Charman 1985; Charman and Walsh 1989) of the irregular aberration of human eyes using this function. According to Walsh and Charman (1985), the terms W_6 to W_9 "loosely" represent coma. The terms W_{10} to W_{14} similarly "loosely" represent spherical aberration.

35.2.1   *Spherical aberration*

None of these terms alone represents the conventional rotationally symmetric spherical aberration, which is a term of the form $(X^2 + Y^2)^2$. Therefore, for the aberration to be a pure conventional spherical aberration, we would need

$$W_{10} = W_{14} = W_{12}/2$$

and all the other aberration terms to be zero. If we wish to determine the equivalent conventional spherical aberration to individual cases, we would need to determine the coefficient $_0W_{4,0}$ of the function

$$W = {}_0W_{4,0}(X^2 + Y^2)^2$$

which best fits equation (35.7), but ignoring the "non-aberration" terms W_0 to W_5Y^2. Application of this principle to the spherical aberration component of the data of Walsh and Charman (1985) gave a least squares best fit value of

$$_0W_{4,0} = 0.000009 \pm 0.000013 \text{ mm}^{-3}$$

This value is about 1/3 the value predicted from the regular aberration measurements given by equation (35.5b), but the distributions ranging from ±1 standard deviations substantially overlap, so there may be no significance in this difference.

35.2.2 Coma

Howland and Howland (1977) and Walsh and Charman (1985) have shown coma-like aberrations are present in the above wave aberration. One reason for the presence of this aberration is the fact that the fovea is off-axis by about 5°. If we look at the values of the coefficients in Table 1 of Walsh and Charman, the values of the coefficients vary from about +0.073 to −0.128, given the pupil coordinates in millimetres and wave aberration in micrometres. We can compare this with the Seidel predictions from Section 35.1.2, where we have

$$_1w_{3,1} = 0.00367 \text{ mm}$$

for an 8 mm diameter pupil. Similar to the rules given by equation (33.75a) for spherical aberration, for this coma we can write that is

$$_1W_{3,1} = {}_1w_{3,1}/\bar{\rho}^3$$

Thus for a pupil diameter of 8 mm (that is $\bar{\rho} = 4$ mm), we have

$$_1W_{3,1} = 0.00367/4^3 = 0.000057 \text{ mm}^{-2} = 0.057\mu\text{m} \cdot \text{mm}^{-3}$$

which is similar in magnitude to the values from Walsh and Charman.

35.3 Chromatic aberrations

The presence of chromatic aberration in the eye can be demonstrated by viewing a small "point" white light source through a red/blue transmitting filter. If the observer is slightly myopic, the centre of the image appears red surrounded by a blue disc. If the observer is slightly hyperopic, the centre is blue with a red

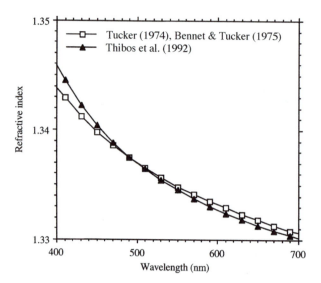

Fig. 35.4: Estimates of the wavelength variation of the refractive index of the ocular media, from equations (35.8a) and (35.8b).

surround. These simple observations predict that an emmetropic eye, focussed correctly say for light about 555 nm, the equivalent power of the eye is slightly higher in blue light and slightly lower in red light.

We can readily model this variation in power because the eye is mostly water and a knowledge of its dispersion allows a theoretical prediction of the equivalent power of the eye as a function of wavelength. Tucker (1974) and Bennett and Tucker (1975) gave the following equation for the refractive index of water, based upon the dispersion equation (1.8d) given in Chapter 1 but truncated to only four terms.

$$n^2(\lambda) = 1.7642 - 1.38 \times 10^{-8}\lambda^2 + 6.12 \times 10^{+3}/\lambda^2$$

$$+ 1.41 \times 10^{+8}/\lambda^4 \tag{35.8a}$$

where here the wavelength must be in nanometres (nm). Thibos et al. (1992) presented a different dispersion equation of the ocular media based upon the experimental chromatic aberration measurements on eight subjects. Their dispersion equation was similar to equation (1.8b) in Chapter 1, but with the value 1.2 of the power index replaced by 1.0 and was

$$n(\lambda) = 1.320535 + 4.685/(\lambda - 214.102) \tag{35.8b}$$

For comparison, these two functions are plotted in Figure 35.4.

35.3.1 V-value

Using the above dispersion equations, we can estimate the V-value for the eye as a whole by calculating the expected values of the refractive index at the appropriate wavelengths and then using equation (1.9) to calculate the V-value. The results of these calculations are shown in Table 35.3 with the V-value of about the low to mid 50's. The Thibos et al. V-value is lower, indicating that it predicts a greater dispersion than the Tucker model.

Table 35.3. The dispersion of the eye and the V-value for the eye as a whole

	λ(nm)	Bennett & Tucker (1975)	Thibos et al. (1992)
n_d	589.3	1.333492	1.333022
n_F	486.1	1.337671	1.337759
n_C	656.3	1.331625	1.331130
V_d		55.15	50.23

35.3.2 Longitudinal chromatic aberration: as a variation of power

More meaningful than Seidel aberrations or the V-value is the power of the eye as a function of wavelength as this can be used to estimate the longitudinal chromatic aberration as a power difference. Tucker (1974) found that a reduced eye model gives an accurate estimate of longitudinal chromatic aberration. The calculation is relatively simple. If we take a reduced eye with a corneal radius of curvature r, the equivalent power $F(\lambda)$ as a function of wavelength is simply

$$F(\lambda) = [n(\lambda) - 1]/r \qquad\qquad (35.9)$$

where $n(\lambda)$ is given by equation (35.8a or b). We have to assign a power to this eye and a convenient power would be an equivalent power of 58.64 m^{-1} (the Gullstrand number 1 relaxed schematic eye value), but this has to be defined at a particular wavelength. We will assume that the middle of the visible spectrum (555 nm) is a reasonable value and therefore this reduced eye is correctly focussed and has a power of 58.64 m^{-1}, for a wavelength of 555 nm. The next step is to determine the corneal radius of curvature r, from the equation

$$r = [n(555) - 1]/0.05864 \text{ mm} \qquad\qquad (35.10)$$

The longitudinal chromatic aberration as a power error $\delta F(\lambda)$ is then given by the equation

$$\delta F(\lambda) = [n(\lambda) - 1]/r - 0.05865 \text{ mm}^{-1} \qquad\qquad (35.11)$$

This power error was calculated using both the Bennett and Tucker and Thibos et al. refractive index distributions. The refractive index data for the corresponding reduced eyes are shown in Table 35.4 and the values of $\delta F(\lambda)$ are plotted in Figure 35.5 as a function of wavelength. It is clear that the longitudinal chromatic aberration leads to a power difference of about 2.5 m^{-1} across the visible spectrum.

However, this is not the value of the chromatic refractive error that is measured in the clinic. For comparisons with clinical measures, we should calculate the chromatic aberration as an equivalent refractive error. A suitable definition of the chromatic longitudinal refractive error $R_E(\lambda)$ is as follows.

For any level of ametropia, it is the difference between the object vergences of the retinal conjugates for a wavelength λ and a reference wavelength $\bar{\lambda}$ for which the eye is emmetropic.

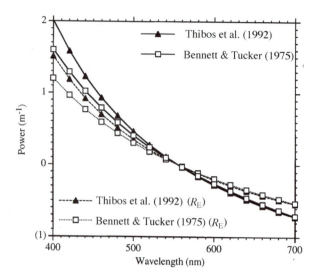

Fig. 35.5: Longitudinal chromatic aberration of a reduced eye with an equivalent power of 58.64 m^{-1}, the same as the Gullstrand number 1 relaxed eye, in terms of the theoretical prediction of the wavelength variation of power of the eye, using equation (35.11) and the equivalent chromatic refractive error equation (35.12).

Table 35.4. *A reduced eye for determining chromatic aberration, designed to have a power of 58.64 m^{-1} at a wavelength of 555 nm*

	Bennett & Tucker (1975)	Thibos et al. (1992)
Refractive index at 555 nm	1.3347	1.3343
Corneal radius of curvature	5.707 mm	5.701 mm
Eye length	22.760 mm	22.754 mm
Distance from nodal point to retina	17.053 mm	17.053 mm

From Thibos et al. (1991), the chromatic refractive error $R_E(\lambda)$ for a reduced model eye with a corneal radius of curvature r, at a wavelength λ relative to the wavelength $\bar{\lambda}$ at which the eye is emmetropic, is given by the equation

$$R_E(\lambda) = \frac{n(\lambda) - n(\bar{\lambda})}{r n(\bar{\lambda})} \tag{35.12}$$

and this curve is plotted in Figure 35.5.

Comparison with spherical aberration

If we compare the amount of spherical aberration in real eyes shown in Figure 35.1 with the longitudinal chromatic aberration shown in Figure 35.5, we may note that the values are similar in magnitude. However, the spherical aberration values are dependent upon pupil size, while the chromatic aberration values are not. Therefore in making a comparison between the two aberrations, we should be aware of the pupil size effect on the spherical aberration.

35.3.3 Transverse chromatic aberration

For a single thin lens with the aperture stop at the lens, the transverse chromatic aberration is zero. Similarly for a more complex system, if the stop is at one of the nodal points, the Seidel transverse chromatic aberration will also be zero. However, in the human eye, since the aperture stop is a few millimetres in front of the front nodal point of the eye, the transverse chromatic aberration is non-zero but small. For a field angle of $5°$, the Gullstrand number 1 relaxed schematic eye has a transverse chromatic aberration of only -0.374% (Appendix 3). The negative value indicates that blue light is focussed closer to the optical axis than red light.

35.4 Aberration compensation

35.4.1 Spherical aberration

There is little evidence that the spherical aberration of the eye is significant for small pupils. For larger pupils where it is greater and if the aberration was rotationally symmetric, it would be possible to compensate for it to some extent by refocussing. Assuming the spherical aberration values given in Section 35.2 are regular we could use equations given in Chapter 33 to calculate the expected amount of refocussing.

We can find the optimum amount of refocussing by minimizing the variance of the wave aberration function, as described in Section 34.2.1.2. According to equation (34.16), if the spherical aberration is purely primary, this defocus term is

$$_0w_{2,0} = -_0w_{4,0} \tag{35.13}$$

If we convert these to $_0W_{2,0}$ and $_0W_{4,0}$ coefficients using equation (33.75), we have

$$_0W_{2,0} = -_0W_{4,0}\bar{\rho}^2 \tag{35.14}$$

From equation (35.5b), the real eyes have an approximate mean value

$$_0W_{4,0} = 0.000025 \text{ mm}^{-3}$$

From equation (35.14), the optimum defocus, which minimizes this effect, is

$$_0W_{2,0} = 0.000025 \times \bar{\rho}^2$$

and from equation (33.89),

$$\delta F = -0.000050 \times \bar{\rho}^2 \text{ mm}^{-1}$$

For a 4 mm diameter pupil, the corresponding defocus would be 0.2 m^{-1} and for an 8 mm diameter pupil the defocus would be 0.8 m^{-1}. A 4 mm diameter pupil is a common diameter in daily activity, for which a refocus of 0.2 m^{-1} is comparable with the depth-of-field and is therefore probably not clinically detectable. For example van Heel (1946) found that correction of the spherical aberration did not improve visual acuity. The failure of such attempts is probably due to the small value of the effect and the complication of the irregularity of the aberration.

35.4.2   Chromatic aberration

The effects of chromatic aberration are attenuated by the unequal spectral response of the eye. The photopic and scotopic spectral responses of the eye, known as the spectral luminous efficiency functions, are shown in Figure 12.1. As the wavelength moves away from the peak values the response of the eye decreases and therefore the effect of any chromatic aberration is decreased.

35.4.2.1   Achromatizing correcting lens

In theory, the longitudinal chromatic aberration of the eye can be corrected by placing a nominally zero power lens in front of the eye that has a longitudinal chromatic aberration equal and opposite to that of the eye. Such lenses have been designed by Bedford and Wyszecki (1957), Fry (1972), Powell (1981) and Lewis et al. (1982). The Bedford and Wyszecki, Fry and Lewis lenses have a symmetric triplet construction. However these lenses, while designed to correct for longitudinal chromatic aberration, cannot correct for the transverse chromatic aberration, and this limits the field-of-view. The Powell lens is an air spaced five element lens (a cemented triplet followed by a cemented doublet). The extra complexity is designed to improve the transverse chromatic aberration and hence give a wider field-of-view than the other lenses.

35.5   The effects of chromatic aberration

35.5.1   Longitudinal chromatic aberration

Longitudinal chromatic aberration, although present in significant amounts, does not seem to greatly reduce visual acuity in white light. Campbell and Gubisch (1967) found that the difference in visual acuity between white and yellow light was only 6%. However in monochromatic light, the chromatic aberration does have a significant effect. It is clear from the preceding results that an emmetropic eye is myopic for blue light and hyperopic for red light. Hence, an emmetrope will have an effective refractive error (approximately -1.5 m^{-1} for the far blue) for distant blue objects, and the effect on visual acuity will become more pronounced as light level decreases and the pupil dilates. On the other hand, the emmetropic eye is hyperopic in red light and can accommodate to clearly see the distant red lights. Myopic eyes will be even more myopic in blue light, but less myopic in red. From Figure 35.5 it can be seen that a 0.5 m^{-1} myope will be approximately emmetropic at 700 nm and -2 m^{-1} myopic in blue.

   The longitudinal chromatic aberration is made use of in the clinical **duo-** or **bi-chrome** test. This is a confirmation test carried out towards the end of refraction. The test chart consists of black letters on a green background and similar letters on a red background. If the refraction is correct the patient should see each set of letters approximately equally clearly. If the green letters are clearer, the eye has a residual hyperopic error and if the red letters are clearer, the eye has a residual myopic error.

35.5.2 Transverse chromatic aberration

Since the optical and visual axes do not coincide, there will be a small amount of transverse chromatic aberration along the visual axis and this is demonstrated by the aberration values given in Appendix 3 for the Gullstrand number 1 schematic eye. Since we have noted that schematic eyes accurately predict chromatic aberration, we can conclude that the level of transverse chromatic aberration in real eyes will be similar to that in the model eye. The schematic eye value is 0.34% at the fovea taken to be 5° off-axis, which corresponds to about 1 minute of arc, which is just on the threshold of conventional resolution.

However, while it is not usually noticed directly, transverse chromatic aberration has an interesting manifestation in the phenomenon of **chromostereopsis**, which is due to a combination of the decentration of the fovea from the optical axis and transverse chromatic aberration. The result of these two factors is to transversely displace the retinal image according to wavelength, which leads to a variation in retinal disparity with wavelength. Because the refractive index with red light is less than that for blue light, the red rays intersect the retina further from the axis than the blue rays. This retinal disparity leads to an apparent longitudinal displacement with binocular vision. Thus, if a red object and a blue object are placed at the same distance, the red object will appear in front of the blue. Chromostereopsis is more easily observed for two colours towards the edges of the spectrum where the retinal image displacement is a maximum.

Not all observers experience this effect. Some experience no chromostereopsis, some see the opposite effect (that is blue in front of red) and others sometimes see red in front of blue and sometimes blue in front of red. These possible observations are supported by theory. Ray tracing through a schematic eye shows that the amount and sign of the dispersion between say the red and a blue beam entering the eye depends where the beam enters the eye. For example if the pupil is progressively decentred towards the nasal side, the disparity decreases, goes through zero for a pupil decentred by about 0.5 mm and then reverses in sign. Once this reversal occurs, blue should be seen in front of red.

35.6 Retinal image quality

Apart from aberrations, diffraction has an effect on retinal image quality. When diffraction is the limiting factor, the Rayleigh criterion and the diffraction limited optical transfer function can be used to assess the image quality in terms of a resolving power. According to the Rayleigh criterion, for a diffraction limited optical system, two points in object space can just be resolved if their angular separation $\Delta\theta$ is given by equation (34.28a), that is

$$\Delta\theta = \frac{1.22\lambda}{D} \quad \text{rad} \tag{35.15}$$

where λ is the wavelength, and D is the diameter of the entrance pupil and in equation (34.28a) the refractive index n is unity. For a wavelength of 550 nm and the entrance pupil diameter (D) ranging from 2 mm at high light levels to 8 mm at low light levels, equation (35.15) gives corresponding resolving powers between $\Delta\theta = 1.15$ and 0.29 minutes of arc, respectively. The optical transfer function theory given in Chapter 34 shows that for a diffraction limited optical system, the spatial frequency limit is given by equation (34.41b), that is

$$\sigma_{max} = \frac{2\tilde{\rho}}{\lambda} \quad c/rad \tag{35.16}$$

where once again the refractive index n of object space is taken as unity. For the pupil range of 2 to 8 mm in diameter ($\tilde{\rho} = 1$ to 4 mm), this equation gives diffraction limited spatial frequency limits of 62.9 c/deg to 252 c/deg.

The above values take no account of aberrations or the retinal neural network. Aberrations must reduce these resolving powers to some extent and we will see that for pupil sizes below about 3 mm, the quality of the retinal image is accurately predicted by diffraction. For pupil diameters above that limit, retinal image quality is worse than predicted by diffraction and thus affected by aberrations.

The quality of the perceived image must also be affected by the neural network of the retina; in particular the size of the retinal cones and rods and the extent to which they are connected at a deeper neural level. According to Polyak (1957), the foveal cone diameter is about 0.0015 mm. Now let us suppose we are to image a sinusoidal or square wave pattern on the retina. The highest spatial frequency that could be resolved is the one in which the bright bars lie on alternating lines of cones. The corresponding angular period of the pattern would be two cone diameters, that is

$$\text{angular period} = 2 \times 0.0015/\mathcal{N}'\mathcal{F}' \quad rad \tag{35.17}$$

where $\mathcal{N}'\mathcal{F}'$ is the distance from the back nodal point to the retina. Now since $\mathcal{N}'\mathcal{F}' \approx 17$ mm (Appendix 3), we have

$$\text{angular period} = 2 \times 0.0015/17 \text{ rad} = 0.6 \text{ minute of arc}$$

which corresponds to about 100 c/deg. The value of 100 c/deg in turn is the diffraction limited resolving power for a pupil diameter of 3.2 mm. While this is approximately the optimum pupil size giving best perceived image quality, the actual resolving power limit of the eye is between 30 and 60 c/deg. Therefore we can conclude that in the fovea, the optics and neural system are closely matched. However a pupil with a 3.3 mm diameter pupil cannot resolve 100 c/deg and the reason is a combination of the presence of aberrations and the fact that a contrast threshold is required before a sinusoidal pattern is detected.

However, once we move away from the fovea and out into the periphery of the retina, the neural network changes and loses its ability to provide good spatial performance. The density of the cones decreases and the density of the rods increases. Both of these related events lead to a reduction in the ability of the neural system to record fine detail in the image (Green 1970; Frisen and Glansholm 1975). Green found that beyond about 5°, the perceived image quality was limited more by the neural system than the optics.

The quality of the retinal image can be investigated either subjectively or objectively. In subjective measurements, the subject is presented with targets of varying levels of fine detail and detection or recognition thresholds can be used to measure the resolution limit. The outcomes of these types of measurements are affected by the level of aberrations and diffraction which affect the retinal image, the quality of the neural network and finally the subject's ability to make appropriate responses. Objective measures can be divided into two types. In one type, simple targets are imaged on the retina and the reflected image observed. This type does not depend on the neural network, although it depends to some extent upon the way the retina reflects incident light. In the other type, electrophysiological recordings can be taken from the skin and therefore this method includes the neural network factors but does not require the subjects to respond. In the remaining part of this chapter, we will look at some of these techniques and their results.

35.6.1   The line and point spread functions

In the objective measures of retinal image quality, the ideal approach would be to image a point source of light onto the retina and record the image using the light reflected back out of the eye. Until the advent of lasers and very sensitive detectors, this approach was not possible because a conventional "point" source of light contained too little light to be detected after it had been reflected from the retina. A suitable and simple alternative was to use a line source. Such a source would emit perhaps a hundred times more luminous flux than a "point" source of the same width.

Using a line source produces a **line spread function** instead of the point spread function and using mathematical manipulation, the line spread function can be processed to give both the corresponding point spread function and the modulation transfer function. However, this method is complicated by the fact that the beam has to traverse the optics of the eye twice and therefore is "doubly" aberrated, but fortunately mathematical processing can be used to restore the observed line spread function to its expected form on the retina.

Krauskopf (1962) measured the line spread functions using white light. His results show that the line spread functions were "bell" shaped with half-widths of 3.3 minutes of arc for pupil diameters between 3 and 6 mm, with little variation in that range. For pupil diameters greater than 6 mm, the half-width increases to about 5.5 minutes of arc for pupil diameters of 8 mm.

Campbell and Gubisch (1966) also measured the line spread functions of the eye for different pupil diameters. They plotted the corresponding diffraction limited line spread function on each of the measured line spread functions. Analysis of the curves shows that for pupil diameters of 2.4 mm in diameter, the measured line spread functions well match the diffraction limited curves. The minimum half-width of about 1 minute of arc occurs for a pupil diameter of 2.4 mm and this value is about one-third of those of Krauskopf.

The above measures of light spread function were done with white light and thus the light spread must be affected by a combination of monochromatic aberrations, chromatic aberrations and diffraction. We have seen that the monochromatic aberration is about the same level as the chromatic aberration but much more variable. Therefore we would expect the chromatic aberration to have a predictable contribution which would vary little from person to person. Van Meeteren (1974), referring to previous studies, stated that the white light point spread function is affected more by longitudinal chromatic aberration

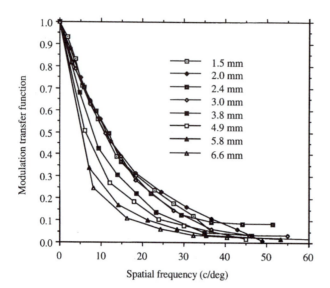

than by spherical aberration and since chromatic aberration varies very little among individuals, unlike spherical aberration, the white light point spread function should be similar from person to person.

The advent of the laser has given us access to a bright point source of light which is most suitable for measurement of the retinal point spread function. Santamaría et al. (1987) recorded the point spread function using monochromatic light from a helium-neon laser and transformed the point spread functions to corresponding modulation transfer functions. Later Artal et al. (1988) discussed a method of transforming the point spread function into wave aberration functions.

35.6.2 The modulation transfer function and the contrast threshold function

The modulation transfer function can be calculated from the line or the point spread function. Campbell and Gubisch (1966) calculated the modulation transfer functions from their measured white light line spread functions. Their mean results for their three subjects for a range of pupil sizes are given in Figure 35.6. Van Meeteren (1974) calculated the expected white light modulation transfer function of the eye using published aberration values and claimed the results were in agreement with experimentally determined values and that chromatic aberration was the most important aberration. As this aberration varies little between individuals van Meeteren claimed that the form of the curves would vary little between individuals. Santamaría et al. (1987) calculated the monochromatic modulation transfer function for two subjects from the point spread functions.

We can also measure the modulation transfer function of the eye by subjective methods; the most common way of doing this is to measure the contrast threshold of sinusoidal patterns for a range of spatial frequencies. However, this measure of contrast is not the modulation transfer function, because the

Fig. 35.7: The contrast threshold functions of the eye from Campbell and Green (1965).

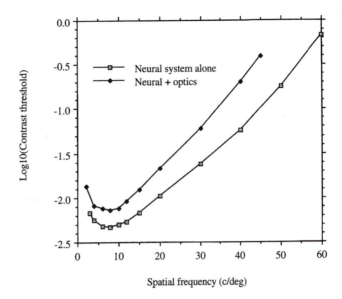

outcome is influenced by the neural system. Such a measure is called the **contrast threshold function**.

The contrast threshold function is defined as the contrast at which a sinusoidal pattern is just visible. This contrast varies with spatial frequency. There are two measures of the contrast threshold function. Probably the most common method of measuring the contrast threshold function is to present sinusoidal patterns on an oscilloscope screen and by some suitable psychophysical procedure vary the contrast until the subject decides that it is just visible. In this case, the final result is affected by the aberrations and diffraction properties of the eye and neural factors and in this book we will call this quantity the (total) contrast threshold function. Alternatively, we can use an interferometer to project sinusoidal patterns on the retina. This can be done using an interferometer similar to the Twyman–Green interferometer described in Chapter 25. In this case, the results are not affected by aberrations or diffraction. In this book, we will call this measure the (neural) contrast threshold function. Examples of these two contrast threshold functions, from Figure 9 of Campbell and Green (1965), are shown in Figure 35.7. The Campbell and Green figure shows the reciprocal of these functions which are known as contrast sensitivity functions.

It is possible to estimate the modulation transfer function from these two contrast threshold functions, as was done by Campbell and Green (1965). The modulation transfer function is the ratio of the (neural) contrast threshold function to the (total) contrast threshold function.

Summary of main symbols

Y	actual ray height in the pupil
S_1	Seidel spherical aberration
$_0w_{m,0}$	wave aberration polynomial coefficients for a normalized pupil

$_0W_{m,0}$ wave aberration polynomial coefficients for the actual pupil
$\delta F(\lambda)$ variation of power of the eye with wavelength

References and bibliography

*Artal P., Santamaría J., and Bescós J. (1988). Retrieval of wave aberration of human eyes from actual point spread function data. *J. Opt. Soc. Am. (A)* 5(8), 1201–1206.

Barnden R. (1974). Calculation of axial polychromatic optical transfer function. *Opt. Acta* 21, 981–1003.

Barnden R. (1976). Extra-axial polychromatic optical transfer function. *Opt. Acta* 23, 1–24.

Barton N.P. (1972). Application of the optical transfer function to visual instruments. *Opt. Acta* 19(6), 473–484.

*Bedford R.E. and Wyszecki G. (1957). Axial chromatic aberration of the human eye. *J. Opt. Soc. Am.* 47, 564–565.

*Bennett A.G. and Tucker J. (1975). Correspondence: chromatic aberration of the eye between 200 and 2000 nm. *Br. J. Physiol. Opt.* 29, 132–135.

*Berny F. and Slansky S. (1970). Wavefront determination resulting from Foucault test as applied to human eye and visual instruments. *Optical Instruments and Techniques 1969.* ed. J. Home Dickson. Oriel Press, Newcastle upon Tyne, pp. 375–386.

*Blaker J.W. (1980). Toward an adaptive model of the human eye. *J. Opt. Soc. Am.* 70, 220–223.

*Campbell F.W. and Green D.G. (1965). Optical and retinal factors affecting visual resolution. *J. Physiol. (Lond)* 181, 576–593.

Campbell F.W. and Gregory A.H. (1960). Effect of pupil size of visual acuity. *Nature (Lond.)* 187, 1121–1123.

*Campbell F.W. and Gubisch R.W. (1966). Optical quality of the human eye. *J. Physiol.* 186, 558–578.

*Campbell F.W. and Gubisch R.W. (1967). The effect of chromatic aberration on visual acuity. *J. Physiol.* 192, 345–358.

*Charman W.N., Jennings J.A., and Whitefoot H. (1978). The refraction of the eye in relation to spherical aberration and pupil size. *Br. J. Physiol. Opt.* 32, 78–93.

*Charman W.N. and Walsh G. (1989). Variations in the local refractive correction of the eye across the entrance pupil. *Optom. Vis. Sci.* 66(1), 34–40.

Crawford B.H. (1937). The luminous efficiency of light rays entering the eye pupil at different points and its relation to brightness threshold measurement. *Proc. Roy. Soc. (B)* 124, 81–96.

*Drasdo N. and Fowler C.W. (1974). Non-linear projection of the retinal image in a wide angle schematic eye. *Br. J. Ophthalmol.* 58, 709–714.

Epstein L.I. (1972). The double sliding door effect. *Ophthalmologica* 165, 117–124.

*Ferree C.E., Rand G., and Hardy C. (1931). Refraction for the peripheral field of vision. *Arch. Ophthal.* 5, 717–731.

*Ferree C.E., Rand G., and Hardy C. (1932). Refractive asymmetry in the temporal and nasal halves of the visual field. *Am. J. Ophthal.* 15(6), 513.

*Frisen L. and Glansholm A. (1975). Optical and neural resolution in peripheral vision. *Invest. Ophthal.* 14, 528–536.

*Fry G.A. (1972). Visibility of colour contrast borders. *Am. J. Optom. Arch. Am. Acad. Optom.* 49, 401–406.

Giles M.K. (1979). Grating detectability: A method to evaluate aberration effects in visual instruments. *Opt. Eng.* 18(1), 33–38.

*Green D.G. (1970). Regional variations in the visual acuity for interference fringes on the retina. *J. Physiol. (Lond.)* 207, 351–356.

*Howland B. and Howland H.C. (1976). Subjective measurement of higher order aberrations of the eye. *Science* 193, 580–582.

*Howland H.C. and Howland B. (1977). A subjective method for the measurement of monochromatic aberrations of the eye. *J. Opt. Soc. Am.* 67(11), 1508–1518.

*Ivanoff A. (1956). About the spherical aberration of the eye. *J. Opt. Soc. Am.* 46(10), 901–903.

*Jenkins T.C.A. (1963a). Aberrations of the eye and their effects on vision: part 1. *Br. J. Physiol. Opt.* 20, 59–61.

*Jenkins T.C.A. (1963b). Aberrations of the eye and their effects on vision: part 2. *Br. J. Physiol. Opt.* 20, 161–201.

*Kooijman A.C. (1983). Light distribution on the retina of a wide angle theoretical eye. *J. Opt. Soc. Am.* 73(11), 1544–1550.

*Koomen M., Tousey R., and Scolnik R. (1949). The spherical aberration of the eye. *J. Opt. Soc. Am.* 39(5), 370–376.

*Krauskopf J. (1962). Light distribution in human retinal images. *J. Opt. Soc. Am.* 52, 1046–1050.

*Lewis A.L., Katz M., and Oehrlein C. (1982). A modified achromatizing lens. *Am. J. Optom. Physiol. Opt.* 59(11), 909–911.

*Lotmar W. (1971). Theoretical eye model with aspheric surfaces. *J. Opt. Soc. Am.* 61(11), 1522–1529.

*Millodot M., Johnson C.A., Lamont A., and Leibowitz H.W. (1975). Effect of dioptrics on peripheral vision. *Vis. Res.* 15, 1357–1362.

*Millodot M. and Lamont A. (1974). Refraction of the periphery of the eye. *J. Opt. Soc. Am.* 64, 110–111.

*Millodot M. and Sivak J. (1979). Contribution of the cornea and lens to spherical aberration of the eye. *Vis. Res.* 19, 685–687.

*Navarro R., Santamaría J., and Bescós J. (1985). Accommodation dependent model of the human eye with asphaerics. *J. Opt. Soc. Am.* (A) 2(8), 1273–1281.

*Polyak S.L. (1957). *The Vertebrate Visual System.* University of Chicago Press, Chicago, 211.

*Powell I. (1981). Lenses for correcting chromatic aberration of the eye. *Appl. Opt.* 20, 4152–4155.

*Rempt F., Hoogerheide J., and Hoogenboom W.P.H. (1971). Peripheral retinoscopy and the skiagram. *Ophthalmologica* 162(1), 1–10.

*Rempt F., Hoogerheide J., and Hoogenboom W.P.H. (1976). Influence of correction of peripheral refractive errors on peripheral static vision. *Ophthalmologica* 173, 128–135.

*Santamaría J., Artal P., and Bescós J. (1987). Determination of the point spread function of the human eyes using a hybrid optical-digital method. *J. Opt. Soc. Am.* (A) 4(6), 1109–1114.

*Schober H., Munker H., and Zolleis F. (1968). Die Aberration des menschlichen Auges und ihre Messung. *Opt. Acta* 15, 47–57.

*Smirnov M.S. (1961). Measurement of wave aberration in the human eye. *Biophysics* 6, 52–65.

*Smith G., Millodot M., and McBrien N. (1988). The effect of accommodation on oblique astigmatism and field curvature of the human eye. *Clin. Exp. Optom.* 71(4), 119–125.

*Thibos L.N., Bradley A., and Zhang X. (1991). Effect of ocular chromatic aberration on monocular visual performance. *Optom. Vis. Sci.* 68(8), 599–607.

*Thibos L.N., Ye M., Zhang X., and Bradley A. (1992). The chromatic eye: a new reduced-eye model of ocular chromatic aberration in humans. *Appl. Opt.* 31(19), 3594–3600.

*Tucker J. (1974). The chromatic aberration of the eye between wavelengths 200 nm and 2000 nm: some theoretical considerations. *Br. J. Physiol. Opt.* 29, 118–125.

*Van den Brink. G. (1962). Measurements of the geometrical aberrations of the human eye. *Vis. Res.* 2, 233–244.

*Van Heel A.C.S. (1946). Correcting the spherical and chromatic aberrations of the eye. *J. Opt. Soc. Am.* 36, 237–239.

*Van Meeteren A. (1974). Calculations on the optical modulation transfer function of the human eye for white light. *Opt. Acta* 21(5), 395–412.

*Walsh G. and Charman W.N. (1985). Measurement of the axial wavefront aberration of the human eye. *Ophthal. Physiol. Opt.* 5(1), 23–31.

*Walsh G., Charman W.N., and Howland H.C. (1984). Objective technique for determining the monochromatic aberrations of the human eye. *J. Opt. Soc. Am.* 1(9), 987–992.

*Welford W.T. (1986). *Aberrations of Optical Systems.* Adam Hilger, Bristol.

Westheimer G. (1955). Spherical aberration of the eye. *Opt. Acta* 2, 151–152.

Westheimer G. (1963). Optical and motor factors in the formation of the retinal image. *J. Opt. Soc. Am.* 53(1), 86–93.

Westheimer G. (1970). Image quality in the human eye. *Opt. Acta* 17(9), 641–658.

Westheimer G. and Campbell F.W. (1962). Light distribution in the image formed by the living human eye. *J. Opt. Soc. Am.* 52(9), 1040–1045.

Part VI

Visual ergonomics

36

Visual ergonomics of monocular systems

36.0   Introduction

The visual performance of a user viewing an image through an optical instrument depends upon a complex interaction between the eye and the instrument. Thus one cannot fully assess an optical instrument without a general understanding of visual optics and occasionally a knowledge of the visual capabilities of the individual user, which must include any anomalies of his or her visual system.

Many visual optical instruments are monocular and some people suffer visual discomfort when using monocular visual instruments for any length of time. Most of this probably stems from the tendency to keep one eye closed while viewing through the instrument. A binocular equivalent instrument will eliminate this problem and offers the possibility of a stereoscopic image. However, binocular instruments may lead to other problems. In this chapter, we will only discuss the ergonomics of monocular instruments and leave the ergonomics of binocular instruments and potential problems until the next chapter.

36.1   Instrument focussing

36.1.1   Image vergence

If an instrument is to be correctly focussed, the image must be within the accommodation range of the user. Classically, the ideal image vergence of visual optical instruments is taken as zero, that is the image is at infinity. The reasoning justifying this situation is that it was traditionally believed that the eye prefers viewing with relaxed accommodation. Unfortunately this approach neglects **instrument accommodation** and any refractive error of the user. The nature of instrument accommodation will be discussed later in Section 36.1.2. The different types of refractive errors and their distribution among the adult population has already been discussed in Chapter 13 and some representative values are shown in Figure 13.9. That diagram shows that only a small proportion of the population is emmetropic.

The image vergence of an instrument must be within the accommodation range of the user and the design of the instrument must cater for the expected range of refractive errors. In most instruments, the instrument can be refocussed at the eyepiece, and a typical focussing range is about ± 5 m^{-1}. From the data in Figure 13.9, it would appear that only a small fraction of the population is outside this range. In some other instruments (e.g. microscopes), the image vergence can be altered by refocussing the eyepiece or by changing the working distance. However with both of these methods, only spherical refractive errors can be corrected. They cannot be used to correct for astigmatic refractive errors and thus users with high levels of astigmatism must use their ophthalmic corrections when viewing; otherwise the uncorrected astigmatic error will lead to some loss in visual resolution and performance.

Fixed focus systems

A number of visual optical instruments are fixed focus, in that the vergence of the image cannot be adjusted to suit the refractive error or instrument accommodation of the user. In these cases, users with a myopic refractive error will not be able to focus the image, while hyperopes will be able to focus the image by accommodating. Probably the most common example is the single lens reflex camera, where the image to be formed on the film plane is directed by a mirror as described in Chapter 21 (cameras) onto a viewing screen. The image on this screen is viewed by a fixed focus simple magnifier. The image vergence from this viewing lens is usually set at a zero vergence but sometimes about -1.0 m^{-1}. An investigation into a suitable optimum value has been reported by Ohzu and Shimojima (1972) and this has been discussed further in Chapter 21.

36.1.1.1   Graticules or cross-hairs

Eyepieces sometimes contain a cross-hair or graticule that can be used for alignment, or control of accommodation and the measurements of image sizes. They can also be used to focus the image by parallax, particularly when the instrument has a large depth-of-field. But to use parallax effectively, the exit pupil of the instrument must be large enough to allow transverse movement of the eye.

In some instruments, the graticule or cross-hair can be focussed independently of the main image. In this case, it should be focussed before the main image is focussed. Some eyepieces have fixed cross-hairs or graticules and these are sometimes placed in the front focal point of the eye lens. In these cases, while the user may be able to focus the eyepiece for the primary image, he or she may not be able to focus the eyepiece for the graticule or cross-hair.

36.1.1.2   Spectacle and contact lens wearers

The above problems are greatly reduced if the ametropic users are able to use their spectacles or contact lenses, particularly for fixed images at infinity and astigmatic refractive errors. Correction with contact lenses poses no further ergonomical problems. However with spectacle lenses, the eye must be placed further back from the eyepiece and most instruments do not have a sufficiently long eye relief to allow pupil matching under these conditions. If the eye is moved back from the instrument exit pupil there will be some reduction in field-of-view. Eye relief and pupil matching for spectacle wearers is discussed further in Section 36.4.

36.1.2   Instrument accommodation

It is now well established that when people look through optical instruments, they accommodate and their amplitude of accommodation also appears to decrease. This phenomenon is sometimes called **instrument myopia** but in fact is more an instrument presbyopia since the accommodation range also appears to decrease. The amount of accommodation varies between individuals with a population mean of about $-1.5\,\mathrm{m}^{-1}$, but with individual values being as high as $-5\,\mathrm{m}^{-1}$. Schober et al. (1970) found a mean value of $-1.75\,\mathrm{m}^{-1}$. Home and Poole (1977) investigated instrument accommodation at two luminance levels (10^3 and 10^{-3} cd/m^2). At the higher level, they found that the mean setting was $-0.75\,\mathrm{m}^{-1}$ and at the lower light level the mean value was $-1.0\,\mathrm{m}^{-1}$.

Schober et al. (1970) and Hennessy (1975) have shown that the design features of the instrument have little influence on the individual levels of instrument accommodation, though there is some evidence that psychological factors play some part, for example a knowledge of the distance of the object. However the main cause is now attributed to the natural or preferred resting point of accommodation being about $-1.5\,\mathrm{m}^{-1}$ in from the far point. Much of the evidence of this belief is that investigations into instrument accommodation, empty field accommodation and low light level accommodation are highly correlated (Leibowitz and Owens 1975; Smith 1983) and therefore most probably have a common link; this is called the **intermediate resting point of accommodation**.

36.1.3   Variation of image vergence with object position and accommodation demands

Most optical instruments are designed for a certain working distance or object position. However there is always some expectation that other working distances will be used. For any given instrument, a change in working or object distance leads to a change in image distance, which in turn leads to a change in the accommodation demands of the observer.

While the focussing eyepiece allows the observer to reduce his or her accommodation demands arising from different working distances, it is worthwhile examining the effect of a change in object or working distance on the image vergence when there is no refocussing of the eyepiece.

In the case of a focal system, if the object is a small distance δl from the front focal point, the resulting image vergence $\delta L'$ at the back principal plane will, from equation (3.43), be

$$\delta L' \approx \delta l F^2/n \quad \text{(focal systems)} \tag{36.1}$$

where F is the equivalent power and n is the refractive index of object space. For an afocal two lens system with an object vergence L_v at the objective, the

Table 36.1. The image space field-of-view diameter of a number of visual optical instruments and eyepieces

	deg
Dioptric telescope	53
Telescope (Keplerian)	17
Telescope (Keplerian) 10×, 20	70
Telescope (Galilean)	17
Eyepieces	
Magnification	
5	24
7	26
10	36
15	36
10 wide angle	42
Micrometer eyepiece	22

image vergence L'_v at the eye lens is given by equation (17.6), that is

$$L'_v = \frac{M^2 L_v}{(1 - dML_v)} \quad \text{(afocal two-lens systems)} \tag{36.2}$$

where M is the magnification and d is the lens separation or system length.

These equations can be used in particular situations to estimate accommodation demands when the instrument is being used at different working distances. We have already done some calculations for simple magnifiers in Chapter 15. However, because the values of $\delta L'$ and L'_v are not the vergences at the eye, they will only give the approximate accommodation demand due to the change in object distance. Apart from using these equations to estimate accommodation demands, they can be used to estimate the depth-of-field of instruments and this will be done in Section 35.3.

36.2   Field-of-view

The field-of-view of an optical instrument is governed by the amount of vignetting or the presence of a field stop. It can be increased by increasing the aperture width of any vignetting component. If this is a lens, the maximum possible width is limited by the $|\rho F|$ condition discussed in Section 6.1.1.5. In many instruments, the eye lens tends to be a vignetting component so that the eyepieces of wide angle systems have a wider aperture than average.

The size of the field-of-view of particular visual instruments can be specified in object or image space. The image space fields-of-view of a sample of instruments and eyepieces are given in Table 36.1. From these data, we can see that typical fields-of-view range from 17° to 70° in diameter, with the mean being 32° in diameter. The instrument with the largest value of 70° would be regarded as a wide angle system.

36.3   Depth-of-field

In any optical system, the ultimate precision in focussing is set by the ability to detect errors in focus. Various methods of focussing an instrument are discussed in Chapter 10, but not all of these are relevant to general viewing through optical instruments. However, each of those methods has a threshold defocus level, below which defocus cannot be detected. We call this limit the depth-of-field. One method of focussing that was not discussed in that chapter was the use of a graticule or cross-hair in the eyepiece and the use of parallax. When the depth-of-field is large, parallax is often a more accurate method of focussing, but to use parallax effectively the exit pupil of the instrument must be large enough to allow transverse movement of the eye.

In this section, we will explore the concept of depth-of-field and develop equations that can be used to calculate the depth-of-field in particular situations. We will restrict the discussion to the case where the object is coherently coupled to the image, as in the case of microscopes and telescopes. The term "coherence" here does not mean the same as the term "coherent light" used in Chapter 25. In this context, coherent image coupling means that if the object is defocussed, the image can be refocussed and this for example applies to simple magnifiers, microscopes and telescopes, but does not apply to optical fibre systems and cameras. In these two examples, the imaging is incoherently coupled and if the object is defocussed, the image cannot be brought back into focus by adjusting the eyepiece. It is left as an exercise for the reader to determine which particular properties of these two devices causes them to have incoherently coupled image formation. To give a hint, it is related to specific properties of the intermediate image and the paths of single rays from object to image.

There are two related approaches to examining the depth-of-field of visual optical instruments. One uses the vergence values of depth-of-field of the eye given in Chapter 13 and the other uses the defocus blur disc approach that was introduced in Chapter 10.

Vergence values of ocular depth-of-field: From the data in Section 13.5, the depth-of-field of the eye varies between subjects and pupil diameter. Figure 13.9 shows a typical mean of the depth-of-field and its variation, which is due to intersubject variability. For a 4 mm diameter pupil the mean depth-of-field is about $\pm 0.25 \, \text{m}^{-1}$. Thus if the image vergence of an optical instrument varies by more than this value, it should be detectable under appropriate conditions. In the following discussion, we will assume that this is the depth-of-field and we will write the threshold vergence as

$$\Delta L'_{\text{th}} = \pm 0.25 \, \text{m}^{-1} \tag{36.3}$$

Defocus blur disc model: The defocus blur disc model is based on geometrical optics and assumes a simple model of ocular defocus. In this model, we assume that the retinal image of an out-of-focus object point is a blur disc, which is a projection of the exit pupil of the eye through the in-focus image point on to the retina. Alternatively, the observed blur disc is the projection of the entrance pupil through the out-of-focus object point on to the in-focus conjugate object plane. Smith (1982) has shown that both approaches, within certain approximations, lead to the same equation for estimating the angular size of the defocus blur disc. He has also shown that the angular diameter ω of the defocus blur disc of an out-of-focus object point is related to the diameter D_e of the entrance pupil of the eye and the difference in vergence ΔL between the in-focus and out-of-focus object points, by the equation

$$\omega = D_e \Delta L \quad \text{radians} \tag{36.4}$$

The approximations made in deriving this equation lead to an error of only about 5% for vergence differences of about $10 \, \text{m}^{-1}$. In the depth-of-field estimates to be investigated here, the equivalent refractive error will be much less than this and hence this equation will be sufficiently accurate for our purposes.

However, we should add that the above equation is based purely upon paraxial optics and neglects the effects of aberrations and diffraction. Neglecting aberrations can be justified on the grounds that the aberrations in an optical system are very variable from system to system and field point to field point and therefore to consider them would make a general study of depth-of-field impossible. However, diffraction is much more predictable. Indirectly, by using experimentally measured depths-of-field of the eye, we will take both of these into account.

The diameter of the defocus blur disc can be used to estimate the depth-of-field as follows. We assume that there is a threshold size ω_{th} of this blur disc, below which the blur disc cannot be distinguished from a point. We could assume that the magnitude of this threshold is approximately the minimum diameter of the point spread function of the eye. This in turn is set by aberrations of the eye and diffraction effects. From the data reviewed in Chapter 35, the minimum half-width of the ocular point spread function is about 1 minute of arc. This will give a total width of about 2 to 3 minutes of arc. However, we could estimate its value from known depths-of-field of the eye, measured in terms of power errors. If we relate the depth-of-field value given by equation (36.3) and assume it is for a 4 mm diameter pupil, then we can substitute these values into equation (36.4), and get $\omega = 0.001$ rad or 3.4 minutes of arc, so set the threshold value of ω_{th} as

$$\omega_{th} = 0.001 \, \text{rad} \tag{36.5}$$

We will now proceed to suppose that the observer is focussed on an object with a fixed level of accommodation and we wish to determine the depth of the object field that appears to be in focus. The solution to this problem depends upon whether the viewing system is focal or afocal and we will begin by looking at focal systems.

36.3.1   *Focal systems, object in the front focal plane*

We can begin by recalling equation (36.1), which we will write in the form

$$\Delta l_{th} = \pm \frac{n \Delta L'_{th}}{F^2} \tag{36.6a}$$

This equation relates the depth-of-field Δl_{th} in object space to the corresponding depth-of-field threshold of the eye $\Delta L'_{th}$, when looking through an optical instrument of equivalent power F. By using equation (36.4), we can write this in terms of the defocus blur disc threshold size

$$\Delta l_{\text{th}} = \pm \frac{n\omega_{\text{th}}}{DF^2} \qquad (36.6b)$$

where D is the smaller value of the entrance pupil of the eye or the exit pupil of the instrument. For natural pupils with a diameter of about 4 mm or more, equation (36.6a) is the more direct, but for smaller or larger pupils equation (36.6b) may be more appropriate. Let us look at some examples.

36.3.1.1 The simple magnifier and microscope

If we assume the optical system is a simple magnifier or microscope, we can replace the power F by the magnification M, using the $F/4$ rule given by equation (15.6), to give

$$\Delta l_{\text{th}} = \pm \frac{n\Delta L'_{\text{th}}}{16M^2} \qquad (36.7a)$$

or

$$\Delta l_{\text{th}} = \pm \frac{n\omega_{\text{th}}}{16DM^2} \qquad (36.7b)$$

Example 36.1: Let us estimate the depth-of-field of a simple magnifier that has a conventional magnification of 5.

Solution: Since simple magnifiers have no intrinsic pupil, the natural pupil of the eye will be the limiting aperture, which can be taken to be about 4 mm in diameter, so let us use equation (36.7a) with $\Delta L'_{\text{th}} = 0.25\,\text{m}^{-1}$ and $n = 1$ since the medium is air. Substituting into this equation gives

$$\Delta l_{\text{th}} = \pm \frac{1 \times 0.25}{16 \times 25} = \pm 0.63\,\text{mm}$$

Example 36.2: Calculate the object space depth-of-field of a microscope with a magnification of 200 and an exit pupil with a diameter of 1 mm. Assume the object space is air.

Solution: Since in this situation, the pupil is only 1 mm in diameter, let us use equation (36.7b). In this equation, we put $n = 1$, $\omega_{\text{th}} = 0.001$, $D = 0.001\,\text{m}$ and $M = 200$. Thus

$$\Delta l_{\text{th}} = \pm \frac{1 \times 0.001}{16 \times 0.001 \times 200^2} = \pm 0.0016\,\text{mm}$$

Now we will proceed to look at the depth-of-field of afocal systems.

36.3.2 Afocal systems

In the case of afocal systems, we can use equation (36.2) but in the approximate form, that is assuming $dML_v \ll 1$, but expressed as

$$L_v = L_v'/M^2 \tag{36.8a}$$

which should be sufficiently accurate because we are only assuming that the defocus or change in vergence L_v is small. Replacing the image vergence by the defocus blur disc, from equation (36.4) this gives us

$$L_v = \pm\omega_{th}/(M^2 D) \tag{36.8b}$$

where once again, the pupil diameter D is the smaller of the eye entrance pupil diameter and the exit pupil diameter of the instrument.

> **Example 36.3:** Calculate the closest distance of clear viewing of a Galilean telescope with a magnification of 3.5, if the pupil size of the observer is 4 mm in diameter.
>
> **Solution:** In this case, we are using a natural, medium diameter pupil, so let us use equation (36.8a). In this equation, $n = 1$, $L_v' = L_{th}' = \pm 0.25 \, \text{m}^{-1}$ and so
>
> $$|L_v| < 0.25/3.5^2 = 0.0204 \, \text{m}^{-1}$$
>
> Thus, the closest distance of distinct vision would be about 49 m.

> **Example 36.4:** For a focimeter with a collimator objective aperture diameter of 10 mm and a telescope magnification of 5, calculate the depth-of-field or precision of the instrument.
>
> **Solution:** In this case, the telescope is working with an entrance pupil diameter of 10 mm and since its magnification is 5, the exit pupil diameter is 2 mm. Since this is a small pupil, we can use equation (36.8b) to estimate the precision of the instrument. In this equation we put $\omega_{th} = 0.001$ and $n = 1$ and so have
>
> $$L_v = \pm 0.001/(5^2 \times 0.002) = \pm 0.02 \, \text{m}^{-1}$$
>
> In practice, the precision may be worse than this because of the spherical aberration of the ophthalmic lens being measured and this precision will worsen with increase in ophthalmic lens power.

36.4 Pupil matching

When visual instruments containing intrinsic aperture stops are used, the viewer automatically places his or her eye in the correct position to give maximum field-of-view; that is the viewer positions the instrument exit pupil to coincide with his or her own entrance pupil. If the two pupils do not match in position, there will in general be some loss in field-of-view. A study of the effect of eye

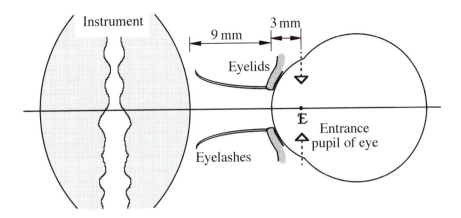

Fig. 36.1: The minimum eye relief for visual optical instruments.

position on field-of-view has been done by Smith (1977). This requires eye reliefs of about 10–15 mm for common instruments. For spectacle wearers, eye reliefs should be at least 20–22 mm. Since very few instruments are designed with these eye reliefs, the spectacle wearers have either to keep their spectacles on and suffer a reduced field-of-view or take the spectacles off and thus may have to refocus the eyepiece to compensate for their refractive error. However, refocussing cannot cope with cylindrical refractive errors. Eye reliefs can be increased by use of an erecting lens as described in Section 17.4.4.1 but are reduced by use of field lenses as shown in Section 17.4.2.

36.4.1 Calculation of the minimum eye relief

The minimum eye relief for unaided viewing can be calculated from a simple anatomical study of the eye as shown in Figure 36.1 and assuming the entrance pupil of the eye is about 3 mm inside the eye from the cornea (Section 13.3). The eyelids and lashes require about an extra 9 mm. Thus the minimum desirable eye relief is about 12 mm.

However the above value assumes that the user is not wearing any spectacles. Once spectacles are worn, the situation becomes a little more complicated, depending upon spectacle lens power, thickness and vertex distance. Instruments designed for spectacle wearers require eye reliefs of at least 20–22 mm and this value is derived as follows. Given that the entrance pupil of the eye is about 3 mm inside the eye, the spectacle vertex distance is about 15 mm and the lens thickness is 3 mm, the entrance pupil of the eye is 21 mm from the front surface of the lens. However, we have shown in Section 14.1.5 that a spectacle lens modifies the position and size of the entrance pupil. For negative power lenses, it comes closer and smaller and for positive lenses it is larger and farther away from the lens.

Some instruments such as simple magnifiers, Galilean telescopes and camera viewfinders do not have intrinsic aperture stops, and therefore do not have exit pupils. In practice, the aperture stop (iris) of the eye acts as the aperture stop of the combined system. Usually, the closer the eye to the instrument, the larger the field-of-view. In the assessment of these systems, it is important to put an artificial stop where the entrance pupil of the eye would normally be placed and this stop should have about the same diameter as that of the entrance pupil of the eye.

36.4.1.1   Long eye reliefs

For normal optical systems, the eye relief needs to be 12 mm or a little longer for the unaided eye and about 21 mm for spectacle wearers. However, in a number of instruments there is a need for much longer eye reliefs. Perhaps the most common example here are the gunsights used on recoiling guns. In these cases, the eye reliefs are well in excess of 21 mm and may be as long as 6–8 cm. Long eye reliefs are usually achieved by using an erecting lens optical system.

Long eye reliefs lead to some minor but preventable problems. For example the positioning of the eye and maintaining visual alignment become more critical and difficult. In conventional instruments, the eyepiece mount or eyepiece cup is used for this purpose. In the case of long eye reliefs, the eye cup must be extended and one of the most convenient ways of doing this is by the use of folding rubber cups. The extendable cup also decreases the probability of glare from ambient bright sources.

36.4.2   Instrument exit pupil size

The size of the exit pupil of a visual optical system depends to some extent on the conditions of use, such as ambient light level, the intended task and whether the eye will be stationary while viewing or be required to rotate to fixate on peripheral objects in the field-of-view.

36.4.2.1   The stationary eye

For the stationary eye, the exit pupil of the instrument need be no larger than the entrance pupil of the eye under the conditions of use. Therefore for high light levels, the exit pupil need only have a diameter of between 2 and 3 mm and for very low light levels, the exit pupil should have a diameter of about 6–8 mm. The effect of ambient light level on entrance pupil diameter of the eye is discussed in Section 13.3.2.

In general terms it may be concluded that an optical system such as a telescope used at low light levels should have an exit pupil size similar to the pupil size of the eye under the same conditions. However, a larger exit pupil requires a proportionally larger objective with a corresponding increase in weight. The gain in night-time performance from a larger exit pupil may not be as great as theoretically predicted for two reasons: (a) aberrations of the eye are much greater at larger pupils and thus reduce acuity, (b) the Stiles–Crawford effect occurs and (c) many people do not have pupils that open up to 7 or 8 mm at very low light levels.

36.4.2.2   The rotating eye

Conventional visual optical instruments have image fields-of-view from about 20° to 50° in diameter. However as optical system design techniques improve and production costs decrease, the field-of-view of many systems is slowly increasing. The larger the angular field-of-view, the greater the tendency for the eye to rotate to view peripheral points of interest rather than rotate the head and instrument. When this is the case, pupil matching becomes more complex.

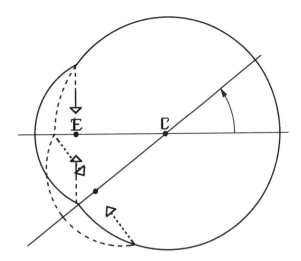

Consider the situation as shown in Figure 36.2. Assume the pupils are perfectly matched in position and size, but only when the eye is stationary and looking along the optical axis of the instrument. Now as soon as the eye rotates to observe an off-axis point, the pupils shear and there is no longer pupil matching in position. As the eye rotates, there is an increasing vignetting effect and the amount of light entering the eye decreases. When vignetting becomes total, the whole field goes dark and nothing is seen. In other words when the eye rotates to look at a peripheral object, the whole field may disappear. When the eye returns to axial viewing, the field returns.

This problem can be reduced or overcome by increasing the diameter of the instrument exit pupil. The minimum diameter of this pupil will depend upon the entrance pupil diameter of the eye and the field-of-view of the instrument. The diameter can be found by using simple trigonometry as follows. Consider the situation as shown in Figure 36.3 and a given field-of-view angular radius ψ. Let the entrance pupil of the eye have a radius $\bar{\rho}_e$ and the desired minimum radius of the instrument exit pupil be $\bar{\rho}_i'$. On eye rotation, vignetting becomes total when the top edge of the eye entrance pupil coincides with the bottom edge of the instrument exit pupil and when the eye will have rotated by an angle ψ. From the diagram it follows that

$$\tan(\phi) = \bar{\rho}_i'/\varepsilon c$$

and

$$\tan(\gamma) = \bar{\rho}_e/\varepsilon c$$

where c is the centre of rotation of the eye, and εc is about 12 mm in the standard schematic eye. We also have from the diagram

$$\psi = \phi + \gamma$$

Since εc, $\bar{\rho}_e$ and ψ are immediately known and $\bar{\rho}_i'$ is the quantity required, it

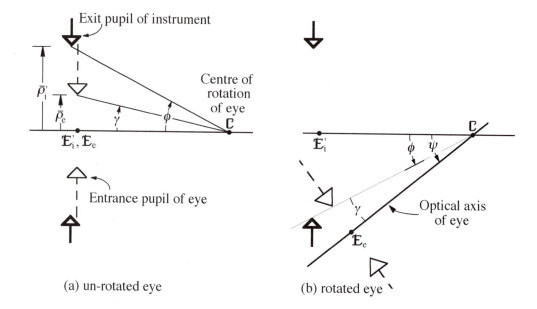

(a) un-rotated eye   (b) rotated eye

can be found by solving the following equations, in order

Fig. 36.3: Calculation of instrument exit pupil size for a rotating eye.

$$\gamma = \arctan(\bar{\rho}_e/\varepsilon c) \qquad (36.9a)$$

$$\phi = \psi - \gamma \qquad (36.9b)$$

and finally

$$\bar{\rho}_i' = \varepsilon c \times \tan(\phi) \qquad (36.9c)$$

Example 36.5: Calculate the required exit pupil diameter of a wide angle optical system with an image field-of-view of 70°.

Solution: From the above data, $\psi = 35°$. If the system is designed for daytime use, the eye entrance pupil will have a diameter of approximately 4 mm, that is $\bar{\rho}_e = 2$ mm. Now the distance from the entrance pupil to centre of rotation of the eye is about 12 mm, that is $\varepsilon c = 12$ mm. Thus solving for $\bar{\rho}_i'$

$$\gamma = \arctan(2/12) = 9.5°$$

$$\phi = 35° - 9.5° = 25.5°$$

Therefore

$$\bar{\rho}_i' = 12 \times \tan(25.5°) = 5.7 \, \text{mm}$$

That is the instrument exit pupil should have a minimum diameter of 11.5 mm, which is considerably larger than the 4 mm value required for a stationary eye.

36.5   Image luminance

36.5.1   Extended sources

When we view an object of luminance L, through an optical instrument, we would like to know what is the perceived luminance. We can find the answer to this problem by the following reasoning. For any optical system, the image illuminance E', of an object of finite size and luminance L, is given by equation (12.21a), that is

$$E' = \tau \pi L [n' \sin(\alpha')]^2 / n^2 \qquad (36.10)$$

where τ is the transmittance of the system, α' is the angular radius of its exit pupil measured at the Gaussian image, n is the refractive index of object space and n' is the refractive index of the image space medium. Now equation (36.10) applies to any optical system, no matter how complex and therefore applies to the eye alone and a combination of the eye and a visual optical system. The only parameters in this equation that can change between these two applications are the transmittance τ and α' the angular radius of the effective exit pupil of the eye. Thus providing the instrument exit pupil is not smaller than the entrance pupil of the eye, the angle α' is the same in both cases and therefore apart from the change in value of the transmittance τ, the retinal illuminance of the object is the same, whether viewing unaided or through the instrument. If the retinal illuminance is the same for both cases, the perceived luminances will be the same.

This argument also shows that a visual optical instrument cannot increase the retinal illuminance (that is the apparent luminance) of an object. In fact, the perceived luminance will always be less because of light losses in the instrument and occasionally because the instrument has an exit pupil smaller than the entrance pupil of the eye. Let us now look at some particular and common instruments.

36.5.1.1   Spectacle lenses

It was shown in Chapter 14 that spectacle lenses change the effective size of the entrance pupil and therefore change the amount of light gathered from any object. Positive power lenses increase the effective pupil size and therefore increase the amount of light collected from the object and negative power lenses decrease the effective size of the pupil. Therefore, we may conclude that positive power lenses must lead to an increase in apparent brightness. But it cannot, for the reasons given above. We can explain this contradiction, by noting that ophthalmic lenses also change the size of the retinal image (spectacle magnification) by a proportional amount. Therefore for any object of finite dimensions, the change in luminous flux entering the eye as a result of the change in entrance pupil diameter exactly matches the change in retinal image size and so the retinal image illuminance remains unchanged, apart from the transmittances losses in the lens.

36.5.1.2   Microscopes

With microscopes, it is important that as magnification is changed, the image luminance remains reasonably constant and this requires the exit pupil of the microscope to remain constant with magnification. We have seen in Section 16.6.3, that the exit pupil diameter D' is proportional to the numerical aperture (NA) of the objective and related by equation (16.19b), that is

$$D' = \frac{NA}{2M} \qquad\qquad (36.11)$$

where M is the overall magnification of the microscope. In terms of the individual objective magnification M_o and the eyepiece magnification M_e, we can use equation (16.6) to write this as

$$D' = \frac{NA}{2M_oM_e} \qquad\qquad (36.12)$$

This equation shows that if the apparent image brightness is to remain constant as the magnification is increased, then the objective numerical aperture is proportional to the magnification of the objective, that is

$$NA = (2D'M_e)M_o \qquad\qquad (36.13)$$

Therefore the objectives must have a higher numerical aperture as the magnification increases. However, while optical designers try to satisfy this requirement, it is not always possible. For example, the numerical aperture in air cannot be greater than 1.0 and this makes it difficult to satisfy the requirement for high power objectives. In these cases, this limit is overcome to some extent by immersing the sample in a liquid with a high refractive index, which now allows the object space numerical aperture to be greater than 1.0 and approach the index value of the liquid. These objectives are called oil immersion objectives. However, constant image illuminance in a microscope can, to some extent, be maintained by increasing the illuminance of the object using an auxiliary light source and therefore this facility does ease the above requirement of numerical aperture. However, as we saw in Chapter 34, higher numerical apertures are also required for good image quality.

36.5.1.3   Telescopes

The diameter of the entrance pupil of a telescope is usually much larger than the entrance pupil of the eye and therefore can gather much more light from an object than can the eye alone. However, as stated above, this does not make the image brighter. This is because the retinal image is enlarged by the same proportion. On the other hand, the telescope can increase the brightness of point sources and this makes it very useful in astronomy, in which most of the objects are effectively point sources. Let us see how this comes about.

36.5.2 *Point sources*

36.5.2.1 *Keplerian telescopes*

For a nominal point source such as a star, the brightness of the source is described by its luminous intensity I and not a luminance L and surface area δA. For a point source at a distance d, the luminous flux ($Flux$) accepted by the telescope is given by equation (12.13c), that is

$$Flux = \pi I D_t^2/(4d^2) \tag{36.14}$$

where D_t is the diameter of the entrance pupil of the telescope. As before, taking into account the losses within the system, the luminous flux ($Flux'$) leaving the telescope is

$$Flux' = \tau \pi I D_t^2/(4d^2) \tag{36.15}$$

If the telescope is coupled to the eye and we assume perfect pupil matching and also assume the combined system is aberration free, all of this flux is distributed over the diffraction limited point spread function. If the source is at a great distance, the peak illuminance $E'(0)$ in this distribution is given by equation (26.36a), that is

$$E'(0) = \tau I \left(\frac{\pi F}{4d\lambda n'}\right)^2 D_t^4 \tag{36.16}$$

where F is the equivalent power of the combined system and not the power of the eye alone. If an afocal system of magnification M is coupled to a focal system, the equivalent power of the combined system is the equivalent power of the focal system divided by M and it is left as a problem for the reader to prove this result. Using this fact, the above equation can be written as

$$E'(0) = \tau I \left(\frac{\pi F_e}{4d\lambda n'M}\right)^2 D_t^4 \tag{36.17}$$

where F_e is the equivalent power of the eye and M is the magnification of the telescope. Let us look at several different situations.

(a) *Perfect pupil matching*: If there is perfect pupil matching then we have

$$D_t = MD_e$$

where D_e is the diameter of the entrance pupil of the eye. Equation (36.17) becomes

$$E'(0) = \tau I \left(\frac{\pi F_e}{4d\lambda n'}\right)^2 M^2 D_e^4 \tag{36.18}$$

Therefore if there is perfect pupil matching, a telescope increases the brightness of point sources proportional to the square of the magnification.

(b) *Exit pupil of the telescope larger than the entrance pupil of the eye*: In this
     case, the iris of the eye acts as the limiting aperture stop of the combined
     system and the effective diameter D_t of the entrance pupil of the telescope
     is reduced to simply $M D_e$, that is

$$D_t = M D_e$$

and therefore equation (36.18) also applies in this case.

(c) *Exit pupil of the telescope smaller than that of the eye*: In this case, the
     aperture stop of the telescope becomes the limiting aperture of the com-
     bined system and the effective entrance pupil diameter of the telescope
     remains as D_t and therefore equation (36.17) applies.

36.5.2.2 Galilean telescopes

Galilean telescopes do not have intrinsic pupils. Instead the iris of the eye acts
as the aperture stop of the total system. Therefore there is always perfect pupil
matching and the same comments apply as to the Keplerian telescope with
perfect pupil matching.

36.5.2.3 Photometric gain

For point sources, imaged by a diffraction limited system, we have seen that
the luminous flux in the image is proportional to the square of the diameter of
the effective entrance pupil. For small diffraction limited sources, the threshold
of detection is proportional to the luminous flux in the image (Ricco's law) and
therefore the probability of seeing increases as the square of the effective pupil
diameter. At super-threshold light levels, we can assume that the perceived
brightness is proportional to the luminous flux in the image.

We can quantify the change in image brightness by introducing a quantity
that we will call the **photometric gain**, denote this quantity by the symbol G
and define it as the luminous flux accepted by the effective entrance pupil of the
telescope compared with that seen with the unaided eye. From equation (36.13)
we have

$$G = \frac{D_t^2}{D_e^2} \tag{36.19}$$

where D_t must be the effective entrance pupil diameter of the telescope, that is
it satisfies the following conditions.

(a) If the exit pupil of the telescope is larger than the entrance pupil of the eye,

$$D_t = M D_e \tag{36.20}$$

(b) If the exit pupil of the telescope is smaller than the entrance pupil of the
     eye, then D_t remains unchanged.

If there is perfect pupil matching then $D_t = M D_e$ and equation (36.19) shows
that the photometric gain is simply the square of the telescope magnification.

Let us look at a numerical example.

> **Example 36.6:** Compare the photometric gains of $10\times$, 20 and $7\times$, 50 telescopes for an eye with a pupil of 4 mm in diameter.
>
> For the $10\times$, 20 telescope, the exit pupil is $20/10 = 2\,\mathrm{mm}$ in diameter, which is smaller than the eye entrance pupil value of 4 mm. Therefore we use equation (36.19) directly and substitute $D_t = 20$ and $D_e = 4$, giving a gain of 25.
>
> For the $7\times$, 50 telescope, the exit pupil is $50/7$ mm, which is greater than the entrance pupil of the eye; therefore using equation (36.20) we reset the effective entrance pupil to 28 mm. Then we substitute $D_t = 28$ and $D_e = 4$ into equation (36.19), giving a gain of 49.
>
> These results show that the $7\times$, 50 telescope, while having a smaller magnification, gives an image that is about twice as bright as the $10\times$, 20 telescope.

36.5.3 The Stiles–Crawford effect

The above discussion neglects the Stiles–Crawford effect (Stiles and Crawford 1933), which describes the decrease in visual efficiency for light passing into the eye through the edge of the pupil. Thus increasing pupil size to increase retinal illuminance does not always lead to the expected increase in visual performance. The Stiles–Crawford effect could be taken into account by regarding the entrance pupil of the eye as being apodized by a suitable factor. Various photometric aspects of the Stiles–Crawford effect are discussed in Section 13.7.1.

36.6 Visual performance

In this context, we define the visual performance of an instrument as the measure of its ability to increase the visibility of detail that is otherwise not visible to the eye. The detail we wish to see may not always be high contrast and may sometimes be almost at threshold contrast. We can investigate the potential visual performance of particular instruments using the image quality criteria that we have met in Chapter 34. Of particular usefulness are veiling glare and the optical transfer function. However, the effects of veiling glare are difficult to quantify because in any particular instrument, the amount of veiling glare is dependent upon the environment, for example the presence of bright lights within or near the field-of-view, their position, brightness and number. The performance of many instruments can be seriously degraded by veiling glare particularly under low light levels, as a result of the presence of bright lights in the vicinity.

Scattering by atmospheric dust and water can seriously reduce the contrast of distant scenes, particularly along a horizontal line of sight. For a very turbid atmosphere, the scattering can reduce the contrast of many distant objects to below threshold level.

In this section, we will use the concept of optical transfer function and its visual relation, the contrast threshold function, to investigate the ability of visual optical instruments to improve the visibility of periodic and non-periodic objects or targets, beginning with periodic targets.

36.6.1 Periodic targets

Let us look at the effect of magnification on sinusoidal periodic targets. The resolution limit of the eye is about 30 to 60 c/deg, so that targets with higher frequencies will not be detected and can only be detected if their frequencies are made less than the above resolution limit. An optical system changes the spatial frequency and contrast of a periodic pattern. If the system has a magnification M, the image spatial frequency σ' and the object spatial frequency σ are related by the equation

$$\sigma' = \sigma/M \qquad\qquad\qquad (36.21)$$

Thus to lower the spatial frequencies of a periodic pattern, the magnification has to be greater than unity.

> **Example 36.7:** What is the minimum magnification needed to make visible a sinusoidal target of frequency 500 c/deg?
>
> **Solution:** If we set a resolution limit of say 30 c/deg, the magnification required to convert 500 c/deg to 30 c/deg is, from equation (36.20),
>
> $$M = 500/30 = 16.7$$

However, the above model is oversimplified, because we have neglected the effect of the optical transfer function in the imaging process and the **neural contrast threshold function** of the eye. If we extend our modelling of the process to include these quantities, we must distinguish two types of image coupling to the eye, (a) coherent and (b) incoherent. The image is coherently coupled if the system does not form any intermediate images on a diffusing surface such as a viewing screen. An example of coherent coupling is the viewing through simple magnifiers, microscopes and telescopes. In incoherent coupling, an intermediate image is formed on a viewing screen which scatters the incident light. The only two examples of incoherently coupled imaging discussed in this book are the viewing system on a single lens reflex camera and fibre optical relays. We will look at these in turn, starting with coherent coupling.

36.6.1.1 Coherent coupling

If there is coherent coupling, the optical system/eye combination can be regarded as a single system for ray tracing purposes and one must consider the instrument and eye as a single unit with the smaller of the instrument exit pupil and eye entrance pupil being the effective pupil of the combined system. The optical transfer function has to be evaluated for the system as a single unit. In most cases, we can ignore the phase component, since this only shifts the pattern laterally, with no effect on its visibility. Therefore in the following discussion, we need only refer to the modulation transfer function part of the optical transfer function. However, the phase effect is very important if we extend the discussion to non-sinusoidal periodic patterns because the spatial frequency phase shift effects can lead to distortions in the shape of the pattern.

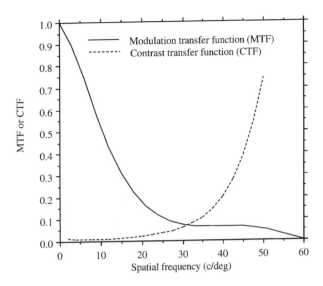

Fig. 36.4: Neural contrast
threshold function (CTF)
and instrument/eye
modulation transfer
function (MTF).

Figure 36.4 shows a possible modulation transfer function of a combination
of a particular instrument and eye. The curve gives the resulting retinal image
contrast of a sinusoidal pattern of unit contrast due to the combined effects
of aberrations and diffraction. To decide whether a sinusoidal pattern with a
particular spatial frequency and contrast can be seen, we also need to plot on
this diagram the neural contrast threshold component of the retina. The dotted
curve on the diagram is a typical neural threshold curve similar to the one
plotted in Figure 35.7. However, we should note that the contrast threshold
curves vary between individuals and therefore that of a particular individual
could be different from that shown in the diagram. However, here we are more
concerned with general principles and we will show how to apply such a curve.
For a periodic pattern to be seen, the retinal image contrast must lie above the
neural contrast threshold function curve.

Example 36.8: Suppose that the modulation transfer function shown
in Figure 36.4 were for an instrument with a magnification of 4. Let
us estimate whether we could detect a sinusoidal pattern of 80 c/deg
but with a contrast of 0.7.

Solution: From equation (36.20), the image spatial frequency will
be 20 c/deg. From the curve shown in Figure 36.4, the modulation
transfer function value of 20 c/deg in image space is 0.18. Therefore
the contrast in the image is $0.18 \times 0.7 = 0.13$. If we compare this
with the neural contrast threshold value of 0.1 at this frequency, the
image contrast is higher than this threshold value and therefore should
be seen.

We will now argue that magnification does not always improve visibility and
use Figure 35.7 for this purpose. Suppose that the unaided eye is looking at a
20 c/deg sinusoidal pattern with a contrast of 0.01. If we plot the corresponding
point on the diagram, we would see that it is below the neural + optics contrast
threshold curve and therefore the pattern would not be visible. Now suppose

we observe this pattern with an aberration free 4 × magnifying system, pupil matched to the eye. The modulation transfer function of the magnifier/eye system would be the same as that of the eye alone and therefore the pattern would be imaged on the retina with a frequency of 5 c/deg but with no change in contrast. The corresponding point on the diagram would now be above the contrast threshold curve and therefore the pattern would be visible. However, if the magnification had been 20 ×, the corresponding point would once more be below the curve and thus the pattern would be invisible.

36.6.1.2   *Incoherent coupling*

Now if the instrument forms an intermediate image on a diffusing screen (e.g. viewing screen in a camera), the coupling is incoherent. In this case, there are three modulation transfer functions: those of the front end of the instrument, the diffusing screen and the optical system used to view the screen and eye combination. The overall effective modulation transfer function is the product of these three functions. Once this overall modulation transfer function is known, the probability of seeing a particular pattern can be determined by the same procedure used in the case of coherent coupling.

36.6.2   *Non-periodic targets*

For non-periodic objects, the threshold contrast of an object decreases as the area of the object increases. The actual relationship will depend upon the shape of the object. For circular objects, Blackwell and Taylor (1969) have shown that the threshold contrast decreases with increase in size in an approximately hyperbolic manner, but for diameters greater than about 30 minutes of arc, the threshold does not appear to decrease much further with increase in size. Therefore we may conclude that providing veiling glare, aberrations and diffraction do not reduce the contrast of the object, the probability of seeing a small low contrast object always increases with magnification. However, once the disc image is greater than 30 minutes of arc, the extra magnification does not offer much more information.

Let us conclude this section by looking at the magnification of an edge. Magnifying an edge does not change its size and therefore we cannot apply the above rule. If we neglect the effects of veiling glare, aberrations and diffraction, the image of an edge is exactly the same as the object. Therefore magnifying it cannot improve its visibility. If we now include the effect of veiling glare, this will reduce the contrast. Aberrations and diffraction will blur the edge to some extent. All of these effects will reduce the probability of seeing the edge. Therefore, magnifying an edge cannot offer any improvement in its visibility and will probably decrease it.

36.7   Magnification limits and empty magnification

The role of magnification in increasing the visibility of fine detail has already been discussed in Section 36.6.1. We showed that magnification did not always improve the visibility of the detail. Whether a certain magnification improves the visibility of a target depends upon the nature of the target, its contrast, whether it is periodic such as a sinusoidal pattern and if so at what frequency and contrast or whether it is a simple pattern such as a disc or an edge. It also depends upon whether the visual task is one of detection or resolution (that is recognition). We argued that magnification could not increase the visibility of an edge. If the target were a simple disc, above a certain upper limit magnification, further magnification did not improve visibility. If the target were a sinusoidal pattern too much magnification could make the target even less visible. These observations lead us to the concept of **empty magnification**. This is defined as the magnification above which no more extra detail can be seen. From the foregoing discussion, it follows that the region of empty magnification depends upon the nature of the target being observed, but it also depends upon the particular instrument. Let us look at the microscope, which has an upper limit of magnification of about 2000.

36.7.1 The microscope

From the above discussion, it is clear that the upper limit of magnification depends upon the nature of the target. Westheimer (1972) argued that for the detection of low contrast objects, the upper limit corresponds to the Airy disc subtending about 5 minutes of arc at the eye. We can find the corresponding upper limit of magnification as follows. We take equation (34.28a) applied to the exit pupil of the microscope, and this gives an Airy disc radius (θ)

$$\theta = \frac{1.22\lambda}{D'_m} \tag{36.22}$$

where D'_m is the diameter of the exit pupil of the microscope. This diameter is related to the objective numerical aperture (NA) and microscope magnification, by equation (16.19b), that is

$$D'_m = \frac{NA}{2M} \tag{36.23}$$

Eliminating D'_m from these two equations gives us

$$\theta = \frac{2.44\lambda M}{NA} \tag{36.24}$$

where the wavelength must now be in metres since equation (16.19b) assumes metre units. Since θ has a maximum size, the corresponding limit to the magnification is

$$M = \frac{\theta NA}{2.44\lambda} \tag{36.25}$$

If we set the angle θ to 2.5 minutes of arc (half the diameter of 5 minutes of arc), this equation reduces to

$$M = 0.000298 NA/\lambda \tag{36.26}$$

If we take light in the middle of the visible spectrum, then the upper limit of magnification is

$$M \approx 540NA \tag{36.27}$$

Alternatively, we can look at the magnification of sinusoidal patterns, as suggested by Charman (1974). From equation (34.41a), the linear spatial frequency resolution limit σ_L of the objective is given by the equation

$$\sigma_L = 2n\sin(\alpha)/\lambda \quad \text{c/unit distance} \tag{36.28}$$

If F is the equivalent power of the microscope as a whole, the corresponding angular spatial frequency limit σ_A' in image space is given by equation (16.13b), that is

$$\sigma_A' = \sigma_L/(4M) \quad \text{c/rad} \tag{36.29}$$

where M is the magnification of the microscope. Combining the above two equations gives

$$\sigma_A' = \frac{NA}{2\lambda M} \tag{36.30}$$

Let us now choose the magnification that corresponds to some preset spatial frequency σ_A'. The desired magnification is then given as

$$M = \frac{NA}{2\lambda\sigma_A'} \tag{36.31}$$

where λ must be expressed in metres since we have incorporated equation (16.13b) into this equation which is only correct if the spatial frequency σ_L is in c/m. If we take a typical wavelength value of 555 nm, which is at the middle of the visible region, and convert the spatial frequency σ_A' to be in c/deg instead of c/rad, equation (36.28) reduces to

$$M = \frac{15724NA}{\sigma_A'} \quad (\sigma_A' \text{ in c/deg}) \tag{36.32}$$

If the resolving power spatial frequency limit σ_A' of the microscope is placed at the spatial frequency of the eye where the threshold contrast is minimum, that is at about 10 c/deg, the above magnification limit could now be expressed as

$$M = 1600NA \tag{36.33}$$

For a non-immersion objective, $n = 1$, and thus the maximum numerical aperture is 1.00 and the upper limit of useful magnification is 1600. A different value of σ_A' will of course lead to a different value of this upper limit.

36.8   Aberration tolerances

36.8.1   Monochromatic aberrations: general

The ultimate image quality of any visual optical system, such as a telescope, is that of the quality of the retinal image and in this case this image must be affected by the aberrations of the combination of the visual optical system and the optical system of the eye. Normally the aberrations of the combined systems are additive. Only when there is an incoherent intermediate image do the aberrations have to be considered separately.

In visual optical systems, aberrations have an effect on both the image quality and the formation of the entrance and exit pupils. Matters are complicated if the aperture stop position is variable, as the aberrations depend upon this stop. This is particularly relevant to systems such as simple magnifiers and Galilean telescopes which have no intrinsic aperture stop and hence no exit pupil and therefore no well-defined eye position.

Using an aberration generator, Giles (1977) examined the effects of approximately 1 to 2 wavelengths of primary spherical aberration, coma and astigmatism of the resolution of a 3-bar resolution chart and a grating, at different contrasts. The pupil size was restricted to 2.2 mm. He found that even 1 wavelength of these aberrations leads to some loss of resolution, but the eye accommodated to reduce the effects of spherical aberration and astigmatism but not coma. His results could be used to predict the loss of resolving power for these simple targets for such low aberration values, but the results are of limited use because of the small pupil and simple targets. In a later paper, Giles (1979), using the same aberration levels, developed a detectability degradation function to examine the effects of aberrations on the detectability of grating patterns. The results were essentially the same as in the earlier paper.

Charman and Whitefoot (1978) confirmed that some of the astigmatism and field curvature in an instrument could be partly compensated by refocussing of the eye, providing that the images were within the accommodation range of the user.

Burton and Haig (1984) carried out an investigation into the threshold tolerances of the eye to defocus, Seidel spherical aberration, coma and astigmatism for four very different types of scenes, using an image processor and simulated aberrations. They concluded that at the 75% detection level, the tolerance to defocus and spherical aberration were about equal at 0.213λ each, and 0.406λ for coma and 0.303λ for astigmatism.

The values of Burton and Haig are much less than those of Giles and most likely are due to the difference in criteria used. The importance of the choice of criterion has been confirmed by Mouroulis and Zhang (1992), who examined the effect of coma and astigmatism singly and combined on subjective measures of image quality and on various image quality criteria. They found that the area under the modulation transfer function from 5 to 24 cycles/deg and the radius of the circle that encloses 84% of the energy in the point spread function were well correlated with the subjective responses and the Strehl intensity ratio, and the variance of the wave aberration function was not well correlated with radius.

36.8.2   Chromatic aberrations

A system in monochromatic light has no effective chromatic aberration, no matter what the computed level of the Seidel aberrations C_L and C_T. Therefore the tolerance to residual levels of the chromatic aberrations will depend upon the spectral band-width of the light source used. Let us look at the worst case, that of using white light.

From our knowledge of the width of the point spread function of the eye, we may expect that the tolerance to chromatic aberration will not be noticeable if the angular width of the colour fringing is less than the angular width of the point spread function. However there will be a difference between the value that is just noticeable and the value that just causes a detectable reduction in visual performance. Furthermore, in the latter case, the value will depend upon the visual performance criteria used.

Mouroulis et al. (1993) reported the findings of an earlier study by Mouroulis and Woo (1988), which showed that adding three wavelengths of longitudinal chromatic aberration to that already in the eye did not affect resolving power for three subjects. From the results of their own study on transverse colour aberration, they found that a transverse angular aberration of about 2.5 minutes of arc was a reasonable tolerance to this aberration.

Prisms

We can extend this result to determine the maximum power of a prism free of noticeable transverse chromatic aberration. The deviation angle θ for a thin prism is given by equation (8.5), that is

$$\theta = \beta(\mu - 1)$$

A change $\delta\mu$ in μ leads to a change $\delta\theta$ in θ given by the equation

$$\delta\theta = \beta\delta\mu$$

If we eliminate β from this equation by using equation (8.7), we have

$$\delta\theta = F_p\delta\mu/[100(\mu - 1)]$$

where F_p is the power of the prism. If we set the variation of index as the variation from the F and C spectral lines, we can replace the term $(\mu - 1)/\delta\mu$ by V_d from equation (1.9), and setting a limit $\delta\theta$ of 2.5 minutes of arc in radians, we have the tolerance limit of

$$F_p < 0.073V_d \quad \text{prism dioptres} \tag{36.34}$$

where V_d is the V-value of the prism material. Since the V-values range from about 30 to about 70, the maximum prism power, free of noticeable colour fringing, would range from 2.2 to 5 prism dioptres, respectively.

36.9   Maxwellian view

In a number of instruments, a small light source is used for providing a uniform field-of-view. When the eye is placed so that the image of the source coincides with the entrance pupil of the eye the field-of-view has maximum width and maximum uniformity of luminance, and this optical arrangement is called **Maxwellian view**. Figure 36.5 shows the arrangement. The source s is imaged

Fig. 36.5: Maxwellian
view.

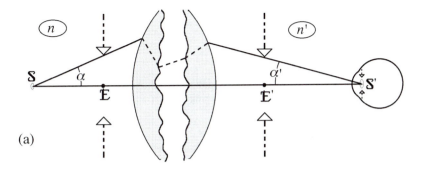

(a)

to s', which is co-incident with the entrance pupil of the eye. The exit pupil of the instrument is then uniformly illuminated and subtends an angular radius α' to the eye. The eye is approximately focussed on this exit pupil.

In some interpretations, the source s may be regarded as the aperture stop and hence entrance pupil of the optical system, and the conventional exit pupil of the optical system becomes the object that is viewed by the eye. The image of the source s' is then the new exit pupil of the system, and this is placed in the entrance pupil of the eye, leading to pupil matching in position if not in size.

Maxwellian view is very useful in photometry and colorimetry, where fluctuations in pupil size would normally affect retinal illuminances. If the Maxwellian source is much smaller than the pupil of the eye, pupil fluctuations will not affect retinal illuminance. It is also useful if small eye movements may affect the image brightness. For example, if the source image were about the same size as the eye entrance pupil, small eye movements would shear the two pupils, leading to some vignetting and loss of brightness. This can be overcome by either having a very large source image size that is much greater than the eye pupil size or having a very small source image size.

36.9.1 Luminance of illuminated field

In Maxwellian view, the eye sees a field of even illumination with an angular radius α' as shown in Figure 36.5. We need to know the luminance of this field. For an extended source of luminance L, the illuminance in the plane of the image is given by equation (12.21a). Denoting this illuminance as E_1, then

$$E_1 = \pi \tau L \sin^2(\alpha') \tag{36.35}$$

If the source has area A_s and the image has area A_s', the amount of luminous flux F_1 falling on an area of magnitude A_s' is given by the equation

$$F_1 = E_1 A_s' = \pi \tau L \sin^2(\alpha') A_s' \tag{36.36}$$

Let us assume that the image of the source is smaller than the entrance pupil of the eye, that is $A_s' < A_p$, where A_p is the area of the entrance pupil of the eye. This flux will fall on the retina and the exit pupil of the Maxwellian system will appear to have a certain luminance, which we will denote as L'.

Now suppose the observed bright field of the instrument exit pupil were replaced by a Lambertian source of the same luminance L'. The illuminance

E_2 in the plane of the pupil will be

$$E_2 = \pi L' \sin^2(\alpha') \tag{36.37}$$

and since the source is Lambertian, it will completely fill the pupil of the eye, and therefore the luminous flux F_2 entering the eye is

$$F_2 = E_2 A_p = \pi L' \sin^2(\alpha') A_p \tag{36.38}$$

Now the area of the retina illuminated is the same in both cases and therefore the ratio of retinal illuminances and hence luminances is the ratio of the fluxes entering the eye. Since the two luminances are equal, the ratio of fluxes must be equal and so

$$\pi L' \sin^2(\alpha') A_p = \pi \tau L \sin^2(\alpha') A'_s$$

Therefore

$$L' = \tau L A'_s / A_p \tag{36.39}$$

with the restriction that the source image is smaller than the entrance pupil of the eye.

36.9.2   *Adapting pupil size*

If the pupil is fully and as usual uniformly illuminated, its size adapts to the light level by a, possibly, negative feed back loop. In the Maxwellian view situation, where the source image is usually smaller than the entrance pupil, this feed back system cannot operate and therefore we could ask the question what happens to the pupil size. Palmer (1966) claimed that the feed back loop was a negative loop and predicted that the pupil diameter should increase. He measured the pupil diameter for the same scene luminance but varied the diameter of the source image. He found that when the source image was smaller than the entrance pupil, the entrance pupil was larger than when it was filled and uniformly illuminated.

Exercises and problems

36.1   Calculate the object space depth-of-field of a microscope with a magnification of 100 and an exit pupil diameter of 2 mm. Assume a threshold defocus blur disc of 2 minutes of arc in diameter.

ANSWER: ± 0.0002 mm

36.2   Consider a wide angle optical system with a field-of-view of 100°. Estimate the desired exit pupil diameter for an eye pupil of 2 mm, so that the eye can rotate to see the edge of the field.

ANSWER: diameter $= 24.2$ mm

Summary of main symbols

D_e entrance pupil diameter of the eye

M magnification of a system

Section 36.1 Instrument focussing

δl focus error

$\delta L'$ vergence focus error

L_v, L'_v object and image vergences for an afocal system (from objective and eye lens)

Section 36.3: Depth-of-field

ω angular diameter of defocus blur disc

ω_{th} threshold value of ω (taken as 2 minutes of arc in most examples)

$\Delta L'_{th}$ vergence depth-of-field of the eye

Δl_{th} corresponding depth-of-field in object space

L_v, L'_v object and image vergences for an afocal system (from objective and eye lens)

Section 36.4: Pupil matching

ε_e centre of entrance pupil of eye

ε'_i centre of exit pupil of the instrument

c centre of rotation of eye

ψ angular radius of field-of-view of an instrument in image space

$\bar{\rho}'_i$ radius of exit pupil of instrument

$\bar{\rho}_e$ radius of entrance pupil of eye

Section 36.5: Image luminance

L luminance of image

E' retinal illuminance

M_o, M_e magnifications of objective and eye lens of a microscope

τ luminous transmittance of an optical system

α' angular radius of exit pupil measured at axial image point

D_t, D'_t entrance and exit pupil diameters of a telescope

d distance of a point source

I luminous intensity of a point source

F_e equivalent power of the eye

G photometric gain of a telescope

Section 36.6: Visual performance

σ, σ' object and image space spatial frequencies

Section 36.7: Magnification limits and empty magnification

D'_m	exit pupil diameter of a microscope
σ_L	object linear spatial frequencies
σ'_A	image angular spatial frequencies (usually c/rad)

Section 36.8: Aberration tolerances

β	apex angle of prism
θ	angle of deviation of a thin prism
μ	refractive index of prism
V_d	V-value of the prism material
F_p	power of the prism

Section 36.9: Maxwellian view

s, s'	source and its image
$\varepsilon, \varepsilon'$	entrance and exit pupils of Maxwellian view optical system
F	luminous flux in the image of the source
L	luminance of image
τ	luminous transmittance of an optical system
α, α'	angular radii of pupils of Maxwellian system
A'_s	area of image of source at eye
A_p	area of pupil of the eye

References and bibliography

*Blackwell H.R. and Taylor J.H. (1969). Survey on laboratory studies of visual detection. *NATO Seminar on "Detection, Recognition and Identification of Line-of-sight Targets"*. The Hague, Netherlands, 25–29 August.

*Burton G.J. and Haig N.D. (1984). Effect of Seidel aberrations on visual target discrimination. *J. Opt. Soc. Am. A* 1(4), 373–385.

*Charman W.N. (1974). Optical magnification for visual microscopy. *J. Opt. Soc. Am.* 64(1), 102–104.

*Charman W.N. and Whitefoot H. (1978). Astigmatism, accommodation and visual instrumentation. *Appl. Opt.* 17, 3903–3910.

*Giles M.K. (1977). Aberration tolerances for visual optical systems. *J. Opt. Soc. Am.* 67(5), 634–643.

*Giles M.K. (1979). Grating detectability: A method to evaluate aberration effects in visual instruments. *Opt. Eng.* 18(1), 33–38.

*Hennessy R.T. (1975). Instrument myopia. *J. Opt. Soc. Am.* 65(10), 1114–1120.

*Home R. and Poole J. (1977). Measurement of the preferred binocular dioptric settings at a high and low light level. *Opt. Acta* 24(1), 97–98.

*Leibowitz H.W. and Owens D.A. (1975). Anomalous myopias and the intermediate dark focus of accommodation. *Science* 189, 646–648.

*Mouroulis P., Kim T.G., and Zhao G. (1993). Transverse color tolerances for visual optical systems. *Appl. Opt.* 32(34), 7089–7094.

*Mouroulis P. and Woo G.C. (1988). Chromatic aberration and accommodation in visual instruments. *Optik* 80, 161–166. Cited by Mouroulis et al. (1993).

*Mouroulis P. and Zhang H. (1992). Visual instrument image quality metric and the effects of coma and astigmatism. *J. Opt. Soc. Am. A* 9(1), 34–42.

*Ohzu H. and Shimojima T. (1972). Optimum dioptre value for a view-finder of photographic camera. *Opt. Acta* 19(5), 343–345.

*Palmer D.A. (1966). The size of the human pupil in viewing through optical instruments. *Vis. Res.* 6, 471–477.

Richards O.W. (1976). Instrument myopia in microscopy. *Am. J. Optom. Physiol. Opt.* 53(10), 658–663.

*Schober H.A.W., Dehler H., and Kassel R. (1970). Accommodation during observations with visual instruments. *J. Opt. Soc. Am.* 60(1), 103–107.

Shimojima T. and Hayamizu Y. (1972). Analysis and synthesis of visual phenomena in microscopic vision – with particular reference to visual acuity. *Opt. Acta* 19(5), 455–458.

*Smith G. (1977). The spectacle wearer and visual optical instruments. *Aust. J. Optom.* 60, 382–386.

*Smith G. (1982). The angular diameter of defocus blur discs. *Am. J. Optom. Physiol. Opt.* 59(11), 885–889.

*Smith G. (1983). The accommodative resting states, instrument accommodation and their measurement. *Opt. Acta*, 30(3), 347–359.

*Stiles W.S. and Crawford B.H. (1933). The luminous efficiency of rays entering the eye at different points. *Proc. Roy. Soc. (B)* 112, 428–450.

*Westheimer G. (1972). Optimal magnification in visual microscopy. *J. Opt. Soc. Am.* 62(12), 1502–1504.

Visual ergonomics of binocular and biocular systems

37.0 Introduction

A wide variety of visual optical instruments are designed for binocular viewing, but probably the most well known are binoculars. Binocular viewing seems to offer a number of advantages over monocular viewing. Monocular viewing usually requires one eye to be closed, which may often lead to fatigue or discomfort, especially over extended viewing periods, although it may be possible for some people to keep both eyes open when viewing monocularly. This is possible if the image of the other eye is suppressed. In comparison, binocular viewing uses both eyes and hence is probably far less fatiguing. Binocular instruments also have the potential for stereoscopic viewing. Thus superficially, binocular viewing seems to be superior to monocular viewing. However a badly designed, badly manufactured or damaged binocular system can lead to significant problems in binocular viewing. Therefore we should be aware of the need for design and constructional tolerances for binocular instruments.

37.1 Stereoscopic and non-stereoscopic constructions

Not all binocular systems provide a stereoscopic image. The production of stereoscopic images requires that the two optical axes at the eyepieces be separated in object space. Let us look at several constructions and see how this requirement may be achieved.

37.1.1 *Non-stereoscopic systems*

Figure 37.1 shows a possible schematic construction of a binocular instrument, in which the two optical axes are identical in object space and hence the eyes will not see a stereoscopic image. In practical systems, the beam-splitting and

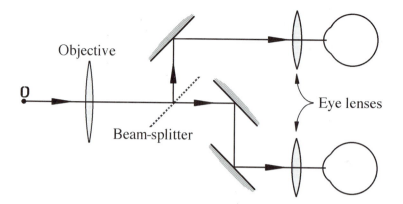

Fig. 37.1: Binocular instrument without a stereoscopic image.

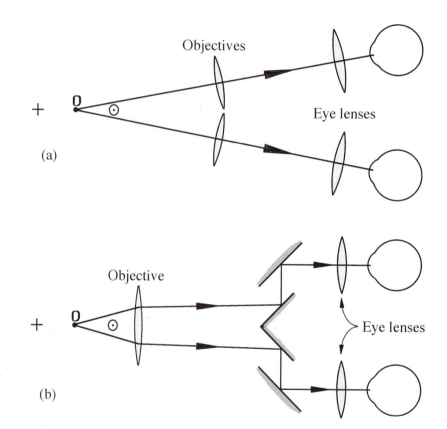

Fig. 37.2: Two possible schematic constructions of a binocular system that provide a stereoscopic image.

deviations are often done with prisms rather than with mirrors as depicted in the diagram.

37.1.2 Stereoscopic systems

In contrast, the construction shown in Figure 37.2 has optical axes that are separated in object space and therefore will give a stereoscopic image. In the

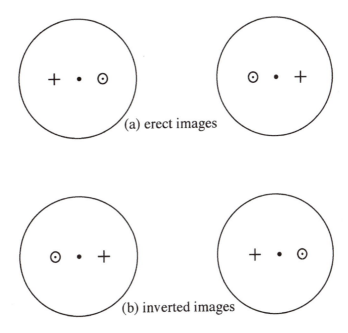

(a) erect images

(b) inverted images

construction shown in Figure 37.2a, the tubes of the system converge to the axial object point and if this is at infinity, the tubes would be parallel. This is the design used in binoculars. It is also useful in some binocular microscopes, but is limited to those of low power. For high power microscopes, the shorter working distance (that is the distance from object to objective) prevents the objectives being placed close together. The only alternative is the use of a single objective as shown in Figure 37.2b. In this arrangement, each eye looks through only one half the objective. Since the objective is usually the aperture stop, the exit pupil for each path is semi-circular, instead of circular.

37.1.2.1 Image orientation and stereopsis

For these constructions, the images must be erect, otherwise reverse stereopsis may be observed. In the constructions shown in Figure 37.2, the observer is viewing the object point o, behind which is a cross and in front is a circle. Let us look at the images formed by these systems, assuming the image is (a) inverted and (b) erect. For both of these two constructions, the retinal image appearances are shown in Figure 37.3. If the images are erect, the left eye will see the cross to the left of the circle, and for the right eye the opposite orientation will be observed, as shown in Figure 37.3a. For the correct stereoscopic effect, this must be the apparent relative orientation of the cross and circle. Now let us look at what happens if the images are inverted. With inverted images, the left eye will observe the cross to the right of the circle as shown in Figure 37.3b and this would be the observation if the cross were in front of the circle. Thus with inverted images, the cross should appear to be in front of the circle instead of behind it, thus giving a reverse stereoscopic effect. For this reason, stereo binocular systems usually have erect images.

However, it should be noted that reverse stereopsis is not always observed with inverted images. If there are other cues to the correct perspective, these

Table 37.1. Range of the eyepiece separation adjustments required to cover certain percentages of the population

	No. of standard deviations (s.d.)	Range (mm)	Percentage of the population
Adult males			
	1	61.0–67.0	68.3
	2	58.0–70.0	95.4
	3	55.0–72.9	99.7
Adult females			
	1	59.0–65.0	68.3
	2	56.0–68.0	95.4
	3	53.0–71.0	99.7

Note: The following mean and standard deviations from Table 13.1 are used: male, mean = 64.0, s.d. = 3.0; female, mean = 62.0, s.d. = 3.0.

may over-ride the above effect. Thus the reverse stereoscopic effect is more likely to occur in situations of reduced perspective cues, such as when the form of the object is unknown or unfamiliar.

37.2 Instrument focussing

The minimum instrument focussing capabilities must be the same as those of monocular systems described in the preceding chapter. This focussing requirement caters for different working distances and the refractive errors of the observer. Binocular systems must have one further refinement. Because of the possibility of a refractive error difference between the two eyes (**anisometropia**), the two eyepieces must be independently focussable. The range of this independent focussing depends upon the population frequency of a given refractive error difference. Rayner (1966) found that 79.8% of refractions had 0.5 m^{-1} or less of anisometropia, 11.8% had from 0.62 m^{-1} to 1.0 m^{-1} of difference, 3.4% had between 1.12 m^{-1} and 1.5 m^{-1} of anisometropia and 5% had 1.62 m^{-1} or greater of anisometropia. According to these data, an independent focussing range of ± 2 m^{-1} will cover at least 95% of the population.

37.3 Convergence of binocular tubes

Binocular optical systems consist of two identical optical paths or "tubes". Examples of these have already been given in Figures 37.1 and 37.2. The optical axes of the tubes may or may not be parallel at the eyepieces. In the example shown in Figures 37.1 and 37.2b, the eyepiece axes are parallel. In binocular systems used for distance viewing the optical axes are usually parallel. However, commercial instruments designed for close work have either parallel eyepiece axes or converging axes. Those that converge have angles of convergences in the range from 11° to 13°. These values correspond to the axes intersecting at a distance of about 300 mm.

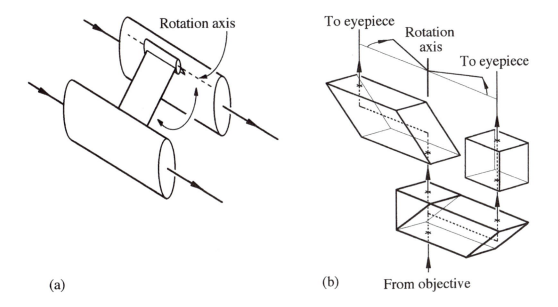

(a)

(b)   From objective

Fig. 37.4: Some possible
optical arrangements for
adjustment of
inter-pupillary distance: (a)
for a binocular in which the
distance between objectives
may also be adjusted and
(b) for systems with a
single objective.

37.4 Inter-pupillary distances and pupil matching

37.4.1 Distant inter-pupillary distance

The distance between the eyepieces of a binocular system must be adjustable
to cover the inter-pupillary range of the majority of users. Data on the distribution of inter-pupillary values can be found in Section 13.8.1 and Table 13.4.
Table 37.1 gives the ranges to cover 1, 2 and 3 standard deviations of the population, using the means and standard deviations taken from Chapter 13.

The eyepiece separation can be achieved by several optical arrangements
depending upon the degree of convergence of the two tubes and whether it
is simultaneously desirable to vary the separation of the objectives. Two of
these are shown in Figure 37.4. The simple example in Figure 37.4a applies
to the common binoculars in which the distance between objectives may also
be varied by the same amount. The two tubes are parallel and connected to a
common axis, which they can swing around. The example shown in Figure 37.4b
applies to those systems which have a single objective. Here the upper prism
arrangement can swing around the central axis and this varies the separation of
the eyepiece axes.

37.4.2 Near inter-pupillary distance settings: converging
binocular axes

When the eyes converge, they turn inwards about the **centre of rotation** of the
eye, thus bringing the pupils close together. From Chapter 13, we take this as
15 mm inside the eye from the corneal vertex. From Appendix 3, the entrance
pupil is about 3 mm in from the corneal vertex and therefore the centre of
rotation is about 12 mm behind the entrance pupil. This situation and distances
are shown in Figure 37.5. One can easily calculate the change in inter-pupillary
distance given the angle of convergence. From the diagram, the change Δp in
inter-pupillary distance p is given by the equation

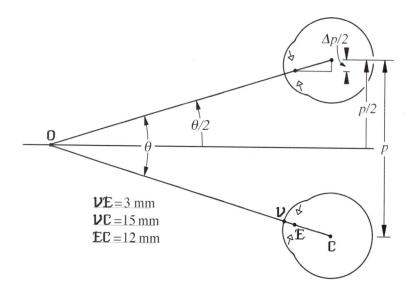

Fig. 37.5:   Effect of
convergence on
inter-pupillary distance.

$$\Delta p = -24\sin(\theta/2) \text{ mm} \tag{37.1}$$

Converging binocular axes seem to have a typical convergence of about 12°.

> **Example 37.1:** Calculate the reduction in inter-pupillary distance if the angle of convergence is 12°.
>
> **Solution:** Here $\theta = 12°$ and therefore from equation (37.1), we have
>
> $$\Delta p = -24\sin(6°) = -2.5 \text{ mm}$$

While the above result is only a small amount relative to the inter-pupillary distance, it is about the same magnitude as the pupil diameter and greater than the exit pupil diameter of a number of high magnification microscopes.

37.5   Accommodation and convergence

Some people have trouble using binocular systems. A typical problem is the inability to see a single clear image. Either the image is sharp but double, blurred but single or worse, blurred and double. These problems arise when there is a conflict between the **accommodation/convergence** demands of the binocular optical system and the accommodation and **accommodative convergence /accommodation** ratio of the individual. The angular convergence of the optical axes of the binocular tubes at the eyepieces is a critical factor in these cases. To understand how this conflict may arise, one can begin by looking at the relationship between the convergence of the axes of the eyes and the accommodation required to focus clearly at the point of convergence.

If the binocular axes of the instrument are converging by an angle θ, as shown in Figure 37.6, the desired angle of convergence of the ocular visual axes has the same value. Otherwise the observer will see two images, which

Fig. 37.6: Relationship
between accommodation
and convergence.

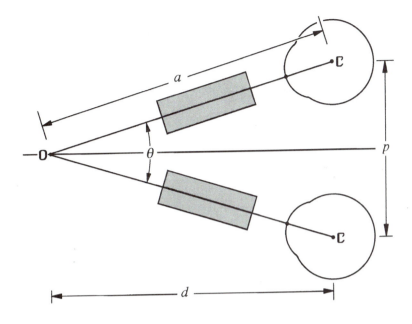

Fig. 37.6: Relationship between accommodation and convergence.

may or may not be in focus. At the same time, a focussed image requires the observer to accommodate appropriately, but the amount of convergence affects the amplitude of accommodation of the observer. Thus the angle of convergence of the binocular eyepiece axes will affect the accommodation response of the observer. This may also be further affected by instrument accommodation. While any accommodation may be compensated for by focussing of the eyepiece, it may not for abnormal or extreme accommodation responses or where the focussing range of the eyepiece is limited.

For a given inter-pupillary distance p and a binocular tube convergence angle of θ, Figure 37.6 can be used to show that the distance d from the centre of rotation to the point of convergence is given by the equation

$$d = p/[2\tan(\theta/2)] \tag{37.2}$$

The equivalent convergence C of the two eyes is often quantified in terms of the power of a prism that would provide the same deviation. Prismatic power is defined in Section 8.1.2.1 and for the general case is defined by the distance y in equation (8.6). Thus from this equation and using the angle θ shown in the diagram, we have

$$C = 200\tan(\theta/2) \quad \text{prism dioptres} \tag{37.3}$$

Again, referring to the diagram, the corresponding required accommodation A is

$$A = 1/a = 2\sin(\theta/2)/p \tag{37.4}$$

where a is the distance from the centre of rotation of the eye to the point of convergence O.

The accommodation A and binocular convergence C can be related by eliminating the angle θ from equations (37.3) and (37.4). After some manipulation, it follows that

$$A = \frac{2C}{p\sqrt{(200^2 + C^2)}} \quad \text{dioptres (accommodation)} \qquad (37.5a)$$

and if we need C in terms of A then

$$C = \frac{200Ap}{\sqrt{(4 - A^2 p^2)}} \quad \text{prism dioptres (convergence)} \qquad (37.5b)$$

The ideal or **stimulus** accommodative convergence/accommodation ratio is C/A. Thus

$$C/A = \frac{200p}{\sqrt{(4 - A^2 p^2)}} \qquad (37.5c)$$

If $A^2 p^2$ is much less than 4, then

$$C/A = 100p \qquad (37.5d)$$

For an average inter-pupillary distance of about ($p =$) 65 mm or 0.065 m, the expected ratio is thus 6.5. This value is significantly greater than actual or **response** ratios found in practice. Studies such as those of Morgan and Peters (1951) have shown that the population mean response ratio value is between 3 and 4. However, Eskridge (1983) found that for young subjects values of about 1.7 were typical. The stimulus ratio cannot change with age, but according to Eskridge, the response ratio increases with age as a result of a decrease in accommodative response with age. As complete presbyopia approaches, the response ratio should approach infinity.

If the observer cannot converge and accommodate by these amounts, he or she will not be able to see a single clear image. However, most instruments have a focussing eyepiece which allows the image vergence to be varied, although only within limits. There are few or no instruments that allow the convergence of the binocular axes to be adjusted to suit the individual. MacLeod and Bannon (1973) state that observers who are **esophoric** prefer converging tubes and **exophoric** users prefer parallel tubes (see Section 32.0.3 and the Glossary in Appendix 4 for a definition of these terms).

For an ideal association, accommodation and convergence must follow the above equations (37.5a) and (37.5b) (the demands), but for a given level of convergence there is a range of possible accommodations within which one sees a single clear image. Outside this range, the observer will see either (i) a single but blurred image, or (ii) a double and clear image, or (iii) a double and blurred image.

The **synoptophore** (see Section 32.1) can be used to examine the convergence/accommodation relationship of individuals. Synoptophore targets with high contrast detail are required, in order to provide accurate accommodation. Negative power lenses are used to stimulate accommodation. The convergence

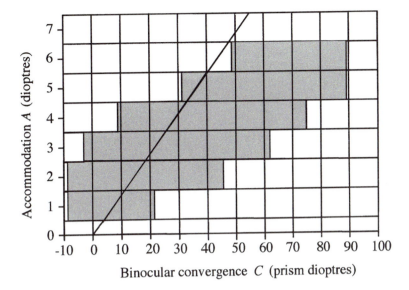

Fig. 37.7: Example of use of the synoptophore to examine the combination of accommodation and convergence (shaded region) that provides a single and clear image.

is varied by rotating the arms of the synoptophore. At each accommodation/convergence level, the subject is asked to report whether the images are single and in focus or some combination of double or blurrred.

An example of a particular response is shown in Figure 37.7. The shaded region is that for a single and clear image. Outside that region, a single clear image could not be seen. Either the image is single and blurred, double and clear or blurred and double. The straight line is the "demand" line determined from equation (37.5d), for that particular observer. The results plotted in the diagram show that the range of accommodation of this observer is very much affected by the binocular convergence and for those instruments with parallel eyepiece axes, the amplitude of accommodation of the observer is only about 3 m^{-1}, which is much less than the full amplitude of accommodation of 6 m^{-1}. These results indicate that this observer would be able to see a single clear image through a binocular system with parallel tubes providing the image vergence is in the range 0.5 to 3.5 m^{-1}. For tubes converging by an angle of $12°$, the corresponding value of C, from equation (37.2), would be approximately 20 prism dioptres and thus, from the diagram, the subject would be able to see a single clear image from this instrument providing the image vergence is now in the range 0.5 to 4.5 m^{-1}.

The ability to converge decreases with age (Pickwell 1985), and this is accompanied by an increase in exophoria with age. Thus, according to MacLeod and Bannon (1973), older users may be better off with parallel binocular tubes.

37.6 Stereoscopic magnification

Binocular vision allows us to perceive depth. If we are looking at two close objects that are a certain distance away, with one in front of the other, the perception of depth allows us to make a judgement of which object is closer. This is known as the stereoscopic effect. The minimum separation that allows us

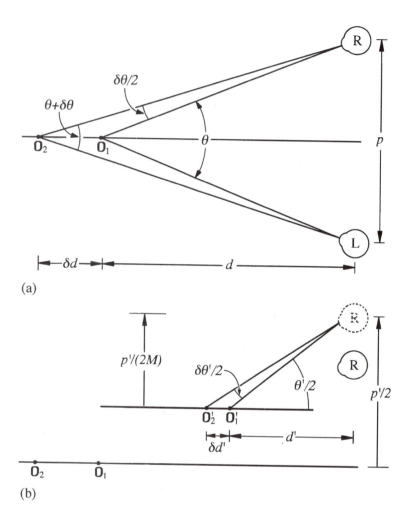

Fig. 37.8: Stereoscopic magnification in an afocal binocular instrument.

to make the judgement is called stereoscopic acuity and its value varies among individuals. Some individuals have no stereopsis and this visual condition is usually associated with **amblyopia**.

Let us examine the optics of stereopsis. Consider a situation as shown in Figure 37.8a. The ability of the eyes to detect a difference in depth or distance between the points o_1 and o_2 is a function of the transverse physical separation of their retinal images. The angle $\delta\theta$ is a measure of this distance. Stereoscopic acuity is defined as the threshold value of this angle, and it may be as small as 5 seconds of arc (Bennett and Rabbetts 1989, 232). From the diagram and assuming the angle θ is small, we have

$$\theta = \frac{p}{d} \quad \text{and} \quad \theta + \delta\theta = \frac{p}{(d + \delta d)} \tag{37.6a\&b}$$

If we solve for $\delta\theta$, and drop its negative sign

$$\delta\theta = \frac{p\delta d}{d(d + \delta d)} \tag{37.7a}$$

If we express this equation in terms of δd, we have

$$\delta d = \frac{\delta \theta d^2}{p - \delta \theta d} \tag{37.7b}$$

At a certain distance d, the corresponding value of δd will be infinite. This occurs when the denominator in equation (37.7b) is zero. We will denote this value of d by the symbol d_{sr} and from equation (37.7b)

$$d_{sr} = p/\delta\theta \tag{37.8}$$

We will call this distance the stereoscopic range and it is the far limit of depth perception. Its value will vary among individuals.

If we return to equation (37.7a) and for a certain distance d, assume that $\delta\theta$ is so small that δd is negligible compared with d, then this equation can be written as

$$\delta\theta = \frac{p\delta d}{d^2} \tag{37.9}$$

This equation shows that the ability to resolve the relative distances between two close points is proportional to the inter-pupillary distance and inversely proportional to the square of the distance. Therefore stereoscopic perception can be increased by increasing the inter-pupillary distance or decreasing the distance to the target.

Binocular systems can enhance the stereoscopic effect. This may be due to (a) a magnification which effectively decreases the apparent distance or (b) an increase in the effective inter-pupillary distance. An example involving both of these is the common pair of binoculars that uses Porro prisms to erect the image. Let us look at the stereoscopic magnification of this type of system.

37.6.1 *Binocular telescopes and stereoscopic magnification*

Figure 37.8b shows a binocular system (but only one side) imaging the points o_1 and o_2 to o_1' and o_2' and the distances d' and $\delta d'$ are as shown. We should note here that these points are not on the axes of the binoculars but displaced by an amount that depends upon inter-pupillary distance and the transverse magnification of the system. Therefore we should now regard the points o_1 and o_2 as representing the edges of off-axis objects of size $p'/2$. In Section 17.3, we showed that such objects are magnified by an amount $1/M$. This places the image points a distance $p'/(2M)$ from the axis. It follows from this diagram and equations (37.6a and b) applied to this situation that

$$\theta' = \frac{p'}{Md'} \quad \text{and} \quad \theta' + \delta\theta' = \frac{p'}{M(d' + \delta d')} \tag{37.10a\&b}$$

If we proceed as for the unaided eye, we can derive the following equation for $\delta\theta'$:

$$\delta\theta' = \frac{p'\delta d'}{Md'^2} \tag{37.11}$$

For large distances, the image distance d' can be related to the object distance d and δd by equation (17.8a), that is

$$d' = d/M^2 \tag{37.12}$$

For small changes in distances, we have

$$\delta d' = \delta d/M^2$$

If we use these two equations to replace d' and $\delta d'$ in equation (37.11) by d and δd, we get

$$\delta\theta' = \frac{p'\delta d M}{d^2} \tag{37.13}$$

If we define the **stereoscopic magnification** M_{st} as

$$M_{st} = \delta\theta'/\delta\theta \tag{37.14}$$

we can combine equations (37.9) and (37.13) to give

$$M_{st} = M(p'/p) \tag{37.15}$$

It is left to the reader to confirm that the new stereoscopic range d'_{sr} is given by the equation

$$d'_{sr} = M_{st}d_{sr} \tag{37.16}$$

These last two equations show that binocular afocal instruments increase stereoscopic effects proportional to the instrument magnification with an additional contribution due to the increased displacement of the optical axes at the objectives. Thus, Keplerian telescopes which use Porro prisms to erect the image and have outwardly displaced objectives also enhance the stereoscopic effect over an equivalent Galilean telescope in which the objectives have the same separation as the user's pupils.

> **Example 37.2:** Calculate the stereoscopic magnification of a 8×, 30 pair of binoculars with the objectives separated by a distance of ($p' = 115$ mm) when used by a viewer with an inter-pupillary distance of 60 mm, that is $p = 60$ mm.
>
> **Solution:** From equation (37.15), the stereoscopic effect is increased by the amount
>
> $$M_{st} = 8 \times 115/60 = 15.33$$

The increase in stereopsis due to the increase in inter-pupillary distance suggests that we can use the simple optical arrangement shown in Figure 37.9a to increase stereopsis, proportional to the increase in separation of the outer mirrors. An interesting modification of this arrangement is the one shown in

Fig. 37.9: (a) Simple
optical arrangement for
increasing the stereoscopic
effect. (b) An arrangement
that may reverse the
stereoscopic effect.

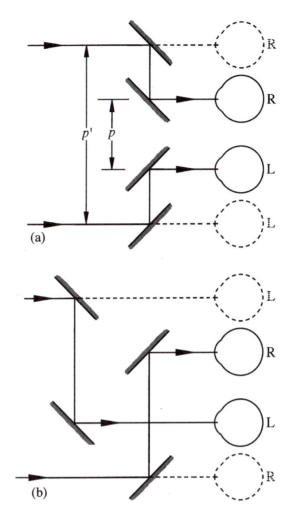

Figure 37.9b, in which the visual axes are crossed over. Optically, this arrangement should give a reversed stereo effect, but whether this is actually observed
depends upon the familiarity with the object.

37.7 Binocular telescopes and binocular convergence

Conventional binocular afocal or telescopic systems have parallel axes and
assume the object is at infinity. In this case, the ocular visual axes are parallel
and the images are at infinity. However, this is not so for objects at a finite
distance. Let us assume that an observer is viewing an object at o_1 a finite
distance, as shown in Figure 37.8. The angle of convergence is the angle θ' in
this diagram. From equations (37.10a) and (37.12), we can write this angle as

$$\theta' = \frac{p'M}{d} \tag{37.17}$$

The required convergence of the eyes is then

$$C = \frac{100p'M}{d} \quad \text{prism dioptres} \tag{37.18}$$

The convergence effect is made use of in the stereoscopic rangefinder described in Section 20.2.2.

Viewing this nearby object may require too much accommodation and this accommodation demand can be eased by refocussing the instrument. We can look into the effect of this refocussing on this convergence by returning to Section 17.7.1. In that section we described a model used to analyse the effect of refocussing on the magnification. We can use this same model and its approximations to show that the change in convergence is proportional to the change in effective magnification. Since the magnification increases on refocussing, the convergence must increase.

37.8 Visual performance

It has been established that with the unaided eye, binocular viewing yields better visual performance than monocular viewing. Campbell and Green (1965) found that contrast thresholds for binocular vision are about $\sqrt{2}$ times lower than for monocular vision, with the actual values not being significantly different statistically from the value of $\sqrt{2}$ which is the theoretical value from noise theory. They also found that the corresponding improvement in visual acuity is about 7%.

We may expect the same to apply to viewing through a binocular system, but providing (i) the two eyes have equal acuity, (ii) there is no conflict between convergence and accommodation demands (Section 37.5), (iii) the instrument is correctly focussed and (iv) there is no magnification difference between the two images. Tolerances on the magnification difference between the two sides of the binocular system are discussed in the next section.

37.9 Optical construction tolerances

As we have just noted in the previous section, to maximize the visual performance of a binocular instrument, the two beam paths should have the same optical properties, such as image vergence and magnification. However, in the construction of a binocular system it is impossible to make each of the binocular paths exactly the same and thus allowances have to be made for manufacturing errors. Depending upon the effects of particular errors on vision, tolerances have to be set for these errors.

37.9.1 Magnification differences

Acceptable magnification differences depend upon the ability of the visual system to cope with binocular differences in retinal image sizes. Davis (1959) states that magnification differences less than 0.75% are not clinically significant; between 1% and 3%, symptoms begin to appear; between 3.25% and 5%, binocular vision begins to be impaired; and over 5% magnification difference, binocular vision is very poor or absent. Reading and Tanlamai (1980) showed

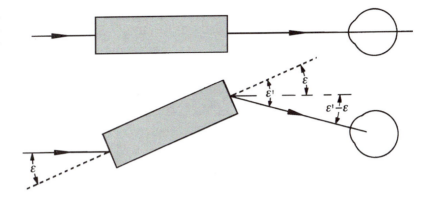

Fig. 37.10: Errors in mis-alignment of the optical axes of nominally parallel binocular tubes.

that stereopsis decreased fairly linearly with magnification difference, with the threshold going from about 5.5 seconds of arc for no magnification difference to about 50 seconds of arc for magnification differences of about 33%. A maximum difference of 2% is suggested in the United States handbook on optical design (MIL-HDBK-141 1962).

37.9.2 Parallelism of optical axes of an afocal system

In nominal afocal systems, the optical axes should be parallel. Any error in parallelism will lead to a loss of parallelism of the visual axes. If one tube axis has an alignment error ϵ as shown in Figure 37.10 and in any direction, an object straight ahead will be displaced by an angle ϵ', where

$$\epsilon' = M\epsilon \tag{37.19}$$

The eye will have to rotate to realign the two images and the angle of rotation θ of the eye will then be given by the equation

$$\theta = \epsilon' - \epsilon = (M - 1)\epsilon \tag{37.20}$$

Because it is easier for eyes to converge than diverge and more difficult for them to mis-align in the vertical direction, the tolerance ϵ_{tol} on the alignment error ϵ depends on the direction of the mis-alignment. If the tolerance on an angular deviation of the eye is θ_{tol}, then

$$\epsilon_{tol} = \theta_{tol}/(M - 1) \tag{37.21}$$

Values of θ_{tol} from a number of sources are shown in Table 37.2.

Mis-alignment of image erecting prisms, particularly Porro prisms, can also lead to a similar mis-aligned image and possible rotation of the image.

37.9.3 Image rotation

It was shown in Section 37.1 that stereoscopic binocular instruments should have erect images; otherwise stereopsis may be reversed. Prisms are commonly used to erect the image, for example in prism binoculars. The prisms in fact

Table 37.2. Tolerances on the deviations θ_{tol} in the alignment of the eyes when using binocular instruments with parallel binocular axes

	Convergence		Divergence		Vertical	
	Minutes of arc	Prism dioptres	Minutes of arc	Prism dioptres	Minutes of arc	Prism dioptres
Johnson (1960)	140	4.0	70	2.0	35	1.0
Martin (1961)	**20–25**	0.58–0.7	**7–8**	0.20–0.23	**7–8**	0.20–0.23
MIL-HDBK-141 (1962, 4–18)	17	**0.5**[a]	17	**0.5**[a]	17	**0.5**

Note: The angle is given in minutes of arc and prism dioptres. The boldface values are those quoted by the source and the non-boldface are derived from those values.
[a] It is stated that the values may be larger than this but larger values would be fatiguing.

rotate the image through 180°. The rotation is achieved in stages by a number of reflections from the internal faces of the prisms. If the prisms are incorrectly set or the angles between the prism faces are in error, the image will not be rotated through exactly 180°. This is a particular problem with the Porro prism assembly, where the two prisms have to be accurately aligned at right angles to give an 180° image rotation. Apart from the alignment errors during construction, severe physical shocks while in use can cause the prisms to move and become mis-aligned.

Small errors in image orientation can be compensated by rotation of one or both of the eyes about their image or visual axes. However, the possible angle of rotation is very small and thus the tolerance on the angular rotation error is much more severe than on the errors in parallelism. However, in contrast to errors in parallelism, tolerance to image rotation is independent of magnification.

37.9.4 Alignment of lenses in stereoscopes

A typical stereoscopic system is shown in Figure 15.13 and shown here in Figure 37.11. If these use high power (short focal length lenses) to give a wide field-of-view as they would in **virtual reality systems**, the system requires careful alignment. The centres of the stereopairs must be accurately aligned with the optical axes of the magnifiers or eye lenses. Any mis-alignment will deviate the image and cause the eyes to go out of alignment, producing unwanted convergence, divergence or superior or inferior mis-alignment.

If one of the pairs is out of centre by a distance δx, as shown in Figure 37.11, and the magnifier has an equivalent focal length f, then the image will be deviated by an angle $\delta\theta$, where

$$\delta\theta = \delta x/f = \delta x F \tag{37.22}$$

This will require the eye to deviate by this amount, and the allowed limits of this angle will be those laid out in Table 37.2 for binocular telescopes.

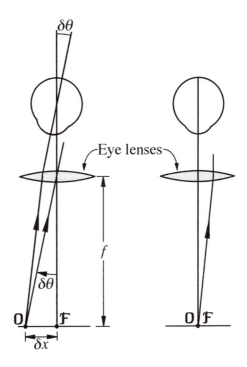

If the images are set at infinity and the aberrations are zero, there is no need to match the separation of the optical axes of the lenses with the inter-pupillary distance of the user accurately. However, if the image is set at some finite distance, for example to allow for myopic users, a mismatch between the user's inter-pupillary distance and eye lens centres will lead to unwanted ocular rotations.

37.9.5   Laboratory technique for checking the binocular axes alignment

A laboratory technique for measuring the axes and image tilts in binocular systems is shown in Figure 37.12, using a pair of binoculars as an example. One tube of the binoculars is aligned with a collimated target as shown and the target image formed through the binocular is observed through an observation telescope which must contain a graticule in its eyepiece. The system is aligned until the target is centred in the observation telescope. The binocular system is then moved sideways until the other tube is aligned with the collimator and the observation telescope. Any alignment or rotation errors in the two tubes will show as a displacement or rotation of the target in the observation telescope. A suitably marked graticule will allow these errors to be measured. This system can also be used for checking the alignment of stereoscopic viewers such as used in virtual reality systems.

37.10   Aberration tolerances

Tolerances to residual aberrations in monocular instruments have already been discussed in the preceding chapter. These same tolerances must also apply to

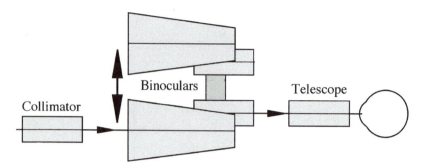

Fig. 37.12: Optical set-up for checking the alignment of the optical axes of a binocular system with parallel axes.

binocular systems. For the specific case of biocular systems such as **head-up displays** other tolerances apply.

37.10.1 Biocular systems

In biocular systems, the two eyes view a single image formed by one single co-axial lens system. As these are effectively a simple magnifier, the irises of the eyes become the aperture stops for the beam entering each eye. Thus this system has two effective pupils that are decentred. In some biocular systems such as head-up displays, it is not usually possible for the eyes to see the whole image from a fixed head position and the head moves to see different parts of the image. As a result, the pupils move and since aberrations depend upon pupil position, the aberrations at any point in the field may change as the head moves. Of particular importance is distortion. Sands (1971) has argued that if the image contains other aberrations, the distortion will depend upon eye position. Haig (1972) has shown that residual levels of these aberrations lead to changes in image vergence and direction, with head movement. Changes in image vergence lead to unequal accommodation demand and changes in direction (mobile distortion) lead to changes in convergence of the eyes. This latter effect may lead to problems in fusion. As a result biocular designs need to have much lower levels of the monochromatic aberrations than one may expect for the small eye pupils. In particular field curvature must be kept low.

The tolerances to the particular aberrations will depend upon the allowed changes in image vergence and image direction and the expected amount of pupil movement. As a guide to the tolerances on changes in vergence, we could use typical values of the depth-of-field of the eye, and set tolerances to changes in image vergence at about ± 0.25 m^{-1}. Tolerances to changes in image direction depend upon two factors: the absolute change in direction and the difference in direction for the two eyes. Limits of this latter effect could be those used to set limits to departure from parallelism in binocular axes given in Table 37.2.

Chromatic aberration may also be a problem. Since the two eyes are decentred from and on opposite sides of the optical axis, there will be some transverse chromatic fringing and this will be in opposite directions for the two eyes. The level of this chromatic aberration will also depend upon pupil position and therefore be a mobile aberration. However this problem may not be so important in some biocular systems such as head-up displays, because the light source is usually a narrow spectral band phosphor.

Exercises and problems

37.1 For a user whose inter-pupillary distance is 63 mm, calculate the stereoscopic magnification for a 8×, 30 pair of binoculars, whose objective centres are 120 mm apart.

ANSWER: $M_{st} = 15.23$

37.2 Using Martin's values set out in Table 37.2, what is the approximate manufacturing tolerance on the amount of divergence of the binocular tubes of an 8× binocular telescope? Take the magnification as $M = +8$.

ANSWER: tolerance = 1.0 minute of arc

37.3 For a typical inter-pupillary distance of 63 mm and a stereoscopic acuity of 20 seconds of arc (from Chapter 13), calculate the stereoscopic range for that observer.

ANSWER: 650 m

Summary of main symbols

θ angle of convergence of the two eyes
C convergence of two eyes (in terms of equivalent prism dioptres)
p inter-pupillary distance of eyes
M magnification of an afocal instrument

Section 37.4: Inter-pupillary distances

Δp change in inter-pupillary distance on convergence

Section 37.5: Accommodation and convergence

c centre of rotation of eye
A accommodation level

Section 37.6: Stereoscopic magnification

M_{st} stereoscopic magnification

Section 37.9: Optical constructional tolerances

θ_{tol} tolerance on the value of θ
ϵ alignment error angle in parallel tube instruments
ϵ_{tol} tolerance on the angle ϵ

References and bibliography

Bedwell C.H. (1972). Viewing stereoscopically through binocular optical systems. *Opt. Acta* 19(5), 341–342.

*Bennett A.G. and Rabbetts R.B. (1989). *Clinical Visual Optics*, 2nd ed. Butterworth-Heinemann, Oxford.

*Campbell F.W. and Green D.G. (1965). Monocular versus binocular visual acuity. *Nature* 208, 191–192.

*Davis R.J. (1959). Empirical corrections for aniseikonia in preschool anisometropes. *Am. J. Optom. Arch. Am. Acad. Optom.* 36(7), 351–364.

Dodwell P.C., Harker G.S., and Behar I. (1968). Pulfrich effect with minimal differential adaptation of the eyes. *Vis. Res.* 8, 1431–1443.

*Eskridge J.B (1983). The AC/A ratio and age – a longitudinal study. *Am. J. Optom. Physiol. Opt.* 60(11), 911–913.

*Haig G.Y. (1972). Visual aberrations of large-pupil systems. *Opt. Acta* 19(6), 543–546.

Hofstetter H.W. (1951). The relationship of proximal convergence to fusional and accommodation convergence. *Am. J. Optom.* 28, 300–308.

Home R. (1976). Binocular summation and its implications in the collimation of binocular instruments. *Assessment of Imaging Systems*, SPIE Vol. 98, Sira, London, 72–78.

*Johnson B.K. (1960). *Optics and Optical Instruments*. Dover Publications, New York.

Knoll H.A. (1959). Proximal factors in convergence: A theoretical consideration. *Am. J. Optom.* 36, 378–381.

*MacLeod D. and Bannon R.E. (1973). Microscopes and eye fatigue. *Industrial Med.* 42(2), 7–9.

*Martin L.C. (1961). *Technical Optics*. Pitman and Sons, London.

Marriot F.H.C. (1972). Visual acuity using binoculars. *Opt. Acta* 19, 385–386.

*MIL-HNDBK-141 (1962). *Military Standardization Handbook: Optical Design*. Defence Supply Agency, Washington D.C.

*Morgan M.W. and Peters H.B. (1951). Accommodation-convergence in presbyopia. *Am. J. Optom. Arch. Am. Acad. Optom.* 28(1), 3–10.

Nyman K.G. (1983). An experimental study in visual strain in microscope work. *Acta Ophthal.* Suppl. 161.

Pickwell L.D. (1985). The increase in convergence inadequacy with age. *Ophthal. Physiol. Opt.* 5(3), 347–348.

*Rayner A.W. (1966). Aniseikonia and magnification in ophthalmic lenses: Problems and solutions. *Am. J. Optom. Arch. Am. Acad. Optom.* 43, 617–632.

*Reading R.W. and Tanlamai T. (1980). The threshold of stereopsis in the presence of differences in magnification of the ocular images. *J. Am. Opt. Assoc.* 51(6), 593–595.

Rogers P.J. (1985). Biocular magnifiers – a review. *1985 International Lens Design Conference*, ed. W.H. Taylor and D.T. Moore. SPIE Vol. 554, Society of Photo-Instrumentation Engineers, Bellingham, 362–370.

*Sands P.J. (1971). Visual aberrations of afocal systems. *Opt. Acta* 18(8), 627–636.

Soderberg I., Calissendorff B., Elofsson S., Knave B., and Nyman K.G. (1983). Investigation of visual strain experienced by microscopy operators at an electronics plant. *Appl. Ergonomics* 14(4), 297–305.

Wallis N.E. (1966). Graphical analysis of accommodation-convergence relationships. *Br. J. Physiol. Opt.* 23(4), 232–241.

Appendices

A1.0 Introduction

This appendix is concerned with more advanced aspects of paraxial ray tracing. For example, we will show that the paraxial ray tracing equations can be expressed in terms of matrices and this allows some interesting properties of optical systems to be investigated. One of these is that there is a linear relationship between the paraxial ray angle and height at one plane in a system and the angle and height for the same ray at any other plane in the system. This linear relationship in turn has some useful applications as we will see. However, before we do this we will first look at some simpler aspects of paraxial rays such as the optical invariant.

A1.1 The optical invariant

Let two distinct rays be traced through an optical system and suppose at the jth surface, the first ray has a height $h_{1,j}$ and angles $u_{1,j}$ and $u'_{1,j}$ to the left and right of this surface, respectively, as shown in Figure A1.1. Let the corresponding values for the second ray be $h_{2,j}$, $u_{2,j}$ and $u'_{2,j}$. It will now be shown that the quantity

$$H = n_j(u_{1,j}h_{2,j} - u_{2,j}h_{1,j}) = n'_j(u'_{1,j}h_{2,j} - u'_{2,j}h_{1,j}) \qquad \text{(A1.1a)}$$

which is calculated on both sides of the surface, has the same numerical value for all surfaces. This quantity H is called the **optical** or **Lagrange invariant**. The proof is simple and utilizes the paraxial refraction and transfer equations. There are two stages of the proof: the first requires equation (A1.1a) to be verified and the second to show that

$$H = n'_j(u'_{1,j}h_{2,j} - u'_{2,j}h_{1,j})$$

$$= n_{j+1}(u_{1,j+1}h_{2,j+1} - u_{2,j+1}h_{1,j+1}) \qquad \text{(A1.1b)}$$

which is evaluated to the right of the jth surface and to the left of the $(j + 1)^{\text{th}}$ surface. To prove equation (A1.1a), take the lefthand side and replace the angles

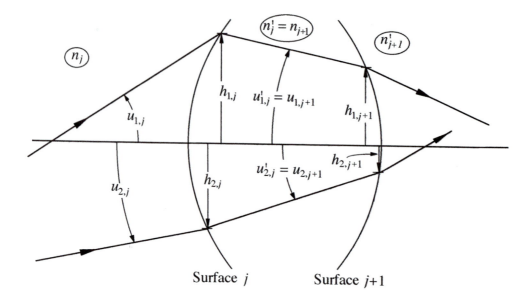

$u_{1,j}$ and $u_{2,j}$ by the corresponding expressions of the form

$$nu = n'u' + hF$$

which is a transformation of the paraxial refraction, equation (2.5b). Thus

$$n_j(u_{1,j}h_{2,j} - u_{2,j}h_{1,j}) = (n'_j u'_{1,j} + h_{1,j}F_j)h_{2,j}$$

$$- (n'_j u'_{2,j} + h_{2,j}F_j)h_{1,j}$$

$$= n'_j u'_{1,j}h_{2,j} - n'_j u'_{2,j}h_{1,j}$$

$$- h_{1,j}h_{2,j}F_j + h_{2,j}h_{1,j}F_j$$

$$= n'_j(u'_{1,j}h_{2,j} - u'_{2,j}h_{1,j})$$

which is the righthand side of equation (A1.1a) and thus completes the first part of the proof. The second part now relies upon a transfer of the two rays from surface j to surface $j + 1$. This is done by replacing $h_{1,j}$ and $h_{2,j}$ with corresponding expressions of the form

$$h = h' - u'd$$

which is a transformation of the paraxial transfer equation, equation (2.12). Thus

$$n'_j(u'_{1,j}h_{2,j} - u'_{2,j}h_{1,j}) = n'_j[u'_{1,j}(h_{2,j+1} - u'_{2,j}d)$$

$$- u'_{2,j}(h_{1,j+1} - u'_{1,j}d)]$$

$$= n'_j(u'_{1,j}h_{2,j+1} - u'_{1,j}u'_{2,j}d$$

$$- u'_{2,j}h_{1,j+1} + u'_{2,j}u_{1,j}d)$$

Since

$$n'_j = n_{j+1}, u'_{1,j} = u_{1,j+1} \quad \text{and} \quad u'_{2,j} = u_{2,j+1},$$

the above expression reduces to

$$n_{j+1}(u_{1,j+1}h_{2,j+1} - u_{2,j+1}h_{1,j+1})$$

which proves equation (A1.1b). Thus the invariance of H has been completely proved.

In the above discussion the two rays were any two distinct rays. However, it is usual in practice to use the paraxial marginal and pupil rays. Using a shortened version of the above invariance equations the optical invariant can be expressed as

$$H = n(\bar{u}h - u\bar{h}) \tag{A1.2}$$

where u and h refer to the marginal ray and $\bar{u}$ and $\bar{h}$ refer to the pupil ray.

Note: The order of the rays in equation (A1.2) is unimportant. For example the optical invariant could have been written

$$H = n(u\bar{h} - \bar{u}h)$$

The only difference is the sign.

A1.2   The equivalent power of a component

It has been established in Chapters 2 and 3 that the paraxial refraction equation of any system or component can be written as

$$n'u' - nu = -hF$$

where h is the height of the ray at the principal planes. If any two distinct rays are traced through this component as shown in Figure A1.2, one can write

$$n'u'_1 - nu_1 = -h_1F$$

and

$$n'u'_2 - nu_2 = -h_2F$$

If the first of these equations is multiplied by u_2, the second by u_1 and then the first is subtracted from the second, one can solve for F to give

$$F = \frac{n'(u'_2u_1 - u'_1u_2)}{(h_1u_2 - h_2u_1)}$$

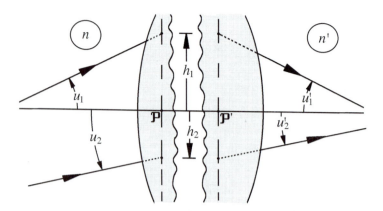

If these two rays were traced through the component, surface by surface, the heights h_1 and h_2 above are the ray heights at the principal planes and not the heights at any particular surface and these values would not be immediately known. However the quantity $h_1 u_2 - h_2 u_1$ is related to the optical invariant H [equation (A1.1a) or (A1.2)]. Here we can write

$$h_1 u_2 - h_2 u_1 = H/n$$

thus

$$F = \frac{nn'(u_2' u_1 - u_1' u_2)}{H} \tag{A1.3}$$

where the optical invariant is here defined as

$$H = n(h_1 u_2 - h_2 u_1) \tag{A1.3a}$$

and the two rays may be the paraxial and pupil rays and H can be evaluated at any surface or plane in terms of quantities directly known from the ray trace. Equation (A1.3) shows that given the paraxial ray trace results of two distinct rays plus the refractive indices, the equivalent power of any component in the system can be conveniently evaluated.

A1.3 Matrix representation

The paraxial refraction and transfer equations given in Chapters 2 and 3 can be expressed in matrix notation. For example, the refraction equation (3.5) at the jth surface can be expressed in the form

$$n_j' u_j' = n_j u_j - h_j F_j \tag{A1.4}$$

where F_j is the power of the surface. This equation can also be expressed in the matrix form

$$\begin{bmatrix} n_j' u_j' \\ h_j' \end{bmatrix} = \begin{bmatrix} 1 & -F_j \\ 0 & 1 \end{bmatrix} \begin{bmatrix} n_j u_j \\ h_j \end{bmatrix} \tag{A1.5}$$

where $h'_j = h_j$ would now be the ray height after refraction, which of course is the same value as before refraction, and therefore the symbol "prime" (') can be omitted if desired. If we now write

$$\mathfrak{p}'_j = \begin{bmatrix} n'_j u'_j \\ h'_j \end{bmatrix}, \quad \mathfrak{R}_j = \begin{bmatrix} 1 & -F_j \\ 0 & 1 \end{bmatrix} \quad \text{and } \mathfrak{p}_j = \begin{bmatrix} n_j u_j \\ h_j \end{bmatrix} \tag{A1.6}$$

we can write a refraction at surface j in the shorthand matrix form

$$\mathfrak{p}'_j = \mathfrak{R}_j \mathfrak{p}_j \tag{A1.7}$$

where $\mathfrak{R}_j$ is the refraction matrix for the jth surface. Thus the "refraction" matrix is a 2×2 matrix and can be seen to have a unit determinant.

Now on the object side of the next surface, surface $(j + 1)$,

$$u_{j+1} = u'_j, \quad n_{j+1} = n'_j$$

and therefore one can write

$$n_{j+1} u_{j+1} = n'_j u'_j$$

The transfer equation between the jth and the $(j + 1)^{\text{th}}$ surfaces can be written

$$h_{j+1} = h_j + u'_j d_j = h'_j + u'_j d_j$$

As before, these equations can be expressed in the matrix form

$$\begin{bmatrix} n_{j+1} u_{j+1} \\ h_{j+1} \end{bmatrix} = \begin{bmatrix} 1 & 0 \\ d_j/n'_j & 1 \end{bmatrix} \begin{bmatrix} n'_j u'_j \\ h'_j \end{bmatrix} \tag{A1.8}$$

We now define the matrix quantities $\mathfrak{p}_{j+1}$ and $\mathfrak{T}_{j,j+1}$ as follows:

$$\mathfrak{p}_{j+1} = \begin{bmatrix} n_{j+1} u_{j+1} \\ h_{j+1} \end{bmatrix} \quad \text{and} \quad \mathfrak{T}_{j,j+1} = \begin{bmatrix} 1 & 0 \\ d_j/n'_j & 1 \end{bmatrix} \tag{A1.9}$$

where $\mathfrak{T}_{j,j+1}$ is the "transfer" matrix for the ray moving from surface j to $j+1$ and also has a unit determinant. The transfer can also be expressed in shorthand matrix form, as follows:

$$\mathfrak{p}_{j+1} = \mathfrak{T}_{j,j+1} \mathfrak{p}'_j \tag{A1.10}$$

It is clearly seen that the determinant of the transfer matrix is also unity. The fact that both the refraction and transfer matrices have a unit determinant will be found useful later.

Using this matrix notation, a refraction and transfer, surface by surface or component by component, then reduces to a multiplication of matrices. If there are k surfaces, the ray tracing can be written as a product of matrices in the form

$$\mathfrak{p}'_k = \mathfrak{R}_k \cdots \mathbb{T}_{3,4}\mathfrak{R}_3\mathbb{T}_{2,3}\mathfrak{R}_2\mathbb{T}_{1,2}\mathfrak{R}_1\mathfrak{p}_1 \tag{A1.11}$$

where

$$\mathfrak{p}'_k = \begin{bmatrix} n'_k u'_k \\ h_k \end{bmatrix} \quad \text{and} \quad \mathfrak{p}_1 = \begin{bmatrix} n_1 u_1 \\ h_1 \end{bmatrix} \tag{A1.11a}$$

and u'_k and u_1 are the image space and object space paraxial angles, respectively, and h_k (or h'_k) and h_1 are the heights of the paraxial ray at the last and first surfaces, respectively.

If we look at the product

$$\mathfrak{R}_k \cdots \mathbb{T}_{3,4}\mathfrak{R}_3\mathbb{T}_{2,3}\mathfrak{R}_2\mathbb{T}_{1,2}\mathfrak{R}_1$$

of the 2×2 matrices in equation (A1.11), this must be reducible to a single 2×2 "refraction" matrix, which we will call the system matrix and denote by the symbol $\mathfrak{S}$. Thus

$$\mathfrak{S} = \mathfrak{R}_k \cdots \mathbb{T}_{3,4}\mathfrak{R}_3\mathbb{T}_{2,3}\mathfrak{R}_2\mathbb{T}_{1,2}\mathfrak{R}_1 \tag{A1.12}$$

Thus $\mathfrak{S}$ can be represented by a matrix of the form

$$\mathfrak{S} = \begin{bmatrix} A & B \\ C & D \end{bmatrix} \tag{A1.13}$$

and since the product of square matrices with unit determinants must also have a unit determinant, it follows that the determinant of $\mathfrak{S}$ must also be unity, that is

$$\det(\mathfrak{S}) = AD - BC = 1 \tag{A1.14}$$

Equation (A1.11) can now be written as

$$\mathfrak{p}'_k = \mathfrak{S}\mathfrak{p}_1 \tag{A1.15a}$$

or

$$\begin{bmatrix} n'_k u'_k \\ h_k \end{bmatrix} = \begin{bmatrix} A & B \\ C & D \end{bmatrix} \begin{bmatrix} n_1 u_1 \\ h_1 \end{bmatrix} \tag{A1.15b}$$

where A, B, C and D depend only on the system constructional parameters.

If the matrices in equation (A1.15b) are multiplied out, we get

$$n'_k u'_k = A n_1 u_1 + B h_1$$

and

$$h_k = C n_1 u_1 + D h_1 \tag{A1.16}$$

If we repeat the above process, but start with the refraction equations (A1.4) in the form

$$u'_j = (n_j/n'_j)u_j - h_j(F_j/n'_j)$$

equations (A1.5) and (A1.8) become

$$\begin{bmatrix} u'_j \\ h'_j \end{bmatrix} = \begin{bmatrix} n_j/n'_j & -F_j/n'_j \\ 0 & 1 \end{bmatrix} \begin{bmatrix} u_j \\ h_j \end{bmatrix} \tag{A1.17}$$

and

$$\begin{bmatrix} u_{j+1} \\ h_{j+1} \end{bmatrix} = \begin{bmatrix} 1 & 0 \\ d_j & 1 \end{bmatrix} \begin{bmatrix} u'_j \\ h'_j \end{bmatrix} \tag{A1.18}$$

and we have as follows

$$\mathfrak{p}'_k = \begin{bmatrix} u'_k \\ h_k \end{bmatrix} \quad \text{and} \quad \mathfrak{p}_1 = \begin{bmatrix} u_1 \\ h_1 \end{bmatrix} \tag{A1.19}$$

$$\mathfrak{R}_j = \begin{bmatrix} n_j/n'_j & -F_j/n'_j \\ 0 & 1 \end{bmatrix} \quad \text{and} \quad \mathfrak{T}_{j,j+1} = \begin{bmatrix} 1 & 0 \\ d_j & 1 \end{bmatrix} \tag{A1.20}$$

Note that the determinant of $\mathfrak{R}_j$ is now n_j/n'_j but that of $\mathfrak{T}_{j,j+1}$ remains 1.0. It now follows that

$$\mathfrak{p}'_k = \mathfrak{S}\mathfrak{p}_1 \tag{A1.21}$$

$$\mathfrak{S} = \mathfrak{R}_k \cdots \mathfrak{T}_{3,4}\mathfrak{R}_3\mathfrak{T}_{2,3}\mathfrak{R}_2\mathfrak{T}_{1,2}\mathfrak{R}_1 \tag{A1.21a}$$

$$\mathfrak{S} = \begin{bmatrix} \mathfrak{A} & \mathfrak{B} \\ \mathfrak{C} & \mathfrak{D} \end{bmatrix} \tag{A1.22}$$

$$\det(\mathfrak{S}) = \mathfrak{A}\mathfrak{D} - \mathfrak{B}\mathfrak{C} = n_1/n'_k \tag{A1.22a}$$

$$\begin{bmatrix} u'_k \\ h_k \end{bmatrix} = \begin{bmatrix} \mathfrak{A} & \mathfrak{B} \\ \mathfrak{C} & \mathfrak{D} \end{bmatrix} \begin{bmatrix} u_1 \\ h_1 \end{bmatrix} \tag{A1.22b}$$

and

$$u'_k = \mathfrak{A}u_1 + \mathfrak{B}h_1$$

$$h_k = \mathfrak{C}u_1 + \mathfrak{D}h_1 \tag{A1.23}$$

This new set of coefficients $\mathfrak{A}$, $\mathfrak{B}$, $\mathfrak{C}$ and $\mathfrak{D}$ is connected to the original set A, B, C and D by the relations

$$\mathfrak{A} = An_1/n_k', \quad \mathfrak{B} = B/n_k', \quad \mathfrak{C} = Cn_1 \quad \text{and} \quad \mathfrak{D} = D \quad \text{(A1.24)}$$

A1.3.1   The A, B, C and D or the $\mathfrak{A}$, $\mathfrak{B}$, $\mathfrak{C}$ and $\mathfrak{D}$ coefficients

The use of coefficients $\mathfrak{A}$, $\mathfrak{B}$, $\mathfrak{C}$ and $\mathfrak{D}$ may be preferred because of the relative simplicity of the pair of equations (A1.23b) compared with equations (A1.16). However, using the set A, B, C and D has the advantage that all the associated matrices have unit determinants, which is a very convenient tool for checking the accuracy of intermediate calculations.

A1.3.2   Reverse ray tracing

Sometimes we may wish to find the object space values, given the image space values. In this case, equations (A1.16) can be written

$$n_1 u_1 = A' n_k' u_k' + B' h_k$$

and

$$h_1 = C' n_k' u_k' + D' h_k \quad \text{(A1.25)}$$

and the matrix equations (A1.15a and b) become

$$\mathfrak{p}_1 = \mathfrak{S}' \mathfrak{p}_k' \quad \text{(A1.26a)}$$

or

$$\begin{bmatrix} n_1 u_1 \\ h_1 \end{bmatrix} = \begin{bmatrix} A' & B' \\ C' & D' \end{bmatrix} \begin{bmatrix} n_k' u_k' \\ h_k \end{bmatrix} \quad \text{(A1.26b)}$$

It is easy to prove that the matrices $\mathfrak{S}$ and $\mathfrak{S}'$ are inverses of each other and therefore their product is the unit matrix. We can show that if $\mathfrak{S}$ is any 2×2 matrix

$$A' = +D/\det(\mathfrak{S}), \quad B' = -B/\det(\mathfrak{S})$$

$$C' = -C/\det(\mathfrak{S}) \quad \text{and} \quad D' = +A/\det(\mathfrak{S}) \quad \text{(A1.27)}$$

where $\det(\mathfrak{S})$ is the determinant value of $\mathfrak{S}$. Since $\det(\mathfrak{S})$ is unity,

$$A' = D, B' = -B, C' = -C \quad \text{and} \quad D' = A \quad \text{(A1.28)}$$

If we use the $\mathfrak{A}$, $\mathfrak{B}$, $\mathfrak{C}$ and $\mathfrak{D}$ coefficients instead, it follows from (A1.27) and (A1.22a) that

$$\mathfrak{A}' = \mathfrak{D}n_k'/n_1, \quad \mathfrak{B}' = -\mathfrak{B}n_k'/n_1$$

$$\mathfrak{C}' = -\mathfrak{C}n_k'/n_1 \quad \text{and} \quad \mathfrak{D}' = \mathfrak{A}n_k'/n_1 \quad \text{(A1.29)}$$

A1.4 Linearity of paraxial ray angles and heights

Equations (A1.23b),

$$u'_k = \mathfrak{A}u_1 + \mathfrak{B}h_1$$

and

$$h_k = \mathfrak{C}u_1 + \mathfrak{D}h_1 \tag{A1.30}$$

derived in the previous section, show that paraxial rays have a linearity property in that the image space angle and ray height are proportional to the object space values. More generally, we can state that the angles and heights at any two planes in an optical system are linearly related and thus we can write

$$u_j^{()} = \mathfrak{A}_{i,j}^{()}u_i^{()} + \mathfrak{B}_{i,j}h_i$$

and

$$h_j = \mathfrak{C}_{i,j}^{()}u_i^{()} + \mathfrak{D}_{i,j}h_i \tag{A1.31}$$

where the i and j subscripts refer to the two planes and the $^{()}$ superscript means that the angles may be specified on either the object or image side of the surface. These equations show that the values of the quantities $\mathfrak{A}_{i,j}$, $\mathfrak{B}_{i,j}$, $\mathfrak{C}_{i,j}$ and $\mathfrak{D}_{i,j}$ depend upon the particular pair of planes. The first and last surfaces or object and image planes are special cases of this general rule.

This linearity property is a very useful property of paraxial rays and has many applications. For example, we used this linearity property in Chapter 2 to show that all points on a plane in object space, which is perpendicular to the optical axis, are paraxially imaged onto a plane in image space which is also perpendicular to the optical axis. We also used the property to show that the transverse magnification in this image plane was independent of object height. In Section A1.6, we will present another application to the calculation of the Gaussian properties of optical systems. For this calculation in special case, we need to know how to find the numerical values of the quantities $\mathfrak{A}$, $\mathfrak{B}$, $\mathfrak{C}$ and $\mathfrak{D}$. Here we will present several methods and give examples for the special case of the pair of surfaces being the first and last surfaces and the angles being specified on the object side for the first surface and the image side for the last surface.

A1.5 Methods for finding the values of (A, B, C, D) or $(\mathfrak{A}, \mathfrak{B}, \mathfrak{C}, \mathfrak{D})$

A1.5.1 By matrix multiplication

In any particular case, the values of A, B, C and D can be found by forming the refraction matrices $\{\mathfrak{R}\}$ and the transfer matrices $\{\mathfrak{T}\}$ and multiplying these together to get the final $\mathfrak{S}$ matrix using equation (A1.12). The elements of this final matrix are the A, B, C and D values. The values of $\mathfrak{A}$, $\mathfrak{B}$, $\mathfrak{C}$ and $\mathfrak{D}$ can then be found from equation (A1.24). Alternatively, we can find $\mathfrak{A}$, $\mathfrak{B}$, $\mathfrak{C}$ and $\mathfrak{D}$ directly with equations (A1.20) and (A1.21a). However, by firstly finding A, B, C and D instead of $\mathfrak{A}$, $\mathfrak{B}$, $\mathfrak{C}$ and $\mathfrak{D}$ we have the advantage that in the

Table A1.1. *The refraction and transfer matrices for the Le Grand theoretical relaxed schematic eye, used in Example A1.1*

$$\Re_1 = \begin{bmatrix} 1 & -0.0483462 \\ 0 & 1 \end{bmatrix} \qquad \mathbb{T}_{1,2} = \begin{bmatrix} 1 & 0 \\ 0.399390 & 1 \end{bmatrix}$$

$$\Re_2 = \begin{bmatrix} 1 & +0.0061077 \\ 0 & 1 \end{bmatrix} \qquad \mathbb{T}_{2,3} = \begin{bmatrix} 1 & 0 \\ 2.280544 & 1 \end{bmatrix}$$

$$\Re_3 = \begin{bmatrix} 1 & -0.0080980 \\ 0 & 1 \end{bmatrix} \qquad \mathbb{T}_{3,4} = \begin{bmatrix} 1 & 0 \\ 2.816901 & 1 \end{bmatrix}$$

$$\Re_4 = \begin{bmatrix} 1 & -0.014 \\ 0 & 1 \end{bmatrix}$$

Final matrix

$$\mathbb{S} = \begin{bmatrix} 0.904420 & -0.059940 \\ 5.448010 & 0.744614 \end{bmatrix}$$

former calculations, the determinants are all unity and this property can be used to check the correctness and accuracy of progressive calculations. Let us look at an example.

> **Example A1.1:** Let us consider the relaxed Le Grand full schematic eye given in Appendix 3. This is a four surface system, so that there are four refraction matrices and three transfer matrices. The elements of the $\{\Re\}$ and $\{\mathbb{T}\}$ matrices were determined from the data given in Appendix 3 and are shown in Table A1.1. The final $\mathbb{S}$ matrix is also shown in this table. The calculations will be done with 6 decimal places. From Table A1.1,
>
> $$A = 0.904420, \; B = -0.059940, \; C = 5.448010 \quad \text{and}$$
>
> $$D = 0.744614$$
>
> For this eye, $n_1 = 1.0$ and $n'_k = n'_4 = 1.336$ and so from equation (A1.24),
>
> $$\mathbb{A} = 0.676961, \; \mathbb{B} = -0.044865, \; \mathbb{C} = 5.448010 \quad \text{and}$$
>
> $$\mathbb{D} = 0.744614$$

For those uncomfortable with matrix multiplication, there are alternative methods that use direct ray tracing. Two ray tracing methods will be described.

A1.5.2   *By numerically tracing one ray*

This method traces a general ray given in terms of an arbitrary angle u_1 in object space and an arbitrary height h_1 at the first surface. By using a numerical example, we will show how this is done.

Example A1.2: Let us consider the relaxed Le Grand full schematic eye used in Example A1.1 and trace a general ray through this eye, with the ray defined only in terms of an arbitrary angle u_1 in object space and an arbitrary height h_1 at the first surface. We will carry out the calculations to six decimal places or at least five significant digits.

Refracting at the first surface, we have

$$n_1' u_1' - n_1 u_1 = -h_1 F_1$$

For the Le Grand eye, $n_1 = 1.0000$, $n_1' = 1.3771$, $C_1 = 1/7.8$ mm^{-1} and therefore

$$F_1 = (1.3771 - 1.0000)/7.8 = 0.048346$$

and so

$$u_1' = u_1/1.3771 - (h_1 \times 0.048346/1.3771)$$

that is

$$u_1' = 0.726164u_1 - 0.035107h_1$$

We now transfer to the second surface with the transfer equation

$$h_2 = h_1 + u_1' d_1$$

with $d_1 = 0.55$. Therefore

$$h_2 = h_1 + (0.726164u_1 - 0.035107h_1)0.55$$

or

$$h_2 = 0.399390u_1 + 0.980691h_1$$

Refracting at the second surface, we have

$$n_2' u_2' - n_2 u_2 = -h_2 F_2$$

For the Le Grand eye, $n_2 = n_1' = 1.3771$, $n_2' = 1.3374$, $C_2 = 1/6.5$ mm^{-1} and therefore

$$F_2 = (1.3374 - 1.3771)/6.5 = -0.0061077$$

and so

$$1.3374u_2' - 1.3771u_2 = -h_2(-0.0061077)$$

$$u_2' = (u_2 1.3771/1.3374) - [h_2(-0.0061077)/1.3374]$$

that is

$$u_2' = 1.029684u_2 + 0.0045668h_2$$

Now $u_2 = u_1'$ and therefore

$$u_2' = 1.029684(0.726164u_1 - 0.035107h_1)$$

$$+ 0.0045668(0.399390u_1 + 0.980691h_1)$$

Therefore

$$u_2' = +0.749543u_1 - 0.031671h_1$$

We now transfer to the third surface with the transfer equation

$$h_3 = h_2 + u_2'd_2$$

with $d_2 = 3.05$. Therefore

$$h_3 = 0.399390u_1 + 0.980691h_1 + (0.749543u_1 - 0.031671h_1)3.05$$

or

$$h_3 = 2.685497u_1 + 0.884095h_1$$

We now refract at the third surface with

$$n_3'u_3' - n_3u_3 = -h_3F_3$$

For the Le Grand eye, $n_3 = n_2' = 1.3374$, $n_3' = 1.42$, $C_3 = 1/10.2$ mm^{-1} and therefore

$$F_3 = (1.42 - 1.3374)/10.2 = +0.0080980$$

and so

$$1.42u_3' - 1.3374u_3 = -h_3(+0.0080980)$$

$$u_3' = (u_31.3374/1.42) - [h_3(+0.0080980)/1.42]$$

that is

$$u_3' = 0.941831u_3 - 0.0057028h_3$$

and since $u_3 = u_2'$ we have

$$u_3' = 0.941831(0.749543u_1 - 0.0316707h_1)$$

$$- 0.0057028(2.685497u_1 + 0.884095h_1)$$

Therefore

$$u_3' = +0.690628u_1 - 0.034870h_1$$

We now transfer to the fourth surface with the transfer equation

$$h_4 = h_3 + u_3'd_3$$

with $d_3 = 4.0$. Therefore

$$h_4 = 2.685497u_1 + 0.884095h_1 + (0.690628u_1 - 0.034870h_1)4.0$$

or

$$h_4 = 5.448010u_1 + 0.744614h_1$$

We now finally refract at the fourth surface with

$$n_4'u_4' - n_4u_4 = -h_4F_4$$

For the Le Grand eye, $n_4 = n_3' = 1.42$, $n_4' = 1.336$, $C_4 = -1/6$ mm^{-1} and therefore

$$F_4 = (1.336 - 1.42)/(-6) = +0.014$$

and so

$$1.336u_4' - 1.42u_4 = -h_4(+014)$$

$$u_4' = (u_41.42/1.336) - (h_40.014/1.336)$$

that is

$$u_4' = 1.062874u_4 - 0.010479h_4$$

Because $u_4 = u_3'$ we have

$$u_4' = 1.062874(0.690628u_1 - 0.034870h_1)$$

$$- 0.010479(5.448010u_1 + 0.744614h_1)$$

Therefore

$$u_4' = +0.676961u_1 - 0.044866h_1$$

In summary we have

$$u_4' = 0.676961u_1 - 0.044866h_1$$

$$h_4 = 5.448010u_1 + 0.744614h_1$$

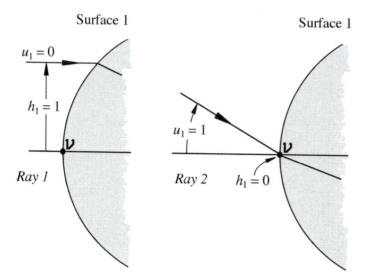

Fig. A1.3: The two rays used to find the values of the coefficients $\mathfrak{A}$, $\mathfrak{B}$, $\mathfrak{C}$ and $\mathfrak{D}$.

and comparing this with equation (A1.25) with $k = 4$, we have

$$\mathfrak{A} = 0.676961, \quad \mathfrak{B} = -0.044866, \quad \mathfrak{C} = 5.448010 \text{ and}$$

$$\mathfrak{D} = 0.744614$$

which, apart from the last digit in $\mathfrak{B}$, agree with the values in Example A1.1.

This numerical example demonstrates the linearity rule described by equations (A1.31).

A1.5.3   *By tracing two distinct rays*

If we look at equations (A1.30), we can find the values of $\mathfrak{A}$, $\mathfrak{B}$, $\mathfrak{C}$ and $\mathfrak{D}$ by tracing any two specific and distinct rays. It is most convenient to trace rays from left to right through any optical system and the most convenient pair of rays are (Ray 1: $u_1 = 0, h_1 = 1$) and (Ray 2: $u_1 = 1, h_1 = 0$), as shown in Figure A1.3.

Ray 1 ($u_1 = 0, h_1 = 1$) will give us final image space values u'_k and h_k and from equation (A1.30)

$$u'_k = \mathfrak{A} \times 0 + \mathfrak{B} \times 1$$

$$h_k = \mathfrak{C} \times 0 + \mathfrak{D} \times 1$$

Therefore for this ray

$$\mathfrak{B} = u'_k \quad \text{and} \quad \mathfrak{D} = h_k$$

For ray 2, $u = 1$ and $h = 0$ and equation (A1.30) gives

$$u'_k = \mathbb{A} \times 1 + \mathbb{B} \times 0$$

$$h_k = \mathbb{C} \times 1 + \mathbb{D} \times 0$$

and so for this ray

$$\mathbb{A} = u'_k \quad \text{and} \quad \mathbb{C} = h_k$$

Thus in summary

$$\mathbb{B} = u'_k \quad \text{and} \quad \mathbb{D} = h_k \qquad \text{from ray 1} \qquad\qquad \text{(A1.32a)}$$

$$\mathbb{A} = u'_k \quad \text{and} \quad \mathbb{C} = h_k \qquad \text{from ray 2} \qquad\qquad \text{(A1.32b)}$$

Example A1.3: Let us apply this method to the Le Grand eye used in Examples A1.1 and A1.2.

Table A1.2 contains the ray trace results for the two rays traced through the Le Grand eye. The data of this eye are given in Appendix 3.

Now for ray 1, $u_1 = 0, h_1 = 1$, and from the ray trace results given in Table A1.2, $u'_4 = -0.044866$ and $h_4 = 0.744614$. Thus from equation (A1.32), we have

$$\mathbb{B} = -0.044865$$

and

$$\mathbb{D} = 0.744614$$

For ray 2, $u_1 = 1, h_1 = 0, u'_4 = 0.676961$ and $h_4 = 5.448010$ and so similarly, from equation (A1.32) we have

$$\mathbb{A} = 0.676961$$

and

$$\mathbb{C} = 5.448010$$

The values of $\mathbb{A}, \mathbb{B}, \mathbb{C}$ and $\mathbb{D}$ are identified in Table A1.2 and are consistent with the values given in Examples A1.1 and A1.2.

In the next section we will show how the values of $\mathbb{A}, \mathbb{B}, \mathbb{C}$ and $\mathbb{D}$ can be used to find some of the Gaussian properties of an optical system, in particular the positions of the cardinal points.

A1.6 Determining the positions of the cardinal points

Equations (A1.30) are very useful for finding the positions of the six cardinal points. However in particular cases we would need to know the numerical values

Table A1.2. *The ray trace results for the Le Grand theoretical relaxed schematic eye and used in Examples A1.3 and A1.4*

Ray 1		Surface	Ray 2	
u	h		u	h
0.000000			1.000000	
	1.000000	1		0.000000
−0.035107			0.726164	
	0.980691	2		0.399390
−0.031671			0.749543	
	0.884095	3		2.685498
−0.034870			0.690628	
	0.744614(= $\mathbb{D}$)	4		5.448010(= $\mathbb{C}$)
−0.044866(= $\mathbb{B}$)			0.676961(= $\mathbb{A}$)	

of $\mathbb{A}$, $\mathbb{B}$, $\mathbb{C}$ and $\mathbb{D}$. Methods for calculating their values were explained in the previous section. In this section, we will derive equations for the positions of the six cardinal points in terms of the values of $\mathbb{A}$, $\mathbb{B}$, $\mathbb{C}$ and $\mathbb{D}$ and then look at a particular example.

A1.6.1 Back principal and focal points

Normally, the back principal and focal points are found by tracing a ray from infinity on the left to image space on the right. This ray has an object space angle $u_1 = 0$ but may have any initial ray height h_1. Substituting this value for u_1 into equations (A1.30) we have

$$u'_k = \mathbb{B}h_1 \tag{A1.33a}$$

and

$$h_k = \mathbb{D}h_1 \tag{A1.33b}$$

Now by definition, the distance $\mathscr{P}'\mathscr{F}'$ between the back principal point $\mathscr{P}'$ and back focal point $\mathscr{F}'$ is

$$\mathscr{P}'\mathscr{F}' = -h_1/u'_k \tag{A1.34}$$

On dividing both sides of equation (A1.33a) by u'_k, we have

$$1 = \mathbb{B}(h_1/u'_k)$$

That is

$$h_1/u'_k = 1/\mathbb{B} \tag{A1.35}$$

and recalling equation (A1.34) we have

$$\wp'\mathcal{F} = -1/\mathbb{B} \tag{A1.36}$$

Similarly by definition, the distance between the back vertex $\mathcal{V}'$ and the back focal point $\mathcal{F}'$ is

$$\mathcal{V}'\mathcal{F} = -h_k/u'_k \tag{A1.37}$$

If we now divide equation (A1.33b) by equation (A1.33a), we have

$$h_k/u'_k = \mathbb{D}/\mathbb{B}$$

Therefore it follows from equation (A1.37) that

$$\mathcal{V}'\mathcal{F} = -\mathbb{D}/\mathbb{B} \tag{A1.38}$$

Now the distance $\mathcal{V}'\wp'$ between the back surface vertex and the back principal point can also be expressed in terms of these coefficients. If we use the relation

$$\mathcal{V}'\wp' = \mathcal{V}'\mathcal{F}' - \wp'\mathcal{F}'$$

then it readily follows from equations (A1.36) and (A1.38) that

$$\mathcal{V}'\wp' = (1 - \mathbb{D})/\mathbb{B} \tag{A1.39}$$

A1.6.2   Front focal and principal points

The front focal and principal points can be found by tracing a ray from infinity on the right to the object space on left. For this ray $u'_k = 0$ and h_k can have any value. Using this ray and the same line of reasoning as before, we can show that

$$\mathcal{V}\mathcal{F} = +\mathbb{A}/\mathbb{B} \tag{A1.40}$$

$$\wp\mathcal{F} = -[\mathbb{C} - (\mathbb{D}\mathbb{A}/\mathbb{B})] \tag{A1.41}$$

and

$$\mathcal{V}\wp = \mathbb{C} + (1 - \mathbb{D})\mathbb{A}/\mathbb{B} \tag{A1.42}$$

A1.6.3   Nodal points

The positions of the nodal points $\mathcal{N}$ and $\mathcal{N}'$ can also be found in terms of the $\mathbb{A}$, $\mathbb{B}$, $\mathbb{C}$ and $\mathbb{D}$ coefficients. The nodal points (one in object space and the other in image space) are defined as points on the optical axis such that any ray passing through them is inclined at the same angle to the axis in both object and image space. Let this angle be u; then

$$u_1 = u'_k = u$$

Thus equation (A1.30) can be rewritten

$$u = \mathfrak{A}u + \mathfrak{B}h_1 \qquad\qquad\qquad (A1.43a)$$

$$h_k = \mathfrak{C}u + \mathfrak{D}h_1 \qquad\qquad\qquad (A1.43b)$$

The distance of the front nodal point $\mathcal{N}$ from the front vertex $\mathcal{V}$ is given by the equation

$$\mathcal{VN} = -h_1/u \qquad\qquad\qquad (A1.44)$$

and for the back nodal point

$$\mathcal{V'N'} = -h_k/u \qquad\qquad\qquad (A1.45)$$

Now returning to equations (A1.43) and dividing both sides by u, we have

$$1 = \mathfrak{A} + \mathfrak{B}(h_1/u)$$

and

$$h_k/u = \mathfrak{C} + \mathfrak{D}(h_1/u)$$

That is, firstly

$$h_1/u = (1 - \mathfrak{A})/\mathfrak{B}$$

It then easily follows that

$$\mathcal{VN} = (\mathfrak{A} - 1)/\mathfrak{B} \qquad\qquad\qquad (A1.46a)$$

and

$$\mathcal{V'N'} = [\mathfrak{D}(\mathfrak{A} - 1)/\mathfrak{B}] - \mathfrak{C} \qquad\qquad\qquad (A1.46b)$$

A1.6.4   Summary

Thus in summary, in terms of the four quantities $\mathfrak{A}$, $\mathfrak{B}$, $\mathfrak{C}$ and $\mathfrak{D}$, the positions of the six cardinal points can be found from the following equations:

$$\mathcal{VF} = \mathfrak{A}/\mathfrak{B}$$

$$\mathcal{PF} = \mathfrak{D}\mathfrak{A}/\mathfrak{B} - \mathfrak{C}$$

$$\mathcal{VP} = \mathfrak{C} + [(1 - \mathfrak{D})\mathfrak{A}/\mathfrak{B}]$$

$$\mathcal{VN} = (\mathfrak{A} - 1)/\mathfrak{B} \qquad\qquad\qquad (A1.47)$$

$$\mathcal{V'F'} = -\mathfrak{D}/\mathfrak{B}$$

$$\mathscr{P}'\mathscr{F}' = -1/\mathfrak{B}$$

$$\mathscr{V}'\mathscr{P}' = (1 - \mathfrak{D})/\mathfrak{B}$$

$$\mathscr{V}'\mathscr{N}' = [\mathfrak{D}(\mathfrak{A} - 1)/\mathfrak{B}] - \mathfrak{C}$$

Let us now use these equations to determine the positions of the six cardinal points of the Le Grand full theoretical schematic eye.

> **Example A1.4:** Using the $\mathfrak{A}$, $\mathfrak{B}$, $\mathfrak{C}$ and $\mathfrak{D}$ coefficients, find the positions of the six cardinal points of the Le Grand full relaxed schematic eye. The values of $\mathfrak{A}$, $\mathfrak{B}$, $\mathfrak{C}$ and $\mathfrak{D}$ are given in Table A1.2. From equations (A1.47)

$$
\begin{aligned}
\mathscr{V}\mathscr{F} &= \mathfrak{A}/\mathfrak{B} = 0.676961/(-0.044866) &&= -15.09 \text{ mm} \\
\mathscr{V}\mathscr{P} &= \mathfrak{C} + [(1 - \mathfrak{D})\mathfrak{A}/\mathfrak{B}] = 5.448010 \\
&\quad + [(1 - 0.744614) \\
&\quad \times 0.676961/(-0.044866)] &&= -1.59 \text{ mm} \\
\mathscr{V}\mathscr{N} &= (\mathfrak{A} - 1)/\mathfrak{B} \\
&= (0.676961 - 1)/(-0.044866) &&= 7.20 \text{ mm} \\
\mathscr{V}'\mathscr{F}' &= -\mathfrak{D}/\mathfrak{B} = -0.744614/(-0.044866) &&= 16.60 \text{ mm} \\
\mathscr{V}'\mathscr{P}' &= (1 - \mathfrak{D})/\mathfrak{B} \\
&= (1 - 0.744614)/(-0.044866) &&= -5.69 \text{ mm} \\
\mathscr{V}'\mathscr{N}' &= [\mathfrak{D}(\mathfrak{A} - 1)/\mathfrak{B}] - \mathfrak{C} \\
&= [0.744614(0.676961 - 1)/(-0.044866)] \\
&\quad - 5.448010 &&= -0.09 \text{ mm}
\end{aligned}
$$

A1.6.5 *Minimum data needed to locate the cardinal points*

We have seen from the above discussion that the cardinal points can be found given the numerical values of the $\mathfrak{A}$, $\mathfrak{B}$, $\mathfrak{C}$ and $\mathfrak{D}$ coefficients. In Section A1.5, we have shown how to calculate their values from the system construction data by matrix methods or by ray tracing one or two rays through the system.

However, let us suppose that we have a finished or constructed system and we do not know its constructional parameters. Without this data we cannot use the matrix or ray trace methods. But using laboratory techniques, we can find the object and image positions of pairs of conjugate planes and the respective magnifications. Let us use this information to determine the $\mathfrak{A}$, $\mathfrak{B}$, $\mathfrak{C}$ and $\mathfrak{D}$ coefficients and we can then use their values to determine the positions of the cardinal points using equations (A1.47).

To find the values of the four quantities $\mathfrak{A}$, $\mathfrak{B}$, $\mathfrak{C}$ and $\mathfrak{D}$ we need four simultaneous equations. Let us suppose that we measure the object distance l_v and image distance l'_v from the front and back vertex planes in the system, for two pairs of conjugate planes denoted by the subscripts A and B. Starting with equations (A1.30), written again here as

$$u'_k = \mathfrak{A}u_1 + \mathfrak{B}h_1 \qquad\qquad (A1.48a)$$

and

$$h_k = \mathfrak{C}u_1 + \mathfrak{D}h_1 \qquad\qquad (A1.48b)$$

we can write

$$\frac{u'_k}{h_k} = \frac{\mathfrak{A} + (\mathfrak{B}h_1/u_1)}{\mathfrak{C} + (\mathfrak{D}h_1/u_1)} \tag{A1.49}$$

Now

$$h_k/u'_k = -l'_\mathrm{v} \quad \text{and} \quad h_1/u_1 = -l_\mathrm{v}$$

Therefore for the two pairs of conjugate planes, we can write equation (A1.49) as

$$\mathfrak{A} - l_{\mathrm{vA}}\mathfrak{B} + l'_{\mathrm{vA}}\mathfrak{C} - l_{\mathrm{vA}}l'_{\mathrm{vA}}\mathfrak{D} = 0 \tag{A1.50a}$$

and

$$\mathfrak{A} - l_{\mathrm{vB}}\mathfrak{B} + l'_{\mathrm{vB}}\mathfrak{C} - l_{\mathrm{vB}}l'_{\mathrm{vB}}\mathfrak{D} = 0 \tag{A1.50b}$$

To find the values of $\mathfrak{A}$, $\mathfrak{B}$, $\mathfrak{C}$ and $\mathfrak{D}$, we need two more equations. However, we know that these quantities are not independent and are connected by equation (A1.22a), namely

$$\mathfrak{A}\mathfrak{D} - \mathfrak{B}\mathfrak{C} = n_1/n'_k \tag{A1.50c}$$

If we know the transverse magnification M of one of these pairs of conjugate planes, we can use equation (A1.48a) in the form

$$u'_k/u_1 = \mathfrak{A} + (\mathfrak{B}h_1/u_1)$$

where

$$M = (n_1 u_1)/(n'_k u'_k)$$

Let us assume we know the magnification M_A for conjugate plane A, then we now have our fourth equation

$$(\mathfrak{A} - \mathfrak{B}l_{\mathrm{vA}}) = n_1/(n'_k M_\mathrm{A}) \tag{A1.50d}$$

Now we have four independent equations and therefore we can solve for the four quantities $\mathfrak{A}$, $\mathfrak{B}$, $\mathfrak{C}$ and $\mathfrak{D}$.

This proves that if the positions of two pairs of conjugate planes and the transverse magnification of one pair are known, then the positions of the cardinal points can be determined.

A1.6.5.1   *Significance to pupil imagery*

Thus it follows from the above results that once the positions of the object and image planes, their magnification and the position of the entrance and exit pupil planes are set, then the magnification of the pupils is set and cannot be used as an independent variable in a design.

A1.7 Some interesting properties of afocal systems

We will now look at the application of this matrix analysis of optical systems to the special case of afocal systems.

Afocal systems are defined as those with zero equivalent power. Thus they have no cardinal points and the general system paraxial refraction equation cannot be applied to an afocal system as a whole. However the processes of refraction and transfer, surface by surface, can be applied; therefore the matrix analysis described in Section A1.3 is very useful for afocal systems and thus we can write the overall refraction of an afocal system in the matrix form given by equation (A1.15a) or (A1.23a). Here we will use the former, that is for the overall system, and we will use equation (A1.15b), that is

$$\begin{bmatrix} n'_k u'_k \\ h_k \end{bmatrix} = \begin{bmatrix} A & B \\ C & D \end{bmatrix} \begin{bmatrix} n_1 u_1 \\ h_1 \end{bmatrix} \tag{A1.51}$$

If the object is at infinity for an afocal system, the image must also be at infinity and if we trace a marginal ray, both u_1 and u'_k must be zero for any value of initial ray height h_1. It then follows from equation (A1.51) that the matrix element B must be zero and equation (A1.51) reduces to

$$\begin{bmatrix} n'_k u'_k \\ h_k \end{bmatrix} = \begin{bmatrix} A & 0 \\ C & D \end{bmatrix} \begin{bmatrix} n_1 u_1 \\ h_1 \end{bmatrix} \tag{A1.52}$$

While we have taken the special case of a marginal ray to show that the B matrix element is zero, this matrix equation applies to any ray.

Suppose now that we trace a pupil ray, that is now regard the angles u'_k and u_1 as pupil ray angles, then we can define the angular magnification M as follows.

$$M = u'_k / u_1$$

It then follows firstly that the matrix element A of the expanded form of (A1.51) has the value

$$A = n'_k M / n_1$$

and secondly, since the system matrix $\mathcal{S}$ must have a unit determinant, that

$$D = 1/A = n_1/(n'_k M)$$

Therefore, the above matrix equation can now be written in the form

$$\begin{bmatrix} n'_k u'_k \\ h_k \end{bmatrix} = \begin{bmatrix} n'_k M / n_1 & 0 \\ C & n_1/(n'_k M) \end{bmatrix} \begin{bmatrix} n_1 u_1 \\ h_1 \end{bmatrix} \tag{A1.53}$$

Now the matrix element C remains the only unknown, but could be found by tracing a ray with initial height $h_1 = 0$ and a non-zero angle angle u_1. If the final ray height h_k is recorded, then it follows from equation (A1.53) that

$$C = h_k/(n_1 u_1) \qquad \text{for } h_1 = 0 \tag{A1.54}$$

As an exercise let us use this matrix development and the above equations to look at the case of an afocal system imaging an object at a finite distance.

A1.7.1 *Conjugates at a finite distance*

Let the object be at a finite distance, say a distance l_v to the left of the first surface. We can now transfer to this surface by inserting the transfer matrix

$$\mathbb{T}_{0,1} = \begin{bmatrix} 1 & 0 \\ -l_v/n_1 & 1 \end{bmatrix}$$

after the system matrix in equation (A1.53). We must also transfer the ray from the last surface to the image plane by a similar matrix placed before the system matrix. Let the distance between the last surface and the image plane be l'_v. The system matrix in equation (A1.53) now has to be modified to

$$\mathbb{S} = \begin{bmatrix} 1 & 0 \\ l'_v/n'_k & 1 \end{bmatrix} \begin{bmatrix} n'_k M/n_1 & 0 \\ C & n_1/(n'_k M) \end{bmatrix} \begin{bmatrix} 1 & 0 \\ -l_v/n_1 & 1 \end{bmatrix}$$

and this can be multiplied out to give

$$\begin{bmatrix} n'_k M/n_1 & 0 \\ (l'_v M/n_1) + C - [l_v/(n'_k M)] & n_1/(n'_k M) \end{bmatrix} \tag{A1.55}$$

Now if we regard the ray being transferred as the marginal ray, then the ray heights at the object and image planes are both zero. Thus we now have

$$\begin{bmatrix} n'_k u'_k \\ 0 \end{bmatrix} = \begin{bmatrix} n'_k M/n_1 & 0 \\ (l'_v M/n_1) + C - [l_v/(n'_k M)] & n_1/(n'_k M) \end{bmatrix} \begin{bmatrix} n_1 u_1 \\ 0 \end{bmatrix}$$

$$\tag{A1.56}$$

Multiplying out this matrix equation, we have

$$n'_k u'_k = (n'_k M/n_1) n_1 u_1$$

and

$$0 = \{(l'_v M/n_1) + C - [l_v/(n'_k M)]\} n_1 u_1$$

Both of these equations reveal interesting special properties of afocal systems. The first can be written as

$$\frac{n_1 u_1}{n'_k u'_k} = \frac{n_1}{n'_k M}$$

The lefthand term is the transverse magnification for the pair of conjugate planes at a finite distance and the righthand side is a constant. Therefore this equation shows that the transverse magnification for conjugate planes at a finite distance is independent of position of the planes. This result is a more general form of equation (17.9) and here allows the refractive indices of object and image to be other than unity.

If we now take this second equation, the lefthand side is zero and u_1 must have a non-zero value since it is the marginal ray. Therefore the term in the brackets is zero. That is

$$(l'_v M / n_1) + C - [l_v / (n'_k M)] = 0 \tag{A1.57}$$

Manipulating this equation and replacing the distances l_v and l'_v by their respective vergences, we have

$$L'_v = \frac{n'^2_k M^2 L_v}{n_1 (n_1 - C n'_k M L_v)} \tag{A1.58}$$

This equation gives the vergence L'_v of an image measured from the eye lens for an afocal system when the image has a vergence L_v measured from the objective. A similar equation [equation (17.6)] was given in Chapter 17 for the specific case of a two lens afocal system. This specific case can be deduced here as follows.

Now if we consider the case of a two thin lens afocal system, returning to equation (A1.52), we have

$$h_2 = C n_1 u_1 + D h_1$$

If we now trace a ray with an angle u_1 and a height h_1 of zero at the first component of our two component system, then

$$h_2 = C n_1 u_1 \tag{A1.59}$$

Let us now trace this ray through the first component. At this component, the paraxial refraction equation is

$$u'_1 u'_1 - n_1 u_1 = 0$$

Transferring to the second component we have

$$h_2 = d u'_1$$

That is

$$h_2 = d n_1 u_1 / n'_1 \tag{A1.60}$$

If we compare equations (A1.59) and (A1.60) then

$$C = d / n'_1 = d / \mu$$

where $n'_1 = \mu$, the refractive index of the medium separating the two lenses. Thus for the special case of a two lens afocal system, equation (A1.58) reduces to

$$L'_v = \frac{n'^2_k M^2 L_v}{n_1 (n_1 - n'_k M d L_v / \mu)} \tag{A1.61}$$

which is an alternative and elegant derivation to the proof of equation (17.6), except for the more general form of the refractive indices for object and image space.

Summary of main symbols

Section A1.2: The equivalent power of a component

u_1, h_1 any paraxial ray
u_2, h_2 a second and distinct paraxial ray

Section A1.6: Determining the Gaussian properties

u_1, h_1 paraxial ray angle and height in object space
u'_k, h_k corresponding angle and height in image space
$\mathfrak{A}, \mathfrak{B}, \mathfrak{C}$ and $\mathfrak{D}$ co-efficients

Appendix 2

Alternative ray trace procedures

A2.0   Introduction

While the paraxial refraction equation (2.14) and the transfer equation (2.17) are a very useful pair of equations for ray tracing through general optical systems, other schemes exist and two such schemes are based upon the "lens" equation (2.11): one using distances and the other using vergences.

A2.1   Ray tracing using distances

To find the image position in Example 2.3, the "lens" equation (2.11) could have been used instead of the paraxial refraction equation, and this would have been the quicker method. Let us extend the lens equation to ray tracing through a system of surfaces.

Let us look at the imaging performed by surface j, as shown in Figure A2.1. The image produced by the preceding surface is at a distance l_j from surface j. This surface re-images this image at a distance l'_j, where from equation (2.11),

$$l'_j = \frac{n'_j l_j}{(n_j + l_j F_j)} \tag{A2.1}$$

Before we can apply this equation to the next surface, we need to determine the new object distance l_{j+1} from the next surface. From the diagram, it follows that this distance is given by the equation

$$l_{j+1} = l'_j - d_j \tag{A2.2}$$

where $d_j =$ the distance between surface j and surface $j+1$. Equations (A2.1) and (A2.2) are analogous to the conventional paraxial refraction and transfer equations, respectively, and repeated use of these equations allows us to locate the position of the image plane for a system of any complexity.

If we need to find the magnification for that pair of conjugate planes, we can apply equations (3.46) and (3.49) at each surface, that is

$$M_j = \frac{\eta'_j}{\eta_j} = \frac{n_j l'_j}{n'_j l_j} \tag{A2.3}$$

where the subscript j has been used to make it clear that this is the magnification of one surface, surface j. Now for a system of k surfaces, the overall magnification is the product of the magnifications of the individual surfaces.

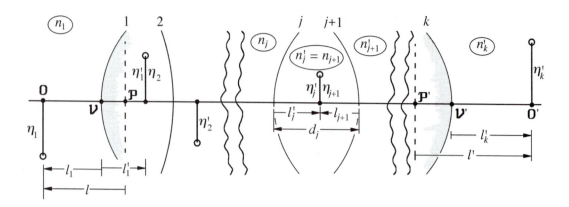

Fig. A2.1:   The general
ray tracing with distances
instead of angles

Therefore since

$$n'_j = n_{j+1}$$

we have

$$M = \frac{n_1 l'_1 l'_2 l'_3 \cdots l'_k}{n'_k l_1 l_2 l_3 \cdots l_k} \qquad (A2.4)$$

If we trace a ray from infinity ($l_1 = \infty$), we can find the equivalent power. To show how this can be done, we recall equation (3.50b) and write this in the form

$$F = (1/M - 1)(n_1/l)$$

for any object distance l, which is the distance from the front principal plane to the object. If we now substitute for M using equation (A2.4), we have

$$F = \frac{n'_k}{n_1} \frac{l_1 l_2 l_3 \cdots l_k}{l'_1 l'_2 l'_3 \cdots l'_k} \frac{n_1}{l} - \frac{n_1}{l}$$

Now the distance l_1 is the distance from the first surface to the object and using Figure A2.1, we can see that this distance is connected to the distance l by the equation

$$l_1 = l + \mathcal{VP}$$

In the limit

$$l(\text{and } l_1) \Rightarrow \infty, \qquad l = l_1$$

and therefore the above expression for F reduces to

$$F = \frac{l_2 l_3 \cdots l_k}{l'_1 l'_2 l'_3 \cdots l'_k} n'_k \qquad (A2.5)$$

A2.2   Ray tracing using vergences

We can also trace rays using the vergence form of the lens equation, that is equation (3.63), but applying this to the jth surface we have

$$L'_j = F_j + L_j \tag{A2.6}$$

where we define the vergences L_j and L'_j by equations (3.62a), that is

$$L_j = n_j/l_j \quad \text{and} \quad L'_j = n'_j/l'_j \tag{A2.7}$$

The corresponding transfer equation (A2.2) becomes

$$L_{j+1} = L'_j/[1 - L'_j(d_j/n_j)] \tag{A2.8}$$

and equations (A2.6) and (A2.8) are used repeatedly through the k surfaces.

It readily follows that the magnification equation (A2.4) can be expressed in terms of vergences as

$$M = \frac{L_1 L_2 \cdots L_k}{L'_1 L'_2 \cdots L'_k} \tag{A2.9}$$

and that from equation (A2.5), the equivalent power equation is

$$F = \frac{L'_1 L'_2 \cdots L'_k}{L_2 \cdots L_k} \tag{A2.10}$$

with $L_1 = 0$.

A2.3   Comparisons with conventional paraxial ray tracing

While the equations just described are popular in ophthalmic optics, they have serious limitations for general use, because they become indeterminate if any intermediate values of l are zero or infinite. Furthermore, in many ray tracing problems, it is often important to know the ray heights at each surface and these are only given by the full paraxial ray trace using the paraxial refraction equation (2.14) and the transfer equation (2.17).

Appendix 3

Schematic eyes

A3.0   Introduction

This appendix lists the constructional data, the Gaussian constants such as the pupil positions and sizes, and the positions of the cardinal points of the most common paraxial schematic eyes. All eyes have an entrance pupil diameter of 8 mm.

The Seidel aberrations have also been evaluated for the Gullstrand number 1 schematic eye and the Emsley reduced eye, using an 8 mm entrance pupil diameter, a semi-field angle of $5°$ and a reference wavelength of 555 nm. The V-values are estimated from the Thibos et al. (1992) data given in Table 35.3. The $5°$ angle was chosen to simulate the effect of the distance of the fovea from the optical axis. With a field angle of $5°$ and a pupil diameter of 8 mm, the optical invariant H in all cases is

$$H = 0.349955$$

A3.0.1   Units

Distances are, unless otherwise specified, in millimetres and measured from the front surface vertex of the cornea.

Powers are expressed in units of dioptres (m^{-1}).

Accommodation levels are measured at the anterior corneal surface vertex.

Aberrations: The Seidel aberrations S_1, S_2, S_3, S_4, C_L and the primary wave aberration coefficients are given in units of wavelengths ($\lambda = 555$ nm), except S_5 and C_T, which are given as percentages.

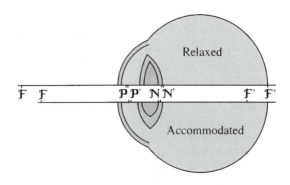

Fig. A3.1: The Gullstrand
number 1 schematic eye,
showing the position of
cardinal points for relaxed
(above) and accommodated
(below) versions.

A3.1   Paraxial schematic eyes

A3.1.1   *Eyes and source of data*

Eye	Accommodation level (m^{-1})	Source of data[a]	Figure
Gullstrand number 1			
relaxed	0	1,2	A3.1
accommodated	10.878	1	A3.1
Le Grand (theoretical)			
relaxed	0	3	A3.2
accommodated	7.053	3	A3.2
Le Grand (simplified)			
relaxed	0	3	
accommodated	7.034	3	
Gullstrand number 2[b]			
relaxed	0	2	A3.3
accommodated	8.599	2	A3.3
Emsley reduced	0	4	A3.4

[a] The constructional data were taken from published references as follows, but the Gaussian and aberration values were determined by the authors.

(1)   A. Gullstrand in Southall (1924, 392).
(2)   H.H. Emsley (1952) Vol.1, 346.
(3)   Y. Le Grand and S.G. Hage (1980, 65–66).
(4)   H.H. Emsley (1952) Vol. 1, 40, 41.

[b] This eye is a modified version of the original as published by Gullstrand (see Southall 1924).

Gullstrand number 1 eye: Relaxed

Please refer to Figure A3.1 for cardinal point positions.

Six surfaces; aperture stop (IRIS) at surface 3.

Surface	*n*	*r*	*d*	*Σd*	*V*	
	1.000					
1		7.700				
	1.376		0.500	0.500	50.23	cornea
2		6.800				
	1.336		3.100	3.600	50.23	aqueous
3		10.000				
	1.386		0.546	4.146	50.23	lens: cortex
4		7.911				
	1.406		2.419	6.565	50.23	lens: core
5		−5.760				
	1.386		0.635	7.200	50.23	lens: cortex
6		−6.000				
	1.336					vitreous

Equivalent powers

Cornea:	front surface	48.831
	back surface	−5.882
	complete	43.053
Lens cortex:	front surface	5.000
Lens core:	front surface	2.528
	back surface	3.472
Lens cortex:	back surface	8.333
Lens:	complete	19.111
Eye:	complete	58.636

Eye length   24.385

Cardinal point positions

	Front	Back
Focal points	−15.706	24.385
Principal points	1.348	1.601
Nodal points	7.078	7.331

$$\mathcal{FP} = \mathcal{N'R'} = \mathcal{N'F} = 17.054 \qquad \mathcal{P'F'} = \mathcal{P'R'} = \mathcal{P'O'} = 22.785$$

Pupils

	Entrance	Exit
Positions	3.047	3.665
Radii	4.000	3.638
Magnifications	1.133	1.031

Marginal and pupil rays

	u	h	$\bar{u}$	$\bar{h}$	
	0.000000		0.087489		
1		4.000000		−0.266593	1
	−0.141951		0.073043		
2		3.929024		−0.230071	2
	−0.128902		0.074217		
3		3.529429		0.000000	3
	−0.136984		0.071539		
4		3.454636		0.039060	4
	−0.141247		0.070451		
5		3.112958		0.209483	5
	−0.151084		0.070943		
6		3.017020		0.254532	6
	−0.175557		0.072011		

Gaussian properties

	Object space	Image space	
N.A.	0.000	0.2345	
νo	∞	17.185	$\mathcal{P}'o' = 22.785$
η'	∞	1.492	

Seidel aberrations: by surface

	$S_1/8$	$S_2/2$	$S_3/4$	$S_4/4$	$\%S_5$	$C_L/2$	$\%C_T$
1	25.0805	10.2096	0.5195	1.9577	−0.0800	14.0130	−0.4523
2	−2.1257	−0.7649	−0.0344	−0.1765	0.0060	−1.6905	0.0482
3	0.1674	0.2218	0.0367	0.1490	−0.0195	0.9474	−0.0995
4	0.2183	0.2229	0.0284	0.0716	−0.0081	0.5148	−0.0417
5	5.5054	−1.1009	0.0275	0.0983	0.0020	1.0703	0.0170
6	12.5019	−2.1811	0.0476	0.2483	0.0041	2.4522	0.0339
Sum	41.3473	6.6073	0.6254	2.3483	−0.0955	17.3071	−0.4944

Primary aberration coefficients

$_0w_{4,0} = 41.347$ $\qquad$ $_1w_{3,1} = 6.607$ $\qquad$ $_2w_{2,0} = 2.974$ $\qquad$ $_2w_{2,2} = 1.251$

Image surface radii of curvature
Sagittal surface   −13.886
Tangential surface   −9.774
Petzval surface   −17.584

Purkinje images

Surface	Relative size	Distance from corneal pole	Relative brightness
1	1.000	3.850	1.0000
2	0.882	3.765	0.00826
3	1.967	10.620	0.0128
6	−0.760	3.979	0.0128

Gullstrand number 1 eye: Accommodated

Please refer to Figure A3.1 for cardinal point positions.

Six surfaces; aperture stop (IRIS) at surface 3.

Surface	n	r	d	Σd	V	
	1.0000					
1		7.700				
	1.3760		0.5000	0.5000	50.23	cornea
2		6.800				
	1.3360		2.7000	3.2000	50.23	aqueous
3		5.333				
	1.3860		0.6725	3.8725	50.23	lens: cortex
4		2.655				
	1.4060		2.6550	6.5275	50.23	lens: core
5		−2.655				
	1.3860		0.6725	7.2000	50.23	lens: cortex
6		−5.333				
	1.3360				50.23	vitreous

Note: Radius of 5.333 was given as 5.3̇ by Gullstrand.

Equivalent powers

Cornea:	front surface	48.831
	back surface	−5.882
	complete	43.053
Lens cortex:	front surface	9.376
Lens core:	front surface	7.533
	back surface	7.533
Lens cortex:	back surface	9.376
Lens:	complete	33.057
Eye:	complete	70.576

Cardinal point positions

	Front	Back
Focal points	−12.397	21.016
Principal points	1.772	2.086
Nodal points	6.533	6.847

$\mathcal{F}'\mathcal{R}' = 3.370 \qquad \mathcal{N}'\mathcal{R}' = 17.539 \qquad \mathcal{P}'\mathcal{R}' = \mathcal{P}'o' = 22.300$

$\mathcal{P}\,\mathcal{F} = -14.169 \qquad \mathcal{P}'\mathcal{F}' = 18.930$

Near point position $= -92.00$ mm or 10.870 m^{-1} of accommodation.

Pupils

	Entrance	Exit
Positions	2.668	3.212
Radii	4.000	3.762
Magnifications	1.177	1.051

Marginal and pupil rays

u		h		$\bar{u}$		$\bar{h}$	
+0.042255				0.087489			
	1	3.887259				−0.233432	1
−0.107242				0.071866			
	2	3.833638				−0.197499	2
−0.093573				0.073148			
	3	3.580991				−0.000000	3
−0.114421				0.070509			
	4	3.504043				+0.047417	4
−0.131567				0.069252			
	5	3.154732				+0.231282	5
−0.150612				0.068994			
	6	3.053445				+0.277680	6
−0.177677				0.069628			

Gaussian properties

	Object space	Image space	
N.A.	0.0423	0.2374	
$\mathcal{V}\,o$	−92.000	17.185	$\mathcal{P}'o' = 22.299$
η	−8.282	1.474	

Fig. A3.2: The Le Grand full theoretical schematic eye, showing the position of cardinal points for relaxed (above) and accommodated (below) versions.

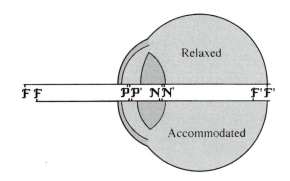

Seidel aberrations: by surface

	$S_1/8$	$S_2/2$	$S_3/4$	$S_4/4$	$\%S_5$	$C_L/2$	$\%C_T$
1	31.4963	13.1658	0.6879	1.9577	−0.0877	14.3419	−0.4754
2	−2.6909	−1.0096	−0.0473	−0.1765	0.0067	−1.7277	0.0514
3	6.0170	3.0464	0.1928	0.2793	−0.0190	2.4794	−0.0995
4	24.2749	7.1186	0.2609	0.2132	−0.0110	2.0999	−0.0488
5	36.9213	1.9985	0.0135	0.2132	−0.0010	2.0999	−0.0090
6	16.8060	−1.5734	0.0184	0.2793	0.0022	2.7446	0.0204
Sum	112.8247	22.7463	1.1263	2.7663	−0.1098	22.0380	−0.5610

Primary aberration coefficients
$$_0w_{4,0} = 112.825 \quad _1w_{3,1} = 22.746 \quad _2w_{2,0} = 3.893 \quad _2w_{2,2} = 2.253$$

Image surface radii of curvature
Sagittal surface   −10.608
Tangential surface   −6.719
Petzval surface   −14.927

Purkinje images

Surface	Relative size	Distance from corneal pole	Relative brightness
1	1.000	3.850	1.000
2	0.882	3.765	0.00826
3	0.800	5.750	0.0128
6	−0.702	4.313	0.0128

Le Grand full theoretical eye: Relaxed

Please refer to Figure A3.2 for cardinal point positions.

Four surfaces: aperture stop (IRIS) at surface 3.

Surface	n	r	d	Σd	
	1.0000				
1		7.800			
	1.3771		0.5500	0.5500	cornea
2		6.500			
	1.3374		3.0500	3.6000	aqueous
3		10.200			
	1.4200		4.0000	7.6000	lens
4		−6.000			
	1.3360				vitreous

Equivalent powers

Cornea:	front surface	48.346
	back surface	−6.108
	complete	42.356
Lens:	front surface	8.098
	back surface	14.000
	complete	21.779
Eye:	complete	59.940
Eye length		24.197

Cardinal point positions

	Front	Back
Focal points	−15.089	24.197
Principal points	1.595	1.908
Nodal points	7.200	7.513

$\mathcal{FP} = \mathcal{N'R'} = \mathcal{N'F} = 16.683$ $\mathcal{P'F} = \mathcal{P'R'} = \mathcal{P'O'} = 22.289$

Pupils

	Entrance	Exit
Positions	3.038	3.682
Radii	4.000	3.682
Magnifications	1.131	1.041

Le Grand full theoretical eye: Accommodated

Please refer to Figure A3.2 for cardinal point positions.

Four surfaces: aperture stop (IRIS) at surface 3.

Surface	n	r	d	Σd	
	1.0000				
1		7.800			
	1.3771		0.5500	0.5500	cornea
2		6.500			
	1.3374		2.6500	3.2000	aqueous
3		6.000			
	1.4270		4.5000	7.7000	lens
4		−5.500			
	1.3360				vitreous

Equivalent powers

Cornea:	front surface	48.346
	back surface	−6.108
	complete	42.356
Lens:	front surface	14.933
	back surface	16.545
	complete	30.700
Eye:	complete	67.677

Cardinal point positions

	Front	Back
Focal points	−12.957	21.932
Principal points	1.819	2.192
Nodal points	6.784	7.156

$$\mathcal{F}'\mathcal{R}' = 2.265 \qquad \mathcal{N}'\mathcal{R}' = 17.040 \qquad \mathcal{P}'\mathcal{R}' = \mathcal{P}'O' = 22.005$$

$$\mathcal{P}\mathcal{F} = -14.776 \qquad \mathcal{P}'\mathcal{F}' = 19.741$$

Near point position $= -141.793$ mm ≈ 7.053 m^{-1} of accommodation.

Pupils

	Entrance	Exit
Positions	2.660	3.255
Radii	4.000	3.785
Magnifications	1.115	1.055

Le Grand: Simplified relaxed

Four surfaces: aperture stop (IRIS) at surface 3.

Surface	n	r	d	Σd	
	1.0000				
1		8.000			
	1.3360		3.6000	3.6000	aqueous
2		0.000			
	1.3360		2.7700	6.3700	aqueous
3		10.200			
	1.4208		0.0000	6.3700	lens
4		−6.000			
	1.3360				vitreous

Note: In this eye, the lens has zero thickness.

Equivalent powers

Cornea:	complete	42.000
Lens:	front surface	8.314
	back surface	14.133
	complete	22.447
Eye:	complete	59.952
Eye length		24.192

Cardinal point positions

	Front	Back
Focal points	−14.895	24.192
Principal points	1.785	1.907
Nodal points	7.390	7.512
$\mathcal{FP} = \mathcal{N'R'} = \mathcal{N'F'} = 16.680$		$\mathcal{P'F'} = \mathcal{P'R'} = \mathcal{P'O'} = 22.285$

Pupils In this eye, the aperture stop or iris is placed so as to give the same entrance pupil positions as Le Grand's exact eye, i.e., the aperture stop is placed 3.60 mm from the cornea.

	Entrance	Exit
Positions	3.038	3.465
Radii	4.000	3.720
Magnifications	1.128	1.049

Le Grand: Simplified accommodated

Four surfaces: aperture stop (IRIS) at surface 3.

Surface	n	r	d	Σd	
	1.0000				
1		8.000			
	1.3360		3.2000	3.2000	aqueous
2		0.000			
	1.3360		2.5800	5.7800	aqueous
3		6.000			
	1.4260		0.0000	5.7800	lens
4		−5.500			
	1.3360				vitreous

Equivalent powers

Cornea:	complete	42.000
Lens:	front surface	15.000
	back surface	16.364
	complete	31.364
Eye:	complete	67.665

Eye length   24.192

Cardinal point positions

	Front	Back
Focal points	−12.773	21.937
Principal points	2.005	2.192
Nodal points	6.971	7.158

$\mathcal{F}'\mathcal{R}' = 2.255$ $\mathcal{N}'\mathcal{R}' = 17.034$ $\mathcal{P}'\mathcal{R}' = \mathcal{P}'\mathcal{O}' = 22.000$

$\mathcal{P}\,\mathcal{F} = -14.779$ $\mathcal{P}'\mathcal{F}' = 19.744$

Near point position $= -142.162$ mm ≈ 7.034 m^{-1} of accommodation.

Pupils

	Entrance	Exit
Positions	2.663	3.034
Radii	4.000	3.830
Magnifications	1.112	1.064

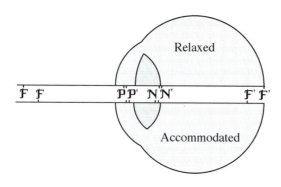

Fig. A3.3: The Gullstrand number 2 schematic eye, showing the position of cardinal points for relaxed (above) and accommodated (below) versions.

Gullstrand number 2 simplified eye: Relaxed

Please refer to Figure A3.3 for cardinal point positions.

Three surfaces: aperture stop (IRIS) at surface 2.

Surface	n	r	d	Σd	
	1.0000				
1		7.800			
	4/3		3.6000	3.6000	aqueous
2		10.000			
	1.4160		3.6000	7.2000	lens
3		−6.000			
	4/3				vitreous

Equivalent powers

Cornea:	complete	42.735
Lens:	front surface	8.267
	back surface	13.778
	complete	21.755
Eye:	complete	60.483

Eye length   23.896

Cardinal point positions

	Front	Back
Focal points	−14.983	23.896
Principal points	1.550	1.851
Nodal points	7.062	7.363

$\mathcal{FP} = \mathcal{N'R'} = \mathcal{N'F} = 16.534$ $\mathcal{P'F} = \mathcal{P'R'} = \mathcal{P'O'} = 22.045$

Pupils

	Entrance	Exit
Positions	3.052	3.687
Radii	4.000	3.667
Magnifications	1.130	1.036

Gullstrand number 2 simplified eye: Accommodated

Please refer to Figure A3.3 for cardinal point positions.

Three surfaces: aperture stop (IRIS) at surface 3.

Surface	n	r	d	Σd	
	1.0000				
1		7.800			
	4/3		3.2000	3.2000	aqueous
2		5.000			
	1.4160		4.0000	7.2000	lens
3		−5.000			
	4/3				vitreous

Equivalent powers

Cornea:	complete	42.735
Lens:	front surface	16.533
	back surface	16.533
	complete	32.295
Eye:	complete	69.721

Cardinal point positions

	Front	Back
Focal points	−12.561	21.252
Principal points	1.782	2.128
Nodal points	6.562	6.909

$$\mathcal{F}'\mathcal{R}' = 2.644 \qquad \mathcal{N}'\mathcal{R}' = 16.987 \qquad \mathcal{P}'\mathcal{R}' = \mathcal{P}'\mathcal{O}' = 21.768$$
$$\mathcal{P}\,\mathcal{F} = -14.343 \qquad \mathcal{P}'\mathcal{F}' = 19.124$$

Near point position $= -116.298$ mm ≈ 8.599 m^{-1} of accommodation.

Pupils

	Entrance	Exit
Positions	2.674	3.249
Radii	4.000	3.766
Magnifications	1.114	1.049

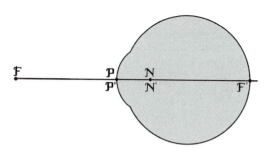

Fig. A3.4:   The Emsley
reduced schematic eye,
showing the position of
cardinal points.

Reduced eye (Emsley 1952)

Please refer to Figure A3.4 for cardinal point positions. This eye is designed to
have an equivalent power of exactly 60 m^{-1}.

One surface: aperture stop (IRIS) at surface 1.

Surface	n	r	V	
	1.0000			
1		50/9		
	4/3		50.23	vitreous

Equivalent power
Eye: complete   60.000

Eye length   22.222

Cardinal point positions

	Front	Back
Focal points	−16.667	22.222
Principal points	0.000	0.000
Nodal points	5.556	5.556

$$\mathcal{F}\mathcal{P} = \mathcal{N}'\mathcal{R}' = \mathcal{N}'\mathcal{F} = 16.667 \qquad \mathcal{P}'\mathcal{F}' = \mathcal{P}'\mathcal{R}' = \mathcal{P}'o' = 22.222$$

Pupils

	Entrance	Exit
Positions	0.000	0.000
Radii	4.000	4.000
Magnifications	1.000	1.000

Marginal and pupil rays

	u	h	$\bar{u}$	$\bar{h}$	
	+0.000000		0.087489		
1		4.000000		0.000000	1
	−0.180000		0.077613		

Gaussian properties

	Object space	Image space
N.A.	0.000	0.240
ν_O	∞	22.222
η'	∞	1.458

$\mathcal{P}'o' = 22.222$

Seidel aberrations: by surface

	$S_1/8$	$S_2/2$	$S_3/4$	$S_4/4$	$\%S_5$	$C_L/2$	$\%C_T$
1	63.0487	30.6448	1.8619	2.4825	−0.1674	17.2181	−0.6636
Sum	63.0487	30.6448	1.8619	2.4825	−0.1674	17.2181	−0.6636

Primary aberration coefficients

$$_0w_{4,0} = 63.207 \qquad _1w_{3,1} = 30.645 \qquad _2w_{2,0} = 4.344 \qquad _2w_{2,2} = 3.724$$

Image surface radii of curvature

Sagittal surface	−9.505
Tangential surface	−5.118
Petzval surface	−16.633

A3.2 Summary of some useful data

A3.2.1 *General*

	Length
Gullstrand number 1	24.385
Le Grand (theoretical)	24.197
Le Grand (simplified)	24.192
Gullstrand number 2	23.896
Emsley reduced	22.222

A3.2.2 *Relaxed Eyes*

	$F\,(\mathrm{m}^{-1})$	$\mathcal{N}'\mathcal{F}'$	$\nu\varepsilon$	$_0w_{4,0}$
Gullstrand number 1	58.636	17.054	3.047	41.347
Le Grand (theoretical)	59.940	16.683	3.038	
Le Grand (simplified)	59.952	16.680	3.038	
Gullstrand number 2	60.483	6.534	3.052	
Emsley reduced	60.000	16.667	—	63.049

A3.2.3   Accommodated eyes

	$F\,(\mathrm{m}^{-1})$	Accomm. (m^{-1})	$\mathcal{F}'\mathcal{R}'$	$_0w_{4,0}$
Gullstrand number 1	70.584	10.878	3.371	112.865
Le Grand (theoretical)	67.677	7.053	2.265	
Le Grand (simplified)	67.665	7.034	2.255	
Gullstrand number 2	69.721	8.599	2.644	

Summary of symbols

n	refractive indices
r	radius of curvature
d	surface separations
Σd	distances from corneal pole
V	V-value
$\mathcal{R},\,\mathcal{R}'$	retinal conjugates
$u,\,h$	marginal ray angle and height
$\bar{u},\,\bar{h}$	pupil ray angle and height
H	optical invariant
η'	retinal image size for a 5 degree object space field
$_0w_{4,0}$	spherical aberration
$_1w_{3,1}$	coma
$_2w_{2,2}$	astigmatism
$_2w_{2,0}$	field curvature
S_1	Seidel spherical aberration
S_2	Seidel coma
S_3	Seidel astigmatism
S_4	Seidel (Petzval) curvature
S_5	Seidel distortion
C_L	Seidel longitudinal chromatic aberration
C_T	Seidel transverse chromatic aberration

References and bibliography

Cowan A. (1928). *An Introductory Course in Ophthalmic Optics*. F.A. Davis, Philadelphia.

*Emsley H.H. (1952). *Visual Optics*, 5th ed., Vol. 1, Butterworths, London.

*Le Grand Y. and Hage S.G. (1980). *Physiological Optics*. Springer-Verlag, Berlin.

*Southall J.P.C. (1924). *Helmholtz's Treatise on Physiological Optics*, Vol. 1. Optical Society of America, Merasha, Wisc., Appendices by A. Gullstrand.

Stine G.H. (1934). Tables for accurate retinal localization. *Am. J. Ophthal.* 17, 314–324.

Appendix 4

Glossary of terms

A4.0   International, national and alternative terms

Some ophthalmic instruments and terms have different names, sometimes depending upon the country. The following material gives some alternatives.

Microscope	compound microscope
Telescope	Keplerian, astronomical telescope; terrestrial telescope (if image is erect)
Macroscope	tele-microscope; near point telescope
Focimeters	vertometers (Australia); lensometer (USA)
Keratometers	ophthalmometers (ISO)
Optometers	refractometers (ISO)
Optical invariant	Lagrange invariant
Pupil ray	chief ray, principal ray, reference ray
Vertex focal lengths	focal distances

Note: ISO = International Standards Organization.

A4.1   Glossary

Aberration   An optical phenomenon due to the departure of real rays from their ideal (paraxial) paths. That is, a beam of real rays from a given object point are not concurrent at the corresponding Gaussian image point. Aberrations are usually classified as monochromatic (spherical aberration, coma, astigmatism, field curvature and distortion) and chromatic (longitudinal and transverse). The presence of aberrations leads to a decrease in image quality.

Achromatic doublet   A combination of two lenses, one with a positive power and one with a negative power and made of different glasses. The combination is usually designed to have zero longitudinal chromatic (Seidel C_L) aberration.

Aerial   Relates to an image that is real and can be projected onto a screen.

Afocal   Describes an optical system with zero equivalent power, but usually having a transverse or angular magnification.

Airy disc   The central disc of light of the monochromatic diffraction limited point spread function. Its diameter is inversely proportional to the diameter of the aperture stop. This disc does not have a uniform illuminance and is surrounded by an infinite number of concentric rings of decreasing illuminance.

Amblyopia   A visual condition in which one eye has reduced visual acuity without any obvious pathology.

Ametropia   A refractive error of the eye, in which the conjugate to the retina is not at infinity for the relaxed eye.

Aniseikonia   The condition in which the two eyes have different ocular image sizes when viewing the same object.

Aperture   The opening in an optical component that allows the light to pass through.

Aperture radius   The radius of the aperture of an optical component. It has the symbol ρ. This is not the radius of curvature of a refracting or reflecting surface.

Aperture stop   The surface in an optical system that limits the width of the beam from an axial object point.

Astigmatism   An optical aberration in which a point object is imaged as two separated and perpendicular lines, separated in space along the pupil ray.

Barlow lens   A negative power lens placed in front of the back focal point of a positive power lens in order to increase the equivalent focal length of the positive lens.

Binocular viewing   The situation in which an optical system forms two separate images of the object and each eye views only one of these images. (See "Biocular viewing" for comparison.)

Biocular viewing   The viewing of the same or a single image by both eyes simultaneously. For example, the viewing of the image formed by a single wide simple magnifier by both eyes at the same time. (See "Binocular viewing" for comparison.)

Bravais points   Self-conjugate points, that is points where the object and image coincide, though not necessarily with unit magnification. For a two thin lenses system in air, Bravais points exist if the lens separation d and the focal lengths f_1 and f_2 have the relationship $d^2 > 4 f_1 f_2$.

Cardinal points   The six important points of an optical system: namely two each of focal point, principal point and nodal point (see these separately).

Catadioptric   An optical system that contains both reflecting and refracting components.

Catoptric   An optical system that only contains reflecting components.

Chromatic aberration   The variation in the path of a ray due to a change in the refractive index with wavelength.

Coherence   A property of a light source in which all the photons have a fixed phase relationship with each other. That is the light beam appears as a long continuous sinusoidal wave train with no sudden changes or phase.

Coherent imaging   (1) Imaging with coherent light. (2) Imaging by an optical system, in which each wavefront from a point object retains its continuity as it passes through the system. In contrast, with incoherent imaging, each of the original wavefronts becomes jumbled or destroyed and new wavefronts are formed.

Coherent light   Light that has a high degree of coherence.

Collimation   The placing of a target or light source in the front focal plane of an optical system so that the image is at infinity. Collimated light is sometimes referred to as "parallel" light. This is misleading because all targets and sources have a finite size and thus the beam emerging from the collimating lens must diverge.

Collimator or collimating lens   A lens used for collimating targets and sources and usually corrected for spherical aberration, coma and chromatic aberration.

Coma   An optical aberration due to the variation of magnification with aperture.

Condenser (lens)   A positive power lens used to image an illuminating source into the entrance pupil of an imaging lens. A condenser lens is a special type of field lens.

Conjugate   The relation between an object and its image. The conjugate of the object is the image and the conjugate of the image is the object.

Contrast   The ratio between two light levels (usually luminances) in an image. It is numerically defined in different ways. The two most common are

$$\frac{\Delta L}{L} \quad \text{and} \quad \frac{L_{max} - L_{min}}{L_{max} + L_{min}}$$

where L is a luminance. In the lefthand definition, L in the denominator is the higher of the two luminances.

Cornea   The front shiny transparent surface of the eye.

Critical angle   When light is passing from a medium of higher refractive index to one of lower index, the angle of incidence at which the refracted angle is just 90°. For angles greater than the critical angle, the light is totally internally reflected.

Curvature   The reciprocal of the radius of curvature of a cylindrical or spherical surface.

Density (neutral)   Symbol D. The logarithm to the base 10 of the reciprocal of the transmittance (T) for a given light source. Thus

$$D = \log_{10}(1/T) \quad \text{that is} \quad D = -\log_{10}(T)$$

Dielectric   A material that is a very good electric insulator. A typical example in optical applications is glass.

Diffraction   A phenomenon of physical optics in which light does not propagate as expected from geometrical ray theory. In effect, it is the bending of light around corners and edges.

Diffraction – Fraunhofer   The diffraction pattern at the focus of a beam.

Diffraction – Fresnel   The diffraction pattern at any finite distance from an aperture, except at the focus of the beam.

Diffuse   The ability of a surface to scatter incident light in a wide range of directions. Synonyms are matte and rough.

Dioptre   The ophthalmic unit of lens power in the metric system. The symbol is D and it is used in place of m^{-1}. However, in this text we will use the notation m^{-1} instead of D, because we need to reduce confusion since D is used as the pupil diameter.

Diplopia   Double vision.

Dispersion   The variation of refractive index with wavelength. With all optical media in the visible region the refractive index decreases with increase in wavelength.

Dissociation   The condition in which the stimulus to the two eyes is so altered as to suspend normal ocular positions.

Distortion   An optical aberration in which the transverse magnification is a function of field angle. The aberration may be positive (pincushion) or negative (barrel).

Emmetropia   The ideal refractive state of the eye, in which the conjugate to the retina is at infinity for the relaxed eye.

Entrance pupil   The image of the aperture stop in object space.

Esophoria   The tendency for an eye to turn inwards relative to its fellow eye.

Exit pupil   The image of the aperture stop in image space.

Exophoria   Tendency for eye to turn outwards relative to its fellow eye.

Exposure   A measure of the amount of light received at the image plane. It is the product of the illuminance and exposure time. The units are lux.s.

Eye lens   The lens nearest the eye in many visual optical instruments. It usually operates as a simple magnifier and magnifies an intermediate image formed inside the instrument.

Eyepiece   A type of simple magnifier designed to view aerial images. It usually is more complex than a simple magnifier in that it consists of the main magnifying lens (the eye lens) and a field lens. There are a number of different optical designs. (See Section 15.4.)

Eye relief   The distance between the back vertex of the eye lens or eyepiece and the exit pupil.

F-number   The ratio of the equivalent focal length of a lens and the diameter of its entrance pupil. Strictly, it is only a valid quantity if the object or image is at infinity.

F-stops   The markings or settings on a camera lens, indicating the *F-number* setting of the lens.

Far point   The object space conjugate of the retina when the eye is completely relaxed, that is when the power of the eye is minimum.

Field curvature An optical aberration that causes an object plane to be imaged point-to-point on a curved image surface.

Field lens A positive power lens placed in an optical system, usually at or close to the object or an intermediate image plane, with the purpose of pupil matching and thus maximizing the field-of-view.

Field stop An aperture placed at an image plane to control the size of the field-of-view.

Focal Relates to an optical system having a non-zero equivalent power.

Focal length The distance from some defined plane to a focal point. (See Vertex focal length and Equivalent focal length.)

Focal point The image point of an axial object at infinity and also the object point for the axial image point at infinity.

Focimeter An instrument for measuring the vertex power of a lens, usually an ophthalmic lens. In Australia, it is usually called a vertometer. In the USA, it is usually called a lensometer.

Focus (verb) To bring together in image space rays from a point in the object space, that is to make the rays concurrent at one point in image space.

Gaussian optics The application of paraxial optics outside or beyond the paraxial region.

Geometrical optics Optics in which all physical optics effects are neglected; that is light can be considered as travelling along rays which only change direction by refraction or reflection.

Goniometer Instrument for mounting a device so that it can be rotated about one or more perpendicular axes. In one implementation, it may have only one axis of rotation and a collimator and telescope can be rotated about this axis with their optical axes intersecting the axis of rotation.

Graticule A plate of optical quality marked with some type of scale and placed in an optical system in either the object or some intermediate image plane to allow measurements to be made of object size.

Homogeneous Having the same physical properties throughout the medium.

Illuminance The amount of light (luminous flux) falling on a surface per unit area. Unit: lumen/m^2 (lm/m^2) or lux.

Isotropic Having radiating or electrical properties that are the same in all directions.

Keratometer An instrument for measuring the radius of curvature or the surface power of the anterior corneal surface.

Lambertian A source or surface whose luminance is independent of viewing direction.

Lens (thin) A single lens consisting of two refracting surfaces and of zero thickness.

Lens – eye or eye lens   The lens of many instruments which is closest to the eye. It usually acts as a simple magnifier looking at an aerial image.

Lens – inverse or reverse telephoto   An optical system that has a short equivalent focal length but a long back vertex focal length.

Lens – objective   The leading lens in many instruments, such as telescopes. It is the lens closest to the object.

Lens – telephoto   An optical system that has a long equivalent focal length and a short back focal length and system length.

Line spread function (LSF)   The light distribution in the image of a line object or source.

Longitudinal   In the direction of the optical axis or of travel.

Lumen   The unit for luminous flux or the quantity of light.

Luminance   The objective measure of brightness of an extended source or surface. Unit: candela/m^2 (cd/m^2) or lm/sr/m^2.

Luminous flux   The quantity of light. Unit: lumen or lm.

Luminous intensity   The luminous flux in a cone per unit solid angle. Unit: candela (cd) or lumen/steradian (lm/sr). It is useful as a measure of the brightness of "point" sources.

Macroscope   An instrument for magnifying objects at intermediate distances. In other texts it is sometimes called a "tele-microscope" or a "near point telescope".

Magnification – longitudinal   Ratio of the image plane movement to the object plane movement along the optical axis.

Magnification – transverse (or lateral)   Ratio of the image and object sizes, with the sizes measured in a plane perpendicular to the optical axis.

Marginal ray   A ray travelling at or near the edge of a lens or pupil.

Microscope   An instrument for magnifying close objects. The magnification is achieved by a two stage magnification process.

Modulation transfer function (MTF)   The ratio of the image to object contrasts for pure sinusoidal patterns. It is the optical transfer function without the phase factor.

Monocular   Viewing with only one eye.

Near point   The conjugate point to the retina when the eye is maximally accommodated, that is when the power of the eye is maximum.

Near point cap   A positive power auxiliary lens placed over either the objective or the eye lens to modify a telescope for viewing objects at a finite distance.

Nodal points   Two axial points, one in object space and the other in image space, through which a ray passes while subtending the same angle with the axis in both spaces. They are cardinal points.

Numerical aperture   The product of the refractive index and the sine of the angular radius of the pupil, measured from the axial object or image point.

Nystagmus   Abnormal involuntary rapid eye movements that prevent adequate fixation.

Objective (lens)   The lens nearest the object in many optical instruments, such as telescopes and microscopes.

Optical axis   The line joining the centres of curvatures of surfaces in centred optical systems.

Optical centre   The actual point of intersection with the axis of a nodal ray. It is the same point for the ray at any height or angle.

Optical element   Something that is designed to modify a light beam in any way, that is refract, reflect, scatter, deviate, absorb, modify the spectral content or alter the size, such as lens, mirror, diffuser, filter or aperture. Optical elements are primarily designed as components of an optical instrument or system.

Optical instrument or system   A construction containing one or more optical elements which has some image forming or flux gathering purpose.

Optical path length   The product of the physical distance travelled and the refractive index.

Optical transfer function   Similar to the modulation transfer function but includes the effects of changes in phase due to aberrations.

Optical tube length   Applies to microscopes. The distance between the back focal point of the objective and the front focal point of the eye lens. It has the symbol t and usually has a value of 160 mm.

Paraxial optics   The optics of rays that are limitingly close to the optical axis.

Phoria   A binocular vision term describing the tendency of one eye to move out of alignment with the other.

Physical optics   Optics that cover the wave nature of light and include diffraction and interference phenomena.

Point spread function (PSF)   The light distribution in the image of a point object or source.

Polarization   Relates to the direction of the plane of the electric vector of electromagnetic radiation.

Power – equivalent   The refractive power of an equivalent thin lens placed at the principal planes.

Power – surface   The refractive power of a surface, given by the equation $F = C(n' - n)$.

Power – vertex   The power of a system measured at a surface vertex, that is the power of a thin lens placed at the surface vertex and having the same back focal point position.

Principal planes   Conjugate planes in a focal system having positive unit transverse magnification. They are also planes at which the system can be replaced by thin lenses of the same equivalent power under certain conditions.

Ray   The geometrical optical representation of the path of a small portion of a light beam. In isotropic materials, rays are perpendicular to the wavefronts.

Ray – finite   An actual ray as opposed to a paraxial ray. Also known as a real or exact ray.

Ray – meridional   Any ray intersecting the optical axis. In a rotationally symmetric optical system, meridional rays stay in the same plane (which contains the optical axis) as they are refracted or reflected through the system. Ray tracing of meridional rays is a two dimensional exercise. (For contrast see Ray – skew.)

Ray – paraxial marginal   A paraxial ray traced from the axial object point, through the edges of the entrance pupil, aperture stop and exit pupil, and meeting the optical axis at the paraxial axial image point.

Ray – paraxial pupil   A ray traced from the edge of the object field, through the centres of the entrance pupil, aperture stop and exit pupil, and meeting the paraxial image plane at the edge of the field. This ray is also known as the chief, reference or principal ray. The pupil ray may be regarded as the central ray of the beam.

Ray – real   An actual ray, as opposed to a paraxial ray. Also known as a finite ray.

Ray – skew   A ray that does not intersect the optical axis, and as a result does not stay in the same plane as it passes through the system. Ray tracing of skew rays is a three dimensional exercise.

Real image   An image which can be projected onto a screen.

Refractive index   The ratio of the speed (or velocity) of light in vacuum to the speed (or velocity) of light in a particular medium. Relative refractive index is the ratio using the speed in air instead of in a vacuum.

Resolution limit   A measure of the dimensions of the smallest detail that can be resolved by an optical system such as the highest spatial frequency of a periodic pattern or the separation of two point sources.

Reticle   Similar to graticule. (See graticule.)

Retina   The light sensitive tissue of cells on the inside surface of the back of the eye.

Retroreflection   The reflection of a ray or beam back in the direction it came from but not necessarily along the same path, that is it travels back along a parallel path.

Spectrometer   A device that is designed for analysis of the spectral content of light beams. It usually consists of a table to hold a dispersive device such as a prism or diffraction grating and can be rotated about an axis. A collimator and telescope are attached but free to rotate about the same axis with their optical axes intersecting the table axis.

Specular   A surface that reflects incident light according to Snell's law. Synonyms are glossy, mirror-like, regular.

Spherical aberration   A rotationally symmetric optical aberration in which the effective power of the system depends upon the ray height in the pupil.

Spurious resolution   The apparent resolution of a sinusoidal or square wave grating at a frequency greater than the (conventional) resolution limit, that is greater than the frequency at which the optical transfer function is first zero. The contrast of spuriously resolved patterns may be reversed.

Stigmatoscopy   Measurement of the refractive error of the eye using a point source as the target or stimulus.

Telecentric   The situation in which the aperture stop is placed at one of the focal points of an optical system. This puts one of the pupils at infinity.

Telephoto lens   A lens system in which the system length plus back focal length is less than the equivalent focal length.

Telephoto ratio   The ratio of the sum of the system length and back focal length to the equivalent focal length.

Telescope   An afocal instrument for magnifying distant objects.

Throw   The distance between the object and image planes.

Total internal reflection   A phenomenon which occurs when light is passing from a medium of higher index to one of lower index. For angles of incidence greater than the critical angle all the incident energy is totally reflected back into the incident medium.

Translucent   Transmitting light, allowing the shape of the object but not its fine detail to be seen.

Transparent   Allowing the passage of light.

Transverse   In a direction perpendicular to the optical axis or direction of travel.

Tropia   The breakdown of the alignment of the two eyes, associated with absence of fusion of the images presented to each eye.

Vergence   The product of a refractive index and the reciprocal of a distance. Sometimes known as reduced vergence.

Vertometer   An instrument for measuring the vertex powers of a lens, usually an ophthalmic lens. This term is used in Australia. It is also known as a focimeter (United Kingdom) and lensometer (USA).

Vignetting   The blocking of rays by a surface other than the aperture or field stop. Usually occurs towards the edge of the field.

Virtual   An image that cannot be projected onto a screen. The image formed by a visual optical instrument is usually virtual.

Visual acuity   A measure of the smallest detail that can be detected or resolved by the eye.

Visual space   The projection of the retinal image back out into object space.

Wave aberration function   The optical path difference between the actual aberrated wavefront and the ideal unaberrated wavefront (the reference sphere) and measured along the ray. It is expressed as a function of the co-ordinates in the entrance or exit pupil of the intersection of the ray with the reference sphere.

Wave aberration polynomial   Representation of the wave aberration function by a polynomial function, such as a power series in radius and azimuth of a point in the pupil, or by Zernike polynomials.

Wavefront   Usually the crest or trough of a sinusoidal wave motion. More generally, it can be defined as the surface of constant phase.

Working distance   Distance from the eye to the object or from an instrument to the object.

Appendix 5

Resolution and visual acuity charts

A5.0   Introduction

There are a number of different methods for analysing the image quality of an optical system. One of the simplest is to measure the resolving power, which is a measure of the smallest detail that can be resolved in the image. But even with this simple concept there are several different ways of defining resolving power in terms of the type of detail and test method. Some of the most common are listed below.

(1)   The detail may be in terms of point objects. In this case the resolving power is a measure of the separation at which two points can just be resolved. This definition is used extensively in optical astronomy since the most common requirement here is to be able to resolve two very close stars.

(2)   An alternative approach is to measure the resolving power of a periodic target, such as a sinusoidal or square wave pattern. The resolving power is the highest spatial frequency of the periodic pattern that can just be resolved.

In practice, the resolution limit of an optical system is most easily measured using a resolving power chart, which consists of a set of patterns containing detail as described above, but repeated at a number of different levels of detail size.

A5.0.1   The Rayleigh criterion

The Rayleigh criterion assumes diffraction limited optics and specifies the resolution limit in terms of the minimum separation of point sources. In a diffraction limited optical system, the image of a point source object is the diffraction limited point spread function discussed in Chapters 26 and 34. The Rayleigh criterion states that two point sources can just be resolved if the maximum of one point spread function falls on the first zero of the other. For sources at a great distance, the angular separation θ, in object space, is related only to the wavelength λ and entrance pupil diameter D by the simple equation

$$\theta = 1.22\lambda/D \tag{A5.1}$$

As this equation applies to a single wavelength, care should be exercised in applying it to a polychromatic point source.

A5.0.2   *Sinusoidal or square wave patterns*

One advantage of using sinusoidal patterns is that the image quality can be assessed mathematically using image quality criteria such as the optical transfer function, which expresses the reduction in contrast of a sinusoidal pattern as a function of its spatial frequency. The calculation of the optical transfer function involves the aberrations of the optical system and diffraction effects. If the system is aberration free, the optical transfer function is only affected by diffraction and has the simple form given by equation (34.40), that is

$$G(s) = [\beta - \sin(\beta)]/\pi \tag{A5.2a}$$

where

$$\beta = 2\arccos(s/2) \tag{A5.2b}$$

where $G(s)$ is the reduction in contrast and s is a normalized spatial frequency related to the actual spatial frequency σ by equation (34.37a or b), that is

$$\sigma = sn\sin(\alpha)/\lambda \quad \text{(c/unit distance) finite conjugate} \tag{A5.3a}$$

or

$$\sigma = sn/\lambda \quad \text{(c/rad) infinite conjugate} \tag{A5.3b}$$

where λ is the wavelength, n is the refractive index of the space, α is the angular radius of the pupil and $\bar{\rho}$ is its physical radius. The term "$n\sin(\alpha)$" is in fact the numerical aperture. The diffraction limited optical transfer function $G(s)$ is plotted in Figure 34.8.

A5.0.2.1   *Effect of defocus on the optical transfer function*

The contrast in the image is affected by the level of any aberrations and defocussing. The effect of aberrations is usually complex and will not be examined here, apart from the effects of astigmatism.

As one increases the level of defocus, the contrast in the image decreases as shown in Figure A5.1, but it does not decrease steadily to zero and stay zero. As shown in the diagram, beyond the defocus at which the contrast goes to zero, the contrast increases again but in a negative direction. Thus the image reappears. This is known as spurious resolution, and is more obvious in well corrected and diffraction limited optical systems.

A5.0.2.2   *Effect of astigmatism*

Of all the aberrations, astigmatism has the most interesting effect on resolution. Astigmatism was introduced in Chapter 5 and discussed further in Chapter 33. It has a similar effect to defocus, but with the level of defocus differing between the tangential and sagittal sections and with a magnitude that depends upon the distance from the axis. Thus the contrast reduction for gratings lined up in the tangential section will be different from that in the sagittal section. This effect is observed best when the gratings are orientated as shown in Figure A5.2.

Fig. A5.1:   Effect of
progressive defocus on the
contrast of a sinusoidal
target, showing regions of
spurious resolution.

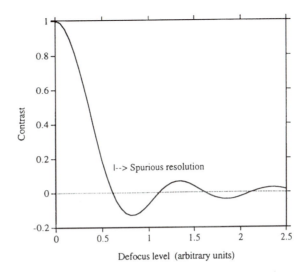

Defocus level (arbitrary units)

Fig. A5.2:   The sagittal
and tangential orientations.

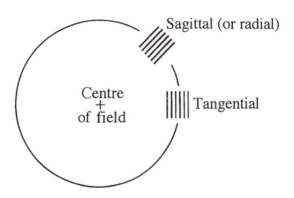

A5.0.3   *Effect of detector acuity*

In most cases, the measurement of resolving power depends somewhat on the
detector, because detectors themselves also have a resolving power which must
be much higher than that expected from the optical system being investigated.
Let us consider the example of the eye. The eye has a resolving power limit
of about 60 c/deg. Fortunately when the resolving power of the image under
investigation is greater than this, an optical aid such as a microscope or a
telescope can be used effectively to increase the resolving power of the eye.
However, once again the resolving power of this optical aid must be much
higher than the system or image being examined.

   While resolving power as described above is used extensively in instrumental
optics, it is not often used in assessing the quality of the retinal image. In this case
the term "visual acuity" is used instead; it is measured using letters rather than
point objects or periodic targets. However, periodic targets are becoming more
frequently used in the assessment of retinal image quality, and the resolving
power of the eye.

A5.1   Resolution charts

Because of the difficulty in making sinusoidal gratings, square wave gratings are usually used instead. Resolution charts are made up of sets of black and white equally spaced bars, that is a square wave pattern, over a wide range of spatial frequencies. In practice, at a given spatial frequency, the bars are arranged in sets of three or five bars. The frequencies are usually based upon a sequence described by the following equation:

$$\sigma(i) = \sigma(1)r^{i-1} \quad \text{cycles/unit distance} \tag{A5.4}$$

where $\sigma(i)$ is the frequency of the ith pattern, $\sigma(1)$ is the spatial frequency of the first pattern and r is the size ratio.

At any given spatial frequency, the patterns consist of orthogonal pairs which are used as a check for the presence of astigmatism either in the image or detector. The presence of astigmatism in the detector can be investigated by rotating the detector (e.g. eye) through 90°. Any change in contrast of the pattern indicates astigmatism. To check for astigmatism in the image, the pattern should be oriented so that one set of bars lies along the line joining the pattern to the centre of the image field, that is radially. The other pattern will then lie along tangents to circles concentric to the field centre, as shown in Figure A5.2. The pattern in the radial direction will be at maximum contrast if the astigmatic sagittal image lies in the image plane. The other pattern will have maximum contrast if the astigmatic tangential image lies in this plane. These rules can be confirmed by observing the sagittal and tangential ray patterns shown in Figure 5.6.

A5.1.1   *The NBS chart or ISO Test Chart Number 2*

The National Bureau of Standards (NBS) chart, denoted as "Standard Reference Material 1010a", is shown in Figure A5.3. The patterns are black bars on a white background, the length of each bar is 14 times the width, and the bars and spaces are of equal width. The chart is printed on glossy photographic paper.

The spatial frequency range is from 1.0 c/mm to 18 c/mm. Each pattern is made up of two orthogonal groups of five parallel bars. The number written next to each pattern is the corresponding spatial frequency (in c/mm). The spatial frequencies in this chart are given by equation (A5.4), with $\sigma(1) = 1.0$ c/mm and $r = 1.122$ and the frequencies rounded (most times) to the one decimal place.

A5.1.2   *The USAAF chart*

The United States of America Air Force (USAAF) resolution chart is shown in Figure A5.4. This chart is similar to the NBS chart above but each pattern is composed of only three and not five bars or 2.5 cycles and contains a much wider range of spatial frequencies. The highest spatial frequencies cannot be reproduced in our diagram. The boundary of each triad of lines is a square and thus the bars are 2.5 cycles or periods in length. This chart is usually produced on either high resolution photographic film or on a thin metal film but both are placed on a glass substrate.

The frequencies of this chart also follow equation (A5.4) with $\sigma(1) = 1.0$ c/mm and $r = 1.12246$. The number next to each set of bars is not a

Fig. A5.3: The NBS resolution chart (not to scale).

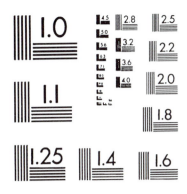

MICROCOPY RESOLUTION TEST CHART
NATIONAL BUREAU OF STANDARDS
STANDARD REFERENCE MATERIAL 1010a
(ANSI and ISO TEST CHART No. 2)

Fig. A5.4: The form of the USAAF resolution chart (not to scale).

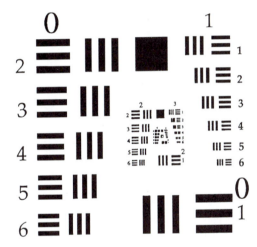

direct measure of the spatial frequency. Table A5.1 gives the corresponding spatial frequency (σ) and block width for each pattern. The highest spatial frequency is 228 c/mm. Since the block is a square and the width is 2.5 cycles in length, the width w or length of each bar is given by the equation

$$w = 2.5/\sigma \tag{A5.5}$$

A5.1.3 The Siemen's star

Siemen's star is a radial chart of the type shown in Figure A5.5. This chart has the special property that the spatial frequency, along a circular arc around the centre of the chart, is constant along the arc but varies with distance from the centre. This chart has been used in Chapter 34 to demonstrate spurious resolution.

Table A5.1. The spatial frequency (σ) and block width or bar length of the triad of bars in the USAAF resolving power chart

Pattern		σ c/mm	Width mm	Pattern		σ c/mm	Width mm
0	1	1.00	2.500	4	1	16.0	0.1563
0	2	1.122	2.227	4	2	17.96	0.1392
0	3	1.260	1.984	4	3	20.16	0.1240
0	4	1.414	1.768	4	4	22.62	0.1105
0	5	1.587	1.575	4	5	25.39	0.09845
0	6	1.782	1.403	4	6	28.50	0.08771
1	1	2.00	1.25	5	1	32.0	0.07814
1	2	2.245	1.114	5	2	35.91	0.06962
1	3	2.520	0.9922	5	3	40.31	0.06202
1	4	2.828	0.8839	5	4	45.25	0.05525
1	5	3.175	0.7875	5	5	50.79	0.04923
1	6	3.563	0.7016	5	6	57.00	0.04386
2	1	4.00	0.6250	6	1	64.0	0.03907
2	2	4.489	0.5569	6	2	71.82	0.03481
2	3	5.039	0.4961	6	3	80.62	0.03101
2	4	5.656	0.4420	6	4	90.49	0.02763
2	5	6.349	0.3938	6	5	101.6	0.02461
2	6	7.126	0.3508	6	6	114.0	0.02193
3	1	8.00	0.3125	7	1	128.0	0.01954
3	2	8.979	0.2784	7	2	143.6	0.01741
3	3	10.07	0.2481	7	3	161.2	0.01551
3	4	11.31	0.2210	7	4	181.0	0.01381
3	5	12.70	0.1969	7	5	203.1	0.01231
3	6	14.25	0.1754	7	6	228.0	0.01096

A5.2   Visual acuity charts

Visual acuity (VA) is usually measured using a recognition task rather than just a detection or resolution task. Since recognition involves a higher cognitive process, the results are often not easy to compare with a simple resolving power test.

A5.2.1   The Snellen chart

The Snellen chart consists of letters of the alphabet. The letters are arranged in rows, with each row a different size and corresponding to a different level of acuity. The acuity rating of each row is the distance at which a person with a standard "normal" acuity would just be able to recognize the letters of that row. The acuity level is often specified as a fraction defined as

$$\text{Snellen VA} = \frac{\text{testing distance } (m)}{\text{distance } (m) \text{ at which test line letters subtend 5 minutes of arc}}$$

(A5.6)

Fig. A5.5: The Siemen's
           star chart.

The 6/6 acuity line is the "normal" line and contains letters that are 5 minutes of arc high. A typical letter such as E, which consists of three horizontal bars, can be thought of as consisting of 2.5 cycles of a square wave pattern with a spatial frequency of 0.5 c/minutes of arc or 30 c/deg.

All the letters on the same line have the same size and the number of letters on each line decreases in general as the letters get larger. Often the line with the largest letters, usually the 6/60 line, has only one letter.

A5.2.2   The Bailey–Lovie chart

The Bailey–Lovie chart is a modified Snellen chart (Bailey and Lovie 1976) with an equal number of letters per line and the size difference between adjacent lines varies with a constant ratio of $10^{+0.1}$ (or $10^{-0.1}$).

Summary of main symbols

D	diameter of pupil
λ	wavelength
$G(s)$	diffraction limited optical or modulation transfer function
s	reduced spatial frequency
σ	linear spatial frequency (c/unit length)
$\sigma(i)$	spatial frequency of the ith pattern in a resolution chart
r	geometrical ratio of adjacent test patterns in resolution charts

References and bibliography

*Bailey I.L. and Lovie J.E. (1976). New design principles for acuity letter charts. *Am. J. Optom. Physiol. Opt.* 53(11), 740–745.

Index